Fine Particles Science and Technology

NATO ASI Series

Advanced Science Institutes Series

A Series presenting the results of activities sponsored by the NATO Science Committee, which aims at the dissemination of advanced scientific and technological knowledge, with a view to strengthening links between scientific communities.

The Series is published by an international board of publishers in conjunction with the NATO Scientific Affairs Division

A Life Sciences **B Physics**	Plenum Publishing Corporation London and New York
C Mathematical and Physical Sciences **D Behavioural and Social Sciences** **E Applied Sciences**	Kluwer Academic Publishers Dordrecht, Boston and London
F Computer and Systems Sciences **G Ecological Sciences** **H Cell Biology** **I Global Environmental Change**	Springer-Verlag Berlin, Heidelberg, New York, London, Paris and Tokyo

PARTNERSHIP SUB-SERIES

1. Disarmament Technologies	Kluwer Academic Publishers
2. Environment	Springer-Verlag / Kluwer Academic Publishers
3. High Technology	Kluwer Academic Publishers
4. Science and Technology Policy	Kluwer Academic Publishers
5. Computer Networking	Kluwer Academic Publishers

The Partnership Sub-Series incorporates activities undertaken in collaboration with NATO's Cooperation Partners, the countries of the CIS and Central and Eastern Europe, in Priority Areas of concern to those countries.

NATO-PCO-DATA BASE

The electronic index to the NATO ASI Series provides full bibliographical references (with keywords and/or abstracts) to more than 50000 contributions from international scientists published in all sections of the NATO ASI Series.
Access to the NATO-PCO-DATA BASE is possible in two ways:

– via online FILE 128 (NATO-PCO-DATA BASE) hosted by ESRIN, Via Galileo Galilei, I-00044 Frascati, Italy.

– via CD-ROM "NATO-PCO-DATA BASE" with user-friendly retrieval software in English, French and German (© WTV GmbH and DATAWARE Technologies Inc. 1989).

The CD-ROM can be ordered through any member of the Board of Publishers or through NATO-PCO, Overijse, Belgium.

Series 3: High Technology – Vol. 12

Fine Particles Science and Technology

From Micro to Nanoparticles

edited by

Ezio Pelizzetti
Dipartimento di Chimica Analitica,
Università di Torino,
Torino, Italy

Kluwer Academic Publishers

Dordrecht / Boston / London

Published in cooperation with NATO Scientific Affairs Division

Proceedings of the NATO Advanced Research Workshop on
Fine Particles Science and Technology:
From Micro to Nanoparticles
Acquafredda di Maratea, Italy
July 15–21, 1995

A C.I.P. Catalogue record for this book is available from the Library of Congress.

DOI: 10.1007/978-94-009-0259-6

Published by Kluwer Academic Publishers,
P.O. Box 17, 3300 AA Dordrecht, The Netherlands.

Kluwer Academic Publishers incorporates the publishing programmes of
D. Reidel, Martinus Nijhoff, Dr W. Junk and MTP Press.

Printed on acid-free paper

TABLE OF CONTENTS

PREFACE

The NATO Advanced Research Workshop on *Fine Particles Science and Technology. From Micro to Nanoparticles* was held at the Hotel Villa del Mare, Acquafredda di Maratea, Italy, on July 15-21, 1995. It was extremely stimulating, with the participation of 56 speakers and 40 researchers, including world-wide leading scientists in the field from academia, research institutes and industries.

The so-called monodisperse microparticle originally attracted the interest of researchers essentially for quantitative interpretation of surface interactions as a function of morphology and size. In the recent years significant progress has been made in the preparation of a large number of inorganic colloids consisting of particles of different chemical composition, shape and modal size, especially by precipitation. These materials have then been used in attempts to elucidate some crucial aspects of surface science such as heterocoagulation, particle adhesion, corrosion to name a few.

Presently, the importance of well-defined fine particles has been recognized in numerous applications, as ceramics, catalysis, pigments, recording materials, medical diagnostic and many others. For example, the role of fine particles in various ceramic applications has become evident since the quality and reproducibility can be more readily achieved by starting with a well-defined powder. Processing (e.g. compaction, sintering, etc.) can in many cases be done more efficiently and at a lower temperature if uniform particles are used. Different properties of matter will be needed when dealing with electro-optical, capacitor, piezoelectric, magnetic composite, superconducting or other ceramic materials. Similar aspects are present in the request of pigments and inks for high-speed printing and the diagnostic and novel disease treatments in medicine.

Semiconductor particles in the nanometer size have also been prepared and investigated in the last decade. Surface boundary processes such as photocatalytic reactions on semiconductors are not largely understood and nanometer particles can be used to elucidate these mechanisms. The particles are so small that the colloidal solutions are completely transparent, so stationary and time-resolved optical spectroscopy can be employed to study the reaction processes. The use of colloidal particles both in solution and in thin layers for solar energy conversion, synthesis of organic compounds and destruction of contaminants have attracted large interest.

An important aspect is the influence of the particle dimension and surface properties on the energies of the electronic levels and on the reactivity and kinetics of the charge carriers. In the range of nanometer size, a gradual transition from solid state to molecular structure occurs as the size decrease. Consequently, a splitting of the energy bands into quantized levels occurs. Particles that exhibit this phenomenon are called "Q-particles". The optical, electronic and catalytic properties of this quantized material drastically differ from those of the corresponding macrocrystalline substances or micrometer particles. Q-particles of many materials have been recently prepared in form of colloidal solutions or embedded in porous matrices and are stable over a long period of time. In sandwich colloids, in which Q-particles of different materials are coupled as well as in porous semiconductor electrodes containing Q-particles in the pores, very efficient primary charge separation is observed. Enhanced photocatalytic activity and electrodes with high photocurrents can be achieved.

The purposes of the workshop were to bring together two communities that are traditionally isolated, the colloidal chemistry and the nanoparticles, quantum-size effect, community and to cover issues ranging from very fundamental physics and chemistry to technologically applied problems.

Five intensive days were planned for this NATO ARW covering the major areas of interest including: i) the understanding of the formation mechanism of uniform fine particles and the search of new processes, besides precipitation, for the generation of well-defined colloids; ii) the elucidation of the formation mechanism of fine particles by precipitation (actually the most used technique) which requires the knowledge of the relationship between the complex chemistry of solutions in which the solid is formed, and the composition, structure and other characteristics of the resulting products; iii) the sequence of events leading to the formation of monodisperse colloids.

Significant fundamental advances related to nanoparticles were presented at the workshop, and many open questions have been highlighted. Among these issues are: i) factors determining the rate of electron transfer from solution or adsorbed species to the particles or among the particles themselves. Data were presented showing that the reorganization energy for electron transfer is mostly due to the reorganization of the medium, implying very small reorganization energy in the solid; ii) strategies to induce new crystallographic structures in nanoparticles; iii) spectroscopic properties of nanoparticles. Size dependent dielectric function could account for same observation in metallic clusters and small particles.

It was reported that the majority of uniform particles are internally composite (i.e. built of much smaller sub-units). At present, it is not understood why a huge number of primary particles can aggregate to much larger identical colloids. This evidence naturally leads to the much discussed possibility that some of the quantum size properties should be retained in the larger particles. It turns out that even the frontier systems of nanoparticles are fast moving toward possible applications.

At the interface between micro and nanoparticles is the growth mechanism in constrained systems (e.g. inverted micelles, cavities of clays or zeolites, Langmuir-Blodgett films and similar microreactors). Observations were presented that the microreactor does control the growth and limits the size of the resultant particles. However, the particles often growth to sizes much larger than the ones dictated by the size of the microenvironment.

In the book the following topics: microparticles, nanoparticles, applications to materials preparation, synthesis, and properties, applications to environment and other areas are sequentially presented. The hope is that this book will stimulate the research in the fundamental and applied aspects in the micro to nanometer sized particles.

I would like to express here my gratitude to the co-director of the NATO ARW, Prof. Jerzy Szczypa, who contributed considerably to the success of the workshop. Furthermore, I would also like to express my sincere tanks to the members of the Scientific Advisory Committee, who helped in the difficult task of selecting the topics and the lecturers. My special tanks must also be addressed to the Local Organizing Committee, who contribute with enthusiasm to the success of the meeting.

The workshop would not have been the same without the dedicated help of the staff of the Hotel Villa del Mare, to whom we owe the creation of a very warm and pleasant atmosphere, conducive to scientific and social interactions among the participants.

Finally, I wish to thank the NATO Scientific Affairs Division who supported financially this Conference. As well, I would also like to gratefully acknowledge the financial contribution of Center for Advanced Materials Processing (Potsdam, USA), Italian National Research Council (CNR, Italy), IACIS, and Regione Basilicata.

Ezio Pelizzetti
Dipartimento di Chimica Analitica
Università di Torino

Torino, October 18, 1995

Scientific Committee

E. Pelizzetti, University of Torino, Italy
J. Szczypa, Marie Curie University, Poland
D. Horn, BASF, Germany
E. Matijevic, Clarkson University, USA
D. Meisel, Argonne National Lab., USA
T. Sugimoto, Tohoku University, Japan

Organizing Committee

E. Pelizzetti, University of Torino, Italy
C. Minero, University of Torino, Italy
E.Pramauro, University of Torino, Italy
V. Maurino, University of Torino, Italy

sponsored by

Center for Advanced Materials Processing, Potsdam, USA
CNR, Italy
IACIS
NATO Scientific Affairs Division
Regione Basilicata

LIST OF CONTRIBUTORS

Dr. Z. **Adamczyk**
Institute of Catalysis
Polish Academy of Sciences
ul. Niezapominajek
Krakow 30239
POLAND
FAX # 0048-12-251923

Prof. Mufit **Akinc**
Material Science and Engineering
Dept.
Iowa State University
3053 Gilman Hall
Ames - IA 50011-3110
USA
FAX # 001-515-2945444

Prof. P. **Alivisatos**
Dept. of Chemistry
University of California
Berkeley, CA 94720
USA
FAX # 001-510-6426911

Prof. Mark **Anderson**
Water Chemistry Program
University of Wisconsin
660, North Park Street
Madison - WI 53706-1484
USA
FAX # 001-608-2620454

Dr. Detlef **Bahnemann**
Institut fur Solarenergieforschung
Sokelant Strasse, 5
30165 Hannover
GERMANY
FAX # 0049-511-3585010

Dr. Miguel **Blesa**
Departamento Quimica de Reactores
Comision Nacional de Energia
Atomica
Avda del Libertador 8250
1429 Buenos Aires
ARGENTINA
FAX # 0054-1-7041164

Prof. E. **Blums**
Institute of Physics
Latvian Academy of Sciences
229021 Salaspils-1
LETTONIA
FAX # 00371-88-21153

Dr. Louis **Brus**
Bell Labs.
AT&T
600, Mountain Av. - P.O.Box 636
Murray Hill, NJ 07974-0636
USA
FAX # 001-908-5823958

Prof. Imre **Dekany**
Colloid Institute
University of Szeged
6720 Szeged
HUNGARY
FAX # 0036-62-321347

Prof. S.S. **Dukhin**
PENKEM InC.
Adams St. 341
Bedford Hills - NY 10507
USA
FAX # 001-914-241482

Prof. Janos H. **Fendler**
Dept. of Chemistry
Center for Science and Technology
Syracuse University
Syracuse - NY 13244-4100
USA
FAX # 001-315-4434869

Dr. Michal **Freedhoff**
Dept. of Chemistry
University of Rochester
914 Hutchinson Hall
Rochester NY 14627 0216
USA
FAX # 001-716-4736889

Prof. Kunio **Furusawa**
Dept. of Chemistry
The University of Tsukuba
Sakaramura, Tennohdai
Ibaraki 305
JAPAN
FAX # 0081-298-536504

Prof. Fernando **Galembeck**
Universidade Estadual de Campinas
Caixa Postal 6154
13081 Campinas
BRASIL
FAX # 0055-192-393805

Prof. Dante **Gatteschi**
Dipartimento di Chimica
Università di Firenze
via Maragliano 77
50144 FIRENZE
ITALY
FAX # 0039-55-354845

Prof. Michael **Graetzel**
Institute de Chimie Physique
EPFL
Ecublens
1015 Lausanne
SWITZERLAND
FAX # 0041-21-6936100

Prof. Franz **Grieser**
School of Chemistry
University of Melbourne
Parkville
Victoria 3052
AUSTRALIA
FAX # 0061-3-3446233

Dr. Michael L. **Hair**
Xerox Research Center of Canada
2660, Speakinon Drive
Mississauga, Ontario L5K2L1
CANADA
FAX # 001-905-8227022

Prof. Shuichi **Hamada**
Dept. of Applied Chemistry
Science University of Tokyo
Kagurazaka 1-3, Shinjuku-ku, Tokyo
JAPAN
FAX # 0081-3-32352214

Dr. Gary **Hodes**
Dept. of Materials and Interfaces
Weizmann Institute of Science
76100 Rehovot
ISRAEL
FAX # 00972-8-344137

Dr. Dieter **Horn**
Polymer and Solid-State Physics
BASF Aktien Gesellschaft
67056 Ludwigshafen
GERMANY
FAX # 0049-621-6092281

Prof. Tatsuo **Ishikawa**
School of Chemistry
Osaka University of Education
Asahigaoka, Kashiwara-shi
Osaka 582
JAPAN
FAX # 0081-729-763269

Prof. W. **Janusz**
Dept. of Chemistry
Maria Curie Sklodowska University
Pl. Marii Curie Sklodowkiej, 3
20031 Lublin
POLAND
FAX # 0048-81-32811

Prof. Barry **Jennings**
English China Clays Int.
1015 Arlington Business Park -
Theale
Reading RG7 4SA
GREAT BRITAIN
FAX # 0044-734-309500

Prof. C. **Kaiser**
Institut fur Anorganische Chemie
J. Gutemberg Universitat
Postfach 3980
55099 Mainz
GERMANY
FAX # 0049-6131-392710

Prof. N. **Kallay**
Dept. of Chemistry
University of Zagreb
P.O.BOX 163
41001 Zagreb
CROATIA
FAX # 00385-41-446528

Dr. Prashant **Kamat**
Radiation Laboratory
University of Notre Dame
Notre Dame, IN 46556-0579
USA
FAX # 001-219-6318068

Prof. Kenneth **Klabunde**
Dept. of Chemistry
Kansas State University
Willard Hall
Manhattan, KS 66506-3701
USA
FAX # 001-913-5326666

Dr. I. **Kobal**
Josef Stefan Institute
Universiy of Ljubljana
P.O.BOX 100
Ljubljana
SLOVENIA
FAX # 00386-61-219385

Prof. Kijiro **Kon-no**
Dept. of Industrial Chemistry - Faculty of Engineering
Science University of Tokyo
Kagurazaka 1-3, Shinjuku-ku
Tokyo
JAPAN
FAX # 0081-3-32604294

Dr. M. **Kosmulski**
Institute of Catalysis
Polish Academy of Sciences
Lublin
POLAND
FAX # 0048-81-32811

Prof. Uwe **Kreibig**
Physical Institute
RWTH
Sommerfeld Strasse 14
52074 Aachen
GERMANY
FAX # 0049-241-8888

Prof. Boris **Levy**
Dept. of Chemistry
Boston University
590, Commonwealth Av.
Boston, MA 02215
USA
FAX # 001-617-3536466

Dr. Paul A. **Liberti**
Immunikon Corp.
1310 Masons Mill II
Huntingdon Valley PA 19006
USA
FAX # 001-215-9380437

Prof. Ray **Mackay**
CAMP
Clarkson University
Potsdam, NY 13699-5665
USA
FAX # 001-315-2687615

Prof. Egon **Matijevic**
CAMP
Clarkson University
Potsdam, NY 13699-5665
USA
FAX # 001-315-2686656

Dr. Dani **Meisel**
Chemistry Division
Argonne National Laboratory
Argonne, IL 60439
USA
FAX # 001-708-2524993

Dr. Olga **Micic**
National Renewable Energy Lab.
1617 Cole Blvd.
Golden, CO 80401
USA
FAX # 001-303-3846150

Prof. Claudio **Minero**
Dipartimento di Chimica Analitica
Università di Torino
via Pietro Giuria 5
10125 TORINO
ITALY
FAX # 0039-11-6707615

Dr. Marcel **Minor**
Agric. University
Dept.Phys. Colloid Chem.
P.O. BOX 8038
Wageningen
HOLLAND
FAX # 0031-8370-83777

Prof. Claudio **Morterra**
Dipartimento di Chimica Fisica
Università di Torino
via Pietro Giuria 7
10125 TORINO
ITALY
FAX # 0039-11-6707589

Dr. Arthur **Nozik**
National Renewable Energy Lab.
1617 Cole Blvd.
Golden, CO 80401
USA
FAX # 001-303-4995707

Prof. Ezio **Pelizzetti**
Dipartimento di Chimica Analitica
Università di Torino
via Pietro Giuria 5
10125 TORINO
ITALY
FAX # 0039-11-6707615

Prof. Maria Paula **Pileni**
Laboratoire SRSI
Université Pierre et Marie Curie
Bat. F74 - 4 Place Jussieu
75230 Paris Cedex
FRANCE
FAX # 0033-44-272515

Prof. Terry A. **Ring**
Dept. of Chemical Engineering
University of Utah
Salt Lake City, UT 84112
USA
FAX # 001-801-5818692

Prof. Helmut **Schmidt**
Institut fur neue Materialien
Universitat des Saarlandes
66041 Saarbrucken
GERMANY
FAX # 0049-681-3025223

Dr. Carlos J. **Serna**
Area de Ciencia y Technologia de Materiales
CSIC
Calle Serrano 115
28006 Madrid
SPAIN
FAX # 0034-1-5624526

Prof. Nick **Serpone**
Dept. of Chemistry
Concordia University
1455 de Maisonneuve Blvd. W.
Montreal QUEBEC H3G1M8
CANADA
FAX # 001-514-8482868

Prof. E.D. **Shchukin**
Dept. of Geography and
Environmental Engineering
John Hopkins University
Charles and 34th Streets
Baltimore MD 21218
USA
FAX # 001-410-5168996

Prof. Johan **Sjoblom**
Dept. of Chemistry
University of Bergen
Allegt 41
5007 Bergen
NORWEY
FAX # 0047-55-322446

Dr. Gro **Stakkestad**
Prototech A/S
Fantoftvegen 38 - P.O.BOX 12
N-5036 Fantoft, Bergen
NORWEY
FAX # 0047-55-574114

Prof. Tadao **Sugimoto**
Institut for Advanced Material
Processing
Tohoku University
Katahira 2-1-1
Aobaku, Sendai 980
JAPAN
FAX # 0081-22-2610938

Prof. Kerry J. **Thomas**
Dept. of Chemistry
University of Notre Dame
Notre Dame - IN 46556-0579
USA
FAX # 001-219-6318068

Prof. Micha **Tomkiewicz**
Dept. of Physics
Brooklyn College of CUNY
Brooklyn, N.Y. 11210
USA
FAX # 001-718-9514645

Prof. Naoki **Toshima**
Dept. of Industrial Chemistry
The University of Tokyo
Hongo 7-3, Bunkyo-ku
Tokyo
JAPAN
FAX # 0081-3-38129254

Dr. Brian **Vincent**
School of Chemistry
University of Bristol
Bristol BS81TS
GREAT BRITAIN
FAX # 0044-272-250612

Prof. Agienus **Vrij**
Rijksuniversiteit
Vant'Hoff Laboratorium
Padualaan 8
3584 Utrecht 2506
HOLLAND
FAX # 0031-30-533870

Prof. Frank **Willig**
Fritz-Haber Institute
Faradayweg 4-6
1000 Berlin 33
GERMANY
FAX # 0049-30-8305520

LIST OF OTHER PARTICIPANTS

Dr. Ralf **Anselmann**
Postfach 4119
E. Merck
Frankfurter Strasse 250
6100 DARSTADT 1
GERMANY
FAX # 0049-6151-726649

Dr. Helmut **Auweter**
Dept. ZKM - Bldg. G201
BASF AG
D-67056 Ludwigshafen
GERMANY
FAX # 0049-621-6021863

Prof. Suna **Balci**
G.U.M.M.F.
Kinya Muh. Bol - Maltepe
Ankara
TURKEY
FAX # 0090-312-2308434

Dr. Nazan **Bergizadi**
Eczacivik Fakultesi
Istambul Universitesi
34452 Beyazit - Istambul
TURKEY
FAX # 0090-212-5190812

Dr. Magali **Boutonnet Kizling**
Chemical Eng. & Tech.
Royal Institute of Technology
10044 Stockholm
SWEDEN
FAX # 0046-8-108579

Dr. **Camara**
ZDH/ZDP M311
BASF AG
D-67056 Ludwigshafen
GERMANY
FAX # 0049-621-6079398

Dr. T. **Chopin**
Centre de Recherches d'Aubervilliers
Rhone Poulenc Research
52 Rue de la Haie-Coq
93308 Aubervilliers CEDEX
FRANCE
FAX # 0033-1-49376100

Dr. Felipe **Costa**
Dept. de Fisica
Universidade Do Minho
Largo do Paco
4709 Braga Codex
PORTUGAL
FAX # 00351-53-604339

Prof. Shlomo **Efrima**
Dept. of Chemistry
Ben-Gurion University
P.O.BOX 653
Beer-Sheva 84105
ISRAEL
FAX # 00972-7-274359

Dr. Valentin **Erochenko**
DLD Int.
135 Boulevard Hassmann
75008 PARIS
FRANCE
FAX # 0033-1-45621515

Dr. Sasha **Gorer**
Dept. of Materials
Weizmann Institute of Science
Rehovot 76100
ISRAEL
FAX # 00972-8-344137

Dr. D. **Gulamova**
3/4 Pushkin Str.
Taskent
UZBEKISTAN, CIS
FAX # 007-3712-367880

Prof. H. **Hofmann**
Dept. des Materiaux
EPFL
MX-D Ecublens
CH-1015 Lausanne
SWITZERLAND
FAX # 0041-21-6933089

Dr. H. **Jachow**
ZAA/I-J653
BASF AG
D-6700 Ludwigshafen
GERMANY
FAX # 0049-621-6072794

Dr. Manfred **Kuehnle**
XMX Corp.
46 Manning Road
Billerica, MA 01821
USA
FAX # 001-508-6708448

Dr. Henri **Lehn**
DLD Int.
135 Boulevard Hassmann
75008 PARIS
FRANCE
FAX # 0033-1-45621515

Dr. Hartmann **Leube**
ZKD/S-B1
BASF AG
D-67056 Ludwigshafen
GERMANY
FAX # 0049-621-6040855

Dr. Valter **Maurino**
Dipartimento di Chimica Analitica
Università di Torino
via Pietro Giuria 5
10125 TORINO
ITALY
FAX # 0039-11-6707615

Dr. Mary D. **Meza**
Bangs Laboratories, Inc.
979 Keystone Way
Carmel, IN 46032-2823
USA
FAX # 001-317-5758801

Dr. Herbert **Neumeister**
Forschungszentrum Julich GMBH
Inst. Angewandte Physikalische
Chemie
52425 Julich
GERMANY
FAX # 0049-2461-612493

Dr. Elena **Obraztsova**
General Physics Institute
Russian Academy of Sciences
38 Vavilov Str.
117942 Moscow
RUSSIA
FAX # 007-095-1350270

Dr. Sefer Cem **Okumus**
Esentepe Kampusu Serdivan
Sakarya Universitesi
54040 Adapazari
TURKEY
FAX # 0090-264-3431275

Dr. Gilles **Picard**
TecnoBiochip
via della Marina
57030 Marciana Elba (LI)
ITALY
FAX # 0039-565-901136

Dr. Valeria **Porcarelli**
Istituto di Metallurgia
Università di Bologna
Vilae Risorgimento 4
40136 Bologna
ITALY
FAX # 0039-51-6443467

Dr. Francesca **Porta**
Dip. Chimica Inorganica e
Metalloorganica
Università di Milano
via Venezian 21
20133 MILANO
iTALY
FAX # 0039-2-2362748

Dr. Frank **Runge**
Dept. ZKM - Bldg. G201
BASF AG
D-67056 Ludwigshafen
GERMANY
FAX # 0049-621-6021863

Prof. Brian **Scarlett**
Dept. of Chemical Process Technology
Delft University
P.O.BOX 5045
2600 GA Delft
HOLLAND
FAX # 0031-15-784452

Dr. Jean-Christoph **Simon**
Centre d'Etudes du Ripault
B.P.16
37260 Monts
FRANCE
FAX # 0033-47-345104

Dr. Barbara **Sultzberger**
EAWAG, ETH
Ueberland Strasse 133
8600 Dubendorf - Zurich
SWITZERLAND
FAX # 0041-1-8235028

Prof. Jerzy **Szczypa**
Institut of Chemistry
Maria Curie-Sklodowska University
Pl. Marii Curie-Sklodowskiej 3
20031 LUBLIN
POLAND
FAX # 0048-8132811

Dr. Isobel **Tjedor-Tjedor**
Water Chemistry Program
University of Wisconsin
660 North Park Street
Madison, WI 53706-1484
USA
FAX # 001-608-2620454

Dr. Gabrielle **Walker**
New Scientist
Kings Reach Tower - Stamford Street
London SE1 9LS
ENGLAND
FAX # 0044- 171-2615134

Dr. Robertino **Zanoni**
Dip. di Chimica
Università La Sapienza
P.le A. Moro 5
00185 Roma
ITALY
FAX # 0039-6-490631

INTERNALLY AND EXTERNALLY COMPOSITE MONODISPERSED COLLOID PARTICLES

EGON MATIJEVIĆ
Center for Advanced Materials Processing
Clarkson University, Potsdam, NY 13699–5814, USA

1. Introduction

Composite colloidal particles represent systems of interest in numerous applications and they also offer challenging problems, which have yet to be resolved. The availability of uniform dispersions is an obvious prerequisite in order to quantify phenomena related to the functions and properties of such materials.

Internally composite particles may consist of compounds of two or more cations and the same anion, or of one cation and a combination of anions. The content of constituent species in these solids may vary, often depending on the method of preparation. Alternately, composite particles can be stoichiometrically precisely defined. Structurally, these dispersions may consist of amorphous matter, single crystals, or solids that show X-ray characteristics of known crystalline materials, but are actually built of a large number of much smaller subunits.

Externally composite particles consist essentially of cores covered with a distinct coating of a different chemical composition. The two parts of these solids can be both inorganic, or either one can be organic and the other inorganic. Obviously, many properties of such particles will depend on the composition and the size of cores and shells. They may also be altered by post-treatment, such as calcination at elevated temperatures.

A special case of such dispersions deals with hollow particles in which the core is a spherical void.

The preparation of a variety of uniform composite particles of different shapes and some of their properties are described below.

2. Internally Composite Colloidal Particles

2.1 PREPARATION

Many different approaches may yield finely dispersed particles of internally mixed composition. Only precipitation in homogeneous solutions and the aerosol techniques will be dealt with in this article.

E. Pelizzetti (ed.), Fine Particles Science and Technology, 1–12.

In the precipitation method composite particles can form by simply mixing convenient reactants, such as – for example – acidic solutions of an aluminum salt and alkaline solutions of sodium silicate to produce colloidal alumosilicates [1]. Such procedures lead, as a rule, to ill defined dispersions in composition and morphology. Much better results in terms of particle uniformity can be achieved by controlled double or triple jet precipitation (DJCP or TJCP) processes [2].

To generate "monodispersed" systems, methods based on kinetically controlled chemical reactions in homogeneous solutions, which result in supersaturation of the desired products, have been proven most successful [3,4]. In order to produce composite particles, the solutions must initially contain, or generate *in situ*, a mixture of cations or anions expected to be present in the resulting particles.

In the aerosol technique droplets of a mixed composition are contacted with a gaseous co–reactant, converting them into solid spheres [5,6]. The size distribution of the latter is determined by the original liquid aerosol.

2.2. PRECIPITATION METHOD

2.2.1. *Particles of Stoichiometric Composition*

In some instances, the least soluble solid product, resulting from chemical reactions in homogeneous solutions, can be a stoichiometrically well defined compound. An example of such a process is the formation of ferric basic sulfates, which have the composition

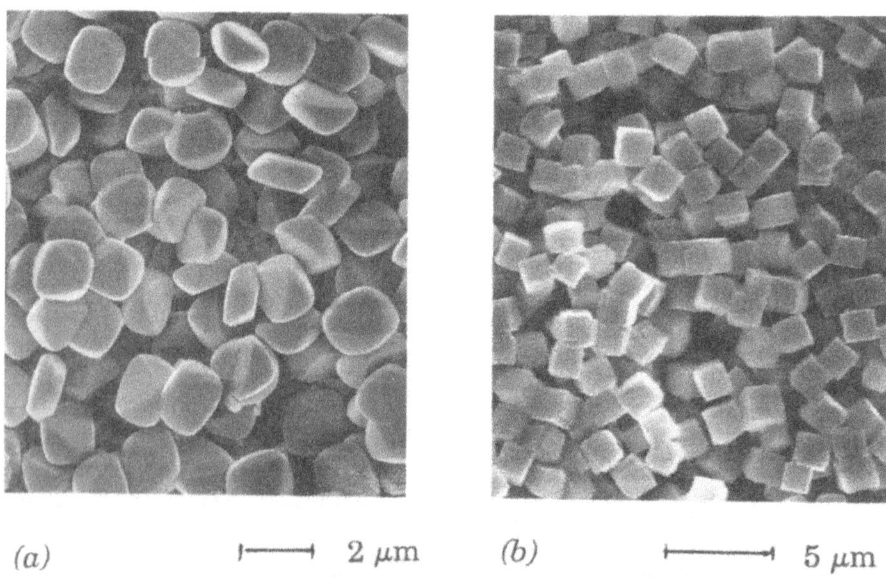

Figure 1.(a) Scanning electron micrograph (SEM) of hexagonal alunite, $Fe_3(SO_4)_2(OH)_5 \cdot 2H_2O$, particles obtained by aging at 80°C for 1.5 h a solution 0.18 mol dm^{-3} in $Fe(NO_3)_3$ and 0.53 mol dm^{-3} in $(NH_4)_2SO_4$ [7]. *(b)* SEM of neighborite, $NaMgF_3$, particles obtained by aging at 80°C for 3 h a solution 1.6 x10^{-1} mol dm^{-3} in NaF and 4.0 x 10^{-2} mol dm^{-3} in $MgCl_2$.

of alunites (jarosites), precipitated by forced hydrolysis of iron(III) salt solutions in the presence of sulfate ions. Figure 1a illustrates the so produced particles of $Fe_3(SO_4)_2(OH)_5 \cdot 2H_2O$ [7]. Another example is offered in Figure 1b, which shows the scanning electron micrograph (SEM) of uniform colloidal crystals of neighborite, $NaMgF_3$.

Obviously, in the first case the compound includes two different anions and in the second case it contains two different cations in fixed stoichiometric ratios.

The CDJP process was employed to generate spherical pure or doped barium titanate particles of narrow size distribution, as illustrated in Figure 2a [2]. Figure 2b compares sintered densities of two commercial barium titanates with that prepared by the CDJP technique. It is quite apparent that the latter sample reaches theoretical density at a much lower temperature than the other materials. The dielectric properties of these solids are very sensitive to the content in cations and to their microstructure. The fractured pellets produced from powders illustrated in Figure 2 show a fairly uniform small grain size (1–2 μm) favorable for the manufacture of multilayered capacitors. Indeed, the relative dielectric constant of 5000 at 20°C of this sintered $BaTiO_3$ is the highest value reported in the literature [2].

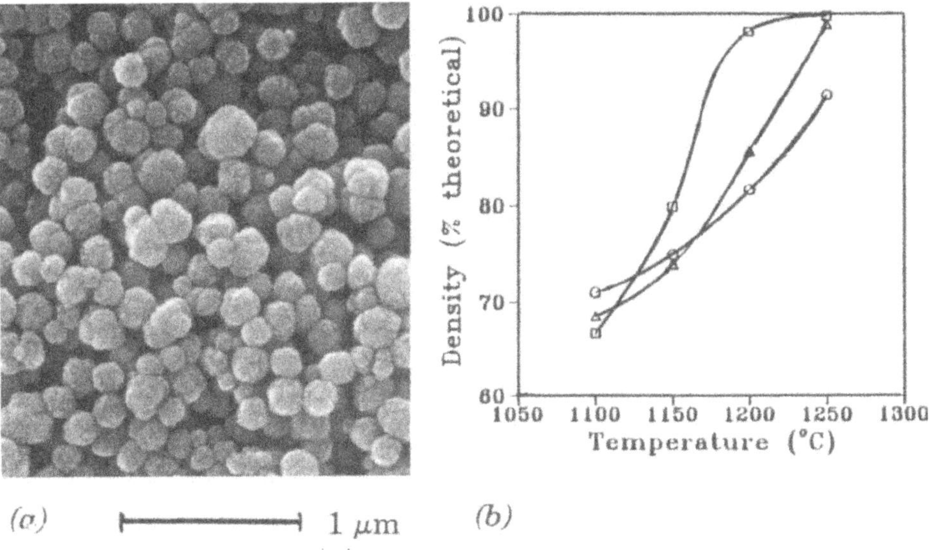

Figure 2.(a) SEM of $BaTiO_3$ particles prepared by the CDJP process, by reacting $BaCl_2$, $TiCl_4$, and NaOH at 85°C for 8 min. *(b)* The plot of the density as a function of sintering temperature of pellets made from this powder (□), and of two different commercial powders (Δ,O) [2].

2.2.2. *Particles of Variable Composition*

Precipitations from solutions of mixed metal salts or organometallic compounds most frequently yield particles, which differ in molar ratios of constituent cations from those in the original media. Indeed, the composition of such solids varies with the reaction time, which leads to the conclusion that the resulting particles, although quite uniform in shape and size, are internally chemically inhomogeneous.

Figure 3(a) is the SEM of a dispersion obtained by aging a solution of aluminum *tert*-butoxide, $Al(t\text{-}OBu)_3$, and tetraethyl orthosilicate (TEOS) in a *tert*-butoxide/water mixture containing sulfate ions [8]. A variation of experimental conditions in terms of reactant concentrations yields spherical silica/alumina particles, but their composition changes with growth and it differs in Al and Si content from that in the solution [8]. The internal inhomogeneity is further documented by the electrokinetic properties. Figure 3b shows that particles prepared in media having a fourfold excess of TEOS (an excess albeit different from that of silica in the solids) have mobilities characteristics of pure $Al(OH)_3$ (with an isoelectric point of 9.2). Particles precipitated from solutions of the initial molar ratio $[TEOS]/[Al(t\text{-}OBu)_3] = 8$ remain negatively charged over the entire pH range, as one would expect for pure silica [8].

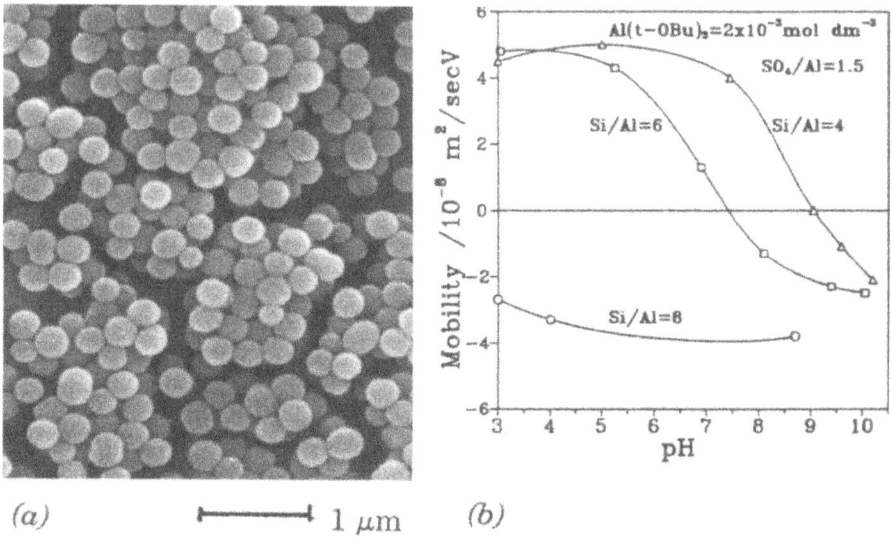

Figure 3(a) SEM of mixed silica–alumina particles obtained by aging at 98 °C for 18 h a system consisting of 2×10^{-3} mol dm^{-3} $Al(t\text{-}OBu)_3$, 8×10^{-3} mol dm^{-3} TEOS, and 3×10^{-3} mol dm^{-3} $(NH_4)_2SO_4$. *3(b)* Electrophoretic mobilities as a function of the pH of dispersions prepared from solutions of initial molar concentration ratios $[TEOS]/[Al(t\text{-}OBu)_3]$ of 4 (Δ), 6 (□), and 8 (O), at a constant concentration of $[Al(t\text{-}OBu)_3] = 2 \times 10^{-3}$ mol dm^{-3} and the ratio $[SO_4]/[Al] = 1.5$, all aged at 98°C for 24 h [8].

The same effects in terms of the variable internal composition have been established with a number of other systems, including different metal (hydrous) oxides [9–11], or metal phosphates [12]. For example, the huge change in the composition with reaction time (which is equivalent to the particle growth) of spherical particles consisting of aluminum basic sulfate and yttrium basic carbonate, is illustrated in Figure 4a [11], and of nickel basic phosphate and cadmium phosphate particles in Figure 4b [12]. The observed inhomogeneities are obviously caused by different rates of precipitation of the constituent species, which depend on their relative solubility and the concentration of reactants in solutions; the latter condition changes as the reaction progresses with time. All above described dispersions consist of spherical amorphous particles. However, on calcination these solids may crystallize into stoichiometrically defined mixed metal oxides, as is the case with Y/Cu- and La/Cu-oxide systems [9].

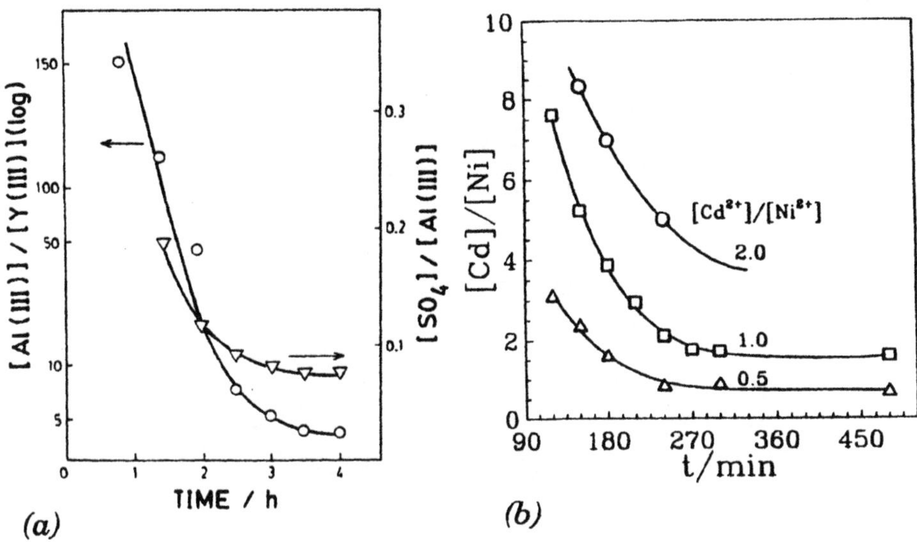

Figure 4(a) Changes of molar ratios [Al(III)]/[Y(III)] and $[SO_4]/[Al(III)]$ in solids obtained by aging at 90°C for increasing periods of time acidic solutions containing 5×10^{-3} mol dm^{-3} Al^{3+} and 2×10^{-3} mol dm^{-3} Y^{3+} ions, urea, and PVP [11]. *(b)* The change in molar ratios of mixed nickel basic phosphate and cadmium phosphate particles, precipitated in solutions of initial molar ratios $[Cd^{2+}]/[Ni^{2+}]$ = 2.0, 1.0, and 0.5, respectively, aged at 80°C [12].

The difference in bulk and surface compositions has been also found in precipitated particles that exhibit crystallinity by X-ray diffraction. For example, the dependence of electrophoretic mobility on the pH was the same for several nickel ferrites of varying content in nickel [13].

2.3. AEROSOL METHOD

Two approaches may be used to generate aerosol droplets of mixed compositions. In the first method the vapors of two (or more) volatile compounds are co-condensed, while in the second method a solution of these compounds is nebulized. There are advantages and disadvantages in either technique. The condensation process can yield droplets of very narrow size distributions, but their composition is difficult to control. In contrast, droplets obtained by mechanical means tend to be polydisperse, but their composition is fixed.

Figure 5 illustrates particles obtained by nebulizing a liquid mixture of TEOS and titanium(IV) ethoxide, $Ti(OEt)_4$, and hydrolyzing the droplets with water vapor [6]. The overall atomic ratio in these particles was found to be Ti:Al=0.035:1.0, while the ratio of the same atoms in the surface layer, determined by ESCA, was Ti:Al=0.48:1 [14]. Obviously, the so prepared mixed oxides were internally inhomogeneous.

Similar results were observed with particles obtained by co-condensation of vapors of TEOS and Al(*sec*-OBu)$_3$ and subsequent exposure to water vapor.

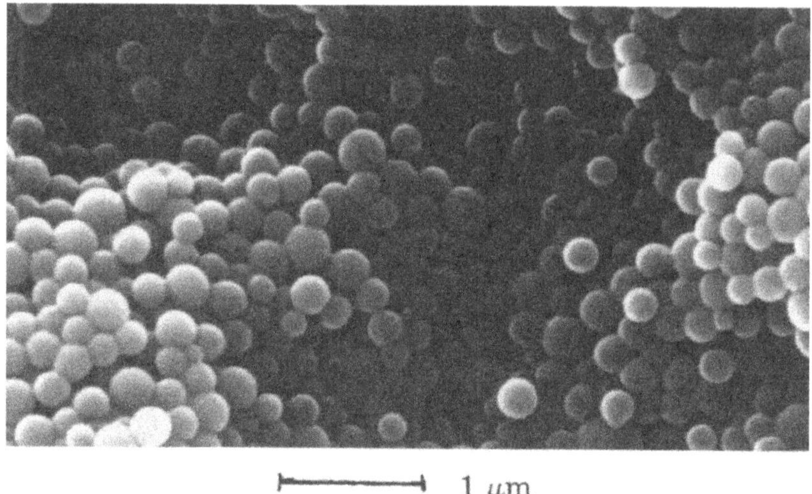

1 μm

Figure 5 SEM of TiO_2/SiO_2 particles obtained by hydrolysis of aerosol droplets consisting of a mixture of TEOS and $Ti(OEt)_4$ [6].

3. Coated Particles

There are many reasons for covering particles with a layer of different composition. In doing so one may affect their reactivity, as well as optical, magnetic, adsorptive, conductive, and other properties. The coating can also be carried out to protect the cores from possible adverse effects of the environment, or to control the stability of their dispersions.

In principle, coatings can be achieved by depositing small particles on larger ones of a different kind [15]. Such layers are, as a rule, inhomogeneous and their thickness is difficult to control. In the other method the outer layer is formed *in situ* on preformed particles, which can be both accomplished by precipitation and by chemical reactions in aerosols. These techniques have resulted in very uniform coatings, the width of which could be controlled over a reasonably broad range. It is noteworthy, that both inorganic and organic compounds can be used in such preparations, and no chemical interaction between the core and coating materials is required in order to achieve the deposition. It appears that the cores simply act as "seeds" for the precipitation of outer layers.

3.1 INORGANIC COATINGS ON INORGANIC OR ORGANIC CORES

To coat particles by precipitation, cores are dispersed in a solution containing the electrolyte or an organometallic compound, that will yield the outer layer, and the system is then treated under the similar conditions as if no preformed solids were present. In order to produce only coated particles, a balance must be established in terms of the amount of the core material and the solution concentration of the compound that forms the outer layer.

Figure 6 shows silica (a) and a polymer latex (b) covered with yttrium basic carbonate. Despite chemically different cores, the coatings are achieved under similar experimental conditions [16,17].

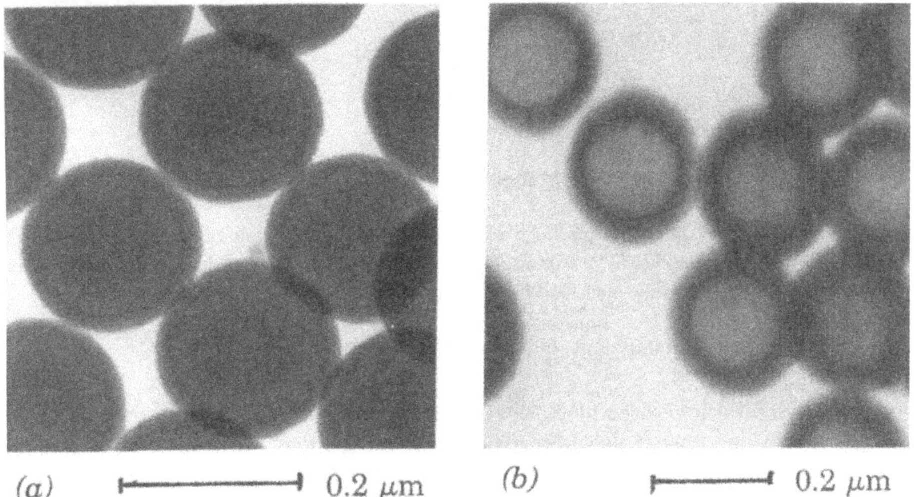

Figure 6.(a) TEM of silica particles coated with yttrium basic carbonate by aging at 90°C an aqueous dispersion of SiO_2 cores containing $Y(NO_3)_3$ and urea [16]. *(b)* TEM of a polystyrene latex coated with yttrium basic carbonate under similar conditions as in *(a)* [17].

Figure 7(a) illustrates ellipsoidal hematite ($\alpha-Fe_2O_3$) particles with a homogeneous layer of silica [19]. The same kind of dispersions can then be calcined at elevated temperatures in the presence of hydrogen. This treatment reduces the iron oxide core to iron and seals the silica shell, resulting in stable magnetic particles [20], shown in Figure 7(b).

The natures of other coated particles can be altered by post-treatments. For example, on heating, the shell of yttrium basic carbonate on iron oxide is converted to Y_2O_3 [20], while the same coating on silica yields on calcination yttrium silicate. It was demonstrated that ceramic bodies made from uniform coated particles of silica on yttria or vice versa greatly differed in their properties, which also depended on the relative volumes of the external and internal parts of the precursor material [16].

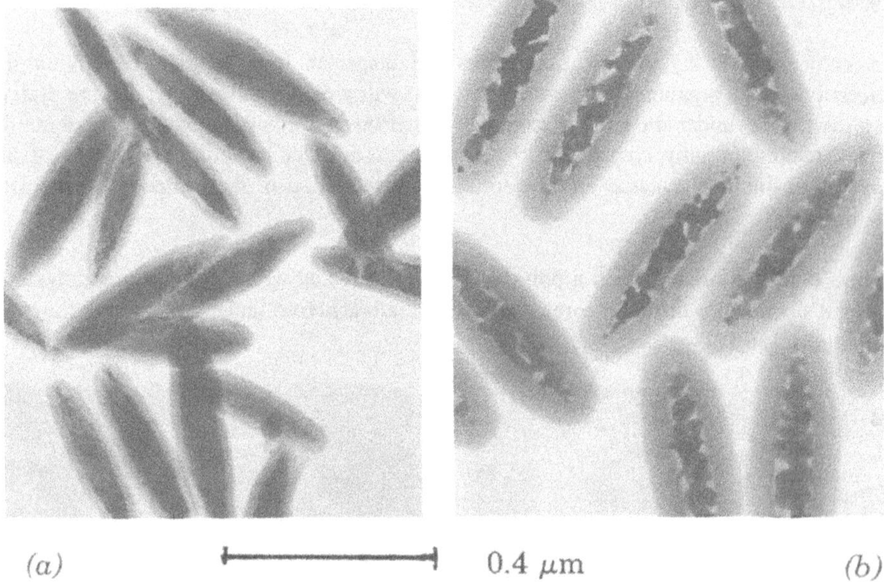

Figure 7. (a) TEM of hematite particles coated with silica by aging at 40°C a dispersion of α–Fe_2O_3 in the presence TEOS, propanol and ammonia [18]. *(b)* TEM of iron coated with silica, obtained by heating powders, such as illustrated in *(a)*, at 450°C in hydrogen [19].

3.2 ORGANIC COATINGS ON INORGANIC CORES

It is equally useful to coat particles with polymers, which can be achieved by adsorption of macromolecular species (such as proteins) on inorganic cores [21], or by adsorption of monomers with subsequent polymerization. The latter stage normally requires a catalyst, which may be the carrier particle itself [23], as in coating of iron, cerium, or copper oxides with polypyrrole [22, 23], or of copper oxide with polyaniline [24]. In some other cases a catalyst needs to be introduced into the reaction system, as was done in forming a shell of poly(divinylbenzene) on $SiO_2/Al(OH)_3$ particles in the presence of vinyl pyridine and benzoyl peroxide [25]. In some cases oxygen can enhance the polymerization process [23,24].

Figure 8(a) illustrates CuO cores coated with polypyrrole [23]. While in this example the polymer shell is nonconductive, the polypyrrole layers on iron oxide were conductive [22].

The polymer coatings can also be produced by the aerosol technique. The TEM in Figure 8(b) shows such particles consisting of titania cores covered with polyurea [26]. The procedures involves first generating titania cores from droplets of Ti(IV)–ethoxide, $Ti(OEt)_4$, by exposing them to water vapor, followed by condensation of hexamethylene-diisocyanate on the solids so formed, and finally reacting the particles, covered with the monomer film, with ethylenediamine vapor [26]. The entire production takes place continuously in an inert gas stream with a reaction time measured in seconds.

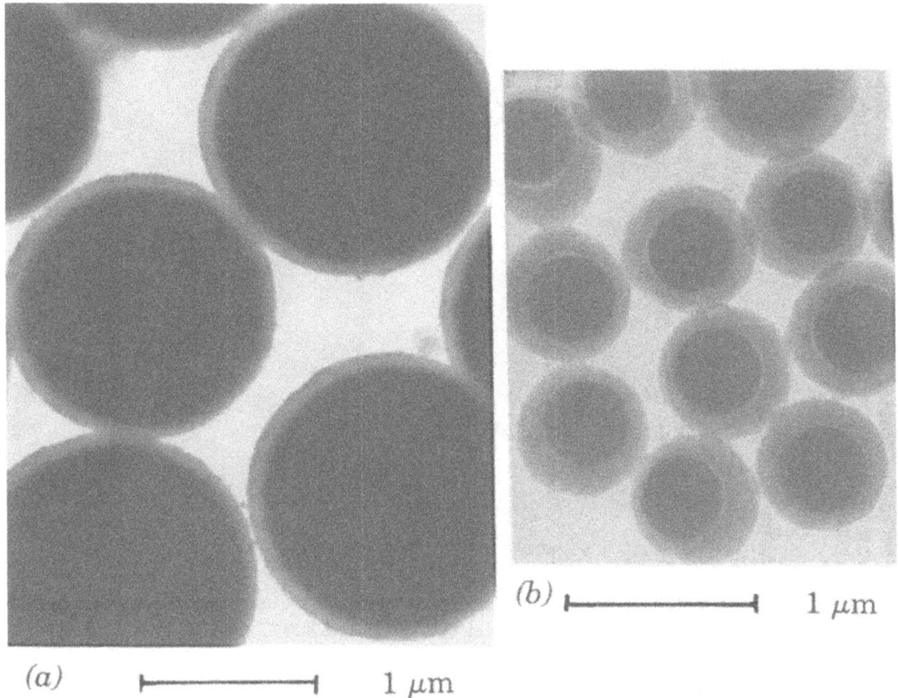

Figure 8.(a) TEM of CuO particles coated with polypyrrole by adsorbing pyrrole on the cores from 5% ethanol solutions in the presence of polyvinyl alcohol (PVA) and aging the system at 100°C for 24h [23]. *(b)* TEM of TiO_2 particles coated with polyurea by the aerosol technique as described in the text [26].

4. Hollow Particles

Hollow particles represent a special case of composite systems, since the core consists of a void. Such dispersions are characterized by low density and unique optical and other properties.

Uniform spherical shells of yttrium and zirconium oxides were obtained by calcination at elevated temperatures in air polystyrene latex particles coated with the respective metal compounds [27,28]. Under proper conditions the polymer was burnt away, leaving behind hollow spheres, the dimensions of which depended on the diameter of latex part-icles and the thickness of the shells used as precursors.

Figure 9 displays such particles of yttrium oxide [27]. On dissolving these hollow particles, no polymer could be detected in the resulting solutions.

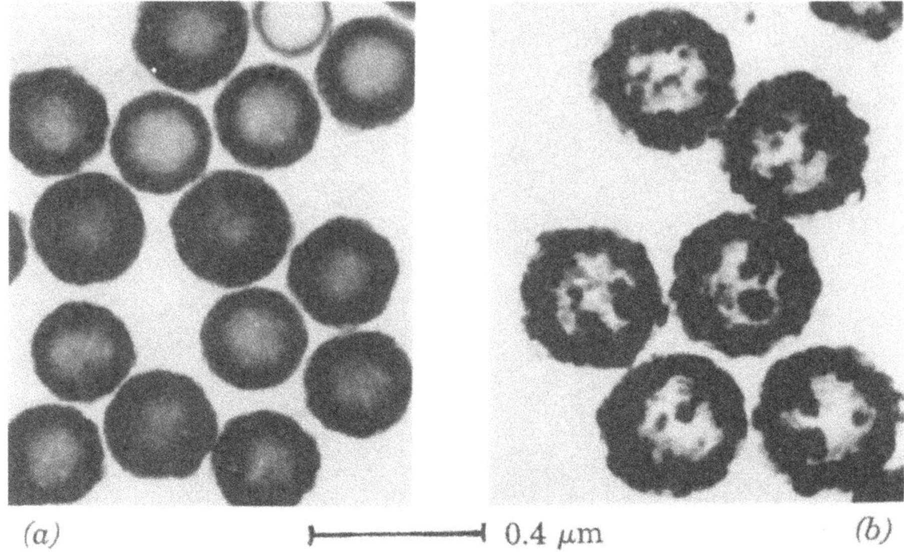

Figure 9. TEM of hollow spheres of yttrium oxide obtained by calcining polystyrene particles coated with yttrium basic carbonate for 3h at 600°C (a) and 800°C (b) [27].

5. Concluding Remarks

This short review should clearly demonstrate that the described techniques make it possible to generate in a reproducible manner a large number of internally and externally composite particles of uniform size and shape. By creating such systems, different properties of dispersed matter can be altered to fit specific needs. However, it is fair to state that the described developments represent only a beginning of an area of colloid science with countless opportunities to follow.

6. References

1. Brace, R. and Matijević, E. (1977) Coprecipitation of Silica with Aluminum Hydroxide, *Colloid & Polymer Science* **255**, 153–161.

2. Her, Y.-S., Matijević, E., and Chon, M.C. (submitted) Preparation of Well Defined Colloidal Barium Titanate Crystals by the Controlled Double Jet Precipitation, *Journal of Materials Research*

3. Matijević, E. (1993) Preparation and Properties of Uniform Size Colloids, *Chemistry of Materials* **5**, 412–426.

4. Matijević, E. (1994) Uniform Colloid Dispersions – Achievements and Challenges, *Langmuir* **10**, 8–16.

5. Visca, M. and Matijević, E. (1979) Preparation of Uniform Colloidal Dispersions by Chemical Reaction in Aerosols. I. Spherical Particles of Titanium Dioxide, *Journal of Colloid and Interface Science* **68**, 308–319.

6. Matijević, E., Zhong, Q., and Partch, R.E. (1995) Preparation of Uniform Colloidal Dispersions by Chemical Reaction in Aerosols. VI. Silica/Titania Composite Particles, *Aerosol Science and Technology* **22**, 162–171.

7. Matijević, E., Sapieszko, R.S., and Melville, J.B. (1975) Ferric Hydrous Oxide Sols. I. Monodispersed Basic Iron(III) Sulfate Particles, *Journal of Colloid and Interface Science* **50**, 567–581.

8. Nishikawa, S. and Matijević, E. (1994) Preparation of Uniform Monodispersed Spherical Silica–Alumina Particles by Hydrolysis of Mixed Alkoxides, *Journal of Colloid and Interface Science* **165**, 141–147.

9. Ribot, F., Kratohvil, S., and Matijević, E. (1989) Preparation and Properties of Uniform Mixed Colloidal Particles. VI. Copper(II)–Yttrium(III) and Copper(II)–Lanthanum(III) Compounds, *Journal of Materials Research* **4**, 1123–1131.

10. Aiken, B., Hsu, W.P., and Matijević, E. (1988) Preparation and Properties of Monodispersed Colloidal Particles of Lanthanide Compounds. III. Y(III) and Mixed Y(III)/Ce(III) Systems, *Journal of the American Ceramic Society* **71**, 845–853.

11. Hsu, W.P., Wang, G., and Matijević, E. (1991) Preparation and Properties of Uniform Colloidal Particles of Mixed Composition. VII. Aluminum and Yttrium Compounds, *Colloids and Surfaces* **61**, 255–267.

12. Quibén, J. and Matijević, E. (1994) Preparation and Properties of Uniform Colloidal Particles of Mixed Composition. VIII. Cadmium and Nickel Phosphates, *Colloids and Surfaces* **82**, 237–246.

13. Regazzoni, A.E. and Matijević, E. (1982) Formation of Spherical Colloidal Nickel Ferrite Particles as Model Corrosion Products, *Corrosion* **38**, 212–218.

14. Ingebrethsen, B.J., Matijević, E., and Partch, R.E. (1983) Preparation of Uniform Colloidal Dispersions by Chemical Reactions in Aerosols. III. Mixed Titania/Alumina Colloidal Spheres, *Journal of Colloid and Interface Science* **95**, 228–239.

15. Bleier, A. and Matijević, E. (1978) Heterocoagulation. III. Interactions of Polyvinyl Chloride Latex with Ludox Silica, *Journal of the Chemical Society, Faraday Transactions I* **74**, 1346–1359.

16. Giesche, H. and Matijević, E. (1994) Preparation, Characterization, and Sinterability of Well Defined Silica/Yttria Powders, *Journal of Materials Research* **9**, 436–450.

17. Kawahashi, N. and Matijević, E. (1990) Preparation and Properties of Uniform Coated Colloidal Particles. V. Yttrium Basic Carbonate on Polystyrene Latex, *Journal of Colloid and Interface Science* **138**, 534–542

18. Ohmori, M. and Matijević, E. (1992) Preparation and Properties of Uniform Coated Inorganic Colloidal Particles. VII. Silica on Hematite, *Journal of Colloid and Interface Science* **150**, 594–598.

19. Ohmori, M. and Matijević, E. (1993) Preparation and Properties of Uniform Coated Inorganic Particles. 8. Silica on Iron, *Journal of Colloid and Interface Science* **160**, 288–292.

20. Aiken, B. and Matijević, E. (1988) Preparation and Properties of Uniform Coated Colloidal Particles. IV. Yttrium Basic Carbonate and Oxide on Hematite, *Journal of Colloid and Interface Science* **126**, 645–649.

21. Johnson, J.E. and Matijević, E. (1992) Interactions of Proteins with Uniform Colloidal Hematite and Chromium Hydroxide Particles. I. Adsorption, *Colloid & Polymer Science* **270**, 353–363.

22. Partch, R., Gangolli, S., Matijević, E., Cai, W., and Arajs, S. (1991) Conducting Polymer Composites. I. Surface Induced Polymerization of Pyrrole on Iron and Cerium(IV) Oxide Particles, *Journal of Colloid and Interface Science* **144**, 27–35.

23. Huang, C.-L. and Matijević, E. (1995) Coating of Uniform Inorganic Particles with Polymers. III. Polypyrrole on Different Metal Oxides, *Journal of Materials Research*, **10**, 1327–1336.

24. Huang, C.-L., Partch, R.E. and Matijević, E. (1995) Coating of Uniform Inorganic Particles with Polymers. II. Polyaniline on Copper Oxide, *Journal of Colloid and Interface Science* **170**, 275–283.

25. Oyama, H.T., Sprycha, R., Xie, Y., Partch, R.E. and Matijević, E. (1993) Coating of Uniform Inorganic Particles with Polymers. I, *Journal of Colloid and Interface Science* **160**, 298–303.

26. Mayville, F.C., Partch, R.E. and Matijević, E. (1987) Preparation of Uniform Spherical Titania Particles Coated with Polyurea by the Aerosol Technique, *Journal of Colloid and Interface Science* **120**, 135–139.

27. Kawahashi, N. and Matijević, E. (1991) Preparation of Hollow Spherical Particles of Yttrium Compounds, *Journal of Colloid and Interface Science* **143**, 103–110.

28. Kawahashi, N. Persson, C. and Matijević, E. (1991) Zirconium Compounds as Coatings on Polystyrene Latex and as Hollow Spheres, *Journal of Materials Chemistry* **1**, 577–582.

is the characteristic particle dimension and

$$\varkappa^{-1} = \mathrm{Le} = \left(\varepsilon kT/8\pi e^2 I \right) \quad (1)$$

is the Debye screening length, ε is the dielectric constant, k the Boltzmann constant, T the absolute temperature, e the elementary charge, I the ionic strength of the solution.

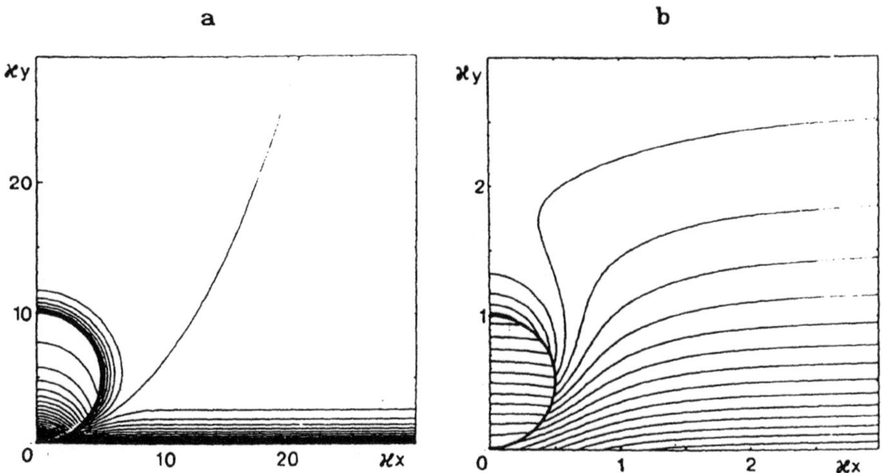

Figure 1. Electrostatic potential distribution within and outside a sphere in the vicinity of a planar interface calulated by solving the nonlinear PB equation in bispherical coordinates for $\varepsilon = 78$, $\varepsilon_p = 2.5$ (polystyrene), $\psi_1 e/kT = 3$, $\psi_2 e/kT = -1.5$: *Part a:* $\varkappa a = 5$ (microparticles) *Part b:* $\varkappa a = 0.5$ (nanoparticles).

For micro- and larger particles $\varkappa a$ assumes values larger than unity whereas for nanoparticles $\varkappa a$ may be comparable with unity and smaller for dilute electrolytes. This affects considerably the electrostatic field distribution in the vicinity of adsorbed particles as shown in *Figure 1* (the results were obtained from an exact numerical solution of the governing Poisson-Boltzmann equation in bipolar coordinate system). As can be observed, for microparticles the electrostatic field of the interface penetrates very little the adsorbed sphere. Hence, adsorption of further particles due to lateral repulsion is expected to occur at interface only, at distances exceeding Le. By contrast, for nanoparticles, the field from the interface may well cause an inversion of the surface potential of the adsorbed particles, especially for larger asymmetry of the particle and interface potentials. In this case, adsorption can occur at particles sitting at the interface, hence a formation of bi- or multi-layer coverage is expected. These effect can be described quantitatively if the interaction energy for various particle configurations is determined. This seems prohibitive in the general case due to the inherent many-body problem. One can formulate, however, approximate models of the EDL interactions the most useful being the linear superposition approach (LSA) and the Derjaguin summation methods. When combined with the additivity principle these models allow one to calculate interaction energy for complex configurations of spherical and anisotropic particles.

The great advantage of the Derjaguin model [4-7] is that it enables one to calculate the interactions between bodies of complicated geometry, e.g., spheres, cylinders, spheroids (for which exact solutions are often inaccessible) in terms of the very well studied cases of plate interactions. Thus, for arbitrary convex bodies one can formulate the simple expression for interaction energy ϕ in which the geometrical and electrostatic effects are separated from each other, i.e.,

$$\phi(h_o) = 2\pi G \int_{h_o}^{\infty} \phi_p(h)\, dh \qquad (2)$$

where h_o is the closest separation between particle surfaces, ϕ_p is the interaction energy between two flat plates and the geometrical factor G is expressed as [5,7]

$$G = \sqrt{\frac{R_1' R_1'' R_2' R_2''}{(R_1' + R_2')(R_1'' + R_2'')}} \qquad (3)$$

where R_1', R_1'', R_2', R_2'' are the four principal radii of curvature of the interacting bodies evaluated at the minimum separation point (some explicit expressions for G in the case of spheroidal particles can be found in [8]).

As far as ϕ_p is concerned one can choose from a variety of available results in the form of either tables [9], elliptic integrals [10], exact solutions [7,14], or approximate solutions valid for short or large separations [11]. The linear model of Hogg, Healy and Furstenau (HHF)

[12] valid for low zeta potentials is also very often used.

Especially useful expressions valid for arbitrary zeta potentials and larger distances between particles can be derived when exploiting the LSA. Thus, for example, the electrostatic interaction energy of two spherical particles of radius a_1 and a_2 immersed in an electrolyte of arbitrary composition can be approximated by the Yukawa-type potential [13]

$$\phi = \frac{1}{2} \varepsilon \frac{(kT)^2}{e^2} Y_1 Y_2 \frac{2 a_1 a_2}{a_1 + a_2 + h_o} e^{-\kappa h_o} \qquad (4)$$

where Y_1, Y_2 are the dimensionless correction functions fulfilling the inequality:

$$\frac{1}{4} \tanh \frac{\psi e}{4 kT} < Y < \frac{\psi e}{kT} \qquad (5)$$

In practice, for $\kappa a > 1$, one can assume $Y = \frac{1}{4} \tanh \frac{\psi e}{kT}$ with an accuracy of a few percent .

It was shown in [18] that Eq.(4) can also be used for spheroidal particles if a_1, a_2 are replaced by the mean radii of curvature R_1 and R_2 of the interacting spheroid pair evaluated at the outermost points of their surfaces. Calculating the exact position of these points for arbitrary spheroid orientations and consequently h_o is rather difficult and can only be performed numerically.

It should be mentioned that at present, except LSA (additivity), there is no theoretical method for describing the true many body electrostatic interactions at interfaces, e.g., including the simplest case when two particles are approaching the surface.

In *Figure 2* the results for ϕ stemming from the Derjaguin and LSA models are confronted with the exact numerical solutions obtained for a typical parameter set, i.e., $\kappa a = 5$, $\psi_1 e/kT = 3$ (interface), $\psi_2 e/kT = -1.5$ (particle). Both the particle/particle (upper part of *Figure 2*) and the particle/interface interactions are plotted. As can be seen for $\kappa h_o > 0.5$ the LSA model is in a good agreement with the exact solution, whereas the linear HHF model shows a tendency to overestimate the interaction energy at all separations (this is especially evident when the constant charge boundary conditions are applied). Much larger differences occur at closer separations when $h_o < 0.5$.

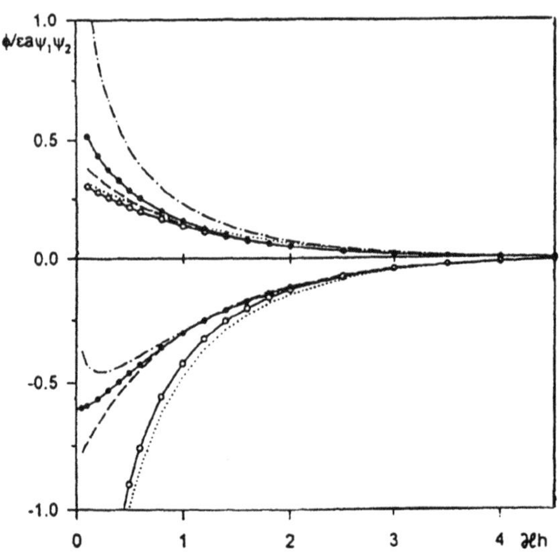

Figure 2. The electrostatic interaction energy between two particles (upper part) and particle interface (lower part) calculated from various models for the set of parameters identical as for *Figure 1*, -•- exact solution for constant charge, -o- exact solution for constant potential, --- LSA, ... HHF constant potential, -.- HHF constant charge. All curves for $\kappa a = 5$.

However, from a practical viewpoint the exact value of ϕ at close separations is of a minor importance (except for the barrier controlled adsorption regime) because the energy assumes large values either positive (particle-particle) or negative (particle interface). In both cases a relatively large uncertainty of ϕ can be tolerated . Additionally, the van der Waals interactions not considered explicitly in our discussion, dominate at close separations.

2.2. THE EFFECTIVE HARD-PARTICLE APPROXIMATION

As demonstrated in our previous works [2-3,18] for interacting spherical particles, adsorption kinetics as well as structure of the adsorbed layer can well be reflected by applying the effective hard particle (EHP) concept introduced originally by Barker and Henderson [14]. According to this approach the true electrostatic interaction potential discussed above can be replaced by the effective hard particle (square wall) potential, i.e.,

$$\phi = \infty \quad \text{for } h_o < h^* \text{ and } \phi = 0 \text{ otherwise}$$

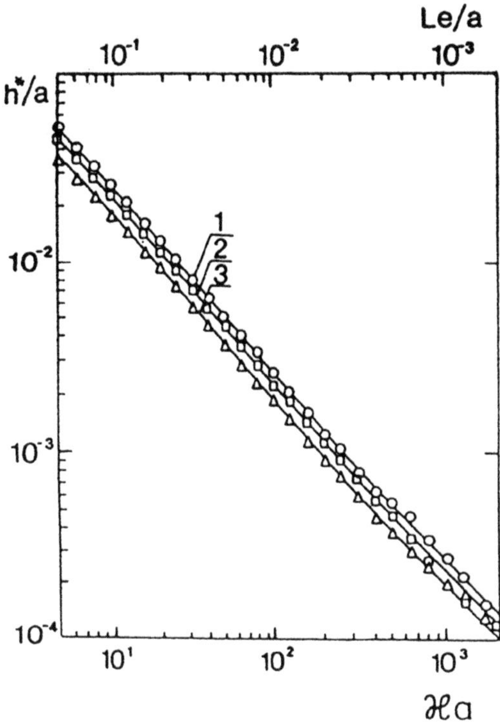

Figure 3. The effective interaction range h^*/a vs $\varkappa a$ calculated numerically for spheroidal particles (Derjaguin model of EDL) characterized by various elongation parameter A (including the case A = 1, i.e., spheres), ϕ = 100 kT, the lines denote the results calculated from Eq.(6).
1. A = 1; 2. A = 0.5; 3. A = 0.2.

Physically this means that the interacting particles can be treated as hard (noninteracting) ones having the equivalent dimension (radius) a^* increased over the geometrical dimension by h^*, i.e., the effective interaction range. Thus, the effective cross-section of the spherical particle becomes simply $S^* = \pi \ (a+h^*)^2$. Instead of analyzing the complicated interaction energy profiles one can use a^* to predict adsorption kinetics. This concept is applicable for both spherical and anisotropic (spheroidal) particles as shown in *Figure 3*. The comparison of the exact MC results with the approximate derived from the above Derjaguin and LSA approximation shows that h^* depends linearly on Le with

$$h^* = \xi \, Le \tag{6}$$

with

$$\xi = \frac{1}{2} \ln \left[\frac{\phi_o}{\phi^*} \right] \tag{7}$$

$\phi_o = \varepsilon \, a \, Y_1 Y_2 \left(\frac{\varepsilon}{kT} \right)^2$ and ϕ^* assuming in the Derjaguin model values close to 1 kT unit. As discussed in due course the EHP concept can be useful for describing particle adsorption kinetics of colloid particles.

3. Particle Adsorption Kinetics

3.1. THEORETICAL MODELS

One of the most widespread ways of accounting for the nonlinear particle adsorption regimes is to introduce the surface blocking parameter $B(\theta)$ (often referred to as the available surface function [15-18]) as a natural extension of the Langmuir approach derived for gas adsorption [19]. Thus, in this quasi-Langmuir approach, $B(\theta)$ is calculated from the simple relationship

$$B(\theta) = \frac{N_{free}}{N_{mx}} = \frac{N_{mx} - N_{occ}}{N_{mx}} = 1 - \frac{\theta}{\theta_L} \tag{8}$$

where N_{free} is the concentration of the "active centers" available, N_{mx} is the total number of active centers and N_{occ} is the number of active centers occupied by adsorbed particles, $\theta_L = S_g N_{mx}$, $\theta = S_g N_{occ}$; S_g is the geometrical cross-section area of a particle. It should be noted that θ_L is an empirical parameter determined *a posteriori* from experimental data.

Despite its limited validity the Langmuir approach is very commonly used for describing the blocking effects of proteins [20-22], colloids [23-25], and bacteria [26].

A more realistic theoretical determination of B for colloidal particles can be achieved using recent theoretical models the most widespread being the random sequential adsorption (RSA) approach developed in [27,1-2,15-16]. The surface blocking parameter $B(\theta)$ in the RSA method can be calculated from the equation

$$B(\theta) = \frac{j}{j_o} = \frac{p}{p_o} = \frac{N_{succ}}{N_t} \tag{9}$$

where j is the adsorption flux for a given configuration of particles at the interface, j_o is the stationary value of the adsorption flux (at uncovered surface), p is the probability of adsorbing a particle on a covered target, p_o that on uncovered surface, N_{succ} is the number of successful trials and N_t is the number of trials (for a given surface concentration θ).

Despite of the simplicity of the underlaying assumptions the topology of particle distributions generated in RSA processes becomes complex for higher surface concentrations and a proper description of the RSA process in 2D can only be achieved in terms of time-consuming Monte-Carlo type computer simulations [1-3, 15-18]. However, for low and moderate surface coverage range $B(\theta)$ in the RSA sequences can be approximated by the simple polynomial expression [3]

$$B(\theta) = 1 - C_1 \theta + C_2 \theta^2 + 0 (\theta^3) \tag{10}$$

where the C_1 and C_2 constants (independent of θ) for spheroidal particles are [8]

$$C_1 = \frac{4 (\pi^2-4)}{\pi^2} + \frac{8}{\pi^2} \left(A + \frac{1}{A}\right) + 4\left(1+ \frac{1}{A}\right) H^* + \frac{4}{A} H^{*2}$$

$$C_2 = q(A)\ C_1^2 \tag{11}$$

$$q(A) = 0.126 + 0.181\ (A+\frac{1}{A})^4 \ / \ (0.448+A+ \frac{1}{A}\ )^4$$

where $H^* = h^*/a$ is the dimensionless effective interaction range discussed above and $A = a/b$ is the ratio of the spheroid axes.

Using Eq.(10) one can derive by integration the following analytical expression describing particle adsorption kinetics [3,8]

$$\theta(\tau) = \theta_{st} \frac{1 - \exp(-p\, C_1 \theta_{ch}\, \tau)}{1 - \frac{\theta_{st}}{\theta_2} \exp(-p\, C_1 \theta_{ch}\, \tau)} \tag{12}$$

where

$$\theta_{st} = \frac{C_1}{2\, C_2} \left[1 - p\right]; \quad \theta_2 = \frac{C_1}{2\, C_2} \left[1 - p\right]; \; p = \left(1 - \frac{4\, C_2}{C_1^2}\right)^{1/2}$$

and $\tau = t/t_{ch}$ is the dimensionless adsorption time, $t_{ch} = \theta_\infty S_g \bar{j}_o n_b$ is the characteristic adsorption time, $\theta_\infty = 0.546$, n_b is the particle number concentration in the bulk.

It is interesting to mention that Eq.(10) can also be applied when the flow-induced blocking effects occur [28] (referred to as the hydrodynamic scattering effect HSE). The C_1 and C_2 constants assume then the form [28]

$$C_1 = C_1^o + C_h \bar{G} \; ; \qquad C_2 = \frac{1}{8} C_1^2 \tag{13}$$

where C_1^o, C_h are the dimensionless constants and $\bar{G}$ is the dimensionless shear rate at the interface.

Considering Eq.(13) one can express θ_{st} for the flow controlled adsorption in the form,

$$\frac{1}{\theta_{st}} = \frac{1}{\theta_o} + C_h' \bar{G} \tag{14}$$

where θ_o, Ch' are the dimensionless constants.

In the general case, for arbitrary surface concentrations and various transport conditions, $B(\theta)$ and particle adsorption kinetics can be quantitatively predicted using the sequential brownian dynamics (SBD) method [28-30].

The force fields stemming from interface and all previously adsorbed particles are taken into account via the additivity rule.

3.2. ILLUSTRATIVE EXPERIMENTAL RESULTS.

The experimental results discussed hereafter were obtained in the stagnation point flow cell by using the *in situ* microscope observation method [1-3,28]. Monodisperse suspensions of polystyrene latex were used as model colloids prepared according to the surfactant free polymerization procedure. The zeta potential of the latex varied between -40 and -60 mV for the ionic strength ranging from I = 10^{-3} to 10^{-5} M. The adsorbent surface was made of freshly cleaved mica sheets whose surface charge was converted to positive by adsorption of silanes which assured barrierless adsorption regime. The zeta potential of the treated mica varied between +60 ÷ +100 mV (for the same ionic strength range) as determined by the streaming potential measurements.

The flow rate in the cell was regulated by changing the hydrostatic pressure gradient. Under the flow conditions used in the our experiments particle adsorption was irreversible. Adsorption kinetics was determined via the image analyzing system with a possibility to measure individual particle coordinates and their number over given surface area. The pair correlation function was obtained using the micrograph technique combined with the digitizer which gives the resolution of 4600 per 2700 pixels.

In *Figure 4* some typical kinetic runs are presented for various bulk suspension concentrations of the 0.88 μm diameter latex ($I=10^{-4}$ M, Reynolds number Re = 8). As can be seen in this *Figure* the MC RSA simulations as well as the limiting analytical solution derived from Eq.(12) describe very well the experimental results whereas the Langmuir model (with B(θ) calculated from Eq.(8)) proved inadequate.

One may, therefore, conclude that under certain combination of flow and diffusion effects the conditions for the RSA mechanism are fulfilled. This is further confirmed by the results shown in *Figure 5* where adsorption kinetics was measured for various ionic strength of the latex suspension. As can be seen the increase in the ionic strength resulted in the increase in adsorption rate in accordance with the MC RSA simulations and the limiting theoretical predictions (cf. Eq.(12) with B(θ) calculated from Eq.(10)). It is interesting to mention that the RSA model proved also adequate for packed bed filtration as confirmed by precise experiments performed recently by Johnson and Elimelech [31].

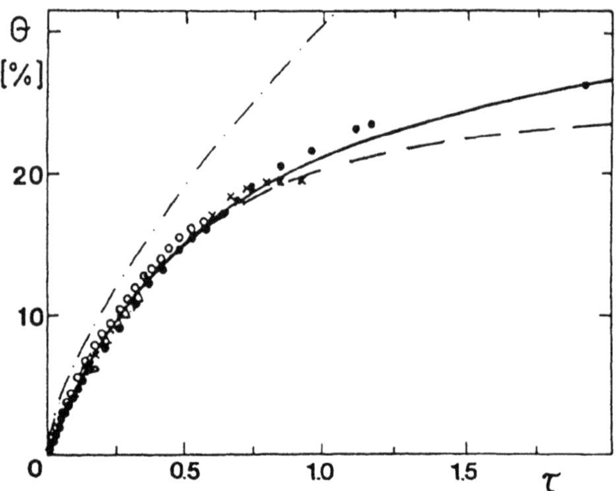

Figure 4. Adsorption kinetics of negatively charged latex on modified mica (a = 0.44 μm, I = 10^{-4} M, Re = 8) expressed as the dependence of the dimensionless surface concentration θ on the dimensionless time τ. The continuous lines denote the theoretical results obtained from the MC RSA simulations, the broken line shows the results calculated from Eq.(12) and the dashed/dotted line the Langmuir model.

The results shown in *Figures 4-5* suggest that the EHP concept in which interacting particles are replaced by hard spheres works well for low and moderate surface concentrations. Note that the surface coverages attained for low ionic strength (κa = 5) for relatively long adsorption times remain much smaller that the maximum hexagonal packing (θ = 91%) or the "random" monolayer for hard spheres equal 54.6% [1-3,15-16]. This may sometime lead to misinterpretation of experimental data concerning microparticle adsorption. The apparent kinetic saturation observed in long-lasting adsorption runs was often treated as thermodynamic equilibrium whereas in fact particle adsorption remained partially or totally irreversible.

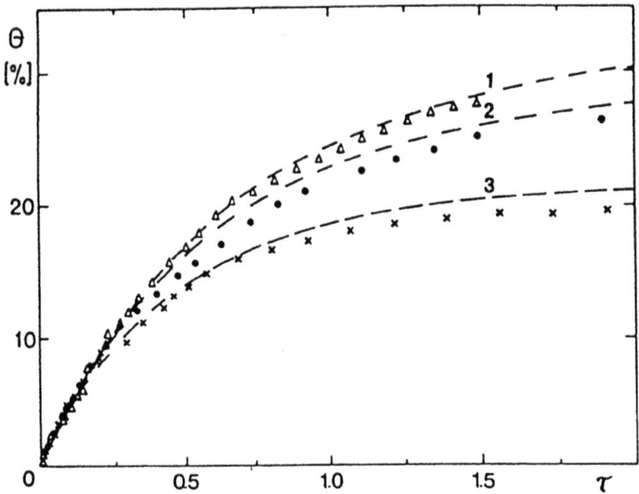

Figure 5. Measured dependence of θ on τ for negative latex on mica, a = 0.5 μm, Re = 8; 1. for I = 10^{-3} M, 2. I = 10^{-4} M, 3. I = 10^{-5} M; broken lines denote analytical approximations given by Eq.(12).

Besides the ionic strength also the suspension flow rate (shear rate at the interface) plays an important role in adsorption of microparticles. This is shown in *Figure 6* where kinetic runs performed for Re = 2, 8 and 16 are plotted (I =10^{-4} M, a = 0.5 μm). In contrast to common expectations the increase in the flow rate (which increased the initial adsorption flux $\bar{j}_0$) resulted in a considerable decrease in adsorption rate for longer times. For higher Re the quasi-stationary value of θ_{st} was established, much smaller than the RSA predictions. Note that the results shown in *Figure 6* can well be accounted for by the SBD numerical simulations with the electrostatic forces calculated using the LSA method (Eq.(4)) and the additivity principle. Also, the analytical results stemming from Eq.(12) approximate well the experimental results for higher Re values.

As shown in *Figure 7*, the experimentally measured dependence of $1/\theta_{st}$ on $\bar{G}$ (the circular and the slot impinging jet cells [28]) can be reflected by the linear relationship as predicted by Eq.(14).

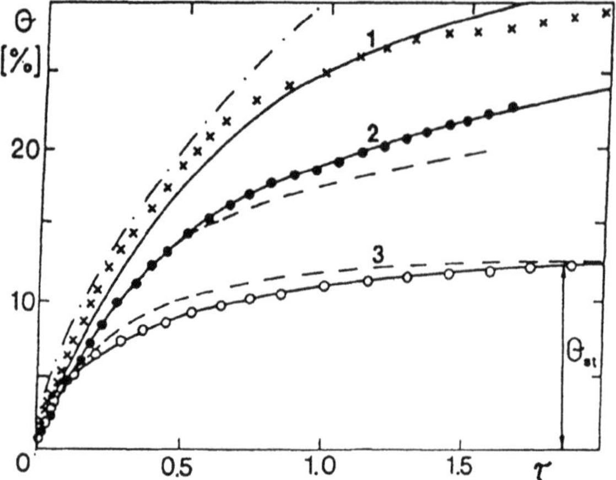

Figure 6. Measured dependence of θ on τ for various Re numbers (at the distance of 150 μm from the center of cell); negative latex a = 0.5 μm, I = 10^{-4} M, 1. Re = 2, 2. Re = 8, 3. Re = 16; continuous lines-SBD numerical simulations, broken lines analytical approximation from Eq.(12), dashed/dotted line shows the results predicted by the Langmuir model.

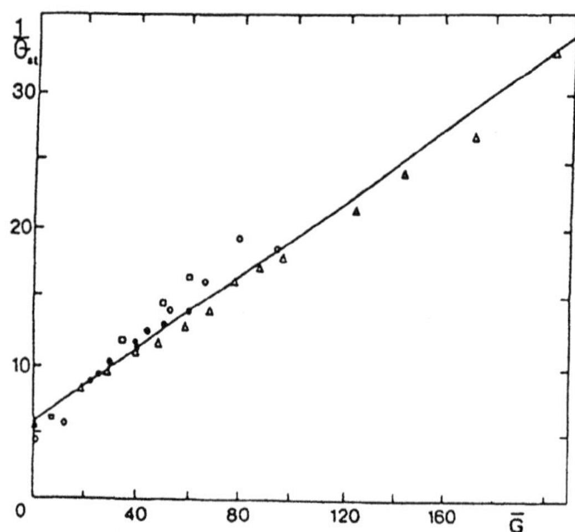

Figure 7. The universal plot showing the dependence of the quasi-stationary surface concentration θ_{st} on the dimension-less shear rate $\bar{G}$. The points represent experimental results obtained for various particle sizes and Re values and the straight line denotes the theoretical predictions calculated from Eq.(14) with C'_h = 0.16.

The results shown in *Figures 6-7* suggest that in adsorption experiments involving larger particles (bacteria) the increase in mixing intensity can result in the decrease of the adsorption rate for longer time and consequently in the quasi-stationary surface coverages as well. This phenomenon analyzed in some detail in [28] was attributed to the hydrodynamic scattering effect HSA occurring due to the coupling between hydrodynamic and electrostatic force fields.

3. Structure of Adsorbed Layers

The structure of the adsorbed layer can be quantitatively characterized in terms of the 2D pair correlation function g_{12} which can be interpreted as the averaged normalized (to the uniform bulk value) surface concentration expressed relative the local coordinate system with its center at a particle. The use of the micro-sized latex suspensions opened the possibility for a direct determination of g_{12} from micrographs of adsorbed particles.

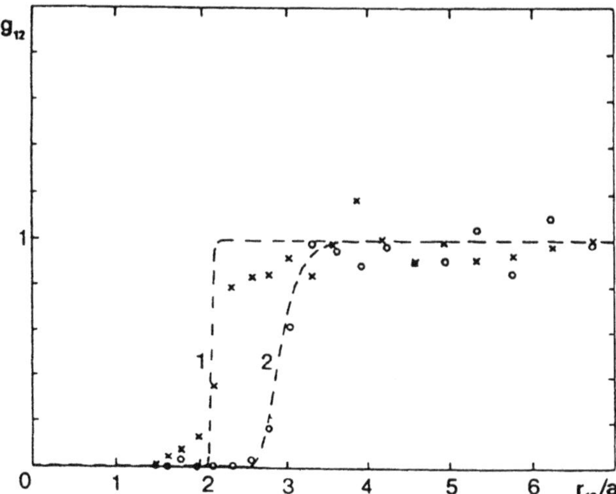

Figure 8. The pair correlation function g_{12} determined for negative latex on mica ($a = 0.5$ μm, Re = 8, θ = 4%, center of the cell); 1. I = 10^{-3} M, 2. I = 10^{-5} M, broken lines denote Boltzmann distribution $g_{12} = \exp(-\phi_2/kT)$.

In *Figure 8* we present g_{12} obtained for low surface concentration of latex particles on mica ($\theta = 4\%$) for two different ionic strength $I = 10^{-3}$ M ($\kappa a = 50$) and $I = 10^{-5}$ M ($\kappa a = 5$). It was found that the experimental g_{12} shown in *Figure 8* can well be reflected by the Boltzmann distribution, i.e.,

$$g_{12}(H) = e^{-\phi_2(H)/kT} \tag{15}$$

However, the mean pair interaction energy $\phi_2(H)$ which fitted best the experimental data was found smaller than the electrostatic interaction energy predicted from previously discussed Eq.(4). This effect can be explained either by the nonlinear induction effect of the interface not accounted for by the LSA or by the charge leakage from the adsorbed particles promoted by the large field in the gap between the adsorbed particle and the interface.

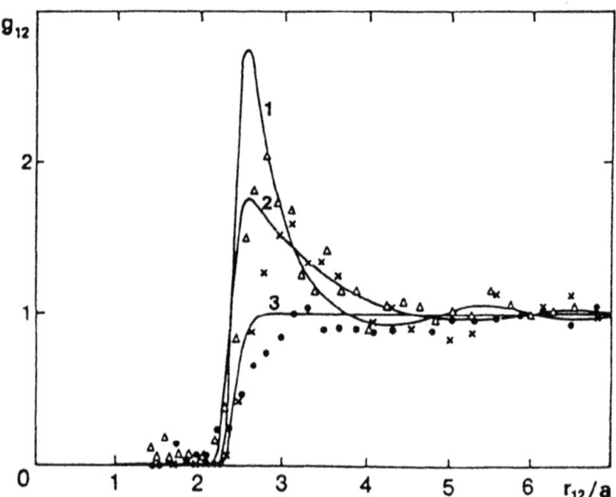

Figure 9. g_{12} determined for various transport mechanisms (constant $\kappa a = 12$, $\theta = 11\%$) 1. sedimentation controlled adsorption (melamine latex a = 1 μm), 2. diffusion controlled adsorption (polystyrene latex a = 0.74 μm, Re = 0.6) 3. flow controlled adsorption (polystyrene latex a = 0.74 μm, Re=16)

It seems therefore that the experimental g_{12} measurements pose the unique possibility of determining the true many body interactions under dynamic conditions since by inverting Eq.(15) ϕ_2 can be calculated.

For higher surface concentrations (cf. *Figure 9*) the g_{12} function clearly indicates that a short-range ordering of the irreversibly adsorbed particle phase occurs, analogous to a 2D liquid or a glassy structure.

It is also interesting to note that the transport mechanism of bringing particles to interface exerted a significant role in determining the structure of the adsorbed layer (cf. *Figure 9*). The hydrodynamic forces, due to previously mentioned HSE decrease the peak observed on the g_{12} vs distance relationship. By contrast, external forces such as gravity (playing a large role for denser melamine latex particles) increase considerably the short-range order which is reflected by the increase in the peak height. However, the gravity force alone is to weak to overcome the interparticle EDL forces and provoke a true 2D crystallization. This can only be accomplished by the capillary forces [32], e.g., upon drying the adsorbed layer as demonstrated spectacularly in [33].

Concluding Remarks

Kinetic measurements as well as the analysis of the structure of adsorbed layer revealed that the electrostatic double-layer forces exerted a significant influence on adsorption phenomena of microparticles. For smaller particles and moderate surface concentrations this can well be accounted for by the RSA model using the equivalent hard-particle concept. For larger particles and high intensity flows ($\bar{G} >> 1$) the lateral electrostatic interactions play even more significant role provoking the hydrodynamic scattering effect which diminished particle adsorption kinetics. The apparent kinetic saturation of interfaces was observed in this case with θ_{st} inversely proportional to $\bar{G}$.

Theoretical considerations, especially the equivalent hard particle concept, suggest that the experimental results obtained for spheres are likely to retain their validity for other particle geometries as well, e.g., spheroids.

Acknowledgement
This work was supported by the KBN Grant 2 P203 039 05.

References

1. Adamczyk, Z., Zembala, M., Siwek, B. and Warszyński, P. (1990) Structure and ordering in localized of particles, J.Colloid Interface Sci. **140**, 123-137.
2. Adamczyk, Z., Siwek, B. and Zembala, M. (1992) Kinetics of localized adsorption of particles on homogeneous surfaces, J.Colloid Interface Sci. **151**, 351-369.
3. Adamczyk, Z., Siwek, B., Zembala, M. and Belouschek, P. (1994) Kinetics of localized adsorption of colloid particles, Adv. Colloid Interface Sci. **48**, 151-280.
4. Derjaguin B.V.(1940) On the repulsive forces between charged colloid particles and on the theory of slow coagulation and stability of lyophobic sols. Trans. Faraday Soc. **36**, 203.
5. White L.R.(1983) On the Deryaguin Approximation for the interaction of macrobodies. J.Colloid Interface Sci. **95**, 286-288.
6. Adamczyk, Z., Belouschek, P. and Lorenz, D. (1990) Electrostatic interactions of bodies bearing thin double layers. I. General Formulation. Ber.Bunsenges.Phys.Chem. **94**, 1483-1492.
7. Adamczyk, Z., Belouschek, P., and Lorenz, D., (1991) Electrostatic Interactions of Bodies Bearing Thin Double-Layers. Bull.Pol.Ac.Chem. **39**, **423-448**.
8. Adamczyk, Z. and Weroński P., (1995) Kinetics of irreversible adsorption of spheroidal particles. LAngmuir submitted.
9. Devereux, O.F. and de Bruyn P.L. (1963) "Interactions of Plane-Parallel Layers. M.I.T. Press, Cambridge, Massachusttts.
10. Mc.Cormak, D., Carnie S.L. and Chan D.Y.C. (1995) Calculation of electric double-layer force and interaction free energy between dissimilar surfaces. J.Colloid Interface Sci. **169**, 177-196.
11. Adamczyk, Z., Belouschek, P. and Lorenz, D. (1990) Electrostatic interactions of bodies bearing thin double layers. II. Exact Numerical Solutions (1990) Ber.Bunsenges.Phys.Chem. **94**, 1492-1499.
12. Hogg R., Healy, T.W. and Furstenau D.W., (1966) Mutual coagulation of colloidal dispersion. Trans.Faraday.Soc. **62**, 1638-1651.
13. Bell, G.M., Levine, S. and McCartney, L.N. (1970) Approximate methods of determining the double-layer free energy of interaction between two charged colloidal spheres. J.Colloid Interface Sci. **33**, 335-359.

14. Barker J.A. and Henderson D. (1967) Perturbation theory and equation of state for fluids. II. A successful theory of liquids. J.Chem.Phys. **47**, 4714-4721.
15. Schaaf P. and J.Talbot, J. (1989) Surface exclusion effects in adsorption processes. J.Chem.Phys. **91**, 4401-4409.
16. Schaaf, P. and albot (1989) Kinetics of random sequential adsorption. J. Phys.Rev.Lett. **62**, 175-177.
17. Viot, P., Tarjus, G., Ricci, S.M. and Talbot.J. (1992) Random sequential adsorption of anisotropic particles. I. Jamming limit and asymptotic behavior. J.Chem.Phys. **97**, 5212-5219.
18. Ricci, S.M., Talbot, J. Tarjus, G.and Viot, P. (1992) Random sequential adsorption of anisotropic particles. II. Low coverage kinetics. J.Chem.Phys. **97**, 5219-5228.
19. Langmuir, I. (1918) The adsorption of gases an plane surfaces of glass. mica and platinium. J.Am.Chem.Soc. **40**, 1361.
20. Feder, J. and Giaever, I. (1980) Adsorption of ferritin. J.Colloid Interface Sci. 78, 144-154.
21. Young, B.R., Pitt, W.G. and Cooper, S.l. (1988) Protein adsorption on polymeric biomaterials I. Adsorption isotherms. J.Colloid Interface Sci. **124**, 28-43.
22. Aptel, J.D., Voegel, J.C. and Schmitt, A. (1988) Adsorption kinetics of proteins onto solid surfaces in the limit of the interfacial interaction control. Colloids Surf. **29**, 359-371.
23. Vincent, B., Young, C.A. and Tadros, Th.F. (1980) Adsorption of small, positive particles onto Large, negative particles in presence of polymer JCS Faraday I **76**, 665-673.
24. Kallay, N., Tomic, M., Biskup, B., Kunjasic, I. and Matijevic, E. (1987) Particle adhesion and removal in model systems. XI. Kinetics of attachment and detachment for hematite-glass systems. Colloids Surf. **28**, 185-197.
25. Dąbroś, T. and van de Ven, T.G.M. (1983) A direct method for studying particle deposition onto solid surfaces. Colloid Polymer Sci. **261**, 694-707.
26. Gibbons, R.J., Moreno, E.C. and Etherden I. (1983) Concentration-dependent multiple binding sites on saliva-treated hydroxyapatite for streptococcus sanginus. Infect. Immun. **39**, 280-289.
27. Hinrichsen, E.L., Feder, J. and Jossang T.(1986) Geometry of random sequential adsorption. J.Stat.Phys. **11**, 793-827.
28. Adamczyk, Z., Siwek, B. and Szyk, L. (1995) Flow-induced surface blocking effects in adsorption

of colloid particles. J.Colloid Interface Sci. in press.
29. Ansell, G.S.and Dickinson (1986) Sediment formation by Brownian dynamics simulation: Effect of Colloidal and hydrodynamic interactions on the sediment structure. J.Chem.Phys. **83**, 4079-4086.
30. Adamczyk, Z., Siwek, B. and Zembala M. (1993) Influence of transport mechanism on adsorption of interacting colloid particles. Colloids Surf. A, **76**, 115-124.
31. Johnson, P.R., Elimelech, M., (1995) Dynamics of colloid deposition in porous media: blocking based on random sequential adsorption. Langmuir, **11**, 801.
32. Kralchevsky P.A. and Nagayama K. (1994) Capillary forces between Colloid Particles. Langmuir, **10**, 23-37.
33. Dushkin C.D., Nagayama K., Miwa T., Kralchevsky P.A. (1993) Colored multilayers from transparent submicrometers spheres. Langmuir, **9**, 3695-3701.

POSSIBILITIES AND LIMITATIONS IN THE GROWTH OF MIXED METAL OXIDE PARTICLES FROM AQUEOUS MEDIA

MIGUEL A. BLESA, GALO J. de A.A. SOLER-ILLIA,
ROBERTO J. CANDAL AND ALBERTO E. REGAZZONI
Comisión Nacional de Energía Atómica,
Avenida del Libertador 8250,
(1429) Buenos Aires, REPUBLICA ARGENTINA, and
INQUIMAE, Facultad de Ciencias Exactas y Naturales,
Universidad de Buenos Aires,
Pabellón II, Ciudad Universitaria,
(1428) Buenos Aires, REPUBLICA ARGENTINA.

1. Abstract

Synthesis of precursors for mixed metal oxides can be achieved by coprecipitation from aqueous media. This procedure offers clear advantages over traditional solid state reactions. There are however important limitations:

(i) The nucleation and growth kinetics usually lead to inhomogeneous in-depth profiles of the concentration of the metal ions.

(ii) The mixing of a sparingly soluble metal oxide with a readily soluble one requires either the coprecipitation of basic salts (carbonates or oxalates), or the alteration of the solubility behavior. Changes in solubility can be achieved by using oxidative or reductive precipitation or by using complexants, including hydroxide ions.

(iii) Mixed oxides in contact with water may be corroded selectively, through the leaching of the more soluble component.

Examples of the above problems shall be given; the chosen systems are mixed Cu-Zn oxides, $YBa_2Cu_3O_{7-x}$ and barium hexaferrite.

Mixed copper - zinc oxides: Precursors were synthesized by homogeneous alkalinization with urea. The nature of the solids depends on the ratio Cu:Zn; only when this ratio is larger than 10:1 the precursors evolve smoothly to Zn-substituted tenorite. Good spherical monodisperse solids can be prepared. Other detected phases are aurichalcite and Zn-substituted malachites.

$YBa_2Cu_3O_{7-x}$: In contact with water, barium and yttrium are selectively

E. Pelizzetti (ed.), Fine Particles Science and Technology, 33–45.

leached, whereas copper oxide is reprecipitated. The kinetics of the evolution depend on the composition of the aqueous medium.

BaO.6Fe₂O₃: Precursors for this oxide can be prepared from strongly alkaline ferrate(VI) solutions, by reductive coprecipitation of Fe_2O_3 + $Ba(OH)_2$.

2. Introduction

The search of particulate metal oxides of adequate particle size and morphology has driven the studies of wet synthesis, based on the controlled hydrolysis of adequate salt solutions. Several procedures to achieve controlled hydrolysis have been explored, and a wide variety of particulate metal oxides have been synthesized in an amazing array of sizes and shapes [1, 2]. In very selected cases, the procedure has been shown to be suitable for scaling up, at least, to bench scale [3, 4], and the use of cheap chemicals and mild conditions makes the economic prospects promising. Of course, the process presents also some important limitations, that must be clearly understood; a comparison with the other important soft chemical route, the alkoxide synthesis, is given in Table 1. In fact, the alkoxide route also implies controlled hydrolysis; in this case, however, the modulation of the hydrolysis is focused on the alkoxide moiety, whereas in aqueous media either the

TABLE 1. Comparison of the characteristics of controlled hydrolysis in aqueous media *vs*. the alkoxide route

CONTROLLED HYDROLYSIS	ALKOXIDE ROUTE
crystalline solids may be obtained	in general, amorphous solids are formed, that require a further thermal treatment
very adequate for mildly acidic metal cations, e.g. M(II) and M(III)	very adequate for highly acidic metal ions, e.g. M(IV) and M(V).
simple experimental setup	water must be carefully excluded or controlled
coprecipitation of two or more metal cations depends crucially on solubility behaviour	very adequate to coprecipitate two or more metal ions from heterometallic homogeneous alkoxides
synthesis of unstable phases possible (depending on the necessity of thermal treatments)	synthesis of unstable phases possible (depending on the characteristics of the thermal treatment)

aqueous ion or a complex formed with a hard complexant is hydrolyzed essentially by changing, directly or indirectly, the pH of the medium.

In what follows we shall address the problems encountered in the attempts to coprecipitate two or more metal cations by controlled hydrolysis of aqueous or complexed metal ions in water solutions.

3. Thermodynamic constraints

Let us consider a very typical system, for which successful preparative methods have been developed: barium titanate (see e.g. [5]). This compound crystallizes in a distorted perovskite structure, the stable polymorph at ambient temperature being the tetragonal variety. The sequence

$$\textit{trigonal} \xrightleftharpoons{173\text{ K}} \textit{orthorhombic} \xrightleftharpoons{273\text{ K}} \textit{tetragonal} \xrightleftharpoons{393\text{ K}} \textit{cubic}$$

shows the temperatures at which phase transitions are expected to occur. The tetragonal polymorph is the most important material because of its ferroelectric properties, which are associated with an elongated *c*-axis and the lack of a symmetry centre (space group P4mm). High temperature solid state synthesis, or thermal teatment of precursors obtained by soft procedures, should lead to the cubic variety. The formation of tetragonal $BaTiO_3$ upon lowering of the temperature may however be impeded by a series of factors [6]. For instance, in a study of the hydrothermal synthesis [6], stabilization of the cubic polymorph was observed when chloride ions were absent fron the reacting system, a fact that was interpreted in terms of the observed influence of Cl^- on particle size.

The direct formation of barium titanate from aqueous solutions at room temperature is thermodynamically possible because it is appreciably less soluble than $Ba(OH)_2.8H_2O$. *Figure 1* shows the solubility of both phases as a function of pH at 298 K; since dissolution of TiO_2 can be neglected, the solubility equilibrium for $BaTiO_3$ is written as

$$Ba^{2+} + 2OH^- + TiO_2 = BaTiO_3 + H_2O \qquad (1)$$

Thus, whenever Ba^{2+} precipitation is possible, barium titanate shall form. The absolute solubility of the mixed oxide is however high enough to prevent its formation at pH values lower than ca. 10.5.

A very important related aspect, is the behaviour of $BaTiO_3$ when immersed in pure water (or, in general, in aqueous media of $[Ba^{2+}]$ and pH lower than the equilibrium values shown in *Figure 1*). Very detailed studies of the corrosion of perovskites have been performed in the context of the project to choose adequate hosts for the final storage of nuclear wastes [7, 8]. The solid particles are known to evolve, and the operation of a dissolution reprecipitation mechanism was shown to produce a TiO_2 layer on top of the perovskite particles [9].

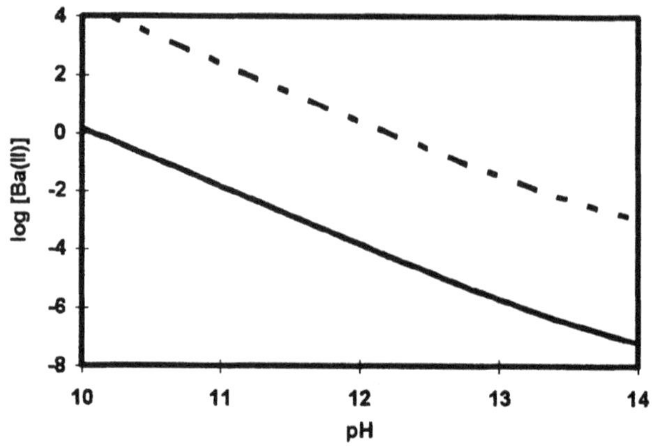

Figure 1. Solubility of $BaTiO_3$ (solid line) and $Ba(OH)_2.8H_2O$ (dashed line) as a function of pH; from data compilled in Refs. [10] and [11].

4. The attack of aqueous solutions on $YBa_2Cu_3O_{7-x}$

A similar case, recently studied in our laboratories, is the evolution of $YBa_2Cu_3O_{7-x}$ particles immersed in aqueous media [12]. The available thermodynamic information is more limited, but the mixed oxide is certainly totally soluble in acidic media:

$$YBa_2Cu_3O_{7-x} + 13\,H^+ \rightarrow Y^{3+} + 2\,Ba^{2+} + 3\,Cu^{2+} + \frac{0.5-x}{2}\,O_2 + 6.5\,H_2O \qquad (2)$$

Figure 2 shows the time evolution of the concentration of barium, yttrium and copper when the solid is put in contact with solutions of pH 2 and 3, in NaCl 1 mol dm^{-3} media. The kinetics follow the contracting volume rate law, and the values of the rate constants indicate surface reaction rate control. It is interesting to attempt a description of this attack in terms of the surface complexation approach [13]. For a mixed oxide surface, an array of sites of varying acidities is present. In our case, neglecting for the moment the problem of the formal oxidation state of copper, the possible surface protonation sites are copper, yttrium and barium ions. The surface equilibria can be written as:

$$\equiv Cu\text{-}OH + H^+ = \equiv Cu\text{-}OH_2^+ \tag{3}$$

$$\equiv Y\text{-}OH + H^+ = \equiv Y\text{-}OH_2^+ \tag{4}$$

$$\equiv Ba\text{-}OH + H^+ = \equiv Ba\text{-}OH_2^+ \tag{5}$$

The order of proton affinities is Ba > Y > Cu. In fact, the surface species ≡Ba-OH is highly unlikely, except in very strong alkaline media. The protonated species $\equiv Ba\text{-}OH_2^+$ should be very labile; the order of lability from the surface is expected to be the same as the order of basicity, the dissolution of barium being diffusion controlled. Thus, dissolution in these acidic media may be visualized as controlled by the removal of protonated copper surface complexes of the type depicted in equation (3), from a surface highly enriched in copper. Dissolution of barium and yttrium may proceed by proton diffusion directly from bulk solution, or by surface diffusion from adjacent protonated copper sites.

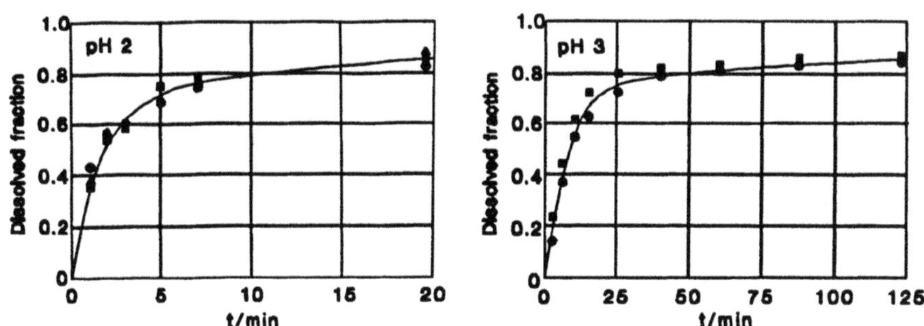

Figure 2. Time dependence of the fractions of Ba (▲), Y (■) and Cu (●) dissolved from $YBa_2Cu_3O_{6.8}$ in 1 mol dm^{-3} NaCl solutions of different pH; T = 298 K; initial mass of solid: 75 mg; solution volume: 0.1 dm^3.

At pH 2 or 3, the surface copper enrichment cannot be observed directly. In contrast, at pH 5, the limitation to the attack of the solid by copper dissolution becomes apparent, as may be seen in *Figure 3*. However, the passivation mechanism involves in fact dissolution and reprecipitation of a sparingly soluble basic copper salt, as demonstrated by the presence of a chloride peak in the EDS spectrum of a copper enriched surface from a sample partially dissolved at pH 5 in NaCl media, and by the detailed analysis of the kinetic behaviour. Studies of the leaching of several mixed oxides demonstrate that leaching of one of the component oxides in mixed oxide phases is at most restricted to a few monolayers, and the most usual dissolution mechanism seems to involve congruent steady state dissolution followed by reprecipitation of the most sparingly soluble simple oxide [9, 14]. The resulting kinetic behaviour is therefore parabolic.

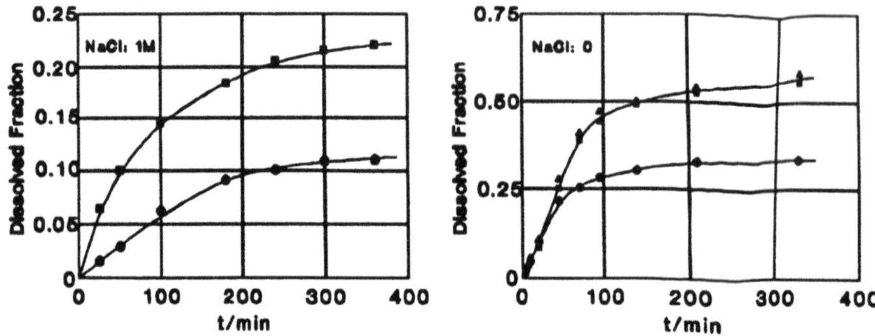

Figure 3. Time dependence of the fractions of Ba (▲), Y (■) and Cu (●) dissolved from $YBa_2Cu_3O_{6.8}$ in solutions of different NaCl concentration at pH 5.0; $T = 298$ K; initial mass of solid: 75 mg; solution volume: 0.1 dm^3.

For phases like barium titanate or ilmenite, the process reverse to that depicted in equation (1) is expected to take place to a degree defined by solution pH, and the surface of mixed oxide particles immersed in aqueous media should in general be expected to be enriched in the less soluble oxide. This characteristic poses a limitation to the purity of the particles obtained by ageing in aqueous media, but, on the other hand, it suggests a potentially useful way to generate, during preparation, passivating layers of the type required to stabilize the many oxides that are not very stable in water (and even in air).

As a final comment on this system, the degree of oxidation of the solid influences strongly its corrosivity. Copper(III) ion is known to be more labile than Cu(II), and the higher its concentration, the higher the dissolution rate [15].

5. Mixed copper-zinc oxides

Zinc-substituted tenorite, $Zn_xCu_{1-x}O$, is an important material for heterogeneous catalysis. However, the actual existence of this material as homogeneous phase is under question [16]. The crystal chemistry of zinc and copper are very diverse; zinc oxide crystallizes in the wurtzite structure, with tetracoordinated Zn(II), whereas the copper oxide structure reflects the typical 4+2 coordination environment associated with d^9 electronic configuration, the Cu(II) coordination being close to square planar.

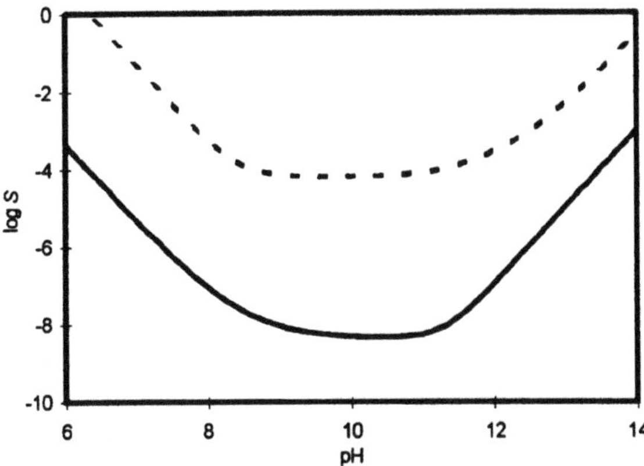

Figure 4. Comparison of the solubilities of $Cu(OH)_2$ (solid line) and $Zn(OH)_2$ *(am)* (dashed line) as a function of pH at 298 K; from [17].

Figure 4 shows the solubility/pH profiles of copper and zinc hydroxides. The higher solubility of $Zn(OH)_2$ is essentially due to the larger contribution of $Zn(OH)_2(aq)$ to the total solubility. The absolute solubility of the latter limits severely the pH window in which coprecipitation from dilute solutions is efficient ($9 \leq pH \leq 11$). Also, control of the simultaneous precipitation of Cu(II) and Zn(II) as hydroxides is complicated by their different solubilities. Therefore, much more attention has been devoted to the formation of basic mixed carbonates [18-20]. *Figure 5* shows superimposed the phase diagrams at 298 K for the individual metal ions in the form of log C_T vs pH plots (where C_T is the total carbonate concentration) for a metal ion concentration of 2.5×10^{-3} mol dm^{-3}. In mixed systems, further to the solids indicated in *Figure 5*, a mixed compound, aurichalcite, $(Zn,Cu)_5(OH)_6(CO_3)_2$ must be taken into account, together with the incorporation of zinc in malachite. The structure of aurichalcite is related to that of hydrozincite. Malachite may accept zinc by incorporating it in tetrahedral environments. The properties of catalysts obtained by thermal treatment of both compounds have shown that malachite is a better precursor [19], a fact probably related with phase segregation in the mixed solids formed when the ratio Zn:Cu is relatively high.

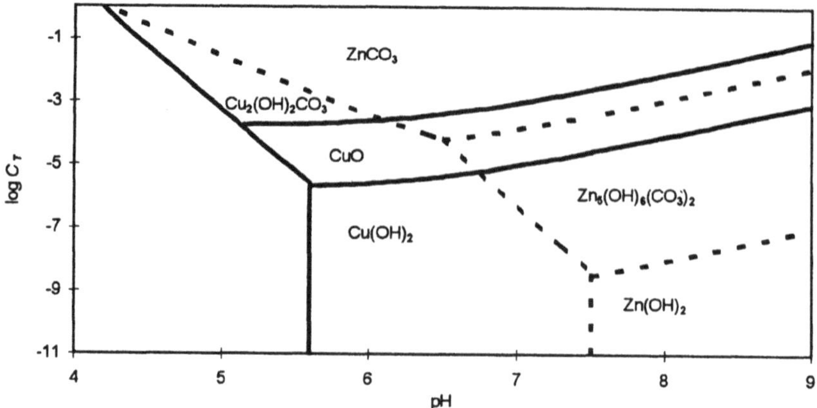

Figure 5. Cu^{2+}-H_2O-CO_2 (solid lines) and Zn^{2+}-H_2O-CO_2 (dashed lines) phase diagrams; constructed from data compilled in Refs. [10] and [17].

We have conducted a systematic study of the controlled precipitation of mixtures of copper and zinc from solutions alkalinized by the hydrolysis of urea [21]. Previous studies [22, 23] had shown that the sequence of formation of solids as the ageing time (and the solution pH) increased was:

amorphous copper basic carbonate → malachite → tenorite

Within certain limits, the addition of zinc does not produce significant changes in the nature of the obtained solids, that apparently incorporate zinc substitutionally, without giving rise to new phases. The limits of this 'solubility' may be set at around 25% [atomic ratio 100 x Zn/(Cu+Zn)] for the amorphous basic carbonate, 20% for malachite, and 4% for tenorite. At higher zinc concentrations, aurichalcite formation is always observed, either as a mixture with one of the above phases, or as a pure solid at high Zn percentages.

Figure 6 shows the time evolution of copper and zinc concentrations in media containing either the individual ions or a mixture of both. It can readily be seen that, during the precipitation of the amorphous basic carbonate, copper is removed from solution much faster than zinc. *Figure 7* shows the time evolution of pH; as expected, during the removal of metal ions pH does not change much, and the rate of solid formation is governed by the rate of urea hydrolysis. The higher solubility of zinc is reflected in the need of higher carbonate and/or pH to precipitate it. The mixture of cations behaves to a first approximation as expected from the overlapping of the two pure systems. However, some copper redissolution is observed when zinc is precipitated, a phenomenon probably related to a nucleation and growth kinetic effect during zinc precipitation; after nucleation, supersaturation is rapidly removed, and the solution variables change to determine an enhanced copper solubility. A further effect is the possible copper complexation by ammonia at the relatively high pH reached during zinc precipitation.

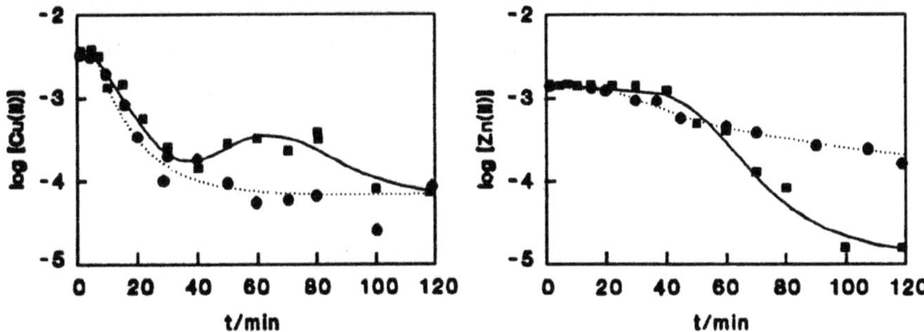

Figure 6. Concentrations of metal ions remaining in solution during the ageing of systems containing 3.5 x 10^{-3} mol dm^{-3} $Cu(NO_3)_2$ and/or 1.5 x 10^{-3} mol dm^{-3} $Zn(NO_3)_2$ and 0.5 mol dm^{-3} urea at 363 K; (●) individual systems; (■) mixed system.

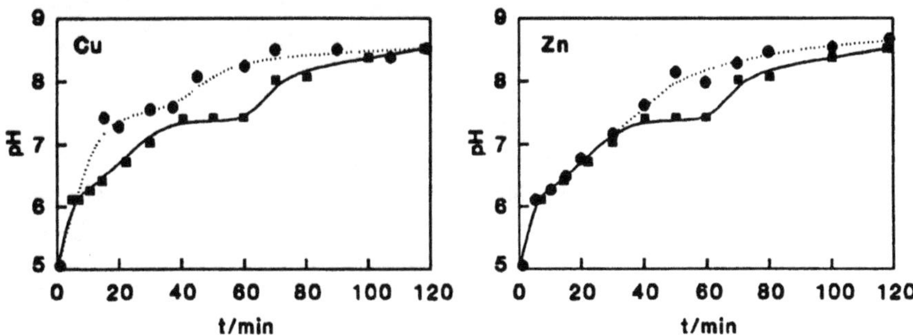

Figure 7. Changes in solution pH during the ageing of systems containing 3.5 x 10^{-3} mol dm^{-3} $Cu(NO_3)_2$ and/or 1.5 x 10^{-3} mol dm^{-3} $Zn(NO_3)_2$ and 0.5 mol dm^{-3} urea at 363 K; (●) individual systems; (■) mixed system.

This behaviour leads to amorphous basic carbonate particles of inhomogeneous composition; the inhomogeneity is partially relieved during ageing, until malachite is generated from the amorphous material; at this point, the solubility of zinc in malachite shall determine whether homogenous particles or mixtures of phases form; as stated, the limit is around 20 % Zn. Further ageing leads to tenorite, and in this phase the solubility limits are very stringent; our results indicate that only up to 4% Zn may be homogeneously incorporated. This limit is in agreement with the values quoted earlier in the literature.

In summary, the complex aqueous chemistry involved determines the feasibility of obtaining homogeneously mixed solids; however, once these factors are adequately recognized and handled, it is possible to prepare good spherical particles of narrow size distribution of all the involved phases. *Figure 8* shows some electron micrographs of typical preparations, together with the conditions of synthesis.

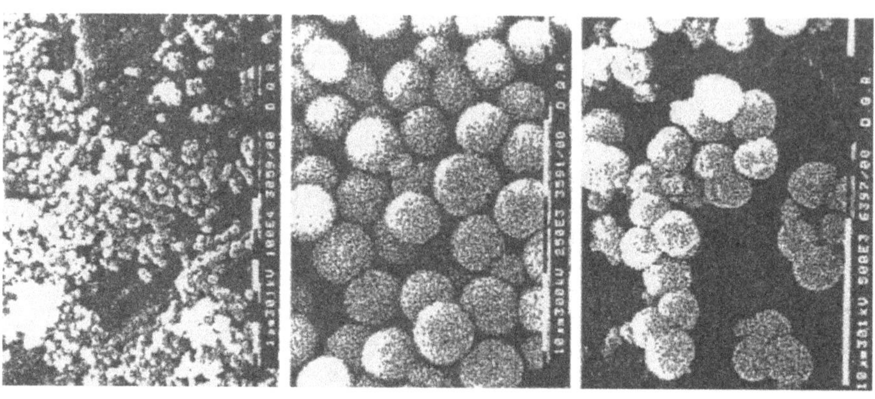

Figure 8. SEM photographs of: (left) Zn-containing copper amorphous basic carbonate particles precipitated form a 8.5×10^{-4} mol dm^{-3} $Zn(NO_3)_2$, 4.0×10^{-3} mol dm^{-3} $Cu(NO_3)_2$ and 0.5 mol dm^{-3} urea solution aged at 363 K for 30 minutes; (centre) aurichalcite particles precipitated form a 2.6×10^{-3} mol dm^{-3} $Zn(NO_3)_2$, 2.2×10^{-3} mol dm^{-3} $Cu(NO_3)_2$ and 0.5 mol dm^{-3} urea solution aged at 363 K for 50 minutes; (right) Zn-doped tenorite particles precipitated form a 3.0×10^{-4} mol dm^{-3} $Zn(NO_3)_2$, 5.0×10^{-3} mol dm^{-3} $Cu(NO_3)_2$ and 0.5 mol dm^{-3} urea solution aged at 363 K for 210 minutes.

6. Barium hexaferrite

Whereas the difficulties in the synthesis of zinc-doped tenorite reflect the thermodynamic constraints, the problems encountered in the synthesis of barium hexaferrite, $BaO.6Fe_2O_3$, are due to kinetic factors. These kinetic factors are also related to the differing crystal chemistry of the involved solids. Cubic ferrites, such as the spinels $NiFe_2O_4$ and $CoFe_2O_4$, can be prepared by isomorphous substitution during the oxidative formation of magnetite from ferrous hydroxide slurries [24, 25]. Oxidative formation of ferric compounds in alkaline media is known to direct the structure to the γ (cubic) varieties, due to the influence of ferrous ions during the early stages of crystallization [26-28]. Divalent cations of sizes not very different from that of Fe(II) can be accommodated in the growing crystal, and good mixed oxides can be obtained (although in this case also, the in-depth composition is variable [24, 29]). The bulky Ba(II) forms mixed oxides with iron that are structurally more related to the hexagonal α-polymorphs. Accordingly, in the presence of Fe(II), the cubic phases formed do not incorporate Ba(II) [30]. We have found a simple way to circumvent this difficulty [31], by avoiding the presence of Fe(II) and synthesizing an adequate precursor in a strongly oxidizing alkaline medium. For this purpose, strongly alkaline solutions of potassium ferrate(VI) were prepared by oxidation of ferric salts with potassium hypochlorite. These solutions were mixed with barium chloride, in order to obtain a mixed solid. The reaction proceeds with strong release of oxygen, and this precursor is not barium ferrate (a well known compound), but rather an intimate mixture of ferrihydrite and barium hydroxide (and/or peroxide). A rather mild heat treatment, at ca. 700 °C, suffices to produce well crystallized barium hexaferrite. Again, the crucial factor to control is the amount of iron and barium incorporated in the precursor, in order to obtain later the pure desired product, and our experiments demonstrate that an excess of iron is required in the mother liquor, because not all is incorporated in the solid precursor.

7. Acknowledgments

This work was partially supported by grants from Universidad de Buenos Aires and Fundación Antorchas, to which the authors are indebt. MAB and AER are members of CONICET.

8. References

1. Matijević, E. (1990) Preparation of well defined ceramic powders, in D.P. Vskokovic, H. Palmour, III, and R.M. Spriggs (eds.), *Science of sintering*, Plenum Press, New York, pp. 101-115.
2. Matijević, E. (1994) Uniform inorganic colloid dispersions. Achievements and challenges, *Langmuir* 10, 8-16.
3. Kallay, N., Fischer, I, and Matijević, E. (1985) A method for continuous preparation of uniform colloidal hematite particles, *Colloids Surf.* 13, 145-149.
4. Her, Y.S., Matijević, E. and Wilcox, W.R. (1990) Continuous precipitation of monodispersed yttrium basic carbonate powders, *Powder Technol.* 61, 173-177.

5. Gherardi, P. and Matijević, E. (1988) Homogeneous precipitation of spherical colloidal barium titanate particles, *Colloids Surf.* 32, 257-274.
6. Dutta, P.K. and Gregg, J.R. (1992) Hydrothermal synthesis of tetragonal barium titanate, *Chem. Mater.* 4, 843-846.
7. Segall, R.L., Myhra, S., Smart, R.St.C. and Turner, P.S. (1984) *Evaluation of critical properties of synroc for disposal of high level radioactive waste*, NERDDP EG84/285, Australian Govt. Publ. Office, Camberra.
8. Ringwood, A.E., Kesson, S.E., Reeve, K.D., Levins, D.M. and Ramm, E.J. (1988) Synroc, in W. Lutze and R.C. Ewing (eds.), *Radioactive waste forms for the future*, North-Holland Physics Publishing, Amsterdam, pp. 223-334.
9. Turner, P.S, Jones, C.F., Myhra, S., Neall, F.B., Pham, D.K. and Smart, R. St.C. (1989) Dissolution mechanisms of oxides and titanate ceramics - Electron microscope and surface analytical studies, in L.-C. Dufour, C. Monty and G. Petot-Ervas (eds.), *Surfaces and interfaces of ceramic materials*, Kluwer Academic Publishers, Dordrecht, pp. 663-690.
10. Robie, R.A., Hemingway, B.S. and Fisher, J.R. (1978) Thermodynamic properties of minerals and related substances at 298.15 K and 1 Bar (10^5 Pascals) pressure and at higher temperatures, *Geological Survey Bulletin* 1452, U.S. Govt. Publ. Office, Washington DC.
11. Bard, A.J., Parsons, R. and Jordan, J. (1985) *Standard potentials in aqueous solutions*, Marcel Dekker, New York.
12. Candal, R.J., Blesa, M.A. and Regazzoni, A.E. (1995) The attack of $YBa_2Cu_3O_{6.8}$ by acidic aqueous solutions, submitted.
13. Blesa, M.A., Morando, P.J. and Regazzoni, A.E. (1994) *Chemical dissolution of metal oxides*, CRC Press, Boca Raton, Florida.
14. Pham, D.K., Ru-Peng, Z., Fielding, P.E., Myhra, S., and Turner, P.S. (1991) The degradation of $YBa_2Cu_3O_7$ resulting from exposure to wet and dry steam, *J. Mater. Res.* 9, 1148-1155.
15. Zhou, P.-J., Riley, D.R. and McDevitt, J.T. (1993) Relative corrosion reactivity and surface microstructure of $YBa_2Cu_3O_{7-x}$ samples with different oxygen contents, *Chem. Mater.* 5, 361-365.
16. Stacey, M.H. and Shannon, M.D. (1985) The decomposition of Cu-Zn hydroxycarbonate solid solutions, in P. Barret and L.-C. Dufour (eds.), *Reactivity of solids*, Elsevier, Amsterdam, pp. 713-718.
17. Baes, Jr., C.F. and Mesmer, R.E. (1976) *The hydrolysis of cations*, John Wiley & Sons, New York.
18. Porta, P., De Rossi, S., Ferraris, G., Lo Jacono, M., Minelli, G. and Moretti, G. (1988) Structural characterization of malachite-like coprecipitated precursors of binary CuO-ZnO catalysts, *J. Catal.* 109, 367-377.
19. Waller, D. Stirling, D. Stone, F.S. and Spencer, M.S. (1989) Copper-zinc oxide catalysts: activity in relation to precursor structure and morphology, *Faraday Discuss. Chem. Soc.* 87, 107-120.
20. Pollard, A.M., Spencer, M.S., Thomas, R.G. and Williams, P.A. (1992) Georgeite and azurite as precursors in the preparation of co-precipitated copper/zinc oxide catalysts, *Appl. Catal. A* 85, 1-11.

21. Soler-Illia, G.J. de A.A., Candal, R.J., Blesa, M.A. and Regazzoni, A.E. (1994) Coprecipitation of mixed copper-zinc hydroxycarbonate particles, IX Argentine Meeting of Chemical Physics, San Luis, Argentina.
22. Candal, R.J., Regazzoni, A.E. and Blesa, M.A. (1992) Precipitation of copper(II) hydrous oxides and copper(II) basic salts, *J. Mater. Chem.* 2, 657-661.
23. Candal, R.J., Regazzoni, A.E. and Blesa, M.A. (1992) Precipitation of monodispersed mixed copper-gadolinium basic carbonate particles, *Colloids Surf. A* 79, 191-198.
24. Regazzoni, A.E. and Matijević, E. (1982) Formation of spherical colloidal nickel ferrite particles as model corrosion products, *Corrosion* 38, 212-218.
25. Tamura, H. and Matijević, E. (1982) Precipitation of cobalt ferrites, *J. Colloid Interface Sci.* 90, 100-109.
26. Domingo-Pascual, C., Rodríguez-Clemente, R. and Blesa, M.A. (1991) The pathways to spinel iron oxides by oxidation of iron(II) in basic media, *Mat. Res. Bull.* 26, 47-55.
27. Domingo-Pascual, C., Rodríguez-Clemente, R. and Blesa, M.A. (1993) Nature and reactivity of intermediates in the autooxidation of iron(II) in aqueous acid media, *Solids State Ionics* 59, 187-195.
28. Domingo-Pascual, C., Rodríguez-Clemente, R. and Blesa, M.A. (1993) Kinetics of oxidative precipitation of iron oxide particles, *Colloids Surf. A* 79, 177-189.
29. Regazzoni, A.E. and Matijević, E. (1984) Interactions of metal hydrous oxides with chelating agents. IV Dissolution of nickel ferrite in EDTA solutions, *Corrosion* 40, 257-261.
30. Domingo-Pascual, C. (1992) Synthesis and morphological characterization of magnetite (Fe_3O_4), goethite (α-FeOOH) and lepidocrosite (γ-FeOOH) particles obtained by oxidation of Fe(II), Ph.D. thesis, Universidad de Barcelona, Spain.
31. Jacobo, S. and Blesa, M.A. (1995) work in progress.

ELECTROCHEMICAL CHARACTERIZATION OF SUBMICRON PARTICLE SURFACE INCLUDING SURFACE CHEMISTRY AND STRUCTURE

S.S. DUKHIN
Institute of Colloid and Water Chemistry
National Academy of Sciences of Ukraine
Vernadskogo Ave., 42, Kiev, 252680, Ukraine

ABSTRACT. The advantage of characterization of submicron particles is that the measurement of the specific surface conductivity (SSC) and therefore the determination of the mobile charge and Stern potential in addition to the electrokinetic charge and potential is possible, which is not feasible with large particles. The ion adsorption and the stability ratio depend on Stern potential exponentially. A substitution of zeta potential for it in the above mentioned exponents can result in a large error. Consequently Stern potential measurement (instead of using zeta potential) is a necessary condition to quantify the theory of ion-particle and particle-particle interactions. Low-frequency dielectric dispersion provides the best opportunity for its determination due to high sensitivity to SSC. Recent development in quantifying the particle-particle interaction based on three variants of nonstandard electrokinetic model is characterized. The combined measurements of mobile and electrokinetic charges enable us to discriminate between the models and to choose the variant of theory of particle-particle interaction. SSC and zeta potential investigation in a wide range of electrolyte concentration and pH enable us to specify the values of invariant parameters of the models.

1. Introduction

General laws of ion-particle and particle-particle interaction are established in colloid science. However their application to technological processes is extremely restricted because information about actually occurring colloid particles is usually either absent or insufficient. The disperse system must be characterized to enable prediction of colloid particle behavior in technological processes on the basis of general colloid science.

The characterization of the surface of a dispersed particle [1, 2] comprises: the nature and surface concentration of active sites (ionogenic groups, ion adsorption active sites, etc.); ion adsorption binding characteristics such as adsorption potentials or ion binding constants for inorganic and organic ions; microgeometric characteristics of the surface, its porosity and

E. Pelizzetti (ed.), Fine Particles Science and Technology, 47–59.

roughness, its contamination or coverage by the adsorption layer; surface heterogeneity [3].

A charge of the surface is determined by means of titration. Electrokinetic measurements enable the diffuse layer charge (so-called electrokinetic charge σ_ζ) to be estimated. The formula determining it follows from the Gouy-Chapman theory which establishes the relation between the charge and potential distribution in the diffuse part of the double layer (DL)

$$\sigma_\zeta = -\frac{4n_0 ze}{\kappa}\sinh\left(\frac{ze\zeta}{2kT}\right), \qquad (1)$$

where κ^{-1} is Debye length, z is counter-ion valency, k is Boltzmann constant, T is temperature, e is proton charge, n_0 is electrolyte concentration, ζ is electrokinetic potential.

In many colloid systems, the titrable charge essentially exceeds the electrokinetic charge [1, 2].

This has resulted in development of three alternative models:

1. Zeta potential is measured some distance apart from the outer Helmholtz plane (OHP). There is a part of diffuse layer charge between the slip plane and the surface which does not contribute to the electroosmotic slip.

2. Zeta potential is measured in the OHP, and surface charge is mostly screened by the counter-ions of the Stern layer. Their flow along the Stern layer is possible.

3. The porous surface model. The charge is also distributed over the inner surface while counter-ions are localized partly in pores and partly in the portion of the diffuse layer that is associated with electrokinetic phenomena.

The models radically differ from the standard electrokinetic model (SEM) [4] and can be called the non-standard ones. Within the framework of SEM, the absence of roughness or porosity of the surface is postulated, which means that it should be smooth on the molecular level. In addition, the surface should be impermeable for ions, i.e. their flows through the surface or along it over the Stern layer should be absent.

These model representations can be used as the basis of the electrochemical characterization of many disperse systems. In general two stages should be differentiated in the methodology of electrochemical characterization: the former is the assignment of the given disperse system to the standard or non-standard electrokinetic models and the latter is the individualizing of the parameters of this model.

2. Specific Surface Conductance and Electrokinetic Models

Development of electrokinetic investigations in the 70s [2, 5] and in the 80 - 90s [6 - 10] has shown that along with titration and determination of zeta potential, measurement of specific surface conductance κ^σ is very important for the electrochemical characterization of a surface.

The well-known Bikerman formula was the first for characterizing the relationship between κ^σ and zeta potential

$$\kappa^{\sigma} = \frac{2F^2c}{RT}\left[D^{+}\left(e^{-\frac{\tilde{\zeta}}{2}} - 1\right)\left(1 + 3m^{+}\right) + D^{-}\left(e^{\frac{\tilde{\zeta}}{2}} - 1\right)\left(1 + 3m^{-}\right)\right], \tag{2}$$

where dimensionless criteria $m^{\pm}$ [4] characterize relative contributions of the electroosmosis component to the surface conductance, R is gas constant per mole, F is Faraday number, $\tilde{\zeta} = F\zeta/RT$, $D^{\pm}$ are diffusivities of cation and anion.

Thus, the criterion of applicability of SEM is the equality of κ^{σ} as measured directly and as calculated on the basis of the measured zeta potential. Information content of this criterion is determined by the fact that its test is based on two *independent* measurements [2]. *Two streams arise* as a result of superposition of electric field on a disperse system with their measurement making it possible to obtain information about DL: *the stream of matter* (the volume velocity of electrolyte) and *the stream of charge* (the electric current).

Information is presented in the vast majority of literature about electrokinetic investigations which are restricted to either the measurement of the stream of matter (electroosmosis, electrophoresis) or to the stream of charge caused by the stream of matter (streaming potential or streaming current). This information can be characterized no more than qualitative (the sign of the charge) or semi-quantitative (a slightly or a strongly charged surface). The exceptions are individual papers in which the measurements of the surface charge were conducted by titration or by the adsorption method in parallel with the measurement of zeta potential. Coincidence of these charges and σ_{ζ} is indicative of the correspondence of given disperse system to SEM.

In most cases (except for the case of applicability of SEM), a measurement of zeta potential contains information only about a part of the screening charge, the measurement of κ^{σ} gives information about the screening charge as a whole. Therefore, the electrochemical characterization of a particle surface can be quantified on the basis of a combination of electrokinetic investigations including the measurements of both the stream of matter and that of charge (κ^{σ} measurement).

3. The Additional Surface Conductivity

On the basis of experimental material, Bikerman [11] arrived at the conclusion that gels represent a much greater obstacle to the motion of a liquid than to the passage of a current. The same conclusion is valid for charged surfaces covered with an adsorbed layer of macromolecules, or even for uncharged (or charged) surfaces with an adsorbed layer of polyelectrolyte. In the last few years much attention has been paid to electrokinetic phenomena as applied to such system and a method describing electroosmotic flow through this type of gel was also developed [10, 12]. The hydrodynamic resistance which fragments of macromolecules exhibit to the electroosmotic flow of liquid greatly surpasses the viscous resistance of free liquid.

A charge distributed on the surface of a particle or on segments of an adsorbed macroion is compensated by mobile counter-ions partly distributed inside the gel and partly in the diffuse layer beyond it.

Additional surface conductivity (ASC) is caused by ion flow along that portion of the diffuse layer that is confined between the surface of the particle and the slip plane [2, 5, 6, 8].

DL is described by at least two parameters: zeta potential and the additional surface conductivity κ^{σ}_{a}. The absence of κ^{σ}_{a} corresponds to SEM. The nonstandard electrokinetic model corresponds to the κ^{σ}_{a} presence.

Thus a distinction between the contribution from ionic motion in the tangential electric field (σ_m) and the convection in electroosmotic fluid flow (σ_ζ) in κ^{σ} is important

$$\kappa^{\sigma} = \left[\sigma^{+}_{m}\Lambda_{+} - \sigma^{-}_{m}\Lambda_{-} + \frac{2\varepsilon RT\sigma^{+}_{\zeta}}{\eta Z^{+}} - \frac{2\varepsilon RT\sigma^{-}_{\zeta}}{\eta Z^{-}}\right]F^{-1}, \tag{3}$$

where $\Lambda^{\pm}$ and $Z^{\pm}$ are equivalent conductivities and the valencies of the counter-ion and coion, respectively. If the diffuse layer charge is high, the contribution to the charge by negative adsorption will be small as compared to that by the positive adsorption, so the equation for mobile charge can be simplified

$$\sigma_m = \left[\kappa^{\sigma}F - \frac{2\varepsilon RT\sigma_{\zeta}}{\eta Z^{+}}\right]\frac{1}{\Lambda^{+}} . \tag{4}$$

The simultaneous measurements of mobile σ_m and electrokinetic charges enable us to discriminate between standard and nonstandard models. For the SEM, σ_s and σ_m charges coincide. The mobile charge exceeds the electrokinetic one if the nonstandard model is valid.

The ASC model is simpler [13] if the porous layer or gel layer does not bear a charge (model 1). Then the charge is presumed to be present only in the form of an infinitely thin layer on the inner boundary of the porous layer with the non-porous substance of the particle. The potential of this layer is an analogue of the outer Helmholtz plane potential Ψ_d and the usual description of the diffuse part of DL retains its significance when the porosity is considerable. As the size of the pores decreases, the diffusivity of counter-ions in the pores $D^{\pm}_{\delta}$ can also be lowered [2]. Therefore, when setting up an expression for surface conductance, one should take into account the fact that tangential ion flows within the ASC layer (ζ, Ψ_d) are characterized by diffusivities $D^{\pm}_{\delta}$ whereas volumetric values are preserved in the hydrodynamically mobile part of DL. On the basis of such a procedure, we have generalized Bikerman's formula (Eq. (166b) in [2]):

$$\kappa^{\sigma} = \frac{2F^2 n}{RT\kappa}\left[D_{\delta}^{+}\left(e^{-\frac{\Psi_2}{2}} - e^{-\frac{\tilde{\zeta}}{2}}\right) + D_{\sigma}^{-}\left(e^{\frac{\Psi_2}{2}} - e^{\frac{\tilde{\zeta}}{2}}\right) + \right.$$
$$\left. + D^{+}\left(e^{-\frac{\tilde{\zeta}}{2}} - 1\right)\left(1 + 3m^{+}\right) + D^{-}\left(e^{\frac{\tilde{\zeta}}{2}} - 1\right)\left(1 + 3m^{-}\right)\right] \tag{5}$$

4. Advantages of Electrochemical Characterization of Submicron Particles

The change of conductivity of a suspension caused by surface conductivity is noticeable if the ratio, which we have labeled as Rel [2]

$$\mathrm{Rel} = \frac{\kappa^{\sigma}}{\mathrm{K}a} \tag{6}$$

is not small. We have proposed [2] to distinguish among two trends and two sections in the theory of electrokinetic phenomena:

1. Equilibrium double layer and electrokinetic phenomena.
2. Polarized double layer and electrokinetic phenomena.

If Rel is small, the influence of the stream of charge on the stream of matter is weak and DL deviation from equilibrium can be neglected. The theory of the equilibrium DL is sufficient to describe the electrokinetic phenomena, as Smoluchowski has done. However measurement of κ^{σ} is impossible. As result even discrimination between electrokinetic models is very complicated.

The polarization state $(\mathrm{Rel} \geq 1)$ enables, in principle, the measurement of a stream of charge along with a stream of matter to be carried out. In addition, the stream of charge shows up by its effect on the velocity of the particle. Thus, the condition $(\mathrm{Rel} \geq 1)$ offers an opportunity for the electrochemical characterization of a surface. The smaller particle the bigger is the range of electrolyte concentration and surface charge within which Rel is not very small and κ^{σ} measurement is possible. The smaller particle, the higher is the characteristic frequency of low-frequency dielectric dispersion (LFDD) [5, 14] $\left(a^2/D\right)$ that is favourable for its measurement. However for very small particle the condition of thin DL $(\kappa a >> 1)$ is violated and LFDD weakens.

5. Complex Electrochemical Characterization of Spherical Particles

Measurements of electric conductance of dilute latices as a function of volume fraction [15, 16], measurement of high-frequency conductance of concentrated latices [6, 7], measurement of LFDD on latices [7] made it possible to calculate Rel and then κ^{σ} and σ_m. The latter is found to be close to σ_t. Such agreement points to the validity of the theoretical basis actually used in interpreting these experimental data. This basis is the Maxwell-Wagner theory extended to charged particles by O'Konski as applied to high-frequency dispersion and by Dukhin and Shilov [5] as applied to low-frequency dispersion (concentration polarization of DL).

Moreover, the smallness of σ_ζ as compared to σ_m established in this work confirms that κ^σ is the most important characteristic of these surfaces. The surface charge is determined by σ_m, i.e. by κ^σ measurement, σ_ζ is not important. Its smallness means that the investigated systems correspond to nonstandard model. A small difference between σ_t and σ_m means that nonstandard model 1 is valid.

Determination of zeta potential under polarization conditions is possible by combined measurements of σ_t or κ^σ and of electrokinetic measurements. Calculating the Ψ_d-potential on the basis of data [6, 7, 15, 16] about the mobile charge, or data about σ_t [9] and using then Dukhin-Semenikhin formula [13], the researchers [7, 9, 15, 16] have obtained a new concentration dependence $\zeta(c)$. It is marked by the absence of a maximum.

Discrimination between the first and the second nonstandard electrokinetic models is possible by comparison of σ_t and σ_m.

The efficiency of applying a combination of electrokinetic and conductivity measurements to the study of counter-ion adsorption was demonstrated in papers [8, 16]. Negatively charged latices with low surface charge ($< 7\ \mu C \cdot cm^{-2}$) showed no evidence of Stern layer in simple indifferent electrolytes but do so in the presence of cesium and alkyl ammonium ions. This conclusion follows from the revealed closeness of the mobile and titration charges in simple indifferent electrolytes and a decreased value of σ_m as compared with σ_t for cesium.

6. Mechanism of Surface Charge Generation and κ_a^σ

The most important mechanisms of surface charge generation are :
(a) surface dissociation
(b) ion adsorption from solution.
Discrimination between these mechanisms is the first stage in the electrochemical characterization. To simplify the analysis, we consider binary electrolytes. It is well known that qualitative distinctions of zeta potential dependence on pH or electrolyte concentration arise due to different mechanisms of surface charge generation. The same is valid for κ_a^σ.

At surface dissociation, a part of DL (ionogenic groups) is immobile. All components of double layer generated by ion adsorption are mobile. It causes big distinctions in additional surface conductivity and the possibility of discrimination between mechanisms of surface charge generation. At surface dissociation, κ_a^σ is caused by the lateral flux of the adsorbed counter-ions because ionogenic groups are immobile and coion adsorption is low or negative. For a given electrolyte its concentration increase leads to the growth of the counter-ion adsorption and correspondingly to growth of κ_a^σ.

Simultaneously the charge of diffuse layer σ_d decreases, which leads to the decrease of Bikerman conductivity κ^{σ_d}. The bigger counter-ion adsorbability, the higher is counter-ion adsorption at a given electrolyte concentration. However the surface diffusivity can decrease at the growth of the adsorption energy. It means that both decrease or increase of κ_a^σ can occur at the adsorption energy growth.

If Dl is generated due to the *preferential ion adsorption*, the electrolyte concentration increase leads to growth of κ_a^σ, because both potential determining ion and counter-ion adsorption grow up. It leads to surface charge increase that can be accompanied by the increase of the diffuse layer charge and correspondingly by increase of κ^{σ_d}. It is the distinction from the case of surface dissociation because κ^{σ_d} decreases with increase in concentration, as it was mentioned above.

The increase of adsorbability of potential determining ions at given electrolyte concentration leads to increase of surface charge, κ_a^σ and κ^σ. It is a distinction from the surface dissociation case, because an increase in counterion adsorbability leads to a decrease of κ^σ.

Experimental investigation of the influence of pH and electrolyte concentration on both zeta potential and κ_a^σ makes the procedure of discrimination between different mechanisms of surface charge generation more exact.

Zeta potential calculation is impossible without κ^σ measurement and the application of the theories of κ^σ, double layer polarization, and its influence on electrophoresis. Thus, the specification of the theories of these phenomena in respect to different models of the surface charge generation is necessary for the quantifying of zeta potential determination. The electrophoresis equation for first model is given in [2, 13], for second model in [18]. Some new results for third model are given in the next section.

7. Additional Surface Conductance of Porous Surface Layer

Recently the theory of additional surface conductivity was extended over model 3 in papers [19, 20]. Old investigations where different assumptions were made are discussed in this papers and in [1].

The diversity of these assumptions may be integrated within the framework of one multiparameter model of additional surface conductance. Constitutive parts of this model are: a spatially homogeneous surface layer of finite thickness; an equation of the Langmuir-Stern adsorption isotherm for ions of any type; and finite diffusion coefficients for all types of ions within the surface layer.

Ions can be adsorbed in the porous layer according to equation [3]

$$A_i = \frac{Q_i K_i C_i e^{-z_i \Psi}}{1 + K_i C_i e^{-z_i \Psi}}, \tag{7}$$

where K_i is the constant of adsorption affinity for ion i, Q_i is the concentration of adsorption sites for ion i, and Ψ is the electric potential. Assume that adsorbed ions are characterized by the diffusion coefficient D_{is} which is less than the bulk diffusion coefficient D_i. Then the additional surface conductance is given by the equation

$$K_{SA} = \frac{F^2}{RT} \sum_{i=1}^{N} z_i^2 D_{is} \int_0^{\infty} A_i(x)dx. \tag{8}$$

The key parameter determining the potential profile in the surface layer is the thickness of the electrical double layer in gel: when the surface layer thickness is considerably greater than the thickness of the double layer, almost the entire surface layer has the same potential, known as the Donnan potential. This potential corresponds to electroneutral gel: due to its value and sign the potential hinders the adsorption of strongly adsorbed ions with a high affinity to the surface (potential-determining ions), and promotes the adsorption of ions with a low affinity. Equation for the Donnan potential Ψ_D is

$$\sum z_i A_i(\Psi_D) = 0. \tag{9}$$

Even when the potential inside a surface layer of moderate thickness is less than the Donnan potential (Fig. 1), the latter is the important and convenient parameter determining the surface layer behavior.

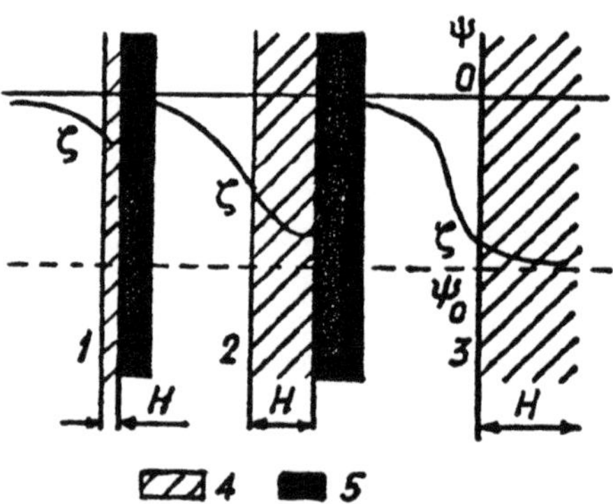

Figure 1. The potential distribution in a surface layer: (1) thin isopotential gel layer, (2) gel layer of intermediate thickness, (3) infinitely thick gel layer, (4) gel layer, (5) the particle bulk impermeable to ions.

When the thickness of the surface layer is much less than the double layer thickness, the potential inside this layer is almost constant. This surface layer may be called isopotential.

Concentration of the potential-determining ions (these ions in the diffuse part of a double layer are coions) exceeds concentration of counter-ions. The difference between the concentrations depends on the ratio of the adsorption affinities of coions and counter-ions. When this ratio is high, zeta potential and the ratio of the concentrations are also high, indicating virtually that, in this case, counter-ions are excluded from the surface layer and its charge depends only on coions. When difference between the affinities is small, zeta potential is also low, and concentrations of coions and counter-ions differ only slightly; i.e., quasi-equivalent adsorption takes place.

Thus, when zeta potential is high, the additional surface conductance is proportional to the charge of the surface layer, but this charge, in turn, equals the electrokinetic charge of the diffuse part of the electrical double layer; this latter charge determines the Bikerman component of total surface conductance. Thus, the additional surface conductance for sufficiently strongly charged particles is expected to be, at least, commensurable with the Bikerman surface conductance. However, more probably, the additional surface conductance will be less than the Bikerman surface conductance because of the low mobility of ions in the surface layer.

When zeta potential is low and quasi-equivalent adsorption is realized, the sum of adsorptions can essentially exceed their difference; i.e. the surface layer charge that equals the electrokinetic charge can constitute only a small part of the total number of the charge carriers in the surface layer. In this case, the additional surface conductance will strongly predominare over the Bikerman conductance, and measurements of surface conductance provide direct information on the surface layer properties.

Thus, in studying surface conductance of the particle with a thin surface layer, we have two different situations for high and low zeta potentials. In the former case, the Bikerman surface conductance predominates over additional surface conductance; in the latter case (low zeta potential), the additional surface conductance can predominate.

These results obtained for thin layer [19] mainly preserve for thicker layer where the assumption of the isopotentiality is not valid (see Fig. 2).

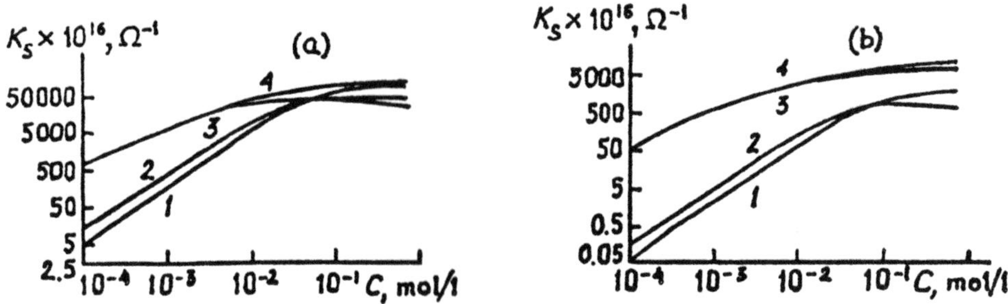

Figure 2. Dependence of (1, 2) the Bikerman and (3,4) the additional surface conductance on the electrolyte concentration: for K^- (1, 3) 1 , (2, 4) 10 l / mol; for H (a) 10, (b) 1 Å.
$Q^+ = Q^- = 0.01$ mol / l; $K^+ = 100$ l / mol.

Specific calculations have been made on the basis of Poisson-Langmuir equation for the particular case of the low Donnan potential of the gel layer [20]. It was found that interfacial potential difference is mostly localized in a medium (gel or solution) where screening is less effective. This fact is responsible for the possible appearance of a maximum of the concentration dependence of a zeta potential.

When only one type of potential-determining ions is present in a solution, a decrease in the electrolyte concentration results in unlimited decreases in adsorption, charge and potential. Addition of ions of the third and subsequent types to the system is revealed in a less pronounced effect of the charge variation with electrolyte concentration.

Strong concentration dependence of surface conductance is helpful in order to distinguish experimentally between the model of the porous surface layer and any other possible models. In addition, the relatively high sensitivity of the surface conductance to the parameters of the porous layer makes it possible to determine the latter.

8. Role of Surface Characterization in Quantifying Ion-Particle and Particle-Particle Interaction

The universally recognized theory of hydrosol stability is closely related to SEM. The important task of the extension of this theory over the real colloids dictates the necessity to generalize DLVO equations in respect to *models 1 - 3*. At some conditions the existing DLVO equations can be used when *model 1* is valid. The total potential drop across diffuse layer Ψ_{ef} as calculated from κ^{σ} measurement according Eq. (5) has to be substituted into DLVO equation instead of Stern potential. This approach can be valid if the distance of the electrostatic barrier to the surface exceeds the thickness of porous layer, estimated on the basis of Ψ_{ef} and zeta potentials.

Dynamic interpretation of colloid stability is necessary in case of *model 2*. The transient disequilibration of Stern layer proceeds during double layer interaction [21].

Surface charge relaxation may take place by the lateral transport of adsorbed ions, that shows the importance of κ_a^{σ}. When the rate of the lateral ion transport is rapid as compared to the rate of particle encounter, the double layer is continuously equilibrated during a collision (quasi-equilibrium interaction) but interesting dynamic features emerge when this condition is not met. (the "dynamic" situation). For the rate of coagulation, substantial deviations from those predicted by the traditional theory of slow coagulation are found. For an experimental cheek of the theory in the transition regime (and to establish the conditions for the limiting cases) one requires information such as ion surface diffusivities at given potential for isolated particle.

Treatment of the electrostatic interaction in terms of the extended model of the porous double layer (*model 3*) is of special importance for the analysis of biocolloid stability.

Analyzing the effect of gel layer on stability, Ohshima [10] considered the case of constant density of a fixed spatial charge, which is typical only for very acidic or very basic ionites. In general, the charge density is dependent on the electrical double layer potential. This evidence significantly complicates the problem, since it involves variation in the potential inside of gel layer, thus necessitating the relevant calculation of charge distribution. A general approach to the derivation of the theory of ion-electrostatic interaction of the particles with charged gel layers, and analysis of simple limiting cases of the general formulas is given in [22]. Without writing cumbersome formulas, we will point out the main result (see Fig. 3).

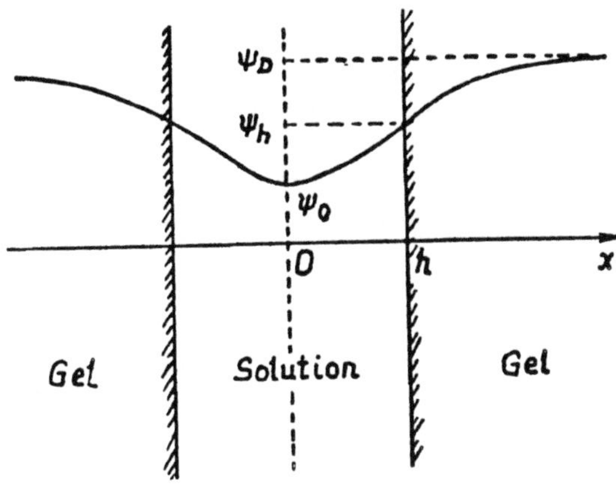

Figure 3. Coordinate system and notations to figures.

Let us assume that the thickness of gel layer l exceeds the internal Debye radius κ_i^{-1} since the effective Stern potential equals the Donnan potential Ψ_D, and hence does not change as the particles come closer together. Let the particles be identical, i.e. the normal component of the electric field in the symmetry plane equals to zero. As the particles approach, the symmetry plane comes closer to the surface of the particle so that in the limit $h \to 0$ electric field on the surface tends to zero. Taking into consideration the fact that this condition is also met inside the thin gel layer, it may be concluded that at $h \to 0$ the gel layer is isopotential and zeta potential tends to Ψ_D. The important conclusion ensues that particle interaction at small distances proceeds under conditions of the almost constant potential equal to Ψ_D.

This conclusion is true also in the case of a thin gel layer whose thickness is less than κ_i^{-1}. As the particles approach, the electrokinetic potential grows due to the overlapping of the diffuse layers, which causes also a growth of Ψ_{ef}. Thus Ψ_{ef} which for an isolated particle at $\kappa_i^{-1} < h$ is less than Ψ_D, grows when the particles approach and approximates Ψ_D. The fact that as a result of this growth Ψ_{ef} and zeta potential attain values equal to Ψ_D follows from the isopotentiality of the gel layer drawn together to the limit $h \to 0$.

Nevertheless, the interaction does not proceed at Ψ_D but at a somewhat lower potential Ψ_{max} corresponding to the position of the energy barrier which does not coincide with the surface of the particle and does not correspond to $h \to 0$. The thinner the gel layer, the greater is the difference between Ψ_{max} and Ψ_D at the same value of the Hamaker constant.

The main parameter controlling the electrostatic component of the disjoining pressure of the particles covered with the gel layer is the Donnan potential of the gel. This value is beyond the reach of direct measurement. However, it may be calculated by complex electrosurface measurements, including the detection of the surface charge by titration and the measurement of surface electrical conductivity [22]. When the Donnan potential is replaced by zeta potential, which can be measured directly using electrophoresis, significant error may arise during the calculation of the disjoining pressure.

The author is grateful to the International Association for the Promotion of Cooperation with Scientists from the Independent States of the Former Soviet Union grant No. (1) 93-3372 and to Soros Foundation grant No. UAF000

References

1. Hunter, R.J. (1987) ***Foundation of Colloid Science* 1**, Oxford University Press, Oxford.
2. Dukhin, S.S. and Derjaguin, B.V. (1974) ***Electrokinetic Phenomena,*** in: E. Matijevic (ed.), Surface and Colloid Science **7**, Wiley, N.Y., Toronto.
3. Koopal, L.K. (1993) Adsorption of ions and surfactants, in B. Dobias (ed.), ***Coagulation and Flocculation***, Marcel Decker, New York, pp. 101-127.
4. Overbeek, J.Th.G. (1950) Electrophoretic velocity of colloids, ***Adv. Coll. Sci.*3**, 97-135.
5. Dukhin, S.S. and Shilov, V.N. (1974) ***Dielectric Phenomena and the Double Layer in Disperse Systems and Polyelectrolytes***, John Willey, N.Y., Toronto.
6. Midmore, B.R. and Hunter, R.J. (1988) The effect of electrolyte concentration and co-ion type on the ζ -potential of polystyrene latices, ***J. Colloid Interface Sci.* 122**, 521-529.
7. Midmore, B.R., Diggins, D., and Hunter, R.J. (1989) The effect of temperature and ion type on the effective charge in polystyren latices, ***J. Colloid Interface Sci.* 129**, 153-161.
8. Van der Linde, A.J. and Bijsterbosch, B.H. (1989) Zeta potentials as calculated from streaming potential and conductivity data, ***Symposium Proceeding, Electrokinetic Phenomena 89,*** Germany, Dresden, pp. 265-275.
9. Fernandez-Barbero, A., Martin-Rodriguez, A., Callejas-Fernandez, J., and Hidalgo-Alvarez, R. (1994) On the calculation of electrokinetic potential and Hamaker constant of model colloids, ***J. Colloid Interface Sci.* 162**, 257-260.
10. Hiroyuki Ohshima (1994) Electrophoretic mobility of soft particles, ***J. Colloid Interface Sci.* 163**, 474-483.
11. Bikerman, J.J. (1970) ***Physical Surfaces***, Acad. Press, NY, Ch. 8.
12. Donath, E. and Voigt, A. (1985) Elektrokinetik an strukturen Grenzflachen, ***Symposium Proceeding, Elektrokinetische Erscheinungen - 1985***, Germany, Dresden, pp. 77-84.
13. Dukhin, S.S. and Semenikhin, N.M. (1970) Calculation of electrophoretic and diffusiophoretic mobility of solid spherical particles, ***Kolloidn. Zh.,* 32**, 360-368.
14. Mandel, M and Odijk, T. (1984) Dielectric properties of polyelectrolyte solutions, ***Ann. Rev. Phys. Chem.* 35**, 75-108.
15. Baran A.A., Dudkina, L.M., and Soboleva, N.M. (1981) Investigation of the structure of a double layer of polysterene latexes, ***Kolloidn. Zh.* 43**, 211-219.
16. Shubin, V.E., Sidorova, M.P., Chechik, O.S., and Sacharova, N.A. (1991) Electric surface properties of polystyrene latices in electrolyte solutions and the structure of the double layer on the latex surface, ***Kolloidn. Zh.* 53**, 187-190.
17. Lyklema, J., Springer, M.M., Shilov, V.N., and Dukhin, S.S. (1986) Relaxation of the double layer around colloidal particles and the low-frequency dielectric dispersion, ***J. Electroanal Chem. Interfacial Electrochem.*198**, 20-26.
18. Simonova, T.S. and Shilov, V.N. (1986) Influence of the mobility of ions of the dense portion of double layer on electrophoresis and electrical conductivity of disperse systems, ***Kolloidn. Zh.*48**, 370-373.
19. Zharkikh, N.I., Pendze, H., and Dukhin, S.S. (1994) The additional surface conductance of the thin porous layer of a colloidal particle, ***Colloid Journal* 56**, 648-654.
20. Zharkikh, N.I. and Dukhin, S.S. (1994) Additional surface conductance of the weakly charged porous surface layer of a colloidal particle, ***Colloid Journal* 56**, 668-673.
21. Dukhin, S.S., Shulepov, Yu.V., and Lyklema, J. (1994) Criteria of applicability of the approximations of constant charge and quasi-equilibrium, ***Colloid Journal* 56**, 566-572.
22. Zharkikh, N.I. and Dukhin, S.S. (1994) Electrostatic interaction between two charged gel layers, ***Colloid Journal* 56**, 680-686.

A NEW METHOD IN PREPARATION OF MONODISPERSE PARTICLES

T. SUGIMOTO
Institute for Advanced Materials Processing
Tohoku University
Katahira 2-1-1, Aobaku, Sendai 980-77, Japan

1. Introduction

It seems needless to say that monodispersed particles perfectly controlled in size, shape and internal structure are the ultimate goals of general advanced materials such as electromagnetic devices, modern ceramics, catalysts, photosensitive materials, medicines, pigments, cosmetics, etc., since the properties of these materials decisively depend on their size, shape and structure. However, when we try to produce monodispersed particles as ideal materials on industrial scale, we have to be confronted with essential problems of their exceedingly low productivity or high costs, because most of them have been prepared in extremely dilute systems (mostly of the order of 10^{-3} – 10^{-2} mole dm^{-3}) to avoid the tremendous coagulation or need costly metal alkoxides as starting materials. In addition, when we employ the sol–gel process for the preparation of amorphous metal oxides starting from metal alkoxides, we usually suffer from a problem of its rather poor reproducibility. Still more difficult issue is the perfect control of their mean size, shape and internal structure, and it is hardly achieved in most monodispersed systems. Thus, it has appeared almost impossible in principle to realize the dream of producing monodispersed particles from ordinary metal salts in highly condensed systems, not to mention the precise control of the above properties.

Recently, however, we have succeeded in preparation of monodispersed hematite(α-Fe_2O_3) particles in a highly condensed system with yield of almost 100% starting from an inexpensive iron salt, ferric chloride.[1-4] The simple process has been explored on an idea that if we make use of highly condensed ferric hydroxide gel as a matrix as well as a reservoir of ferric ions for growing hematite particles, it may be possible to avoid the coagulation of the hematite particles even in the presence of a highly concentrated electrolyte by fixing them in the gel matrix. In addition, the gel matrix is expected to disappear completely at the end of the reaction. Since we prepare the sol of hematite from hydroxide gel, the procedure has been named "gel–sol method" as a reversal of the sol–gel process, though there are almost no common aspects between them. And, the gel–sol method showed excellent reproducibility. Thus, the gel–sol method is a kind of general process of high potentialities for the synthesis of general monodisperse particles and may give us a clue to solve the essential problem of low productivity. Moreover, the gel–sol method has turned out to be a quite promising procedure for the precise control of the mean size, shape, and internal structure of the products.

E. Pelizzetti (ed.), Fine Particles Science and Technology, 61–70.

2. Preparation of α–Fe_3O_3 Particles by the Gel–Sol Method

2.1 THE STANDARD PROCEDURE AND GROWTH MECHANISM

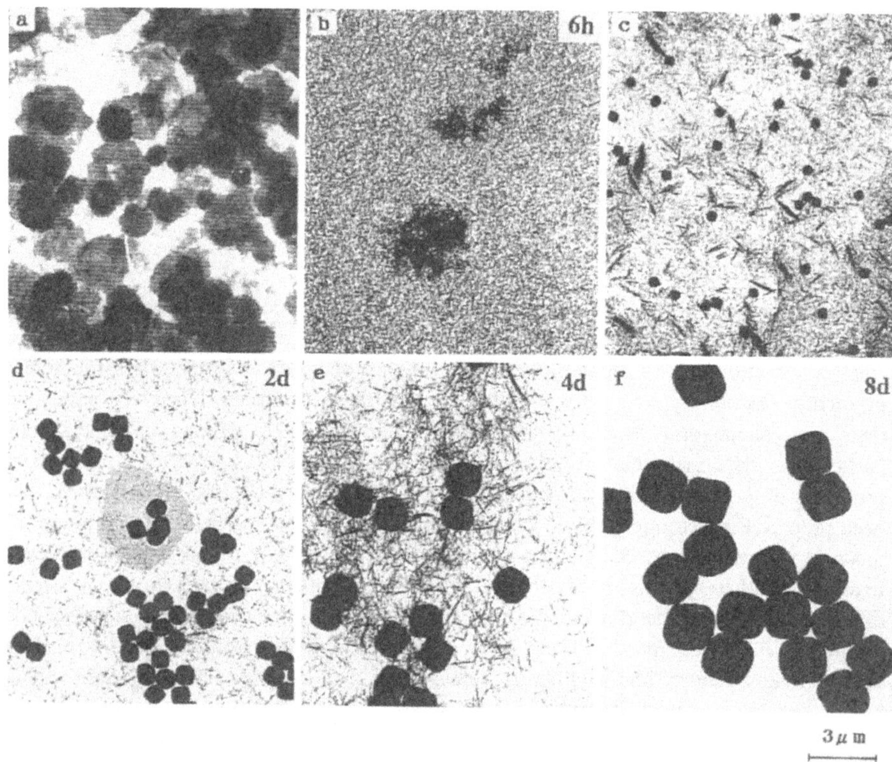

Figure 1. The phase transformation and the growth of pseudocubic hematite particles with aging time under the standard conditiions: (a)0 h, (b)6 h, (c)1 day, (d)2 days, (e)4 days, and (f)8 days.

The standard procedure for the preparation of monodisperse α–Fe_2O_3 particles is as follows.[1] One hundred ml of 5.4 mol dm^{-3} NaOH is added to the same volume of 2 mol dm^{-3} $FeCl_3$ in 10 min at room temperature under agitation. The resulting highly viscous 0.9 mol dm^{-3} $Fe(OH)_3$ gel with an excess of 0.1 mol dm^{-3} Fe^{3+} in the solution phase was aged at 100 °C in an oven for 8 days. As shown in the transmission electron micrographs in Fig. 1, the amorphous ferric hydroxide gel in Fig. 1a was transformed into very fine β–FeOOH particles (b) after 6 h of aging at 100 °C, as identified by XRD. After 24 h, monodispersed pseudocubic particles were observed with the great amount of β–FeOOH. Then the hematite particles continued to grow at the expense of the β–FeOOH and finally the β–FeOOH particles were completely dissolved after 8 days leaving only the grown α–Fe_2O_3 particles. The yield of the hematite was almost 100%. Thus, the hematite particles were formed through a two–step phase transformation of the solid phase, which was found to proceed by dissolution–recrystallization mechanism via solution phase. In this process, the extremely viscous suspension of β–FeOOH also worked as a matrix of α–Fe_2O_3. Fig. 2 shows the compositional changes in the solution phase and in the solid phase as well as the change of pH measured at room temperature. As obvious from this figure, the $Fe(OH)_3$ gel were completely dissolved in a few hours at 100 °C but, instead, β–FeOOH particles were formed at the same time, and the proportion of the β–FeOOH gradually declined to be replaced by α–Fe_2O_3 with the progress of aging. Another important aspect of this figure is the changing behavior of the concentration of the supernatant iron and pH. Namely, the initial rapid drops of the supernatant iron concentration and pH are due to the formation of β–FeOOH of the relatively lower solubility product than $Fe(OH)_3$ and it serves for the formation of monodispersed particles of hematite by reducing the supersaturation for the renucleation of the hematite. Besides, the lowering of pH is especially effective for reducing the concentration of the precursory complex of hematite such as $Fe(OH)_2^+$ which is thought to be specifically responsible for the nucleation as well as the growth of hematite. The intermediate β–FeOOH plays the key role in the monodispersed particle formation of hematite by satisfying the most important two requirements for the monodispersed particle formation: *i.e.*, 1) Inhibition of coagulation of growing particles and 2)Distinct separation between nucleation and growth periods.[5]

The nucleation of the hematite particles occurs during the addition of sodium hydroxide and the early part of aging in the oven kept at 100 °C for the formation of β–FeOOH, as is obvious from Fig. 3 in which the excess concentrations of Fe^{3+} before and after aging are shown to have a strong effect on the final size of the hematite. Also, in a different experiment, it has been

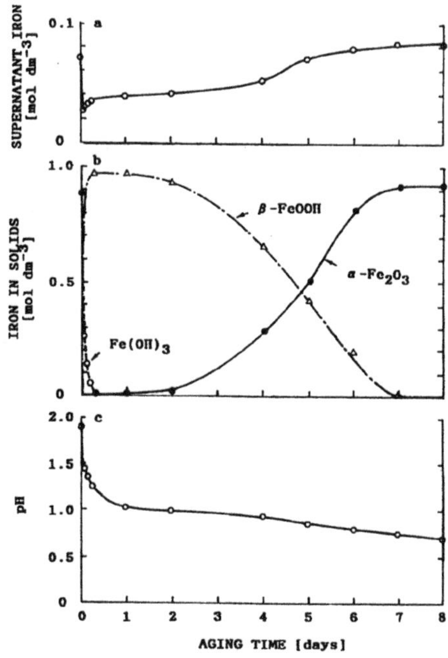

Figure 2. Changes of the composition of iron species and pH with aging time, where all measurements were conducted at room temperature: (a)supernatant iron, (b)iron in solid, and (c)pH.

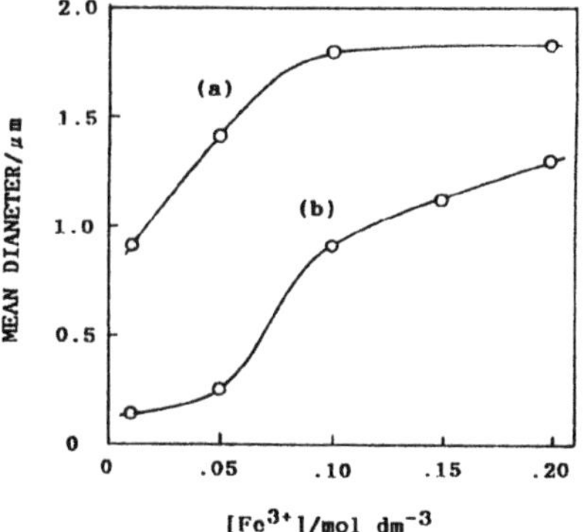

Figure 3. (a)Effect of $[Fe^{3+}]$ before the start of aging at 100 °C on the final mean size of the hematite, where all $[Fe^{3+}]$ values at start of aging were adjusted to 0.1 mol dm^{-3}. (b)Effect of $[Fe^{3+}]$ at the start of aging at 100 °C on the final mean size of the hematite, where all $[Fe^{3+}]$ values were fixed at 0.01 mol dm^{-3} until their variation just before the start of aging.

confirmed that the nucleation of hematite is negligible when the addition of 5.4 mol dm^{-3} NaOH is conducted at 3 °C for 30 min under the otherwise standard conditions.

2.2 SHAPE CONTROL

When we added some adsorptives for hematite such as sulfates,[3,4] inorganic phosphates,[3] organic phosphate, hydroquinones,[6,7] EDTA,[6,7] and nitrilotriacetic acid (NTA)[6,7] converted the morphology of the hematite particles into ellipsoidal particles. Particularly, as the concentrations of sulfate and inorganic phosphate were increased, the ellipsoids turned into peanut-type particles.[3,4] Fig. 4 shows the typical effect of sulfate ions on the

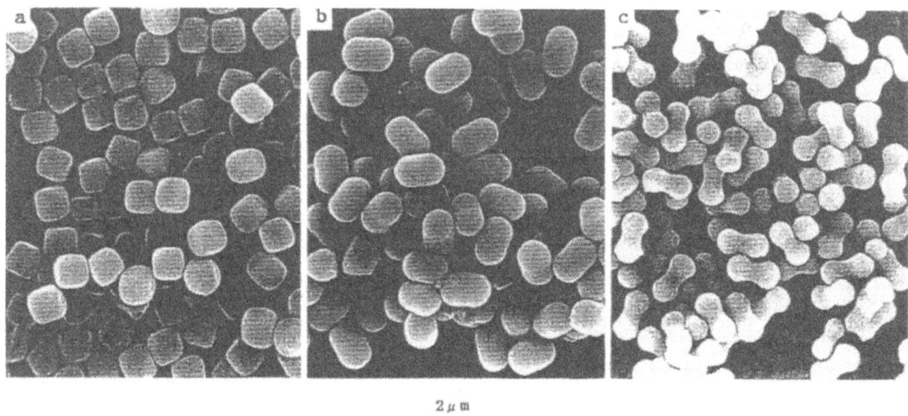

Figure 4. Scanning electron micrographs depicting the effect of sulfate ions on the morphology of monodisperse hematite particles prepared by the gel-sol method. The concentrations of Na_2SO_4 in the gels are (a)0, (b)$10^{-2}M$, and (c)$3\times10^{-2}M$.

morphology of the hematite particles. The revolutional axis of these ellipsoidal and peanut-type particles was found to agree with the c-axis of the hexagonal crystal system of hematite and these particles were all polycrystals consisting of much smaller elongated crystallites oriented in regular directions. At least for sulfate, inorganic phosphate, and hydroquinone, the most parts, say, ca. 90 %, of the used additives were observed to be incorporated into the resulting hematite particles. The mechanism of the shape control has been explained in terms of the the anisotropic growth of the crystallites of the hematite particles due to the selective adsortption of the additives to the crystal planes parallel to the c-axis of each crystallite. Since only irregular shaped particle were produced when we used ferric nitrate or ferric perchlorate as the starting iron salt, the highly concentrated chloride ions themselves must be responsible for the formation of the pseudocubic particles bound by {012} planes, as identified by OPML-XRD method.[8] Presumably, the chloride ions may strongly be adsorbed to the {012} planes and retard the growth in the direction normal to {012} planes. On the other hand, we have also succeeded in preparation of monodispersed platelet hematite particles of hexagonal shape for the first time, as shown in Fig. 5, which were prepared by the gel-sol method starting from β–FeOOH at an extremely high pH.[9] The platelet particles were disclosed to be bound by the well developed {001} basal planes and {012} side planes by the OPML–XRD with the measurement of the angle between the basal plane and the side planes from the SEM. Hence, it seems that hydroxide ions most strongly suppress the growth of the basal planes of hamatite and thus enhance the anisotropic growth in the direction normal to the c–axis.

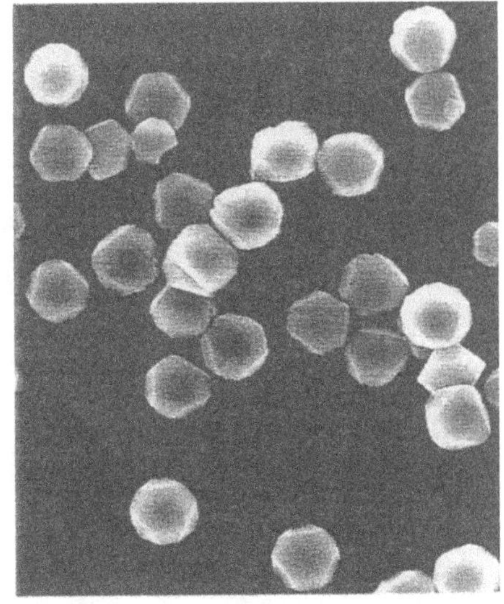

3 μm

Figure 5. Scanning electron micrograph of monodisperse platelet–type hematite prepared by the gel–sol method.

2.3 SIZE CONTROL

As has been shown in Fig. 3, the control of the final size of the hematite can be achieved by variation of the concentration of excess Fe^{3+} ions, or pH, in the preparation of ferric hydroxide gel. If we control the temperature during the precipitation of ferric hydroxide gel, the same purpose can be attained. However, these methods are not satisfactory for the size control, since considerable growth occurs concurrently with nucleation so that the size distribution becomes rather broad. Thus, we ground pseudocubic hematite particles using a planetary mill, resupended the ground powder, retrieved the supernatant suspension by centrifugation at 10,000 rpm for 30 min and used it as a seed suspension to be added to ferric hydroxide gel.[10,11] The average size of the seeds were ca. 80 **Å**. Fig. 6 shows the effect of seeding, in which the volume of the seed suspension was varied and a fixed amount of sodium sulfate, 3.0×10^{-2} mol dm^{-3} in final concentration, was added. Obviously, systematic control of the size of the hematite with the same aspect ratio (length ratio of the longer axis to the shorter) of each particle is possible by this procedure and we have found that it was also possible to change the aspect ratio with a fixed particle volume by this method. The seeding technique provides us with exceedingly uniform particles precisely controlled in size and shape, because we can grow only the seeds without spontaneous nucleation in this system.

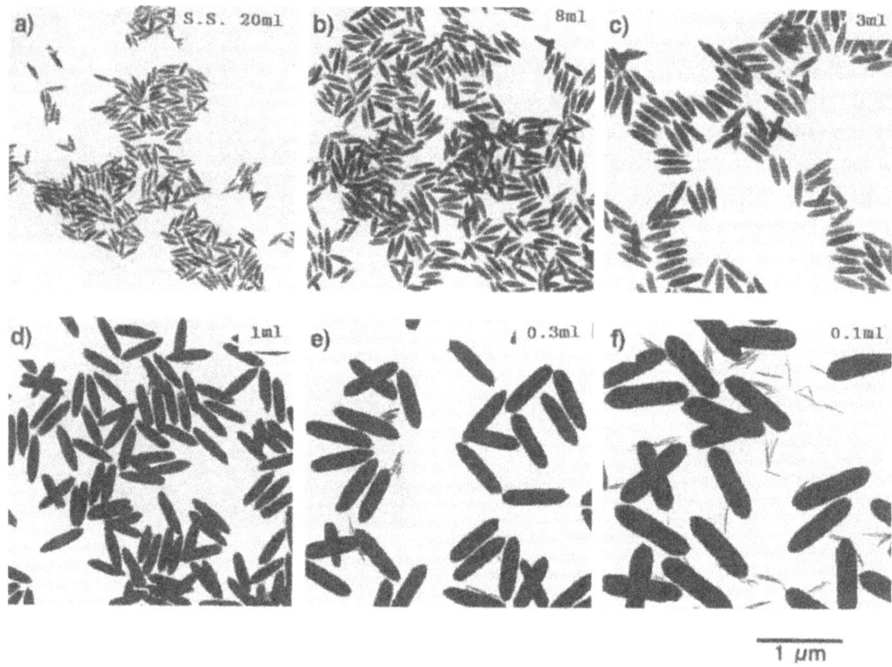

Figure 6. Size control of ellipsoidal hematite particles by changing the added amount of the seed suspension to the ferric hydroxide gels at a fixed concentration of sodium sulfate, $3.0\times10^{-2}M$, where the gels were aged for 48 h at 100 °C.

2.4 INTERNAL STRUCTURE CONTROL

The shape controllers including anions such as chloride, sulfate and phosphate, and organic compounds such as organic phosphate, hydroquinones, EDTA, and NTA have also a strong effect on the internal structure of the product. For instance, Fig. 7 shows TEM's of slices of the pseudocubic and peanut–type particles prepared with an ultramicrotome.[12-14] Obviously, these particles are polycrystals consisting of much smaller subcrystals radially developed from the center or revolutional axis of each particle. The average sizes of the subcrystals were ca. 25 nm for the pseudocubic particles, while 5–10 nm in width and 70 nm in length for the peanut–type particles. Since we found that, when we used ferric nitrate or ferric perchlorate as the starting salt, the resulting hematite particles were single crystals, the polycrystalline structure of the pseudocubic particles seems to be due to the strong adsorption of the highly concentrated chloride ions (3.0 mol dm^{-3} in the solution phase) to each subcrystal by inhibiting the lateral fusion of the two–dimensional nuclei on the growing surfaces. Thus, it is expected that the considerable amount of chloride ions may remain in the grain boundaries of the subcryatals. Actually, the atomic ratio of the internal chloride to Fe in a hematite particle prepared under the standard conditions was found to be ca. 0.02

from EDX measurement on a sliced pseudocubic particle.[14] The elongated subcrystals in the peanut-type particles are thought to be brought about by the selective adsorption of the sulfate ions to the crystal planes of each subcrystal parallel to the c-axis. The selective adsorption of sulfate and phosphate has been explained in terms of the fitting between the interatomic distances of O–O of these anions and Fe–Fe of the crystal planes of hematite.[4] It was also confirmed that about 90 % of the total sulfate ions were incorporated in the peanut-type particles, as measured by ion-chromatography on the supernatant solution and EDX on a sliced peanut-type particle. Furthermore, we found a considerable difference between the effects of sulfate and phosphate on the internal structures of ellipsoidal particles when they were grown from a β-FeOOH suspension at 140 °C for three days with these anions in the presence of much reduced concentration of chloride ions (0.1 mol dm^{-3}). Namely, while the particles prepared with sulfate ions (3.0×10^{-2} mol dm^{-3}) remained typical polycrystals, those with phosphate ions (3.0×10^{-3} mol dm^{-3}) were found to be almost single crystals, as disclosed by transmission electron microscopy and electron beam diffractometry on the respective slices[15] (Fig. 8). By using the same technique, the platelet-type particles were found to be single crystals.[16]

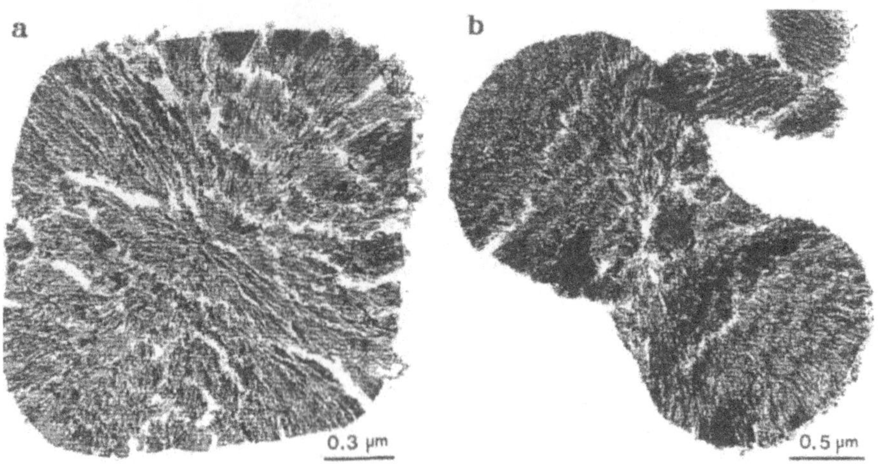

Figure 7. Transmission electron micrographs of sliced pseudocubic and peanut-type hematite particles with an ultramicrotome.

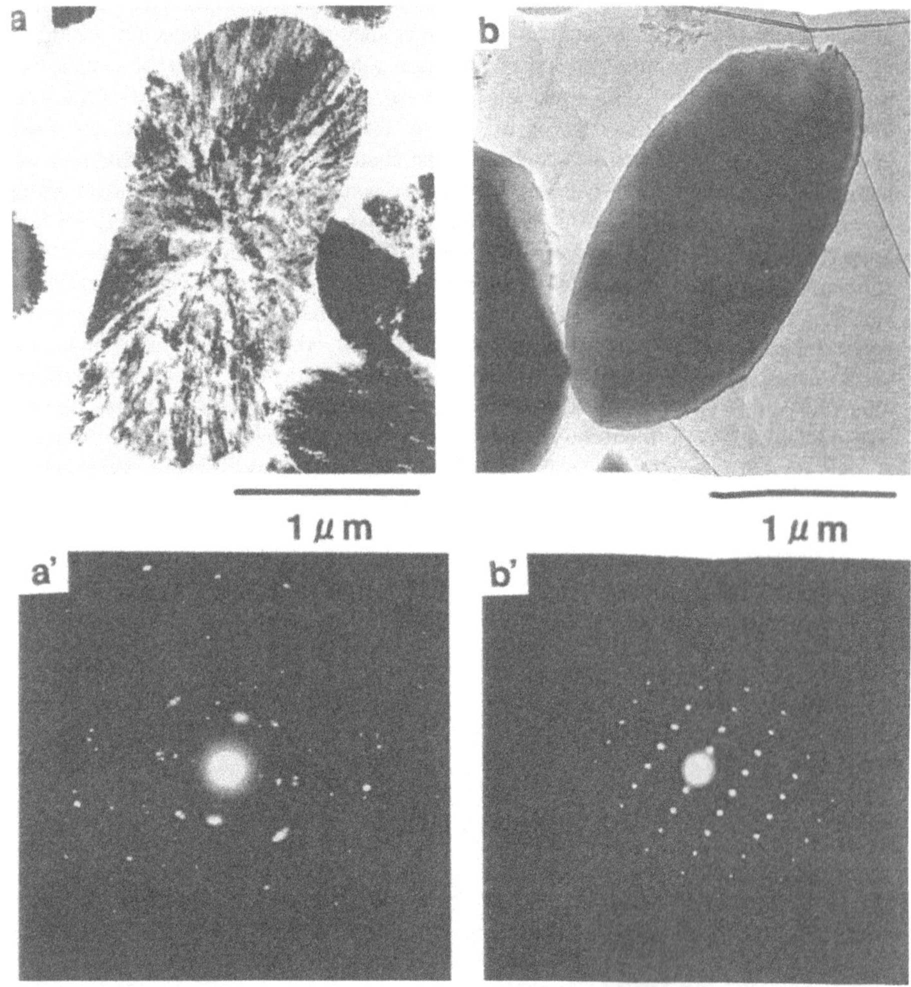

Figure 8. Transmission electron micrographs and electron beam diffractions of sliced ellipsoidal hematite particles prepared in the presence of $3.0\times10^{-2}M$ sulfate ions (a) and $3.0\times10^{-3}M$ phosphate ions (b). Both were prepared by aging β-FeOOH suspensions at 140 °C for 3 days in the presence of 0.1 *M* HCl.

3. Applications of the Gel-Sol Method

We are now developing the gel-sol method for different particles and we have so far prepared monodispersed particles of CdS,[17-19] Cu_2O,[20] basic aluminum sulfate, etc., in large quantities from highly condensed hydroxides of these metals. In the case of CdS, 1 mol dm^{-3} crystalline $Cd(OH)_2$ was completely converted into monodispersed spherical particles of the order of 40 nm with 1.1 mol dm^{-3} thioacetamide (TAA) at 20 °C and pH 8.5 in a pH buffer system of NH_3/NH_4^+ of gelatin solution. The reaction was so rapid as to be finished within 2 min and, surprisingly, TAA was found to be converted into acetonitrile by this reaction. In this system, gelatin (1 wt%) was indispensable to inhibit the direct attack of TAA to $Cd(OH)_2$ and coagulation of the CdS particles. In addition gelatin also worked as a pH buffer. On the other hand, the spherical Cu_2O particles were prepared from CuO crystals (0.5 – 2.0 mol dm^{-3}) with hydrazine at 30 °C and pH 9.3 in the presence of gelatin. The gelatin played again the important multiple roles similar to the case of CdS. In the standard conditions of the preparation of the basic aluminum sulfate, 9 ml of 4.8 mol dm^{-3} NaOH was added to 15 ml of mixed solution of 0.352 mol dm^{-3} in $Al_2(SO_4)_3$ and 0.896 mol dm^{-3} in $Al(NO_3)_3$, followed by aging at 100 °C for 3 days. SEM's of these particles are shown in Fig. 9.

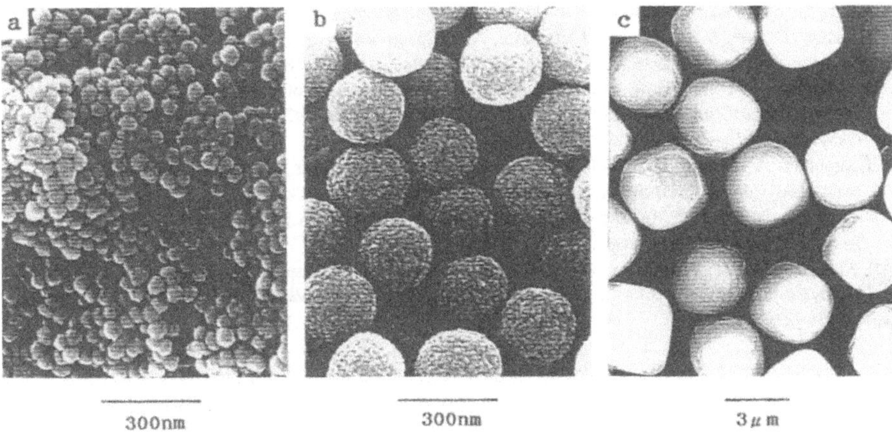

Figure 9. Monodispersed particles of CdS, Cu_2O, and basic aluminum sulfate prepared from highly condensed hydroxides of the corresponding metals: (a)CdS, (b)Cu_2O, and (c)basic aluminum sulfate.

References

1. Sugimoto, T. and Sakata, K. (1992) Preparation of monodisperse pseudocubic α–Fe_2O_3 particles from condensed ferric hydroxide gel, *J. Colloid Interface Sci.* **152**, 587–590.
2. Sugimoto, T., Sakata, K., and Muramatsu, A. (1993) Formation mechanism of monodisperse pseudocubic α–Fe_2O_3 particles from condensed ferric hydroxide gel, *J. Colloid Interface Sci.* **159**, 372–382.
3. Sugimoto, T., Khan, M. M., and Muramatsu, A. (1993) Preparation of monodisperse peanut–type α–Fe_2O_3 particles from condensed ferric hydroxide gel, *Colloids Surf. A* **70**, 167–169.
4. Sugimoto, T., Khan, M. M., Muramatsu, A., and Itoh, H. (1993) Formation mechanism of monodisperse peanut–type α–Fe_2O_3 particles from condensed ferric hydroxide gel, *Colloids Surf. A* **79**, 233–247.
5. Sugimoto, T. (1987) Preparation of monodispersed colloidal particles, *Adv. Colloid Interface Sci.* **28**, 65–108.
6. Itoh, H., Mochida, T., and Sugimoto, T. (1992) Formation mechanism of monodisperse ellipsoidal hematite particles by gel–sol method, *Proc. 64th Annual Meeting Chem. Soc. Japan*, Niigata, vol.1, pp.110.
7. Itoh, H., Mochida, T., and Sugimoto, T. (1993) Effect of added organic compounds on the morphology of hematite colloid prepared by gel–sol method, *Proc. 65th Annual Meeting Chem. Soc. Japan*, Tokyo, vol.1, pp.305.
8. Sugimoto, T., Muramatsu, A., Sakata, K., and Shindo, D. (1993) Characterization of hematite particles of different shapes, *J. Colloid Interface Sci.* **158**, 420–428.

9. Sugimoto, T., Waki, S., and Itoh, H. (1995) Preparation of monodisperse platelet-type hematite particles by gel-sol method, *Proc. 69th Annual Meeting Chem. Soc. Japan*, Kyoto, I, pp.254.
10. Sugimoto, T., Ndei, M. K., and Itoh, H. (1993) Shape and structure control of monodisperse ellipsoidal α-Fe_2O_3 particles by gel-sol method, *Proc. 65th Annual Meeting Chem. Soc. Japan*, Tokyo, vol.1, pp.305.
11. Sugimoto, T., Sakata, K., and Itoh, H. (1993) Shape and structure control of monodisperse spindle-type hematite particles by gel-sol method, *Proc. 46th Symp. Colloid Interface Chem.*, Tokyo, pp.306-307.
12. Shindo, D., Aita, S., Park, G., and Sugimoto, T. (1993) High-voltage high-resolution electron microscopy of monodispersed pseudocubic and peanut-type hematite particles, *Mater. Trans., JIM* **34**, 1226-1228.
13. Shindo, D., Park, G., Waseda, Y., and Sugimoto, T. (1994) Internal structure analysis of monodispersed peanut-type hematite particles produced by the gel-sol method, *J. Colloid Interface Sci.* **168**, 478-484.
14. Park, G., Shindo, D., Waseda, Y., and Sugimoto, T. (in press) Internal structure and analysis of monodispersed pseudocubic hematite particles by electron microscopy, *J. Colloid Interface Sci.*
15. Sugimoto, T. and Sakata, K. (1994) Shape and internal structure control of monodispersed hematite particles by gel-sol method, *Proc. 47th Symp. Colloid Interface Chem.*, Okayama, pp.240-241.
16. Shindo, D., Lee, B., Waseda, Y., Muramatsu, A., and Sugimoto, T. (1993) Crystallography of platelet-type hematite particles by electron microscopy, *Mater. Trans., JIM* **34**, 580-585.
17. Sugimoto, T., Dirige, E. G., and Muramatsu, A. (1994) Synthesis of uniform cadmium sulfide (CdS) particles from a condensed cadmium hydroxide suspension, *Proc. 47th Symp. Colloid Interface Chem.*, Okayama, pp.236-237.
18. Sugimoto, T, Dirige, E. G., and Muramatsu, A. (in press) Synthesis of uniform CdS particles from condensed $Cd(OH)_2$ suspension, *J. Colloid Interface Sci.*
19. Sugimoto, T, Dirige, E. G., and Muramatsu, A. (in press) Formation mechanism of uniform CdS particles from condensed $Cd(OH)_2$ suspension, *J. Colloid Interface Sci.*
20. Muramatsu, A. and Sugimoto, T. (1994) New synthetic method of spherical uniform Cu_2O particles with highly condensed CuO sol as a source of Cu, *Proc. 47th Symp. Colloid Interface Chem.*, Okayama, pp.238-239.

NONPOROUS SILICA MICROSPHERES IN THE MICRON AND SUBMICRON SIZE RANGE:

Manufacture, characterization and application

CH. KAISER[1], M. HANSON[1], H. GIESCHE[2], J. KINKEL[3] AND K.K. UNGER[1]

1) *Institut für Anorganische Chemie und Analytische Chemie, Johannes Gutenberg-Universität, Becherweg 24, D-55099 Mainz, Germany*
2) *School of Ceramic Engineering and Science, New York State College of Ceramics, Alfred University N.Y. 14802-1296, USA*
3) *E. Merck, Research and Development, P.O. Box 4119, D-64293 Darmstadt, Germany*

1. Introduction

Uniform nonporous silica microspheres in the micrometer and submicrometer size range are formed by a process which involves reacting alkoxysilanes with water in a short chain alcohol using dissolved ammonia as a morphological catalyst.

The particles are widely used in different areas. Beside the classical priority fields as support materials for chromatographic applications and polymer fillers further innovative applications appeared in the last years. To improve the slipperiness of polymeric films or fibers fine silica particles are inserted into the polymeric material. The particles are also used as precursors for ceramic materials. Moreover, cosmetic industries utilize the spheres in their products.

2. Synthetic routes for producing uniform nonporous silica microspheres

2.1. SYNTHESIS OF PARTICLES BETWEEN 0.05 AND 2 μm

In 1956 Kolbe [1] reported on the formation of silica particles by the reaction of tetraethoxysilane with water and bases in an alcoholic solution. The experimental conditions for producing uniform silica spheres in a defined size range between 0.05 μm to 2 μm have been described first by Stöber, Fink and Bohn in 1968 [2]. The Stöber method involves an alkoxysilane as the silica source, water as hydrolyzing agent, ammonia as catalyst and short chain alcohol as solvent. The growth process involves two reactions. The alkoxysilane is hydrolyzed in a few seconds followed by condensation of the silicic acids to polysilicic acid and spherical silica hydrogel.

E. Pelizzetti (ed.), Fine Particles Science and Technology, 71–84.

The reactions of the Stöber process are summarized as follows:

Hydrolysis

$$\equiv Si\text{-}OR + H_2O \underset{\text{(Re-esterifcation)}}{\rightleftharpoons} \equiv Si\text{-}OH + R\text{-}OH$$

Condensation

$$\equiv Si\text{-}OR + HO\text{-}Si\equiv \underset{\text{(Alcoholysis)}}{\rightleftharpoons} \equiv Si - O - Si\equiv + R\text{-}OH$$

$$\equiv Si\text{-}OH + HO\text{-}Si\equiv \underset{\text{(Hydrolysis)}}{\rightleftharpoons} \equiv Si - O - Si\equiv + H_2O$$

The surface of the composed silica spheres is covered with hydroxyl groups. Ammonia is not only used as a morphological catalyst but it also provides the silica particles with a negative, stabilizing surface charge. The final particle dispersion is monodisperse with a standard deviation of the mean particle diameter of less than 10%. The hydrogel particles are subjected to calcination to form dense spherical particles.

Kaiser[3] introduced a continuous process for the synthesis of Stöber particles by using a tube reactor (Fig. 1). The precursor solution A (tetraalkoxysilane, alcohol) and solution B (alcohol, aqueous ammonia) are pumped into a static mixer where the mixing takes only seconds. The injection of inert gas bubbles behind the static mixer divides the solvent into segments with a volume of about 5 ml, thus producing so-called micro batches. The gas bubble partition avoids a laminar flow profile in the tube reactor which would lead to polydisperse particle sizes due to the concentration gradient in the reactor.

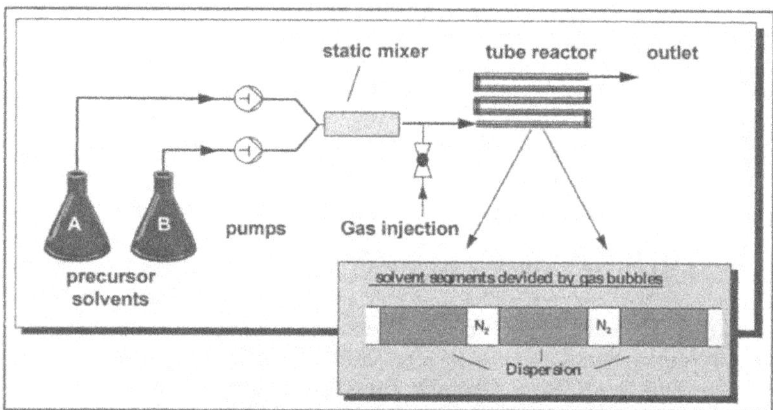

Figure 1: Continuous synthesis of silica spheres

The condensation of the silicic acid monomers starts after a few minutes depending on the reaction conditions ie. the temperature employed and the concentration of the reagents. The process can be easily followed by observing the increasing opalescence of the particle dispersion caused by the change of the optical density of the growing silica spheres. The standard deviation of the mean particle diameter is about 10% and, thus, comparable with the products obtained from the batch process under the same conditions.

Investigations of the surface by helium adsorption [4] - and Small Angle X-Ray Scattering (SAXS) [5] measurements indicates a microporosity of the particles. After calcination of the particles at 1000°C these micropores will be closed and a smooth surface is obtained corresponding to the geometrical surface of the spherical particles.

In order to observe the formation of the silica spheres after nucleation the light scattering technique can be used also when the optical density of the silica spheres is similar to the optical density of the solvent [6 ,7]. By means of this technique an overall rate of the growth process can be determined. An indirect way of observing the growing process of the particles is the detection of the resulting hydrolyzed ethanol concentration and the ethoxy groups of the tetraethoxysilane by dynamic ^{13}C nuclear magnetic resonance (NMR) technique [8] as a function of time when the synthesis is performed in propanol (Fig. 2). These measurements also allow to assess the reaction kinetics.

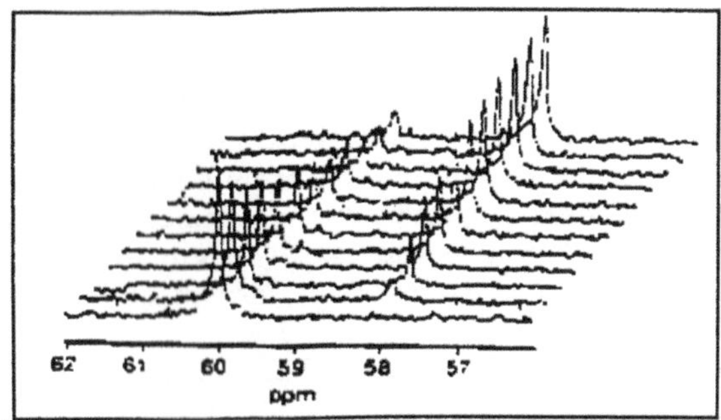

Figure 2: Ethoxy- and ethanol concentration during the forming process in propanol

In some investigations it is shown that there is a number of parameters responsible for the particle size and aggregation behaviour. By increasing the amount of water and ammonia in the synthesis procedure in a certain concentration range the particle grow to maximum size observed under constant alkoxysilane concentration and reaction temperature [9]. After the maximum size is reached the addition of more water or ammonia will lead to a decrease of the particle size (Fig. 3).
Another possibility to alter the silica particle sizes is to vary the reaction temperature and the type of alcohol as solvent [10]. Regardless of the alcohol used as solvent the particle size increases with decreasing temperatures (Fig. 4).

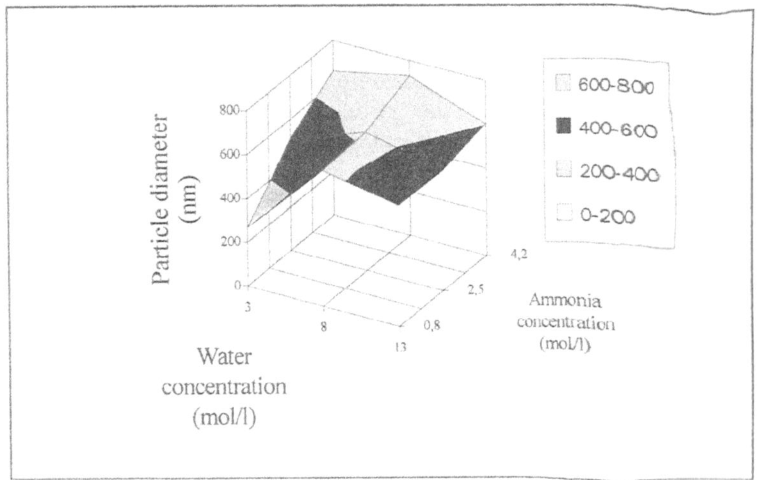

Figure 3: Particle sizes by varying the water and ammonia concentration at room temperature

The alcohol employed as solvent affects the condensation since it reacts with the partially hydrolyzed alkoxysilane leading to a reesterification. The dependence of the particle size on the different C1 to C4 alcohols cannot be described by simple rules. Above room temperature the particle size increases in direct correlation to the chain length of the alcohol while at lower temperatures the effect is reversed, except in the case of methanol (Fig.: 4).

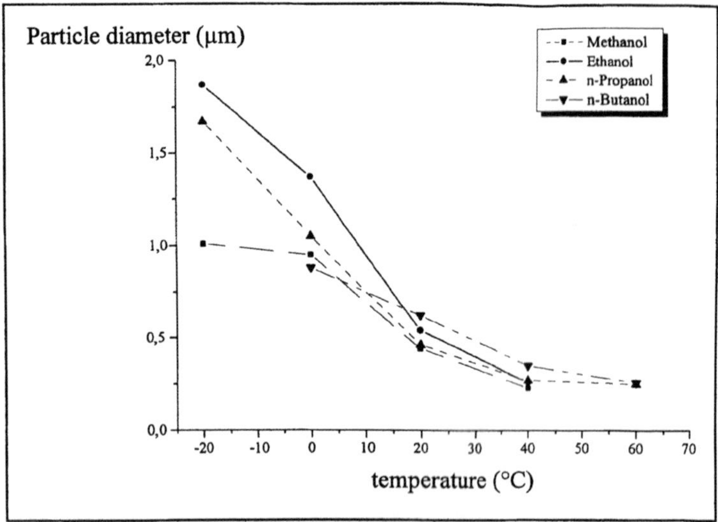

Figure 4 Dependence of the particle size on the reaction temperature and the type of alcohol [10]

The dependence of the particle size on the alcoholic solvent can partly be explained partly by the reduced condensation rate which leads to a smaller number of nuclei in the first part of the reaction. As a consequence the particle size increases.

2.2. SYNTHESIS OF SILICA PARTICLES BETWEEN 2 μm AND 10 μm

Above a certain size of the silica particles formed by the Stöber method agglomeration of the particles will occur together with a further increase of size. The repulsion forces of the negative charged surfaces will not prevent agglomeration. Thus, the particle size is limited to a maximum of 2 μm. During the last decade several synthetic routes have been developed to make particles up to a size of 10 μm [11 ,12 ,13 ,14]. The formation process of larger particles is in most cases based on the hydrolysis of alkoxysilanes and a deposition of silica on seed particles.

Either Stöber particles or commercially available Ludox particles can be used as seeds. Ludox particles consist of nearly compact polysilicic acid spheres with particle diameters between 20 and 30 nm. The suspension in water is stabilized either by sodium-ions or ammonia.

Unger and Giesche [11] used Stöber spheres as seed particles for a layer-by-layer growth by continuous addition of an alkoxysilane/alcohol solution and aqueous ammonia to compemsate for dilution. To avoid a second nucleation that would lead to a polydisperse particle size distribution the concentration of the solutions and the mixing conditions have to be chosen carefully. With this method it is possible to reach particle sizes of up to 10 μm.

Kovats, Jelinek and Erbacher [12] invented a method to form powders of compact, spherical silica particles based on Ludox particles as seed spheres. Ludox seed particles are dispersed in a starting mixture of water, ammonia and ethanol. The continuous addition of tetraalkoxysilane, ammonia and water lead to a further growth of the seed particles by polycondensation of silicic acid. The final particles are dried under vacuum and calcinated to obtain a smooth and non-porous surface.

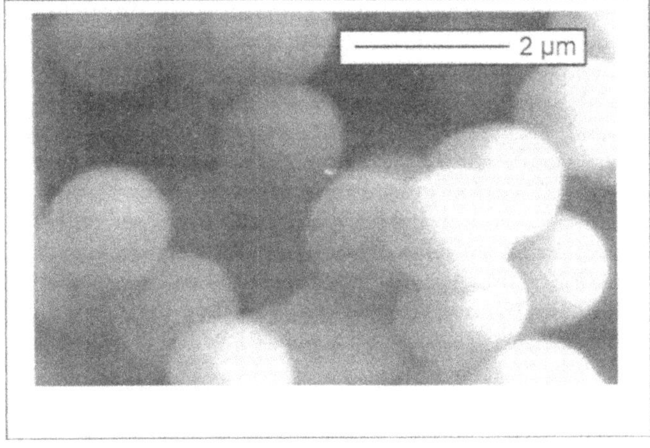

SEM-image of Silica spheres made by a layer-by-layer growth on Stöber spheres

Barder and Dubois [14] described a different process to achieve uniform silica microspheres up to particle diameters of 10 µm based on the hydrolysis of alkoxysilanes. The process is characterized by employing precursor solutions and feed rates which initially yield a two-phase reaction mixture. The feed solution containing 25 to 30 % w/w alkoxysilane, an alcohol and an aqueous ammonia solution is composed to form a two-phase mixture because of the insolubility of the silane in water (Fig. 5). As the hydrolyzis reaction proceeds the two-phase liquid mixture converts into a single phase due to the additional alcohol produced during hydrolization. When the solution becomes homogeneous the formation of new particles is completed.

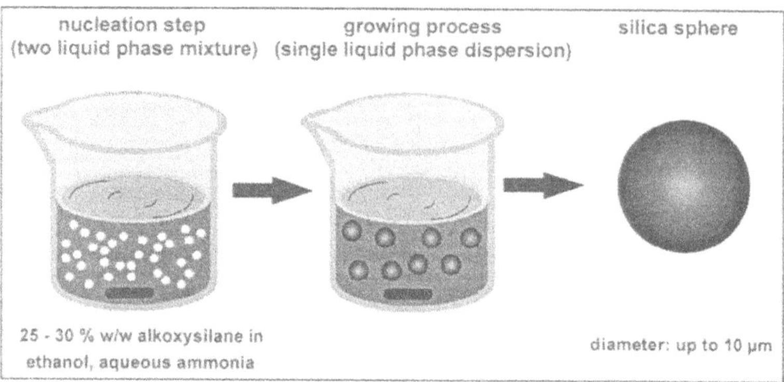

Figure 5: Synthetic route to produce silica spheres of up to 10 µm

2.3. GROWTH MECHANISM

The growth mechanism of the Stöber particles is discussed in the literature very controversive during the last decade [15 ,16]. The first mechanism proposed a growing process of the spheres was suggested by Stöber et al [2]. They took over the theory for the formation of monodispersed hydrosols proposed by LaMer and Dinegar [17] in 1950. In this model the silicic acid is formed until a critical supersaturation is reached and nucleation occurs. The particles grow continuously due to the addition of monomeric silicic acid and, thus, decrease the concentration of monomers in the solution. After the concentration of the silicic acid drops below the critical level, nucleation stops and the particles grow due to monomeric addition until the equilibrium concentration of the precipitating species is reached (Fig. 6a).

Matsoukas and Gulari [18 ,19 ,20] suggested a growth model where nucleation is the result of the reaction between two reactive monomers. Studies by means of light scattering and Raman spectroscopy showed that particle formation is characterized by an incubation period after which no significant nucleation takes place. The model for particle growth is based on a monomer addition process in the presence of a slow initiation step (hydrolysis). Thus the overall growth rate is limited by the rate of hydrolysis which is a first order rate in excess of water (Fig. 6b).

A different growth model was proposed by Bogush and Zukoski [21 ,22]. The critical concentration of monomeric silicic acid is higher than the value required for particle nucleation also in the advanced course of the reaction. Thus, it is assumed that the growth mechanism involves reactions in solution which do not depend on the presence of particles and result in the formation of silica nuclei which aggregate to form particles. The uniform particle size distribution is determined by the size dependent aggregation rates. These assumptions are proved by conductivity and volume measurements.

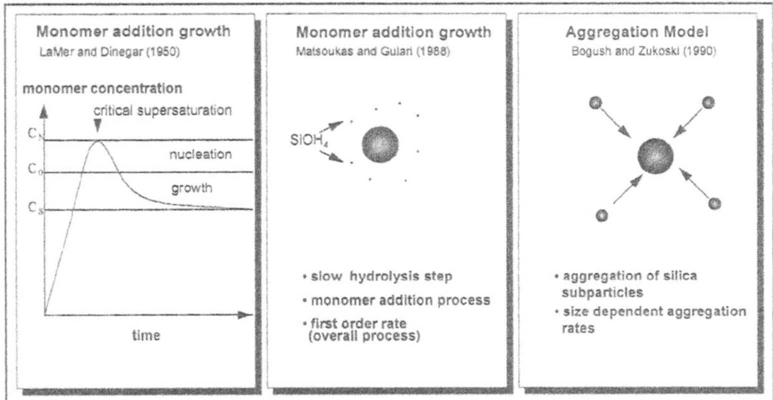

Figure 6: Growth models for the synthesis of silica spheres

Other theoretical models are located between these two extreme positions.

3. Characterization of the uniform silica spheres

The properties of uniform nonporous silica microspheres are studied by various characterization methods.

3.1. SURFACE PROPERTIES

The external surface of the silica spheres is covered with hydroxyl groups with a density of about 8 $\mu mol/m^2$ after drying. Because of the hydrophilic properties of the silica surface water will be easily adsorbed. A way of determining the degree of surface coverage with hydroxyl groups and adsorbed water is ^{1}H-nuclear-magnetic-resonance (NMR) spectroscopy [23 ,24]. Using deuterated trifluoroacetic acid is a popular method to assess the concentration of hydroxyl groups. The knowledge of the surface concentration of hydroxyl groups is needed when a modification of the silica surface with chemical binding species is desired.

The surface of the silica particle in this hydrated state is not nonporous. Several measurements by helium adsorption [25] or SAXS showed a microporosity of the surface in a pore size range of several Ångstroms. To achieve nonporous particles calcination has to performed following the preparation.

Physisorbed water is removed by drying of the silica spheres at 150°C. Total dehydration of the silica surface is achieved by calcination at 1000°C. The hydroxyl groups on the surface are converted into siloxane bonds leaving water. The surface is nonporous and hydrophobic after the calcination (Fig. 7). To receive a reactive hydroxyl covered surface the calcined particles will be treated either by water again or by mineral acids at temperatures of about 75 - 95°C [12].

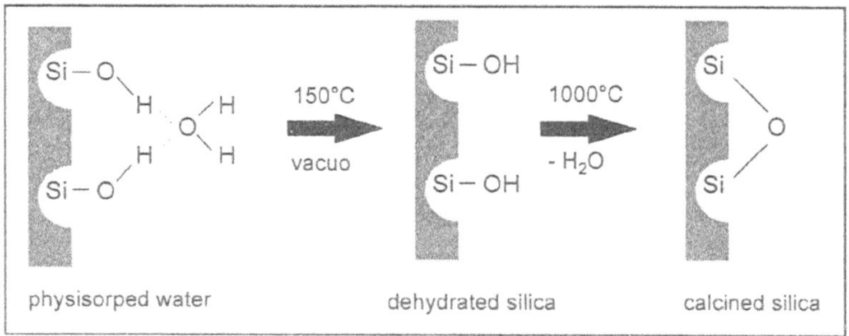

Figure 7: Drying and calcination processes

3.2. PARTICLE PROPERTIES

The most important property of silica particles prepared via alkoxysilane hydrolysis in a basic medium is their uniform shape and the narrow size distribution. These characteristics are measured either by light scattering techniques [26] or by electron microscopy images.

During the calcination process the weight loss of the particles is not only caused by losses of water or included ammonia but also by enclosed alkoxy groups the latter remain from an incomplete hydrolysis of the alkoxysilane. The silica particles formed by the 'Stöber' method are not fully condensed and contain only about 60% fully hydrolyzed silicic monomers [27]. The rest of the partly hydrolyzed silica species contains various amounts of ethoxy groups that lead to a weight loss of about 4 % when heating the particles (Fig. 8). The results of thermogravimetrical measurements are confirmed by elemental analysis [27].

It can be concluded from transmission electron microscopy images that before drying the surface of the silica spheres made according to the Stöber method exhibits a surface roughness that is probably caused by a difference in the molecular weight of the building silicon moieties reacting with the growing particle [27].

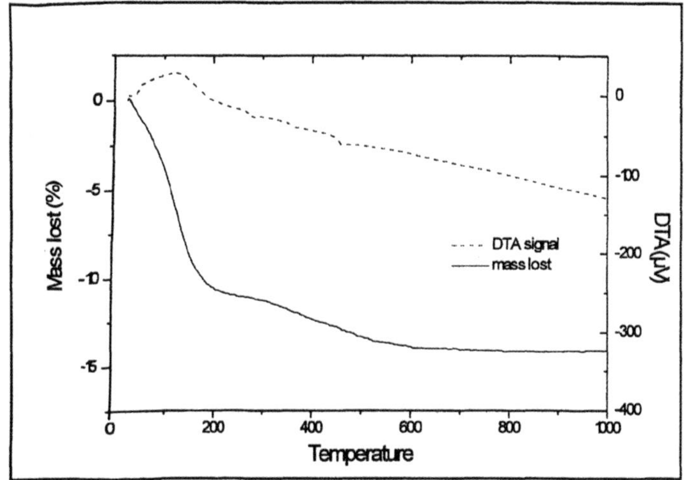

Figure 8: TG/DTA measurement of uniform nonporous silica spheres

The density of the final calcined particles is determined to about 2 ± 0.2 g/ml measured by drying a known volume of a concentrated dispersion in ethanol and weighing the residue [26].

4. Modification of the Silica Surface

Modified nonporous silica spheres are applied in chromatography and in material science as composite materials, fillers, insert in magnetic tapes, etc. The surface is modified either by silanization of the silanol groups (reaction of surface hydroxyl groups with organo-chlorosilanes) or by grafting and adsorbing polymers.

4.1. MODIFICATION BY SILANIZATION

Nowadays the silanization of the silica surface is a convenient method to modify the surface properties or to introduce functional groups to the silica support material [28 ,29 ,30] (Fig. 9). A simple method of linking aminopropyl-, mercaptopropyl- and octadecyltrimethoxysilane is described by Badley et al. [31]. The modified silica surfaces can be characterized by thermogravimetrical analysis, elemental analysis, fourier transform infrared spectroscopy (FTIR) and ^{13}C and ^{29}Si cross-polarization magic angle spinning NMR [32 ,33].

Compared to pure silica particles the densities of the modified silica spheres decrease from 2 g/ml to values between 1,58 g/ml (octadecyl-modified) and 1,90 g/ml (mercaptopropyl-modified) [31]. The carbon content of the particles increase up to 15% [31]. For many applications the surface coverage the silica support is crucial. The average surface coverage amounts to about 3 $\mu mol/m^2$.

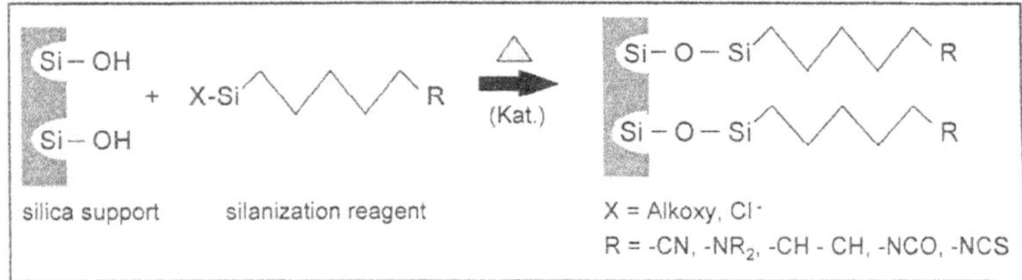

Figure 9: Scheme of the Silanization of the silica surfaces

4.2. MODIFICATION BY POLYMER COATINGS

The immobilization of the polymers on the silica support can be performed by various methods. The polymer can be linked to silanol groups on the surface via functional or reactive groups of the polymer which will provide a chemical bonding to the silica surface (Fig. 10). Either monomers or oligomers are used as precursors and physisorbed or grafted to the surface prior to immobilization Also trimethoxysilyl-terminated polymers can react with silanol groups [34 ,35].

Another way of modification is the direct esterification of poly(ethylene oxide) [36] or through urethane linkage [37]. After grafting poly(methyl methacrylate) chains can grow from the silica surface by anionic polymerization [38].

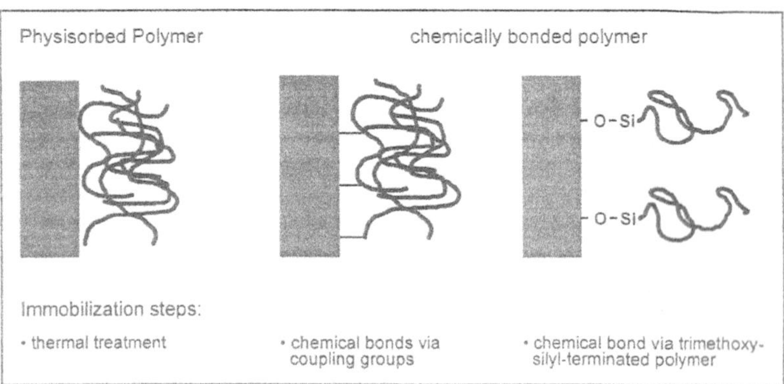

Figure 10: Polymer coatings on silica surfaces

The physisorption of reactive prepolymers on the silica surface is another way of modification [39 ,40 ,41 ,42]. The adsorbed prepolymer is immobilized on the silica surface by thermal treatment or γ radiation (Fig. 10).

5. Applications

5.1. CHROMATOGRAPHIC SUPPORT

Nonporous silica spheres of 1.5 μm diameter were first introduced by Anspach et al. [43] in 1984 as supports in High Performance Affinity Chromatography. The aim was to avoid restricted diffusion, slow adsorption-desorption kinetics, inaccessibility of ligands at the surface and losses in bioactivity of biopolymers to be separated and purified. In the following years this type of support was successfully applied in fast high resolution separations of biopolymers by means of Reversed Phase [44], Hydrophobic Interaction [45] and Ion Exchange [46] High Performance Liquid Chromatography (HPLC). According to the variant of HPLC the surface of nonporous silica particles was functionalized with hydrophobic (Reversed Phase and Hydrophobic Interaction Chromatography) ionic groups (Ion-Exchange Chromatography) following the silanization and polymerization procedure as discussed previously. The low surface area of the support of about 5 m^2/cc of column volume was not a critical parameter in these separation techniques because the biopolymer exhibited high distribution coefficients on these packings and were separated by means of gradient elution. The modified supports were packed into short columns of 30 mm in length and 4 to 8 mm inner diameter. The short column length was mandatory to keep the column pressure drop below 400 bar. Due to the small particle size and the absence of an interparticle porosity the separations were extremely fast and could be achieved in less than one minute (Fig. 11).

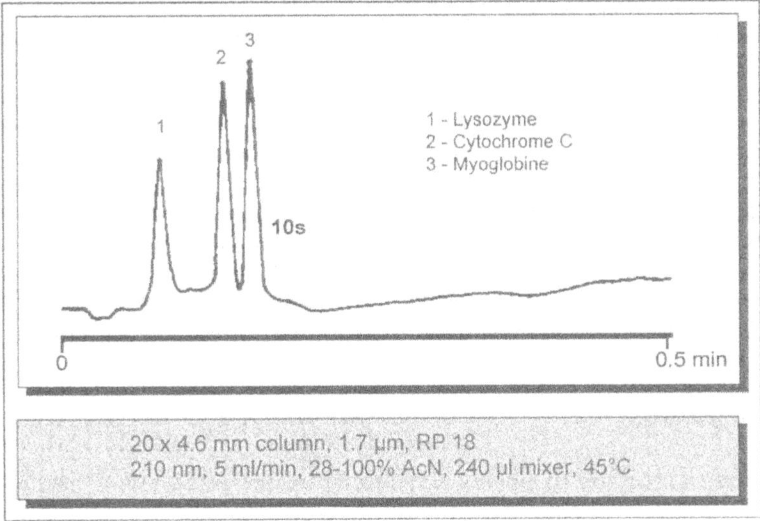

Figure 11: Separation of LCM Proteins

In order to further extend the use of nonporous silica spheres of 1.5 μm to the separation of low molecular weight analytes the packing technology and column bed structure was examined in depth by Giesche et al. [47] with respect to column performance under isocratic elution conditions. The use of low molecular weight analytes for the separation on short columns with modified nonporous silica spheres has some distinct consequences as compared to columns packed with modified porous silica spheres as

- The specific surface of nonporous silica spheres is by a factor of 100 lower than for porous silicas. As a consequence the retention of analytes is about 100 times lower on columns with porous particles at constant mobile phase composition. The reduced retention on columns with nonporous silica can be enlarged by using a mobile phase with a lower solvent strength. In Reversed Phase Chromatography e.g. the water content of the eluent has to be increased.
- The loadability of column packed with modified nonporous silica also diminishes by a factor of 100 sample amounts of 1μg and less should be injected to avoid an overloading.
- Due to the low column dead volume (a column of 30 x 4.6 mm posses a column dead volume of about 200 μl) the injection volume, the column connection and the volume of the detector cell must be smaller than 20 μl, otherwise the column performance decreases drastically. This means that columns with modified nonporous 1.5 μm silica spheres require a specially designed low dead volume HPLC equipment different from the traditional systems by which columns with 5 μm particles are operated. The gain is in analysis time and in the number of analyses per unit time. This concept has now been commercialized by MICRA Scientific, Inc. Northbrook, IL. USA.

5.2. MODELLING

Nonporous silica particles of particle diameters between 0.1 and 2 μm have been subjected to agglomeration by means of the isostatic pressure technique, gravity sedimentation, centrifugation the slurry technique [48] and by means of sintering [49]. The compact aim was to achieve highly regular and uniform agglomerates which can be compared with modeled compacts of uniform spheres. The coordination, porosity and pore size of agglomerates and aggregates were examined by electron scanning microscopy, nitrogen adsorption and mercury intrusion and extrusion techniques [50].
Columns packed with modified nonporous 1.5 μm spheres were also employed in HPLC of biopolymers and Affinity Chromatography [51] to assess the mass transfer kinetics by means of frontal analysis and to fit the experimental results with model calculations.

5.3. ADDITIVES

Nonporous silica microspheres are employed as additives to pigments to make them fluffy. Modified with polyethyleneglycols the beads are incorporated into polyester foils to improve the adhesion behaviour.

[1] G. Kolbe, Dissertation (1956) Jena, Germany
[2] W. Stöber, A. Fink and E. Bohn, (1968), Controlled Growth of Monodisperse Silica Spheres in the Micron Range, J. Colloid Interface Sci., **26**, 62
[3] C. Kaiser, Diploma thesis, 1993, Mainz, Germany
[4] H. Reichert, Diploma thesis, 1986,Universität Mainz, Germany
[5] A.K. Van Helden, J.W. Jansen and A. Vrij, (1980) Preparation and Characterization of Spherical Monodisperse Silica Dispersions in Nonaqueous Solvents, **81**, 354
[6] T. Matsoukas and E. Gulari, (1988), Dynamics of Growth of Silica Particles from Ammonia-Catalyzed Hydrolysis of Tetra-ethyl-orthosilicateJ. Colloid Interface Sci., **124**, 252
[7] A.P. Philips and A. Vrij, (1987), Polydispersity probed by Light Scattering of Secondary particles in controlled growth experiments of Silica SpheresJ. Chem. Phys., **87**, 5634
[8] A. van Blaaderen, J. van Geest and A. Vrij, (1992), Monodisperse Colloidal Silica Spheres from Tetraalkoxysilanes: Particle Formation and Growth Mechanism, J. Colloid Interface Sci. **154**, 481
[9] H. Giesche, Dissertation, 1987, Mainz, Germany
[10] C.G. Tan, B.D. Bowen., N. Epstein, (1987), Production of monodisperse Silica Spheres: Effect of Temperature, J. Colloid Interface Sci., **118**, 290-293
[11] K.K.Unger and H. Giesche, US patent 4,775,520
[12] E. sz. Kovats, L. Jelinek, C. Erbacher, Europian patent EP0.574.642 A1
[13] S. Coenen and C.G. de Kruif, (1988), Synthesis and Growth of Colloidal Silica Particles, J. Colloid Interface Sci., **124**, 104-110
[14] T. J. Barder, P.D. Dubois, US patent 4.983.369
[15] A.P. Philipse, (1988), Quantitative aspects of the growth of (charged) silica spheres, Colloid Polym. Sci., **266**, 1174
[16] E. Matijevic, (1981), Monodispersed Metal (Hydrous) Oxide - A Fascinating Field of Colloid Science, Acc, Chem. Res., **14**, 22-29
[17] V. LaMer and R.H. Dinegar, (1950), Theory, Production and Mechanism of Formation of Monodispersed Hydrosols, J. Am. Chem. Soc., **72**, 4847-4854
[18] T. Matsoukas and E. Gulari, (1989) Monomer-Addition Growth with a Slow InitiationStep: A Growth Model for Silica Particles from Alkoxides, J. Colloid Interface Sci., **132**, 13-21
[19] T. Matsoukas and E. Gulari, (1988), Dynamics of Growth of Silica Particles from Ammonia-Catalyzed Hydrolysis of Tetra-ethyl-orthosilicate, J. Colloid Interface Sci., **124**, 252-261
[20] T. Matsoukas and E. Gulari, (1991), Self SharpeningDistributions Revisited - Polydispersity in Growth by Monomer Addition, J. Colloid Interface Sci., **145**, 557-562
[21] G.H. Bogush, C.F. Zukoski IV, (1991), Studies of the kinetics of the precipitation of uniform silica particles through the Hydrolysis and Condensation of silicon alkoxides, J. Colloid Interface Sci., **142**, 1-18
[22] G.H. Bogush, C.F. Zukoski IV, (1991), Uniform Silica Particle Precipitation: An Aggregative Growth Model, J. Colloid Interface Sci., **142**, 19
[23] M. Holik, 2nd NMR Seminar, Prague, 1974
[24] M. Holik and B. Matejková, (1981), Determination of Hydroxyl Groups and Water Content in Silica by Nuclear Magnetic Resonance Spectroscopy, J. Chromatogr., **213**, 33-39
[25] H. Reichert, Diploma thesis 198.., Universität Mainz, Germany
[26] H.G. Barth and R.B. Flippen, Anal. Chem. **67** (1995) 257R
[27] A. van Blaaderen and A.P.M. Kentgens, (1992), Particle morphology and chemical microstructure of colloidal silica spheres made from alkoxysilanes, J. Non-Cryst. Solids **149**, 161
[28] K.C. Vrancken, L. De Coster, P. Van Der Voort, P.J. Grobert etc. J. Colloid Interface Sci **170** (1995) 71
[29] A. van Blaaderen and J. Vrij, (1992), Synthesis and Characterization of Monodisperse Colloidal Organo-silica spheres, J. Colloid Interface Sci. **156** 1
[30] J. N. Kinkel, Dissertation, Universität Mainz, 1984, Germany
[31] R.D. Badley, W.T. Ford, F.J. McEnroe and R.A. Assink, Surface Modification of Colloidal Silica, Langmuir **6** (1990) 792
[32] J.W. de Haan, H.M. Van den Bogart, J.J. Ponjeé and L.J.M. van de Ven, (1986), Characterization of Modified Silica Powders by Fourier Transform Infrared Spectroscopy and Cross Polarization Magic Angle Spinning NMR, J. Colloid Interface Sci. **110** 591-600
[33] K.C. Vrancken, L. De Coster, P. Van Der Voort, P.J. Grobert etc. J. Colloid Interface Sci **170** (1995) 71
[34] C. Apel, R. Jordan, K.K. Unger, HPLC 1995, Insbruck, Austria

[35] J. Edwards, S. Lenon, A.F. Toussaint, B. Vincent, ACS Symp Ser. **240** (1984) 281
[36] H. Ben Ouda, H. Hommel, A.P. Legrand, H. Balard, E. Papirer, J. Colloid Interface Sci. **122** (1988) 441
[37] K.- Bridger, B. Vincent, Eur. Polym. J. **16** (1980) 1017
[38] E. Schomaker, A.-J. Zwarteveen, G. Challa, M. Capka, Polym. Commun. **29** (1988) 158
[39] G. Schomburg, LC-GC, **6** (1988) 37
[40] M. Hanson, K.K. Unger and G. Schomburg, J. Chromatogr. **517** (1990) 269
[41] U. Bien-Vogelsang, A. Deege, H. Figge, J. Köhler, G. Schomburg, Chromatographia **19** (1984) 170
[42] H. Figge, A.Deege, J. Köhler, G. Schomburg, J. Chromatogr. **351** (1986) 393
[43] K.K. Unger, J.N. Kinkel, B.Anspach and H.Giesche, Evaluation of Advanced Silica Packings for the Separation of Biopolymers by High Performance Liquid Chromatography, I. Design and Properties of Parent Silicas, J. Chrom., **296** (1984) 3-14
[44] K.K. Unger, G.Jilge, J.N. Kinkel and M.T.W. Hearn, (1986), Evaluation of Advanced Silica Packings for the Separation of Biopolymers by High-Performance Liquid Chromatography II. Performance of Non-Porous Monodisperse 1.5 µm Silica Beads in the Separation of Proteins by Reversd-Phase Gradient Elution High Performance Liquid Chromatography, J. Chromatogr., **359**, 61-72
[45] R. Janzen, K.K. Unger, H. Giesche, J.N. Kinkel and M.T.W Hearn, (1987), Evaluation of Advanced Silica Packings for the Separation of Biopolymers by High-Performance Liquid Chromatography V. Performance of Non-Porous Monodisperse 1.5 µm Bonded Silicas in the Separation of Proteins by Hydrophobic-Interaction Chromatography, J. Chromatogr., **397**, 91-97
[46] G. Jilge, K.K. Unger, U.Esser, H.-J. Schäfer, G. Rathgeber and W. Müller, (1989) Design, Chromatographic Performance and Apllication of Non-Porous Silica Based Anion Exchangers, J. Chromatogr., **476**, 135-145
[47] H. Giesche, K.K. Unger, U. Esser, B. Eray, U. Trüdinger and J.N. Kinkel, (1989) Packing Technology, Column Bed Structure and Chromatographic Performance of 1-2 µm Non-Porous Silicas in High Performance Liquid Chromatography, J. Chromatogr., **465**, 39-57
[48] H. Giesche, Dissertation, 1989, Mainz, Germany
[49] H. Giesche and K.K. Unger (1989) in H. Hauser and G.L. Messing (editors): Ceramic Powder Processing Science, 2nd Int. Conference, Berchtesgaden, Germany 1988, Deutsche Keramische Gesellschaft, Köln
[50] H. Giesche, K.K. Unger and U. Esser, (1989), Hysteresis in Nitrogen Sorption and Mercury Porosimetry on Mesoporous Model Adsorbents Made of Aggregated Monodisperse Silica Spheres, **37**, 93-113
[51] A.I. Liapis, B. Anspach, M.E. Findley, J. Davies, M.T.W. Hearn and K.K. Unger, (1989) Biospecific Adsorption of Lysozyme onto Monoclonal Antibody Ligand Immobilized on Non-Porous Particles, Biotechnology and Bioengineering, **34**, 467-477

ADSORPTION OF CATIONS ON COLLOIDAL OXIDE PARTICLES. COBALT IONS ON HEMATITE

N. OGRINC[1], I. KOBAL[1], E. MATIJEVIĆ[2], and N. KALLAY[3]

[1] *Physical Chemistry Group, J. Stefan Institute, P.O.Box 100, 61111 Ljubljana, Slovenia*

[2] *Center for Advanced Materials Processing, Clarkson University, Potsdam, New York 13699-5814, USA*

[3] *Laboratory of Physical Chemistry, Faculty of Science, University of Zagreb, P.O. Box 163, 10001 Zagreb, Croatia*

1. Introduction

Adsorption of ions in colloidal suspensions plays an important role in science, technology, and every day life. In the cooling water system of a nuclear power plant, depending on the physico-chemical conditions, radioactive cobalt is either dissolved or sorbed on colloidal magnetite and hematite corrosion products particles [1]. The extent of the contamination of surface waters due to a uranium mine and mill greatly depends on the content of natural colloids in these waters to which uranium, radium, and other members in the decay chain are attached [2]. Release of chromium from leather industry is subject of similar phenomena [3]. There is no need to mention that the transport of other pollutants and their uptake by living organisms is governed by sorption equilibria at suspended particles of naturally occurring colloids, mainly being hydrated oxides [4].

It is not surprising that these systems have been extensively investigated both experimentally and theoretically. Since the real problems are usually too complex, investigators often take model systems and study them under well defined conditions. Understandably, colloids of metal oxides attract the major attention.

Adsorption of a number of cations and anions onto hydrous ferric oxides has been already systematically investigated and the results were theoretically interpreted in the terms of the "surface complexation" or "site binding" model.

E. Pelizzetti (ed.), Fine Particles Science and Technology, 85–96.

Several modifications have been introduced into the former model, which differ in the structure of the electrical interfacial layer, as well as mechanisms and stoichiometry of binding of ions [5].

The aim of this paper is to interpret the adsorption of cobalt ions on suspended hematite particles in aqueous solutions, using the electrical triple layer model within the "2-pK" concept [6].

2. Experimental

The experimental data to be dealt with in this paper were reported elsewhere [7]. The adsorption of Co^{2+} ions on spherical hematite particles, with the diameter of 125 ± 10 nm and a specific surface area of 13.3 m^2 g^{-1}, suspended in aqueous $NaNO_3$ solutions was studied as a function of the pH, ionic strength, and initial cobalt concentration. In the adsorption process, on average 1.4 H^+ ions are released per one Co^{2+} ion adsorbed. The adsorption curves shifted to higher pH values with increasing initial cobalt concentration, and no significant dependence of the uptake of cobalt ions on the ionic strength was observed.

Potentiometric pH-titration in the absence of Co^{2+} was carried out with NaOH (3.4 mol dm^{-3}) using 105 cm^3 of a suspension containing 36.7 g dm^{-3} of hematite in 0.10 mol dm^{-3} $NaNO_3$ under nitrogen atmosphere [8].

3. Surface complexation model

In the interpretation, the "2-pK" concept was used assuming the electrical triple layer model (TLM). The following surface equilibria with the corresponding equilibrium constants were taken into consideration [5,9,10]:

$$FeOH_2^+ \rightleftharpoons FeOH + H^+ \qquad Ka_1^{int} = \frac{[FeOH][H^+]}{[FeOH_2^+]} e^{-\frac{F\psi_0}{RT}} \qquad (1)$$

$$FeOH \rightleftharpoons FeO^- + H^+ \qquad Ka_2^{int} = \frac{[FeO^-][H^+]}{[FeOH]} e^{-\frac{F\psi_0}{RT}} \qquad (2)$$

$$FeOH + Na^+ \rightleftharpoons FeO^-Na^+ + H^+ \qquad K_{Na}^{int} = \frac{[FeONa][H^+]}{[FeOH][Na^+]} e^{-\frac{F}{RT}(\psi_0-\psi_i)} \qquad (3)$$

$$FeOH_2^+NO_3^- \rightleftharpoons FeOH + NO_3^- + H^+ \qquad K_{NO_3}^{int} = \frac{[FeOH][NO_3^-][H^+]}{[FeOH_2NO_3]} e^{-\frac{F}{RT}(\psi_0-\psi_i)} \qquad (4)$$

$$FeOH + Co^{2+} \rightleftharpoons FeOCo^+ + H^+ \qquad K_{Co,1}^{int} = \frac{[FeOCo^+][H^+]}{[FeOH][Co^{2+}]} e^{-\frac{F}{RT}(\psi_0-2\psi_i)} \qquad (5)$$

$$2FeOH + Co^{2+} \rightleftharpoons (FeO)_2Co + 2H^+ \qquad K_{Co,2}^{int} = \frac{[(FeO)_2Co][H^+]^2}{[FeOH]^2[Co^{2+}]} e^{-\frac{2F}{RT}(\psi_0-\psi_i)} \qquad (6)$$

$$FeOH + Co^{2+} + H_2O \rightleftharpoons FeO^-CoOH^+ + 2H^+ \qquad K_{Co,3}^{int} = \frac{[FeO^-CoOH^+][H^+]^2}{[FeOH][Co^{2+}]} e^{-\frac{2F}{RT}(\psi_0-\psi_i)} \qquad (7)$$

where ψ is the surface potential and subscript i is either 0 or β, representing the inner and outer plane, respectively, while the other symbols have the usual meaning. The equilibrium constants were defined according to Ref [11]. The surface concentration of OH groups, N_s, is given as:

$$\begin{aligned} N_s = [FeOH]_T &= [FeOH] + [FeOH_2^+] + [FeO^-] + [FeONa] + \\ &+ [FeOH_2NO_3] + [FeOCo^+] + [(FeO)_2Co] + \\ &+ [FeOCoOH^+] \end{aligned}$$

and the surface concentration of cobalt adsorbed, Γ_{Co}, as:

$$\Gamma_{Co} = [FeOCo^+] + [(FeO)_2Co] + [FeOCoOH^+]$$

As commonly done, FeO^- and $FeOH_2^+$ were considered to be at the 0 plane, while charges resulting from the surface complexes formed by association of Na^+, Co^{2+} and NO_3^- ions are allowed to be located either at the 0 or at β plane of the TLM.

In calculations the following triple layer models were used: (*1a*): $i = 0$ in Eq (3) to (7); (*2a*): $i = 0$ in Eq (3) and (4) and $i = \beta$ in Eq (5) to (7); (*3a*): $i = 0$ in Eq (3) and (4), $i = 0$ in Eq (5), and $i = \beta$ in Eq (6) and (7); (*1b*): $i = \beta$ in Eq (3) and (4), and $i = 0$ in Eq (5) to (7); (*2b*): $i = \beta$ in Eq (3) to (7), and (*3b*): $i = \beta$ in Eq (3) and (4), $i = 0$ in Eq (5), and $i = \beta$ in Eq (6) and (7).

Surface potentials ψ_0, ψ_β, ψ_d are related to surface charge densities σ_0, σ_β, σ_d, through inner- and outer-layer capacitances, (C_1, C_2), by well known relationships [12]:

$$\begin{aligned} C_1 &= \frac{\sigma_0}{\psi_0 - \psi_\beta} \\ C_2 &= \frac{\sigma_d}{\psi_d - \psi_\beta} \end{aligned}$$

The diffuse layer potential ψ_d was calculated using the Gouy-Chapman expression:

$$\psi_d = \frac{2RT}{F} ln(\frac{\sigma_d}{2\sqrt{2RT\epsilon I}} + \sqrt{\frac{\sigma_d^2}{8RT\epsilon I} + 1})$$

where ϵ is the permitivity and I is the effective ionic strength, corrected for the adsorption of ions from the bulk. The condition $\sigma_0 + \sigma_\beta + \sigma_d = 0$ is fulfilled.

There are several computer programs for modeling adsorption in terms of the surface complexation model [13-16]. In this study SCPLOT [16] was used, which allows for a simultaneous fitting of potentiometric titration and adsorption data at different pH, ionic strength, and initial Co^{2+} concentration. However it was found to be more convenient to first fit the potentiometric titration, and then the cobalt adsorption experimental points. No parameters required in the TLM model were experimentally determinated, instead N_s, Ka_1^{int}, Ka_2^{int}, C_1, C_2, K_{Na}^{int}, $K_{NO_3}^{int}$, $K_{Co,1}^{int}$, $K_{Co,2}^{int}$ and $K_{Co,3}^{int}$ were considered adjustable in fitting the experimental results. The success of a model was judged either by observing the calculated curve plotted through the experimental points on the screen, or by searching for the minimum in the sum of squares of differences (ssq) between the calculated and experimental values of the pH in the potentiometric titration, and of the percentage of adsorbed cobalt in cobalt adsorption.

4. Results and discussion

For the interpretation of the potentiometric titration data of the hematite suspension the *1a* and *1b* TLM were used (with no Co^{2+} association included). The starting values of adjustable parameters were selected from the literature for similar systems. Since for hematite, N_s values between 5 and 10 OH groups per nm^2 were reported [17], 12 groups per 1 nm^2 were initially chosen. For the amorphous iron oxyhydroxide in $NaNO_3$ the data in the literature are given as $Ka_1^{int} = 7.9{\cdot}10^{-6}$ mol dm^{-3}, $Ka_2^{int} = 2.0{\cdot}10^{-11}$ mol dm^{-3}, $K_{Na}^{int} = 1.0{\cdot}10^{-9}$, and $K_{NO_3}^{int} = 2.5{\cdot}10^{-7}$ (mol dm^{-3})^2 [18] and were used as starting values. For capacitance of the inner-layer capacitor a value between 0.1 and 2.0 Fm^{-2} is expected [18], thus, 1.5 Fm^{-2} was selected, while the outer-layer capacitance should be $\sim$ 0.2 Fm^{-2} [11] which is taken as our starting value.

For metal oxides Δ pK_{com}^{int} = pK_{Na}^{int} - p$K_{NO_3}^{int}$ is expected to be from 3 to 6 [19,20] and Δ pKa = pKa_1^{int} - pKa_2^{int} should be < 3 [20,21] as matched by the *1b*, but not by the *1a* model. The ssq is 0.17 for the *1b* and is ten times higher for the *1a* model. The *1b* model seems to be more appropriate therefore its best-fit values were used as starting points for the analysis of the adsorption data, as well as for the calculated titration curve, shown together with experimental points in Figure 1. Model *2b* is inferior as compared with *1a* in the capacitance C_2, which is higher than expected (0.2 Fm^{-2} in *1a* TLM).

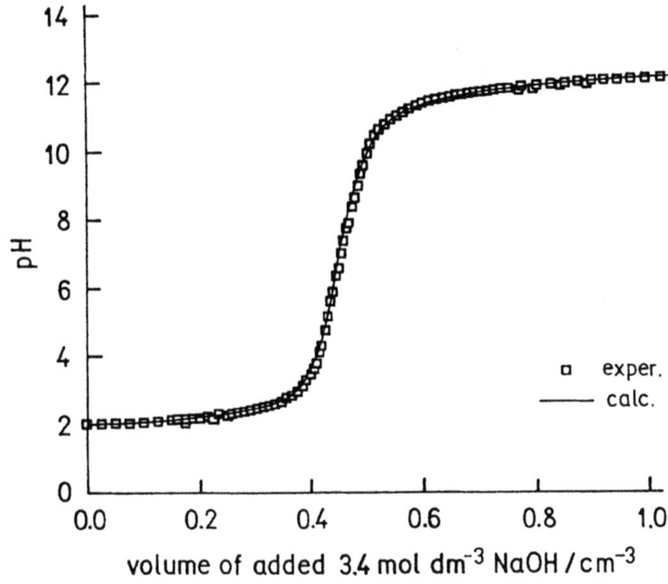

Figure 1. Potentiometric titration of hematite in an aqueous $NaNO_3$ suspension, I= 0.1 mol dm^{-3}, T = 298 K.

For the adsorption analysis, cases *1b*, *2b* and *3b* (see above) for the TLM were considerd. The three cobalt assocation equilibrium constants were taken as adjustable parameters: $K^{int}_{Co,1}$ (-), $K^{int}_{Co,2}$ (-), $K^{int}_{Co,3}$ (mol dm^{-3}). Although all experimental points could be fitted simultaneously, the calculation was done in two steps. First the entire pH range was covered for the initial cobalt concentration $C_{Co,in} = 1.7 \cdot 10^{-7}$ mol dm^{-3}, and different ionic strength, I, (69 points), and then the I = 10^{-4} mol dm^{-3} was selected and the analysis was carried out for different initial cobalt concentrations (33 points).

With *1b* and *2b* the analysis was initiated with only one cobalt association equilibrium constant different from zero. Figures 2a and 2b show that the latter does not influence the result. In the next step different values of a selected cobalt association equilibrium constant were chosen, while one of

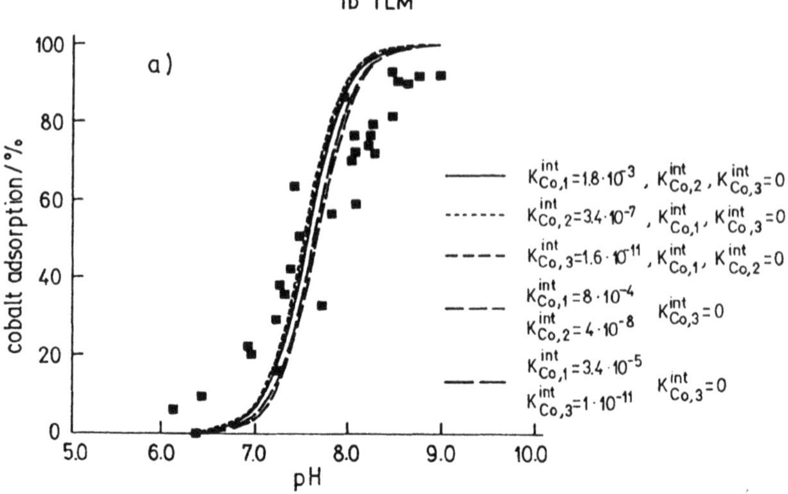

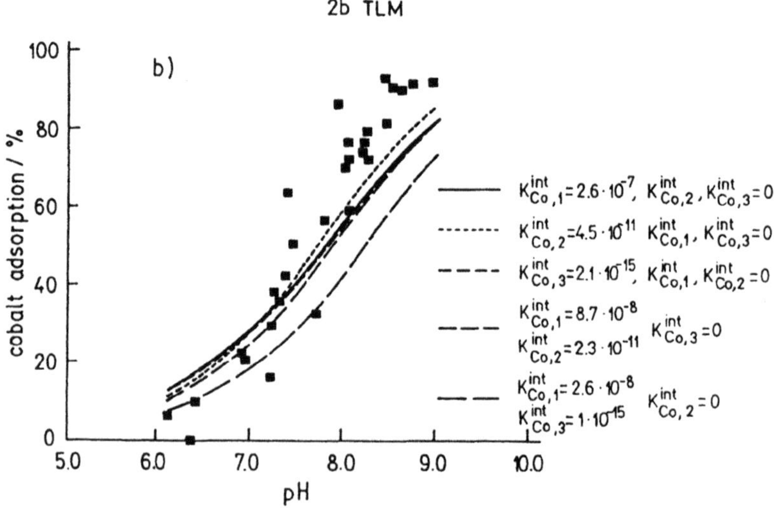

Figure 2. Experimental data and calculated adsorption curves for various selections of non-zero single, or a pair of cobalt association equilibrium constants for different TLM, $I = 10^{-4}$ mol dm^{-3}, $C_{Co,in} = 1.7\cdot10^{-7}$ mol dm^{-3}, T = 298 K.

the other two was set to zero and the optimum value of the remaining one was obtained by seeking the minimum in the ssq. Surprisingly the quality of the fit is pretty much the same for only one, as for a pair of non-zero cobalt association constants (Figures 2a and 2b). It was also found that the calculated adsorption curves for case *1b* are only slightly dependent on the ionic strength, in agreement with the experimental results, but the curves are too steep (Figure 2a), while the slope of the curves for *2b* is too low (Figure 2b) and the adsorption significantly depended on the ionic strength.

Since the slope of calculated adsorption curves with the *1b* could not be altered, the case *3b* was tested, being a hybrid between *1b* and *2b*, which was proven successful in the adsorption of Pb and Cd on geothite [11]. First the dependence of the ssq on the adjustable parameters for the potentiometric titration and adsorption was assessed. The ssq minima are essentially at the same values of N_s and C_1 for titration and adsorption, but do not coincide for Ka_2^{int}, K_{Na}^{int} and $K_{NO_3}^{int}$. In these cases the titration values are preferred because they are based on a larger number of more accurate experimental points. The best-fit adsorption value was only taken for Ka_1^{int}, because the titration ssq minimum is less pronounced in titration than in adsorption.

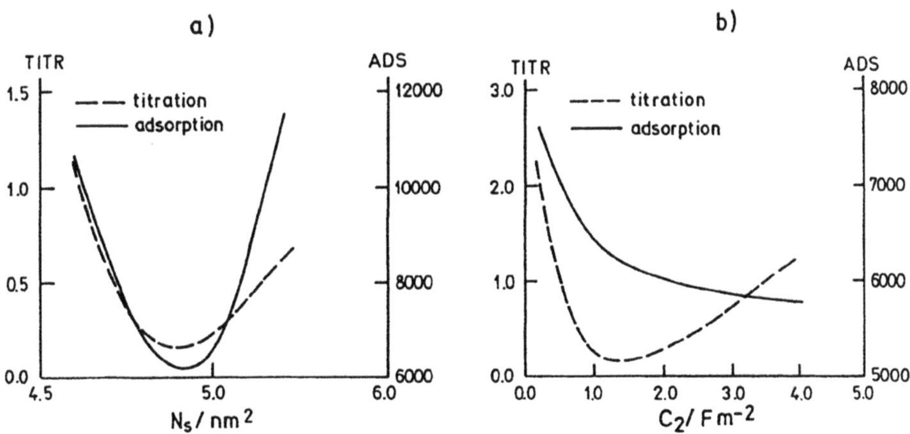

Figure 3. Dependence of the sum of squares (ssq) of differences between the calculated and experimental percent of cobalt adsorption (ADS), and pH values (TITR) on adjustable parameters a) N_s and b) C_2 case.

For sake of saving space, only two curves are shown in Figure 3: a) N_s case with coincidence of ssq minima, and b) C_2 case with only titration ssq minimum. Thus the best-fit values as obtained from both titration and adsorption are as follows: N_s = 4.8 nm^{-2}, Ka_1^{int} = 9.6·10^{-9} mol dm^{-3}, Ka_2^{int} = 7.8·10^{-9} mol dm^{-3}, C_1 = 1.9 Fm^{-2}, C_2 = 1.4 Fm^{-2}, K_{Na}^{int} = 1.1·10^{-9}, $K_{NO_3}^{int}$ = 1.9·10^{-12} (mol dm^{-3})^2.

The calculations for *3b* case were then carried out as with the additional case in which one selected constant was fixed and the optimum values for the other two were sought. Again pairs ($K_{Co,1}^{int}$, $K_{Co,2}^{int}$) and ($K_{Co,1}^{int}$, $K_{Co,3}^{int}$) yielded satisfactory results, but not the pair ($K_{Co,2}^{int}$, $K_{Co,3}^{int}$), in agreement with 1.4 H^+ ions released per 1 Co^{2+} ion adsorbed. The explanation may be found in Figure 4, presenting the ssq for different selections of constants. Because there is no minimum in Figure 4c, a low (or zero) value of $K_{Co,3}^{int}$ is preferred. The *3b* TLM actually gives a better slope of the adsorption curve than either *1b* or *2b* as seen on Figure 5a for both pairs of non-zero cobalt association constants. The ionic strength dependence is slightly worse but the agreement for I = 10^{-4} mol dm^{-3} is almost perfect with the lowest ssq value obtained with $K_{Co,1}^{int}$ = 3.2·10^{-4} and $K_{Co,2}^{int}$ = 2.2·10^{-11} or with the combination $K_{Co,1}^{int}$ = 7.8·10^{-4} and $K_{Co,3}^{int}$ = 3.5·10^{-16} mol dm^{-3}. Based on one of these most acceptable combinations the adsorption curves for different initial cobalt concentrations are shown in Figure 5b, which indicate satisfactory agreement.

5. Conclusions

The experimental potentiometric titration and adsorption data were analyzed within the framework of the surface complexation model using the TLM. The titration data suggest, that Na^+ and NO_3^- ions are located at the β plane, as is the common assumption with the TLM for the background electrolyte. Even though three different cobalt surface association equilibria were assumed and the *3b* TLM (two cobalt complexes at β and one at 0 plane) is slightly more successful in fitting the experimental data, this complex system does not seem to be justified on the basis of the experimental data. The finding that each one of the three cobalt association equilibrium constants, the other two being zero, may give almost the same fit, and that introduction of an additional constant caused no improvement, was disappointing. Intuitively, one may assume that the three cobalt association equilibria considered in this study indeed took place at the hematite particle surface, but this case could not be confirmed by analysis. It also appears that more accurate experimental data could not

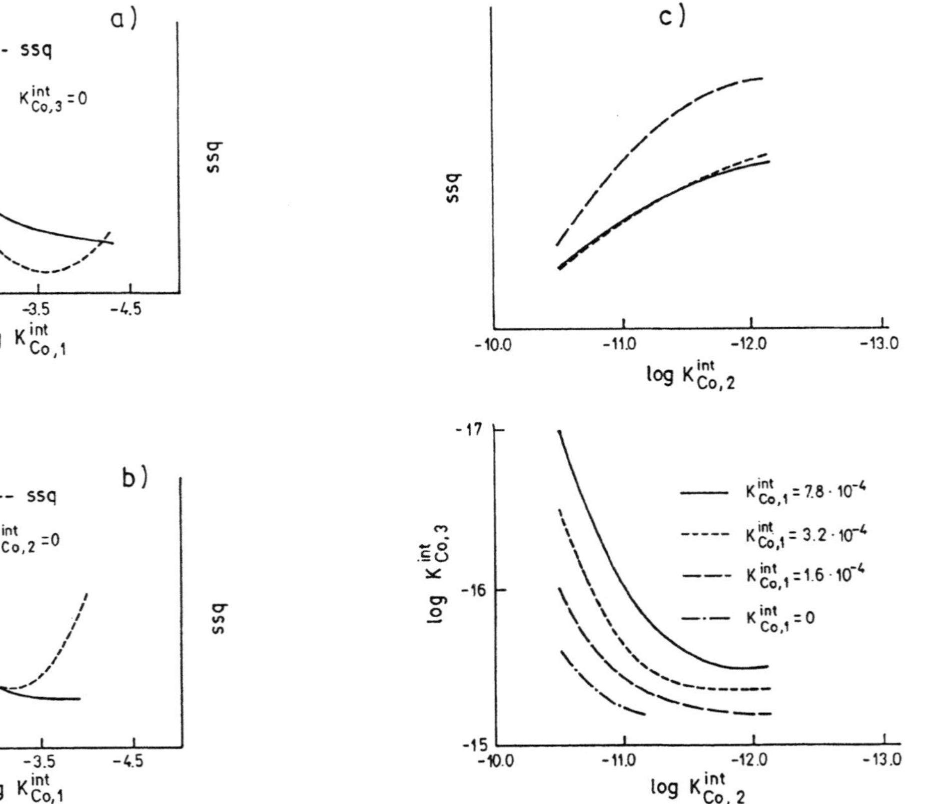

Figure 4. Optimal combinations of cobalt association equilibrium constants

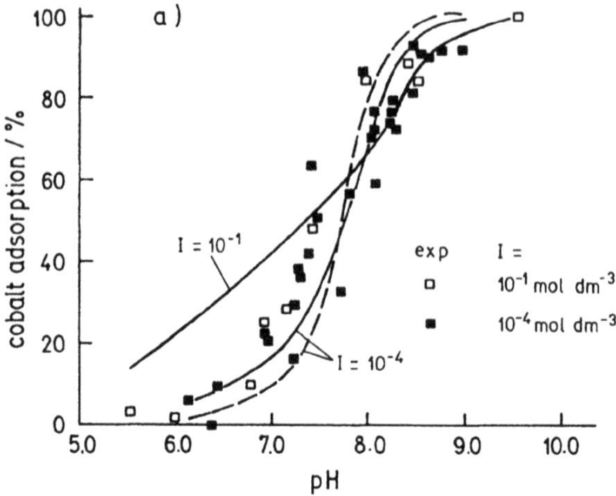

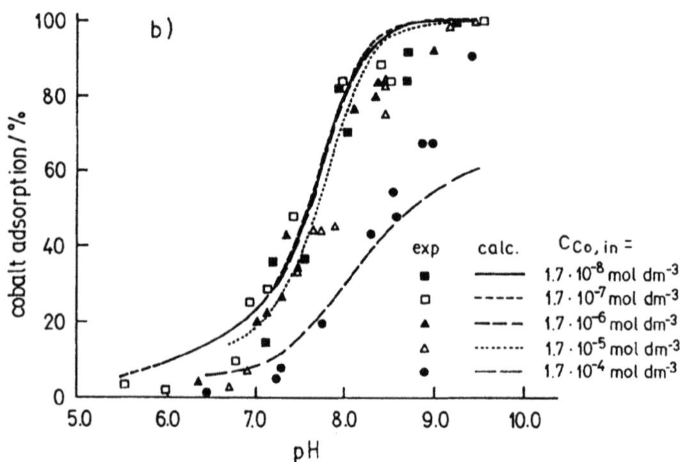

Figure 5. Adsorption of cobalt ions on hematite, *3b* TLM, T = 298 K: N_s = 4.8 nm^{-2}, $Ka_1^{int} = 9.6 \cdot 10^{-9}$ mol dm^{-3}, $Ka_2^{int} = 7.8 \cdot 10^{-9}$ mol dm^{-3}, C_1 = 1.9 Fm^{-2}, C_2 = 1.4 Fm^{-2}, $K_{Na}^{int} = 1.1 \cdot 10^{-9}$, $K_{NO_3}^{int} = 1.9 \cdot 10^{-12}$ $(mol\ dm^{-3})^2$:
a) — $K_{Co,1}^{int} = 3.2 \cdot 10^{-4}$, $K_{Co,2}^{int} = 2.2 \cdot 10^{-11}$, I = 0.1 and 10^{-4} mol dm^{-3};
- - - $K_{Co,1}^{int} = 7.8 \cdot 10^{-4}$, $K_{Co,3}^{int} = 3.5 \cdot 10^{-16}$ mol dm^{-3}, I = 10^{-4} mol dm^{-3}
b) $K_{Co,1}^{int} = 3.2 \cdot 10^{-4}$, $K_{Co,2}^{int} = 2.2 \cdot 10^{-11}$.

significantly improve the results. Consequently, the following values for the equilibrium constants using the *3b* TLM: $K^{int}_{Co,1} = 3.2 \cdot 10^{-4}$, $K^{int}_{Co,2} = 2.2 \cdot 10^{-11}$ or $K^{int}_{Co,1} = 7.8 \cdot 10^{-4}$, $K^{int}_{Co,3} = 3.5 \cdot 10^{-16}$ mol dm^{-3} are considered reasonable.

As already known, different assumed models may satisfactorily fit adsorption data. Consequently, one is not able to establish the actual mechanism of surface charging and structure of the interfacial layer.

One strategy, described here, is to analyze two different kinds of measurements, i. e. the potentiometric pH-titration to evaluate N_s, Ka^{int}_1, Ka^{int}_2, C_1, C_2, K^{int}_{Na}, $K^{int}_{NO_3}$, and separately the adsorption of Co^{2+} ions, which resulted in less freedom in the fitting procedure.

References

1. Honda, T., Izumiya, M., and Minato A. (1984) Radiation buildup on stainless steel in a boiling water reactor environment, *Nucl. Technol*, **64**, 35-42.
2. Beneš, P., Borovec, Z., and Strejc P. (1985) Interaction of radium with freshwater sediments and their mineral compounds, *J. Radioanal. Nucl. Chem.*, **89**, 339-351.
3. Musić, S., Ristić, M., and Tonković M. (1986) Sorption of chromium(VI) on hydrous iron particles, *Z. Wasser-Abwasser-Forsch.*, **19**, 186-196.
4. Lieser, K.H., Ament, A., Hill, R., Singh, R.N., Stigl, U., and Thybusch, B. (1990) Colloids in groundwater and their influence on migration of trace elements and radionuclides, *Radiochim. Acta*, **49**, 83-100.
5. Dzombak, D.A., and Morel, F.M.M. (1989) *Surface Complexation Modeling*, A Wiley-Interscience Publication, John Wiley & Sons, New York, USA.
6. Kallay, N. and Žalac, S. (1994) Enthalpy of interfacial charging of metal oxide/water systems, *Croat. Chim. Acta* **67**, 467-479.
7. Kobal, I., Hesleitner, P., and Matijević, E. (1988) Adsorption at solid/liquid interfaces. 6. Interactions of Co^{2+} ions with spherical hematite particles, *Colloids Surfaces*, **33**, 167-174.
8. Hesleitner, P., Babić, D., Kallay, N., and Matijević E. (1987) Adsorption at solid/solution interfaces. 3. Surface charge and potential of colloidal hematite, *Langmuir*, **3**, 815.
9. Morel, F.M.M., Yeasted, J.V., and Westall, J.C. (1981) Adsorption models: a mathematical analysis in the framework of general equilibrium calculation in *Adsorption of Inorganics at Solid-Liquid Interfaces* (M.A. Anderson and A.J.Rubin, Eds.), Ann Arbor Science, Michigan, USA, pp.263.

10. Tewari, P.H., Campbell, A.B., and Lee, W. (1972) Adsorption of Co^{2+} by oxides from aqueous solutions, *Can. J. Chem.*, **50**, 1642-1648.

11. Hayes, K.F., and Leckie, J.O. (1987) Modeling ionic strength effects on cation adsorption at hydrous oxide/solution interface, *J. Colloid Interface Sci.* **115**, 564-572.

12. Blesa., M.A., and Kallay, N. (1988) The metal oxide - electrolyte solution interface revisited, *Adv. Colloid Interface Sci.*, **28**, 111-134.

13. Westall, J.C., and Morel, F.M.M. (1977) FITEQL: A general algorithm for the determination of metal-ligand complex stability constants from experimental data, *Technical Note 19*, Ralph M. Parsons Laboratory, Department of Civil Engineering, Massachusetts Institute of Technology, Cambridge, Massachusetts, USA.

14. Westall, J.C., Zachary, J.L., and Morel, F.M.M. (1976) MINEQL: A computer program for the calculation of chemical equilibrium composition of aqueous systems, *Technical Note 18*, Ralph M. Parsons Laboratory, Department of Civil Engineering, Massachusetts Institute of Technology, Cambridge, Massachusetts, USA.

15. Westall, J.C. (1982) FITEQL: A program for the determination of chemical equilibrium constants from experimental data,*Technical Report*, Chemistry Department, Oregon State University, Corvallis, Oregon, USA.

16. Trkov, A., Ogrinc, N., and Kobal, I. (1992) Modeling surface complexation at the colloid/electrolyte interface, *Computers Chem.*, **16**, 341-344.

17. Schindler, P.W. (1981) Surface complexes at the oxide/water interfaces in *Adsorption of Inorganics at Solid-Liquid Interfaces* (M.A. Anderson and A.J.Rubin, Eds.), Ann Arbor Science, Michigan, USA, pp.1.

18. Davis, J.A., and Leckie, J.O. (1978) Surface ionization and complexation at the oxide/water interface. II. Surface properties of amorphous iron oxyhydroxide and adsorption of metal ions, *J.Colloid Interface Sci.*, **67**, 90-107.

19. Yates, D.E., Levine, S., and Healy T.W. (1974) Site-binding model of the electrical double layer at the oxide/water interface *J.Chem.Soc. Faraday 1*, **70**, 1807.

20. Davis, J.A., James, R.O., and Leckie, J.O. (1978) Surface ionization and complexation at the oxide/water interface. I. Computation of electrical double layer properties in simple electrolytes *J.Colloid Interface Sci.*, **63**, 480-499.

21. Sprycha, R., Kosmulski, M., and Szczypa, J. (1989) Ionic components of charge on oxides, *J. Colloid Interface Sci.*, **128**, 88-95.

FORMATION OF UNIFORM CRYSTALLINE GERMANIUM(IV) OXIDE PARTICLES

S. HAMADA
Department of Applied Chemistry, Faculty of Science, and Institute of Colloid and Surface Chemistry, Science University of Tokyo
1–3, Kagurazaka, Shinjuku-ku, Tokyo 162, JAPAN

Abstract

Uniform polyhedral particles of crystalline germanium(IV) oxide with an average size of 2.0 μm and a relative standard deviation of 0.08 were prepared by cooling from 50 to 10 °C an ethanolic mixed solution (30% by volume) of 8.4×10^{-2} mol dm^{-3} germanium (IV) and 1.7 mol dm^{-3} hydrochloric acid. The particles grew through a polynuclear layers mechanism as judged from chronomal analysis based on a fractional change of soluble germanium(IV) species remained in a supernatant solution. Uniform cuboidal particles of crystalline germanium(IV) oxide were also prepared by oxidizing at 100 °C a mixed solution of 0.11 mol dm^{-3} tris(oxalato)germanate(IV) complex, 4.0 mol dm^{-3} sulfuric acid, and ethanol of 13% by volume with 2.2 mol dm^{-3} hydrogen peroxide. An average size of the uniform particles thus obtained was estimated to be 3.4 μm with a relative standard deviation of 0.12. Additionally, uniform tetrahedral particles of sodium germanate(IV) ($Na_2Ge_3O_7\cdot7H_2O$) were obtained at 100 °C and an initial pH_i 8.5 from an ethanolic solution (50% by volume) of 3.5×10^{-2} mol dm^{-3} 8–quinolinolatogermanium (IV) complex in the presence of a large amount of sodium ions.

1. Introduction

Many kinds of uniform metal (hydrous) oxide particles [1,2], such as aluminium(III) [3,4], chromium(III) [5], gallium(III) [6,7], etc., have been prepared by forced hydrolysis of corresponding metal ions at elevated temperatures. However, uniform germanium(IV) oxide particles have not been prepared from a homogeneous solution, although monodispersed spherical particles of silica and titania were prepared as quadrivalent compounds [8–12].

In the present work, preparation of monodispersed germanium(IV) oxide particles

E. Pelizzetti (ed.), Fine Particles Science and Technology, 97–107.

were studied through a simple process based on a temperature dependence of the solubility [13,14] of germanium(IV) oxide. In addition, uniform particles of germanium(IV) oxide and corresponding composite oxide were also obtained from tris(oxalato)germanate (IV) and 8–quinolinolatogermanium(IV) solutions, respectively, by an aging method at elevated temperatures.

2. Experimental

2.1 MATERIALS

Germanium(IV) oxide of 99.99% purity was molten with ten times quantity of sodium carbonate. The solid thus obtained was dissolved in a diluted hydrochloric acid solution to yield a germanium(IV) solution acidified at pH 2. Then, sodium ions were removed by passing the solution through an H-type cation exchange resin (Amberlite IR-120B) column. The solution was filtered through a membrane filter, 0.2 μm in pore size, before use as a stock solution of germanium(IV). A total concentration of germanium(IV) in the solution was determined spectrophotometrically by a molybdenum blue method [15–17].

Potassium tris(oxalato)germanate(IV) [18–20] was prepared by refluxing a mixture of 0.064 mol germanium(IV) oxide, 0.064 mol potassium oxalate monohydrate, and 0.128 mol oxalic acid dihydrate in 300 cm^3 of water for about 3 h until the solids were completely dissolved. Then, ethanol was added at 40 °C with stirring to yield crystals of potassium tris(oxalato)germanate(IV) upon further cooling at ~0 °C.

8–Quinolinolatogermanium(IV) solution was prepared by adding a 3.5×10^{-2} mol dm^{-3} 8–quinolinol solution to germanium(IV) solution obtained from alkali fusion of germanium(IV) oxide, as described above, at specified concentration ratios and initial pH_i, but without elimination of sodium ions. The solution thus prepared was filtered through a membrane filter, 0.2 μm in pore size, before use.

2.2 PROCEDURES

In order to examine optimum conditions to get uniform germanium(IV) oxide particles by a simple process based on a temperature dependence of their solubility, an ethanolic mixed solution (30% by volume) of germanium(IV) was prepared at concentrations from 7.1×10^{-2} to 9.0×10^{-2} mol dm^{-3} at 50 °C. An aliquot of the solution (~40 cm^3) thus prepared was quickly cooled to 10 or 25 °C in a thermostated bath and kept for a desired period of time (1 – 10 h). Resulting particles were centrifuged at 3000 rpm and then repeatedly washed with a distilled water/ethanol mixture at room temperature using an ultrasonic bath.

A total concentration of germanium(IV) species remained in a supernatant solution was assayed spectrophotometrically using molybdenum blue method [15–17], after the solids were removed by centrifugation at 3000 rpm, followed by filtration through a membrane filter, 0.1 μm in pore size.

Monodispersed crystalline germanium(IV) oxide particles were also prepared by oxidizing at 100 °C for 15 h a tris(oxalato)germanate(IV) solution at an initial pH_i range between 7.9 and 8.9 with hydrogen peroxide. In addition, uniform crystalline sodium germanate(IV) particles were prepared by aging at 100 °C for 35 min an 8–quinolinolato–germanium(IV) solution at an initial pH_i 9.5 in the presence of a large amount of sodium ions. The pH values in this work are defined as apparent ones, because those obtained empirically slightly shift from real values in a alcohol/water mixed solution [21].

Morphology and size of the particles were observed by scanning electron microscopy. The particles were identified with an X–ray powder diffractometer. Diffraction angles were calibrated using a standard silicon. Differential thermal analysis (DTA) and thermogravimetry (TG) of the particles were carried out using a DTA analyzer.

3. Results and Discussion

3.1 PRECIPITATION OF UNIFORM GERMANIUM(IV) OXIDE PARTICLES FROM GERMANIUM(IV)/HYDROCHLORIC ACID/ETHANOL/WATER MIXED SOLUTIONS

Figure 1 shows uniform polyhedral particles prepared by cooling quickly from 50 to 10 °C a mixed solution of 8.4×10^{-2} mol dm^{-3} germanium(IV), 1.7 mol dm^{-3} hydrochloric

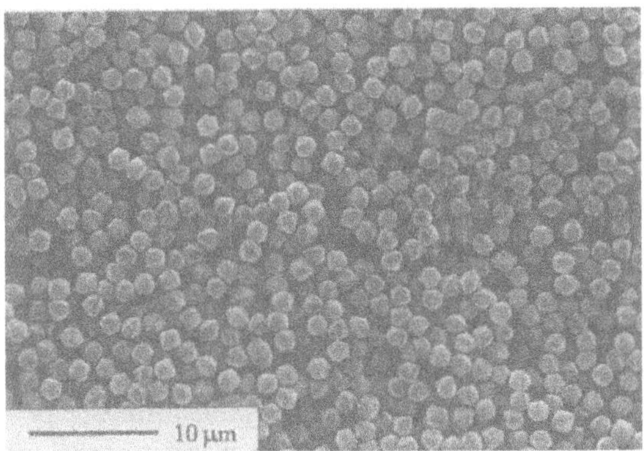

Fig. 1 Scanning electron micrograph of polyhedral GeO_2 particles obtained by coolong at 10 °C for 3 h an ethanolic solution (30 % by volume) of 8.4×10^{-2} mol dm^{-3} $Ge(IV)_t$ and 1.7 mol dm^{-3} HCl.

acid and ethanol of 30% by volume, and keeping for 3 h at the same temperature. An average size of the particle was estimated as 2.0 μm with a relative standard deviation of 0.08. Such the excellent uniform particles could be also prepared even by cooling down to 25 °C. However, any particles did not appear at initial germanium(IV) concentrations less than 7.0×10^{-2} mol dm^{-3} at 10 °C. Thus, an enough supersaturation of germanium(IV) species should be achieved for producing uniform germanium(IV) oxide particles. Solubility [13,14] of germanium(IV) oxide is much greater than those of common trivalent metal oxides, so that optimum concentrations for preparing uniform germanium (IV) oxide particles would be greater as much as about two orders than those for the uniform trivalent metal oxide particles.

Figure 2 illustrates the X-ray powder diffraction pattern of the particles. They were identified to be crystalline hexagonal germanium(IV) oxide. The spacings and relative intensities of the diffraction peaks were agreed well with those in a literature [22].

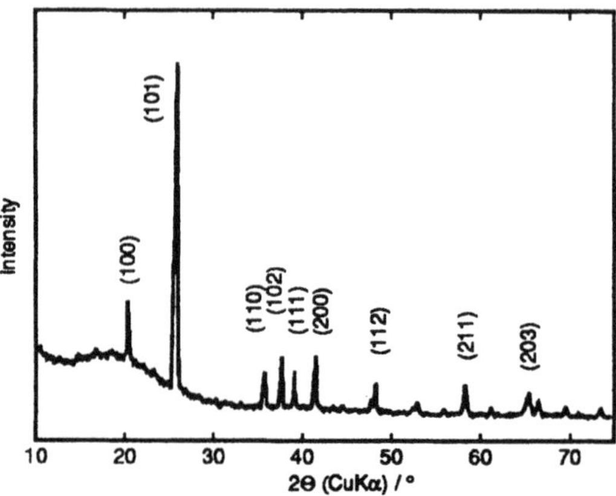

Fig. 2 X-ray powder diffraction pattern of uniform GeO_2 particles obtained under the same conditions shown in Fig. 1.

Figure 3 shows results of differential thermal analysis (DTA) and thermogravimetry (TG) of the monodispersed particles taken at a heating rate of 20 °C min^{-1}. The sample was previously dried in a desiccator for more than one week before DTA/TG measurement. Two endothermic peaks around 330 and 740 °C, accompanied about 3% of weight loss, were observed in contrast to few weight change below 300 °C. These two broad endothermic peaks suggest an existence of two kinds of structural water in the particles, whereas the particles indicated the sharp X-ray powder diffraction peaks as well-crystallized hexagonal germanium(IV) oxide. A sharp endothermic peak arose from fusion of germanium(IV) oxide was also observed at about 1120 °C.

3.2 GROWTH RATE OF UNIFORM GERMANIUM (IV) OXIDE PARTICLES

Figure 4 illustrates features of growth of the uniform germanium(IV) oxide particles during the reaction at 10 and 25 °C under the conditions given in Fig. 1. The particles grew linearly at both 10 and 25 °C, except for early stages of the reaction. An average

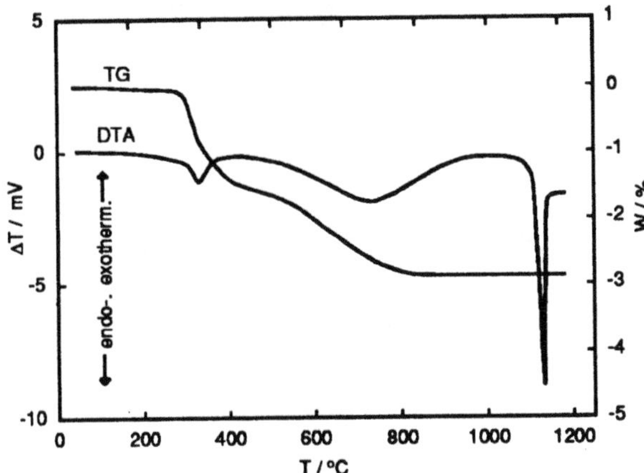

Fig. 3 DTA and TG patterns of the same GeO_2 particles taken at heating rate of 20 °C min^{-1}.

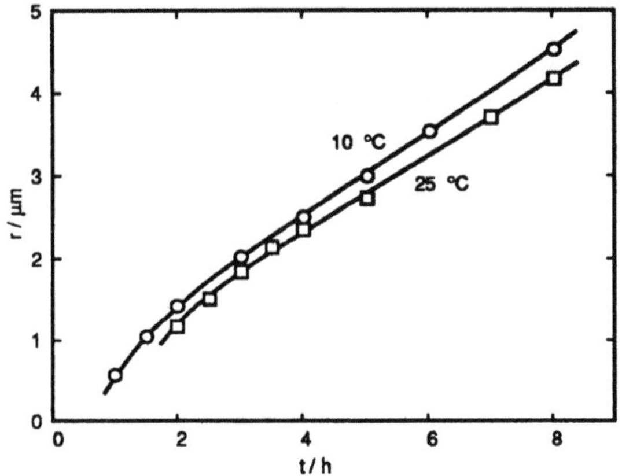

Fig. 4 Growth of uniform GeO_2 particles at 10 and 25 °C during reaction at the same concentrations shown in Fig. 1.

size of the particles obtained at 10 °C was somewhat larger than that produced at 25 °C, depending on degree of supersaturation.

Figure 5 shows an effect of acidity of the solution on the particle growth at initial stages of the reaction under the same conditions shown in Fig. 1 except for the acid concentration. The particle growth was significantly enhanced with increasing the acid concentration from 1.7 to 3.4 mol dm^{-3}. Such the pronounced effect of acid content on the particle growth would arise from a difference in the apparent solubilities of germanium (IV) oxide [13] dependent on acidity, as usually observed in the case of silica [23]. A kind of alcohol in a mixed solution also affected the growth of uniform germanium(IV) oxide particles under the same concentrations. Higher alcohols such as 2–propanol and

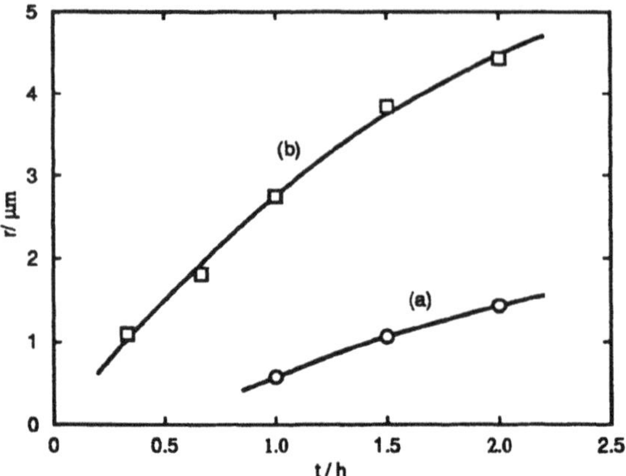

Fig. 5 Effect of acid concentration on particle growth of GeO_2 at an early stage of reaction under the same conditions shown in Fig. 1 except for acid content. Conditions: $[Ge(IV)]_t = 8.4 \times 10^{-2}$ mol dm^{-3}, $[HCl]_t = 1.7$ mol dm^{-3} (a) and 3.4 mol dm^{-3} (b), and ethanol 30% by volume.

t–butanol clearly enhanced growth rate of the germanium(IV) oxide particles than ethanol, whereas morphology of the particles did not depend on the alcohol.

3.3 RATE–DETERMINING STEP FOR GROWTH OF GERMANIUM(IV) OXIDE PARTICLES

Figure 6 displays a concentration change of soluble germanium(IV) species remained in a supernatant solution as a function of reaction time at 10 °C under the same conditions shown in Fig. 1. The concentration finally reached 3.0×10^{-2} mol dm^{-3} (a fraction of 0.36) after 325 h. A similar fractional change was also observed at 25 °C with a final fraction of 0.49 (4.0×10^{-2} mol dm^{-3} $Ge(IV)_t$) after 260 h.

The reaction degree, α, is defined as:

$$\alpha = (C_0 - C)/(C_0 - C_s), \tag{1}$$

where C_0, C_s and C are the initial and final concentrations of germanium(IV) species, and that at reaction time t, respectively. Thus, the modal size, r, of monodispersed particles at t can be expressed as:

$$r = r_f \alpha^{1/3}, \tag{2}$$

where r_f is the final size of the particle.

According to Nielsen [24], the change in the reaction degree, $d\alpha/dt$, is given by the following equation (3) for the growth of particles through a polynuclear layers mechanism:

$$\frac{d\alpha}{dt} = K_p \alpha^{2/3}(1-\alpha)^{(m+2)/3} \tag{3}$$

$$K_p = 1.8\, r_f^{-1} k_m^{1/3} C_0^{(m+2)/3} d^{7/3} D^{2/3}, \tag{4}$$

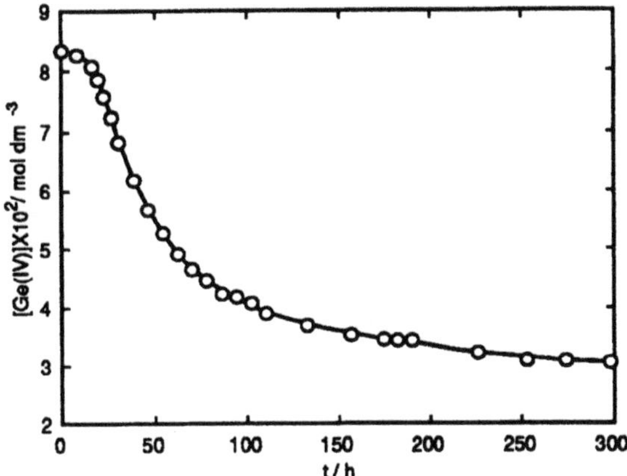

Fig. 6 Change in concentration of soluble Ge(IV) species during reaction at 10 °C under the same conditions shown in Fig. 1.

where k_m and m stand for the apparent rate constant of the two-dimensional surface nucleation and its exponential concentration dependence, d is the height of the surface nuclei, and D is the diffusion coefficient of the species. The parameters K_p and m can be calculated by a least-squares method using logarithmic values of $d\alpha/dt$, α, and $(1 - \alpha)$ obtained from the results shown in Fig. 6, because K_p is constant under the given conditions. The changes in the observed reaction degree were previously smoothened by a polynominal equation. The values of K_p and m were estimated as 13.8, and 3.81 at 10 °C, respectively.

The solid curve in Fig. 7 shows the reaction degrees obtained from numerical integration

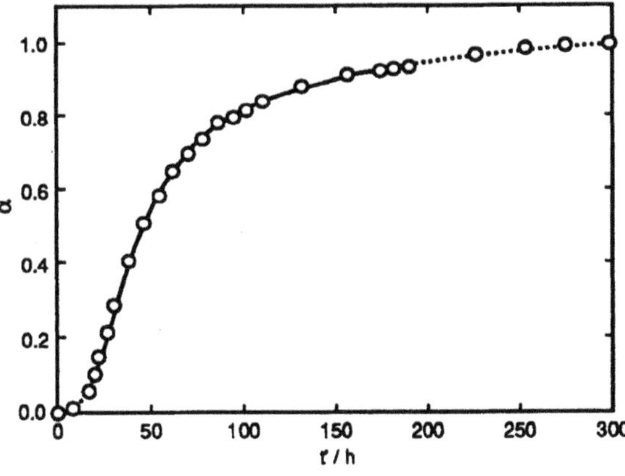

Fig. 7 Plots of reaction degrees of GeO_2 particles at 10 °C obtained from results in Fig. 7. Reaction time, t', in the figure is normalized in terms of induction period.

of Eq. (3) by the Runge–Kutta method using the parameters obtained above, which agreed well with the empirical ones over a wide range of the reaction time. Similar numerical treatment was also made assuming a diffusion mechanism [24]. Calculated values, however, did not fit the experimental data. Thus, it would appear that the polyhedral particles of germanium(IV) oxide grow through the polynuclear layers mechanism including dehydration process of germanium(IV) hydroxocomplexes even at low temperatures.

3.4 FORMATION OF GERMANIUM(IV) OXIDE PARTICLES FROM TRIS–(OXALATO)GERMANATE(IV) SOLUTIONS

Free metal ions would be released from the corresponding metal complexes at elevated temperatures, if the complex formation is exothermic. An advantage of such the complexes is that supersaturation of actual precursors of the particles would be appropriately reduced for producing uniform metal (hydrous) oxide particles [4, 25–27] through complexation under a limited set of conditions, not only as a reservoir of the metal ion species. Therefore, a tris(oxalato)germanate(IV) complex solution was aged at 100 °C to prepare uniform particles over a concentration range from 4.3×10^{-2} to 5.4×10^{-2} mol dm^{-3} $Ge(IV)_t$ and an initial pH_i range from 7.9 to 8.9. However, any particle did not appear, because the oxalato complex is so stable to release an enough amount of free germanium(IV) hydroxocomplexes in spite of the exothermic complexation [15-17].

Thus, tris(oxalato)germanate(IV) complex were degraded by oxidizing oxalate anions with hydrogen peroxide, leading to germanium(IV) hydroxoccomplexes. Figure 8 shows cuboidal particles, with an average size of 3.4 μm and a relative standard deviation of 0.12, obtained by aging at 100 °C for 15 h a mixed solution of 0.11 mol dm^{-3} tris(oxalato)–

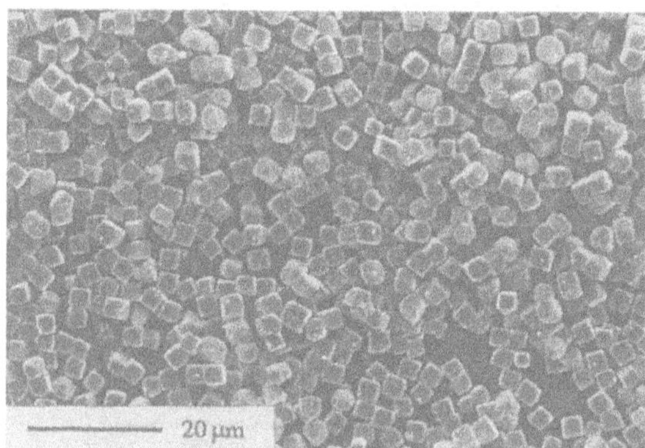

Fig. 8 Scanning electron micrograph of cuboidal GeO_2 particles precipitated from tris(oxalato)germanate(IV) solution by aging at 100 °C for 15 h. Conditions: $[Ge(IV)]_t = 1.1 \times 10^{-1}$ mol dm^{-3}, $[H_2O_2]_t = 2.2$ mol dm^{-3} , $[H_2SO_4] = 4.0$ mol dm^{-3}, and ethanol 13% by volume.

germanate(IV), 4.0 mol dm^{-3} sulfuric acid, 2.2 mol dm^{-3} hydrogen peroxide, and ethanol 13% by volume. The particles thus obtained were identified as well–crystallized hexagonal germanium(IV) oxides [22] by X–ray powder diffractometry. However, this procedure was somewhat disadvantageous due to disturbance of homogeneous reaction by carbon dioxide bubbles generated at the oxidation of the oxalate anions.

3.5 PRECIPITATION OF SODIUM GERMANATE(IV) PARTICLES FROM 8–QUINOLINOLATOGERMANIUM(IV) SOLUTIONS

The enthalpy change of the formation of 8–quinolinolatogermanium(IV) complex [28] has been reported to be –76.0 kJ mol^{-1}. Thus, preparation of germanium(IV) compound particles was attempted by aging the 8–quinolinolato complex solution at elevated temperatures. Figure 9 illustrates tetrahedral particles obtained by aging at 100 °C for 35 min a solution of 3.5×10^{-2} mol dm^{-3} $Ge(IV)_t$ at a concentration ratio, $[C_9H_7NO]_t$ / $[Ge(IV)]_t$, of 2.0, and ethanol of 30% by volume at the pH_i of 8.5 in an oil bath. An average size of the particles thus obtained was estimated to be 2.6 µm with a relative

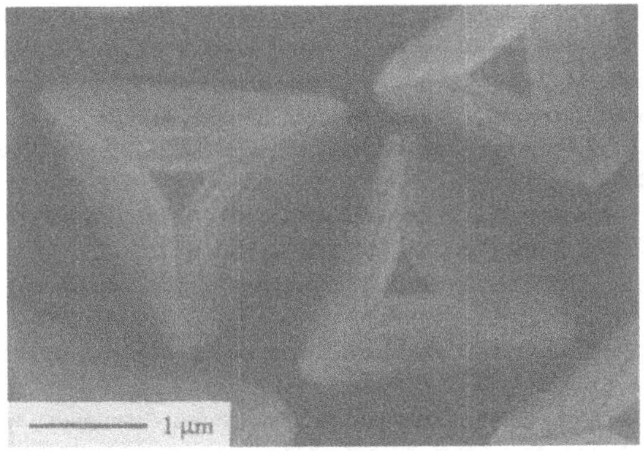

Fig. 9 Scanning electron micrograph of $Na_2Ge_3O_7 \cdot 7H_2O$ particles prepared from 8–quiolinolatogermanium(IV) solution in the presence of sodium ions by aging at 100 °C and pH_i 8.5 for 35 min. Conditions: $[Ge(IV)]_t = 3.5 \times 10^{-2}$ mol dm^{-3}, $[C_9H_7NO]_t$ / $[Ge(IV)]_t = 2.0$, and ethanol 50% by volume.

standard deviation of 0.17. The particles thus obtained were identified as $Na_2Ge_3O_7 \cdot H_2O$ by X-ray powder diffractometry [29]. However, morphology and uniformity of the particles were strongly dependent on the preparing conditions such as germanium(IV) concentration, initial pH_i, and alcohol content.

4. References

1. Matijevic, E. (1981) Monodispersed Metal (Hydrous) Oxides – A Fascinating Field of Colloid Science,

Acc. Chem. Res., **14**, 22 – 29.

2. Matijevic, E. (1985) Production of Monodispersed Colloidal Particles, *Ann. Rev. Mater. Sci.*, **15**, 483 – 516.
3. Brace, R. and Matijevic, E. (1973) Aluminum Hydrous Oxide Sols. I. Spherical Particles of Narrow Size Distribution, *J. Inorg. Nucl. Chem.*, **35**, 3691 – 3705.
4. Hamada, S., Kudo, Y., and Hasegawa, S. (1991) Preparation of uniform spherical aluminum hydrous oxide particles from acetylacetonatoaluminum solutions, *Colloid Polym. Sci.*, **269**, 290 – 294.
5. Matijevic, E., Lindsay, A. D., Kratohvil, S., Jones, M. E., Larson, R. I., and Cayey, N. W. (1971) Characterzation and Stability of Chromium Hydroxide Sols of Narrow Size Distribution, *J. Colloid Interface Sci.*, **36**, 273 – 281.
6. Hamada, S., Bando, K., and Kudo Y. (1984) Formation of Monodispersed Gallium Hydrous Oxide Particles by Hydrolysis at Elevated Temperatures, *Nippon Kagaku Kaishi*, **1984**, 1068 – 1071.
7. Hamada, S., Bando, K., and Kudo Y. (1986) The Formation Process of Hydrous Gallium(III) Oxide Particles Obtained by Hydrolysis at Elevated Temperatures, *Bull. Chem. Soc. Jpn.*, **59**, 2063 – 2069.
8. Stöber, W., Fink A., and Bohn, E. (1987) Controlled Growth of Monodisperse Silica Spheres *J. Colloid Interface Sci.*, **26**, 62 – 69.
9. Tan, C. G., Bowen B. D., and Epstein, N. (1987) Production of Monodisperse Colloidal Silica Spheres: Effect of Temperature, *J. Colloid Interface Sci.*, **118**, 290 – 293.
10. Shimohira, T. and H. Ishijima, H. (1981) Synthesis of Monodispersed Spherical Silica, *Nippon Kagakukaishi* **1981**, 1503 – 1505.
11. Matijevic, E., Budnik, E. M., and Meites, L. (1977) Preparation and Mechanism of Formation of Titanium Dioxide Hydrosols of Narrow Size Distribution, *J. Colloid Interface Sci.*, **61**, 302 – 311.
12. Santacesaria, E., Tonello, M., Storti, G., Pace, R. C., and Carra, S. (1986) Kinetics of Titanium Dioxide Precipitation by Thermal Hydrolysis, *J. Colloid Interface Sci.*, **111**, 44 – 53.
13. Linke, W. F. (1958) *Solubilities: Inorganic and Metal–Organic Compounds*, Vol. I, Am. Chem. Soc., Washington DC, p. 1074.
14. Laubengayer, A. W. and Morton, D. S. (1932) GERMANIUM. XXXIX. THE POLYMORPHISM OF GERMANIUM DIOXIDE, *J. Am. Chem. Soc.*, **54**, 2303 – 2320.
15. Erdey, L. and Bodor, A. (1951) Colorimetrische Milkrobestimmung des Germaniums, *Z. Anal. Chem.*, **134**, 81 – 85.
16. Kitson, R. E. and Mello, M. G. (1944) Colorimetric Determination of Germanium as Molybdigermanic Acid, *Ind. Eng. Chem.*, **16**, 128 – 130.
17. Boltz, D. F. and Mellon, M. G. (1947) Determination of Phosphorus, Germanium, Silicon, and Arsenic by the Heteropoly Blue Method, *Anal. Chem.*, **19**, 873 – 877.
18. H. F. Holtzclaw, Jr. (ed) (1966) *Inorg. Synth.*, McGraw–Hill, New York, Vol. VIII, p. 34.
19. Tchakirian A. (1939) La Chimie du Germanium, *Ann. Chim.* (Paris), **12**, 415 – 428.
20. Moeller, T. and Nielsen, N. C. (1953) Partial Resolution of the Tris–(oxalato)–germanate(IV) Ion, *J. Am. Chem. Soc.*, **75**, 5106 – 5107.
21. Bates, R. G. (1964) *Determination of pH*, John Wiley, New York, p.222.
22. McMurdie, H. (1986) *Powder Diffraction*, Vol. **1**, p. 69.
23. Linke, W. F. (1965) *Solubilities: Inorganic and Metal–Organic Compounds*, Vol. II, Am. Chem. Soc., Washington D.C., p. 1452.
24. Nielsen, A. V. (1964) *Kinetics of Precipitation*, Pergamon Press, New York, Chaps. 3 and 4.
25. Hamada, S., Niizeki, S., and Kudo, Y. (1986) The Precipitation of Monodispersed α–Iron(III) Oxide Particles from Iron(III) Chloride–Glycine System in Aqueous and 2–Propanol/Water Media, *Bull. Chem. Soc. Jpn.*, **59**, 3443–3450.
26. Hamada, S., Kudo, Y., and Matsumoto, T. (1989) The Preparation of Monodispersed α–Iron(III) Oxide Particles from 2,2'–Bipyridineiron(III) Complex, *Bull. Chem, Soc. Jpn.*, **62**, 1017–1022.
27. Hamada. S,., Kudo. Y., and Kobayashi, T. (1993) Precipitation of Uniform Indium Hydroxide Particles from Indium 2–Aminobutyrato Complex Solutions, *Colloid & Surfaces A, Physicochemical and Engineering*

Aspects, 79, 227–232.

28. Tsau, J., Matsouo, S., Clerc, P., and Benoit, R. (1967) Chélates du germanium(IV) avec l'hydroxy–8–quinoléine et des dérives substitués, *Bull. Soc. Chim. Fra.*, **1967**, 1039 – 1043.
29. Knyazev, E. A. and Borisova, S. V. (1967) THE GeO_2–Na_2O–H_2O SYSTEM AT 25 °C, *Russ. J. Inorg. Chem.*, **12**, 1469 – 1471.

MODEL PARTICLES FOR CONCENTRATED COLLOIDAL DISPERSIONS: FROM HARD SPHERES TO SOFT RODS.

Fifteen years of colloid research in Utrecht: particle synthesis and physics of suspensions

A. VRIJ AND A.P. PHILIPSE
Van 't Hoff Laboratory for Physical and Colloid Chemistry, Utrecht University
Padualaan 8, 3584 CH Utrecht, The Netherlands

Abstract

In Utrecht we study 'fine' particles in the colloidal size range in connection with (concentrated) dispersions of such particles in a liquid carrier. Originally the particles were spherical in shape (e.g. silica spheres) but more recently also elongated particles (e.g. boehmite rods) are synthesized and studied. It is very important that the particles are non-clustered during the synthesis and thereafter. The particles are coated with a surface layer of chain molecules or silane coupling agents to make them soluble and stable against coagulation in organic solvents. In that case the particles/solvent system behaves as a 'Complex Fluid' in which the particles show 'hard' or 'soft' repulsion forces as revealed by light or (small-angle) neutron scattering and other techniques like sedimentation, diffusion, rheology and colloidal filtration.
In more recent studies 'compound' particles were synthesized and studied, consisting e.g. of silica spheres labeled with a fluorescent core, silica spheres with a magnetic core, and silica rods with a core of boehmite, all in aqueous and organic solvents.
Spheres with a fluorescent core are used as labels in optical studies like Confocal Scanning Laser Microscopy. The particles with a magnetic core show very interesting behaviour in an external magnetic field and the elongated particles show double refraction e.g. in a shear field. More over, also phase transitions from isotropic to liquid-crystal like suspensions have been observed.

A brief overview will be given of the synthesis and physical experimentation augmented by references, allowing the interested reader to pursue a particular topic in more detail.

1. Introduction

In this contribution to the workshop we would like to emphasize our approach to the theme on 'Fine Particles, from Micro to Nano'. In Utrecht we study 'fine' particles in a

E. Pelizzetti (ed.), Fine Particles Science and Technology, 109–124.

liquid medium. Such systems are called: Colloidal Suspensions or Colloidal Dispersions.
Particles are in the colloidal size range when having the following physical properties. Particles must be large enough to behave as (quasi) macroscopic from an atomic point of view ($a > 1$ nm) and at the same time must be small enough to possess sufficient Brownian motion to remain suspended ($a < 1$ μm). In many instances such systems can be viewed as 'supra-molecular liquids', i.e fluids containing 'molecules' with a much inflated size.
The term 'Colloid' (greek: glue-like) reminishes the property of such particles to easily stick to each other and to many kinds of surfaces. This is due to the fundamental reason that a particle of colloidal size has a (very) large surface (interaction) area compared to an atom, whereas the thermal energy of translation is the same (kT)

The managing of colloidal particles during the process of synthesis and thereafter is therefore a delicate task. All the time the interaction forces between the particles must be kept sufficiently repulsive. If not, the particles will clump together often forming clusters that are not redispersable any more.
Our emphasis is to prepare and study model systems of non-clustered particles for fundamental research. But it is also found that model systems can sometimes be developed further for specific applications. In this respect the compound particles, and the particles with fluorescent or magnetic labels may be interesting.

In this paper we will describe in a brief overview the research we have done with inorganic colloidal particles in the last fifteen years at the Van 't Hoff-laboratory. After this we will give some recent results. For the experimental and theoretical details we refer to the listed references. The colloidal particle systems comprising microemulsions [1a-b] and latices [1c] will not be reviewed here.

The paper is organized as follows. First, spherical particles with a silica core and several surface coatings are treated. The synthesis and physico-chemical properties of (concentrated) dispersions will be described. Second, elongated particles and particles with compound cores are mentioned.

2. Preparation of Organophylic Silica Particles.

Silica dispersions are prepared in three steps. First, spherical silica cores are synthesized [2] in an ethanol-ammonia mixture by the polymerization of tetraethoxysilane (TES), following a method originally developed by Stöber et al. [3]. The schematical reaction is:

$$nSi(OC_2H_5)_4 + 2nH_2O \xrightarrow[NH_3]{C_2H_5OH} nSiO_2 + 4nC_2H_5OH$$

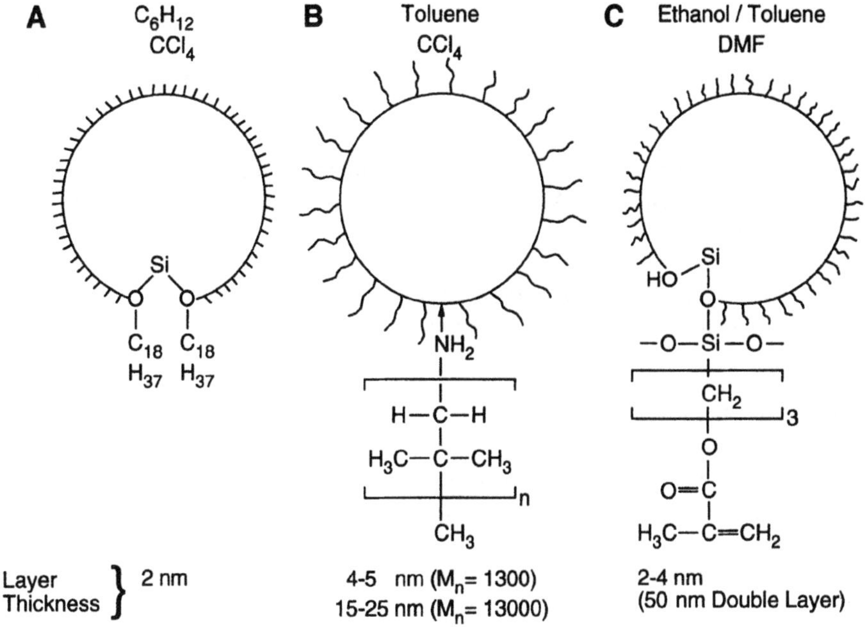

Figure 1. Silica spheres with different surface coatings.

The second step is a further growth in size of the silica spheres in the alcosol by the addition of TES [4]. The reaction vessels are kept at a constant temperature. This problably helps to avoid secondary nucleation. The mechanism of particle formation and growth were also studied in our laboratory [5].

The third step is a coating of the spheres. Three types of coating were studied by us and are schematically depicted in Fig. 1. In case A the NH_3 and ethanol are gradually removed by destillation while octadecanol is gradually added and the temperature ultimately reaches a value of about 190 degrees centigrade, and were two more hours are allowed to complete the etherification of the surface silanol groups [2a]. The gradual change of solvent environment helps to prevent the particles from clustering. The octadecyl-coated spheres are soluble in non-polar solvents like cyclohexane. They can even be redispersed after (gently) drying (in a nitrogen stream).
In case B the particles are lyophylized with poly-isobutene chains carrying functionalized end groups with a NH_2-function [6]. Also grafting reactions with poly-hydroxy-stearic acid have been developed [6c]
In case C the surface silanol groups are reacted with the silane coupling agent γ-methacryloxypropyl-trimethoxy-silane (TPM). TPM-oligomers adsorb on the particle

surface and condense with surface silanol [4] to a thin organophylic layer, which makes the particle soluble in polar organic solvents.
The particles are electrically charged in the original alcosol and in sols of type C. They are uncharged in the cases A and B.

For more details also about the characterization with techniques like static and dynamic light scattering, SANS and SAXS, EM, solid state NMR, see the original litterature [2b], [5a-d], [7a-b], [9].

Time consuming sedimentation procedures are often required in the purification steps in order to avoid unwanted clustering of particles.

3. Interparticle Forces

An important aim of study is monitoring the interaction potential between the particles in the suspending liquid. This may range from steeply repulsive to deeply attractive, as depicted in Fig. 2I,II.
In Fig.2I, we show a schematic picture of ranges of repulsive forces, scaling from steep(A) to soft(C). This is the trend found in the dispersions given in Fig. 1. We have in A, stearyl chains in cyclohexane; in B, PIB chains in cyclohexane and in C, TPM-coating in an toluene/ethanol mixture. The steepness is correlated with the "interpenetration depth" of the overlapping surface layers. In C the softness of the outer part is mainly determined by the thickness of the electrical double layer.

In all these surface coating/solvent couples the minimum requirement for repulsive interaction is a 'good solvation' of the surface layers by the solvent molecules. Good solvation means that molecular segments of the surface layers are preferentially attracted by neighboring solvent molecules *over* neighboring segments.

Attractive interactions come into play when the solvation is less good. This means that molecular surface layer segments are preferentially attracted by other, neighboring segments rather than by solvent. Attractions can be invoked in case A by replacing cyclohexane by toluene or benzene and in case C by increasing the fraction of toluene. The attraction can sometimes be fine-tuned by adjusting temperature or pressure [8].

3.1. WHERE ARE THE LONDON-VAN DER WAALS FORCES?

An often asked question. Attractive van der Waals forces are universal and are therefore also present here. If they will be effective depends on the energy that can be gained by replacing solvent (material) by particle (material). With silica cores in organic solvent the refractive index difference is small. (For lightscattering experiments it is often even nearly matched, see section 4). This implies that the (macroscopic) London-van der Waals forces between the particle cores are small. They may usually be neglected, at least for not too large particles [6b].

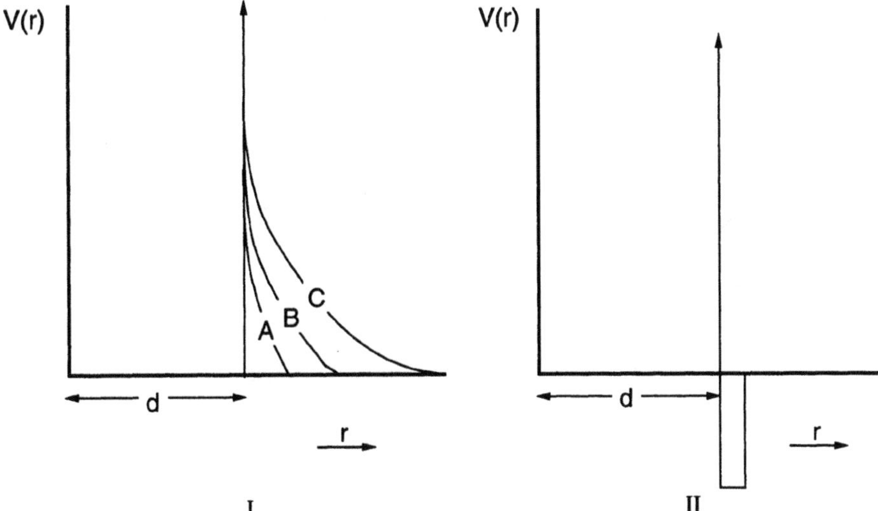

Figure 2. Repulsive, I, and attractive, II, interparticle potentials.

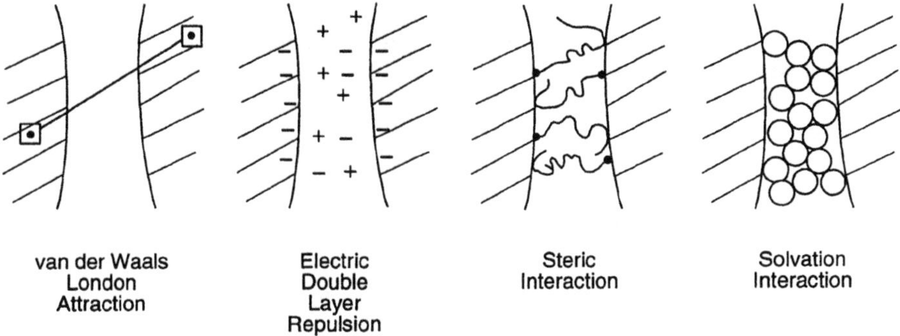

Figure 3. Several different types of interparticle forces.

But what about the London-van der Waals forces between the surface layers? Here the same principle applies, but it is not appropriate to use a macroscopic concept of the London-Van der Waals forces for molecular distances.

For the contribution of the surface layers a careful consideration of the (free) energy of the 'overlapping' surface layers is required. In the process of two approaching surface layers, solvent molecules are replaced by surface molecular segments. The free energy of this exchange process has an energy component (molecular London-van der Waals forces) but also an entropy component. The details are often unknown but qualitatively

Structure Factor

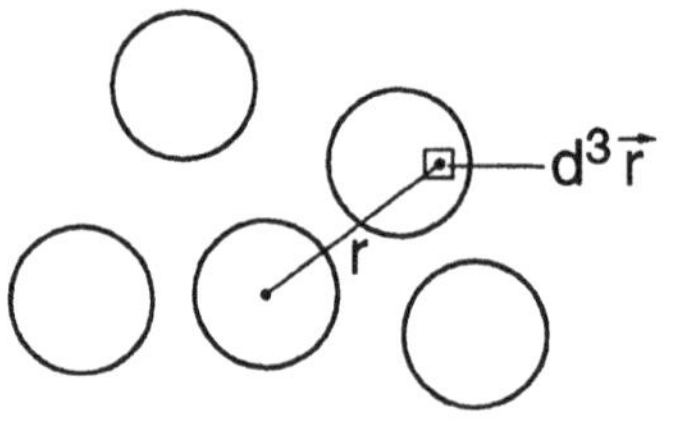

$$S(K) = 1 + 4\pi\rho \int_0^\infty [g(r)-1] \frac{\sin Kr}{Kr} r^2 dr$$

$d^3\vec{r}$

r

g(r)= Radial Distribution Function

$$K = \frac{2\pi}{\lambda} \sin\frac{\theta}{2}$$

Colloidal Fluid

θ= Scattering Angle

Figure 4. Radial distribution function and structure factor of a colloidal fluid.

this is expressed as "good solvation", when it is difficult to remove solvent molecules from the surface layers and "poor solvation" when it is relatively easy.
The several types of interactions are depicted in Fig. 3.

4. Mesoscopic Structures, Scattering and Thermodynamic Properties in Colloidal Suspensions.

Statistical physical theories and computer simulations can be used to connect interparticle forces (section 3) with mesoscopic and macroscopic properties of suspensions containing the colloidal particles. Quite often these are copies of atomic or low-molecular fluid theories.

In Fig 4, a schematic picture of a repulsive particle fluid is depicted. Its structure can be characterized by the pair distribution function g(r), which gives the probability to find a particle on a distance r from a given particle.

The light or small-angle neutron (resp. -X-ray) scattering angular spectrum S(K) is a measurable function connected with g(r) via a Fourier transform. An example is shown in Fig. 5 for suspension C (see Fig. 1) at several particle concentrations.
Higher concentrations correlate with more pronounced structure functions S(K). The plots in Fig. 5 compare favorably with soft-sphere models of liquid state physics. Similar results were found for system A, which particles behave as 'hard spheres'. Also polydispersity effects [9b] are considered in these studies. See references [7a, 9a-b] for further details.

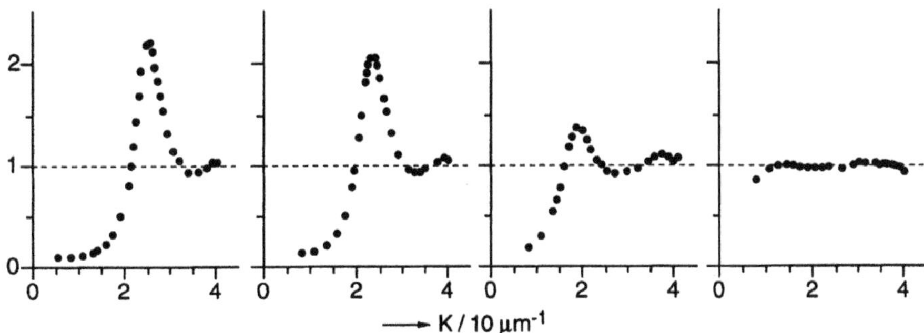

Figure 5. Structure factor of silica spheres with TPM-coating, Left to right: concentration (g/ml) is 0.16, 0.13, 0.069, 0.015. [9a]

An important feature of our colloidal silica systems is that the refractive index difference of the suspended particles and the suspending solvent is rather small and that it can be varied. Thus one can make the light scattering intensity 'tunable' and much smaller than usual. In this way multiple scattering can be avoided and concentrated systems can be studied. This is not possible in most aqueous systems, except fluorbased latex which has a refractive index near 1.35 [10]. Also London-van der Waals forces between particle cores become very weak.

5. Colloidal Particle Suspensions - Instabilities and Phase Separations.

The suspensions as shown in Fig 2I and in Fig. 6 are stable. Instability may be invoked in two different ways.

1) inducing an attraction well in the interparticle potential, see Fig. 2II, or
2) by making the particle concentration very high (while keeping the interparticle forces repulsive).

The first case is the common one. It can be invoked by changing e.g. the solvent composition, the temperature, or the pressure. It usually leads to flocculation/coagulation (see Fig. 6). In our particle systems - and this is different from many hydrophobic colloids in water - the flocculation is usually reversible, i.e. it can be reversed by reversing the 'inducing agens' [11a,d].

A special, but important case is a stable colloidal suspension to which non-adsorbing polymer molecules are added. This may lead to a genuine fluid, separated phase, instead of a flocculated system or gel [11b]. A similar phase separation may take place when small colloids are added [6d]. The particle, just as non-adsorbing polymers, are

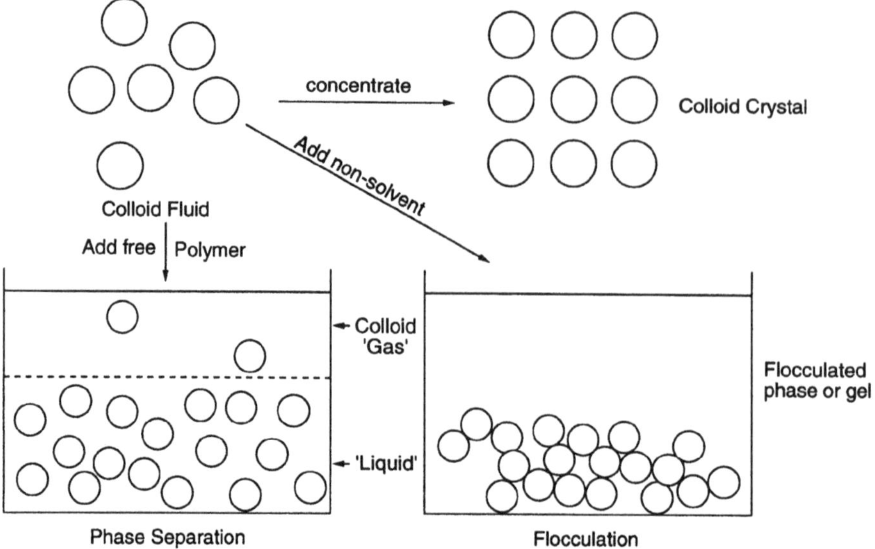

Figure 6. Induced phase transformations of a colloidal fluid.

excluded from the region between the large colloids. This gives rise to an (effective) attractive depletion force [11b, 11d].

In the second case of instability, a stable system with repulsive interparticle forces leads to a "colloidal crystal" exhibiting Bragg-reflections with visible light. This disorder-order transition is depicted in Fig. 6, [11c-d].

6. Diffusion, Rheology, Permeation and Sedimentation.

Our model particles are good candidates for dynamical studies of concentrated dispersions as well. In Fig. 7, we show schematically the phenomenon of diffusion in a concentration gradient. This type of diffusion called 'collective diffusion' is driven by the osmotic pressure gradient:

$$D_c = B_c \left(\frac{\partial \Pi}{\partial \rho} \right)_{\mu_0} \tag{1}$$

Here D_c and B_c are the collective diffusion coefficient and collective mobility and Π and ρ are the osmotic pressure and the number density of the dispersion. Further μ_0 is the chemical potential of the solvent.

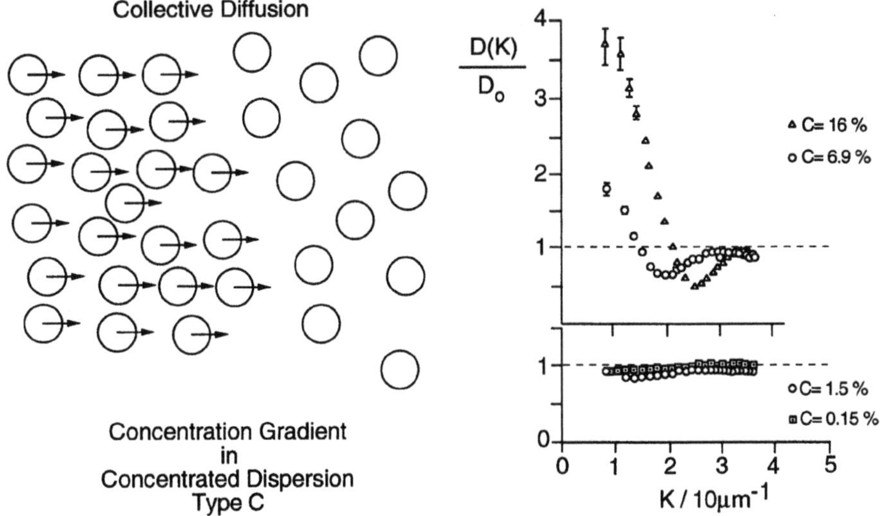

Figure 7. Collective diffusion of a colloidal fluid.

In our systems with nearly refractive index-matched silica spheres D_C can be measured with dynamic light scattering. In dynamic light scattering the relaxation time of a sinusoidal (Brownian) concentration wave with a wavelength $\Gamma=2\pi/K$ is measured, where K is defined in Fig.4.

The plots in Fig. 7 clearly show that whereas for low particle concentration the diffusion coefficient is not a function of K, it becomes a function of K at high volume fractions. This peculiar behavior is caused by the fact that appreciable "structure" is present in the dispersion (see Fig.5) on a length-scale $\Lambda=2\pi/K$. This requires a generalized $\partial\Pi/\partial\rho$ and B_C which are K-dependent. Where a K-dependence of the driving force was already known, our experiments also require a K-dependent mobility. For a recent theory we refer to papers of R. Klein and co-workers [12].

A different type of mobility property is shown in the permeability and sedimentation experiments of Fig. 8. Solvent is pumped through a sedimented bed of particles or particles are sedimented in a centrifugal field [13a,b]. The normalized velocity $v(\phi)/v_0$ is plotted in Fig. 8 as a function of the particle volume fraction. For identical systems $v(\phi)$ is related to B_C,

$$B_C = \frac{D_0}{kT}\frac{v(\phi)}{v_0} \tag{2}$$

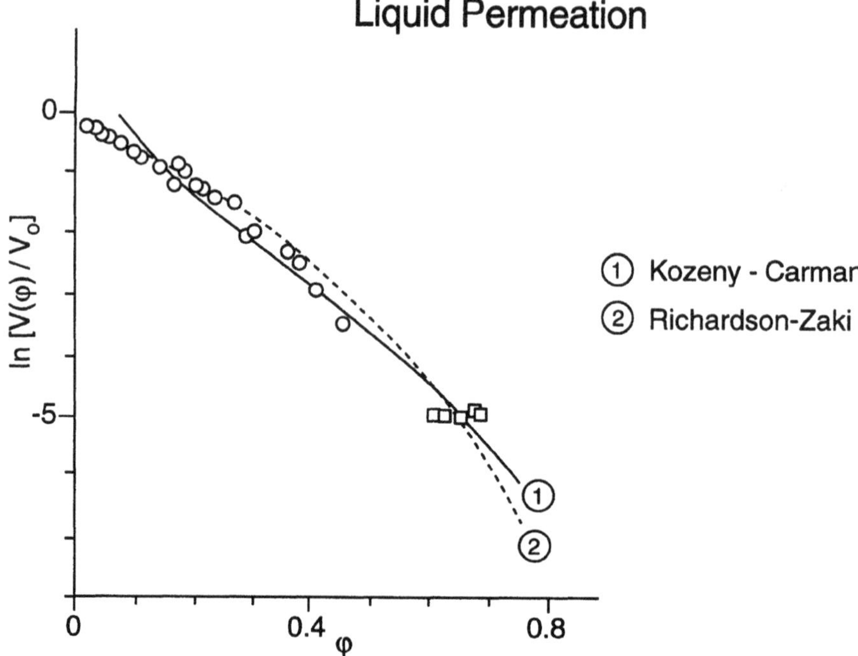

Figure 8. Liquid permeation of (❑) a packed bed and sedimentation (○) of colloidal spheres

The drawn lines are calculated with two different models (Kozeny-Carman and Richardson-Zaki)

Dispersions of (either repulsive or attractive) silica particles are also valuable model systems for rheological studies. For example, shear-rate and concentration dependence has been investigated [13d,e], as well as the influence of shear on colloidal phase transitions [13f,g]

7. Compound Particles. Silica Particles with Fluorescent, Elongated and Magnetic Cores.

In our laboratory particles are prepared and studied which have a layer of silica grown on a core of different material. The periferal silica layer can be covered subsequently with a layer of octadecyl or PIB or TPM to make the compound particles soluble in several solvents.

The periferal silica layer can serve different objectives, i.e., refractive index matching, screening of van der Waals' or magnetic dipole forces, equalizing solvation forces in mixtures of particles.

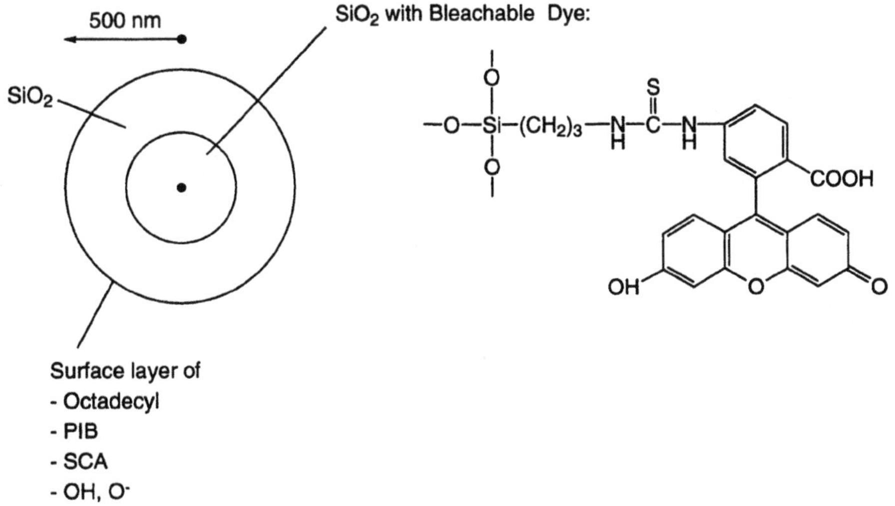

Figure 9. Compound silica particle with bleachable dye in the core. [14]

7.1 SILICA PARTICLES WITH A FLUORESCENT CORE.

Van Blaaderen in our laboratory synthesized stable colloidal dispersions of fluorescent silica spheres [14a]. The dye is first coupled to a silane coupling agent ((3-aminopropyl)-triethoxy-silane) after which it is mixed with tetraethoxysilane and incorporated in a Stöber synthesis. These spheres serve as a particle core on which a subsequent silica layer (without dye) is grown to a compounded particle, see Fig. 9. In the thus optically-labelled spheres the dye is fixed and protected from the outer solvent medium.

The particles are used in a special type of dynamic laser light scattering, i.e., fluorescence recovery after photo-bleaching (FRAP) and forced Rayleigh scattering (FRS). With FRAP a strong laser beam bleaches a diffraction pattern in the solution which decays in time because of the (self) diffusion of bleached colloidal spheres. This makes it possible to measure "long-time" self diffusion of colloidal spheres in a *concentrated* suspension [14b].

Another application is observation of particles in a concentrated dispersion by means of a new type of microscopy: confocal scanning laser spectroscopy (CSLM). The labelled particles can be index-matched so the turbidity of the system is very weak. This allows to look deeply inside the suspension, which is not possible for non index-matched systems, e.g. latex in water, which are milky-white [14c-d]. The weak fluorescence of the core makes location of the particles visible (see Fig. 10)

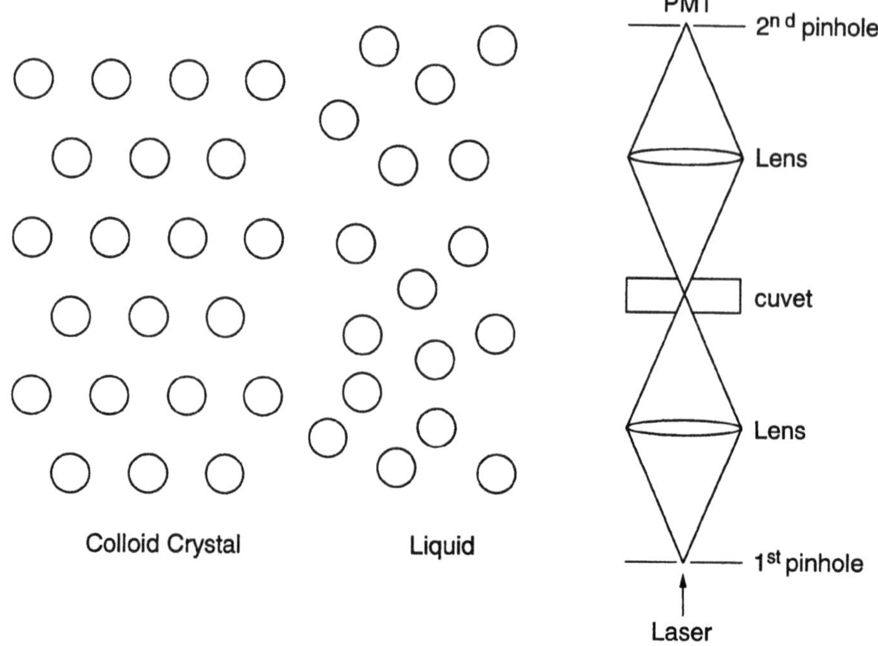

Figure 10. Colloid crystal/liquid interface and the principle of the CSLM-technique.

7.2. SILICA RODS WITH A BOEHMITE-NEEDLE CORE.

Well defined rods with controllable sizes and interparticle interactions are important in the study of physical properties of (model) suspensions [15d].

Recently Buining et al [15a-b] developed an alkoxide route for the synthesis of fairly well-defined boehmite needles in non-aqueous solvents. These systems show streaming birefringence. With these colloids the first experimental demonstration was given for the isotropic-nematic phase separation in suspensions of inorganic rods [15e, 15d]. Philipse and collaborators used these boehmite needles as heterogeneous nuclei for silica-growth [15c]. The resulting rod-like silica particles can be chemically modified with organic groups in a similar manner as for silica spheres (see Fig. 1). The so obtained rods, coated with TPM or with octadecanol, are soluble in polar respectively non-polar organic solvents, see Fig.11. A fluorescent layer can be applied as well [15c]. Phase behavior and reology of these rods are currently being investigated in relation to interparticle attractions or repulsions.

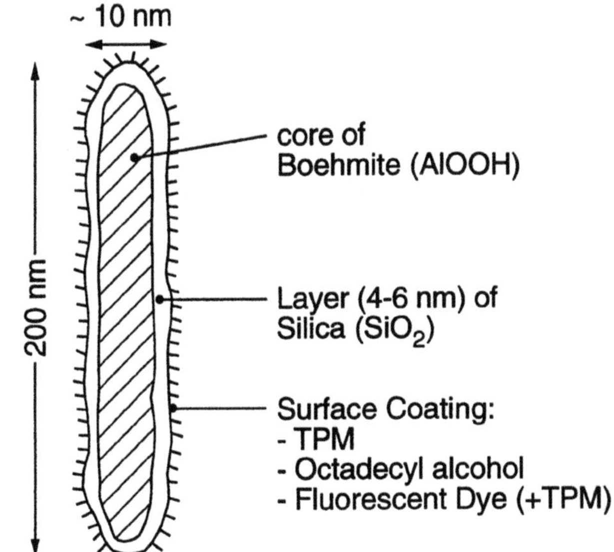

Figure 11. Compound particle with boehmite core and silica layer.

The range of aspect ratios which can be achieved in the boehmite synthesis is fairly restricted. Silica ellipsoids have been employed for studying lower aspect ratios [15f].Imogolite fibers, which have very high aspect ratio's, were found to be very effective stabilizers of metallic nano-particles [15g].

7.3 SILICA SPHERES WITH A MAGNETIC CORE.

Magnetic, colloidal particles are the main constituent of so-called 'ferrofluids', which show very interesting properties in a magnetic field (an example is found in Fig. 12, where the magnetic field, induced by an electric current, gives the fluid meniscus a conical shape). For a recent review see reference [16a].

The particles are usually very small (~10 nm) and have strong interactions because of the magnetic dipole forces. Recently we were able to prepare compound particles around a magnetic core. The magnetic cores are synthesized according to a method of Massart [16b], which produces single-domain (super paramagnetic) particles. Subsequently [16c] a layer of silica is grown and a surface layer of octadecanol or TPM is attached to make the particles soluble in organic solvent, see Fig.12. The underlying idea is to 'tailor' the silica shell thickness such that the colloids are stable spheres with isotropic interactions, whereas an external magnetic field produces weak dipolar attractions and consequently reversible anisotropic structures.

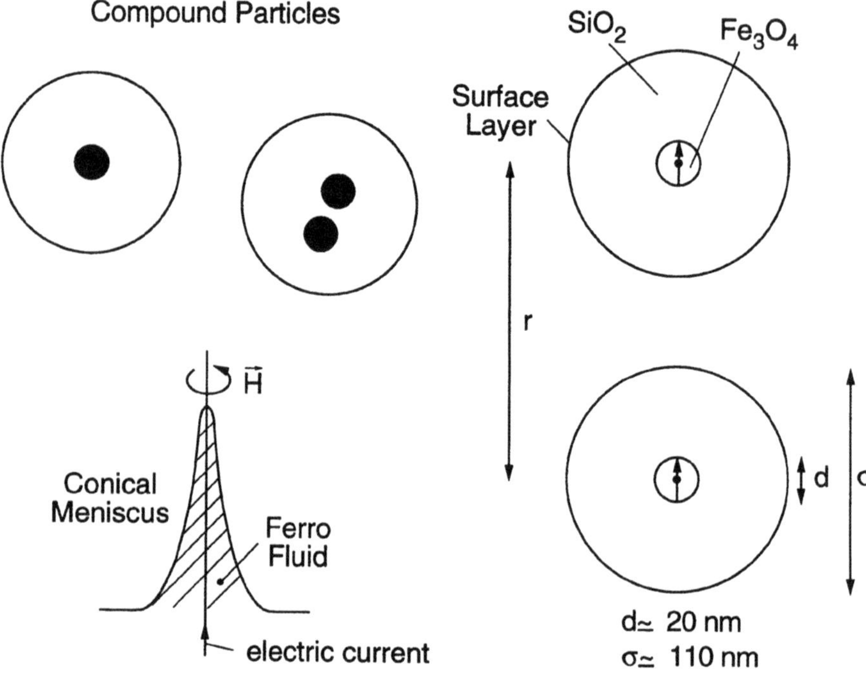

Figure 12. Compound particle with magnetic core and ferrofluid meniscus in a magnetic field.

For two magnetic moments in a head-to-tail configuration (see Fig. 12) the interaction is attractive and proportional to $\mu^2\ r^{-3}$, where μ is the dipole moment and r is the distance between the particle centers.

8. Final Remark

The research on particle (dispersion) synthesis and physical experimentation covers a broad range in both time and scope. For further information we suggest readers to contact A. Vrij or A.P. Philipse for details on subjects related (roughly) to the Figures (1-7), respectively (7-12).

References

1a) Overbeek, J.Th.G., Verhoeckx, G.J., de Bruyn, P.L., and Lekkerkerker, H.N.W. (1987) *J. Colloid Interface Sci.* **119**, 422. b) Kegel, W.K., van Aken, G.A., Bouts, M.N., Lekkerkerker, H.N.W., Overbeek, J.Th.G., and de Bruyn, P.L. (1993) *Langmuir* **9**, 252. c) Nieuwenhuis, E.A., Pathmamanoharan, C., and Vrij, A. (1981) *J. Colloid Interface Sci.* **81**, 196.

2a) Van Helden, A.K., Jansen, J.W., and Vrij, A. (1981) *J. Colloid Interface Sci.* **81**, 354. b) Van Helden, A.K.; Vrij, A. (1980) *J. Colloid Interface Sci.* **78**, 312.

3. Stöber, W., Fink, A., Bohn, E.J. (1968) *J. Colloid Interface Sci.* **26**, 62.

4. Philipse, A.P., Vrij, A. (1989) *J. Colloid Interface Sci.* **128**, 121.

5a) Van Blaaderen, A., van Geest J., and Vrij, A. (1992) *J. Colloid Interface Sci.* **154**, 481. b) Van Blaaderen, A., Kentgens, A.P.M. (1992) *J. Non-Cryst.Solids* **149**, 161. c) Van Blaaderen A., and Vrij A. (1994) in Bergna (ed.) *Colloid Chemistry of Silica*, American Chemical Society, 83-11. d) Van Blaaderen, A., and Vrij, A. (1993) *J. Colloid Interface Sci.* **156**, 1.

6a) Pathamamanoharan, C. (1988) *Colloids Surf.* **34**, 81. b) Philipse, A.P. and Pathamanoharan, C. (1993) *J. Colloid Interface Sci.* **159**, 96. c) Pathamanoharan, C. (1990) *Colloids and Surfaces* **50**, 1. d) Van Duijneveldt, J.S., Heinen, A.W., and Lekkerkerker, H.N.W. (1993) *Europhys. Lett.* **21**, 369.

7a) Duits, M.H.G., May, R.P., Vrij, A., and de Kruif, C.G. (1991) *J. Chem. Phys.* **94**, 4521. b) Penders, M.H.G.M., and Vrij, A., *J. Colloid Polymer Sci.* **268**, 823.

8a) Jansen, J.W., de Kruif, C.G., and Vrij, A. (1986) *J. Colloid Interface Sci.* **114**, 481. b) Rouw, P.W., Vrij, A., and de Kruif, C.G. (1988) *Colloids and Surfaces* **31**, 299. c) Duits, M.H.G., de Kruif, C.G., and Vrij, A. (1992) *Colloid and Polymer Sci.* **270**, 154. d) De Kruif, C.G., and Schouten, J.A. (1990) *J. Chem. Phys.* **92**, 6098.

9a) Philipse, A.P., and Vrij, A. (1988) *J. Chem. Phys.* **88**, 6459. b) Duits, M.H.G., May, R.P., Vrij, A., and de Kruif, C.G. (1991) *J. Chem. Phys.* **94**, 4521. c) Van Beurten, P., and Vrij, A. (1981) *J. Chem. Phys.* **74**, 2744.

10) Degiorgio, V., Piazza, R., Bellini, T., and Visca, M. (1994) *Advances Colloid Interface Sci.* **48**, 61.

11a) Vrij, A., Penders, M.H.G.M., Rouw, P.W., de Kruif, C.G., Dhont, J.K.G., Smits, C., and Lekkerkerker, H.N.W. (1991) *Faraday Disc. Chem. Soc.* **90**, 31. b) Vrij, A., and de Hek, H. (1981) *J. Colloid Interface Sci.* **84**, 409. c) De Kruif, C.G., Rouw, P.W., Jansen, J.W., and Vrij, A. (1985) *J. de Physique* C3, suppl. au no. 3, 295. d) Lekkerkerker, H.N.W., Dhont, J.K.G., Verduin, H., Smits, C., and van Duijneveldt, J.S. (1995) *Physica A* **213**, 18.

12a) Nägele, G., Kellerbauer, O., Krause, R., and Klein, R., (1993) *Phys. Rev. E* **47**, 2562. b) Nägele, G., Sterninger, B., Genz, U., and Klein, R. (1994) *Physica Scripta* **T55**, 119.

13a) Philipse, A.P. and Pathamanoharan, C. (1993) *J. Colloid Interface Sci.* **159**, 96. b) Thies-Weesie, D.M.E., and Philipse, A.P. (1994) *J. Colloid Interface Sci.* **162**. 470. c) Kops-Werkhoven, M.M., and Fijnaut, H.M. (1982) *J. Chem. Phys.* **77**, 2242; (1981) **74**, 1618. d) Dhont, J.K.G., van der Werff, J.C., and de Kruif, C.G. (1989) *Physica A* **160**, 195. e) Van der Werff, J.C., de Kruif, C.G., and Dhont J.K.G. (1989) *Physica A* **160**, 205. f) Dhont J.K.G., Duyndam, A.F., and Ackerson, B.J. (1992) *Langmuir* **8**, 2907. g) Dhont J.K.G., and Verduin, H. (1994) *J. Chem. Phys.* **101**, 6193

14a) Van Blaaderen, A., and Vrij, A. (1992) *Langmuir* **8**, 2921 b) Imhof, A., van Blaaderen, A., Maret, G., Mellema, J., and Dhont, J.K.G. (1994) *J. Chem. Phys.* **100**, 2170. c) Van Blaaderen, A., Imhof, A., Hage, W., and Vrij, A.. (1992) *Langmuir* **8**, 1514. d) Verhaegh, N.A.M., and van Blaaderen, A. (1994) *Langmuir* **10**, 1427.

15a) Buining, P.A., Pathamanoharan, C., Bosboom, M., Jansen, G.B.H., and Lekkerkerker, H.N.W. (1990) *J. Am. Ceram. Soc.* **73**, 2358. b) Buining, P.A., Pathmamanoharan, C., Jansen, G.B.H., and Lekkerkerker, H.N.W. (1991) *J. Am. Ceram. Soc.* **74**, 1303. c) Philipse, A.P., Nechifor, A-M., and Pathamanoharan, C. (1994) *Langmuir* **10**, 4451. d) Vroege, G.J., and Lekkerkerker, H.N.W. (1992) *Rep. Prog. Phys.* **55**, 1241. e) Buining, P.A., and Lekkerkerker, H.N.W. (1993) *J. Phys. Chem.* **97**, 11510. f) Thies-Weesie, D.M.E. (1995) *Sedimentation and Liquid Permeation of Inorganic Colloids,* Thesis, Utrecht University. g) Liz Marzan, L., and Philipse, A.P., (1994) *Colloids Surfaces* **90**, 95.

16a) Cabuil, V.L., Bacri, J.C., Perzynski, R. (1993) *J. Magn. Magn. Mater.* **122.** b) Massart, R. (1981) *IEE Trans. Magn.* **MAG-17**, 1247. c) Philipse, A.P., van Bruggen, M.P.B., and Pathamanoharan, C. (1994) *Langmuir* **10**, 92.

THERMODYNAMICS OF THE METAL OXIDE-ELECTROLYTE INTERFACE CHARGING

N. KALLAY, S. ŽALAC AND D. KOVAČEVIĆ
Laboratory of Physical Chemistry, Faculty of Science,
University of Zagreb, P.O. Box 163, 10001 Zagreb, Croatia

1. Introduction

Most of the properties of colloidal dispersions are determined by equilibria in the electrical interfacial layer. For example, the stability of a colloidal system or adhesion of particles depends on the electrostatic potential at the onset of diffuse region of the interfacial layer. This potential is a result of several interactions of ions with active surface groups. Important systems in nature and technology are aqueous dispersions of metal oxides. Amphoteric surface groups interact with H^+ ions causing development of interfacial charge and electrostatic potential. Several methods are used for characterization of these systems; among them, adsorption measurement and electrokinetics are found to be useful tools [1]. More recently, the equilibria in the interfacial layer were investigated by calorimetry [2-11]. An attempt was made to evaluate the enthalpy of reactions leading to surface charge. The problems in doing so are related to several simultaneous reactions taking place at the interface, so one can hardly distinguish different contributions to the overall enthalpy. An additional problem is the electrostatic effect on the enthalpy.

Recently, standard reaction enthalpies of specified surface reactions were obtained by introducing a "symmetric" calorimetry experiment in which an acid (or base) are added to a basic (or acidic) metal oxide suspension in such a way that point of zero charge is

E. Pelizzetti (ed.), Fine Particles Science and Technology, 125–140.

the mean value of the initial and final pH [12-14]. In this case electrostatic effects cancel and so one obtains the standard reaction enthalpies of surface reactions. The electrostatic effect on the enthalpy of surface charging was investigated by calorimetry titrations [15]. Data interpretation yields the dependence of surface potential on pH.

The purpose of this article is to present a solution to the problems related to the calorimetric experiments and to discuss the physical meaning of enthalpy values with respect to the assumed mechanisms of surface reactions.

2. Mechanism and stoichiometry of surface charging

The basic process at the metal oxide aqueous interface is charging due to the interaction of the surface with H^+ and OH^- ions. Two different approaches are used to describe the charging process of an oxide surface, *i.e.* the adsorption model and the surface complexation model. According to the surface complexation model, charging could be interpreted by the "1-pK" and "2-pK" concepts [17,18]. More advanced models consider the heterogeneity of the surface [19]. Regardless of the actual mechanism taking place at the interface, one is in principle able to apply different stoichiometric equations to describe surface charging. The values of the physical quantities (standard reaction enthalpy, standard equilibrium constant) depend directly on the way in which the reaction equations are written.

2.1. "2-pK" MODEL

One approach to the surface reactions at aqueous metal oxide interface is to consider the hydrated surface sites (MOH) as amphoteric groups involved in protonation and deprotonation reactions (reactions **p** and **d**). This model is sometimes called the "2-pK" model [16,17]. Accordingly, the following equilibria are assumed

$$\mathrm{MOH} + \mathrm{H}^+ \rightarrow \mathrm{MOH_2}^+ ; K_\mathrm{p}^\mathrm{o} = \exp(F\varphi_0/RT)\,\frac{\Gamma(\mathrm{MOH_2}^+)}{a(\mathrm{H}^+)\,\Gamma(\mathrm{MOH})} \tag{1}$$

$$\mathrm{MOH} \rightarrow \mathrm{MO}^- + \mathrm{H}^+ ; K_\mathrm{d}^\mathrm{o} = \exp(-F\varphi_0/RT)\,\frac{a(\mathrm{H}^+)\,\Gamma(\mathrm{MO}^-)}{\Gamma(\mathrm{MOH})} \tag{2}$$

where K_r^o is the standard equilibrium constant of a surface reaction (corresponding to zero electrostatic potential in the interfacial layer), φ_0 is the electrostatic potential at the 0-plane in which charged surface groups are located and Γ_i is the surface concentration of component i, *i.e.* the amount of surface groups divided by the surface area. Symbol M denotes the metal atom at the solid surface.

2.2. "1-pK" MODEL

Another approach is based on the "1-pK" concept [18]. Stoichiometry of the reaction corresponding to the "1-pK" concept depends on the type of oxide, but in all cases it involves binding of one H^+ ion to an active surface site (reaction **h**)

$$M(OH)_n^{z} + H^+ \rightarrow M(OH)_nH^{z+1}\ ;\ K_h^o = \exp(F\varphi_0/RT)\,\frac{\Gamma(M(OH)_nH^{z+1})}{a(H^+)\,\Gamma(M(OH)_nH^{z})} \quad (3)$$

The standard equilibrium constant of the reaction **h**, K_h^o, is related to the K_p^o and K_d^o by

$$K_h^o = (K_p^o/K_d^o)^{1/2} \quad (4)$$

2.3. ADSORPTION MODEL

Alternatively, the binding of H^+ ions, *i.e.* protons (reaction **bp**) and OH^- (hydroxide) ions (reaction **bh**) may be assumed to represent reactions with certain surface sites S, or simply as adsorption [14,20]

$$S + H^+ \rightarrow\ SH^+\ ;\ K_{bp}^o = \exp(-F\varphi_0/RT)\,\frac{\Gamma(SH^+)}{a(H^+)\,\Gamma(S)} \quad (5)$$

$$S + OH^- \rightarrow\ SOH^-\ ;\ K_{bh}^o = \exp(F\varphi_0/RT)\,\frac{\Gamma(SOH^-)}{a(OH^-)\,\Gamma(S)} \quad (6)$$

This formalism yields equivalent results with respect to that associated with the "2-pK" model. If one takes the surface site S identical to the surface group MOH, then the binding of protons (reaction **bp**) becomes equivalent to the surface protonation (reaction **p**) so that

$$K^{o}_{bp} = K^{o}_{p} \tag{7}$$

Addition of the neutralization reaction **n**

$$OH^{-} + H^{+} \rightarrow H_2O;\ K^{o}_{n} = 1/K^{o}_{w} \tag{8}$$

with the equilibrium constant, K^{o}_{n}, (related to the "ionic product" of water), K^{o}_{w}, to the surface deprotonation (reaction **d**) yields reaction **bh** (6) so that

$$K^{o}_{bh} = K^{o}_{d}\, K^{o}_{n} \tag{9}$$

2.4. ASSOCIATION OF COUNTERIONS

Surface charge is simultaneously reduced by association of counterions (cations and anions) from the solution with surface charged groups. The counterion association reactions are independent of the assumptions on the mechanism of surface charging, *i.e.* it is always the binding of one counterion to the oppositely charged surface site. The contribution of the counterion association may be neglected if the experiment is performed at low ionic strength and/or in the vicinity of the pH_{pzc} [21-23].

3. Enthalpy of surface reaction

Generally, the reaction enthalpy is defined as the ratio of the reaction heat (measured at a constant pressure and temperature), Q, and the extent of specified reaction, $\Delta\xi_r$

$$\Delta_r H = Q/\Delta\xi_r \tag{10}$$

The enthalpy of the surface reaction, $\Delta_r H$, can be separated into a "chemical" (standard), $\Delta_r H^o$, and "electrostatic" term, $\Delta_r H_{el}$

$$\Delta_r H = \Delta_r H^o + \Delta_r H_{el} \tag{11}$$

The measured change of enthalpy is a sum of contributions of all reactions taking place in the calorimeter. It is equal to the sum of the products of particular reaction enthalpies and the corresponding extents of reactions. This fact makes the interpretation of calorimetry data relatively complicated. For example, the contribution of the neutralization should be subtracted. A better accuracy of the results requires minimization of the contribution due to neutralization. As it will be shown later, it is sometimes suitable to add acid to a slightly acidic dispersion. The effect of counterion association could be minimized by performing experiments at low ionic strength and/or in the vicinity of point of zero charge. The second step in the interpretation depends on the assumed mechanism of surface charging. Application of the "1-pK" model is simpler, while the "2-pK" model requires a special design of the experiment.

4. Standard enthalpy of surface reactions

4.1. POINT OF ZERO CHARGE DEPENDENCE ON TEMPERATURE

In the vicinity of the point of zero charge, no specific adsorption and only a negligible counterion association may be expected, provided the concentration of neutral electrolyte is low enough [21-23]. At the point of zero charge (pH_{pzc}), one can take $\varphi_0 = 0$ and $\Gamma(MOH_2^+) = \Gamma(MO^-)$ (or $\Gamma(H^+) = \Gamma(OH^-)$), so that

$$pH_{pzc} = 0.5 \lg (K_p^o/K_d^o) = \lg K_h^o = 0.5 \lg (K_{bp}^o K_n^o/K_{bh}^o) \tag{12}$$

The relation between the standard thermodynamic functions (Gibbs energy, $\Delta_r G^o$, enthalpy $\Delta_r H^o$, entropy $\Delta_r S^o$) and the corresponding equilibrium constant of surface reaction is given by

$$-RT\ln K_r^o = \Delta_r G^o = \Delta_r H^o - T\Delta_r S^o \tag{13}$$

so that eqs. (12) and (13) yield the temperature dependence of the point of zero charge for any assumed reaction mechanism

$$pH_{pzc} = \frac{\Delta_d H^o - \Delta_p H^o}{2RT\ln 10} - \frac{\Delta_d S^o - \Delta_d S^o}{2R\ln 10} = -\frac{\Delta_h H^o}{RT\ln 10} + \frac{\Delta_h S^o}{R\ln 10}$$

$$= \frac{\Delta_{bp} H^o + \Delta_n H^o - \Delta_{ph} H^o}{2RT\ln 10} - \frac{\Delta_{bp} S^o + \Delta_n S^o - \Delta_{bh} S^o}{2R\ln 10} \tag{14}$$

The above analysis shows that p.z.c. measurements yield the standard enthalpies and entropies of surface reactions with defined stoichiometries, which do not include electrostatic contribution. The values would depend on the assumed mechanism of surface charging.

In order to obtain reliable values, one needs to avoid specific adsorption and the counterion association which requires the medium of low concentration of "neutral" electrolyte. Another problem is the accuracy of the p.z.c. determination. For p.z.c. determination, potentiometry is commonly applied [1,24,25]. Two different techniques are in use. The first approach is to measure the surface charge at different "neutral" electrolyte concentrations and to obtain the p.z.c. from the intersection of these curves [26]. The slopes of these functions are not markedly different, so the accuracy of the intersection point location is rather poor. There is another, better approach, *i.e.* to add the electrolyte to suspensions at different pH values. The pH of the system where no pH change is seen upon the electrolyte addition is taken to be equal to pH_{pzc}. However, in both methods relatively high electrolyte concentrations are applied, so the requirement of negligible counterion association is jeopardized. In addition, the change in the diffusion junction potential of reference electrode affects the results.

Recently, a new method for p.z.c. determination has been introduced. The method was named "mass titration". It was originally proposed by Noh and Schwarz [27] for pure colloidal systems, and it was later refined by Žalac and Kallay [28] so that it can be also used for contaminated samples. The underlying principle of the method is simple. If the concentration of the metal oxide is high enough, the equilibrium pH corresponds to the pH_{pzc}. Therefore, one can simply prepare a concentrated suspension and measure the pH at different temperatures. The advantage of this method is that a low electrolyte concentration can be used and the accuracy of pH_{pzc} is the same as that of the pH measurements.

The adhesion method, which is suitable for conductive surfaces [29], is not applicable to studies of the temperature dependence of p.z.c. due to its relatively low accuracy.

The disadvantage of electrokinetic methods [1], which yield the isoelectric point, is that pH cannot be measured *in situ*, but either before or after the zeta-potential determination. Temperature control may also be a problem.

4.2. CALORIMETRY

A direct approach to the evaluation of surface reaction enthalpies is calorimetry. In view of the "2-pK" model, the measured heat is given by

$$Q = \Delta\xi_p \, \Delta_p H + \Delta\xi_d \, \Delta_d H + \Delta\xi_n \, \Delta_n H \tag{15}$$

The contribution of neutralization can be deduced since the extent of that reaction may be evaluated from pH measurements and the $\Delta_n H$ value can be taken from the literature. However, contributions of protonation and deprotonation of hydrated surface sites cannot be simply distinguished. The relationships between the extents of surface reactions **p** (1) and **d** (2), and the amounts of the added strong acid, such as HNO_3 ($\Delta n(HNO_3)$), and/or base, such as NaOH ($\Delta n(NaOH)$), are

$$\Delta n(\mathrm{H}^+) = \Delta n(\mathrm{HNO_3}) - \Delta\xi_\mathrm{p} + \Delta\xi_\mathrm{d} - \Delta\xi_\mathrm{n} \tag{16}$$

$$\Delta n(\mathrm{OH}^-) = \Delta n(\mathrm{NaOH}) - \Delta\xi_\mathrm{n} \tag{17}$$

where $\Delta n(\mathrm{H}^+)$ and $\Delta n(\mathrm{OH}^-)$ are the changes in the equilibrium amounts of H^+ and OH^- ions, respectively. By definition, the extents of surface protonation and deprotonation reactions ($\Delta\xi_\mathrm{p}$ and $\Delta\xi_\mathrm{d}$) are

$$\Delta\xi_\mathrm{p} = A\,\Delta\Gamma(\mathrm{MOH_2}^+) \quad \text{and} \quad \Delta\xi_\mathrm{d} = A\,\Delta\Gamma(\mathrm{MO}^-) \tag{18}$$

where A is the surface area of solid oxide. According to eqs. (1), (2) and (18) the ratio of the extents of surface reactions **p** and **d** is given by

$$\frac{\Delta\xi_\mathrm{p}}{\Delta\xi_\mathrm{d}} = -\frac{K_\mathrm{p}^\circ}{K_\mathrm{d}^\circ}\,\frac{a_1(\mathrm{H}^+)\,a_2(\mathrm{H}^+)}{\exp[(\varphi_{0,1} + \varphi_{0,2})\,F/RT]} \tag{19}$$

Subscripts $_1$ and $_2$ denote the initial and final states of the calorimetry experiment, respectively. The surface potential, φ_0, may be approximated by the Nernstian equation

$$\varphi_0 = \frac{RT\ln 10}{F}\,(\mathrm{pH_{pzc}} - \mathrm{pH})\,\alpha \tag{20}$$

where α represents deviation from the ideal slope ($\alpha \leq 1$). The ratio $K_\mathrm{p}^\circ/K_\mathrm{d}^\circ$ is related to the pH_{pzc} by relationship (12) so that a combination of the above equation with eqs. (12) and (19) yields

$$D = \frac{\Delta\xi_\mathrm{p}}{\Delta\xi_\mathrm{d}} = -10^{(1-\alpha)\,(2\mathrm{pH_{pzc}} - \mathrm{pH_2} - \mathrm{pH_1})} \tag{21}$$

If the system fully obeys the Nernst equation ($\alpha = 1$), or if the initial and final pH values are related by

$$0.5\,(\mathrm{pH_1} + \mathrm{pH_2}) = \mathrm{pH_{pzc}} \tag{22}$$

the following relation holds

$$\Delta\xi_p = -\,\Delta\xi_d \tag{23}$$

According to (16), (17) and (23), the extents of surface reactions **p** and **d** are

$$\Delta\xi_p = -\,\Delta\xi_p = 0.5\,\left[\Delta n(\mathrm{HNO_3}) - \Delta\xi_n - c^{\circ}\left(\frac{V_2\,10^{-\mathrm{pH}_2}}{y_2} - \frac{V_1\,10^{-\mathrm{pH}_1}}{y_1}\right)\right] \tag{24}$$

where V is the volume of liquid medium and y the activity coefficient given by the Debye-Hückel equation. The extent of neutralization reaction, $\Delta\xi_n$, can be calculated from pH measurements by

$$\Delta\xi_n = \Delta n(\mathrm{NaOH}) - K_w^{\circ} c^{\circ}\left(\frac{V_2\,10^{\mathrm{pH}_2}}{y_2} - \frac{V_1\,10^{\mathrm{pH}_1}}{y_1}\right) \tag{25}$$

If the difference between the initial pH and $\mathrm{pH_{pzc}}$ is equal to the difference between $\mathrm{pH_{pzc}}$ and the final pH value, the initial and final potentials, $\varphi_{0,1}$ and $\varphi_{0,2}$, are equal in magnitude but opposite in sign. Accordingly, the electrostatic contribution to the partial molar enthalpies of surface species in the initial and final states are also equal in magnitude but opposite in sign so that $\Delta_r H_{el} = 0$. In such a case, one can obtain the difference between the enthalpies of deprotonation and protonation reactions, $\Delta_{ch}H^{\circ}$

$$\Delta_{ch}H^{\circ} = \Delta_d H^{\circ} - \Delta_p H^{\circ} = \frac{Q - \Delta_n H^{\circ}\Delta\xi_n}{\Delta\xi_d} \tag{26}$$

This value is the difference in the standard reaction enthalpies because the electrostatic contributions cancel. Cancellation of two electrostatic effects is based solely on the assumption that function $\varphi_0(\mathrm{pH})$ is differentiable and "symmetrical" around p.z.c., without any need to assume the Nernstian behavior. Since the electrostatic contribution to the total enthalpy exhibits the same behavior as a derivative of the potential, the compensation of electrostatic contribution to the enthalpies takes place. The $\Delta_{ch}H^{\circ}$ value may be recalculated to values associated with other assumed stoichiometries.

According to the "1-pK" model, the measured heat Q is the sum of heats of the surface reaction **h** and neutralization **n** so that

$$Q = \Delta_h H \, \Delta\xi_h + \Delta_n H \, \Delta\xi_n \tag{27}$$

The enthalpy of reaction **h** can be evaluated from

$$\Delta_h H = \frac{Q - \Delta_n H \, \Delta\xi_n}{\Delta\xi_h} \tag{28}$$

The evaluation of the extent of reaction based on the "1-pK" model is simple since the unique reaction equation implies only binding of one hydronium ion. The changes in the equilibrium amounts of hydronium, $\Delta n(H^+)$, and hydroxide ions, $\Delta n(OH^-)$, are

$$\Delta n(H^+) = \Delta n(HNO_3) - \Delta\xi_h - \Delta\xi_n \tag{29}$$

$$\Delta n(OH^-) = \Delta n(NaOH) - \Delta\xi_n \tag{30}$$

The above equations yield the change of the extent of surface reaction **h**, $\Delta\xi_h$

$$\Delta\xi_h = \Delta n(HNO_3) - \Delta\xi_n - c^o\left(\frac{V_2\, 10^{-pH_2}}{y_2} - \frac{V_1\, 10^{-pH_1}}{y_1}\right) \tag{31}$$

which depends not only on the amounts of added acid ($\Delta n(HNO_3)$) or base ($\Delta n(NaOH)$) but also on the extent of neutralization, $\Delta\xi_n$.

Interpretation of experimental results based on the "1-pK" model may also yield the standard reaction enthalpy if the experiment is performed so that the p.z.c. is located midway between the final and initial pH. In such a case, the $\Delta_h H^o$ value is directly related to the difference ($\Delta_d H^o - \Delta_p H^o = \Delta_{ch} H^o$) based on the "2-pK" model by

$$\Delta_h H^o = -\Delta_{ch} H^o/2 \tag{32}$$

The important advantage of the "1-pK" concept lies in the fact that the interpretation does not require the Nernstian behavior.

According to the above analysis, standard enthalpies of surface reaction could be obtained by adding a strong acid (or a base) to the suspension, in such a way that the difference between the initial pH and the pH_{pzc} equals the difference between the pH_{pzc} and the final pH value. Therefore, the calorimetric experiment should be accompanied with simultaneous potentiometric measurements. In order to increase the contribution of surface reactions with respect to the neutralization one, the solid concentration should be kept as high as possible, so it is desirable to use small particles with a large specific surface area.

5. Electrostatic effect on the enthalpy

In the experiments outside the pH_{pzc} region, the electrostatic contribution dominates and evaluation of the extents of surface reactions is more complicated. The effect of electrostatic contribution may be evaluated by subsequently adding small portions of the acid or base to the suspension. In this case, the obtained values of enthalpy correspond to the mean pH value.

When the "2-pK" model is employed, one can assume that the electrostatic contributions to the enthalpies of protonation and deprotonation reactions are equal in magnitude but opposite in sign ($\Delta_d H_{el} = -\Delta_p H_{el}$) because the initial and final electrostatic surface potentials are the same for all surface species, but the change in the charge number is opposite in sign and same in magnitude ($z_p = -z_d$). Accordingly,

$$Q = \Delta\xi_p\,(\Delta_p H^o + \Delta_p H_{el}) + \Delta\xi_d\,(\Delta_{ch} H^o + \Delta_p H^o - \Delta_p H_{el}) + \Delta\xi_n\,\Delta_n H \qquad (33)$$

The above equation could be used for interpretation of calorimetric experiments outside the p.z.c. region. The value of the $\Delta_{ch}H^o$ could be determined separately by the "symmetric" experiment [14,15] or by measuring the temperature dependence of pH_{pzc} [30-32]. The following problem is related to the calculation of the extent of a particular surface reaction.

In the case when acid is added to an acidic, or base to a basic suspension, the extent of neutralization can be neglected and the measured heat is given by

$$Q = (\Delta_p H^o + \Delta_p H_{el})\, \Delta\xi_p + (\Delta_{ch} H^o + \Delta_p H^o - \Delta_p H_{el})\, \frac{\Delta\xi_p}{D} \tag{34}$$

The following equation is obtained by rearranging the above expression

$$Q = \{\frac{\Delta_{ch} H^o}{Y} + (1 + \frac{1}{Y})\, \Delta_p H^o + (1 - \frac{1}{Y})\, (pH_{pzc} - pH)\, \alpha RT \ln 10\}\, \Delta\xi_p \tag{35}$$

where function Y is defined as

$$Y = -10^{2(1-\alpha)\,(pH_{pzc} - pH)} \tag{36}$$

and pH is the mean value between the initial and final pH; $pH = (pH_1 + pH_2)/2$.

The presented analysis makes it possible to reach two interesting targets. One is evaluation of the singular value of the standard reaction enthalpy of protonation and the other is determination of coefficient α.

The extent of protonation reaction, $\Delta\xi_p$, can be calculated as

$$\Delta\xi_p = \frac{\Delta n(HNO_3) - c^o(V_2/y_2 10^{pH2} - V_1/y_1 10^{pH1})}{1 - 1/D} \tag{37}$$

where $\Delta n(HNO_3)$ is the amount of added acid.

Application of the "1-pK" model leads to a simpler situation. The enthalpy of reaction h is defined as

$$\Delta_h H = \Delta_h H^o + zF\varphi = \Delta_h H^o + \alpha RT (pH_{pzc} - pH) \ln 10 \tag{38}$$

Consequently,

$$\frac{d\varphi}{d(pH)} = \frac{d(\Delta_h H)}{d(pH)}\, \frac{1}{F} \tag{39}$$

In this case, one could determine the value of α from the slope in the presentation of $\Delta_h H$ *vs* pH.

6. Conclusion

Different assumptions on the mechanism of surface reactions resulting in surface charge could be hardly verified by performing adsorption and/or electrokinetic measurements. It was shown that several different concepts may be satisfactorily used for the interpretation of experimental data. However, the above analysis of calorimetry data may serve this purpose. In the first step, one should obtain the standard enthalpies of surface reactions on the basis of both "1-pK" and "2-pK" models. For this purpose, one may either use measurements of the temperature dependence of pH_{pzc} or perform the "symmetric" calorimetry experiments, as described above. In the next step, one performs calorimetric titration of a slightly acidic (or basic) dispersion by adding subsequent portions of acid (or base). Data interpretation by the "1-pK" and "2-pK" models yields the dependence of the reaction enthalpy on pH. Since the enthalpy consists of "chemical" (standard) and "electrostatic" contributions, one may compare the results with the Nerstian equation. The results for hematite aqueous dispersion [33] showed that, in the case of the "2-pK" model, the slope is about 150 mV while for the "1-pK" the slope is realistic, *i.e.* about 52 mV. According to these results, one may conclude that the "1-pK" model describes better the situation at the surface.

7. References

1. Kallay, N., Hlady, V., Jednačak-Biščan, J. and Milonjić, S. (1993) Techniques for the Study of Adsorption from Solutions, Chap. 2 in B.W. Rossiter and R.C. Baetzold (eds.), *Investigation of Surfaces and Interfaces*, Part A, Vol. 9 of *Physical Methods in Chemistry*, Interscience Publishers, A Wiley-Interscience Publication, John Wiley & Sons, New York
2. Machesky, M.L. and Anderson, M.A. (1986) Calorimetric acid-base titrations of aqueous goethite and rutile suspensions, *Langmuir* 2, 582-587.
3. Machesky, M.L. and Jacobs, P.F. (1991) Titration calorimetry of aqueous alumina suspensions. I. Results and comparison with similar studies, *Colloids Surf.* 53, 297-314.

4. Wierer, K. A. and Dobiaš, B. (1988) Exchange enthalpies of H^+ and OH^- adsorption on minerals with different characters of potential-determining ions, *J. Colloid Interface Sci.* **122**, 171-177.

5. de Keizer, A., Fokkink, L.G.J. and Lyklema, J. (1990) Thermodynamics of proton charge formation on oxides. Microcalorimetry, *Colloids Surf.* **49**, 149-163.

6. Machesky, M.L. and Jacobs, P.F. (1991) Titration calorimetry of aqueous alumina suspensions. II. Discussion of enthalpy changes with pH and ionic strength, *Colloids Surf.* **53**, 315-328.

7. Denoyel, R., Durand, G., Lafuma, F. and Audebert, R. (1989) Adsorption of cationic polyelectrolytes onto montmorillonite and silica: A microcalorimetric study of their conformation, *J. Colloid Interface Sci.* **139**, 281-290.

8. Partyka, S., Lindheimer, M., Zaini, S., Keh, E. and Brun, B. (1986) Improved calorimetric method to investigate adsorption processes from solutions onto solid surfaces, *Langmuir* **2**, 101-105.

9. Partyka, S., Rudzinski, W., Brun, B. and Clint, J.H. (1989) Calorimetric studies of adsorption of anionic surfactants onto alumina, *Langmuir* **5**, 297-304.

10. Noll, L.A. (1987) Adsorption calorimetry of surfactant interaction with minerals, *Colloids Surf.* **26**, 43-54.

11. Machesky, M.L., Bischoff, B.L. and Anderson, M.A. (1989) Calorimetric investigation of anion adsorption onto goethite, *Environ. Sci. Tehnol.* **23**, 580- 587.

12. Kallay, N., Žalac, S. and Štefanić, G. (1993) Enthalpies of Reactions at Metal Oxide Aqueous Interface, *Langmuir* **9**, 3457-3460.

13. Kallay, N., Žalac, S., Čulin, J., Bieger, U., Pohlmeier, A. and Narres, H.D. (1994) Thermochemistry and adsorption equilibria at hematite-water interface, *Progr. Colloid & Polymer Sci.* **95**, 108-112.

14. Kallay, N. and Žalac, S. (1994) Enthalpy of interfacial charging of metal oxide/water systems, *Croat. Chem. Acta* **67**, 467-479.

15. Žalac, S. and Kallay, N. (submitted) Determination of electrostatic contribution to enthalpy of charging at metal oxide/electrolyte interface, *Croat. Chem. Acta*

16. Yates, D. E., Levine, S. and Healy, T.W. (1974) Site-binding model of the electrical double layer at the oxide/water interface, *J. Chem. Soc. Faraday Trans* I **70**, 1807-1818.

17. Dzombak, D.A. and Morel, F.M.M. (1990) *Surface Complexation Modeling*, A Wiley-Interscience Publication, John Wiley & Sons, New York.

18. Bolt, G.H. and van Riemsdijk, V.H. (1982) B: Physico-Chemical Models, in G.H. Bolt (ed.), *Soil Chemistry*, Elsevier, Amstrerdam.

19. van Riemsdijk, V.H., Bolt, G.H., Koopal, L.K. and Blaakmeer, J. (1986) Electrolyte adsorption on heterogeneous surfaces: Adsorption models, *J. Colloid Interface Sci.* 109, 219-228.

20. Kallay, N. and Babić, D. (1986) Adsorption at solid/solution interfaces II. Surface charge and potential of spherical colloidal titania, *Colloids Surf.* 19, 375-386.

21. Kallay, N. and Tomić, M. (1988) Association of counterions with adsorbed potential-determining ions at a solid/solution interface. 1. Theoretical analysis, *Langmuir* 4, 559-564.

22. Tomić, M. and Kallay, N. (1988) Association of counterions with adsorbed potential-determining ions at a solid/solution interface. 2. Double layer equilibria metal oxide interface, *Langmuir* 4, 565-569.

23. Kallay, N. Sprycha, R. Tomić, M. Žalac, S. and Torbić, Ž. (1990) Some controversies in the understanding of equilibria in electrical interfacial layer, *Croat. Chem. Acta.* 63, 467-487.

24. Hesleitner, P., Babić, D., Kallay, N. and Matijević, E. (1987) Adsorption at solid/solution interfaces. 3. Surface charge and potential of colloidal hematite, *Langmuir* 3, 815-820.

25. Lyklema, J. (1984) Points of zero charge in the presence of specific adsorption, *J. Colloid Interface Sci.* 99, 109-117.

26. Hall, D.G. (1988) A thermodynamic analysis of common intersection points in potentiometric titration studies of solid surfaces, *J. Chem. Soc. Faraday Trans* I 84, 2227-2240.

27. Noh, J.S. and Schwarz, J.A. (1989) Estimation of the point of zero charge of simple oxides by mass titration, *J. Colloid Interface Sci.* 130, 157-164.

28. Žalac, S. and Kallay, N. (1992) Application of mass titration to the point of zero charge determination, *J. Colloid Interface Sci.* 149, 233-240.

29. Kallay, N., Torbić Ž., Barouch, E. and Jednačak-Bišćan, J. (1987) The determination of isoelectric point for metallic surfaces, *J. Colloid Interfac Sci.* 118, 431-435.

30. Berube, Y.G. and de Bruyn, P.L. (1968) Adsorption at the rutile-solution interface. I. Thermodynamic and experimental study, *J. Colloid Interface Sci.* 27, 305-318.

31. Fokkink, L.G.J., de Keizer, A. and Lyklema, J. (1989) Temperature dependance of the electrical double layer on oxides: Rutile and hematite, *J. Colloid Interface Sci.* 127, 116-131.

32. Fokkink, L.G.J., de Keizer, A. and Lyklema, J. (1990) Temperature dependance of cadmium adsorption on oxides. I. Experimental observation and model analysis, *J. Colloid Interface Sci.* **135**, 118-131.

33. Kallay, N., Žalac, S, Kovačević, D., Pohlneier, A. and Schwuger, M., to be published

Fundamentals of Forced Hydrolysis of Indium Hydroxide

Terry A. Ring#, Yoshiyuki Kudo*
*Department of Applied Chemistry
Science University of Tokyo
Tokyo, Japan
#(Speaker)Department of Chemical Engineering
University of Utah
Salt Lake City, Utah 84122

Abstract
During forced hydrolysis an acidified metal salt solution with its solution complexes is placed in a sealed test tube and heated at a specific rate to a high temperature <100°C. By heating the solution equilibria are altered, the hydroxide concentration is increased, and often a metal hydroxide is precipitated from solution as a narrow particle size distribution. This paper presents a theory that links solution complexation equilibria with a population balance model for precipitation predicting the particle size distribution. This model uses classical nucleation theory and growth rates by various rate limiting steps for the growth of the crystals. This theory is compared to forced hydrolysis experiments where dilute Indium nitrate solutions, acidified with nitric acid, were placed in a water bath at 80°C for various periods of time. The experiments produced cubic particles of Indium hydroxide. The experiments were monitored for temperature, pH, turbidity, and particle size distribution, all as a function of time for comparison with this model.

Introduction

Forced hydrolysis is a common way to produce metal hydroxides from an acidified metal salt solution. To induce precipitation, this solution, initially below its solubility and is stable with respect to precipitation, is heated to an elevated temperature in either a forced convection oven or a water bath. Upon heating the solution the concentration of OH- ion increases forcing the precipitation of a metal hydroxide. These metal hydroxides can be precipitated under controlled conditions as monodisperse colloids and as such are of interest for many applications including ceramic powders and pigments. Professor Matijevic[1] over his career has been able to produce monodisperse hydroxides of nearly all the metals in the periodic table. Over the year his work has been expanded by many researchers around the world.

In this work, we develop a model of forced hydrolysis which predicts both the particle size distribution of the precipitate and the concentration of all ions in solution as the precipitation takes place. This is done by performing a mass balance on the species in solution that precipitate and using these concentrations to predict the concentration of the other ions in solution assuming that the solution is in equilibrium with respect to all other species. The results of this model are compared to the experimental system[2,3,4] where a dilute $In(NO_3)_3$ solution acidified with HNO_3 is heated from room temperature to 80°C in a water bath causing the precipitation of cubic submicron $In(OH)_3(s)$. This experimental system

E. Pelizzetti (ed.), Fine Particles Science and Technology, 141–159.

was chosen because it is simple due to 1) a reduced suite of ions and 2) the precipitation of a crystal with a known geometry.

The mass balances are performed for the Indium metal ions, M^{+m}, and OH^- ions and is given by the generalized differential equations given below :

$$\frac{d[OH^-]}{dt} = \{\text{release of } OH^- \text{ from dissociation of } H_2O\}$$

-{Loss of OH^- due to complexation with metal}

-{Loss of OH^- due to nucleation}

-{Loss of OH^- due to crystal growth} (1)

$$\frac{d[M^{+m}]}{dt} = -\{\text{Loss of } M^{+m} \text{ due to complexation with various anions in solution}\}$$

-{Loss of M^{+m} due to nucleation}

-{Loss of M^{+m} due to crystal growth} (2)

where [] symbolizes concentration.The initial conditions for these differential equations are:

$[OH-] = C_2$, at $t = 0$, $[M^{+m}] = C_3$ at $t = 0$ and $[H^+] = C_1$ at all time.

All C_i values are constants. Since we are in dilute solution we will assume that activity and concentration are equivalent. This is not totally accurate, however, to add activity coefficients, which are a function of the ionic strength, and their calculation to these two non-linear simultaneous differential equations would render them much more difficult to evaluate. Each of the terms of the above differential equations is discussed below after a discussion of the reactions that take place as an acidified metal salt solution is heated. To resolve these two coupled differential equations, a finite difference technique was used. This finite difference technique was found to be stable if time steps less than 1 sec. were used.

Reactions in Solution

Metal complexation

Solution speciation reactions with anions $A[= NO_3]$:

$$iM^{+m} + jA^{-a} \underset{}{\overset{K_{Aij}}{\rightleftharpoons}} M_iA_j^{+(im-ja)} \tag{3}$$

$$K_{Aij} = \frac{[M_iA_j]^{+(im-ja)}}{[M^{+m}]^i[A^{-a}]^j}$$

For this work, we will use the NO_3 complexes consisting of i:j of 1:1 and 1:2[5]. Solution speciation reactions with OH^- ions[6] also occur in solution :

$$iM^{+m} + jOH^- \underset{}{\overset{K_{OHij}}{\rightleftharpoons}} M_i(OH)_j^{+(im-j)} \tag{4}$$

$$K_{OHij} = \frac{[M_i(OH)_j]^{+(im-j)}}{[M^{+m}]^i[OH^-]^j}$$

For this work, we will use Biedermann's model[7,8] of solution speciation of Indium hydroxide complexes. These include complexes i:j , 1:1, 1:2, 2:2, 3:4. For these complexes, we have the equilibrium constant, K_{OHij}, and the ΔH_{OHij} of reaction[9].given in Table 1. The reaction kinetics between the 1:1 and the 2:2 complexes have been studied by temperature jump experiments[10].

$$2\ In(OH)^{+2} \rightarrow In_2(OH)_2^{+4}$$

The forward rate constant for this reaction was found to be $4.1 \pm 0.4 \times 10^5\ M^{-1}\ sec^{-1}$. Thus this reaction is fast, occurring within a msec for the solution concentration typically used in forced hydrolysis. Other forward and backward reactions given in equation (4) have not been studied as to their reaction rates as evidence of the lack of information in the litterature.

In general, other solution speciation reactions can take place with other anions (or cations) B[=??] :

$$iM^{+m} + jB^{-b} \underset{}{\overset{K_{Bij}}{\rightleftharpoons}} M_iB_j^{+(im-jb)} \tag{5}$$

$$K_{Bij} = \frac{[M_iB_j]^{+(im-jb)}}{[M^{+m}]^i[B^{-b}]^j}$$

However, in this experimental system, there are no other ions in solution for the metal to complex with but this not always the case. When ammonium ions are in solution many metals are complexed e.g. Cu. Solution speciation reactions with anions A^{-a}, B^{-b} and OH^- lower the concentration of M^{+m} in solution.

Acid Dissociation

Acids present in solution dissociate. For Nitric acid the reaction is :

$$H^+ + NO_3^- \overset{K_1}{\rightleftharpoons} HNO_3 \qquad (6)$$

$$K_1 = \frac{[HNO_3]}{[H^+][NO_3^-]} = 10^{-1.44} \text{ @25°C and } 10^{-1.18} \text{ @70°C, } \Delta H_{K_1} = 3.3 \text{ Kcal/mole}^{11}$$

Water Ionization

Water ionization equilibrium is also necessary to describe the experimental system:

$$H_2O \overset{K_w}{\rightleftharpoons} H^+ + OH^- \qquad (7)$$

$$K_w = [H^+][OH^-] = 10^{-13.999} \text{ @ 25°C and } 10^{-12.422} \text{ @ 90°C}^{12,13}$$

In an acidified metal salt solution $[0H^-]$ is very low. For a solution with a fixed $[H^+]$, the $[0H^-]$ is increased by an increase in the temperature of the solution, since the value of K_w increases with temperature.

The predictions for $[H^+]$ and [OH-] for a 0.00078 M HNO_3 solution at various temperatures between 25°C and 100°C are given in Figure 1. Here we see that the $[OH^-] = K_w/[H+]$ increases from 10^{-11} M to 10^{-9} M while the [H+] and $[NO_3^-]$ are essentially unchanged. Experiments performed on this solution show a pH change of only 0.05 after temperature compensation. This change is similar to the noise level in these experiments. We will take advantage of this insignificant change in [H+] and $[NO_3^-]$ as temperature changes to simplify the necessary differential equations for the mass balance, i.e. no balance is necessary for H+ or for NO_3^-.

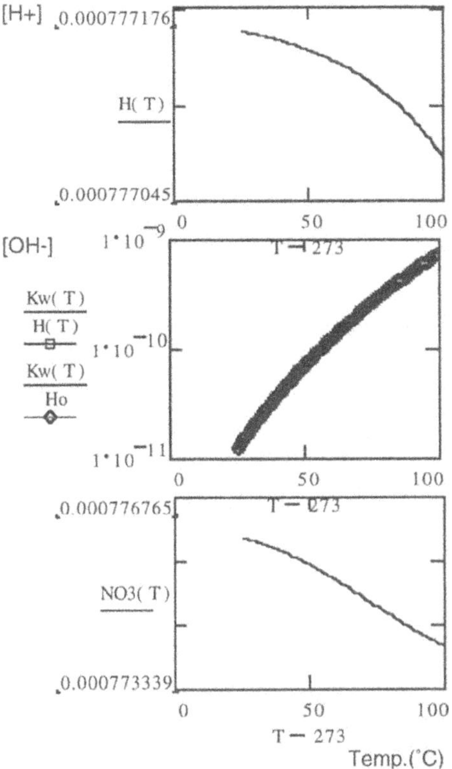

Figure 1. Prediction of $[NO_3^-]$, $[H^+]$ and $[OH^-]$ as a function of temperature between 25 and 100 °C. Please note that $[NO_3^-]$ and $[H^+]$ do not change by more than 0.4% and 0.04%, respectively, in this temperature range. Furthermore, if we assume that the $[H^+]$ is constant the error in $[OH^-]$ is less than 0.04%. This proves that NO_3^- complexes and $[H^+]$ do not vary significantly with temperature.

Precipitation reaction of a metal hydroxide :

$$M^{+m} + mOH^- \underset{}{\overset{K_{sp}}{\rightleftharpoons}} M(OH)_m(s) \qquad (8)$$

$$K_{sp} = [M^{+m}][OH^-]^m$$

where for Indium hydroxide K_{sp} is $10^{-36.9}$ M^3 at 25°C and 10^{-44} M^3 at 90°C(predicted value using ΔH_{sp}=24,789cal/mole[14]).

The supersaturation ratio, S, is given by :

$$S = \frac{[M^{+m}][OH^-]^m}{K_{sp}(T)} \qquad (9)$$

The rate of precipitation will dictate the loss rate of OH^- and In^{+3} ions due to the steps of nucleation and growth of the particles, discussed below.

Kinetics of Reactions and Precipitation

Release of OH^- due to ionization of H_2O

Rate of precipitation is controlled by the release rate of $[OH^-]$. We will make the assumption that the forward and reverse reactions responsible for the dissociation of H_2O are very fast (μsec.) compared to the rate at which heat can be transfered to the solution (min.). Thus the release rate of $[OH^-]$ is limited by the change in temperature of the solution[15], $\frac{dT}{dt}$:

$$\frac{d[OH^-]}{dt} = \frac{d\left(\frac{K_w(T)}{[H+]_0}\right)}{dT}\frac{dT}{dt} = \frac{K_w(T)}{[H+]_0}\left(\frac{-\Delta H_w}{R_g T^2}\right)\frac{dT}{dt} \tag{10}$$

assuming that the $[H^+]_0$ is fixed during the forced hydrolysis experiment. Note, R_g is the gas constant. As a result, the release of $[OH^-]$ is controlled by the heating rate, $\frac{dT}{dt}$, of the solution.

The temperature of the solution is altered by placing a test tube containing the solution into either a water bath or a forced convection oven. In either case, the test tube contents starts at a constant temperature, T_0, often room temperature and is immersed in a fluid with a constant temperature, T_∞, assumed to be an infinite reservoir. Heat is transfered to the solution by convection (natural or forced) through the walls of the test tube. We will make the assumption that the solution inside the test tube is all the same temperature and that the heat transfer is given by[16]:

$$V \rho C_p \frac{dT}{dt} = h_0 (T_\infty - T) A \tag{11}$$

where V is the volume, ρ is the density and C_p is the heat capacity of the solution in the test tube, A is the surface area of the test tube exposed to the heat reservoir and h_o is the overall heat transfer coefficient for both the inside and outside boundary layer, as well as, the heat conduction for the glass wall of the test tube. The solution to this heat balance on the solution in the test tube is given by :

$$T = T_\infty + (T_o - T_\infty) \exp\left[-\left(\frac{h_o A}{V \rho C_p}\right) t\right] \tag{12}$$

if we use the initial condition that the solution is at a constant temperature, T_o. Test tubes filled with the same volume of solution to be used in forced hydrolysis experiments have been subject to heating tests. The temperature in the test tube was measured with a thermocouple and plotted as a function of time. A plot of $(T_o-T_\infty)/(T_o-T_\infty)$ was plotted as a function of time and an exponential curve was used as a best fit (r^2=0.999). Analysis of the slope gives the parameter group $\left(\frac{h_o A}{V \rho C_p}\right)$ [= 0.775 min^{-1}] which is all that is needed to determine the temperature versus time curve for any initial, T_o, and bath, T_∞.

The temperature change is rather slow in the water bath taking more than 5 minutes to reach 80°C. This equation gives a heating rate, $\frac{dT}{dt}$, for the solution in the test tube of :

$$\frac{dT}{dt} = -\left(\frac{h_o A}{V \rho C_p}\right) (T_o - T_\infty) \exp\left[-\left(\frac{h_o A}{V \rho C_p}\right) t\right] \tag{13}$$

This heating rate can be used in the equation describing the release rate of $[OH^-]$, equation (10), completing a theoretical description of the $[OH^-]$ release rate. As a result of the heating rate, the concentration of OH^- can be calculated if losses due to nucleation and growth are accounted for as discussed below.

Loss of M^{+m} due to complexation with anions in solution

The heating rate will also release M^{+m} ions due to the altering of the solution speciation equilibria of the metal with A^{-a} and OH^- ions as given by[17] :

$$\frac{d[M^{+m}]}{dt} = \frac{dT}{dt}\left(\sum_i \frac{1}{i}\sum_j \frac{\{\frac{\Delta H_{OHij} + j\Delta H_w}{R_g T^2}\}}{\frac{1}{K_{OHij}\,[M^{+m}]^i[OH^-]^j} + \frac{1}{[M^{+m}]}}\right) +$$

$$\frac{dT}{dt}\left(\sum_i \frac{1}{i}\sum_j \frac{\{\frac{\Delta H_{Aij}}{R_g T^2}\}}{\frac{1}{K_{Aij}\,[M^{+m}]^i[A^{-a}]^j} + \frac{1}{[M^{+m}]}}\right) \tag{14}$$

where $\frac{dT}{dt}$ is again the heating rate of the solution. This expression assumes that $[A^{-a}]$ is not dependant upon temperature which is the case if it is an independent ion added with the metal salt used in the initial formulation of the solution and not complexing anything but the metal. Other solution speciation, i.e. $[M_iB_j]$ can also be evaluated in the very same way giving and additional term like the M_iA_j term. For the Indium hydroxide system under investigation the equilibrium constants and the enthalpies are given in Table 1.

Table 1 Indium complexation reactions - Equilibrium constants.

i:j	log(K_{OHij}) @25°C	ΔH (Kcal/mole)
Reactions of the type $iM^{+m}+jOH^- \overset{K_{OHij}}{\rightleftharpoons} M_i(OH)_j^{+(im-j)}$		
1:1	9.58	-8.2
1:2	19.67	-13.0
2:2	22.78	-16.0
3:4	46.08	--
Reactions of the type $iM^{+m}+jNO_3^- \overset{K_{NO3ij}}{\rightleftharpoons} M_i(NO_3)_j^{+(im-j)}$		
i:j	**log(K_{NO3ij}) @25°C**	**ΔH (Kcal/mole)**
1:1	-0.43	--
1:2	-0.3	--

Loss of $[OH^-]$ and $[M^{+m}]$ due to Nucleation

With precipitation, we need a mass balance on $[OH^-]$ and $[M^{+m}]$ since they are changing with the formation of a metal hydroxide. The loss of is due to two processes ; nucleation and growth. For nucleation, we have :

$$\frac{d[OH^-]}{dt} = -\frac{3\rho}{M_w} J(S,T)\, R^*(S,T)^3 \tag{15}$$

$$\frac{d[M^{+m}]}{dt} = -\frac{\rho}{M_w} J(S,T)\, R^*(S,T)^3 \tag{16}$$

where ρ is the density and M_w is the molecule weight of the solid, $J[=J_{max} \exp\{\frac{-32\gamma^3 (\rho/M_w)^2}{(R_gT)^3 (\ln S)^2}\}$[18]] is the nucleation rate[19,20,21] which is a function of time (i.e. S and T are functions of time) and $R^*[=\frac{12\gamma\rho/M_w}{3R_gT \ln S}]$ is the size of the critical nucleii[22], assuming nucleii and crystals are cubic. R^* is also a function of time.

Loss of $[OH^-]$ and $[M^{+m}]$ due to Crystal Growth

For crystal growth, we will grow crystal fro the time they are nucleated until time t. The loss of $[OH^-]$ and $[M^{+m}]$ due to crystal growth is therefore given by :

$$\frac{d[OH^-]}{dt} = -\frac{3\rho}{Mw} \int J(S(t),T(t))\, dt \,\Big|_{t_{nucleated}} \int_{t_{nucleated}}^{t} R^2 \frac{dR}{dt} dt \tag{17}$$

$$\frac{d[M^{+m}]}{dt} = -\frac{\rho}{M_w} \int J(S(t),T(t))\, dt \,\Big|_{t_{nucleated}} \int_{t_{nucleated}}^{t} R^2 \frac{dR}{dt} dt \tag{18}$$

where $\frac{dR}{dt}$ is the growth rate of the crystal from the time it is nucleated until time, t. Generally, the growth rate has the form $\frac{dR}{dt} = C f(S) R^n$. Parameters in the various crystal growth rate models[23] are given in Table 2. All growth rates are a function of the saturation ratio, S, which is intern a function of time.

Table 2 Crystal Growth Rate

$dR/dt = C*f(S)*R^n$

Growth Mechanism	C	f(S)	n	Ref
Diffusion Bulk	$\hat{V}DC_{eq}$	S-1	-1	24
Mono Surface Nucleation	$\beta_A D\, d^{-3}$	$\exp[\Delta G^*_s/k_BT]$§	2	25
Poly Surface Nucleation	$\frac{(Dd/3)}{(N_A C_{eq})^{2/3}}$	$(S-1)^{2/3}\exp[\Delta G^*_s/k_BT]$§	0	26
Screw Dislocation	$D_s n_{se}\beta/(y_o{}^2\rho)$	$S^2/S_1 \tanh(S_1/S)$§§	0	27
Heat Conduction	$\hat{V}k_H R\, T^2/\Delta H_f$	lnS §§§	-1	-
Chemical Reaction	$\eta\hat{V}DC_{eq}$	S-1	-1	-

§$\Delta G_s^* = \beta_L{}^2\, \gamma_e{}^2\, d^2/(1\beta_A\, k_BT \ln S)$

§§$S_1 = (y_o/y_s)S$

§§§$\ln S = {}_T\int^T \Delta H_f/(RT^2)\, dT$

$\hat{V}$ is the molar volume, N_{Av} is Advogadro's number, C_{eq} is the equilibrium concentration, D is the diffusion coefficient, sub-s surface, ΔH_f is the heat of fusionn, η is the Damkohler Number, k_H is the thermal conductivity, β_A is the area shape factor for surface nucleii, yo is the distance between steps, n_{se} is the equilibrium surface concentration, $\beta = 1-\sigma_s/S$ is one minus the maxiumum surface supersaturation divided by the solution supersaturation, ρ is the density.

Results of Experiments

Experiments similar to those performed by Hamada, et. al.[28] giving submicron cubic crystalline $In(OH)_3$ particles were performed using a solution $3.96x10^{-4}$ M $In(NO_3)_3$ and $4.04x10^{-4}$ M HNO_3 , $pH_{initial}$ =3.32. Using chronomal analysis[29] on the average particle size, this group found that the growth limited by a poly surface nucleation mechanism. A test tube with 50 ml of this solution was prepared and filtered through a 0.2 μm membrane filter and let stand for 24 hrs at room temperature (22.0±0.5°C). Test tubes filled with this solution were then placed in a Water bath (filled with a ethylene glycol solution to slow evaporation) at 80.0±0.1°C for various times. Test tubes were removed periodically, rapidly quenched to room temperature and measured as to their pH, absorbance and were centrifuged at 700 G and washed repeatedly with distilled water using ultrasonic agitation[30]. The particle size distribution as measured taking a samples on an Al stub for observation with a Cambridge Steroscan 240 Scanning Electron Microscope. The particle size distribution was measured by counting particles of different size on the SEM pictures. The arithmatic mean size and standard deviation were determined from the particle size distribution. All experimental data is presented in Figure 2.

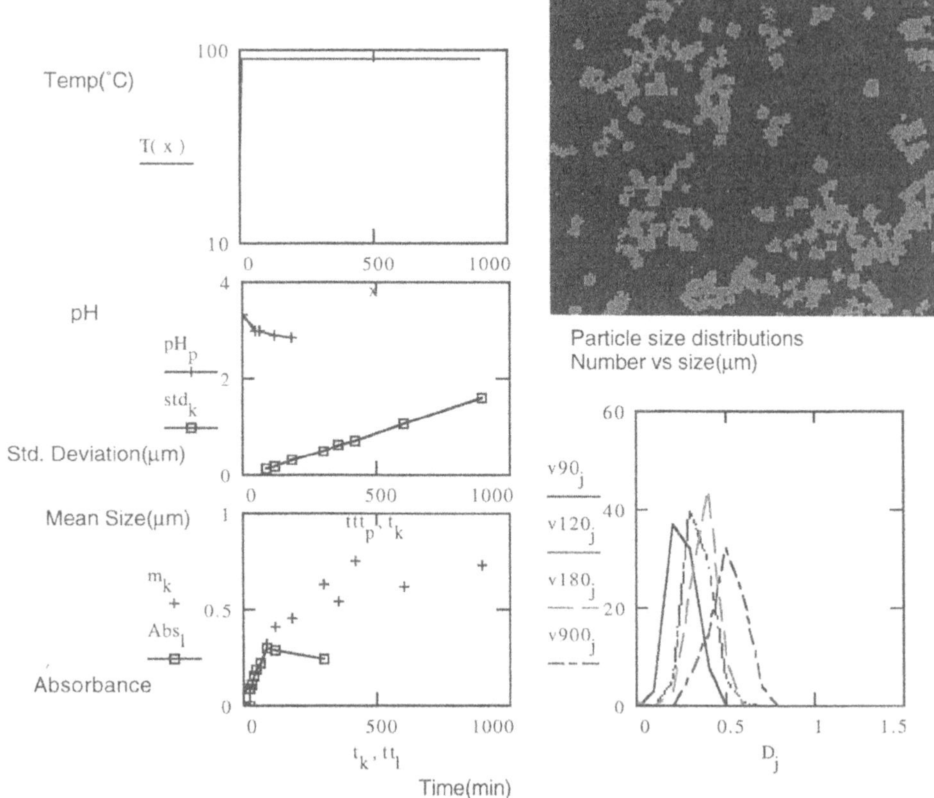

Figure 2. Plot of experimental data for an experiment performed with a 3.96×10^{-4}M $In(NO_3)_3$, 4.04×10^{-4}M HNO_3, pHi=3.32, A) temperature versus time, B) Mean size and Absorbance at λ=390nm (1 cm cell) versus time, C) Standard Deviation of particle size distributions versus time D) Scanning Electron Micrographs of particles produced at 4 hrs. (bar = 10 μm) and E) particle size distributions at various times v90= 90 min., v120= 120 min., etc size = μm.

The measured pH (temperature compensated) decreases approximately 0.30±0.15 pH units after the temperature of the solution is raised. It is not known if the change in pH is real. The change in pH for a 4.044×10^{-4}M HNO_3 solution heated in the same way was less than 0.05 pH units. The solution absorbance is near zero for nearly one hour then increases drastically initially and reaches a maximum at 90 min. As a result, we can presume that the particles are nucleated rather quickly (note, R* is ≈50Å for these values of S and T) and take time to grow to the a size that is sufficiently large to scatter light. The maximum value of absorbance is due to settling which decreases the particle number density as time progresses. The particle size increases with a decreasing rate and the standard deviation of the particle size distribution increases with growth time. This suggests that n >0 in the growth function $\frac{dR}{dt} \alpha R^n$ of Table 2.

Comparison with Calculation

The utilization of this model allows the prediction of the values of $[In^{+3}]$, $[OH^-]$, S and all of the species of the type $In_i(OH^-)_j^{+(im-j)}$ or $In_i(NO_3^-)_j^{+(im-j)}$, assuming that $[NO_3^-]$ is constant. [H+] is also assumed to be constant throughout the experiment. These assumptions are some cause of errors < 0.4% as discussed above, however, these assumptions drastically simplify the calculation and make the problem mathematically tractable. These result are highly dependent upon the growth model used for calculation and the parameters used in the growth rate theory e.g. the diffusion coefficient, D (10^{-6} cm/sec^2) and the parameters used for the interfacial energy ($\gamma = 0.1$ J/m^2). Changing the growth law alters the shape of the particle size distribution as it matures with time. Results are given in Figure 3.

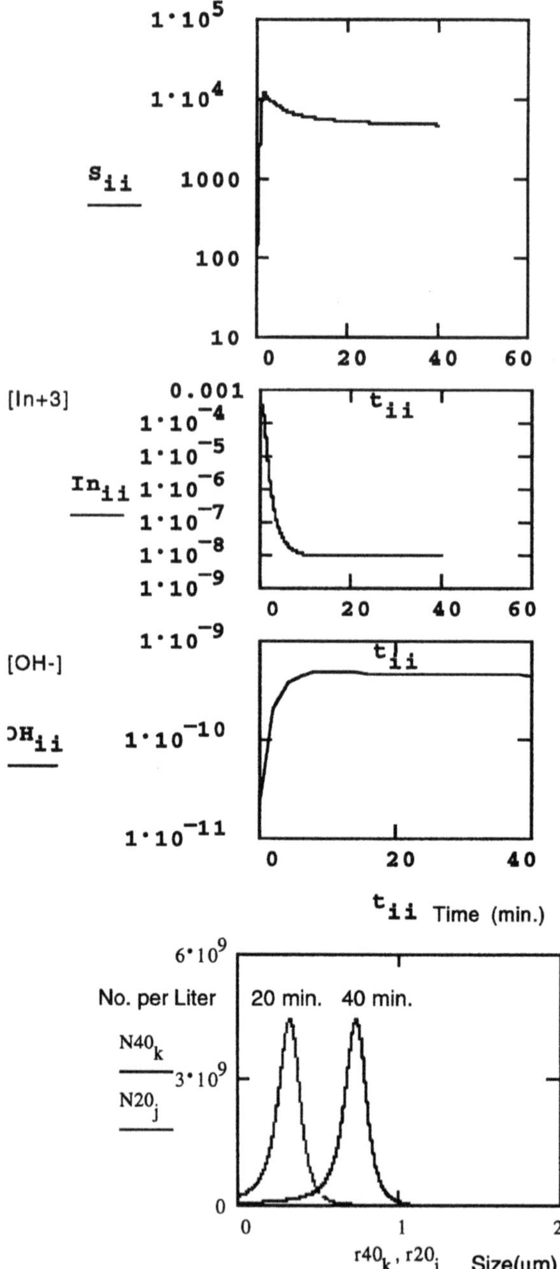

Figure 3 Model results for the Forced Hydrolysis of a solution 3.96×10^{-4}M $In(NO_3)_3$, 4.04×10^{-4}M HNO_3, $pH_i=3.320$. using homogeneous nucleation and poly surface nucleation limited growth. A) Saturation ratio versus time, B) $[In^{+3}]$ versus time, C) [OH-] versus time, D) Particle size distribution at various times.

In Figure 3, we see the values of $[In^{+3}]$, $[OH^-]$ and S as a function of time as the solution is heated from 25° to 80°C. The OH^- concentration increases as the temperature increases. This increase slows as nucleation and growth proceed. The In^{+3} concentration decreases monotonically as nucleation and growth proceeds. Since $S=\frac{[In^{+3}][OH^-]^3}{K_{sp}}$, we see that changes in $[In^{+3}]$ and [OH-] are reflected by the changes in S. S increases from ≈ 1 to ≈ 10,000 as the temperature increases and remains high > 1,000 for time >20 min. allowing growth to proceed for several hours at a very high supersaturation. The particles are nucleated over a specific period of time (10 min. $>t_{nucleation}>$ 5 min.) giving a distribution of nucleii sizes. Separate calculations show that the critical value of S is ≈10,000 thus nucleation will be insignificant for all values of S << 10,000 both before the peak in S and after ≈10 min. The number of particles per unit volume and the particle size as a function of time when compared to experimental values are not accurate. Essentially the number of particles per unit volume has to be increased and the particle size has to be decrease to fit with the experimental data. However, the model trands are similar to the experimental findings. To better taylor the model to experimental different growth laws must be considered and different values for the model parameters which in this case are the diffusion coefficient and the interfacial energy. A slower growth model e.g. poly surface nucleation model moves the model in this direction. This particle size distribution will increase in width during subsequent growth if the the growth law has a value of n >0 , n from Table 2.

The way this model has been formulated gives the ability to predict the changes in the concentration of solution species as the experiment proceeds. This is shown in Figure 4. Here we see that the various $[OH^-]$ complexes initially increase in concentration due to an increase in [OH-] and then decrease in concentration as nucleation and growth proceeds. The concentration of $[In^{+3}]$ and the $In_i(NO_3)_j$ complexes decrease monotonically. The details of this picture are altered if a slower growth rate law is used. When the time for nucleation is increased giving more particles and a broader size distribution for the nucleii at the end of nucleation, the number density of particles is increased. Changing the growth law alters the shape of the particle size distribution as it matures with time.

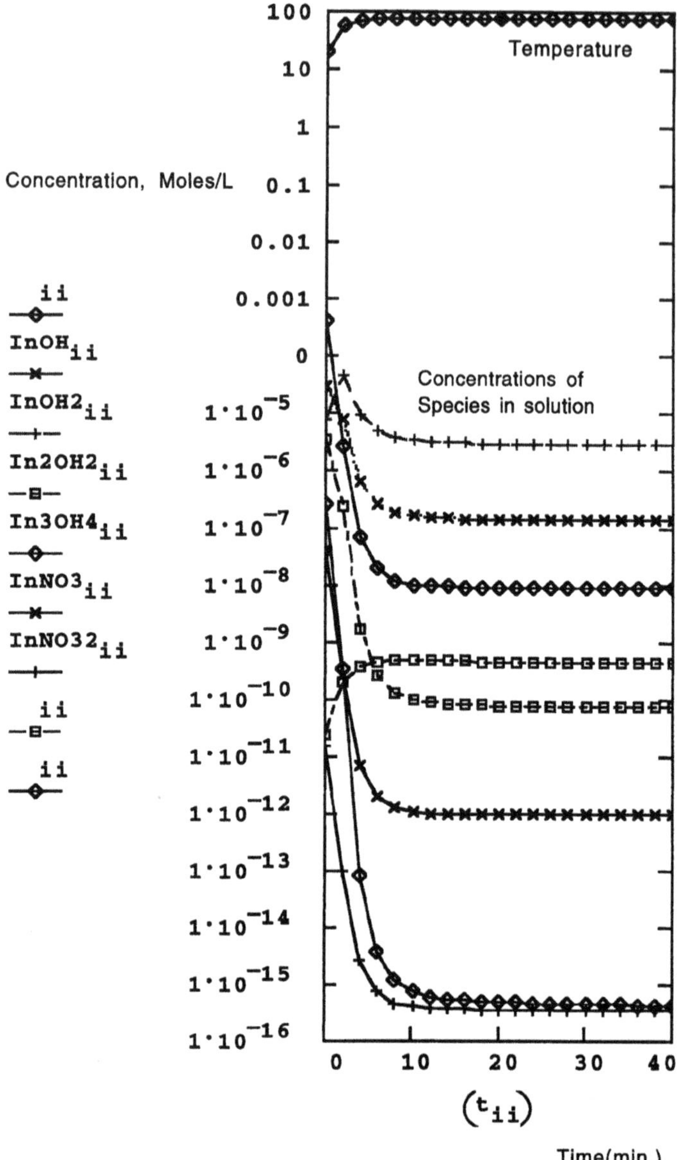

Figure 4. Concentration of solution species versus time of the forced hydrolysis of 3.96×10^{-4}M $In(NO_3)_3$, 4.04×10^{-4}M HNO_3, pHi=3.320. using homogeneous nucleation and diffusion limited growth model.

Conclusions

The hydrolysis of acidified $In(NO_3)_3$ solution has been studied experimentally. The result is a relatively narrow size distribution of submicron cubic $In(OH)_3$ crystals. A model of forced hydrolysis has been developed. This model accounts for the varying concentration of ions in solution as the precipitation proceeds and predicts the precipitated particle size distribution. It assumes that all chemical reactions in solution are fast compared to the change in solution temperature. The formalism of this model is a mass balance on both the metal, In^{+3}, and hydroxide, OH^-, ions in solution which allows calculation of the supersaturation ratio at any time which is the driving force for precipitation. Various crystal growth laws can be incorporated into the model. Comparison between model and experiment shows that a slower crystal grow law better explains the experimental data than the diffusion limited growth model used.

Acknowledgement

The authors would like to thank the Professor S. Hamada and the Science University of Tokyo for providing funding for Dr. Yoshiyuki Kudo to work in the USA on this project. Electron microscopy was provided by Matt Nowell.

References

[1]Distinguished University Professor, Dept. Chemistry, Clarkson University, Potsdam, New York.
[2]Hamada, S. Kudo, Y., and Minagawa, K. Bull. Chem. Soc. Japan 63,102-107(1990).
[3]Yura, K., Fredrikson, K.C. and Matijevic, E., Colloids Surf. 50,281-293(1990).
[4]Yuar, E., and Matijevic, E., J. Colloid Interface Sci.in press.
[5]Ferguson, R.C., Doubud, P. and Tuck, D.G., J. Chem. Soc. (A) 1968, 1058.
[6]Some solution equilibria text books use metal hydryolysis equilibria of the type :

$$iM^{+m} + jH_2O \overset{K_{H2Oij}}{\rightleftharpoons} [M_i(OH)_j]^{im-j} + jH^+$$

To generate the $[M_i(OH)_j]^{im-j}$ ions. This type of equilibria can be converted to solution speciation reactions with the hydroxide ion of the type :

$$iM^{+m} + jOH^- \underset{}{\overset{K_{OHij}}{\rightleftharpoons}} M_i(OH)_j^{+(im-j)}$$

by dividing the hydrolysis equilibium constant by the water dissociation equilibrium constant raised to the j power, as follows :

$$K_{OHij} = \frac{K_{OHij}}{K_w^{\,j}}$$

[7]Biedermann,G., Ark. Kemi.,9,277(1956). and Rec. Trav. Chim., 75,716(1956).
[8]Biedermann, G. and Ferri, D. Acta Chemica Scandinavica A 36,611-22(1982).
[9]Smith, R.M. and Martel, A.E., "Critical Stability Constants, " Vol.4, Inorganic Complexes, Plenum Press, New York, 1976.
[10]Eyring, E.M. and Owen, J.D., J. Phys. Chem.74[9],1825,10970.
[11]Silien, L.G. and Martel, A.E., "Stability Constants of Metal-ion Complexes," 2 nd. ed., Speciation publication, Chemical Society (Great Britain) , No.17, London Chemical Society, 1964.
[12]Silien, L.G. and Martel, A.E., "Stability Constants of Metal-ion Complexes," 2 nd. ed., Speciation publication, Chemical Society (Great Britain) , No.17, London Chemical Society, 1964.
[13]ΔH_w(cal/mole) = -2,6613 + 44.487*T(K) ≈ -13.34 Kcal/mole @ 25°C[13].
[14]determined from a Ksp vs 1/T fit of data from : Lacrois, S., Ann. Chim(France), 4,5(1949), Moelloer, T., J. Amer. Chem. Soc., 63,2625(1941) and Feitknecht, W. and Schindler, P. Pure Appl. Chem., 6,130(1963).
[15]This result was obtained by taking the log of the relationship $[OH^-] = \left(\frac{K_w(T)}{[H+]_o}\right)$ then taking the temperature derivative of both the right and left hand sides, where $[H+]_o$ is a constant. Using the relationship $\frac{d \ln K_w(T)}{dT} = \frac{-\Delta H}{R_g T^2}$ and then expanding the term on the left handside $\frac{d \ln [OH^-]}{dT}$ to $\frac{[H+]_o}{K_w(T)} \frac{d[OH^-]}{dT}$ and rearranging.
[16]Bird.R.B., Stewart, W.E., Lightfoot, E.N., "Transport Phenomenon", John Wiley & Sons, New York, 1960.
[17]This first summation was obtained by rearranging the equilibria expression to :

$$[M_iOH_j]^{+(im-j)} = K_{OHij}\,[M^{+m}]^i[OH^-]^j$$

and taking the derivative with respect to temperature and noting that $\frac{d K_{OHij}}{dT} = K_{OHij}\left\{\frac{-\Delta H_{OHij}}{R_g T^2}\right\}$ and that $\frac{d\{[M_iOH_j]^{+(im-j)}\}}{dT} = -i\,\frac{d\{[M^{+m}]\}}{dT}$ and then rearranging using expressions in ref. 14.

[18]Jmax for aqueous solution is either 10^{33} or 10^{30} #/cm^3/sec depending upon the theory used, see ref. 22.
[19]Becker, R. and Doring, W., Ann. Physik, **24**, 719-752,(1935)
[20]Volmer, M. and Weber, A. , Z. Phys. Chem. **119**,227,(1926)
[21]Nielsen, A.E., Kinetics of Precipitation, Pergamon Press, Oxford, (1964).
[22]Nielsen, A.E., Kinetics of Precipitation, Pergamon Press, Oxford, (1964).
[23]Dirksen,J.A. and Ring, T.A., Chem. Eng. Sci. 46(10),2389-2427(1991).
[24]Volmer, M.M. "Kinetik der Phasenbildung"p.209 Steinkopff, Dresden, Leipzig, 1939.
[25]Nielsen, A.E., Kinetics of Precipitation, Pergamon Press, Oxford, (1964).
[26]ibid

[27]Elwell, D. and Scheel, H.J."Crystal Growth from High-Temperature Solution", Academic Press Inc., London, 1975.
[28]Hamada, S., Kudo, Y. and Minagawa,K., Bulletin of the Chemical Society of Japan, 63[1],102-107(1990).
[29]Nielsen, A.E., Kinetics of Precipitation, Pergamon Press, Oxford, (1964).
[30]Branson 3200 ultrasonic bath.

ADSORPTION OF BACKGROUND ELECTROLYTE IONS AT THE METAL OXIDE/WATER SOLUTION INTERFACE. THEORY AND EXPERIMENT - A COMPARISON

W. JANUSZ
Department of Radiochemistry and Colloid Chemistry,
Faculty of Chemistry, Maria Curie-Sklodowska University,
Pl. M.Curie-Sklodowskiej 3, 20-031 Lublin, Poland.

1. Abstract

Surface complexation models of the electrical double layer at the metal oxide/electrolyte interface are used to describe behavior of such systems dependent on the background electrolyte concentration. When inert electrolyte concentration increases, it well known, that the surface charge increases while potential decreases. However, all of these models predict negligible adsorption of cations from solutions below pHpzc and anions above pHpzc. Numerous adsorption data show detectable adsorption of ions in these pH ranges. According to Johnson's proposal (Johnson E. R. (1984) A Thermodynamic Description of the Double Layer Surrounding Hydrous Oxides *J. Colloid Interface Sci.* 100(2), 540-554) adsorption of ions on neutral sites can take place in addition to the reactions considered by standard complexation model. This adsorption is described by Frumkin isotherm which takes into account lateral interaction with charged groups and adsorbed ions.

2. Introduction

The concentration of electrolyte ions have an influence on the surface charge density of the solid and the distribution of the charge in the interfacial region [1,2]. It is well known that the zeta potential of inorganic oxide decreases and surface charge density increases when inert electrolyte concentration increases. This phenomenon is well described by the site binding theory of the electrical double layer (edl) and by its extended models, like surface complexation model. According to this theory surface charge is created as a result of acid-base dissociation of surface hydroxyl groups:

$$\equiv SOH_2^+ \rightleftharpoons \equiv SOH + H^+ \qquad (1)$$

E. Pelizzetti (ed.), Fine Particles Science and Technology, 161–170.

$$\equiv SOH \rightleftharpoons \equiv SO^{-} + H^{+} \tag{2}$$

and formation of surface complexes at the metal oxide/electrolyte interface:

$$\equiv SOH_2^{+}A^{-} \rightleftharpoons \equiv SOH + A^{-} + H^{+} \tag{3}$$

$$\equiv SOH + K^{+} \rightleftharpoons \equiv SO^{-}K^{+} + H^{+} \tag{4}$$

The role of inert electrolyte ions on the surface charge of oxides is taken into account in the theory of ion association proposed by Kallay and Tomic, assumes that charged groups on the metal oxide surface may associate with counterions [3,4]. Though the above theory differs from the molecular mechanism of adsorption of counterions, it predicts the increase of surface charge with the increase of ionic strength of electrolyte.

The radiotracer technique is very often used as convenient method to determine the ion adsorption at the metal oxide/electrolyte interface [6-12]. The literature data showed the dependence of adsorption density of cations as well as adsorption of anions on pH.

3. Model Calculations vs. Experimental Data

Calculated concentrations of anion and cation complexes vs. pH are depicted on Figs 1 and 2, respectively for three different ionic strengths of the electrolyte. These concentrations were calculated on the basis of the constant capacity model, (Eq. 66-69 in ref .[2]), using the following parameters: $pK_{a1}=4$; $pK_{a2}= 8$; $pK_A=5$; $pK_K= 7$; and the concentration of surface hydroxyl groups $N_s=1*10^{18}$ OH group/m^2. As is seen from Fig. 1, the adsorption of anions decreases when pH of the solutions increases. The opposite situations is observed for cations Fig. 2. However, the disagreement is observed between experimental adsorption data and calculated adsorption densities of background electrolyte ions. The experimental adsorption, of anions from the electrolyte solution for pH > pHpzc, is higher than concentration of $\equiv SOH_2^{+}A^{-}$ form. Similarly, adsorption of cations from the solution for pH < pHpzc is higher then calculated concentration of $\equiv SO^{-}K^{+}$ groups.

The difference between calculated and experimental adsorption data is presumably because of different experimental conditions of acquisition of the surface charge and adsorption data [5]. The measurements of ions adsorption were carried out at high concentration of metal oxide in the electrolyte solution ranging from 3000 to 4500 m^2/dm^3, whereas potentiometric titration was performed at low solids concentration in the suspension i.e. 160-480 m^2/dm^3 [6]. However adsorption and titrations measurements

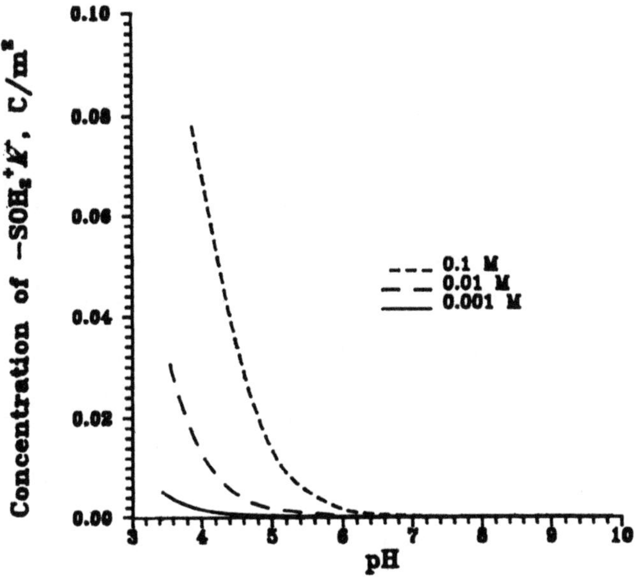

Figure 1. Concentration of $\equiv SOH_2^+A^-$ *groups vs. pH calculated for* $pK_{a1}=4$; $pK_{a2}= 8$; $pK_A=5$; $pK_K= 7$; *concentration of surface hydroxyl groups* $N_s=1*10^{18}$

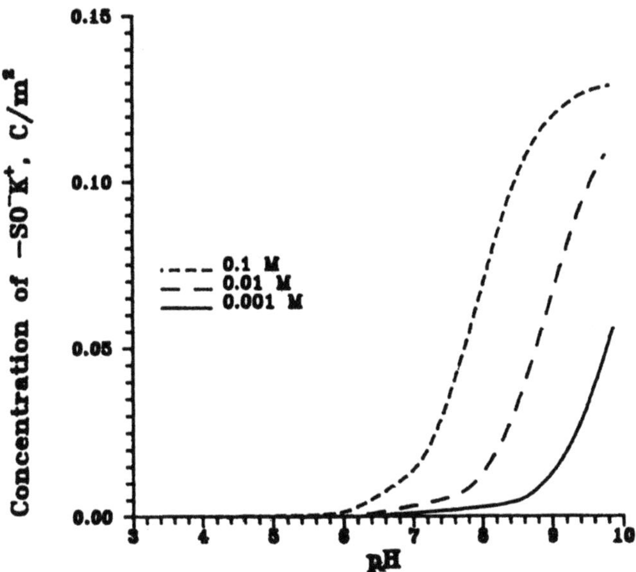

Figure 2. Concentration of $\equiv SO^-K^+$ *groups vs. pH calculated for* $pK_{a1}=4$; $pK_{a2}= 8$; $pK_A=5$; $pK_K= 7$; *concentration of surface hydroxyl groups* $N_s=1*10^{18}$

performed for the same solid:liquid ratio confirmed Sprycha's results [9-12]. The difference between the experimental adsorption data and the concentration complexes formed according to Eqs 3 and 4, may result from the additional interactions in the adsorption layer, not predicted by the site binding theory. For example, Johnson considered adsorption of cations and anions of the inert electrolyte onto neutral sites [13]. Even though H^+ ions do not participate directly, in such model, the ion adsorption is dependent on pH, because of the changes in the concentration of $\equiv$SOH groups and variation of the electrical potential in the edl.

Adsorption of counterions at the neutral surface groups may be a consequence of lateral interactions between ions adsorbed in the inner Helmholtz plane (IHP). The coadsorbed ions could compensate lateral interaction of ions adsorbed in the IHP, and maintain the same value of the surface charge. This event may occur in the case of large adsorption densities of the ions adsorbed in the IHP. Experimentally measured adsorption may be a sum of adsorption in the IHP and coadsorption of ions in the Stern layer. Such an approach may explain the increase of adsorption of sodium with the decrease of pH of the solution ions for pH < pHpzc, reported by Smith et. al. for the alumina/sodium bromide system [14].

Coadsorption of ions in the compact layer of the edl should results in lateral interaction of ion with all charged groups existing on the surface of the metal oxide. The complete physical description of these interactions seems to be very difficult, because of their various nature (interaction of ions and dipole-ions) and a porosity of the surface. Therefore, in this paper Frumkin's isotherm was used to take into account the lateral interactions between charged groups in the adsorbed layer [15].

$$\frac{\theta_{A^-}}{1-\sum\theta_i} EXP\left(\sum\theta_i * f_i\right) = K * a_\pm \tag{5}$$

where

θ_i - fraction of the surface occupied by "i" type of charged group (eg $\equiv SOH_2^+$, $\equiv SOH_2^+A^-$, $\equiv SO^-$, $\equiv SO^-K^+$) or with coadsorbed anions and cations,

f_i - coefficient of interaction between "i" type of charged group($\equiv SOH_2^+$, $\equiv SOH_2^+A^-$, $\equiv SO^-$, $\equiv SO^-K^+$) and with coadsorbed anion or cation,

K - adsorption constants,

a - mean activity of electrolyte ions,

The above equation describes the lateral interactions in the compact layer of the edl, assuming that the number of the surface sites available for the adsortpion, is the same for cations and anions and neglecting the differences between ion sizes.

Metal oxides may contain some amount of impurities which may have an effect on adsorption measurements. The effect of impurities present in the metal oxide on potentiometric titration data was analyzed by Zalac and Kallay[16]. Results of adsorption measurements are also sensitive to the presence of impurities in the investigated system. The results of adsorption measurements of ions, may be affected by isotopic exchange between radiotracer and its stable isotope present on the surface of the oxide as an

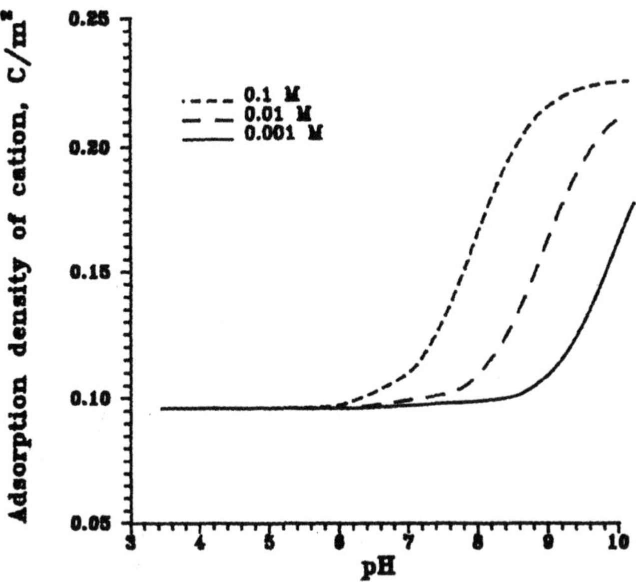

Figure 3. Adsorption density of Na^+ vs. pH calculated for $pK_{a1}=4$; $pK_{a2}= 8$; $pK_A=5$; $pK_K= 7$; concentration of surface hydroxyl groups $N_s=1*10^{18}$ in presence sodium impurities 10μmol/g.

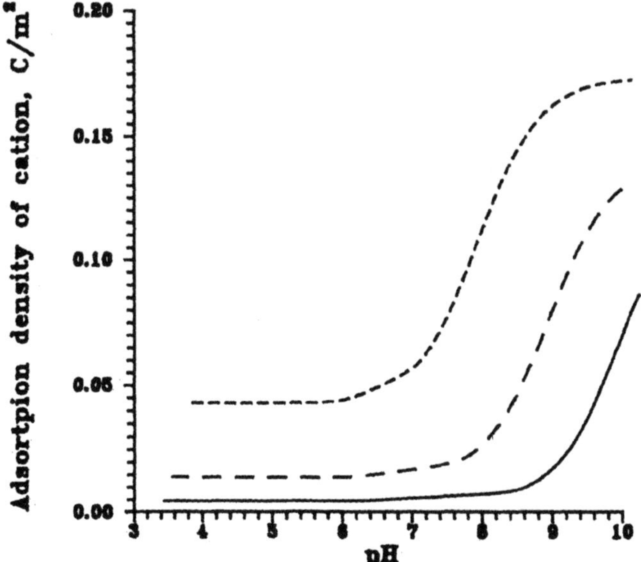

Figure 4. Adsorption density of Na^+ vs. pH calculated for $pK_{a1}=4$; $pK_{a2}= 8$; $pK_A=5$; $pK_K= 7$; concentration of surface hydroxyl groups $N_s=1*10^{18}$ in presence monovalent cations impurities 10μmol/g.

impurity. For instance when adsorption of Na^+ is investigated using Na-22 isotope as an radiotracer then adsorption of this ion, as well as, the isotopic exchange can occur:

$$-Na + {}^*Na^+ \rightleftharpoons -{}^*Na + Na^+ \quad (6)$$

where:

$^*Na$ - radiotracer,

-Na - sodium present in the metal oxide as an impurity,

In such case, uptake of radioactivity of he solution will be greater than that caused by adsorption only (reaction 4). The results of calculations of sodium ions adsorption density vs. pH where isotopic exchange was taken into account, are depicted on Fig 3. Calculations were performed using James and Parks method,(Eqs 66-69 in [2]), on the basis of reactions 1-4 and 6 assuming $pK_{a1}=4$; $pK_{a2}= 8$; $pK_A=5$; $pK_K= 7$ and concentration of surface hydroxyl groups $N_s=1*10^{18}$ of OH group/m^2 and concentration of sodium impurities equal to 10 μmol/g. The specific surface area was equal to $10m^2/g$. As is seen from Fig 3, the presence of sodium impurities increased "the adsorption" of sodium ions for all ionic strengths. In addition, for higher concentrations of hydrogen ions i.e. low pH, the ionic exchange between H^+ ions and impurity ions is also possible.

If, in the system described above, other ions are also present as impurities eg. potassium ions they will undergo ionic exchange with the electrolyte ions as well. In such case, the amount of exchanged ions, will depend not only on the amount of impurities, but also on the exchange constant. For potassium ions present as impurity in metal oxide, dispersed in the electrolyte containing sodium ions, the following ion exchange can take place:

$$-K + Na^+ \rightleftharpoons -Na + K^+ \quad (7)$$

Such ion exchange was taken into account in the model calculations depicted in Fig 4. The electrical double layer parameters were the same as those presented in Fig. 3 except that the constant of ion exchange was assumed equal to 1. As is seen in Fig. 4, the plot of adsorption density of cations is similar to the experimental curve. For pH < pHpzc, adsorption density is not equal to 0, and increases with increasing the concentration of the electrolyte. However, the concentration of $\equiv SO^-K^+$ groups in this range of pH is almost equal to zero. It seems that for pH < pHpzc the process of ion exchange prevails in the system.

The same approach can be used for anions for pH > pHpzc.

4. Optimization of the parameters of the edl for titania(rutile)/solution system.

Some authors pointed out that experimental adsorption densities of background electrolyte ions differ from model calculations based on literature models of the electrical double layer at the metal oxide/electrolyte solution interface [5,9-12,17]. It was shown in previous section that taking into account the ionic exchange between background electrolyte ions and impurities present in the oxide, can significantly improve the agreement between experimental and model calculations. In this section, model parameters were optimized to fit the experimental titration and adsorption data for the rutile electrolyte system.

The adsorption and potentiometric titration measurements were carried out with the same solid: liquid ratio. The preliminary calculation, performed for boundary conditions which may occur in concentrated ($\psi_\beta = 0$) and diluted ($\psi_o = \psi_\beta$) solutions indicated that better agreement was obtained using when $\psi_o >> \psi_\beta$ and $\psi_\beta = 0$ [12]. Parameters of the model were calculated numerically using MINUIT procedure for the optimization of the following function [18]:

$$F = \frac{\sum\left(\frac{\sigma_t - \sigma_o}{\sigma_o}\right)^2 + \sum\left(\frac{[\equiv SOH_2^+A^-] - A_A}{A_A}\right)^2 + \sum\left(\frac{[\equiv SO^-K^+] - A_K}{A_K}\right)^2}{N_p - N_u} \qquad (8)$$

where:

σ_o - surface charge density,

σ_t - surface charge, calculated from the site binding model,

A_A - adsorption density of anion,

A_K - adsorption density of cation,

N_p - number of experimental points,

N_u - number of adjustable parameters,

Optimization of the electrical double layer parameters was carried out for two types of site binding models: constant capacitance model and the model proposed by James and Parks [2]. The best matching was achieved using constant capacitance model in the limited range of pH, pHpzc $\pm$ 1.0 pH unit. This approach is more justified than using constant capacitance model in whole range of pH. Surface charge density calculated, in this way, is very close to the experimental one. Beyond this region, discrepancy between calculated and experimental values becomes greater. To achieve better agreement between theory and experimental data, variable capacity model was applied using the following values of constants $pK_{a1}=2.24$; $pK_{a2}= 8.33$; $pK_A=5$; $pK_K= 7$ (the best fitting values). These constants are very similar to those calculated by Davis et. al. method[12]. The calculated curve of surface charge density vs. pH was in good agreement with experimental potentiometric titration data. However the agreement between model calculations and experimental data of adsorption of background electrolyte ions is less satisfactory. In same cases, experimental adsorption density of given ion was lower than that calculated on the basis of variable capacity model. This may results from the removal of ions from the compact part of the electrical double layer during centrifugation of the suspension to prepare the radiosources.

In same cases experimental adsorption density of a given ion was higher than the calculated one. This may arise from the model simplifications or neglecting of some reactions taking place on the metal oxide/electrolyte interface.

The adsorption of ions under the influence of lateral interactions is easy to describe using Frumkin's isotherm. To use the site binding model for the ion adsorption, under lateral interactions, six additional parameters has to be taken into account in the optimization procedure. Using numerical Minuit procedure the best fitting was obtained for the following values: $pK_{a1}=2,09$; $pK_{a2}=8,08$, $pK_{Cl}=2,74$; $pK_{Na}=7,22$; $pK_{adsCl}=0,95$; $pK_{adsNa}=0,82$; $f_{SOH2}=-0,1$; $f_{SO}=-0,69$; $f_{Na}=-0,8$; $f_{Cl}=0,7$; $f_{SOH2Cl}=-0.18$; $f_{SONa}=1,0$. The values of the interaction coefficients show that the lateral interactions are rather week. Comparison of the adsorption density of sodium and chloride ions with the concentrations of the appropriate surface complexes are depicted in Fig 5. As is seen the agreement between model and experimental curves for sodium is very good. However, the same relationship for chloride ions is poor. This may be result from the assumption of equal lateral interactions for sodium and chloride ions (equal interaction coefficients).

The results of optimization of electrical double layer parameters which take into account the ionic exchange between impurities and electrolyte ions are depicted in the Fig. 6 as a plot of adsorption density vs. pH. Experimental points are also presented for comparison. The best fitting was obtained for the following values: $pK_{a1}=2,32$; $pK_{a2}=8,06$, $pK_{Cl}=2,86$; $pK_{Na}=7,09$; constant of chloride exchange $pK_{exCl}=0,56$; concentration of anion ipurities 29 μmol/g, constant of sodium exchange $pK_{exNa}=2,45$ and concentration of cation impurities 25 μmol/g . As is seen from Fig. 6 good agreement between experimental data and model calculation is observed for sodium ions. However, for chloride ions the experimental adsorption is higher than that calculated sum of adsorption of chloride ions on the basis of the model taking into account the exchange process. This may be the due to the simplifications of the model and neglecting the ion exchange between multivalent and monovalent ions. The measurements of adsorption of chloride ions on very pure metal oxide (Cl^- free) showed that adsorption density for pH > pHpzc is almost equal to zero. The same was observed for sodium ions for pH < pHpzc.

5. Conclusions

The problem of discrepancy between experimental adsorption data of background electrolyte ions at the metal oxide/electrolyte interface and model calculations can be solved taking into account the lateral interaction between ions and surface charge groups in the compact part of the electrical double layer. Adsorption of ions at the metal oxide/electrolyte interface, determined by radiotracer technique, is a sum of real adsorption predicted by site binding theory and isotope exchange and ion exchange between electrolyte solution and metal oxide impurities. More experimental work is needed on impurities free oxide samples to determine the real contribution of surface complexes to the surface charge.

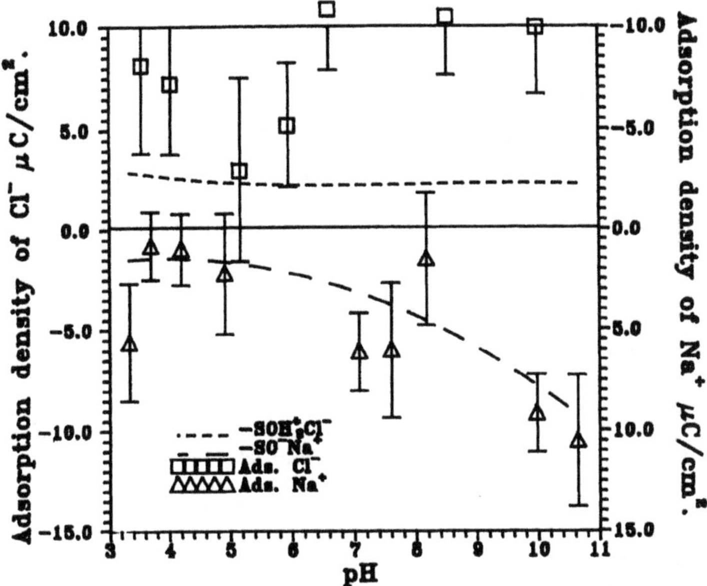

Figure 5. Adsorption density of Na^+ and Cl^- vs. pH for TiO_2(rutile)/0.1 M NaCl system. Dashed lines - model calulation using Frumkin's isotherm .

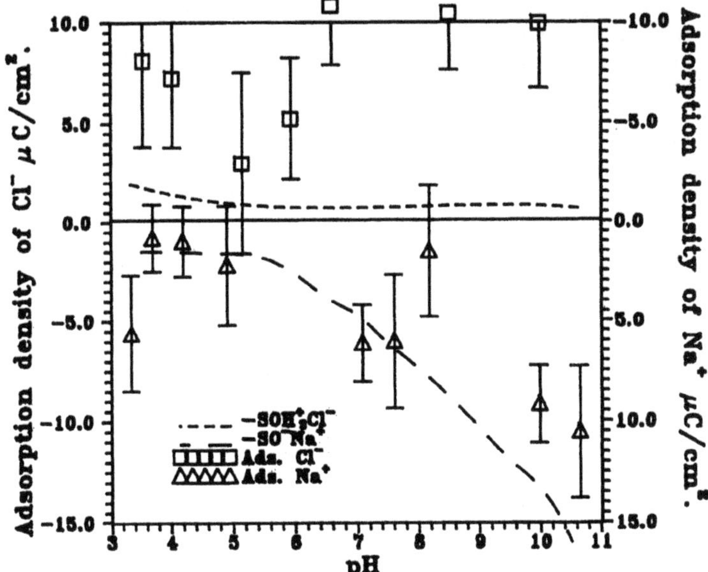

Figure 6. Adsorption density of Na^+ and Cl^- vs. pH for TiO_2(rutile)/0.1 M NaCl system. Dashed lines - model calulation taking into account ionic exchange between electrolyte ions and oxide impurities.

6. References

1. Wiese G.R., James R.O., Yates D.E., and Healy T.W. (1976) Electrochemistry of the Colloid - Water Interface, in ,Bockris J O'M (ed.), *Electrochemistry of Colloid-Water Interface*, Int. Review. Sci. Phys. Chem., Series 2, Butterworths vol. 6, pp. 53-102.
2. James R.O. and Parks G.A. (1982) Characterization of Aqueous Colloids by Their Electrical Double-Layer and Intrinsic Surface Chemical Properties, in Matijević E. (Ed.) *Surface and Colloid Sci.*, Wiley-Interscience, New York. **12**, 119 -216.
3. Kallay N. and Tomić M. (1988) Association of Counterions with Adsorbed Potential-Determining Ions at a Solid/Solution Interface. 1. Theoretical Analysis, *Langmuir*, **4**, 559-564.
4. Tomić M. and Kallay N. (1988) Association of Counterions with Adsorbed Potential-Determining Ions at a Solid/Solution Interface. 2. Double-Layer Equilibria at a Metal Oxide Interface, *Langmuir*, **4**, 565-569,.
5. Hayes K.F., Redden G., Ela W., Leckie J.O. (1991) Surface Complexation Models: An Evaluation of Model Parameter Estimation Using FITEQL and Oxide Mineral Titration Data, *J. Colloid Interface Sci.*, **142**, 448-469.
6. Sprycha R. (1984) Surface Charge and Adsorption of Background Electrolyte Ions at Anatase/Electrolyte Interface, *J. Colloid Interface Sci.*, **102**, 173-185.
7. Smit W., Holten C.L.M. Stein H.N., de Goeij J.J.M., and Theelen H.M.J. (1978) A Radiotracer Determination of the Adsorption of Sodium Ion in the Compact Part of the Double Layer of Vitreous Silica, *J. Colloid Interface Sci.*, **63**, 120-128.
8. Foissy A., Pandou A. M., and Lamarche J. M. (1982) Surface and Diffuse Layer Charge at the TiO_2 - Electrolyte Interface, *Colloids Surf.* **5**, 363-368.
9. Janusz W. and Szczypa J., (1987) Electrical Double Layer at the Alumina/Electrolyte Interface, *Mater. Sci. Forum* **25-26**, 427-430.
10. Janusz W., (1988) Surface Charge and Adsorption of Na^+ and Cl^- Ions at the Zirconium Dioxide/Water Interface, *J. Radioanalytical and Nuclear Chemistry- Articles* **125**, 393-401.
11. Janusz W. (1989) A New Radiotracer Method of Adsorption Studies at Metal Oxide/Aqueous Electrolyte Interface, *J. Radioanalytical and Nuclear Chemistry- Articles* **134**, 193-198.
12. Janusz W. (1989) Adsorption of Sodium and Chloride Ions at the Rutile/Electrolyte Interface - Parameters of the Electric Double Layer, *Mat. Chem. Phys.* **24**, 39-50.
13. Johnson E. R. (1984) A Thermodynamic Description of the Double Layer Surrounding Hydrous Oxides, *J. Colloid Interface Sci.* **100**, 540-554.
14. Smit W. and Van Holten L. M. (1980) Zeta-Potential and Radiotracer Adsorption Measurements on EFG α-Al_2O_3 Single Crystals in NaBr Solutions, *J. Colloid Interface Sci.* **78**, 1-14.
15. Gileadi E. (1967) Adsorption in Electrochemistry in Gileadi E. (ed.) *Electrosorption*, Plenum Press, New York, 13-15.
16. Žalac S. and Kallay N. (1992) Application of Mass Titration to the Point of Zero Charge Determination, *J. Colloid Interface Sci.* **149**, 233-240.
17. Kallay N., Sprycha R., Tomić M., Žalac S. and Torbić Ž. (1990) Some Controversies in the Understanding of Equilibria in Electrical Interfacial Layer, *Croatica Chemica Acta*, **63**, 467-487.
18. James F. and Ross M. (1975) Minuit- A System for Function Minimization and Analysis of the Parameter Errors and Correlations, *Computer Phys. Comm.* **10**, 343-367.

STRUCTURES OF MICROPOROUS HEMATITE PARTICLES PREPARED FROM δ-FeOOH

T. ISHIKAWA
School of Chemistry, Osaka University of Education
4-698-1 Asahigaoka, Kashiwara-shi, Osaka-fu 582, Japan

δ-FeOOH particles have been prepared by oxidation of $Fe(OH)_2$ with H_2O_2. The $Fe(OH)_2$ precursor precipitated from $FeSO_4$ solutions of different concentrations were agied at different temperatures from 40 to 110°C. The δ-FeOOH particles obtained and their thermal decomposition products were characterized by SEM, TEM, XRD, FTIR and N_2 adsorption. The plate-like δ-FeOOH particles with a mean width of 0.1 μm are obtained at a lower $FeSO_4$ concentration. The δ-FeOOH particles formed at a higher $FeSO_4$ concentration are tetradecahedron and the particle size increases from *ca.* 0.1 to 1 μm with elevating the ageing temperature of $Fe(OH)_2$. The tetradecahedron δ-FeOOH particles are less orderly crystalline. The t-plots of N_2 adsorption show the microporous structure of the δ-FeOOH particles obtained by ageing $Fe(OH)_2$ at a higher temperature. The α-Fe_2O_3 particles formed upon outgassing the plate-like δ-FeOOH particles above 200°C have uniform slit-shaped micropores with a 0.8 nm width. On the other hand, the α-Fe_2O_3 particles formed by outgassing tetradecahedron δ-FeOOH particles above 200°C have micropores with the size ranging from 1 to 2 nm.

1. Introduction

δ-FeOOH, one of the four crystalline phases of iron(III) oxide hydroxide, exhibits interesting properties different from those of α-, β- and γ-FeOOH. For instance δ-FeOOH particles show ferromagnetic behavior at room temperature while other phases are antiferromagnetic [1, 2]. Thermal phase transformation from δ-FeOOH to α-Fe_2O_3 is very sensitive to the crystallinity, particles size and heating procedure so that different transformation mechanisms have been proposed [3-6]. δ-FeOOH particles are usually generated in hexagonal plates because they are formed by a topotaxial oxidation reaction of hexagonal plate-like $Fe(OH)_2$ particles [7, 8]. Okamoto found that the Fe^{3+} ions in δ-FeOOH are easily replaced to a fair degree with other divalent metal ions like Co^{2+} and Mn^{2+} ions [1]. This suggests that the oxide or metal particles of iron, which are homogeneously doped with other metal ions or atoms as desired, can be prepared by thermal treatment or reduction of the metal-doped δ-FeOOH particles. It is hard generally to prepare a homogeneous solid solution of metal oxides by means of co-precipitation or solid reaction. Despite the interesting characteristics of δ-FeOOH, the surface and pore structures of this phase and its thermal decomposition products have received less attention compared with those of α-, β- and γ-FeOOH.

E. Pelizzetti (ed.), Fine Particles Science and Technology, 171–183.

Hematite (α-Fe_2O_3) particles are widely used as catalysts, pigments and raw materials of powders for magnetic memory. The α-Fe_2O_3 particles with uniform-sized micropores and a larger pore surface area are suitable as catalysts and adsorbents with high selectivity. The microporous α-Fe_2O_3 particles produced by the thermal decomposition of α- and γ-FeOOH are needle or thin plate-like particles with a submicron size so that these α-Fe_2O_3 particles show a low ratio of micropore surface area to total one. The preparation of α-Fe_2O_3 particles with a high micropore surface area requires the large bulky δ-FeOOH particles as the precursor.

The present study has been aimed at clarifying the surface and pore structures of δ-FeOOH and its thermal decomposition products by various means In this article the pore structures of δ-FeOOH particles and α-Fe_2O_3 particles obtained from the δ-FeOOH particles are discussed based on the results obtained [9, 10].

2. Experimental Section

2.1. MATERIALS

The δ-FeOOH particles with different sizes and shapes were synthesized by rapid oxidation of $Fe(OH)_2$ particles with H_2O_2 [6]. The small plate-like δ-FeOOH particles were prepared as follows. 20 g of $FeSO_4 \cdot 7H_2O$ were dissolved in 200 cm^3 of deionized and distilled water free from O_2. 100 cm^3 of a 10 mol dm^{-3} NaOH solution deoxygenated were added to the $FeSO_4$ solution at 45°C in an N_2 atmosphere. The resulting white $Fe(OH)_2$ precipitates were aged at 45°C for 6 h, after which 50 cm^3 of a 30 % H_2O_2 solution were instantly poured into the $Fe(OH)_2$ suspension. The resulting orange precipitates were filtered off, thoroughly washed with water and vacuum dried at room temperature for one day.

The large tetradecahedron δ-FeOOH particles with different sizes were synthesized by the essentially similar method to the small plate-like particles [9]. Differing from the above-mentioned recipe, more concentrated $FeSO_4$ solutions were used in order to obtain larger particles. $FeSO_4 \cdot 7H_2O$ (44 g) was dissolved in 100 cm^3 of deionized, distilled and deoxygenated water in a Teflon vessel filled with pure N_2. To the $FeSO_4$ solutions was added the NaOH solution prepared by dissolving 60 g of NaOH into 100 cm^3 of water in the similar vessel to the preparation of $FeSO_4$ solutions. The resulted precipitates were agied in the closed reaction vessel filled with pure N_2 at different temperatures in the range 40 - 110 °C for 6 h. To the white $Fe(OH)_2$ suspension formed after this ageing were added 75 cm^3 of a 30 % H_2O_2 solution as quickly as possible. The brawn precipitates formed by the oxidation were washed and dried in the same manner as for the plate-like particles.

The thermal decomposition of δ-FeOOH particles was carried out by outgassing at different temperatures 25 - 300°C for 2 h. The maximum outgassing pressure was kept below 1 Pa by controlling the heating rate, since the pore structure of the thermal decomposition products was influenced by the pressure of H_2O evolved.

2.2. METHODS

The materials so prepared and the thermal decomposition products were examined by the following techniques. The powder X-ray diffraction (XRD) patterns were recorded on a Rigaku diffractometer with nickel-filtered Cu-Kα radiation (30 kV, 15 mA). The IR spectra were taken by KBr method on a Digilab FTIR spectrometer. The electron micrographs were taken with an Akashi scanning electron microscope (SEM) and a Jeol transmission microscope (TEM). The adsorption isotherms of N_2 were measured at the boiling point of liquid N_2 using a computer-aided automatic volumetric equipment assembled in our laboratory. Before N_2 adsorption the samples were outgassed under the similar condition to the thermal decomposition described above.

3. Results and Discussion

3.1. α-Fe_2O_3 PARTICLES PREPARED FROM SMALL PLATE-LIKE δ-FeOOH PARTICLES

To know the thermal transformation of δ-FeOOH to α-Fe_2O_3, were measured the XRD patterns and IR spectra of the samples treated by outgassing at different temperatures. The patterns of the samples outgassed below 100°C had diffraction peaks characteristic of δ-FeOOH. When treated above 125°C two small additional peaks of α-FeOOH were detected. The peaks of α-Fe_2O_3 appeared above 200°C. These results verify that δ-FeOOH transforms to α-FeOOH at 125°C and further to α-Fe_2O_3 at 200°C. However, the main peaks of these three phases are too close to each other to know the exact composition of less crystalline materials such as the present samples.

To support the XRD data we measured the IR spectra of samples treated under the same conditions as the XRD samples, since IR data usually permit us to obtain structural information over a shorter range than the XRD data. The spectrum of the sample before treating had two absorption bands at 1130 and 804 cm^{-1}, near the 1124 and 794 cm^{-1} bands reported for δ-FeOOH [2], although the assignment of these bands remains unclear. The 1130 cm^{-1} band diminished on elevating the outgassing temperature to 175°C, while the 804 cm^{-1} band is detected at 200°C. An additional band appears at 864 cm^{-1} above 125°C and is clearly observed even at 250°C, though shifting to a somewhat higher wavenumber. The 864 and 804 cm^{-1} bands observed for the samples treated above 175°C are close to the 886 and 794 cm^{-1} bands of pure α-FeOOH, which have been assigned to the Fe-O-H bending and Fe-O stretching vibrations of the bulk FeOH groups, respectively [10]. The spectra of the materials treated above 200°C showed three bands at 1328, 890 and 800 cm^{-1}. Along with these bands, two bands due to the skeleton vibration of α-Fe_2O_3 appeared at 572 and 409 cm^{-1}. Therefore, the weak 890 and 800 cm^{-1} bands seem to be due to the Fe-O-H bending and Fe-O stretching vibrations of the FeOH groups, because the α-Fe_2O_3 particles formed by treating above 200°C contain a small amount of OH groups. The 1328 cm^{-1} band cannot be assigned at the moment. As described above, the records of XRD indicate that δ-FeOOH transforms to α-FeOOH at 125°C and then to α-Fe_2O_3 at 200°C. IR spectroscopy proves, however, that the material treated at 125-175°C is a mixture of α- and δ-FeOOH. This discrepancy between the results of XRD and FTIR would result from the difference in the sensitivities of these methods. Accordingly, we can recognize that FTIR is more useful for analyzing less crystalline materials compared with XRD.

To clarify the pore structure of the decomposition products of the plate-like δ-FeOOH particles, we measured the isotherms of N_2 on the samples outgassed at different

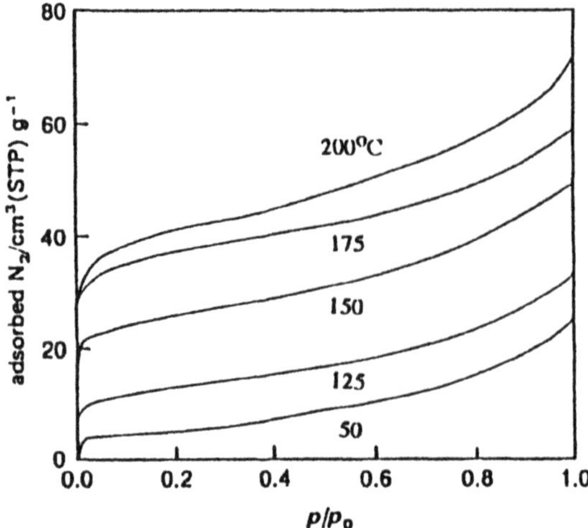

Figure 1. N_2 adsorption isotherms at the boiling point of liquid-N_2 temperature on the samples outgassed at various temperatures.

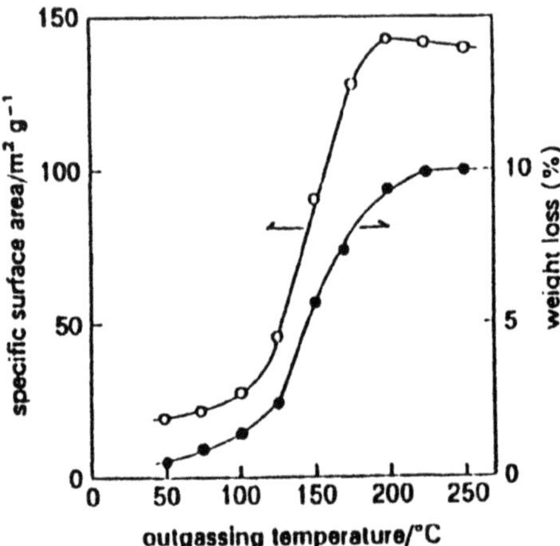

Figure 2. Specific surface area and weight loss of the samples outgassed at different temperatures.

temperatures and illustrated Figure 1. As is seen in Figure 1, the adsorbed amount increases with increasing outgassing temperature and the adsorption isotherms on the samples outgassed at higher temperatures rise steeply at low relative pressures, verifying micropore formation. Figure 2 plots the BET specific surface area determined by employing the BET equation and the weight loss during outgassed against the outgassing temperature by the open and filled circles, respectively. Both the surface area and the weight loss abruptly increase on elevating the outgassing temperature from 100

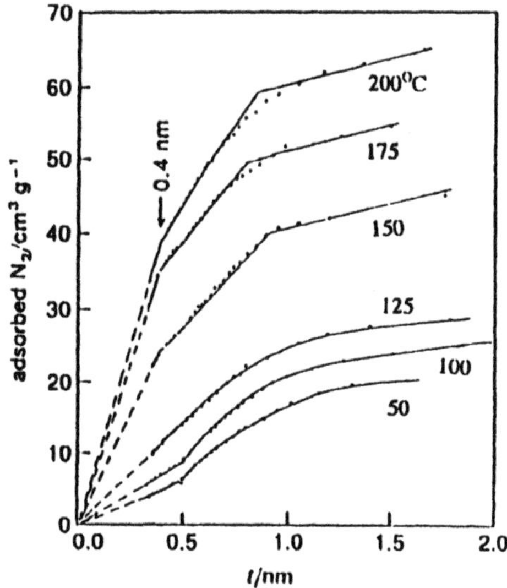

Figure 3. t-plots of the N_2 adsorption isotherm on the particles outgassed at different temperatures.

to 200°C. This suggests that the surface area increases with the desorption of adsorbed H_2O or the release of OH groups from the crystals as H_2O molecules without altering the particle size and morphology as established by TEM. It can be speculated from the N_2 adsorption isotherms in Figure 1 that the sample outgassed at a higher temperature exhibits a higher microporosity. To obtain detailed information on the pore structure of the thermal decomposition products, the t-plots were obtained from the N_2 isotherms of Figure 1 and are depicted in Figure 3. The t-plots for the samples outgassed below 125°C curved upward, meaning that capillary condensation takes place in the slit-like pores between the plate-like particles. The finding that the t-plots of the samples outgassed above 150°C clearly bent at thickness of 0.4 nm indicates the existence of micropores of a 0.8 nm width. The surface area of these micropores of a width of 0.8 nm increases with elevation of the outgassed temperature. Jimenez-Mateos et al. have found pores with *ca.* 2 nm diameter in both δ-FeOOH particles treated in air at 400°C and the particles ground at room temperature, and have concluded that these pores exist between the plate-like particles [12]. However, they have no observed the micropore of a 0.8 nm width as found in this work. It was established in this study that micropores having such homogeneous size distributions can be formed only by treating under vacuum.

Figure 4 displays the TEM pictures of the particles outgassed at 150 and 200°C. The small channels can be clearly observed in each particle. The particles outgassed above 250°C are a little sintered, although the pictures are not shown. The mean width of these channels is 0.8 nm, in surprisingly good agreement with the pore width estimated from the t-plots. From observing the TEM pictures the morphology of the particles is retained even after outgassing at 250°C despite the loss of the inner OH groups. This is also evident from in which the slope of the t-plot in a mesoporous region varies little more than *ca.* 1 nm with the outgassing temperature. Hirokawa et al. have discovered slit-like micropores with *ca.* 1 nm width in the α-Fe_2O_3 particles yielded by outgassing

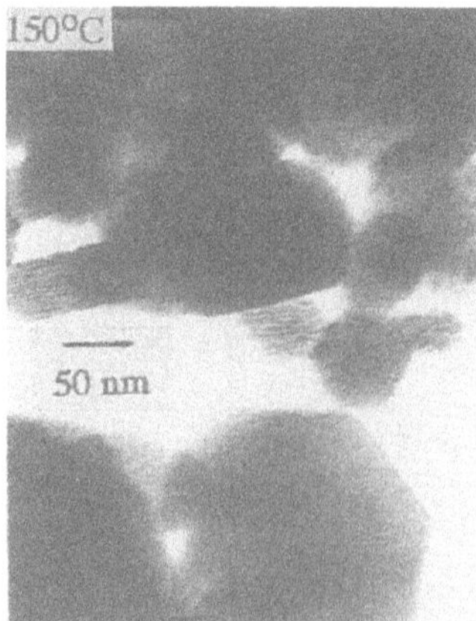

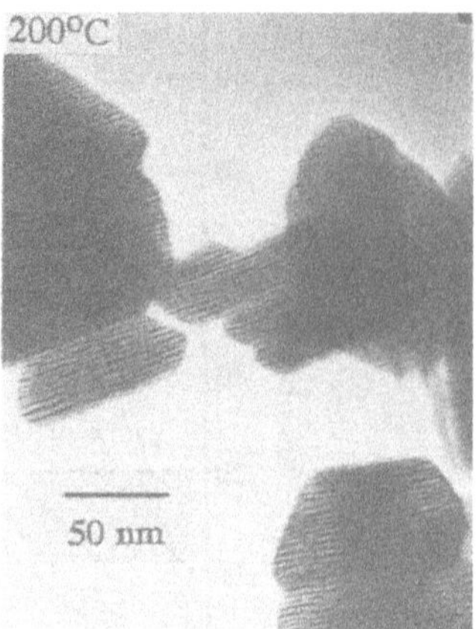

Figure 4. TEM pictures of the samples outgassed at 150 and 200°C.

α-FeOOH [13]. This micropore width is approximately equal to that of the micropore found in the present work, giving support to the proposed thermal transformation of δ-FeOOH to α-Fe_2O_3 *via* the intermediate phase α-FeOOH.

3.2. TEXTURES OF LARGE BULKY δ-FeOOH PARTICLES

The SEM pictures of the particles obtained by ageing $Fe(OH)_2$ for different temperatures are shown in Figure 5. Note that the particle shapes of these samples are tetradecahedron, differing from the plate-like particles described in section 3.1. where the $FeSO_4$ concentration was less than that as for the preparation of the tetradecahedron particles. To our knowledge, no tetradecahedron δ-FeOOH particles have so far been prepared. Figure 6 plots the mean particle size estimated from the SEM pictures against the ageing temperature of $Fe(OH)_2$ (called t_a below) by the triangular marks. The particle size increased from 0.1 to 1 μm with elevating t_a from 40 to 110°C. Okamoto has proposed a formation mechanism of δ-FeOOH that the protons of OH^- ions and the electrons of Fe^{2+} ions in the $Fe(OH)_2$ crystal move to the crystal surface and are oxidized by H_2O_2 [4]. The reaction of δ-FeOOH formation goes on as follows.

$$Fe^{2+} \text{--->} Fe^{3+} + e^-$$
$$2\,OH^- \text{--->} OH^- + O^{2-} + H^+$$
$$1/2\,H_2O_2 \text{--->} 1/2\,H_2O + 1/4\,O_2$$
$$H^+ + e^- + 1/4\,O_2 \text{-->} 1/2\,H_2O$$

$$Fe(OH)_2 + 1/2\,H_2O_2 \text{--->} \delta\text{-}FeOOH + H_2O$$

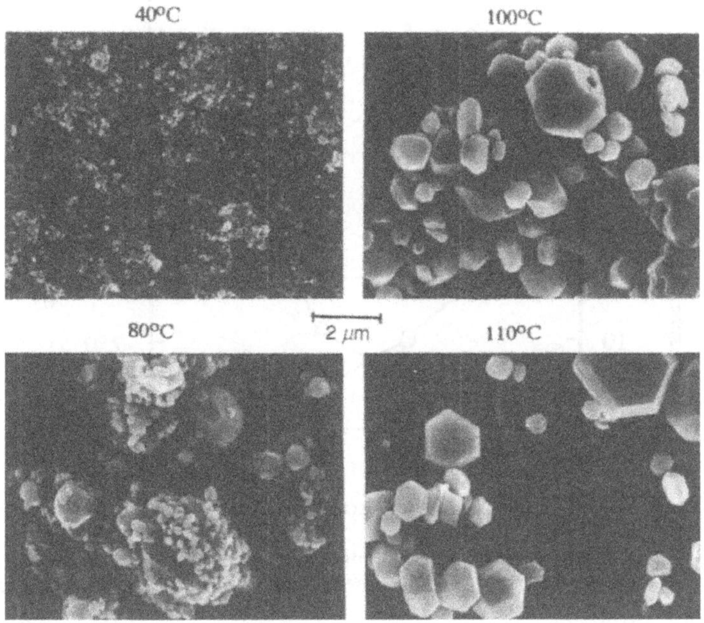

Figure 5. SEM pictures of the materials formed by ageing $Fe(OH)_2$ at different temperatures.

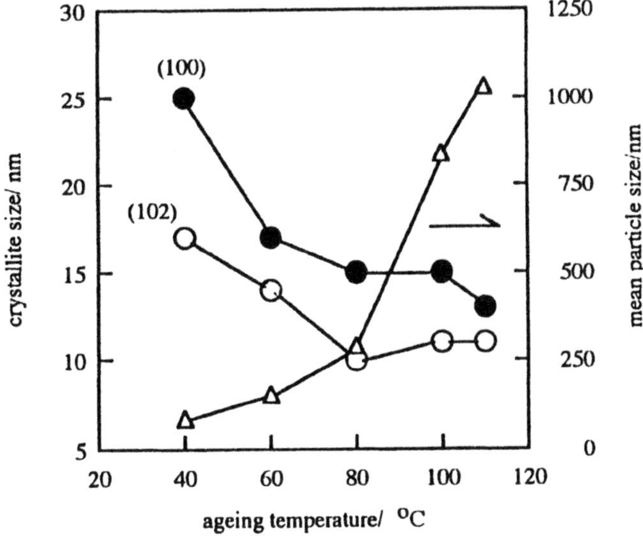

Figure 6. Mean particle and crystallite size of the materials formed by ageing $Fe(OH)_2$ at different temperatures.

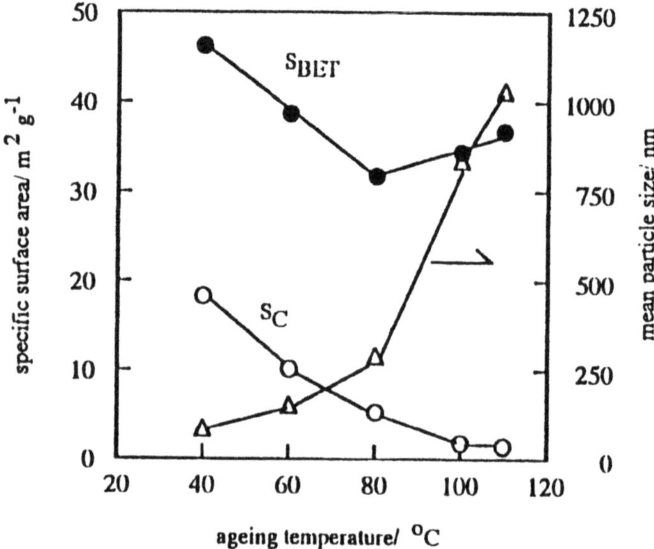

Figure 7. Specific surface areas and mean particle sizes of the materials obtained by ageing $Fe(OH)_2$ at different temperatures. The outgassing temperature before N_2 adsorption was 50°C.

The morphology of $Fe(OH)_2$ is considered to be not changed with these reactions. It seems reasonable therefore that the teradecahedron shape of δ-FeOOH particles originates from the shape of the precursor $Fe(OH)_2$ particles. The growth of $Fe(OH)_2$ particles took place more rapidly at a higher t_a, resulting in the formation of larger δ-FeOOH particles. The XRD patterns of the materials formed at different t_a showed the diffraction peaks characteristic of δ-FeOOH, though being less crystallized. The crystallinity was slightly lowered by elevating t_a. The crystallite sizes of the δ-FeOOH particles, evaluated by the Scherrer equation using the diffraction peaks due to (100) and (102) planes, are plotted as a function of t_a respectively by the filled and open circles in Figure 6. Note that the crystallite sizes are extremely less than the particle size shown by the triangular marks and decrease with the rise of t_a. This means that the δ-FeOOH particles are polycrystalline and become less crystallized with the particle growth. This fact could be explained by the assumption that the inner texture of the precursor $Fe(OH)_2$ particles is changed by the violent oxidation with H_2O_2 without varying the particle morphology, being supported by the XRD result that the precursor $Fe(OH)_2$ is more crystallized as compared with the formed δ-FeOOH. However the exact structure of $Fe(OH)_2$ particles before the oxidation remains unclear for the moment because of a high reactivity to O_2 of this material. Figure 7 shows the change of the BET specific surface area (called S_{BET} below) of the δ-FeOOH particles with t_a by the filled circles. The S_{BET} value was minimum at 80°C of t_a. The open circles in this figure show the specific surface area (noted S_C below) estimated from the mean particle size by approximating the tetradecahedron particles to the spherical ones using the density of δ-FeOOH (3.9 g cm^{-3}). As can be seen in Figure 7, S_{BET} was considerably larger than S_C for all the samples, especially for the powders formed at a high t_a. This

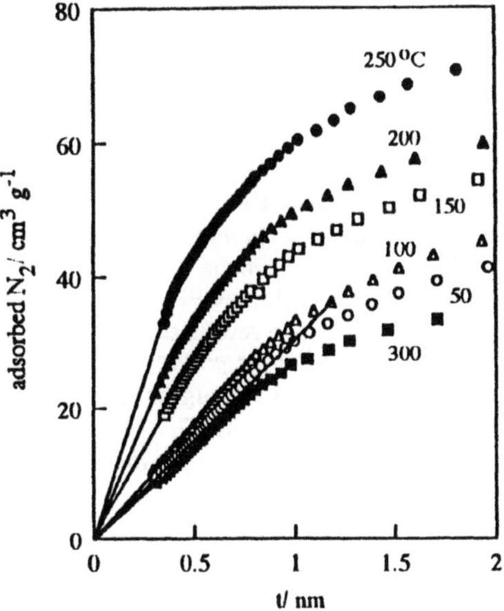

Figure 8. t-plots of N_2 adsorption on the materials formed by outgassing δ-FeOOH at various temperatures. The δ-FeOOH sample was prepared by ageing $Fe(OH)_2$ at 40°C.

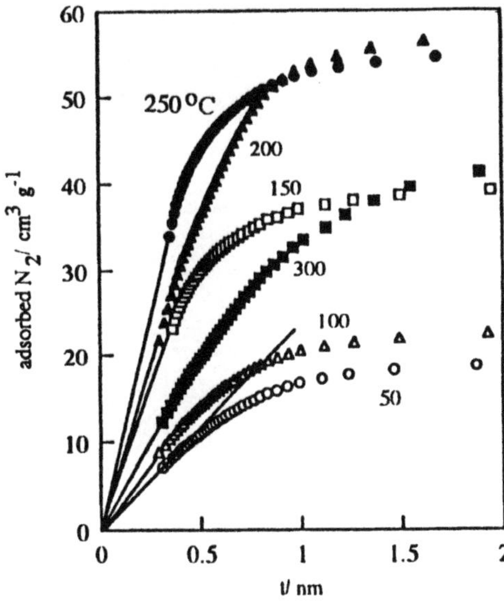

Figure 9. t-plots of N_2 adsorption on the materials formed by outgassing δ-FeOOH at various temperatures. The δ-FeOOH sample was prepared by ageing $Fe(OH)_2$ at 100°C.

finding clearly indicates that these δ-FeOOH particles are porous and the porosity is developed by rising t_a.

To characterize the porous structure of these materials the t-plots of N_2 adsorption isotherms on the samples outgassed at different temperatures 50 and 100°C were obtained and are shown in Figures 8 and 9. The samples outgassed below 100°C were confirmed to be δ-FeOOH by XRD and FTIR. The t-plots for the small δ-FeOOH particles formed at a low t_a of 40°C (Figure 8) indicate that the condensation of N_2 takes place in the voids among the particles. Therefore the pore structure of these small particles can not be discussed from the t-plot. On the other hand the t-plots for the large δ-FeOOH particles obtained at t_a of 100°C (Figure 9) give rise to the existence of the micropores with the width *ca.*1 - 2 nm. These results seem to associate with the fact that the particles formed at a higher t_a are more polycrystalline. Indeed a developed microporosity of the larger particles obtained at a higher t_a is apparent from the larger discrepancy between S_{BET} and S_C as shown in Figure 7. The conclusion may be drawn from the above mentioned results that the micropores of the tetradecahedron δ-FeOOH particles exist between the small crystallites.

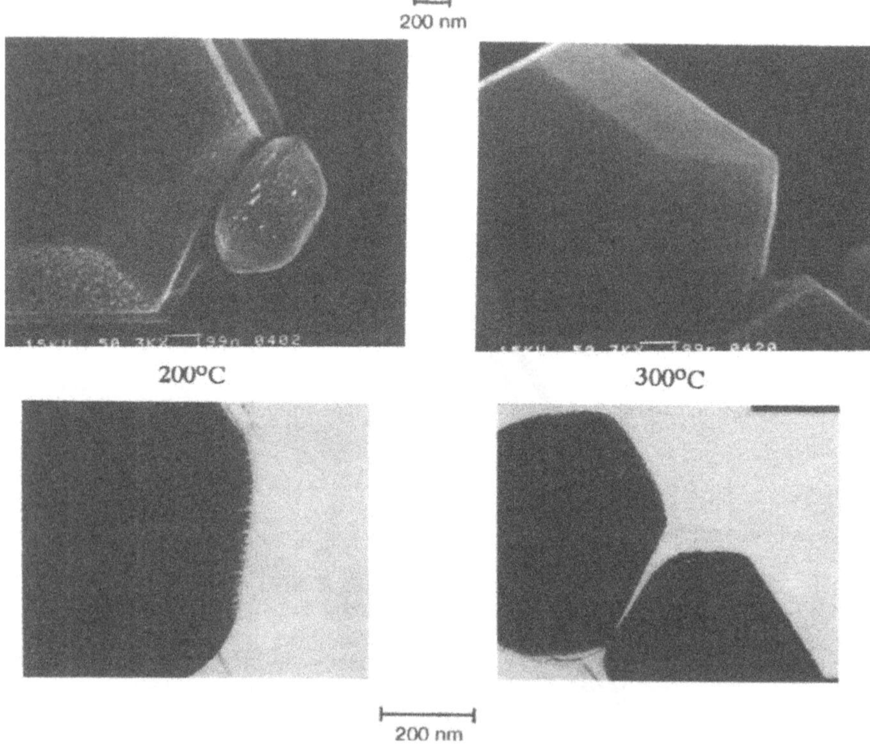

Figure 10. SEM (upper) and TEM (lower) pictures of the particles obtained by outgassing δ-FeOOH at 200 and 300°C. The δ-FeOOH sample was prepared by ageing $Fe(OH)_2$ at 100°C.

3.3. STRUCTURES OF α-Fe_2O_3 PARTICLES PREPARED FROM LARGE BULKY δ-FeOOH PARTICLES

The XRD results showed that the thermal phase transformation of δ-FeOOH takes place as follows; δ-FeOOH transforms into α-FeOOH at the temperature range 150 - 200°C and α-FeOOH turns into α-Fe_2O_3 above 200°C. Figure 10 displays the SEM and TEM pictures of the particles formed at t_a of 100°C and outgassed at 200 and 300°C. From the SEM pictures in the upper side of Figure 10 the morphology of the particles did not essentially vary with the treatment besides the particles were slightly sintered in the case of the sample outgassed above 300°C. On the other hand the TEM pictures in the lower side of Figure 10 indicate that the particles outgassed at 200°C have pores near the particle surface while the pores diminished to some extent when outgassed at 300°C. As mentioned in section 3.1., the small plate-like α-Fe_2O_3 particles have uniform slit-shaped micropores. However, such uniform pores were hardly detected in the tetradecahedron α-Fe_2O_3 particles because of their low transmittance for electron beam.

For understanding of the detailed pore structure of the α-Fe_2O_3 particles obtained from tetradecahedron δ-FeOOH particle, were determined the adsorption isotherms of N_2 on the materials formed by outgassing the δ-FeOOH particles at different temperatures 150 - 300°C. The S_{BET} values are plotted against the outgassing temperature in Figure 11, showing a maximum at 200°C for all the δ-FeOOH samples. The increase of S_{BET} below 200°C is caused by the pore formation with the dehydration of δ-FeOOH to

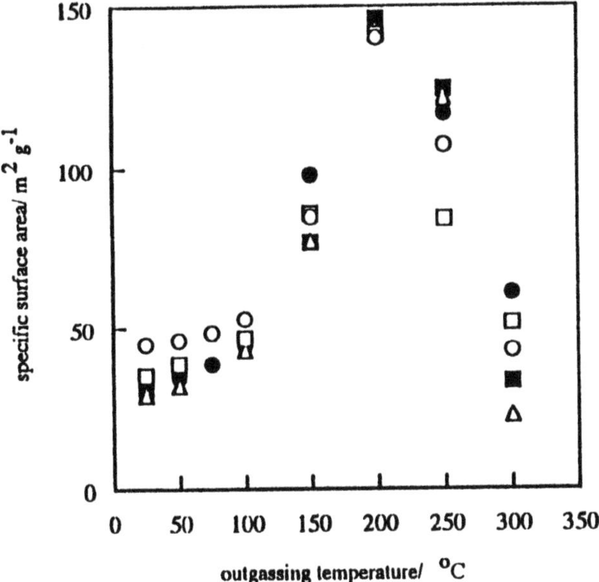

Figure 11. BET N_2 specific surface area of the materials obtained by outgassing the δ-FeOOH particles of different sizes at various temperatures. The δ-FeOOH sample was prepared by ageing $Fe(OH)_2$ at different temperatures. ○, 40; ●, 60; △, 80; ▲, 100; □, 110°C.

α-Fe_2O_3 and the decrease of S_{BET} above 200°C is due to the pore closure by sintering. It is noteworthy that the maximum surface area of all the δ-FeOOH particles with different sizes amounts almost constant value from140 to 146 m^2 g^{-1}. It has been reported that the α-Fe_2O_3 particles obtained by the thermal decomposition of α-FeOOH also show a maximum S_{BET} of *ca.* 140 m^2 g^{-1} at the similar outgassing temperature [14]. This well accordance is reasonable because δ-FeOOH transforms into α-Fe_2O_3 *via* α-FeOOH as mentioned above. These facts may suggest that the α-Fe_2O_3 particles formed by outgassing δ-FeOOH particles at an elevating temperature are the agglomerates of primary plate particles with a constant thickness. The thickness of the primary particles was estimated from the S_{BET} of 140 m^2 g^{-1} and the density of α-Fe_2O_3 (5.21 g cm^{-3}) by assuming that the pore surface is made up of the basal face of the primary plate particles. The thickness thus obtained was 2.7 nm. As described section 3. 1., the small plate-like α-Fe_2O_3 particles also give rise to the maximum S_{BET} of 140 m^2 g^{-1} at an outgassing temperature of 200°C (Figure 2). Therefore, these plate-like particles are also considered to be the agglomerate of the primary plate α-Fe_2O_3 particles. Actually the mean thickness of the primary plate α-Fe_2O_3 particles estimated from the TEM picture shown in Figure 4 amounts to 3 nm close to the calculated value of 2.7 nm noted above.

Further information as to the pore structure of the tetradecahedron α-Fe_2O_3 particles can be obtained from the t-plot analysis of N_2 adsorption isotherm. Figures 8 and 9 show the t-plots for the α-Fe_2O_3 particles formed by outgassing two different δ-FeOOH samples above 200°C. The α-Fe_2O_3 particles obtained from the smaller δ-FeOOH particles synthesized at t_a of 40°C have micropores with the size more than *ca.* 0.8 nm, since the t-plots curve at t >*ca.* 0.4 nm (Figure 8). On the other hand, the α-Fe_2O_3 particles obtained from the larger δ-FeOOH particles synthesized at t_a of 100°C have micropores with the size 0.8 - 2 nm because the t-plots curve at t from 0.4 to 1 nm (Figure 9). The pore structure of the tetradecahedron α-Fe_2O_3 particles is different from that of the small plate-like ones having uniform slit-shaped micropores with a 0.8 nm width shown in Figure 4; the large tetradecahedron particles show a wider pore size distribution as compared to the plate-like ones. This difference may be because the tetradecahedron particles are polycrystalline while the plate-like particles are close to monocrystalline particles. Accordingly the dehydration in the crystallites would give the uniform slit-shaped micropores. In the case of the polycrystallized particles the pores with a wide size distribution form among the crystallites besides the uniform slit-shaped micropores in the crystallites. Figure 12 schematically demonstrates the

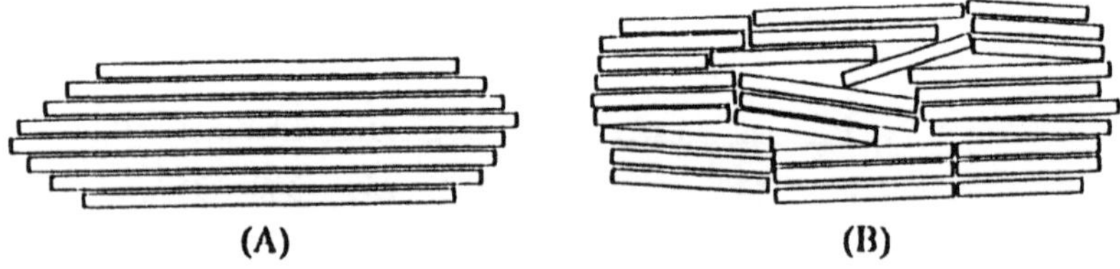

Figure 12. A model of the aggregation of plate-like primary particles with a thickness of 3 nm. (A) and (B) are the microporous α-Fe_2O_3 particles obtained from the small plate-like δ-FeOOH particles and large tetradecahedron ones, respectively.

difference between the structures of α-Fe_2O_3 particles produced from the small plate-like δ-FeOOH particles and large teradecahedron ones. Consequently, it seems reasonable to consider that the microporous α-Fe_2O_3 particles prepared in this study are the agglomelates of the plate-like primary α-Fe_2O_3 particles with a 3 nm thickness and the micropores are voids among the plate-like primary particles. The adsorption of various molecules onto these microporous α-Fe_2O_3 particles was reported in our paper [15].

The author acknowledge Dr. Kazuhiko Kandori, Mrs. Akemi Yasukawa, Mr. Ryuji Orii and Mrs. Wei Yan Cai for their collaborations. This study was supported in part by Nippon Sheet Glass Foundation for Materials Science and Technology and the Science Research Fund of Ministry of Education, Japanese Government.

References

1. Okamoto, S. (1964) Magnetic properties of δ-ferric oxide hydroxides substituted by metal ions, *Kogyo Kagaku Zasshi* **67**, 1855-1860.
2. Okamoto, S. (1968) Structure of δ-FeOOH, *J. Am. Cream. Soc.* **51**, 594-599.
3. Francombe, M. H. and Rooksby, H. P. (1959) Structure transformations effected by the dehydration of diaspore, goethite and delta ferric oxide, *Clay Miner. Bull.* **4**, 1-14.
4. Okamoto, S. (1964) Formation and decomposition of δ-ferric oxide hydroxides, *Kogyo Kagaku Zasshi* **67**, 1850-1855.
5. Loseva, G. V. and Murashko, N. V. (1971) Metastable δ-Fe_2O_3, *Inorg. Mater.* **7**, 1306-1307.
6. Povitskii, V. A., Makarov, E. F., Murashko, N. V. and Salugin, A. N. (1976) Mossbauer study of superparamagnetic δ-FeOOH and its transformation into α-Fe_2O_3, *Phys. Stat. Sol.* (a) **33**, 783-787.
7. Bernal, J. D. , Dasgupta, D. R. and Mackay, A. C. (1959) The oxides and hydroxides of iron and their structural inter-relationships, *Clay Miner. Bull.* **4**, 15-30.
8. Feitknecht, W. (1959) Oxidation of solid hydroxides of iron in aqueous solution, *Z. Electrochem.* **63**, 34-43.
9. Ishikawa, T., Cai, W. Y. and Kandori, K. (1992) Characterization of the thermal decomposition products of δ-FeOOH by Fourier-transformation infrared spectroscopy and N_2 adsorption, *J. Chem. Soc. Faraday Trans.* **88**, 1173-1177.
10. Ishikawa, T., Yasukawa, A., Kandori, K. and Orii, R. (1994) Textures of tetradecahedron δ-FeOOH particles and their thermal decomposition products, *J. Chem. Soc. Faraday Trans.* **90**, 2567-2571.
11. White, W. B. and Roy, R. (1964) Infrared spectra-crystal structure correlations: II. Comparison of simple polymorphic minerals, *Amer. Min.* **49**, 1670-1687.
12. Jimenez-Mateos, J. M., Morales, J. and Tirado, J. L. (1988) Textural evolution of α-Fe_2O_3 obtained by thermal and mechanochemical decomposition of δ-FeOOH, *J. Colloid Interface Sci.* **122**, 507-513.
13. Hirokawa, S., Naito, T. and Yamaguchi, T. (1986) Effect of atmosphere on the geothite decomposition and pore structure of product particles, *J. Colloid Interface Sci.* **112**, 268-273.
14. Ishikawa, T. and Inouye, K. (1972) The structural transformation of ferric oxyhydroxides and their activity to sulfur dioxide, *Bull. Chem. Soc. Japan* **45**, 2350-2354.
15. Ishikawa, T., Cai, W. Y. and Kandori, K. (1993) Adsorption of molecules onto microporous hematite, *Langmuir* **9**, 1125-1128.

ZETA POTENTIAL OF SUBMICRON TITANIA PARTICLES IN MIXED SOLVENTS

M. KOSMULSKI

Institute of Catalysis and Physical Chemistry of Interfaces, Laboratory of Adsorption and Physical Chemistry of Interfaces, Polish Academy of Sciences, Pl.M.C.Skłodowskiej 3, 20031 Lublin 41, Poland

1. Abstract

Anatase particles show negative ζ potentials in mixtures of n-alcohols, DMSO, dioxane and acetone with small (up to 5%) amounts of water. The values of these ζ potentials are not correlated with the empirical parameters characterizing solvent polarity, acidity and basicity. The values of ζ potentials depend not only on the nature of the solvent but also on the impurities, e.g., traces of inorganic salts, present in the solid and liquid phase, whose nature and concentration is beyond the control.

Addition of alkali chlorides causes a reversal of the sign of ζ potential of anatase to positive in the studied mixed solvents. This phenomenon is little sensitive to the nature of the anion. The concentration of a 1-1 salt necessary to reverse the sign of ζ potential of anatase in a given solvent decreases in the series Li>Na>K>Rb>Cs. Thus, in terms of ability to reverse the sign of ζ potential of anatase, cesium is more efficient than the other alkali metal cations.

The concentration of cesium chloride necessary to reverse the sign of ζ potential of anatase c_{rev} increases with the solvent polarity and acidity, represented by the $E_T(30)$ solvent scale, and decreases with the empirical parameter β characterizing solvent basicity. The following linear correlation: $\ln c_{rev} = 0.509\, E_T(30) - 15.29\,\beta - 23.21$ was found. It follows from this equation that in water c_{rev} is very high. Indeed, the sign of ζ potential is not reversed in water over the experimentally available concentration range of CsCl (up to 0.1 mol dm^{-3}).

On the other hand, the above equation predicts $c_{rev} \approx 10^{-3}$mol dm^{-3}) in a completely inert solvent.

E. Pelizzetti (ed.), Fine Particles Science and Technology, 185–196.

2. Introduction

Dealing with fine solid particles, e.g., of oxides dispersed in liquids, one encounters a problem of their stability against coagulation. The stability of such dispersions depends on their ζ potentials[1]. Once the factors affecting the ζ potentials in a given system are known, one can find a way to stabilize or destabilize the system to various practical benefits. Therefore, in spite of experimental and theoretical difficulties, extensive studies of ζ potentials of oxides in various media have been carried out.

In aqueous systems, ζ potentials of oxides depend primarily on the pH value, positive potentials are observed at low, and negative at high pH values, respectively. The pH value at which the sign of ζ potential is reversed is called i.e.p. (isoelectric point). For simple systems containing a pure oxide and aqueous 1-1 inert electrolyte (alkali nitrate, perchlorate), the i.e.p. of a given oxide is independent of the nature of the electrolyte and its concentration so the i.e.p. may be considered as a constant characterizing the acidity of a given oxide.

On the other hand, presence of multivalent cations in aqueous systems usually causes a shift of the i.e.p. toward higher pH values. The shift increases with the concentration of the multivalent cation and depends on the nature of this cation. Multivalent anions shift the i.e.p. toward lower pH values. This phenomenon is termed specific adsorption. Adsorption of other compounds, e.g., ionic surfactants on oxides may be also specific, i.e., lead to a shift of i.e.p.

In this paper the terms *specific* and *inert* are used to distinguish between the ions whose adsorption leads to a shift in i.e.p. (or reversal of sign of ζ potential) and those whose increasing concentration causes a decrease of the absolute value of the ζ potential at a given pH, without a charge reversal. Specific and nonspecific adsorption can be also studied by means of other experimental methods, e.g., direct measurement of ion adsorption and its dependence on various factors. X-ray absorption studies of adsorbed ions enable an insight into the molecular mechanism of ion adsorption. Sometimes, it is not quite clear whether adsorption in a given system is specific and different criteria give contradictory answers [2].

Addition of inert electrolytes leads to a decrease of absolute value of ζ potential and thus to destabilization of the suspension. Using specifically adsorbing counterions, one can destabilize the suspension more effectively, i.e., at a lower electrolyte concentration. However, when the coagulation concentration is exceeded, the sign of ζ potential is reversed and this may lead to re-stabilization.

It is tempting to assume that an inert character of a given electrolyte in aqueous systems can be generalized to other solvent systems, specially water-organic mixtures and polar nonaqueous solvents whose properties are similar to those of water. In some studies, such an assumption is uncritically accepted [3]. In view of problems with pH measurements in nonaqueous or nearly nonaqueous media, it is difficult to verify this assumption directly, i.e., in terms of shift of i.e.p. with the electrolyte concentration. However, some results suggest that the classification of ions into specifically and nonspecifically adsorbing on a given oxide is solvent dependent and that in organic or mixed solvents, alkali metal cations adsorb specifically on oxides, i.e., their adsorption may lead to a reversal of sign of ζ potential.

When water is replaced with a mixed aqueous-organic solvent rich in water, e.g., 30% aqueous ethanol or methanol, the i.e.p. of oxides is shifted toward higher pH values at a sufficiently high concentration of 1-1 electrolytes, e.g., KCl. This means that at certain pH values, negative ζ potentials of oxides are reversed after addition of these salts [4]. The shift of i.e.p. may suggest that cations are specifically adsorbed. However, a directly measured cation adsorption on oxides at a given pH and ionic strength does not increase when water is replaced with a mixed solvent[5,6]. Another explanation of the observed shift of the i.e.p. may be offered, namely, in terms of a shift of the position of the shear plane.

In organic solvents, nonaqueous or containing little water, the reversal of ζ potentials of silica on addition of alkali halides is more pronounced and it occurs at lower salt concentration, as compared with systems rich in water. In view of the lack of standard procedures to measure the pH in such systems, a quantitative representation of the solvent effect in terms of the shift of i.e.p. can be hardly applied. Different systems may be compared in terms of the concentration of a 1-1 salt at which the ζ potential of given oxide is reversed.

Addition of alkali bromides to silica suspensions in 1-butanol leads to a reversal of the ζ potential from negative to positive[7]. This effect is more pronounced for cesium bromide and less for sodium bromide. Addition of alkali chlorides leads to a reversal of ζ potentials of silica in dioxane-water mixtures from negative to positive[8]. In mixtures containing more water, the concentration of cesium chloride required to reverse the ζ potential of silica is higher. The effect is less pronounced for alkali cations of lower molecular weight and little sensitive to the nature of the anions.

The mechanism of reversal of ζ potentials of silica in solutions of alkali chlorides in organic and mixed solvents containing is not clear. In order to explain this phenomenon, more experimental data are desired. It is specially interesting if the phenomenon is unique for silica or typical for oxides and how it is influenced by the nature of organic solvent (co-solvent). To this end, ζ potentials of anatase have been studied as a function of concentration of alkali chlorides in various organic solvents with small admixtures of water.

There are several reasons to study organic solvents with admixtures of water rather than "pure" organic compounds. First, it has been found that in certain systems, traces of water may seriously influence the ζ potentials in nearly anhydrous organic solvents [9]. It is very difficult to avoid moisture completely thus it is easier to get reproducible results controlling the concentration of water by addition of a certain amount, higher than that present in the reagents.

2.1. CHARACTERIZATION OF MIXED SOLVENTS.

A value of solvent dependent physicochemical property XYZ in a solvent j may be expressed as [10,11]:

$$XYZ=(XYZ)_o+aA_j+bB_j+cC_j+ \dots \qquad (1)$$

where $(XYZ)_0$ is XYZ value in the gas phase or in a completely inert solvent, uppercase symbols denote parameters characterizing the solvent j and lowercase symbols denote parameters dependent only on the property XYZ. Certainly, better correlations are obtained with more parameters on the r.h.s. of Eq. (1). On the other hand, for sake of clarity, the number of parameters on the rhs of Eq.(1) should be minimized; one-parameter approach is specially desired.

Physical constants characterizing the solvents like dielectric constant, refractive index or permanent dipole moment or functions thereof, used as parameters of eq.(1) often fail in explaining the solvent effects on equilibrium constants and rate constants of chemical reactions. Therefore, several experimental parameters characterizing solvents have been introduced. The solvent scales may be based on solvatochromic shifts (Reichardt's $E_T(30)$, Kamlet-Abboud-Taft π^*), NMR shifts (Gutman's AN acceptor number) and heats of solvation (Gutman's DN donor number) of certain probe molecule(s) in a series of solvents. Much more examples can be found in the literature [10,11]. Each approach is confined to certain solvents only. Insolubility of probe molecules in some solvents on one hand or too strong solvent-solute interaction inducing a chemical reaction on the other are typical limiting factors.

It is generally agreed that these experimental parameters express combinations of intrinsic solvent properties and specific interactions between the solvent and the probe molecule. Therefore, experimental solvent scales are interrelated. Such correlations make it possible to estimate a value of certain solvent parameter even in a solvent in which this parameter cannot be measured directly because of the mentioned above limitations. A detailed discussion of these correlations can be found in the literature[11].

Dealing with mixed solvents, one must note, that the solvent composition around titania particles (or rather their hydroxyl groups) and around the chromophore in the probe molecule in measurements of experimental solvent parameters is not necessarily the same. When hydroxyl groups of anatase preferentially adsorb one component of binary mixed solvent and, e.g., the solvatochromic dye adsorbs the other component, application of data obtained in a mixed solvent may be misleading. Therefore, in spite of availability of some data (especially $E_T(30)$) for solvent mixtures, the values characterizing pure solvents are preferred in the present study.

3. Experimental

Anatase from Aldrich was washed in order to remove soluble impurities according to the procedure described in [4]. Cesium perchlorate was precipitated from pre-filtered 1 mol dm^{-3} solutions of cesium chloride and sodium perchlorate at 50 °C, the mixture was cooled down to 5 °C and the precipitate was washed with ice-cold (0 °C) water and vacuum dried at 80°C. Other chemicals from Aldrich, Merck and Ubichem were used without purification. The organic-water mixtures were prepared on weight basis. Concentrated solutions of inorganic salts (>1 mmol dm^{-3}) were prepared by dissolving a certain mass of salt in a given volume of mixed solvent while more dilute solutions were prepared by addition of a certain volume of a concentrated (1 or 0.1 mol dm^{-3}) aqueous solution.

The electrophoretic mobilities of anatase were measured by means of Laser Zeemeter 501 (Pen Kem) according to the procedure described in [4]. At least 0.25 dm^3 of suspension was used to wash and fill the cell. In media of $\varepsilon > 15$, the ζ potentials were calculated according to the formula from ref. [13], assuming a complete dissociation of the alkali chlorides and applicability of Walden rule over the studied concentration range. The former assumption seems to be reasonable in view of high dissociation degrees of $LiNO_3$ in pure organic solvents (except for acetone) reported in [14]. The procedure to convert mobilities into ζ potentials in dioxane-water mixtures was adapted from [8]. One has to emphasize here that a change of the ionic strength by less than an order of magnitude has only a small effect on the course of the ζ vs. mobility curves. Therefore, very precise determination of the degree of dissociation is not crucial in calculation of the ζ potentials from the electrophoretic mobilities.

4. Results and Discussion

In all solvent mixtures studied in this paper, anatase assumes a negative ζ potential. This result is not surprising in view of the results by Labib and Williams [15] who have measured ζ potentials of titania in a series or organic solvents. In 1,2-dichloroethane, the ζ potential of titania was found to be positive while in the other organic solvents it was negative and little sensitive to moisture.

4.1. THE EFFECT OF THE NATURE OF THE ORGANIC CO-SOLVENT ON ZETA POTENTIALS

Certainly, the ζ potential of anatase in a given organic solvents depends on the nature of the solvents as well as on the solutes whose nature and concentration is beyond the control. They may originate from the solvent but also from the solid. The effect of salinity on the ζ potentials of anatase in mixed solvents is presented in Table 1. A typical course of the variations of the ζ potential of anatase with [CsCl] is presented in Fig.1.

Reversal of ζ potentials of anatase in water was actually not observed. However, the i.e.p. of silica [12] is shifted toward higher pH values when the concentration of cesium chloride or rubidium chloride is high enough and similar effects may be expected for the other oxides. One must emphasize here that it is extremely difficult to determine an exact position of the i.e.p. at a high ionic strength.

Although systematic studies of the effect of water contents in mixed solvents on the ζ potentials of anatase were not carried out, some random measurements suggest that small variations of water contents (e.g., 1 *vs.* 2%) exert a small effect on the course of ζ ([CsCl]) curves for a given organic co-solvent. Thus, the numbers in Table 1 characterize not only a mixed solvent of certain composition but, generally, mixtures of a given organic co-solvent with small amounts of water.

The measured ζ potentials and the CsCl concentrations necessary to reverse the sign of the ζ potentials of anatase (c_{rev}) in particular solvents are not correlated with their ε values (r=0.75 for ln C_{rev} and ε). Even worse correlation coefficients are obtained for

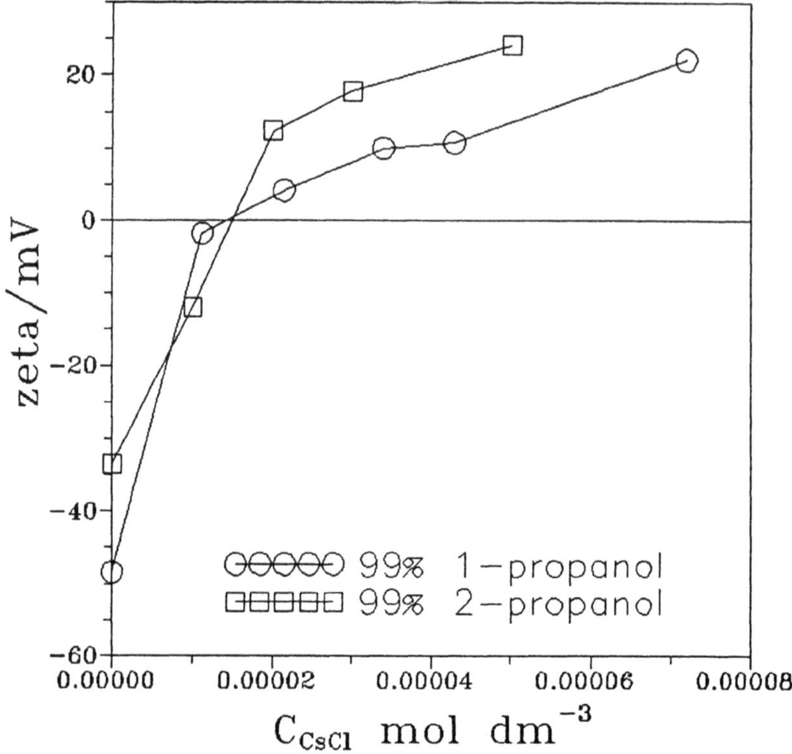

Figure 1. The ζ potentials of anatase in 99% 1-propanol and 2-propanol as the function of CsCl concentration.

the one-parameter Eq.(1) with XYZ equal to ln c_{rev} and A_j equal to one of the following parameters: hydrogen bond donation ability α, hydrogen bond acceptance ability β, polarity/polarizability parameter π^*, $E_T(30)$, DN, AN, Acity, Basity, Kosower's Z and Z' (meaning of these solvent scales and their mutual relations are explained in detail in [10] and [11] and literature quoted therein). It is not surprising when one considers the difference between formation of double layer around titania particles and the processes upon which the above solvent scales are based. This result substantiates the superiority of ε over the above solvent scales in a one-parameter approach to the interpretation of electrokinetic data in various solvents.

However, some two-parameter variants of Eq.(1) with A_j as one and B_j as another of the above parameters, explain the dependence of c_{rev} on the nature of the solvent quite well. No matter what parameter is used as B_j, a in Eq.(1) is always positive when A_j equals to π^* or $E_T(30)$ (c_{rev} increases when π^* or $E_T(30)$ increases). On the other hand, a is usually negative (one exception) when $A_j=\beta$. For the other experimental solvent parameters used as A_j, a random mixture of positive and negative a values is obtained; the sign varies with the parameter used as B_j, thus a relevance of the other solvent scales in the studied phenomena is limited.

Table 1. The effect of cesium on the ζ potential of anatase.

solvent	β	$E_T(30)$	%	ζ of anatase in "pure" mixed solvent (no salt added) mV	[CsCl] at which the sign of ζ is reversed mol dm^{-3}
	pure organic co-solvent, 25°C				
water	0.47	63.1			very high
dioxane	0.37	36.0	95	-40	$2\cdot10^{-5}$
methanol	0.66	55.4	99	-18	$8\cdot10^{-3}$
ethanol	0.75	51.9	96	-11	$1\cdot10^{-4}$
1-propanol	0.90	50.7	99	-48	$2\cdot10^{-5}$
2-propanol	0.84	48.4	99	-34	$2\cdot10^{-5}$
DMSO	0.76	45.1	99	-20	very low ($<10^{-6}$)
acetone	0.43	42.2	99	-89	$4\cdot10^{-4}$ *

* perchlorate

In terms of maximum correlation coefficient, the best equation to describe the dependence of c_{rev} on the nature of the solvent (data from Table 1 excluding water and DMSO) is:

$$\ln c_{rev} = -72.14\,\alpha + 136.18\,Acity - 36.41 \tag{2}$$

In spite of a high correlation coefficient r=0.983, Eq. (2) has two disadvantages. First, c_{rev} in DMSO and water calculated from Eq.(2) are too high to have a physical sense while c_{rev} in DMSO is very low according to the experiment. In addition, the physical significance of Eq.(1) with $A_j=\alpha$ and $B_j=Acity$ is doubtful in view of a values of different sign obtained when α or *Acity* as A_j is combined with another parameter as B_j in Eq.(1).

Therefore, the second best fit (r=0.969), namely the equation:

$$\ln c_{rev} = 0.509\,E_T(30) - 15.29\,\beta - 23.21 \tag{3}$$

is much more promising. The measured c_{rev} values are plotted against those calculated from Eq.(3) in Fig.2. For water and DMSO (full symbols), only calculated values are available but they are in reasonable agreement with the experiment (DMSO) or prediction (water).

Also another equation (r=0.948):

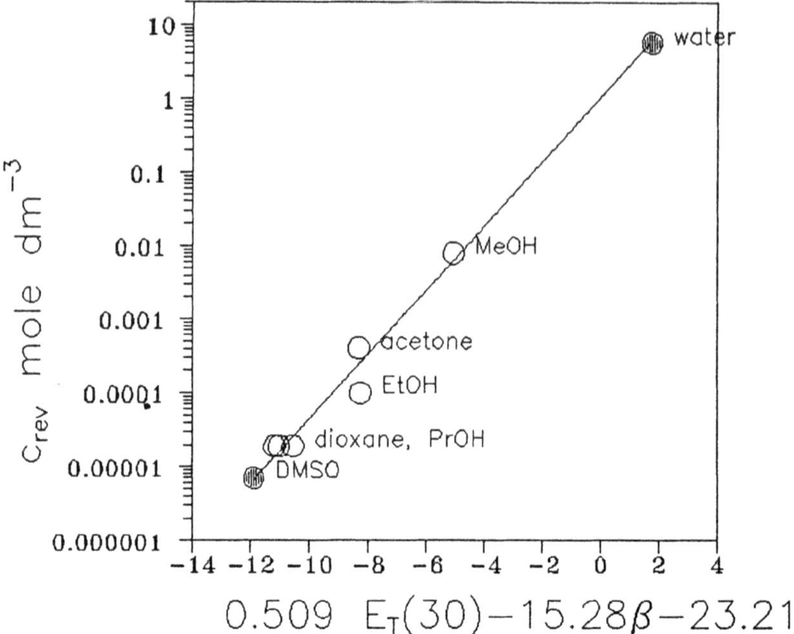

Fig.2. Dependence of the CsCl concentration at which the ζ potential of anatase is reversed on the nature of the organic co-solvent (Eq.(3)).

$$\ln c_{rev} = 0.617\ E_T(30) - 0.419\ DN - 26.93 \quad (4)$$

combines an explanation of the experimental points only slightly worse than Eq.(3) and reasonable prediction for DMSO and water. Here, again, c_{rev} in ethanol stands out of the theoretical line. The validity of Eqs.(3) and (4) is not surprising. It is known [10] that eq.(1) with $E_T(30)$ (representing a blend of solvent polarity and acidity) as A_j and a parameter representing the solvent basicity as B_j is very powerful in interpretation of many solvent dependent properties. Thus, in solvents of high $E_T(30)$ and low basicity (DN or β), CsCl acts as an inert electrolyte and reversal of ζ potential of anatase is possible only at a very high concentration of this salt. In contrast, when the basicity is high and $E_T(30)$ is low, the c_{rev} is low, i.e. the sign of ζ the potential can be easily reversed by addition of this salt. Eqs. (3) and (4) suggest that in a completely inert solvent ($E_T(30)=30$, $\beta=DN=0$), the ζ potential of titania would be reversed at $\approx 10^{-3}$ mol dm^{-3} CsCl. This result can hardly be proved in practice because of limited solubility of CsCl in inert solvents.

An attempt to interpret the values of ζ potential of anatase in seven "pure" mixed solvents (no salt added) listed in Table 1 in terms of one or two parameter versions of Eq.(1) has also been made with rather poor results. Namely all two-parameter combinations yield $r<0.9$ except for one equation ($r=0.936$):

$$\zeta = 52.2\ Z' - 48.0\ Z - 146.9 \qquad (5)$$

However, it does not have any physical sense since Kosower's Z and Z' parameters are known to be proportional, so Eq.(5) is just an accidental coincidence. Therefore, the ζ potentials in "pure" solvents are probably due to some impurities whose concentrations are beyond control.

4.2. ACETONE

As compared with the other solvents studied in this paper, acetone shows some exceptional features. First, potassium, rubidium and cesium chlorides are sparingly soluble in acetone and its mixtures with small amounts of water. Presence of these salts causes a decrease of negative ζ potentials of anatase dispersions in 96-99% acetone but reversal of the sign is not observed over the available concentration range. With lithium chloride, which is much better soluble, no reversal of ζ potentials is observed at concentrations up to 10^{-2} mol dm^{-3}. On the other hand, alkali iodides and perchlorates are soluble in acetone. Since in the previous [8] and present study no essential effect of the nature of the monovalent anions on the reversal of ζ potentials of silica and anatase was detected, cesium perchlorate was used in the studies in acetone. The sign of the ζ potential of anatase is also reversed in 10^{-3}mol dm^{-3} KI.

The negative ζ potential of anatase in acetone-water mixtures is high as compared with other organic co-solvent of similar properties (ε, $E_T(30)$, etc.). This may be due to some specific interaction but also to impurities.

Finally, $LiNO_3$ [13] and probably other inorganic salts, too, behave as relatively weak electrolytes in acetone while in other organic co-solvents of similar properties, they are completely dissociated over the concentration region of interest. Thus, the ionic strength necessary to reverse the sign of ζ potential of anatase in acetone (and dioxane, see [8]) is lower than the analytical concentration of cesium chloride given in Table 1.

4.3. THE EFFECT OF THE NATURE OF THE CATIONS ON ZETA POTENTIALS

In most studied cases (except for DMSO and water), the effect of alkali metal cations on the reversal of ζ potentials of anatase increases from Li to Cs. This means that the concentration of cesium salt required to reverse the ζ potential is lower than the concentration of lithium salt and at the same salt concentration, the ζ potentials increase (become more positive) from Li to Cs, as it is shown in Fig.3.

A similar trend was observed earlier for silica [7,8]. Since in aqueous systems, adsorption on silica at a given pH and ionic strength increases from Li to Cs, the phenomena in mixed solvents seemed to be due to a higher affinity of silica towards cesium. However, in the case of anatase, adsorption of alkali metal cations decreases from Li to Cs (see [16] and references therein). Thus, in spite of an opposite trend in alkali metal cations adsorption from aqueous solution on anatase and on silica, the

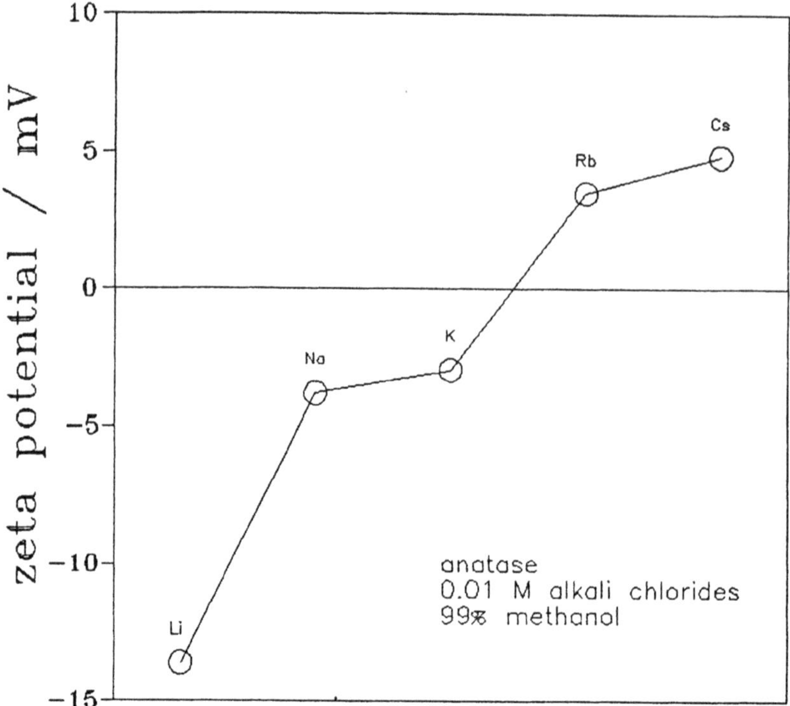

Figure 3. The effect of the nature of the alkali cations on the ζ potential of anatase in 0.01 mol dm^{-3} solutions of alkali chlorides in 99% methanol.

tendency observed for reversal of ζ potentials of these oxides in mixed solvents is identical. This suggests that the difference between particular alkali metal cations is due to the properties of the solution rather than to a special affinity of the oxides towards cesium.

Because of a large size of cesium ions, its salts have a higher degree of dissociation in mixed solvents and at the same analytical concentration they produce a considerably higher ionic strength than corresponding lithium salts.

5. Conclusion

Reversal of the sign of ζ potential from negative to positive in organic solvents with small admixture of water, as an effect of addition of alkali halides, originally found for silica, seems to be a common property of oxides. For silica and anatase, this phenomenon is little sensitive to the nature of the anion but it is much more pronounced for cesium than for lithium. Low polarity and acidity and high basicity of the organic solvent promote this reversal of the sign of ζ potential.

The proposed mechanism of this phenomenon is shown in Fig.4. Dry titania particles have some ions on their surface. Their concentration may be very low but not zero.

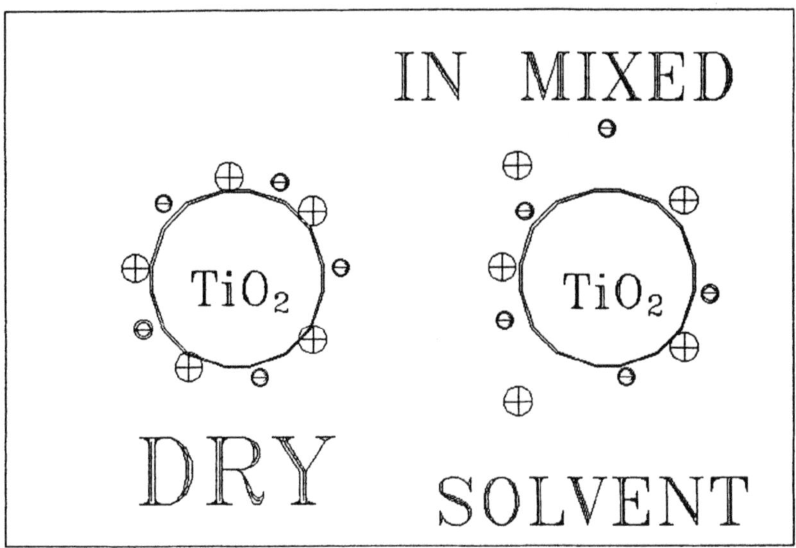

Figure 4. Ions around titania particles.

When such particles are put into a mixed solvent, some of the ions desorb, forming a space charge ("electric double layer") around the particle. Certainly, the tendency to desorb may be different for different ions. It seems that in mixed solvents typical cations present as impurities of titania (e.g. protons) desorb easier than the anions. Therefore, titania particles show negative ζ potentials and there is an excess of cations beyond the shear plane (Fig.4).

When cesium chloride is added into the solution, the trend may be reversed: large cations adsorb stronger than the anions so the net charge within the shear plane turns to positive. Such charge separation is more pronounced in apolar media; in polar solvents like water both anions and cations are strongly solvated and their tendency to adsorb is limited. High basicity of the solvent promotes cation adsorption on and anion desorption from oxides (similar trend in aqueous systems is well known). This explains the physical sense of Eqs. (3) and (4).

6. References

1. Van der Hoeven, Ph.C. and Lyklema, J. (1992) Electrostatic stabilization in nonaqueous media, *Adv. Colloid Interface Sci.* **42**, 205-277.
2. Kosmulski M. (1994) Co-adsorption of mono-and multivalent ions on silica and alumina, *Ber. Bunsenges. Phys. Chem.* **98**, 1062-1067.
3. Kitahara A. (1992) Stability of dispersions in polar media, *Adv. Colloid Interface Sci.* **38** 1-11.
4. Kosmulski, M. and Matijević, E. (1992) Zeta potential of anatase in mixed solvents, *Colloids and Surfaces* **64** 57-65.
5. Kosmulski, M. (1993) Adsorption of methanol and supporting electrolyte on silica and alumina in mixed solvent systems, *J. Colloid Interface Sci.* **156** 305-310.
6. Kosmulski, M. (1993) Oxide/electrolyte interface: Interpretation of supporting electrolyte adsorption data, *Polish Journal of Chemistry* **67** 1831-1839.
7. Zhukov, A.N., Levashova, L.G. and Evstratova, A.Yu (1981) Study of the effect of cations of alkali metals on the electrical and surface properties of quartz in 1-butanol, *Kolloidn. Zh.* **43** 240-245.
8. Kosmulski,M. and Matijević, E. (1991) Microelectrophoresis of silica in mixed solvents of low dielectric constant, *Langmuir* **7** 2066-2071.
9. Logtenberg, E.H.P. and Stein, H.N. (1986) Zeta potential and coagulation of ZnO in alcohols, *Colloids Surfaces* **17** 305-312.
10. Reichardt, Ch. (1994) Solvatochromic dyes as solvent polarity indicators, *Chem.Rev.* **94** 2319-2358.
11. Marcus, Y. (1993) The properties of organic liquids that are relevant to their use as solvating solvents, *Chem.Soc.Rev.* **1993** 409-416.
12. Kosmulski, M. and Matijević, E. (1992) Formation of the surface charge on silica in mixed solvents, *Colloid and Polymer Science* **270** 1046-1048.
13. Ohsawa, K., Murata, M. and Ohshima, H. (1986) Zeta potential and surface charge density of polystyrene latex; comparison with synaptic vesicle and brush border membrane vesicle, *Colloid Polymer Sci.* **264** 1005-1009.
14. De Rooy, N., De Bruyn, P.L. and Overbeek J.T.G. (1980) Stability of dispersions in polar organic media. I. Electrostatic stabilization, *J.Colloid Interface Sci.* **75** 542-554.
15. Labib, M.E. and Williams, R. (1984) The use of ζ-potential measurements in organic solvents to determine the donor-acceptor properties of solid surfaces, *J.Colloid Interface Sci.* **97** 356-366.
16. Sprycha, R. (1984) Surface charge and adsorption of background electrolyte ions at anatase/electrolyte interface, *J.Colloid Interface Sci.* **102** 173-185.

STRUCTURAL AND MAGNETIC PROPERTIES OF FINE γ-Fe_2O_3 PARTICLES

M.P. MORALES, M. OCAÑA, T. GONZALEZ-CARREÑO and
C. J. SERNA
Instituto de Ciencia de Materiales de Madrid, CSIC
Serrano 115 Dpdo. 28006 Madrid (SPAIN)

Abstract

The magnetic properties of γ-Fe_2O_3 uniform particles prepared by two different methods; from α-Fe_2O_3 precursors obtained by homogenous hydrolysis and directly, by spray pyrolysis of iron salt solutions, are analysed and related to their structural characteristics. In the first case, the γ-Fe_2O_3 particles, obtained by reduction-oxidation from the corresponding α-Fe_2O_3 precursors, showed different degrees of order in the vacancies distribution depending on the presence of defects inside the α-Fe_2O_3 particles. It is shown that both the saturation magnetisation and the coercive force decrease with increasing the disorder distribution of the vacancies in γ-Fe_2O_3.
The spray pyrolysis method can produce directly γ-Fe_2O_3 particles with nanometer size and different states of aggregation depending on the used precursor salt. These particles were found to be superparamagnetic with an unusually low blocking temperature. The influence of the size and the geometry of the aggregates is discussed in terms of ac magnetic susceptibility (χ_{ac}).

1. Introduction

In the last twenty years a large number of powders with uniform particle morphology and narrow size distribution have been prepared by different methods [1]. These studies have been carried out not only because of academic interest but also due to the important effect that particle size and shape have on the properties of many products of industrial importance such as ceramics, pigments, recording materials, catalysts, etc.
The great interest of γ-Fe_2O_3 particles in magnetic recording has brought the emergence of new preparation methods that allow to synthesise samples from aggregates to single uniform particles of different morphologies. Thus, spherical and ellipsoidal γ-Fe_2O_3 particles in the micrometer size range can be prepared by reduction-oxidation [2] of uniform α-Fe_2O_3 particles previously obtained by

E. Pelizzetti (ed.), Fine Particles Science and Technology, 197–208.

homogeneous hydrolysis in solution [3-4]. Spherical γ-Fe_2O_3 particles of nanometer size can be also directly obtained by spray pyrolysis of iron salt solutions [5].
The aim of the present work will be to correlate the crystallochemical characteristics of the γ-Fe_2O_3 particles obtained by the above mentioned methods to their magnetic properties. Since the crystallochemical characteristics of the γ-Fe_2O_3 particles were found to depend on the formation mechanism, this question will be analysed in detail.

2. Preparation of α-Fe_2O_3 Precursors

Uniform spherical particles of α-Fe_2O_3 (Fig. 1A) were obtained by hydrolyzing at 100°C a 0.02 M ferric chloride solution [6], and ellipsoidal particles with an axial ratio of 6 (Fig. 1B), by using the same procedure but adding 4.5×10^{-3} M KH_2PO_4 to the above solution [6]. The formation time of the α-Fe_2O_3 particles increases with the particles axial ratio, being two days for spheres and six days for the ellipsoids. It should be noticed the outstanding effect that a small amount of phosphates (Fe/Phosphate atomic ratio ≈ 40) have on the α-Fe_2O_3 particles morphology.

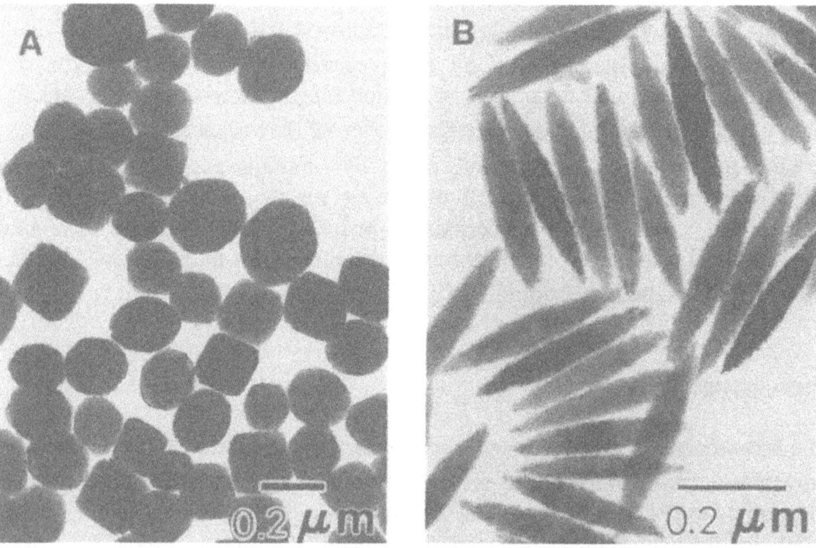

Figure 1. Transmission electron micrographs of the α-Fe_2O_3 samples prepared in the absence (A) and in the presence (B) of phosphate anions

A careful study of the crystalline nature of the particles indicates that, in both cases, they are monocrystalline as it has been shown by electron diffraction patterns carried out on single particles [6]. However, thermogravimetric analyses up to 1000°C (Fig. 2) show a weight loss of about 5% for the ellipsoidal particles due to the release of water molecules and hydroxyl groups, which is not present in the case of spherical particles. In order to elucidate the origin of these water molecules and hydroxyl groups, the mechanism of formation of both, spherical and ellipsoidal α-Fe_2O_3 particles, was studied in detail.

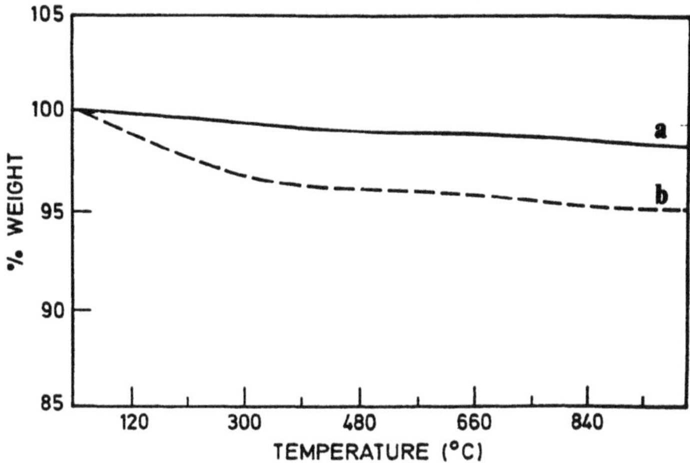

Figure 2. Thermogravimetric analyses of α-Fe_2O_3 samples: a) spheres and b) ellipsoids

3. Mechanism of Formation of α-Fe_2O_3 Particles

The mechanism of particles formation was studied by analyzing samples obtained at different aging times during the formation process by electron microscopy, X-ray diffractometry and infrared spectroscopy.

For the case of spheres (Fig. 3), a broad particle size distribution was observed during most of the aging period [6] indicating that several nucleation events took place. The final particle size uniformity could be achieved through a self-sharpening growth process where the small particles grow more rapidly than the larger ones [6,7].

A quite different mechanism was observed for the ellipsoidal particles; small anisometric particles of α-Fe_2O_3 (primary particles), of about 300 Å were initially formed, which later aggregated in an ordered manner, along their c axes, to give the final ellipsoidal α-Fe_2O_3 particles [6-8]. A schematic diagram of this mechanism can be seen in Fig. 4, which includes the role of the phosphates. According to this mechanism after hematite nucleation, phosphate groups adsorb on the surface plains parallel to the c axis of the nuclei, resulting in the growth of anisometric primary particles, which then aggregate with phosphate desorption producing the final ellipsoidal particles [7-8]. The above results along with some previous observations carried out in other systems [1,9], clearly demonstrate that La Mer approach for the particles formation in solution is only applicable to a limited number of cases whereas a continuous nucleation/aggregation process seems to be a rather common mechanism [7].

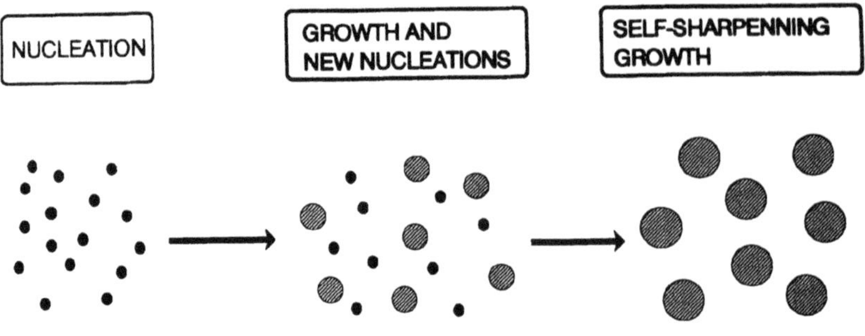

Fig. 3: Steps in the formation of spherical α-Fe_2O_3 particles

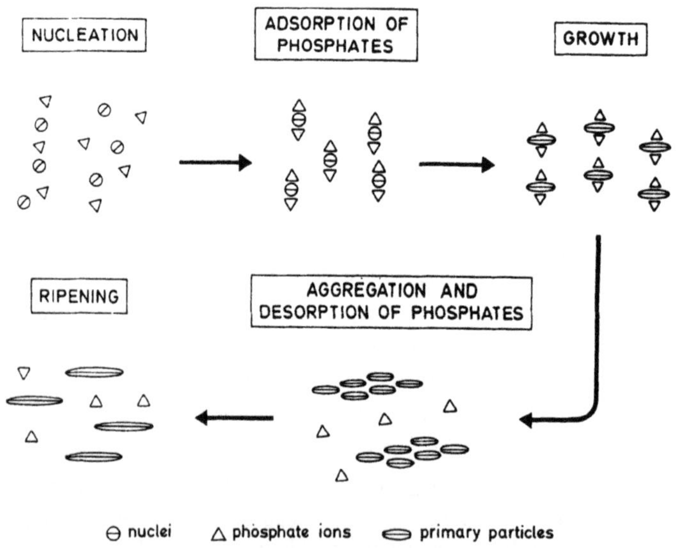

Fig. 4: Steps in the formation of ellipsoidal α-Fe_2O_3 particles

Finally, we must emphasize that the presence of defects inside the α-Fe_2O_3 ellipsoids, as the hydroxyl groups and water molecules associated to the weigh lost detected by thermogravimetric analyses (Fig. 2), is due to the aggregation mechanism operating in their formation. As it will be shown later, the structure of vacancies observed in γ-Fe_2O_3 is a consequence of the defects present in the α-Fe_2O_3 precursors.

4. Preparation of γ-Fe_2O_3 Particles

Uniform γ-Fe_2O_3 microcrystalline particles (> 100 nm) with axial ratios of 1 (sample a) and 6 (sample b) were prepared by reduction-oxidation from the corresponding α-Fe_2O_3 samples (Fig. 1) during which, particle morphology remained unaltered [2]. On the other hand, nanometer size γ-Fe_2O_3 particles (Fig. 5A and B) were obtained directly by spray pyrolysis of iron salt solutions [5]. When using iron (III) nitrate as precursor (sample c), spherical aggregates of small crystallites were obtained (Fig. 5a), whereas hollow spheres (Fig. 5b) resulted when using iron (II) ammonium citrate (sample d). The morphological characteristics of all considered samples are summarized in Table I.

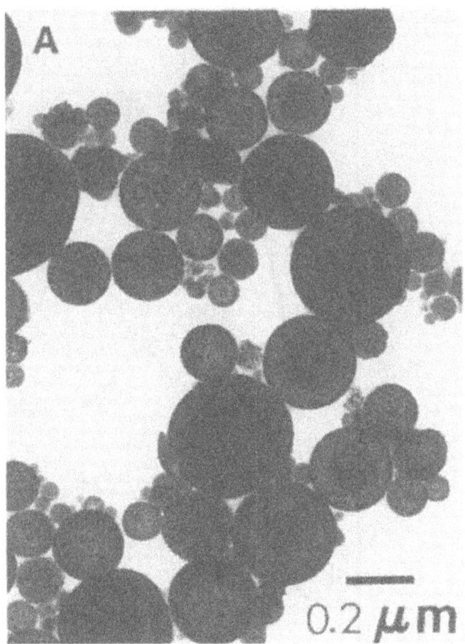

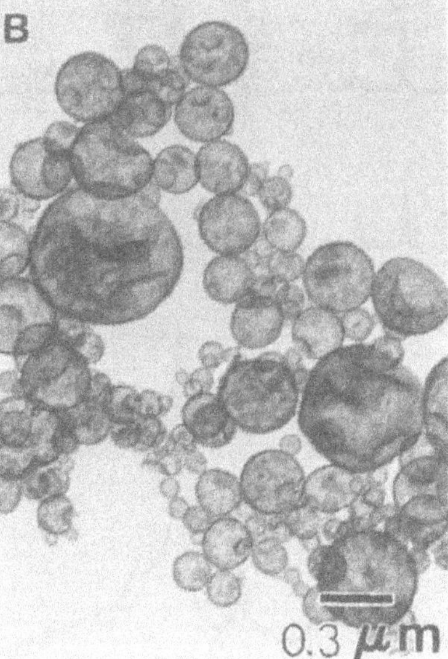

Figure 5. Transmission electron micrograph of γ-Fe_2O_3 particles obtained by spray pyrolysis of solutions of: A) iron (III) nitrate and B) iron (II) ammonium citrate

TABLE 1. Morphological, microstructural and magnetic characteristics of the γ-Fe_2O_3 particles

Sample	Preparation method	Morphology (size in nm)	Crystallinity	Magnetic Properties
a	Reduction-oxidation of α-Fe_2O_3	Monodispersed spheres (150 ± 20)	Single crystals (ordered vacancies)	Ferrimagnetic
b	Reduction-oxidation of α-Fe_2O_3	Monodispersed ellipsoids (540 ± 70 x 90 ± 10)	Single crystals (disordered vacancies)	Ferrimagnetic
c	Spray Pyrolysis (500°C) of Fe (III) nitrate solutions	Compact spherical aggregates (40-500)	Disordered polycrystals (~ 6 nm)	Superparamagnetic
d	Spray Pyrolysis (500°C) of Fe (II) ammonium citrate solutions	Hollow spheres (20-300)	Disordered polycrystals (~ 4 nm)	Superparamagnetic

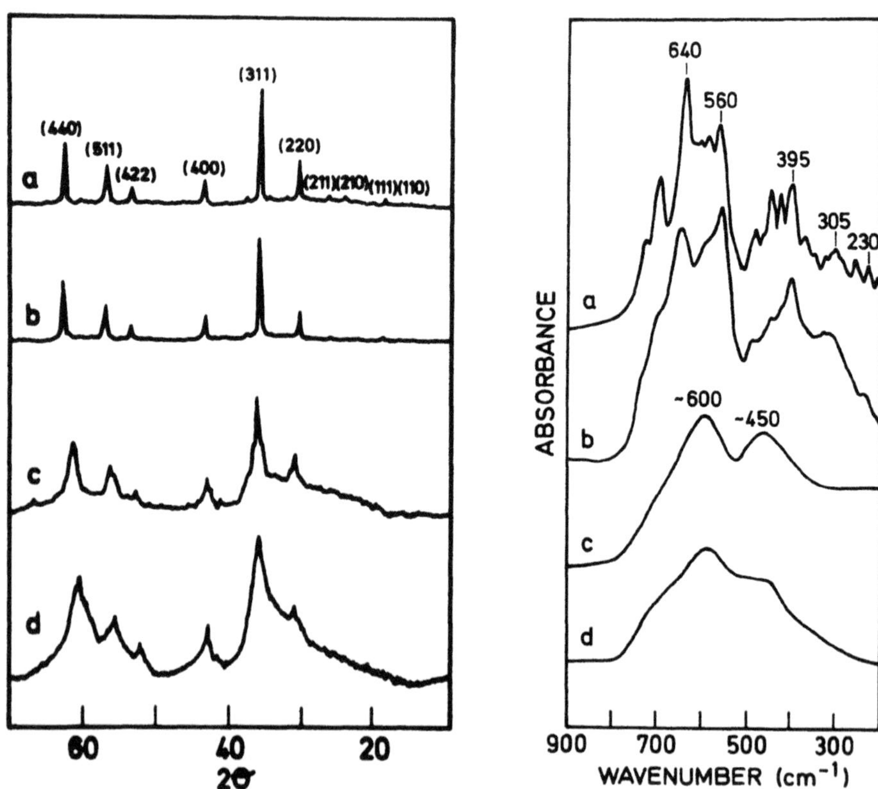

Figure 6. X-ray diffraction patterns (left) and infrared spectra (right) of the γ-Fe_2O_3 samples

5. Microstructural Characteristics of γ-Fe_2O_3 Particles

X-ray powder diffraction patterns of the γ-Fe_2O_3 samples are shown in Figure 6. The spherical particles give reflections (Fig. 6a) that must be indexed on the basis of a tetragonal lattice with a=8.33 Å and c/a=3 [2]. However, the ellipsoidal particles (Fig. 6b) only present the reflections of a disordered spinel structure (space group Fd3m). Therefore, the γ-Fe_2O_3 spheres present an ordered distribution of vacancies, whereas in the ellipsoids, these vacancies are randomly distributed. This lack of vacancies order in the γ-Fe_2O_3 ellipsoidal particles has been associated to the previously mentioned presence of defects in the α-Fe_2O_3 precursors [2]. The γ-Fe_2O_3 nanometer particles (samples c and d) also show the spinel reflections, which in this case are very broad as a consequence of their small particle size (Fig. 6 c and d). These structural differences are also manifested in the infrared powder spectra of the γ-Fe_2O_3 samples (Fig. 6 right). Thus, an increase in the number of absorption bands is observed in going from sample d to sample a, in agreement with the progressive lowering in particle symmetry (from Fd3m to $P4_32_12$) [2].
Lattice images and electron diffraction patterns obtained on isolated γ-Fe_2O_3 particles (Fig. 7) show that in microcrystalline samples (samples a and b), each particle behaves as a single crystal, with the long axis being coincident to the [111] direction for the case of the ellipsoids (Fig. 7 right). However, spot rings characteristics of polycrystalline particles were observed in the γ-Fe_2O_3 samples prepared directly by spray pyrolysis [5]. The mean size of the crystallites composing the particles in these sample, as determine by X-ray diffraction [5], was 6 nm for sample c and 4 nm for sample d (Table I).

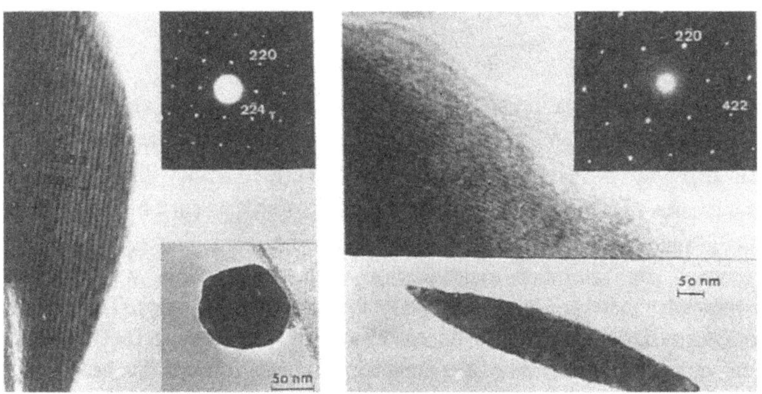

Figure 7. Lattice images and electron diffraction of samples a (left) and b (right)

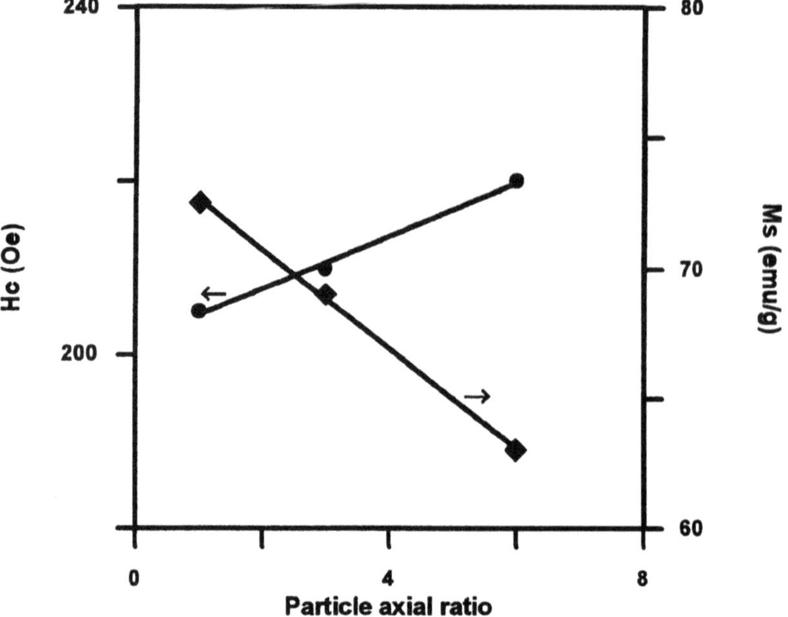

Figure 8. Coercive field (Hc) and saturation magnetization (Ms) of microcrystalline γ-Fe_2O_3 samples as a function of particle axial ratio. Particles with axial ratio of 3 were prepared according to reference [6]

6. Magnetic Properties

The saturation magnetization (Ms) and the coercive force (Hc) found for the microcrystalline samples prepared from α-Fe_2O_3 precursors is plotted versus the particles axial ratio in Figure 8. As we previously mentioned, the vacancy disorder in the γ-Fe_2O_3 particles increases with the density of defects in the α-Fe_2O_3 precursors, i.e. with increasing particle axial ratio, as a consequence of their formation mechanism [2]. As it can be seen from Figure 8, the saturation magnetization decreases regularly with the increase of structural disorder in the particles (increase of the particles axial ratio). This decrease of the saturation magnetization value can be correlated to some canting (not restricted to the surface region) in the directions of the magnetic moments induced through the occurrence of disorder in the distribution of defects. These results are important since they can explain the discrepancies between the saturation magnetization values of the γ-Fe_2O_3 particles published in the literature [10-11].
The expected increase of the coercive force with the particle axial ratio (Fig. 8) appears lower than what it would corresponds when only shape effects are considered to be responsible for the value of that magnitude [10]. This behavior suggests an influence of the

degree of order in the demagnetization processes, as it has been confirmed for samples with the same morphology and different degree of vacancies ordering [11].

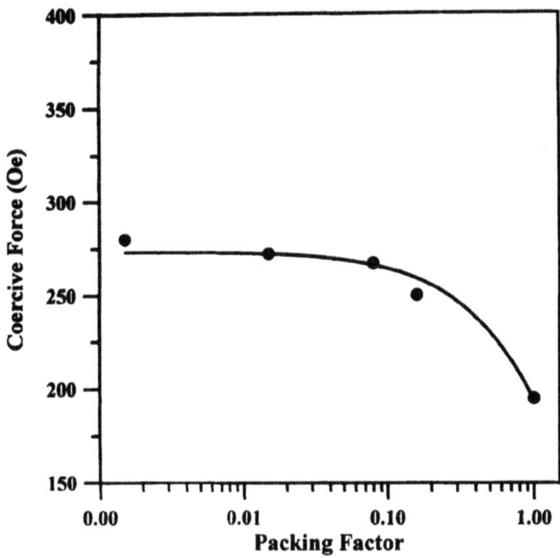

Figure 9. Coercive force (Hc) of sample b dispersed in a silica matrix as a function of the packing factor

It is well known that the value of the coercive force is also influenced by particles interactions since a particle assembly cannot behaves like a collection of isolated particles. To determine the effects of particle dispersion in the monocrystalline samples on Hc, γ-Fe_2O_3 ellipsoidal particles (sample b) were dispersed in a diamagnetic matrix of silica with different degrees of dispersion, defined as a packing factor (γ-Fe_2O_3/SiO_2 volume ratio). The different packing factors were obtained by coating the γ-Fe_2O_3 ellipsoidal particles with silica layers of different thickness using TEOS as silica precursor [12] (packing factors at 0.16 and 0.08) and by dispersing these particles in a Na_2SiO_3 solution which was further gelified at pH=5 (packing factors 0.0145 and 0.0015). An increase in Hc was observed for decreasing packing factors (increasing dispersion) (Fig. 9) indicating a significant reduction in the particles interaction. This is as expected since the magnetic interaction force varies as $1/r^3$ with the distance between particles, so this force decreases quickly as the separation increases.

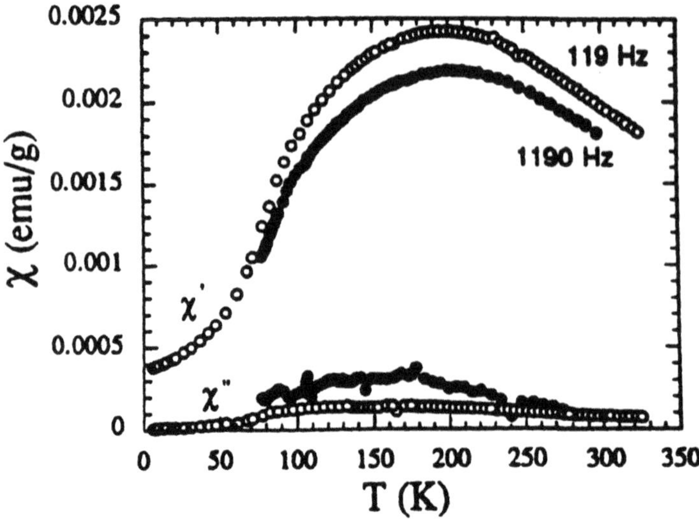

Figure 10. Magnetic susceptibility (χ_{ac}) of sample c as a function of temperature measured at two different frequencies

The magnetic behavior of polycrystalline γ-Fe_2O_3 nanoparticles (samples c and d) is less understood than in the case of isolated particles [13]. The influence of the size and the geometry of the aggregates was analyzed in terms of ac susceptibility, χ_{ac} (T, ω). This magnitude has been measured at 119 and 1190 Hz by an induction method and with an exciting ac field of 10 Oe [14]. For sample c (spherical aggregates) (Fig. 10), the temperature dependence of χ' present a broad maximum that shifts to higher temperatures as frequency increases. In addition, there is a maximum in χ''(T) at slightly lower temperatures. This behavior, also observed in the sample composed by hollow particles (sample d), is characteristic of a transition from the high temperature superparamagnetic state to the low temperature blocked state. Mössbauer spectra of this sample at room temperature and at 18 K confirm the magnetic transition (Fig 11). At room temperature the sample presents a doublet whose Mössbauer parameters are characteristic of Fe^{3+} oxides of small particle size [5]. whereas at 18 K the spectrum is compatible with ferrimagnetic γ-Fe_2O_3 particles. The collapse of the magnetic hyperfine splitting at room temperature due superparamagnetic relaxation has been found in fine particles of γ-Fe_2O_3 with different morphologies [13].

Acknowledgment

This work has been supported by the Spanish CICYT under project MAT93-0789.

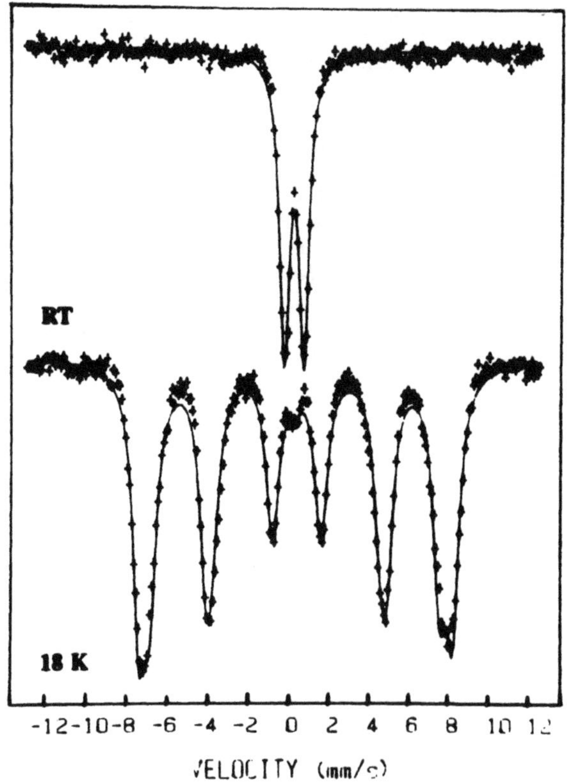

Figure 11. Mössbauer spectra of sample c taken at room temperature (RT) and at 18 K.

References

1. Matijevic, E. (1993) Preparation and properties of uniform size colloids, *Chem. Mater.* **5**, 412
2. Morales, M. P., Pecharromán, C., González-Carreño T. and Serna, C. J. (1994) Structural characteristics of uniform γ-Fe_2O_3 particles with different axial (length/width) ratios, *J. Solid State Chem.* **108**, 158-163
3. Matijevic, E. and Scheiner P. (1978) Ferric hydrous oxide sols. III. Preparation of uniform particles by hydrolysis of Fe (III)-chloride, -nitrate and -perchlorate solutions, *J. Colloid and Interface Sci.* **63**, 509-524
4. Ozaki, M, Kratohvil, S and Matijevic, E. (1984) Formation of monodispersed spindle type hematite particles, *J. Colloid and Interface Sci.* **102**, 146-151
5. González-Carreño, T., Morales, M. P., Gracia, M. and Serna, C. J. (1993) Preparation of uniform γ-Fe_2O_3 particles with nanometer size by spray pyrolysis, *Mater. Letters* **18**, 151-155
6. Morales, M. P., González-Carreño, T. and Serna, C. J. (1992) The formation of α-Fe_2O_3 monodispersed particles in solution, *J. Mater. Research* **7**, 2538-2545
7. Ocaña, M., Rodríguez-Clemente R. and Serna, C. J. (1995) Uniform colloidal particles in solution: Formation mechanism, *Advan. Mater.* **7**, 1006-1011
8. Ocaña, M., Morales M. P. and Serna, C. J. (1995) The growth mechanism of α-Fe_2O_3 ellipsoidal particles in solution, *J. Colloid Interface Sci.* **171**, 85-91

9. Andrés-Vergés, M., Mifsuf A. and Serna, C. J. (1990) Formation of rod-like zinc oxide microcrystals in homogeneous solutions, *J. Chem. Soc. Faraday Trans.* **86**, 959-963
10. Morales, M. P., de Julián, C., González J. M. and. Serna, C. J (1994) The effect of the distribution of vacancies on the magnetic properties of γ-Fe_2O_3 particles, *J. Mater. Res.* **8**, 135-141
11. Morales, M. P., de Julián, C., Serna, C. J and. González J. M. (1994) Magnetic viscosity of uniform γ-Fe_2O_3 particles with different degrees of cationic ordering, *IEEE Trans. Magn.* **30**, 772-774
12. Ohmori M. and Matijevic, E. (1992) Preparation and properties of uniform coated colloidal particles. VII. Silica on hematite, *J. Colloid and Interface Sci.* **150**, 594-598
13. Tronc, E., Prene, P., Jolivet, J. P., d'Orazio, F., Lucari, F., Fiorani, D., Godinho, M., Cherkaoni, R., Nogues, M. and Dormann J. L. (1995) Magnetic behaviour of nanoparticles by Mössbauer spectroscopy and magnetic measurements, *Hyper. Inter.* **95**, 129-148
14. López, F. J. Lázaro, A., González-Carreño, T., Morales, M. P. and Serna, C. J. (1995) Magnetic properties of γ- Fe_2O_3 small particles prepared by spray pyrolysis, *IEEE Trans Magn.* **140-144**, 383-384

WATER, HYDROPHOBES AND SILICA PARTICLES

M. L. HAIR* AND C. P. TRIPP
Xerox Research Centre of Canada,
2660 Speakman Drive, Mississauga, Ontario
L5K 2L1, Canada

1. Abstract

The silanisation of surfaces is important to many of the new particulate technologies. In this paper we discuss specifically the reaction of chlorosilanes with nonporous particles of a pure fumed silica. Infra red (IR) spectroscopy is used to define the surface reactions which occur on surfaces that are completely dry and on those that contain small quantities of adsorbed water.

In the presence of preadsorbed water the chlorosilanes can be fully hydrolysed to the corresponding hydroxysilanes and the the reactions are followed by IR using a thin film technique. It is found that the percentage of the adsorbed methylsilanols that condense with the surface hydroxyl groups is in the order $(CH_3)_3SiOH >$ $(CH_3)_2Si(OH)_2 > CH_3Si(OH)_3$. This is opposite to the expected trend and is explained in terms of a difference between the kinetics of self-condensation and the kinetics of reaction with *surface* hydroxyl groups. Although in solution the tendency to self-condensation increases with the number of hydroxyl groups attached to the silicon atom, the tendency for reaction with the surface hydroxyl groups follows the reverse order.

The effect of curing on the surface bonding can also be investigated. Both the dimethyldisilanol and the trimethylsilanol condense fully and react with the surface silanols at room temperature to form surface Si-O-Si bonds. The surface species is not affected by curing. On the other hand, the methyltrisilanol does not condense with the surface hydroxyl groups at room temperature but is partially crosslinked. Curing completes the crosslinking process and also results in *a few* (less than 5%) attachments to the surface.

2. Introduction

Chlorosilanes are widely used for the surface modification of silicas.[1-4] The surface properties imparted by the silane are defined primarily by the selection of the organo group though the nature of the attachment to the underlying surface is critical. Thus, for instance, it is common practice to covert hydrophilic glass and silica surface into hydrophobic ones by the action of an alkylchlorosilanising agent. The main obstacle that limits the widespread use of chlorosilanes is that there exists a wide variability in the quality of the resulting films. Most of this variability can be traced to the acute sensitivity of the reaction to surface preparation and particularly to the level of surface water.[5-10] This critical dependence on water is avoided using a gas phase reaction at elevated temperature (> 300 °C) where a direct reaction of the chlorosilane with the surface hydroxyl groups (Si_sOH) occurs to form a strong Si_s-O-Si surface bond.[11]

E. Pelizzetti (ed.), Fine Particles Science and Technology, 209–224.

The reaction proceeds both monofunctionally,

$$Si_sOH + R_nSiCl_{4-n} \rightarrow Si_sOSiR_nCl_{3-n} + HCl \quad (1)$$

and difunctionally,

$$2Si_sOH + R_nSiCl_{4-n} \rightarrow (Si_sO)_2SiR_nCl_{2-n} + 2HCl \quad (2)$$

$$Si_s(OH)_2 + R_nSiCl_{4-n} \rightarrow Si_sO_2SiR_nCl_{2-n} + 2HCl \quad (3)$$

where Si_s refers to a surface silicon atom.

Reactions at elevated temperatures limit the number of potential silanising agents to those with sufficient vapour pressure. Thus, silanisation is often done in a nonaqueous solvent near room temperature. In the solution reactions the film quality varies as the water present on the surface is a key component in the reaction. In the absence of surface water, it has been shown that there is no reaction between alkyltrichlorosilane and the silica surface at room temperatures in solution.[5,6] (An exception to this general rule occurs when R contains a fluoro-alkyl group on the gamma carbon atom.[12])

In the solution reaction it has been proposed that the adsorbed chlorosilane is initially hydrolysed to the alkylsilanol by water which is either preadsorbed on the surface or dispersed in the solvent:[12]

$$n\,H_2O + R_{4-n}SiCl_n \rightarrow R_{4-n}Si(OH)_n + n\,HCl \quad (4)$$

The alkylsilanol which is formed then condenses with the surface Si_sOH groups to form a surface Si_s-O-Si bond.

$$Si(OH)_n + R_{4-n}Si(OH)_n \rightarrow Si_sOSiR_{4-n}Si(OH)_{n-1} + H_2O \quad (5)$$

In both steps, the amount of surface water is critical; a thick water layer gives rise to a polymerised silane layer which then sits on top of the surface water layer whereas insufficient water leads to an incomplete monolayer. In the first case, the silane layer can easily be floated off.[9] Thus, it is important that the chlorosilane is securely anchored to the surface through surface Si_s-O-Si linkages if robust films are to be obtained.Recent articles from our laboratory,[5,6,11,12,15] have described in detail the first two steps in the reaction of alkyltrichlorosilanes with high surface area fumed silica at both the solid/liquid and solid/gas interface. Using a thin film IR spectroscopic technique, we were able to obtain direct evidence of surface reaction at the solid/gas interface by following the changes which occurred in bands due to SiOH, Si-O-Si, Si-C, and Si-Cl modes. The ability to obtain spectra in the region 1300 to 200 cm^{-1} was critical. Our findings confirmed that alkyltrichlorosilanes are hydrolysed to alkyltrisilanols by the surface water but that few if any of the adsorbed alkyltrisilanol on the surface reacted with the surface Si_sOH group to form a surface Si_s-O-Si bond. The alkyltrisilanol polymerised in solution and dropped out onto the surface as a thick silane layer. Ellipsometric data of Trau et al.[16] on silica plates and IR studies by Le Grange et al.[17] on oxidised silicon wafers agree with our IR findings.

The lack of reactivity is of some import to discussions on the question of the permanence of the surface film or the electrical properties of a partially reacted surface in a silicon device. The purpose of this paper therefore is to shed light onto the reaction of small particles of silica with mono, di and tri-hydroxy silane and the effect of "curing" of the resultant film.

3. Experimental

A detailed description of the infrared cell, spectrometer and thin film technique has been published previously.[5,11] The fumed silica (Aerosil 380) was obtained from Degussa A.G. and had a measured surface area of 375 m^2g^{-1}. It was dispersed as a thin film (about 0.2 mg/cm^2) on a CsI window. The silica used in these experiments was dehydrated by evacuation at 400 °C for 30 minutes and cooled to room temperature. The silica thus contains only isolated/geminal hydroxyl groups at a density of about 1.4 OH/nm^2.[19]

The absorbance spectra which are shown in the Figures were calculated by reference to a spectrum recorded prior to addition of the silane. For the curing experiments, a thin film of "hydrated" silica was first evacuated at 200 °C and cooled to room temperature. This spectrum was used as the reference. The hydrated silica was then produced by exposing the silica to air for 10 minutes followed by a short evacuation of about 1 minute. This procedure removed approximately 90% of the original amount of adsorbed water. A second reference spectrum was recorded and this was used in the calculation of the absorbance spectrum for the silane addition to a hydrated silica at room temperature. By using spectra of the silica as a reference, the negative bands produced in the absorbance spectra are due to bonds that have been removed from the silica and the positive bands are due to bonds formed on the surface. All spectra were recorded at room temperature.

Trichloromethylsilane (TCMS), dichlorodimethylsilane (DCDMS) trimethylchlorosilane (TMCS) and triethylsilanol (TES) were obtained from Hüls America Inc and were purified by transfer to glass bulbs and several degas / freeze-thaw cycles. IR spectra of the reagents were recorded periodically to monitor purity. An excess amount of silane vapor (about 4 mmole g^{-1}) was added to the hydrated silica reacted for 5 minutes at room temperature, and they evacuated for 5 minutes. During curing, the silica was heated to 200 °C under evacuation for 15 minutes before being cooled to room temperature.

4. Results and Discussion

4.1 TRIETHYLSILANOL (TES)

The simplest reference point for the series of reactions implied by reactions 1-5 is the reaction of triethylsilanol (TES) with the silica surface hydroxyl groups. TES is the intermediate that would be formed by hydrolysis of triethylchlorosilane (Eq. 4). It has the advantage that it is a stable molecule which is fully hydrolysed and could react only monofunctionally with the silica surface. As a side product, only a dimer could form. More importantly, TES is commercially available and has a sufficiently high vapor pressure that it can react from the gas phase at room temperature. The results of adding triethylsilanol to a "hydrated" silica at room temperature are shown in the difference spectra of Figure 1a. From the spectra it is clear that the free surface hydroxyl groups (3747 cm^{-1}) are removed by the reaction. Bands due to the ethyl groups appear around 2900 cm^{-1}. More importantly, however, is the appearance of the band at 1060 cm^{-1} which is indicative of the formation of a surface Si_S-O-Si linkage. In addition, there are a large number of hydrogen bonded triethylsilanol molecules attached to the surface (ca 3400 cm^{-1}) which are removed by (evacuation) at 200 °C. However the number of triethylsilanols that were chemically anchored via a surface Si_S-O-Si bond remained constant -- the band at 1060 cm^{-1} does not change with "curing".

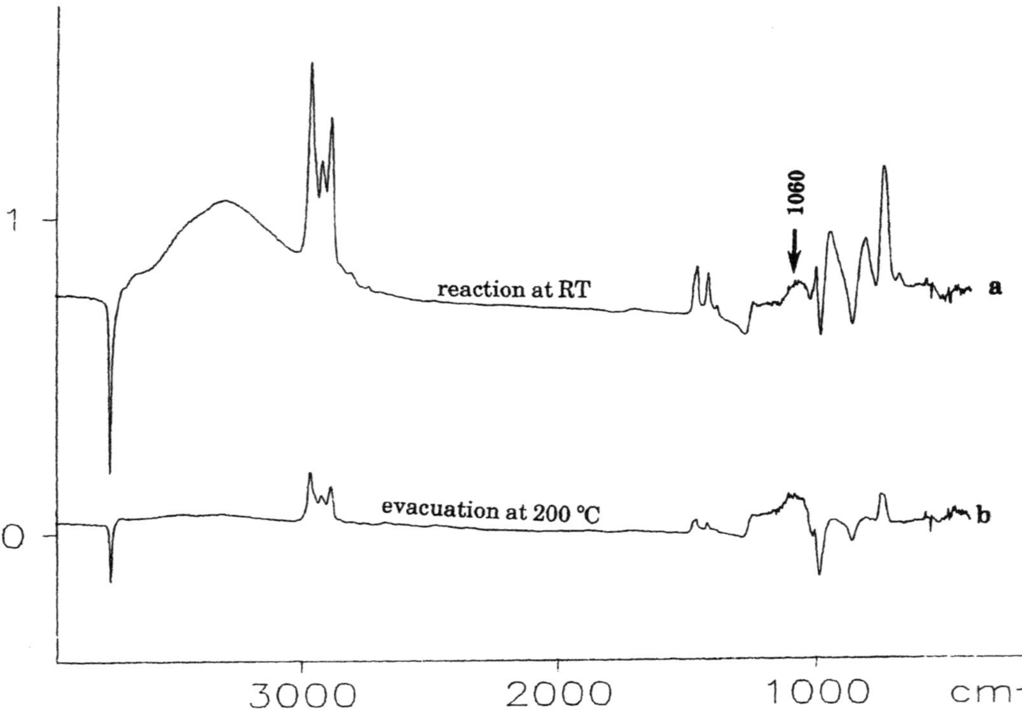

Figure 1. Addition of triethylsilanol to a) a hydrated silica at room temperature followed by b) evacuation at 200 °C.

4.2 TRIMETHYLCHLOROSILANE (TMCS)

The difference spectrum obtained upon adding TMCS to a hydrated silica at room temperature is shown in Figure 2a. Reaction was immediate and there was no change in this spectrum following prolonged evacuation for 2 h. Hydrolysis of the TMCS with the surface water is evidenced by the disappearance of the H_2O deformation mode at 1620 cm^{-1} and formation of HCl (not shown). We have previously shown that no reaction occurs when TMCS is added to a totally dehydrated silica at room temperature.[5] Although the spectrum shown in Figure 2a was recorded after evacuation of the excess TMCS, it is important to note that the band at 1620 cm^{-1} disappeared before the evacuation step. Thus, the surface water was removed by reaction with the TMCS and not during the evacuation of the excess TMCS. The broad band at 3400 cm^{-1} (SiO-H) is due to the alkylsilanol[5] and the weak Si-Cl band from the parent compound (487 cm^{-1}) shows that the attached silane is almost completely hydrolysed. The negative band at 3747 is again due to the SiO-H stretching mode of the isolated/geminal surface silanols and its disappearance shows that there is an interaction of the surface Si_sOH groups and the alkylsilanol. A negative Si_s-OH stretching mode[11] at 973 cm^{-1} accompanies the negative band at 3747 cm^{-1} and is clearly evident in Figure 2c. This negative band at 973 cm^{-1} is not as apparent in Figure 1a because it is superimposed on two positive peaks located at 1007 and 898 cm^{-1}.

The interaction with the surface hydroxyl groups can arise from the physisorption of an alkylsilane species, from hydrogen bonding with the alkylsilanol or from the formation of a surface Si_s-O-Si linkage. These possibilities can be deciphered from examination of the Si-O-Si modes which occur in the region between 1200 and 1000 cm^{-1}. Figure 2a shows that there is one band located at 1060 cm^{-1} in this spectral region. The relative intensities of the 1060/3747 bands can be used to estimate the number of trimethylsilanol that have reacted with the surface Si_S OH groups to form an Si_S-O-Si bond. The gas phase (400 °) reaction of TMCS with the silica surface is known to proceed smoothly and completely and the equivalent spectrum is shown in figure 2C. Thus comparison of the relative intensities (Table 1) leads to the conclusion that about 70% of the trimethylsilanols react with the surface Si_sOH groups to form a Si_s-O-Si bond. To account for the remaining 30% it is noted that there is a band at 3715 cm^{-1} which is attributed to weakly perturbed hydroxyl groups arising from the physisorption of an adsorbate on the surface Si_sOH groups. This band disappears when the sample is "cured" and this is accompanied by a 15% drop in the amount of adsorbed TCMS. After curing, the 1259/3747 ratio was close to the ratio obtained in the 400 °C gas phase reaction. This substantiates that the same reaction occurs at room temperature with a hydrated sample. The curing of this sample simply removes the weakly adsorbed species from the surface.

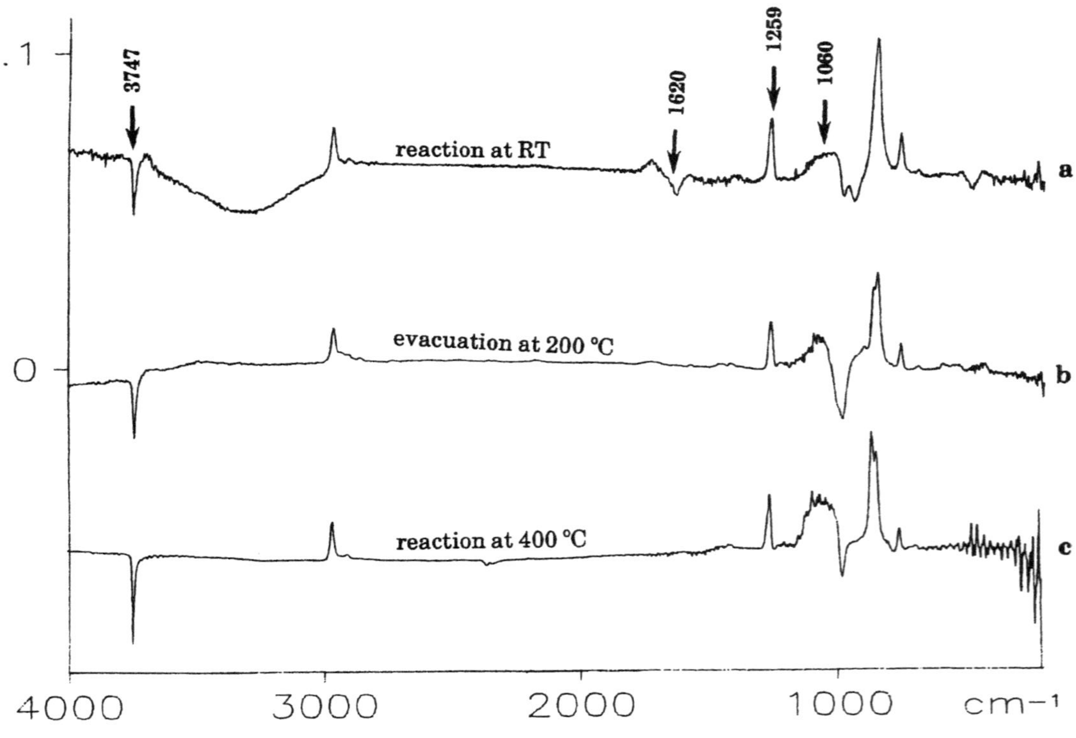

Figure 2. Addition of TMCS to a) a hydrated silica at room temperature followed by b) evacuation at 200 °C. Curve c) is the spectrum obtained after reaction of TMCS at 400 °C with a dehydrated silica.

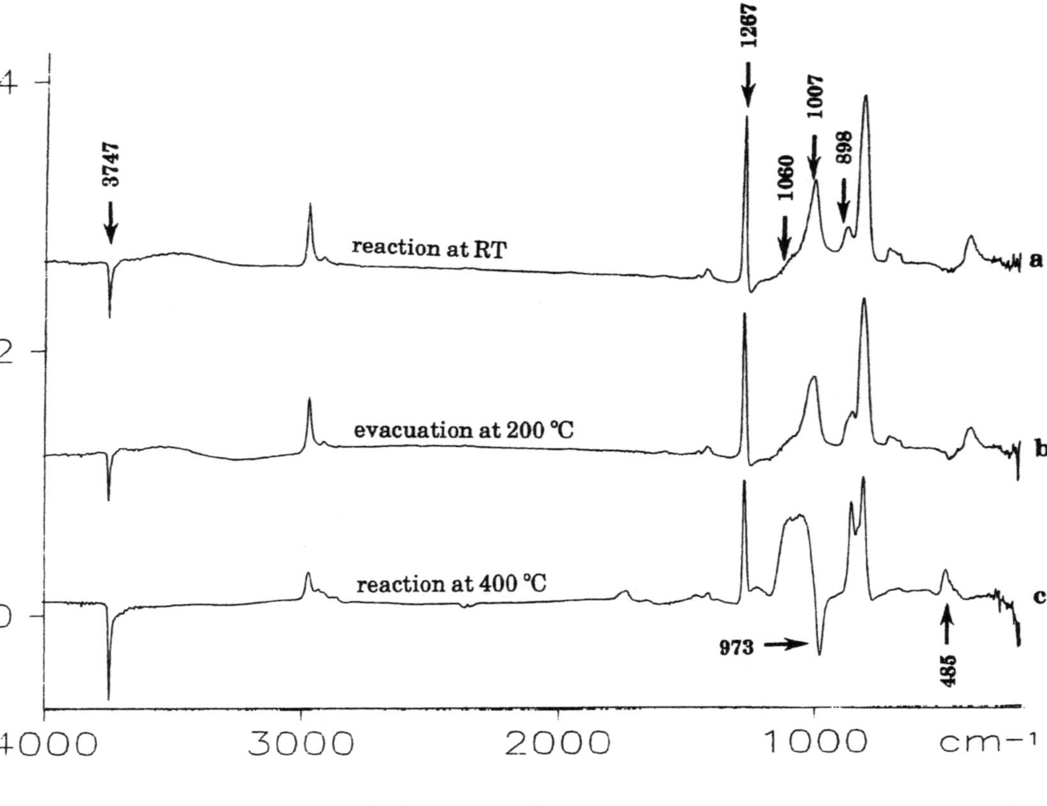

Figure 3. Addition of DCDMS to a) a hydrated silica at room temperature followed by b)evacuation at 200 °C. Curve c) is the spectrum obtained after reaction of DCDMS at 400 °C with a dehydrated silica.

4.3 DICHLORODIMETHYLSILANE (DCDMS)

When DCDMS is exposed to water and is hydrolysed to dihydroxdimethylsilane we can expect to see differences in the rate of condensation with the surface. Compared to the simple monohydroxy compounds the bifunctional molecules can polymerise and form a linear chain. The length of the chain will depend upon the kinetics of condensation relative to the kinetics of the surface interaction. The spectrum obtained upon addition of DCDMS to a hydrated silica is shown in Figure 2a. The absence of a Si-Cl band from the parent compound at 485 cm^{-1} shows that the DCDMS is completely hydrolysed. However, very little of the hydrolysed DCDMS remains on the surface. The weak band at 898 cm^{-1} shows that there are very few dimethylsilanol molecules and the appearance of bands at 1060 cm^{-1} and 1000 cm^{-1} is evidence that a large proportion of the dimethylsilanol molecules have condensed with the surface Si_sOH groups (1060 cm^{-1}) and crosslinked (1000 cm^{-1}) to form both surface and intermolecular Si-O-Si bonds at room temperature. These assignments are discussed fully in the section on TCMS. By comparing the 1060/3747 ratio with that obtained in the high temperature gas phase reaction (see Figure 2c) it can be estimated that about 75% of the surface Si_sOH groups on the hydrated silica are due to Si_s-O-Si formation. Furthermore, the ratio of the 1265/3747 intensities shows that about 2.8 times as many DCDMS molecules are adsorbed per perturbed hydroxyl group on the hydrated silica than was found in the 400 °C gas phase reaction. Since the lateral polymerisation of DCDMS occurs in a linear chain, the molecules bound to the surface through a Si_s-O-Si bond must anchor the silane and terminate the polymer chain. On average therefore about 2 DCDMS molecules are attached through intermolecular Si-O-Si bonds for each DCDMS which is attached to the surface. This oversupply of DCDMS molecules is reflective of the amount of adsorbed water initially present in both samples.

When the above sample is cured by evacuation at 200 °C there is only a small change in the adsorbed amount, and the adsorbed DCDMS species do not undergo any chemical transformation during the curing. This is not surprising since the spectrum in Figure 2a shows that a nearly complete condensation of the dimethylsilanol had already occurred at room temperature.

4.4 TRICHLOROMETHYLSILANE (TCMS)

The difference spectrum obtained upon adding TCMS to a hydrated silica at room temperature is shown in Figure 4a. The reaction was instantaneous and complete and there was no change in this spectrum following prolonged evacuation for 2 h. Hydrolysis of the TCMS with the surface water has occurred and is evidenced by the disappearance of the H_2O deformation mode at 1620 cm^{-1} and formation of HCl (not shown). No reaction occurs when TCMS is added to a totally dehydrated silica at room temperature.[5], so although the spectrum shown in Figure 1a was recorded after evacuation of the excess TCMS, it is again important to note that the band at 1620 cm^{-1} disappeared prior to the evacuation step. The broad bands at 3400 cm^{-1} (SiO-H) and 989 cm^{-1} (Si-OH) are due to the alkylsilanol. The very weak band at 574 cm^{-1} shows that the attached silane is almost completely hydrolysed. The positive band at 3400 cm^{-1} cannot be due to an OH stretching mode of adsorbed water because

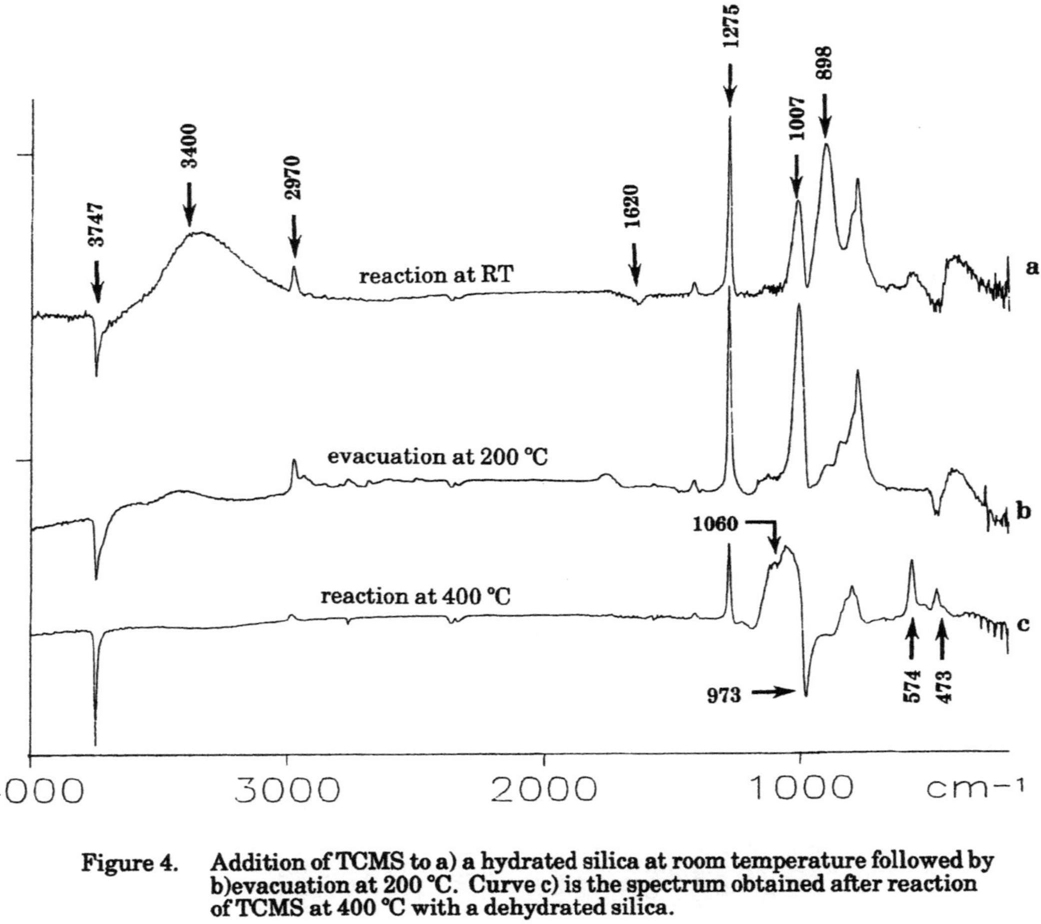

Figure 4. Addition of TCMS to a) a hydrated silica at room temperature followed by b)evacuation at 200 °C. Curve c) is the spectrum obtained after reaction of TCMS at 400 °C with a dehydrated silica.

the H_2O deformation mode at 1620 cm^{-1} is in the negative direction. The negative band at 3747 cm^{-1} is due to the SiO-H stretching mode of the isolated/geminal surface silanols and its disappearance shows that there is an interaction of the surface Si_sOH groups and the alkylsilanol. A negative Si_s-OH stretching mode[11] should accompany the negative band at 3747 cm^{-1} and this band is clearly evident in the spectrum shown in Figure 4c. This negative band at 973 cm^{-1} is not as apparent in Figure 1a because it is superimposed on two positive peaks located at 1007 and 898 cm^{-1}.

The band at 1007 cm^{-1} is of particular significance. Assignment of bands in the Si-O-Si spectral region is difficult because a Si-O-Si bond can arise from several species. A Si-O-Si bond can be due to a reaction with the surface Si_sOH groups, from crosslinking between adjacent alkylsilanols, or from polymerisation products depositing on the surface. The assignment is further complicated in the thin film studies because the bands in this region are superimposed on the intense bulk mode of silica. Subtle changes in the bulk Si-O-Si modes near the adsorbed species could produce spectral changes in the 1200 to 1000 cm^{-1} region. Assignment of the band at 1007 cm^{-1} to a polymerised silane layer can be eliminated based on the spectrum of known polymeric compounds containing $(Si\text{-}O\text{-}Si)_x$. In cyclic compounds where $x=3$ or 4 the Si-O-Si band is split in two components at about 1090 and 1020 cm^{-1} whereas linear compounds show two bands at 1080 and 1020 cm^{-1}.[20,21] On this basis, we rule out deposition of a siloxane polymer on the surface.

Assignment to a surface Si_s-O-Si species can also be discounted after comparing with Figure 4c. That spectrum was produced by the reaction of TCMS at 400 °C with a completely dehydrated silica[11] where the bands have been assigned as follows: negative 3747 cm^{-1} (Si_sO-H stretching), 2970, 2920 cm^{-1} (CH_3 stretching), 1428, 1275 cm^{-1}(CH_3 bending), 1060 cm^{-1}(Si_s-O-Si stretching), negative 973 cm^{-1}(Si_s-OH stretching), 827, 803 cm^{-1}, (CH_3 rocking), 574 cm^{-1} and 473 cm^{-1} (Si-Cl stretching). In the reaction at 400 °C, all adsorbed TCMS molecules are bound to the surface by a Si_s-O-Si bond which is evidenced by the broad band located at 1060 cm^{-1}. Thus, the band at 1007 cm^{-1} cannot be due to a surface Si_s-O-Si bond and thus the removal of surface Si_sOH groups (i.e., the negative bands at 3747 and 973 cm^{-1} in Figure 4a) is not be due to the formation of a surface Si_s-O-Si linkage.

A few of the silanols possibly react to produce an Si_s-O-Si surface bond. However, the relative number of adsorbed TCMS molecules that are either bound to the surface or crosslinked to each other by Si-O-Si bonds can again be estimated by a comparing with spectrum in Figure 4c. The ratio of the integrated intensities of the 1060/1275 cm^{-1} bands (Figure 4) gives an estimate of the relative number of TCMS molecules attached to the surface by Si_s-O-Si bonds in the room temperature reaction with a hydrated silica. Similarly, the 1275/3747 cm^{-1} ratio provides information on the relative number of TCMS molecules per Si_sOH site and the 1060/3747 ratio gives the relative number of perturbed Si_sOH groups converted to Si_s-O-Si bonds. The results are listed in Table 1. (The Si-CH_3 bending mode of TCMS, DCDMS and TMCS are at 1275, 1267 and 1259 cm^{-1} respectively and are generally referred to as 1265 cm^{-1}band.)

A comparison of the 1060/3747 ratio in Figures 1a and 1c shows that the number of surface Si_s-O-Si bonds is below our detection limit which means that a maximum of 5% of the perturbed surface Si_sOH groups can be involved in a condensation reaction with the methyltrisilanol. The adsorption of TCMS must occur mainly by hydrogen bonding interactions of the methylsilanol with the surface Si_sOH groups.

Reactions with Hydrated Silica: Intensity Ratios of Infrared Bands

Reaction	1060/1265	1060/3747	1265/3747
TCMS (RT)	--	--	0.14
TCMS (Curing)	0.10	0.01	0.11
TCMS (400 °C)	1.0	0.056	0.056
DCDMS (RT)	0.18	0.035	0.23
DCMS (Curing)	0.18	0.035	0.23
DCDMS (400 °C)	0.71	0.057	0.08
TMCS (RT)	0.43	0.033	0.077
TMCS (Curing)	0.63	0.032	0.050
TMCS (400 °C)	0.82	0.048	0.058
Et_3SiOH (RT)	0.28	0.01	0.048
Et_3SiOH (Curing)	1.27	0.052	0.041

RT = Reaction at room temperature
Curing = After curing at 200 °C
400 °C = Gas phase reaction at 400 °C

It is possible that the excess TCMS which is adsorbed per surface Si_s-OH group and the resilience of the product to removal is due to self-condensation with formation of an Si-O-Si bond between adjacent alkylsilanols. However, we know that curing the sample (a process known to increase crosslinking) results in an increase in intensity of the band at 1007 cm^{-1} and a decrease in intensity at 3400 and 898 cm^{-1} (The two latter bands are due to the SiO-H and Si-OH vibrations of the alkylsilanol). No other changes occur in the spectrum during the curing step. We also saw earlier that a band near 1007 cm^{-1} is formed when dichlorodimethylsilane (DCDMS) is adsorbed but does not appear with trimethylchlorosilane (TCMS). With TCMS a crosslinked Si-O-Si bond between adsorbed alkylsilanols is not possible and this provides strong support for assigning the 1007 cm^{-1} band to a Si-O-Si bond between adjacent adsorbed alkylsilanols.

The picture that emerges is that the TCMS is fully hydrolysed at room temperature and about 40% of the molecules (from the 1265/3747 ratios)are anchored to the surface by hydrogen bonding interactions with the surface Si_sOH groups. The remaining hydrolysed TCMS are not directly attached to the surface and form a partially polymerised layer through Si-O-Si bonds between molecules. This is similar to Si[29] NMR conclusions reported by Sindorf and Maciel.[24]. Although their experimental protocol was different from ours, Sindorf and Maciel found that the TCMS adsorbed on a hydrated silica was in excess of the number of surface Si_sOH groups and this difference was attributed to polymerisation of silane on the surface.

To study the effect of curing, the sample giving the spectrum shown in Figure 4a was evacuated while the temperature was increased to 200 °C. Changes in the spectrum begin to appear at 100 °C and are complete by 200 °C. The spectrum obtained after evacuation at 200 °C is shown in Figure 4b. The band at 1265 cm^{-1} has not changed in intensity showing that the curing process did not change the amount of silane attached to the surface. There is a 95% reduction in the intensity of the silanol bands at 3400 and 898 cm^{-1} and this is accompanied by a doubling in the intensity of the band at 1007 cm^{-1}. A weak band at 3500 cm^{-1} appears and this could be due to residual hydrogen bonded surface silanols. Evidently, the curing completes the condensation process through the formation of Si-O-Si bonds between adsorbed molecules. There is evidence of a very weak, broad band at 1060 cm^{-1} showing that some of the surface Si_sOH groups condense to form a surface Si_s-O-Si linkage and this is accompanied by about a 25% increase (measured from the negative band at 3747 cm^{-1}) in the number of hydroxyl groups interacting with the silane.

The above results clearly show that curing completes the crosslinking process, reduces the number of available surface Si_sOH groups and causes some condensation between the methylsilanol and the surface Si_sOH groups. However, even after curing, only a few of the adsorbed TCMS molecules are anchored via Si_s-O-Si bonds and a comparison of the 1060/1265 ratios in Figures 1b and 1c suggests that less than 10% of the adsorbed molecules are linked through a surface Si_s-O-Si bond.

Kinetic Trends

The results show that the three methylchlorosilanes, TCMS, DCDMS, and TMCS are fully hydrolysed to methylsilanols upon contact with the surface water on silica and that the tendency for reaction of the adsorbed methylsilanols with the surface Si_sOH groups follow the order;

$$(CH_3)_3SiOH > (CH_3)_2Si(OH)_2 > CH_3Si(OH)_3 \qquad (7)$$

It is well known that the tendency of alkylsilanols to condense increases with the number of hydroxyl groups attached to the silicon atom.[1] Therefore, the tendency to react with the surface hydroxyl group outlined in Equation 5 might have been expected to follow the order:

$$RSi(OH)_3 > R_2Si(OH)_2 > R_3SiOH$$

which is exactly opposite. It is generally accepted that the tendency of organosilanols in solution to self-condensation increases with the number of hydroxyl groups attached to the silicon atom and decreases with the number and size of the organic group.[1] The mechanism for self-condensation is by nucleophilic attack of an O atom on a Si atom of a silane through a pentacoordinate transition state and the attack is favored at a Si atom with a higher number of attached hydroxyl groups.[25,26] However, the self-condensations of alkylsilanols are also extremely sensitive to steric hindrance and it is noted, for instance, that silanediols containing t-butyl groups are extremely resistance to condensation as compared to di-n-alkylsilanediols.[1]

It is our contention that the mechanism for solution condensation is completely different from the mechanism with surface SiOH groups. In solution, a pentacoordinate intermediate is formed in the reaction by attack of the O at 180 or 90 degrees with respect to the leaving group.[25] These geometries are not possible with the attack of O of an alkylsilanol on the Si surface atom and it is energetically improbable that the surface silicon is able to alter its conformation and achieve a pentacoordinate structure.

This argument finds support from previous studies comparing base catalysis in solution to surface reaction.[15] It has been shown that when triethylamine is preadsorbed on surface Si_sOH groups it will promote the room temperature reaction of chlorosilanes with the surface. The triethylamine does not form a pentacoordinate intermediate with the incoming silane but rather forms a strong hydrogen bond with the surface Si_sOH groups. It is this hydrogen bonding that renders the O of the surface Si_sOH group a better nucleophile for attack on a Si atom of the chloroalkylsilane.

An acid-base mechanism can be envisaged for the general reaction between surface silanols and alkylsilanols. In principle, the reaction can occur two ways: the oxygen atom eliminated can come either from the surface

$$Si_sO_sH + HOSiR_3 \rightarrow Si_s-O-SiR_3 + H_2O_s \quad (8)$$

or from the alkylsilanol

$$Si_sO_sH + HOSiR_3 \rightarrow Si_s-O_s-SiR_3 + H_2O \quad (9)$$

The relative acidity of the silanol groups will decide the reaction. The question of surface acidity has been addressed previously and surface pK_a values have been estimated by measuring frequency shifts with a series of adsorbed molecules. It is found for instance that the hydroxyl stretching frequencies for silica[27] (3690 cm^{-1}) and trimethylsilanol[28] (3688 cm^{-1}) in CCl_4 shift 395 cm^{-1} and 238 cm^{-1} respectively by hydrogen bonding to diethylether. The surface Si_sOH group is more acidic than the methylsilanol group and pK_a values of 7.1 and 11 have been obtained.[29] Thus, the surface reaction will proceed according to equation 9 and the water produced will contain a proton from the surface Si_sOH group and an OH from the alkylsilanol. In essence, the condensation reaction involves breaking the O-H bond on the surface and the Si-O bond of the approaching silanol.

If it is assumed that the reaction does proceed by attack of the oxygen atom of the surface Si_sOH group on the silicon atom of the alkylsilanol then a single transition state structure as shown below is plausible.

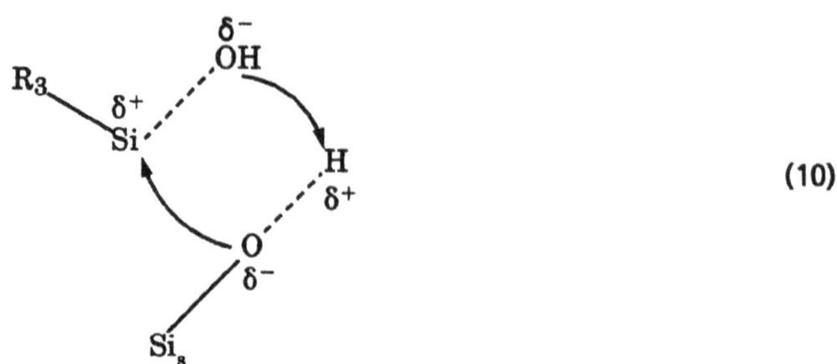

Of the three chlorosilanes, the methyltrisilanol of TCMS is the only adsorbed species that had not fully condensed and reacted with the surface at room temperature. This is a consequence of the competition between self-condensation and surface reaction and it is this difference in kinetics (self-condensation vs reaction with surface) that controls the nature of the adsorbed species. The tendency of self-condensation increases with the number of hydroxyl groups attached to the silicon atom but the tendency for condensation with the surface hydroxyl groups follows the reverse order. Thus, in the first step of the condensation of the methyltrisilanol, self-condensation is favored over surface reaction. One hydroxyl group of the trisilanol reacts with the hydroxyl group of a second trisilanol. Reaction of the second and third hydroxyl groups becomes increasingly difficult[10,14]. A curing step is needed to complete the self-condensation process and give some slight surface reaction. Figure 4 shows that the curing step eliminates the remaining methylsilanols and this effectively doubles the number of intermolecular Si-O-Si bonds.

In interpreting the experimental data in terms of the acid-base characteristics of the surface we become aware of a significant practical conclusion that accounts for the miscellany of results that are often reported for silanising agents. As we are dealing with the acidity of the surface we would expect changing kinetics as the acidity of the surface hydroxyl group is altered. It is well known from the catalytic and molecular sieve literature that introduction of aluminum ions into a silicon framework, for instance, can cause the formation of highly acid Brönsted sites. These would be strong nucleophiles and would accelerate many surface reactions. The introduction of the chemically similar Boron into the silica surface has already been shown to affect the kinetics of gas phase reactions between surface hydroxyls and chlorosilanes.[30]

References

(1) Noll, W. (1968) *Chemistry and Technology of Silicones*; Academic Press: New York.

(2) Plueddemann, E. P. (1982) *Silane Coupling Agents*; Plenum: New York.

(3) Iler, R. K. (1979) *The Chemistry of Silica*; John Wiley and Sons: New York.

(4) Unger, K. K. (1971) *Porous Silica*; Elsevier: Amsterdam.

(5) Tripp, C. P.and Hair, M. L. (1992) *Langmuir*, **8**, 1961.

(6) Tripp, C. P.and Hair, M. L. (1992) *Langmuir*, **8**, 1120.

(7) Brzoska, J. B., Shahidzadeh, N. and Rondelez, F. (1992) *Nature*, **360**, 719.

(8) Angst, D. L.and Simmons, G. W. (1991) *Langmuir*, **7**, 2236.

(9) Silberzan, P., Léger, L., Ausserré, D.and Benattar, J. J. (1991) *Langmuir*, **7**, 1647.

(10) Kessel, C. R.and Granick, S. (1991) *Langmuir*, **7**, 532.

(11) Tripp, C. P.and Hair, M. L. (1991) *Langmuir*, **7**, 923.

(12) Tripp, C. P., Veregin, R. P. N.and Hair, M. L. (1993) *Langmuir*, **9**, 3518.

(13) Sagiv, J. (1980) *J. Am. Chem. Soc.*, **102**, 92.

(14) Osterholtz, F. D., Pohl, E. R. (1992) *J. Adhesion Sci. Technol.*, **6**, 127.

(15) Tripp, C. P.and Hair, M. L. (1993) *J. Phys. Chem.*, **97**, 5693.

(16) Trau, M., Murray, B. S., Grant, K.and Grieser, F. (1992) *J. Colloid Interface Sci.*, **148**, 182.

(17) Le Grange, J. D., Markham, J. L.and Kurkjian, C. R. (1993) *Langmuir*, **9**, 1749.

(18) Caravajal, G. S., Leyden, D. E., Quinting,.G. R.and Maciel, G. E. (1988) *Anal. Chem.*, **60**, 1776.

(19) Morrow, B. A.and McFarlan, A. J. (1991) *Langmuir*, **7**, 1695.

(20) Bellamy, L. J. (1975) *The Infra-red Spectra of Complex Molecules*, 3rd ed.; Chapman and Hall Ltd.: Thetford, UK.

(21) Smith, A. L. (1960) *Spectrochim. Acta*, **16**, 87.

(22) Tripp, C. P.and Hair, M. L. (1993) *Langmuir*, **9**, 3523.

(23) Smith, A. L.; Anderson, D. R. (1984) *Appl. Spectrosc.*, **38**, 822.

(24) Sindorf, D. W.and Maciel, G. E. (1983) *J. Am. Chem. Soc.*, **105**, 3767.

(25) Corriu, R. J. P.and Guerin, C. (1980) *J. Organomet. Chem.*, **198**, 231.

(26) Sommer, L. H. (1965) *Stereochemistry, mechanism and silicon: An introduction to the dynamic stereochemistry and reaction mechanisms of silicon centers; McGraw-Hill Inc.*

(27) Hair, M. L.and Hertl, W. (1969) *J. Phys. Chem.*, **74**, 91.

(28) West, R.and Baney, R. H. (1959) *J. Am. Chem. Soc.*, **81**, 6145.

(29) Hair, M. L. (1967) *Infrared Spectroscopy in Surface Chemistry*; Marcel Dekker: New York.

(30) Hair, M. L.and Hertl, W. (**1973**) *J. Phys. Chem.*, **77** 1965.

DIELECTRIC SPECTROSCOPY OF MODEL COLLOIDS, AND THE ROLE OF CONDUCTION BEHIND THE PLANE OF SHEAR

Theory and experiments

M. MINOR, A. van der WAL, J. LYKLEMA
Department of Physical and Colloid Chemistry
Wageningen Agricultural University
Dreijenplein 6
6703 HB Wageningen
The Netherlands

ABSTRACT. The Standard Electrokinetic Model (SEM) is adapted for additional surface conduction behind the plane of shear. This extended model is applied to dielectric and electrophoretic measurements. As a model system, spherical, homodisperse bacterial cell suspensions are used which exhibit surface conduction behind the plane of shear due to the hairy cell wall. From the combination of these experiment a more realistic ζ-potential as the fraction of counter charge situated behind the plane of shear can be obtained. Because the surface charge is determined from titrating the isolated cell walls, the mobility of counterions inside the cell wall structure is retrieved.

1. Introduction

Although electrokinetic potentials belong to the most common characteristics of charged surfaces, their interpretation in terms of the double layer potentials is not straightforward. Rarely is the ζ-potential identical to the surface potential ψ^0. Mostly ζ is lower, if not much lower. When superequivalent counterion adsorption occurs even the signs of ζ and ψ^0 differ. In these respects ζ rather resembles the Stern potential ψ^d, a correspondence that is supported by the, often observed, correlation between electrokinetic potentials and the stability of electrocratic colloids. Electrokinetic potentials cannot be measured directly but are obtained from experiments by applying an electrokinetic model. The Standard Electrokinetic Model (SEM) is popular in

E. Pelizzetti (ed.), Fine Particles Science and Technology, 225–238.

electrokinetic studies [1-3]. In this model the particles are considered as rigid, impenetrable units, characterised by a slip-plane at which the potential is ζ. An important assumption in SEM is the absence of conduction in that part of the double layer where the solvent is stagnant (no conduction behind the plane of shear). Applying SEM generally results in higher ζ-potentials obtained from d.c. and a.c. conductivity measurements ζ_c compared to the ζ-potential obtained from electrophoretic mobility measurements ζ_e. In other words, the theory is inconsistent. This means that many ζ-potentials reported in literature, have to be revisited, especially those measured at low electrolyte concentrations. The most probable cause for this inconsistency is the neglection of conduction behind the plane of shear [4]. This conduction may be related to surface roughness, porosity, the visco-electric effect [5] or a macromolecular surface layer. Incorporation of this "extra" surface conduction in SEM increases the (complex) conductivity (and ζ_c) and decreases the mobility (and ζ_e). In this way the discrepancy between ζ_c and ζ_e can be taken away to yield ζ_{comb} which may significantly differ from ζ_c and ζ_e especially at low electrolyte concentration. Identifying this ζ_{comb} with the Stern-potential ψ^d may predict the stability of colloidal suspensions more accurately.

Below we present a theory in which such conduction behind the slip plane is incorporated. This theory will be elaborated for the case of alternating fields. If applied to (model) colloids, the surface conductivity behind this plane can be obtained and, if combined with titration data, numbers and mobilities of the charge carriers behind and beyond the shear plane are obtained.

Measurements were performed on a homodisperse bacterial suspension of spherical cells. These particles have a thick cell wall structure of charged macromolecules. Descriptions of the intricacies of these measurements are found in [6,7].

The following theory section is Dukhin and Shilov's famous thin double layer theory [2] applied to spherical particles extended with the incorporation of extra surface conduction behind the plane of shear.

2. Theory

2.1. GOVERNING ELECTROKINETIC EQUATIONS

We consider a spherical particle with radius a in a suspending electrolyte with valences z_i, friction coefficients β_i and number densities n_i. The origin of our spherical coordinate system is located at the centre of the particle. We will identify the position just inside the particle with a^-, the position in the inner part of the diffuse layer, adjacent to the surface, with a, the position where the shear plane is situated with a^ζ

and the position just outside the thin diffuse layer with a^+. An oscillating electrical field with magnitude E_0 and (angular) frequency ω is imposed:

$$\vec{E} = \vec{E}_0 \exp i\omega t \quad , \quad E_0 = \left|\vec{E}_0\right| \tag{1}$$

The electrical potential ψ , the space charge density ρ and the dielectric permittivity $\varepsilon = \varepsilon_0 \varepsilon_r$ are related by Poisson's equation:

$$\vec{\nabla} \cdot \varepsilon \vec{\nabla} \psi = -\rho = -\sum_i z_i e n_i \tag{2}$$

where e is the unit charge, ε_0 the permittivity of vacuum and ε_r the relative permittivity of the medium. The boundary condition at the surface is given by Gauss's law:

$$\left(\varepsilon \frac{\partial \psi}{\partial r} \right) \Bigg|_{a^-}^{a} = -\sigma^0 \tag{3}$$

where σ^0 is the surface charge density. Far away from the particle the electrical field must attain the applied electrical field and the ion concentration must approach its bulk value (superindex ∞):

$$\vec{\nabla} \psi = -\vec{E} \quad \text{as} \quad r \to \infty \tag{4}$$

$$n_i \to n_i^\infty \quad \text{as} \quad r \to \infty \tag{5}$$

The ions are convectively transported by the fluid with the velocity $\vec{v}$ and can move relative to the solvent under the influence of electrical and thermodynamic forces. The flux $\vec{J}_i$ of ions of kind i is given by the Nernst-Planck equation:

$$\vec{J}_i = n_i \vec{v} - \frac{kT}{\beta_i} \left(n_i \frac{z_i e}{kT} \vec{\nabla} \psi + \vec{\nabla} n_i \right) \tag{6}$$

Ions are not able to enter the particle:

$$\vec{J}_i \cdot \vec{e}_r = 0 \quad \text{at} \quad r = a \tag{7}$$

where $\vec{e}_r$ is the normal unit vector on the surface.
The motion of the fluid is described by the Navier-Stokes equations for incompressible fluid flow and, in the frequency domain of interest with the neglection of inertia:

$$\eta \vec{\nabla}^2 \vec{v} - \vec{\nabla} p - \rho \vec{\nabla} \psi = 0 \tag{8}$$

$$\vec{\nabla} \cdot \vec{v} = 0 \tag{9}$$

where η is the viscosity and p the pressure. These equations must be solved subjected to the boundary conditions

$$\begin{aligned} \vec{v} &= 0 \quad \text{at} \quad r = a^\zeta \\ \vec{v} &= -\vec{v}_{ef} \quad \text{as} \quad r \to \infty \end{aligned} \tag{10}$$

The liquid between $r = a$ and $r = a^\zeta$ is stagnant. $\vec{v}_{ef}$ denotes the electrophoretic velocity.
Conservation of ions implies:

$$\frac{\partial n_i}{\partial t} + \vec{\nabla} \cdot \vec{J}_i = 0 \tag{11}$$

When the applied electrical field is small, we can linearize the non-linear equations around the equilibrium values in absence of the imposed field (subscript *eq*). First we introduce small perturbations linear in the applied electrical field:

$$\psi = \psi_{eq} + \delta\psi + O(E_0^2)$$
$$n_i = n_{i,eq} + \delta n_i + O(E_0^2) \tag{12}$$
$$\text{etc.}$$

The velocity is first order in E_0. Substituting these expressions in the Poisson equation and omitting the small non-linear $O(E_0^2)$ terms yields:

$$\vec{\nabla} \cdot \varepsilon \vec{\nabla} \delta\psi = -\delta\rho = -\sum_i z_i e \delta n_i \tag{13}$$

Substituting (12) in the conservation equation, subtracting the zero-flux equilibrium situation and omitting the small non-linear terms gives us the linearized Nernst-Planck equation:

$$\begin{aligned} \vec{J}_i &= n_{i,eq}\vec{v} - \frac{kT}{\beta_i}\left(\frac{z_i e}{kT}\left(n_{i,eq}\vec{\nabla}\delta\psi + \delta n_i \vec{\nabla}\psi_{eq}\right) + \vec{\nabla}\delta n_i\right) = \\ &= n_{i,eq}\vec{v} - n_{i,eq}\frac{kT}{\beta_i}\vec{\nabla}\left[\frac{z_i e}{kT}\delta\psi + \frac{\delta n_i}{n_{i,eq}}\right] \end{aligned} \tag{14}$$

The term between square brackets is just the first-order perturbation in the non-dimensional electrochemical potential.

The linearized Navier-Stokes equation yields:

$$\eta \vec{\nabla}^2 \vec{v} - \vec{\nabla}\delta p - \rho_{eq}\vec{\nabla}\delta\psi - \delta\rho\vec{\nabla}\psi_{eq} = 0 \tag{15}$$

For the linearized conservation equation we obtain:

$$\frac{\partial \delta n_i}{\partial t} + \vec{\nabla} \cdot \vec{J}_i = 0 \tag{16}$$

2.2. EQUATIONS AND FIELDS OUTSIDE THE DOUBLE LAYER

Beyond the thin diffuse double layer with thickness:

$$\kappa^{-1} = \left(\varepsilon^{\infty} kT \Big/ \sum_i z_i^2 e^2 n_i^{\infty}\right)^{1/2} \tag{17}$$

the electrolyte is undisturbed by the double layer

$$\psi_{eq} = 0 \quad , \quad n_{i,eq} = n_i^{\infty} \tag{18}$$

and the set of governing equations simplifies. The solutions to these simplified equations will be designated "far fields" and labelled with the superindex "f".

The Poisson equation reads:

$$\nabla^2 \delta\psi^f = -\frac{1}{\varepsilon^\infty}\sum_i z_i e \delta n_i^f \tag{19}$$

The conservation equations for the species turn into:

$$\frac{\beta_i^\infty}{kT}\frac{\partial \delta n_i^f}{\partial t} = \nabla^2\left(\frac{z_i e}{kT} n_i^\infty \delta\psi^f + \delta n_i^f\right) \tag{20}$$

The convective term cancelled because of incompressibility and the Navier-Stokes equation reduces to Stokes equations. So fluid flow decouples from the potential and densities, which further simplifies the equations outside the double layer.
We will now consider a binary symmetrical electrolyte where the friction coefficients of both ion species are the same. Then from the equations (19) and (20) it can be seen that the perturbations in number densities are equal for both species. These densities obey the ordinary diffusion equation whereas the potential obeys Laplace's equation. Physically this means that the ions diffuse simultaneously; hence there is neither production of net charge, nor generation of associated diffusion potentials outside the double layer:

$$\nabla^2 \delta n^f = \mathrm{i}\omega \frac{\beta^\infty}{kT}\delta n^f \tag{21}$$

$$\nabla^2 \delta\psi^f = 0 \tag{22}$$

The unique solution for δn^f which reduces to zero at infinity is :

$$\delta n^f = \frac{d_n a^3 E_0}{r^2}(1+\lambda r)\exp(-\lambda r)\cos\theta \exp(-\mathrm{i}\omega t) \tag{23}$$

$$\lambda a \equiv (1+\mathrm{i})\sqrt{\omega\tau} \quad , \quad \tau \equiv a^2\beta^\infty/2kT$$

For the potential we find:

$$\delta\psi^f = \left(-E_0 r + \frac{d_\psi a^3 E_0}{r^2}\right)\cos\theta\exp(-\mathrm{i}\omega t) \tag{24}$$

The two polarization characteristics d_n and d_ψ are often called electrical and chemical dipole strength respectively. These can be determined by two additional boundary conditions at the particle surface. In this region the complete set of coupled equations must be solved. However, because we consider the case $\kappa a >> 1$, the double layer can be considered flat and with the assumption of local equilibrium between the diffuse layer and the adjacent electrolyte the two inner boundary conditions will now be derived.

2.3. INNER BOUNDARY CONDITIONS AND THE FAR FIELDS

By subtracting the conservation equation for the far fields from the general conservation equation (16) and integration over the diffuse layer (d), we obtain the integrated flux balance for the perturbations in the excess amounts of ions in the diffuse layer:

$$\frac{\partial}{\partial t}\int_d\left(\delta n_i-\delta n^f\right)dr-\nabla_t^2\left[\frac{z_i e}{kT}\delta\psi^f+\frac{\delta n^f}{n^\infty}\right]_{r=a}\int_d\frac{kT}{\beta^\infty}\left(n_{i,eq}\frac{\beta^\infty}{\beta_i}-n^\infty\right)dr$$
$$-\frac{kT}{\beta^\infty}n^\infty\frac{\partial}{\partial r}\left(\frac{z_i e}{kT}\delta\psi^f+\frac{\delta n^f}{n^\infty}\right)\bigg|_{r=a}+\int_d\vec{\nabla}\cdot\left(n_{i,eq}-n^\infty\right)\vec{v}dr=0 \qquad (25)$$

Because in leading order the net radial flux in the diffuse layer is zero, there is local equilibrium between the diffuse layer and the adjacent electrolyte. Because of this, the electrochemical potential (term between square brackets) could be taken out of the integration and indexed "f". The time derivative can be neglected, as demonstrated by Fixman [8,9]. The flow term, which involves the transport of both ion species, can be solved on the basis of the Navier-Stokes equation and can be expressed in terms of the far fields evaluated at the particle surface.

At not too low surface potentials, the tangential flux of co-ions along the surface is approximately zero due to the relatively low density of these ions. Hence, to a good approximation, (25) reduces to (see also [10]):

$$\frac{\partial}{\partial r}\left[\frac{ze}{kT}\delta\psi^f+\frac{\delta n^f}{n^\infty}\right]_{r=a}=0 \quad , \quad z=\text{valence co-ion} \qquad (26)$$

for co-ions, which states that no co-ions may enter the double layer regime radially because this flow can not be balanced.

The net tangential flux of counterions in the double layer regime is balanced by the flux entering this regime radially from the neighbouring electrolyte:

$$2Rel\,a\,\nabla_t^2\left[\frac{-ze}{kT}\delta\psi^f+\frac{\delta n^f}{n^\infty}\right]_{r=a}+\frac{\partial}{\partial r}\left[\frac{-ze}{kT}\delta\psi^f+\frac{\delta n^f}{n^\infty}\right]_{r=a}=0 \qquad (27)$$

Rel is a dimensionless parameter given by:

$$Rel\equiv\frac{1}{\kappa a}\left[1+\frac{3m}{z^2}+\Theta\right]\left(\exp\left(\frac{ze\zeta}{2kT}\right)-1\right) \qquad (28)$$

where the non-dimensional ionic drag coefficient, m, is given by:

$$m=\frac{2\varepsilon^\infty kT\beta^\infty}{3\eta e^2} \qquad (29)$$

The physical meaning of the non-dimensional *Rel* parameter is a comparison between surface conduction and bulk conduction (surface conductivity times the circumference of the particle divided by the bulk conductivity times the cross-section of the particle). There are three contributions to *Rel*. The first of the three terms between square brackets reflects conduction by the counterions outside the plane of shear. Because the liquid in this region is moving along the particle, additional charge is transported as reflected in the second term (this is the electro-osmotic contribution). The third contribution stems from the conduction behind the plane of shear:

$$\Theta \equiv \frac{\sigma_c^{in}/\overline{\beta}_c^{in}}{\sigma_c^{out}/\beta^{\infty}} \tag{30}$$

with σ_c^{in} the countercharge inside the plane of shear, σ_c^{out} the countercharge outside the plane of shear and $\overline{\beta}_c^{in}$ the average friction coefficient of the counterions inside the stagnant part:

$$\sigma_c^{in}/\overline{\beta}_c^{in} \equiv \int_{stagnant\ part\ d} \rho_c/\beta_c\, dr \tag{31}$$

From these boundary conditions it can be seen that the concentration of neighbouring electrolyte is polarized (called concentration polarization) resulting in $d_n \neq 0$ (see eq. (23)). The current entering the double layer region normally from the adjacent electrolyte is equally carried by co- and counterions. This current is balanced by (an in- or decrease in) the tangential currents in the double layer region. However, these currents are mainly carried by counterions. Hence the double layer will act as a source of neutral electrolyte on one side of the particle and as a sink on the other side. Neutral excesses/depletions will diffuse into the neighbouring electrolyte and spread out over distances of the order of the particle radius. After switching on the external field at t=0, a few times the characteristic time τ is needed to establish a stationary state. In an alternating field the surface is alternatingly acting as a sink or as a source, resulting in the concentration profile given by (23).

Applying these inner boundary conditions to (23) and (24) yields:

$$d_{\psi} = -\frac{1}{2} + \frac{3\,\mathrm{Re}\,l}{2 + 2\,\mathrm{Re}\,l(\gamma+1)} \tag{32}$$

$$\gamma \equiv \frac{1+\lambda a}{1+\lambda a+\frac{1}{2}(\lambda a)^2} = 1 - \frac{\omega\tau\sqrt{\omega\tau}+\mathrm{i}\omega\tau}{(1+\sqrt{\omega\tau})(1+\omega\tau)}$$

or to a good approximation writing d_{ψ} in its real and imaginary parts:

$$d_{\psi} = -\frac{1}{2}\left[\frac{1-Rel}{1+2Rel} - 3\left(\frac{Rel}{1+2Rel}\right)^2 \cdot \left(\frac{\omega\tau\sqrt{\omega\tau}}{(1+\sqrt{\omega\tau})(1+\omega\tau) - Rel\,\omega\tau\sqrt{\omega\tau}/(1+2Rel)} + \frac{\mathrm{i}\,\omega\tau}{(1+\sqrt{\omega\tau})(1+\omega\tau)}\right)\right] \tag{33}$$

2.4. CONDUCTIVITY IN AN ALTERNATING ELECTRICAL FIELD

The complex conductivity $\hat{K}$ of a system is defined by

$$\vec{I}_0 = \hat{K}\vec{E}_0 \tag{34}$$

where $\vec{I} = \vec{I}_0 \exp \mathrm{i}\omega t$ is the macroscopic current density. Formally $\hat{K}$ can be decomposed into its real and complex parts:

$$\hat{K}(\varphi,\omega) \equiv K(\varphi,\omega) + \mathrm{i}\,\omega\varepsilon_0\varepsilon_r(\varphi,\omega) \tag{35}$$

In absence of particles in the cell $\hat{K}$ simply reduces to:

$$\hat{K}(0,\omega) = K^L + \mathrm{i}\,\omega\varepsilon_0\varepsilon_r^L \equiv \hat{K}^L(\omega) \tag{36}$$

where K^L is the conductivity of the electrolyte solution and ε_r^L the relative dielectric permittivity of the electrolyte solution (about 80 for water) and both independent of the frequency in the regime of interest ($\omega/2\pi$<1MHz). For dilute sols of spherical particles with volume fraction φ, $\hat{K}$ can be related to the electrical dipole strength neglecting $O(\varphi^2)$ contributions [2]:

$$\hat{K}(\varphi,\omega) = \hat{K}^L(\omega) + 3\varphi d_\psi \hat{K}^L \tag{37}$$

Using (33) we find to a good approximation [5,11]:

$$K(\varphi,\omega) = K^L + 3\varphi K^L\left[-\frac{1}{2}\frac{1-Rel}{1+2Rel} + \right.$$
$$\left. + \frac{3}{8}\left(\frac{2Rel}{1+2Rel}\right)^2 \frac{\omega\tau\sqrt{\omega\tau}}{(1+\sqrt{\omega\tau})(1+\omega\tau) - Rel\,\omega\tau\sqrt{\omega\tau}/(1+2Rel)}\right] \tag{38}$$

The conductivity of the solution is changed by adding particles. The first contribution between square brackets reflects the static conductivity increment. This contribution is negative for *Rel*<1, positive for *Rel*>1 and zero for *Rel*=1. Upon placing a non-conducting, uncharged particle in an electrolyte solution, the overall static conductivity will decrease because the non-conducting bulk of the particle has taken the place of conducting electrolyte. In general, however, particles are charged and exhibit surface conduction. This gives a positive contribution to the overall conductivity. The two effects cancel at *Rel*=1 which is not surprising given the physical meaning of *Rel.*
The second contribution between brackets reflects the relaxation process. It contributes significantly for $\omega > \tau^{-1}$.
In a similar way we find

$$\varepsilon_r(\varphi,\omega) = \varepsilon_r^L + \varepsilon_r^L \frac{9}{16}\varphi(\kappa a)^2\left(\frac{2Rel}{1+2Rel}\right)^2 \frac{1}{(1+\sqrt{\omega\tau})(1+\omega\tau)} \tag{39}$$

The out-of-phase currents introduced by the relaxation of the neutral charge cloud are experimentally interpreted as a change in the dielectric permittivity. At frequencies $\omega > \tau^{-1}$ these charge clouds cannot fully extend. This gives rise to dispersion around $\omega = \tau^{-1}$.
In dielectric and conductivity studies often increments are considered. They are defined and given by:

$$\Delta K(\varphi,\omega) \equiv K(\varphi,\omega')\Big|_{\omega_{ref,K} \ll \tau^{-1}}^{\omega} =$$
$$= K^L \frac{9}{8}\varphi\left(\frac{2Rel}{1+2Rel}\right)^2 \frac{\omega\tau\sqrt{\omega\tau}}{(1+\sqrt{\omega\tau})(1+\omega\tau) - Rel\,\omega\tau\sqrt{\omega\tau}/(1+2Rel)} \tag{40}$$

and

$$\Delta\varepsilon_r(\varphi,\omega) \equiv \varepsilon_r(\varphi,\omega')\Big|_{\omega_{ref,\varepsilon}>>\tau^{-1}}^{\omega} = \\ = \varepsilon_r^L \frac{9}{16}\varphi(\kappa a)^2\left(\frac{2Rel}{1+2Rel}\right)^2 \frac{1}{(1+\sqrt{\omega\tau})(1+\omega\tau)} \tag{41}$$

2.5. ELECTROPHORETIC MOBILITY

The complete boundary conditions (25) are used by Fixman [8,9] to find the dipole strength, but his analysis results in long and untransparant solutions. This problem is caused by the coupling between the two boundary conditions (one for each ion species) by the osmotic term which is depending on both ion species. Therefore, we prefer the results obtained by the much more simple boundary conditions (26) and (27) and find for the electrophoretic mobility u after some algebra [11]:

$$u\frac{\eta e}{\varepsilon kT} = \frac{e}{kT}\zeta + \frac{2Rel}{1+2Rel}\left\{\frac{2}{z}\ln 2 - \frac{2}{z}\ln\left(1+\exp\left(\frac{ze\zeta}{2kT}\right)\right)\right\} \tag{42}$$

At low Rel values the second part is negligible and the Helmholtz-Smoluchowski result is obtained. This condition is usually met at not too low electrolyte concentration. At higher Rel values ($Rel=O(1)$) the polarization of the double layer must be taken into account. This results in the second, opposite contribution. An important remark is that the conduction (eq. (38) and (39)) only depends on Rel, whereas the mobility is a function of Rel *and* ζ. From the (complex) conductivity experiments Rel can be obtained. Substitution of Rel and the measured mobility in (42) yields ζ_{comb}. Now it is easy to obtain the fraction of conduction behind the plane of shear from (30).

3. Dielectric spectrometer

One of the most characteristic problems in the measurement of low frequency dielectric dispersion is the large ratio between the resistive and capacitive components of the complex admittance $\hat{Y}$ (the out-of-phase currents are small compared to the in-phase currents, especially at low frequencies). Therefore accurate measurements of the components require a powerful phase-sensitive detector. A second, nagging, problem is electrode polarization, which may be particularly troublesome at low frequencies. There are two options to avoid or suppress the latter:

(1) construction of a cell with variable electrode distance, so that the bulk impedance can be separated from the electrode impedance,

(2) working with a four-electrode system, in which the sensor electrodes carry nearly no current which minimizes polarization.

Discussions of measuring cells and electronics can be found in the literature [12,13]. We use the cell from [13] given in figure 1. The cell contains two large current-carrying electrodes b, usually made of platinum. An a.c. voltage is supplied from a generator and the current j through the cell is transformed into a voltage V_i by a current-to-voltage converter. Both V_i and the voltage drop between the sensor electrodes V_s are fed into a detector unit. This unit generates a potential difference $V_s - V_i$ suitable for analysis by a commercial frequency response analyzer which yields the real and imaginary components of the admittance. The cell can be alternately filled with suspension and blank electrolyte via openings c by a pump. The cell is thermostatted.

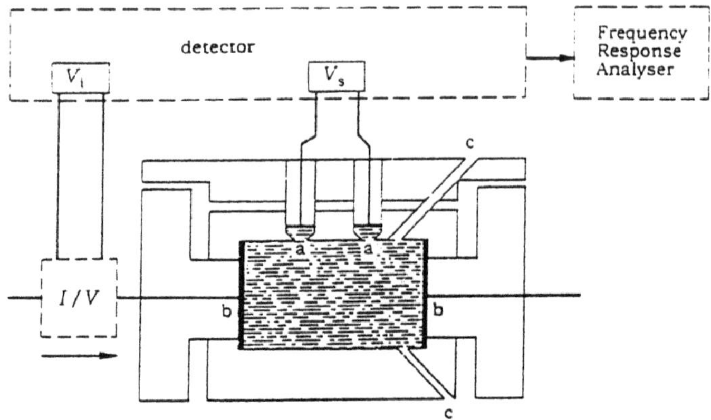

Figure 1. Dielectric spectrometer

4. Discussion of measurements on spherical hairy bacterium cells.

Experiments were performed on living Corynebacterium cells (DSM 44016). At first sight biological systems seem to be far from model systems. But these cells are nearly spherical and we were able to make homodisperse suspensions. The radius of the cells was obtained from dynamic light scattering experiments and E.M. photographs which yielded 570 nm. The cell membrane is impenetrable for ions. We noticed some leakage of ions from the interior of the cells towards the bulk solution, resulting in a slow increase of the overall conduction. However we were able to perform the measurements fast enough to make this artefact neglegibly small [6,7]. Outside the cell membrane the cell wall is situated, consisting of macromolecules. These macromolecules bear acidic and alkaline groups responsible for the surface charge. The cell wall thickness is approximately 40 nm (from E.M. photographs).

All the measurements were done in 1 mM KNO_3 electrolyte. In this situation the Debye screening length is 10 nm which is much smaller than the cell wall thickness. We therefore expect a high fraction of the mobile countercharge to be situated inside the cell wall. The cells are negatively charged at pH=6.5 where all the experiments where carried out. The liquid in this cell wall will be stagnant. A stagnant layer which is conducting may look like a paradox, but the phenomenon is encountered in other places as well. For instance, a few percent of gelatine added to water may hydrodynamically immobilize the liquid completely, without markedly impairing ionic conduction or self-diffusion of dissolved ions.

It follows that such bacterial cells are excellent model systems, displaying surface conduction behind the slip plane. For such cells SEM does not work, but the adapted theory, developed above, can be readily applied.

Experimental results for the permittivity and conductivity increments, measured at 25°C, are given in figure 2.

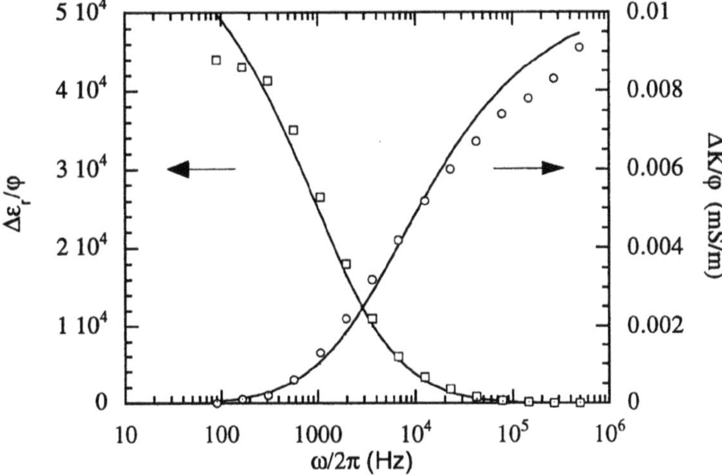

Figure 2. Permittivity and conductivity increment. See text for explanation.

In the frequency range 300 Hz-500 kHz the response showed excellent linear dependency on the volume fraction up to 10%. This means that particle interaction, which was not taken into account by the theory, is indeed negligible. The drawn curves are given by our theory with *Rel* =1.0 and $\tau = 7.6 \cdot 10^{-5}$s. Both curves are explained with the same *Rel* and τ. The observed relaxation frequency for both $\Delta K/\varphi$ and $\Delta\varepsilon_r/\varphi$ agrees well with the ion mobility and cell radius.

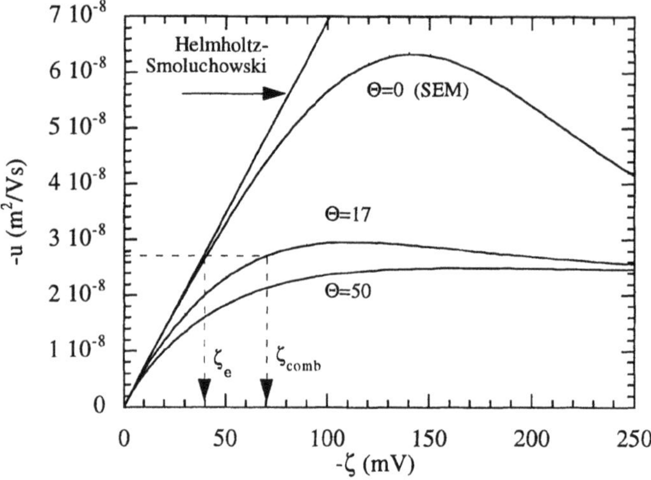

Figure 3. The mobility versus the ζ-potential. See text for explanation.

The mobility of the particles was measured (in 1 mM KNO_3 at 25°C, $E_0 \approx 2500$V/m) with a commercial laser-Doppler velocimetric device (Zetasizer 3, Malvern Instruments, Ltd., Worcestershire, Great Britain). We found u=-2.7·10^{-8}m^2/Vs.
With Rel=1.0 and (42) we obtain: ζ_{comb}=-70 mV , Θ =17.
In figure 3. the electrophoretic mobility versus the ζ-potential is plotted for κa=56, and different Θ-values according to equation (42). ζ_e and ζ_{comb} are shown. If the mobility measurement would have been interpreted under SEM conditions (Θ =0) we would have found ζ_e =-40mV. Interpretation due to Helmholtz-Smoluchowski would also yield $\zeta = -40\text{mV}$. This illustrates the statement made earlier that many literature data of ζ-potentials have to be reconsidered. This present paper indicates how this should be done.

From ζ_{comb} we compute the counterion charge outside the shear plane to be $\sigma_c^{out} = -0.0059\text{C/m}^2$. With (30) we obtain $\sigma_c^{in}\beta^{\infty}/\bar{\beta}_c^{in}$=-0.100C/m^2. Titration of the isolated cell walls yields a surface charge σ^0=-0.35C/m^2. Because Θ is high, nearly all conduction and charge is situated in the stagnant part and we can identify σ_c^{in} with σ^0. From this we find the average diffusivity of the counterions in the cell wall to be 0.29±0.05 times its bulk value. This means that there is approximately 17/0.29=60 times more charge behind the plane of shear than beyond the plane of shear. So application of our model allows interesting additional information to be obtained.

5. Conclusions and perspectives

From (complex) conductivity experiments in combination with electrophoretic measurements ζ_{comb} can be obtained by taking conduction behind the plane of shear into account. This ζ_{comb} may differ significantly from ζ_e and ζ_{con} especially at low electrolyte concentration ($Rel=O(1)$). Then ζ_{comb} is a better choice for the Stern potential ψ^d in stability studies than ζ_e or ζ_{con}.
At higher electrolyte concentration ($Rel<<1$) there will be no significant difference between the electrokinetic potentials ζ_{comb} and ζ_e, even when there is substantial conduction behind the plane of shear. However then, of course, $\zeta_{con}<<\zeta_{comb}\approx\zeta_e$ if ζ_{con} can be determined experimentally at all. From (38) it can be seen that for these low *Rel* values the static conductivity increment is reaching a limiting value which inhibits accurate determination of *Rel* and ζ_{comb}. These problems also occur for the frequency-depending part of the conduction (see (38) and (39)): at low *Rel* these contributions get small.

When it is possible to determine the surface charge from titration, the (average) mobility in the stagnant part of the double layer can be obtained. This determination is not always possible. For instance, in the case of porous Stöber silica, also charge in the bulk of the particle is titrated.
The advantage of complex conductivity measurements over static conductivity measurements is the extra information (the particle radius) and the insensitivity for uncontrolled changes in the background electrolyte conductivity [6,7]. New developments in theory [14] show that exchange of Stern layer charge from one side of the particle to the other side of the particle and exchange from the Stern layer to the bulk solution can influence the complex currents. The kinetics of these effects give rise to new relaxation frequencies. In the future it might be possible to study the kinetics of these effects by alternating current measurements.

Another interesting task is the development of new theory for coupling the observed surface conduction behind the shear-plane to surface roughness, porosity, the visco-electric effect or the presence of a macromolecular surface layer, depending on the kind of system we are studying. Because these theories are not developed yet, trends in Θ can only be interpreted qualitatively.
In the commercial production of colloidal particles, trends in ζ_{comb} and in Θ due to changes in process conditions give information on the changing quality of the product. In this respect this work also is of practical interest.

6. References

(1) Wiersema, P. H., Loeb, A.L., Overbeek, J.Th.G. (1966) Calculation of the electrophoretic mobility of a spherical colloid particle, *J. Colloid Sci.* **22**, 78-99.

(2) Dukhin, S. S., Shilov, V. N. (1974) *Dielectric Phenomena and the Double Layer in Disperse Systems and Polyelectrolytes*, Wiley, New York.

(3) O'Brien, R. W., White, L.R. (1978) Electrophoretic mobility of a spherical colloidal particle, *J. Chem. Soc. Faraday Trans. (II)* **74**, 1607.

(4) Kijlstra, J., van Leeuwen, H.P., Lyklema, J. (1993) Low-frequency dielectric relaxation of hematite and silica sols, *Langmuir* **9**, 1625-1633.

(5) Lyklema (1995) *Fundamentals of Interface and Colloid Science, Solid-liquid interfaces*, Volume II, Academic Press, London.

(6) van der Wal, A., Minor, M. To be published.

(7) van der Wal, A., Minor, M. To be published.

(8) Fixman, M. (1980) Charged macromolecules in external fields. I. The sphere, *J. Chem. Phys.* **72**, 5177-5186.

(9) Fixman, M. (1983) Thin double layer approximation for electrophoresis and dielectric response, *J. Chem. Phys.* **78**, 1483-1491.

(10) Hunter (1989) *Foundation of Colloid Science*, Volume II, Clarendon Press, Oxford.

(11) Minor, M., van der Wal, A. To be published.

(12) Myers, D. F., Saville, D.A. (1989) Dielectric spectroscopy of colloidal suspensions. I. The dielectric spectrometer, *J. Colloid Interface Sci.* **131**, 448-460.

(13) Kijlstra, J., Wegh, R.A.J., van Leeuwen, H.P. (1994) Impedance spectroscopy of colloids, *J. Electroanal. Chem.* **366**, 37-42.

(14) Minor, M. To be published.

SOME COLLOID-CHEMICAL ASPECTS OF THE SMALL PARTICLES CONTACT INTERACTIONS

Eugene D. SHCHUKIN
Institute for Physical Chemistry of the Russian Academy of Sciences, Moscow 117915, Russia, and
The Johns Hopkins University, Baltimore, MD 21218, USA

Depending on the character of the particles' interaction, two main types of the particle contacts (and, respectively, two types of disperse structures) can be considered, namely, coagulational and phase contacts [1-3]. In *coagulational* contacts, disperse phase particle adhesion is limited to simple "touching", either directly or by means of a residual layer of dispersion msdium, mainly due to molecular van der Waals forces. These contacts are not strong (e.g., 10^{-9} - 10^{-7} N) and are mechanically reversible, predetermining the viscoelastic properties of the corresponding thixotropic systems. A *phase* contact, on the other hand, is the result of bridging of particles (by displacing the dispersion medium from an area much bigger than the sizes of elementary cell) due to the same valence bonding as in the bulk of the given phase. These contacts are relatively strong (e.g., 10^{-6} N and higher) and, generally speaking, are mechanically irreversible. Elastic-brittle or elastic-plastic behavior is typical for these structures. Phase contacts determine the material's quality during use. In contrast, properties of coagulational contacts (and corresponding systems) are important during numerous steps of the material production and forming, especially when easy mobility is required. In diluted systems (sols), interactions of particles (energy of adhesion in contact being compared with the kT value) predetermines aggregative stability of a system.

In the Part 1 of this paper, two aspects of coagulational contacts are considered in which small sizes on an object are especially significant: (1.1) conditions of spontaneous transition from an aggregate (coagulational structure) to peptizate (sol) and equilibrium between them in fine disperse systems; (1.2) nano-level processes in contact zone studied by molecular dynamics simulation revealing some definite limits for the thermodynamical (continuous) approach. In both cases, particular attention is paid to the notions of lyophilicity/lyophobicity of a system.

In the Part 2, some results are presented in the experimental studies of the transition form coagulational to phase contacts, allowing one to elucidate arising of the initial bridges between particles as the fluctuational process of the nano-size nuclea formation in contact zone. Such approach is illustrated with two examples: (2.1) phase contacts formation between gypsum crystals (calcium sulphate dihydrate) in supersaturated calcium sulphate solution, during the hydration hardening process; (2.2) bridging of amorphous particles of

E. Pelizzetti (ed.), Fine Particles Science and Technology, 239–253.

polysilicic acid (called below, for simplicity, silica) in supersaturated silicic acid solution, that is, sol-gel transition.

1. Coagulational contact between particles. Spontaneous dispersion

1.1. CONDITIONS OF COAGULATION AND DISAGGREGATION IN FINE DISPERSE SYSTEM

In the colloid-chemical studies of surface phenomena in disperse systems and their stability, one of the decisive factors has been the development of quantitative methods for studying interaction between solid particles in a given medium . The principal thermodynamical characteristic of such interaction - the disjoining pressure $\Pi(h)$ is defined as the force of interaction per unit area between two plane-parallel surfaces separated by a gap h. Its first integral with respect to h gives the work performed upon approach of the surfaces $F(h)$ which is commonly called the specific *free energy of interaction* (like Π, positive for repulsion, and negative for attraction). Direct measurements of this quantity are practically unrealizable for a gap between two solid surfaces. However, in the case of a nonplanar gap, e.g., between two identical spheres with radius R, integration of $\Pi(h)$ over the surface of these particles (double integration) gives the *force* of interaction between them $p_1(h)$. Derjaguin's formula [4] relates these two characteristics by the linear equation: $p_1(h) = \pi RF(\text{h})$; in the case of an equilibrium gap $h = h_0$ (with direct contact), we will write: $p_1(h_0) = p_1 = \pi RF = \pi RF(h_0)$. The quantity F is related to the interfacial energy σ_{12} and to the specific free energy of the film γ as follows: $-F = 2\sigma_{12}\text{-}\gamma$. If the second term is small, i.e., if the contact is perfect, then $\sigma_{12} \sim -F/2$.
Corresponding measurements have been realized using magnetoelectric suspension as the force meter (Fig.1). Electrical current in one direction presses the particles together with a force f_1; current in the reverse direction gives the magnitude of the force of detachment, i.e., the strength of the contact p_1 [1,2,5].These experiments reveal a broad spectrum of the free energy of interaction (and of σ_{12}). In the case of dissimilar phases such as nonpolar surfaces (methilated and fluorinated glass) in an aqueous medium, polar surfaces (glass or quartz) in nonpolar hydrocarbons, or these and other surfaces in air, values of $-F/2$ are measured in tens of ergs/cm^2 (mJ/m^2); this is a lyophobic character of the interface. In contrast, in the case of similar boundary phases such as nonpolar particles in hydrocarbons or a hydroxylated glass surface in water, the interaction is weakened by 3-4 orders of magnitude, down to ~ 0.01 erg/cm^2; this is lyophilicity of the interface. Media of intermediate polarity, their mixtures, and solutions of surfactants make it possible to cover all this enormous interval and thereby control the properties of the system (Fig.2; in this and following figures, F is given by modulus).

These results allowed also to take a new approach to one of the central problems in stability: the possibility of spontaneous dispersion (peptization) of coagulated small particles and formation of a sol in equiblibrium with the coagulate - a thermodynamically stable, lyophilic dispersion with a certain colloidal solubility [2,6,7].

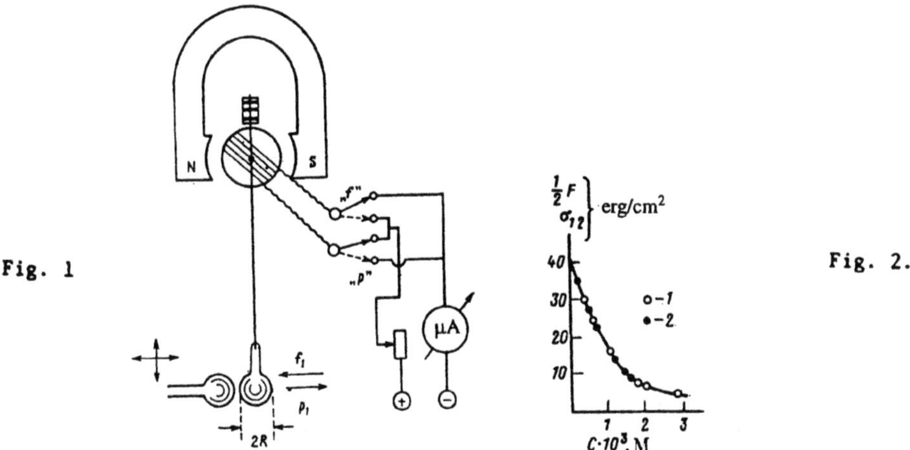

Figure 1. Scheme of instrument for measuring force of adhesion p_1 between two macroscopic particles after they have been pressed together with force f_1 (instrument uses a magnetoelectric galvanometer as the force meter).

Figure 2. Specific free energy of interaction (a half, by modulus) $F/2 = p_1/2\pi R$ (erg/cm^2) of hydrophobized (methylated) surfaces of glass (1), and interfacial energy σ_{12} (erg/cm^2), evaluated from the change in work of wetting $\Delta W_w = RT$ Int($\Gamma d\ln C$) (2) as functions of concentration C of sodium dodecylsulphate in dispersion medium.

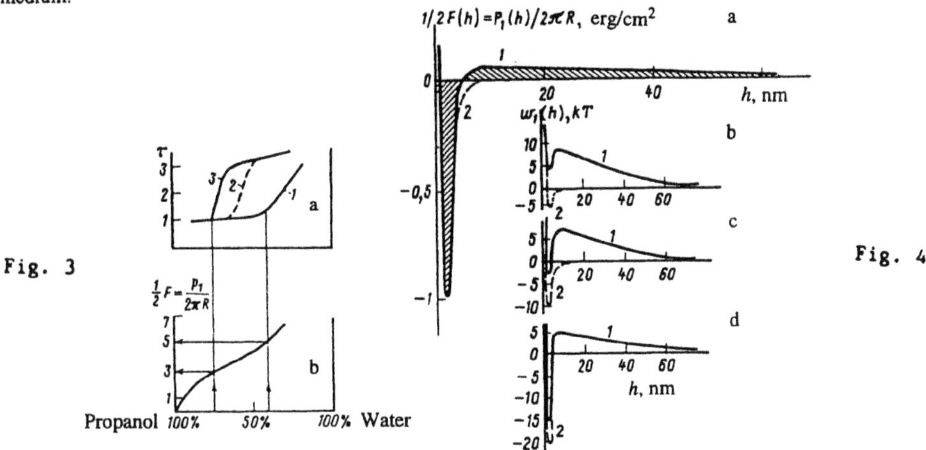

Figure 3. Interaction of hydrophobic (methylated) surfaces in mixtures of propanol and water: increase in turbidity τ (rel. units) of dispersions of aerosil (a), and free energy of interaction F (a half, by modulus; erg/cm^2), according to measurements of the force of adhesion p_1 between particles with radius $R \sim 1$ mm (b), as functions of polarity of medium, without foreign electrolyte (1), and with addition of KCl to a concentration of 10^{-2} (2) or 10^{-1} M (3).

Figure 4. Specific free energy of interaction F (a half; erg/cm^2) from measurements of the force of interaction $p_1(h)$ (dyn) between hydrophylic surfaces of spherical glass particles (a), and energy of interaction u_1, in units of kT, for corresponding nano-size paricles (b-d), as functions of distance h, in aqueous solutions of cetylpiridinium bromide: without foreign electrolyte (1), and in the presence of 10^{-1} M NaCl (2). Concentrations: 10^{-6} M (b), 5.10^{-6} M (c), or 2.10^{-5} M (d).

Such a process is possible if the work of detachment of the particle $^1/_2zw_1$ is compensated by its inclusion in thermal Brownian motion; or, in other words, the increase $\Delta\mathcal{F}$ in the free energy of the system $n(z/2)w_1$ is commensurate with its decrease due to the entropy term $nkT\ln\alpha$, where α is determined by the ratio of free volumes in the peptizate and in the aggregate, or, in a rougher approximation, by the ratio of concentrations of molecules of dispersion medium n_0 and disperse phase particles n: $\Delta\mathcal{F}= n(z/2)w_1-nkT\ln\alpha < 0$. Here, z is the coordination number (number of neighboring particles in the aggregate), $w_1 =]u_1[$ is the work of the bond rupture between two neighbouring particles. The minimum in free energy of the system determines the equilibrium colloidal solubility $n \sim n_0\exp[-(z/2)w_1/kT]$. Thus, of decisive role here is the energy of adhesion in an individual contact $u = -$ Int $p_1(h)dh$ from ∞ to h_0. Direct experimental measurements of this critically important value encounter severe difficulties; we call this "the problem of the third integral", since u_1 is the third integral of the disjoining pressure. Theoretical estimate is simple only for dispersion component of the attraction: $w_1{}^d = AR/12h_0$, where A is the complex Hamaker constant, h_0 is of the order of a few angstroms. For other types of interaction, $w_1 = p_1(h_0)h_{eff}$, and, depending on the conditions, the effective value h_{eff} may amount from angstroms up to hundreds and thousands of angstroms.

In order to introduce determinacy into the situation, we require an additional equation, i.e., an independent experimental method for observing the interaction in question. This is achieved by measuring the parameter $F/2 = p_1/2\pi R$ on macroscopic, millimeter-size spheres and comparing these values with observations of stability, with the same variations of the medium, in analogous colloidally disperse systems - sols of hydrophilic or hydrophobic (methylated) aerosil with individual particle size $2R_a \sim 10$ - 20 nm [5,7]. Upon the addition of water to an initially stable sol of a methylated aerosil in propanol (i.e., upon lyophobization of the system), turbidity sets in at a critical concentration of water C_c amounting to several tens of persent by volume (Fig.3). Measurements of the force of adhesion of similarly methylated spherical samples in a mixture of propanol and water with a concentration C_c give a value $F_c/2 = p_{1c}/2\pi R$ of several ergs/cm^2. Values p_1 and R are proportional; thus, for the aerosil particles, when coagulation sets in, the force of adhesion is $p_{1ca} \sim \pi R_a F_c$.

An independent, approximate evaluation of w_{1c} for aerosil particles is possible from the following consideration. In the formula for colloidal solubility, the value of the exponent can lie only within a relatively narrow interval of values $(z/2)w_{1c}/kT = \beta = \ln(n_0/n) \sim 10$ - 15. It is actually easy to see that even $\beta = 5$ corresponds to an unrealistically high concentration of the disperse phase, and $\beta = 25$ corresponds to a vanishingly small one. Since the number z in a loose coagulate is quite low, and the factor $z/2$ differs little from unity, we can give the following order-of-magnitude estimate: $w_{1c} \sim (10\text{-}15)kT \sim (4\text{-}6).10^{-13}$ erg. This enables evaluate h_{eff}: under these particular conditions $h_{eff} \sim w_{1c}/p_{1ca} \sim w_{1c}/\pi R_a F_c \sim 5$-10 angstroms.which is in agreement with data [8,9], obtained by different methods.

In this same scheme of solution of the "problem of the third integral of $\Pi(h)$", i.e., in evaluation of the energy of the small paricles adhesion, a quantitative analysis has been made of the influence of an electrolyte on the magnitude of w_1, i.e., on the depth of the primary potential well [2,5]. Experiments have shown that the DLVO electrostatic barrier cannot have any significant effect on the value $F/2$; even if this value drops to 1-2 erg/cm^2, it is still far greater than the height of the barrier U^e, which is no higher than 0.05-0.1 erg/cm^2. Therefore, measurements of the force p_1 give practically the same results with or without an added electrolyte (Fig.4). However, integral of $p_1^e(h) = U^e(h)\pi R_a$ across the large distance h_{eff}^e may give a considerable contribution w^e. For example, setting $U^e = 0.1$ erg/cm^2, $R_a = 10$ nm, and $h_{eff}^e = 15$ nm, we find $w_1^e \sim U^e\pi R_a h_{eff}^e \sim 4.10^{-13}$ erg $\sim 10kT$. An analogous estimate of the depth $w_1 \sim]F/2[.2\pi R_a h_{eff}$ (when barrier is suppressed), with $R_a = 10$ nm, $h_{eff} \sim$ a few angstroms, and $-F/2 \sim 2$ erg/cm^2, gives $w_1 \sim (10\text{-}20)kT$.

This means that as long as free energy of interaction is no more than a few erg/cm^2 (which is already very far from "complete lyophilicity" of the surface), the system under consideration is stable, even when an electrolyte is introduced. In the absence of electrolyte, stability is maintained with greater lyophobization, since w^e is deducted from w_1. Let us emphasize once more that, in this basically new formulation of the problem, we are concerned with thermodynamic factors of stability, i.e., not the kinetic role of the DLVO barrier, but rather its influence on the decrease in depth of the primary potential well - the energy of adhesion. With a deep well (tens of kT), the contribution of the DLVO barrier no longer protects against coagulation. This was shown both for sols of a hydrophobic aerosil in water-alcohol mixtures and for a hydrophilic aerosil - in aqueous solutions of cetylpyridinium bromide [2,5].This approach to study of the interaction of disperse phase particles allows to take a new step in solving the fundamental problem of the nature of the transition between lyophilicity and lyophobicity in fine disperse systems.

1.2. MOLECULAR DYNAMICS SIMULATION OF AN ELEMENTARY CONTACT BETWEEN PARTICLES IN WETTING LIQUID

The approach [4,5] mentioned above relates the disjoining pressure integrated over the contact area for smooth coNvex surfaces (i.e., the force between particles $p_1(h)$) to the corresponding integral with respect to the distance (i.e., the free energy of surfaces interaction $F(h)$). But this is correct only if the distance between opposite parts of these two surfaces is proportional to the square of the distance from the center of contact; molecular structure is not taken into account in this approach. However, both experimental and theoretical studies show that molecular structure of the liquid causes disjoining pressure to oscillate with the gap width [8,10]; the finite size of liquid molecules restricts their penetrating the narrow gap where adhesion is overcome during contact rupture.

Computer simulation of the rupture of contact between a nano-size solid particle and a flat surface in a liquid medium has been used by us to understand this detachment process on the molecular level, and find out to what extent the continuous approach is still valid for a small system [11]. Both liquid (l), particle (p) and solid substrate (s) have been described as two-dimensional Lennard-Jones systems, with dimensionless parameters ε, r_0 and m, and corresponding units for time, velocity and force.

Two model particles were considered (Fig.5and 6). The shape of the "large" (convex) particle was chosen to get the gap width proportional to the distance from the center of contact: in two dimensions this gives proportionality between the force of adhesion and the free energy of interaction of two plane-parallel surfaces. The "small" particle has a larger ("flat") contact area with substrate and almost no surface upon which a positive disjoining force from the liquid could act. Parameters of interaction have been chosen to insure the complete lyophilicity of a system: $\varepsilon_{ls} = \varepsilon_{lp} = {}^1/_2(\varepsilon_{ll}+\varepsilon_{sp})$.

Contact rupture has been simulated in two ways. Either the central atom of the particle was moved upward with a constant velocity, and the external force and work were monitored (alongside with other variables of the system), - or a constant force (or no force) was applied to the central atom, and the mean time for the particle to escape was measured in series of computer experiments. Some runs were also performed to monitor averages and fluctuations of the force, and viscous resistance from the liquid.

"Large" (convex) paricle. The first run of experiments was done with no external forrce: to look for a possibility of spontaneous detachment in such lyophilic system. Molecular pictures of the escape process are variable (Fig.5): either the particle escapes - in 60% of the cases, 50 time units each, or it is tilted and "glued" to substrate, with one monolayer of liquid between. Disjoining pressure of the liquid is negative, i.e., it presses particle to substrate; it originates as a rather small difference between large forces on the top and bottom parts. There are large fluctuations in the forces, many times exceeding the average force.

Then, rupture was induced by moving the central atom of a particle with various constant velocities. Only for the lowest one (0.005), the work performed upon the system was less than s/p energy. At the higher velocities, external work increases due to viscous resistance.

The force necessary to move a particle with constant velocity (Fig.7) is at first positive, and becomes negative at a distance 0.20. It anticorrelates with the force from the liquid upon particle. Thus, oscillations of the l/p force for such convex nano-size particle proceed with essentially smaller period than the molecular diameter.

Runs with a constant force did not show any variations of the escape time with forces up to 5 units. This is explained with the predominant role of fluctuations.

"Small" (flat) particle has a wider contact with subsrate while its surface towards the liquid is much less. The total force from the liquid is also negative, and close to that for the large particle; fluctuations are 2.5 times less, and the contact strength - 2 times more. This results in quite different character of the particle detachment. Under constant force, time for rupture obeys the well known linear dependence of ln (t) vs. force. In liquid, the picture for this lyophilic system is qualitatively the same as in vacuum, but the time to escape is longer, i.e., liquid presses particle to substrate.

Under conditions of constant velocity, penetration of liquid molecules into the flat gap between particle and substrate is observed only after displacement more than 0.6 length units (Fig.6). Therefore, external work for small particle does first increase above the energy of p/s interaction, - to start to decrease at the moment of liquid penetration.

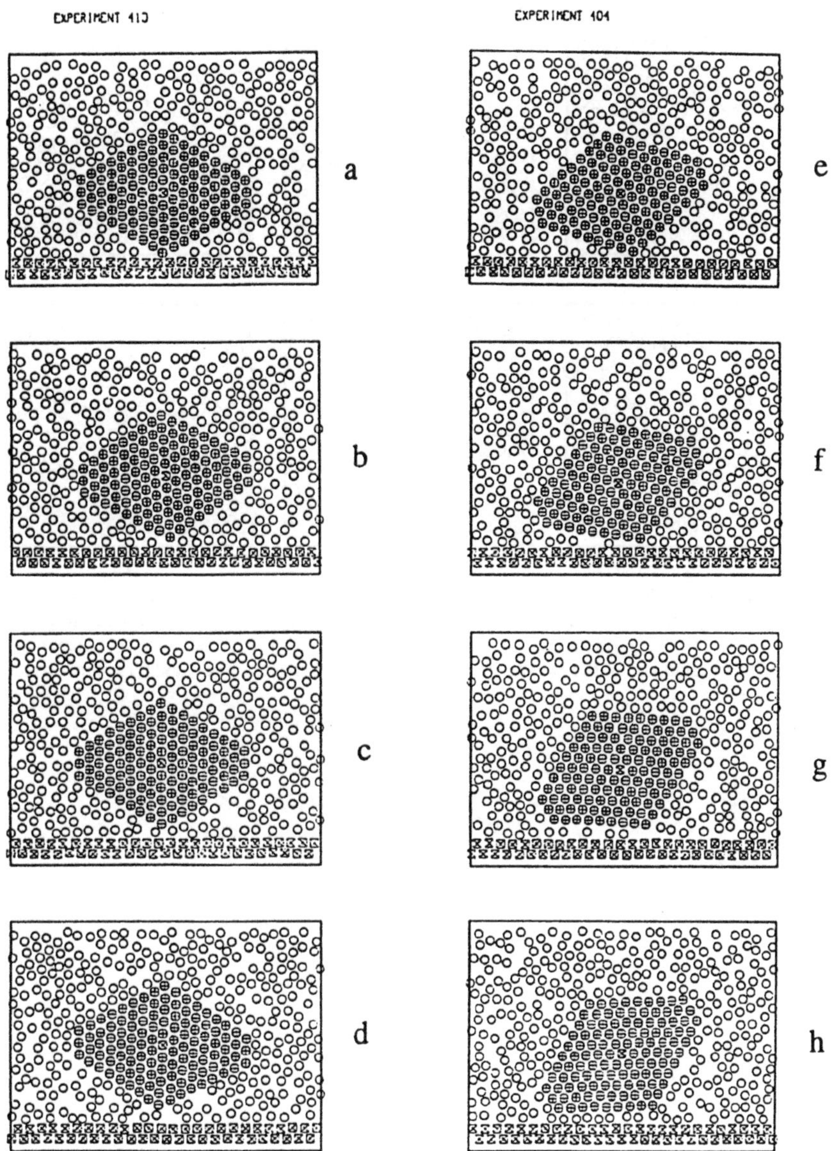

Figure 5. Spontaneous rupture of the "large" (convex) particle/substrate contact: (a) initial state for both runs; (b-d) contact rupture without tilt (25, 30 and 35 time units), (e-h) tilt of the particle (at 10, 20 30 and 40 time units).

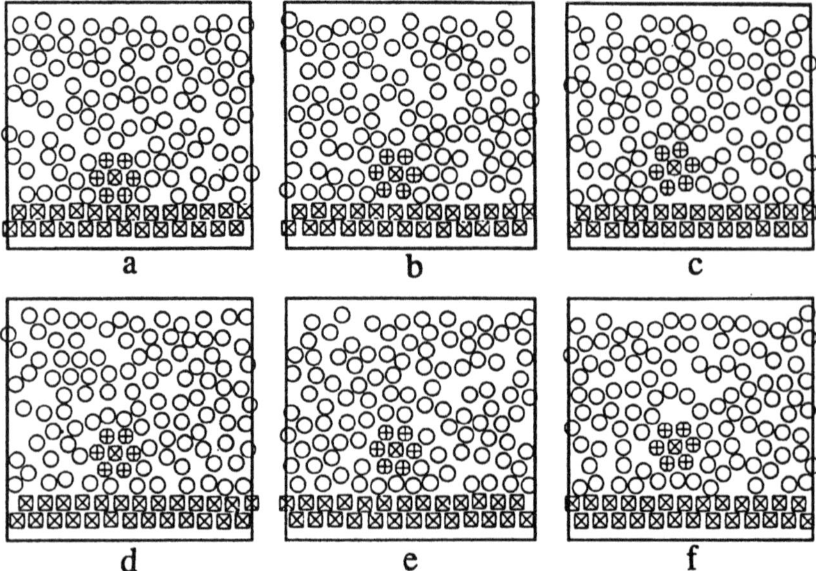

Figure 6. Rupture of the "small" (flat) particle/subsrtate contact when the particle is moved with the constant velocity 0.005. The pictures are at displacements 0, 0.15, 0.5, 0.8, 0.95, and 1.125 from the initial state. They correspond to the following situations: (a) initial state, (b) maximum external force, (c) maximum work performed, (d) first minimum of the external force, (e) first minimum of the work, and (f) second maximum of the external force.

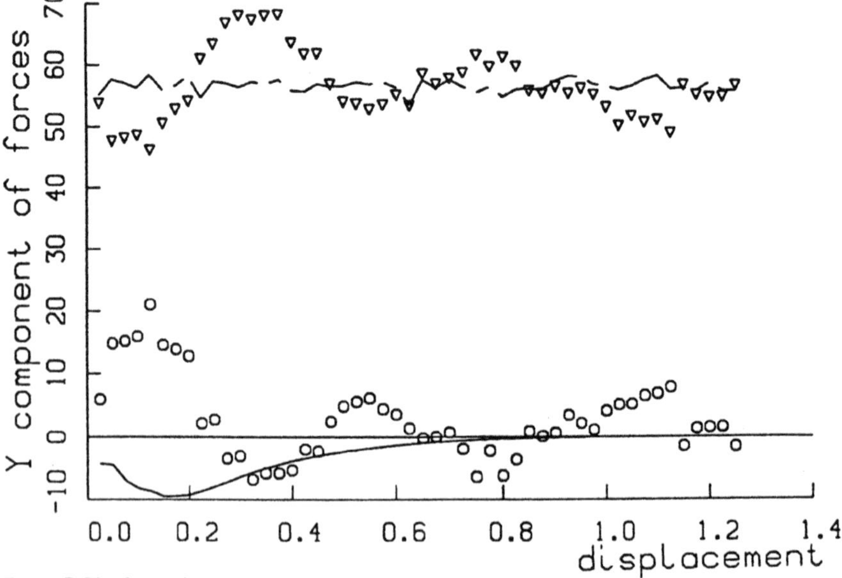

Figure 7. The forces that act upon the "large" (convex) particle in the process of contact rupture with velocity 0.005: external force (o), force from the substrate (solid line) equal to the external force in vacuum, force from the liquid upon the bottom of the particle (▽), force from the liquid upon the top part of the particle (dashed line, by modulus). The total force from the liquid upon the particle (giving disjoining pressure) is easily seen from the last two curves.

Some conclusions. The source of the observed *joining* force is that the liquid is restricted in penetrating in the contact area between particle and subsrate. Hydrostatic pressure acts, therefore, on the larger area on the top than on the bottom of the particle. These effects compete: joining force caused by hydrostatic pressure, and disjoining (positive) force due penetration of the liquid. The question is one of the balance between them. For spheres (in 3-dimensional case), both effects scale in the same way with system size. What determines balance is therefore hydrostatic pressure and geometry of the contact zone: for more sharp configuration the disjoining factor may be predominant, for more flat - the joining one [12]. The less is the angle between surfaces, the less are period and amplitude of mentioned oscillations in forces and energies. So, these effects are specific just for very small objects. In macroscopic systems, the role of asymmetry in hydrostatic pressure can be much less than in our nano-scale model. However, the high enough hydrostatic pressure might be even a way to control surface aggregation of colloidal particles.

Thus, the performed investigation shows that although the macrocopic thermodynamic theory explains a part of the behavior of a molecular system like the one in this study, new phenomena appear that only may be understood by an explicid reference to the molecular structure of the involved components. The influence of the liquid solvent upon the force and the energy of adhesion are qualitatively changed for the nano-size particles. This is caused by the discrete molecular structure of the solid and liquid phases, peculiarities of the microgeometry of the contact zone and the immossibility for liquid molecules to penetrate into the very narrow gaps where the rupture of contact takes place. Attraction and repulsion occur to be distributed in space. In such microscopic system, the role of thermal fluctuations is also increased. The joining fluctuations that press the particle towars the substrate are of less importance for the rubture process, while the disjoining ones may cause the escape of the particle. When a convex particle is moved away from the surface with a constant low velocity, the force of adhesion in not reduced by the fluid, although less work is done on the system than for the case without liquid.

In the models investigated: lyophilic systems with solid particles on flat subsrates, the liquid does insure the spontaneous escape of the convex ("large") particle. Such spontaneous escape is caused by disjoining pressure fluctuations, while the mean force from the liquid presses the particle against the substrate (i.e., the disjoining pressure is negative). For the small particle, there is also a negative disjoining pressure, but here the fluctuations are much smaller. This increases the mean time to escape.

2. Phase contact formation. Nano-size bridge nucleation between particles

Practically all commercially available materials are structures with phase contacts, starting from small ones compared to the particle size (e.g., in porous catalyst supports), to complete particle bridging (in compact metals, some composites). Many ceramic materials, products of powder technology, and structures formed by hydration may be considered to be intermediate cases. Transformations from coagulational to phase contacts are as a rule the main processes during production of materials. Such transfer may be a result of mutual plastic deformation of particles during their compression (powder metallurgy, cold welding), of sintering (ceramics, etc.), isothermal mass transfer through the liquid phase or through its residual menisci (deterioration of hydroscopic powders), and so on. In the mentioned cases, the dispersity (usually relatively coarse) has been predetermined by the initial powder. The finest structures are formed, however, by nucleation (condensation) of the highly disperse new phase in a metastable initial (mother) phase. Various crystalline and amorphous (inorganic, organic, bioorganic) materials, including cementitious structures, correspond to these *condensational* structures.

2.1. BRIDGING OF GYPSUM CRYSTALS IN CALCIUM SULPHATE SOLUTIONS

The technique of study of the conditions for formation of individual phase contacts between particles and their properties based on direct measurements of cohesive forces between two individual solid particles is similar to that described in the Part 1. Two crystals are taken: one of them is fastened to the manipulator, the other is attached to the galvanometer hand. By passing the current of defined direction and value through the circuit of the frame, one of the particles is pressed to the other by force f. After keeping the system for a time t in a given liquid phase, the current is passed in the reverse direction, and the force necessary to separate the particles (i.e., contact strength p_1) is obtained [1,3].

Crystals of calcium sulphate dihydrate (gypsum) were placed into calcium sulphate solutions supersaturated with respect to gypsum (solubilities of hemihydrate and dihydrate in water are 6.2 and 2 g/L, respectively).The following parameters were varied: supersaturation α of solution, crystal contact time t, compression force f and the crystallographic orientation of the gypsum crystal contact edges.

Measurements on individual crystals showed a wide scatter of cohesion forces for the contacts formed under the same conditions. Results of these measurments are given in the form of histograms $\rho = dn/(n_0 d\log p_1)$, where n and n_0 are the current and the total number of measurements (Fig.8).Two types of contacts occurred, essentially differing in strength: "weak" (coagulation) contacts with $p_1<10^{-2}$ dyne, and "strong" phase contacts with $p_1>10^{-1}$ dyne. Probability of bridging, W_b, may be defined as the ratio of the number of "strong" contacts w_t to the total number of contacts w_0: $W_b = w_t/w_0$. It increases with supersaturation value and contact time (Table 1). This dependence permits the initial step of bridging to connect with the fluctuational process of the primary nucleus contact formation, which is critical for the given supersaturation, in the course of hydration: $CaSO_4.{}^1/_2H_2O + {}^3/_2H_2O = CaSO_4.2H_2O$. Once arising, initial bridge spreads immediately over contact zone.

Mathematical treatment of the experimental results gives: $W_b = 1\text{-}\exp(-It)$. According to the fluctuation theory of formation of a new phase nucleus: $I = I_0\exp(-A_c/kT)$; A_c is the work of formation of the critical nucleus contact. The nucleus contact is assumed to have the form of a rectangular prism with a square base a x a in the gap h wide; we consider that the nucleus contact is cooriented with the crystal lattice of only one crystal (Fig.9). The terms σ_1, σ_2, σ_3 are the specific free surface energies of the nucleus/solution, nucleus/crystal, and crystal/solution boundaries. From the condition of an extremum of free energy, we get for critical nucleus contact: $a_c = 2h\sigma_1/[(hkT/V_1)\ln\alpha+\sigma^*]$, where $\sigma^*= 2\sigma_3-\sigma_2$, and, respectively, $A_c = 4h^2\sigma_1{}^2/[(hkT/V_1)\ln\alpha+\sigma^*]$. This essentially differs from

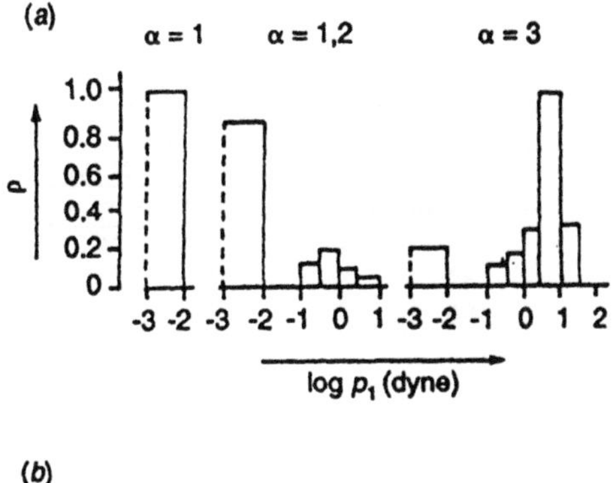

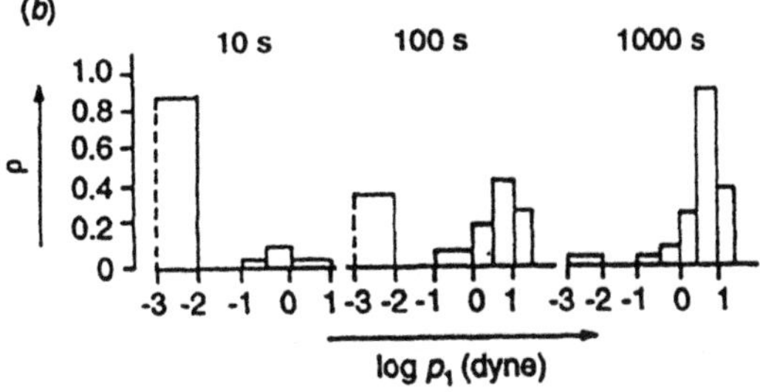

Figure 8. Differential distributions of contact strength p_1 emerging between gypsum dihydrate crystals brought together under force f = 10 dyne in calcium sulphate solutions: (a) for various supersaturations α (t = 10 s); (b) for various time of contact t (α = 1.8); ρ - distribution density.

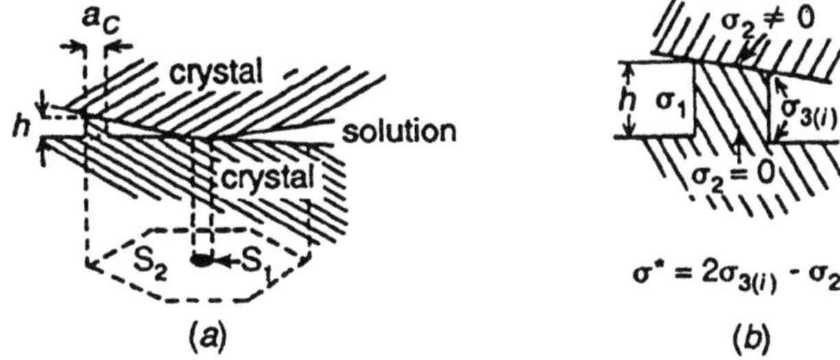

Figure 9. A possible model for the contact zone.

TABLE 1. Probabilities of bridging of gypsum and of amorphous silica particles (W_b) as a function of contact time (t) and supersaturation (α) of solution.

t (s)	W_b (%) Gypsum* $\alpha = 1.2$	$\alpha = 1.8$	$\alpha = 3.0$	Silica (pH 7) $\alpha = 1.2$	$\alpha = 1.4$	$\alpha = 3.0$
10	0	7	38	3	16	32
100	17	49	83	16	54	81
1000	33	96	100	31	64	92

TABLE 2. Probability of bridging (W_b) of $CaSO_4.2H_2O$ crystals in supersaturated calcium sulphate solutions as a function of compressing force (f); α is supersaturation, t is time of contact.

t	W_b (%) $\alpha = 1.2$ f (dyne)			$\alpha = 1.5$ f (dyne)			$\alpha = 1.8$ f (dyne)			$\alpha = 3.0$ f (dyne)		
(s)	0.1	1	10	0.1	1	10	0.1	1	10	0.1	1	10
10	0	0	0	0	1	5	0	2	7	0	7	38
100	0	4	17	0	8	24	1	14	49	4	53	83
1000	4	22	33	7	44	68	10	60	96	33	91	100

TABLE 3. The effect of solution supersaturation (α) on probability of silica particles bridging (W_b) at different pH values; $f = 10$ dyne, $t = 100$ s.

	pH 0.5		5.0		7.0				8.5	
α	3.5	7.0	2.5	5.0	1.2	1.4	1.8	3.0	1.1	2.2
W_b (%)	3.0	22.0	6.0	25.0	16.0	54.0	68.0	81.0	1.0	25.0

the traditional consideration of the two-dimensional and three-dimensional nucleation: when $\alpha=1$, A_c is finit. Graphical treatment of data on $W_b(\alpha,t)$ gives reasonable values A_c~(6-9)kT, h~1 nm, and a_c~(0.6-0.9) nm for this nano-nucleus.

The probability of bridging increases with the force f compressing the crystals (Table 2). To explain this, it can be assumed that the nucleus contact is most likely to appear not in the very zone of the "plastic contact" S_1 where the stresses are high, where solution is extruded, and thus bridge formation is difficult, - rather it occurs in the periphery of the contact zone at the area S_2 (proportional to f), where the stresses are reduced, the gap between the crystals is of suitable width, and the solution is present.

Bridging of gypsum crystals with fluorite, calcite, quartz, as well as bridging of such "alien" crystals with each other in supersaturated calcium sulphate solutions, was also studied and showed principal dependence of the probability W_b upon the conditions. Solutions containing electrolytes and surfactants were used along with pure calcium sulphate solutions.

2.2. BRIDGING OF AMORPHOUS SILICA PARTICLES IN SILICIC ACID DISPERSIONS

Silica cylindrical threads modified in silica sols (pH 7; 0.6 mass% SiO_2) were brought together (normal to each other) in supersaturated (with respect to amorfous silica) "solutions" of silicic acid. Values of pH, α, t, and f were greatly varied. An elementary event of bridging is considered here as the siloxan groups formation due to the condensation reaction: -Si-OH + HO-Si- = -Si-O-Si- + H_2O. Despite the more complicated chemistry of these systems compared with gypsum, such bridging can also be explained by the fluctuational mechanism of nucleus contact formation.

Like in the case of gypsum, all histograms (Fig.10) show a sharp transition from "weak" coagulation contacts ($p_1 < 10^{-2}$ dyne) to "strong" phase ones ($p_1 > 10^{-1}$ dyne). Absence of the intermediate values permits the precise distinguishing of the bridging. Calculations following the algorithm mentioned above for gypsum, and using experimental data on the dependence of W_b for silica samples on α and t, give for the nucleus contact size: h = 1.2 nm, a_c = 0.3-0.8 nm at α = 3.0-1.2; the work of its formation A_c seems to be sufficiently less than that for the dimeric nucleus in traditional consideration.

The higher is the rate of chemical reaction, the lower is the residual concentration of the dissolved silica. Comparision of the data on this residual concentration C_t (1 h after mixing sodium silicate solution and hydrochloric acid) with the probability W_b vs. pH shows that high W_b values occur at pH 7-8, that is, at maximum polycondensation rate (Fig.11). Low W_b values are realized at pH 2-3, i.e., under the conditions of minimum reaction rate. Along with this, at each given pH, W_b increases with α value (Table 3)..

Thus, systematic studies on bridging of individual samples, both crystalline and amorphous, show that the basic issue is the fluctuation process of formation of a critical nucleus contact in the gap between two particles. Once such contact is established, it then spreads rapidly over the whole contact zone. The probability of formation of a nucleus contact is determined by supersaturation (metastability) of the solution, time of contact,

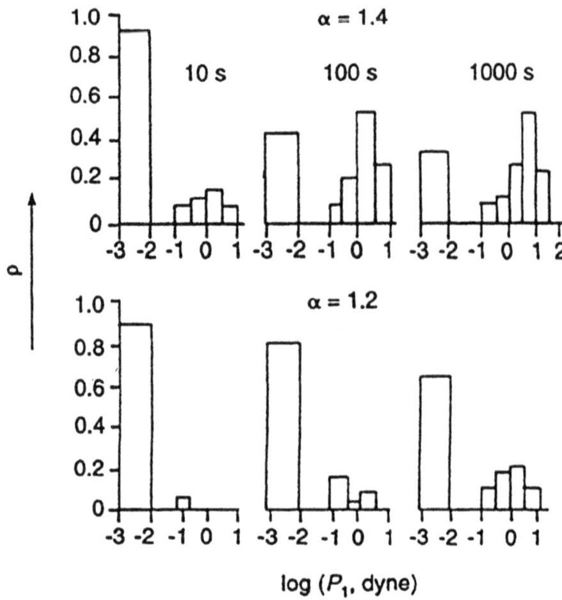

Figure 10. Differencial distributions of sftrength (p_1) of the contacts emerging between silica samples in supersaturated silicic acid solutions; pH 7, f = 10 dyne.

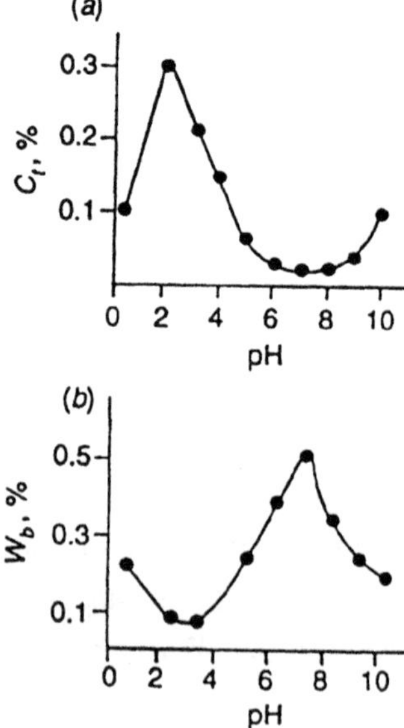

Figure 11. Dependence of (a) residual concentration of dissolved silica (C_t) in sol, and (b) probability of phase contact formation (W_b) on pH values.

and the force pressing the particles to each other; it varies for contacts with particles of different natures (e.g., fillers), and is influenced greatly by the presence of electrolytes, addition of surfactants, and so on.

Acknowledgments

This work is supported in part by the Sandia National Laboratories and the International Science Foundation grant.

References

1. Shchukin, E.D. and Amelina, E.A. (1979) Contact interactions in disperse systems, *Adv. in Colloid and Interface Sci.* **11**, 235-287.
2. Shchukin, E.D. (1990) Development of teaching of P.A.Rehbinder on surface phenomena in disperse systems, *Bull. Acad. Sci. USSR, Chem. Sci.* **10**, 2424-2446.
3. Shchukin, E.D., Amelina, E.A., and Kontorovich, S.I. (1992) Formation of contacts between particles and development of internal stresses during hydration hardening, in Skalny, J. (ed.), *Materials Science of Concrete III*, The American Ceramic Society, Westerwille, OH, pp.1-35.
4. Derjaguin, B.V., Churaev, N.V., and Muller, V.M. (1987) *Surface forces*, Consultant Bureau, New York and London.
5. Shchukin, E.D., and Yaminsky, V.V. (1985) Thermodynamic factors of the sol-gel transition, *Colloids and Surfaces* **32**, 19-32; 33-55.
6. Shchukin, E.D. and Rehbinder, P.A. (1958) New surfaces formation during deformation and fracture of solids in surface-active media, *Kolloidn. Zh.* **20**, 645-654.
7. Shchukin, E.D., Amelina, E.A., and Yaminsky, V.V. (1982) Thermodynamic equilibrium between coagulation and peptization, *J. Colloid and Interface Sci.* **90**, 137 -142.
8. Israelachvili, J.N. (1985) *Intermolecular and surface forces*, Academic Press, London.
9. Luckham, P.F. and de L. Costello, B.A. (1983) Recent developments in the measurement of interparticle forces, *Adv. in Colloid and Interface Sci.* **44**, 183-240.
10. Shchukin, E.D. and Yushchenko, V.S. (1985) Molecular dynamics simulation of surface phenomena in solids, *Abstracts of the 5-th Int. Conf. on Surface and Colloid Sci.*, Potsdam, N.Y., p.7
11. Yushchenko, V.S., Edholm, O., and Shchukin, E.D. (submitted) Molecular dynamics of the particle/substrate contact rupture in a liquid medium, *Colloids and Surfaces.*
12. Yushchenko, V.S., Edholm, O., and Shchukin, E.D. (1992) Influence of hydrostatic pressure on interaction of particles and stability of suspensions, *Colloids and Surfaces* **62**, 131-134.

ELECTRO-OPTIC CHARACTERISATION OF ANISOTROPIC SUB-MICRON PARTICLES

B.R. JENNINGS
English China Clays plc, St. Austell, UK. and
Chemistry Department, University of Bristol, Bristol, UK.

1. Introduction

Optical anisotropy can be manifest for colloidal particles through a number of phenomena [1]. Birefringence, dichroism and light scattering arise from variability of refractive index, absorption coefficient or geometry associated with the different principal axes of the particles. Furthermore, anisotropic particles are generally anisodiametric, and are most conveniently represented by the principal shapes of spheroids of revolution. Whereas many of the theories which account either for colloidal behaviour or for the methods used to characterise the same assume spherical models [2,3], this is far from an adequate description of real systems. Commercially important colloids are continuously being exploited for their needle-like, discotic, tabular or aggregated nature. Furthermore, naturally occurring colloids are seldom of a single monodisperse size species and samples are engineered for specific narrow, broad or truncated size distributions.

The characterisation of polydisperse non-spherical systems is therefore timely and the dependence of electro-optical methods on both anisotropy and particle size is especially relevant. Following a brief resumé of the principles of electrically induced birefringence, the concept is developed as to how the fast transient responses of colloidal suspensions to a single short-duration pulsed field can be analysed to fit a log-normal distribution to the sample heterogenity. When such data are compared with similar estimates obtained using established, alternative methods, such as photon correlation, sedimentation rate, optical diffraction or viscosity, an estimation of the minor dimension can be made. This yields the axial ratio and approximate shape. Finally, the sensitivity of the changes induced in the scattered light intensity upon particle alignment is directly proportional to the anisotropic shape of the particles. This can be exploited as a fine indicator for flocculation and aggregation processes.

2. Basic Principles

Briefly stated, a non-oriented random dispersion of anisotropic particles gives rise to an isotropic medium which is not birefringent. In the presence of an applied electric field (E), particle alignment is partially imposed as the field couples with any permanent or induced dipole

E. Pelizzetti (ed.), Fine Particles Science and Technology, 255–266.

moment associated with the particle geometry. This torque competes with Brownian energy and the resulting orientation function (Φ) defines the degree to which the inherent anisotropy of the particles is manifest. In the case of electrically induced birefringence (Δn), the observed effect for a suspension of volume concentration (c_v) in a medium of refractive index (n) for particles of inherent optical geometry ($g_3 - g_1$), is given by the equation [4],

$$\Delta n = \frac{2\pi c_v}{n} (g_3 - g_1) \Phi \tag{1}$$

where Φ is a function of the electrical properties of the particles, E^2 and $(1/kT)$, with k the Boltzmann constant and T the absolute temperature. At low fields, a direct proportionality between Φ (and hence Δn) and E^2 occurs, whilst at high fields, Φ is unity and hence field independent [5].

When a rectangular, pulsed field is applied, the response is transient in nature and decays away after the passing of the field. It is this decay which is most easily analysed for particle characterisation. With a polydisperse sample, each size species contributes to the decay, which is recorded through the compounded summation by the equation [6],

$$\Delta n_t = \Sigma_i (\Delta n_o)_i \exp(-6D_i t) \tag{2}$$

at any time t after the commencement of the decay, for which t=0. The parameter D_i is the rotary diffusion coefficient for any size species i, which contributes a birefringence Δn_i to the total observed effect.

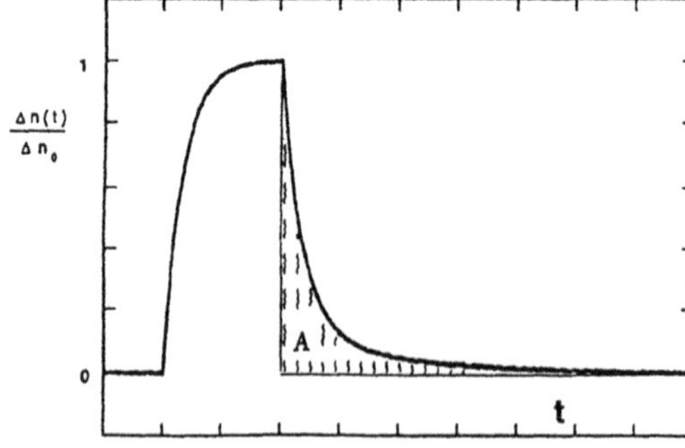

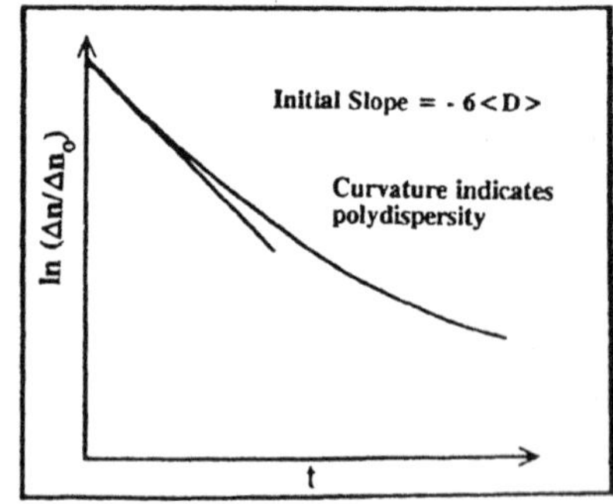

"Figure 1" - Representative transient and semi-log plot of the normalised birefringence decay. The area A and initial slopes yield <1/D> and <D> respectively.

For such a sample, a plot of the $\log_e$ of $(\Delta n/\Delta n_o)$ vs. t is curved (figure 1). Here Δn_o is simply $\Sigma_i(\Delta n_o)_i$. The initial slope (S) defines a specific average rotary diffusion coefficient $<D>$. The area under the decay curve (A) is related to the average parameter $<1/D>$. Equations for these two parameters under the conditions of low field strength (where Φ is proportional to E^2) and high field strength (where Φ is independent of E) have been presented elsewhere [7].

Electrically induced birefringence can be readily measured with the apparatus indicated in figure 2. Details of the setting of the optical components, the manipulation of the data recorded and the design of auxillary apparatus have all been described elsewhere both for general birefringence amplitude measurements [8] and for accurate transient decay capture and analysis [9].

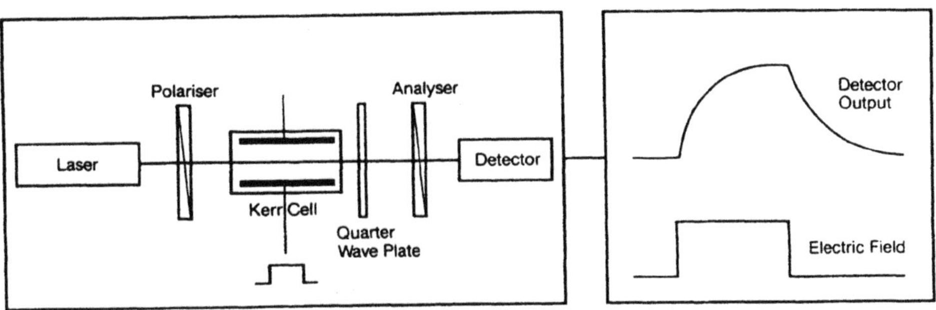

"Figure 2" - Principles of an electrically-induced birefringence system for particle size analysis. The detector output is transient with rates dependent on particle size and shape.

3. Size distributions from single transients

Various attempts have been made in the past to deconvolute transient decay curves in order to isolate the composite species. Often complex mathematical analyses have been produced [10], which have proved difficult to realise experimentally due to the unrealistically high precision required of the related experimental data. Early methods included "peeling" out the major components within the decay [11] and using values of $<D>$ under high $<D_H>$ and low $<D_L>$ field conditions to give data which could be fitted to one or other two parameter distribution functions. One major advantage of methods which measure rotary diffusion coefficients is that $<D>$ can be calculated discreetly for non-spherical shapes [12].

A recent advance[13] has been made, based on the realisation that the two parameters $<D>$ and $<1/D>$ have different weightings on the particle size. They can thus be obtained from a single low (or high) field transient and interpreted to yield information from which the two defining parameters of the median (m) and breadth (σ) of a log-normal distribution of the major particle dimension can be obtained where[14]

$$f(x) = \frac{1}{\sqrt{2\pi}} \cdot \frac{1}{x\sigma} \exp\left\{\frac{-[\ln(x/m)]^2}{2\sigma^2}\right\} \qquad (3)$$

with x the major dimension. For all rigid particles, one may write[13]

$$<D_L> = \frac{Q\,F(\rho)}{m^3} \cdot \exp\{\tfrac{1}{2}\,[9-6(u+s)]\sigma^2\} \qquad (4)$$

$$<1/D_L> = \frac{m^3}{Q\,F(\rho)} \exp\{\tfrac{1}{2}\,[9+6(u+s)]\sigma^2\} \qquad (5)$$

with ρ the axial ratio and $Q = 3kT/\pi\eta$.

In these equations, u and s are the power dependencies of the particle volume V on the major dimension x, i.e. V is proportional to x^u. Similarly, the electrical polarisability anisotropy $\Delta\alpha$ depends upon x to the power s. Values of u and s are given in table 1, along with expressions for the shape factor $F(\rho)$ for the particle morphologies of interest.

The expressions are somewhat simpler for the high field values of $<D_H>$, which can be collated in the expression[13]

$$D_H = \frac{QF(\rho)}{m^3} \cdot \exp\left\{\frac{N\sigma^2}{2}\right\} \qquad (6)$$

where N has the values three, minus three or minus nine for rods, discs and ellipsoids (including spheres) respectively. Practical realisation of the high field orientation limit is neither easily nor unequivocally attained experimentally. Low field data are more reliable.

TABLE 1. Shape dependent parameters for rotary diffusion coefficients[12]

Shape	u	s	F(ρ)
spheres	3	3	Unity
rods	1	2	$\ln(2\rho)-1.57+7\left(\frac{1}{\ln(2\rho)}-0.28\right)^{2}$
prolate ellipsoids	1	2*	$\frac{\rho^4}{2(\rho^4-1)}\left\{\frac{(2\rho^2-1)\ln[\rho+(\rho^2-1)^{1/2}]}{\rho(\rho^2-1)^{1/2}}-1\right\}$
discs	2	2	$\pi/4$
oblate ellipsoids	2	2*	$\frac{\rho^3}{2(1-\rho^4)}\left\{\frac{(2-\rho^2)\tan^{-1}[(\rho^2-1)^{1/2}]}{(\rho^2-1)^{1/2}}-1\right\}$

* Assumed from rod and disc data, with ρ large.

In all cases, the following products hold

$$\langle D_L\rangle \langle 1/D_L\rangle = \langle D_H\rangle \langle 1/D_H\rangle = \exp(9\sigma^2) \qquad (7)$$

so that σ is readily found. When σ is reinserted into any expression for $\langle D\rangle$ or $\langle 1/D\rangle$, the parameter m is obtained. As field pulses of a millisec duration are used, automated capture and analysis of any decay results in the generation of size distribution data within a minute.

Representative results are shown on a sample of near-spherical, prolate, emulsion polymerised particles of polytetrafluoroethylene (PTFE) for which $\langle D_L\rangle = 34\text{s}^{-1}$, $\langle 1/D_L\rangle = 5.24 \times 10^{-2}$s so that, with $\rho=1.5$, $\sigma=0.24$ and m=312nm. An electron micrograph cumulative histogram (figure 3) on some 600 particles demonstrates the validity of the method.

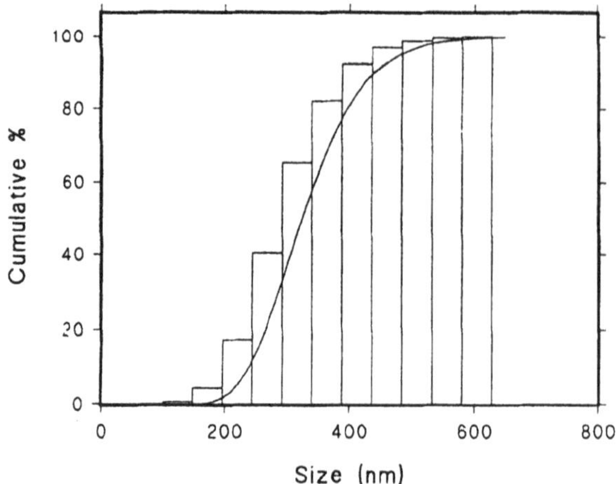

"Figure 3" - Log-normal distribution of a PTFE sample from a single transient (full curve) and from electron micrograph data for some 600 particles.

4. Axial Ratio Determination

The majority of colloidal particles are not spherical. Yet most of the conventional methods for particle size measurement are based upon evaluation of the equivalent spherical diameter (d). Experimentally determined values of this parameter depend upon the physical principle which forms the basis of the method of measurement. Recently [15], equations were derived for the equivalent spherical diameter (esd) of well formed spheroids for each of the common classes of measuring instrument. These are summarised in table 2 for disc particles where the value of d is expressed as a ratio of the true axial dimension (x) for this particle geometry. Each method has a distinctive dependence on the axial ratio. Only if the particles are true spheres do all values of the esd by all methods coincide. Non coincidence thereby *indicates* anisotropy.

Comparative measurement of the esd for a given monodisperse system, by two methods thus allows an estimation to be made of ρ. This is illustrated in figure 4 where the equations have been expressed as a ratio of d_s/d_r and plotted versus the axial ratio for prolate and oblate geometries. With the aid of this graph, measurements of the two esd values enable ρ to be obtained. Knowing x, the secondary dimension is thus evaluated.

TABLE 2. Ratio of the experimental equivalent spherical diameter (d) to the true surface diameter (δ) of discs of axial ratio ρ

Measured Parameter	Experimental Method	d/δ
Projected Area	Fraunhoffer diffraction	$(0.5 + 1/\rho)^{1/2}$
Stokes' diameter	Sedimentation	$\{(3 \arctan \rho)/2\rho\}^{1/2}$
Translatory diffusion	Photon correlation	$(1/\arctan \rho)$
Rotary diffusion	Transient electrooptics	$(4\pi/3)^{1/3}$
Particle volume	) Conductivity) Mie Scattering	$(3/2\rho)^{1/3}$

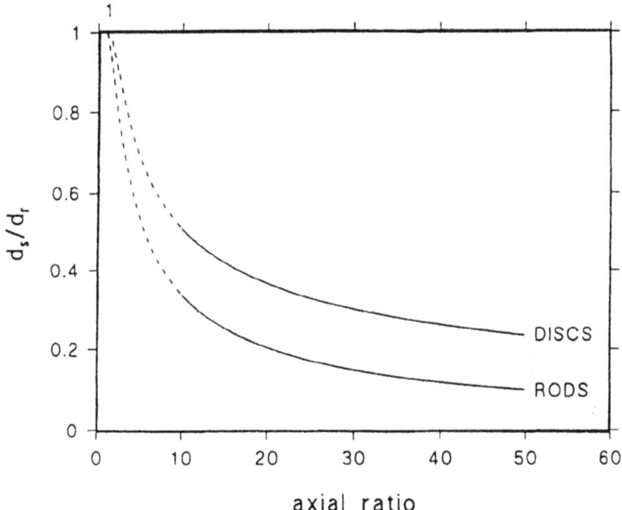

"Figure 4" - Ratio of the two equivalent spherical diameter ratios d_s/d_r from sedimentation and rotary diffusion experiments respectively.

The method is illustrated by measurements on an aqueous suspension of the clay mineral kaolinite. These particles are irregular thin plates which may, to a first approximation, be represented by a disc model. Samples are always highly polydisperse, of indiscreet morphology, and are never spherical. Their commercial uses often depend upon their axial ratio and its control. Sedimentation rate data were recorded using a "Sedigraph" instrument and the relevant cumulative distribution of the esd is shown as the broken curve in figure 5. The equivalent spherical diameter from electric birefringence measurements is shown in the same graph. The two distributions do not coincide, reflecting the high particle non-sphericity. An estimate can be made of the approximate particle thickness by comparing the values of the size of the esd for each distribution at the mid point of the cumulative curves and by assuming that a disc best models the particle shape. In this case, values of 310nm and 640nm respectively were obtained, giving a ratio of 0.494 for d_s/d_r. From figure 4, such a ratio indicates an aspect ratio of 11. Auxillary electron microscopy data demonstrated a ρ of 12. Hence the method is justified. Such analysis could be extended throughout the distribution curve to find the variability in ρ with major dimension.

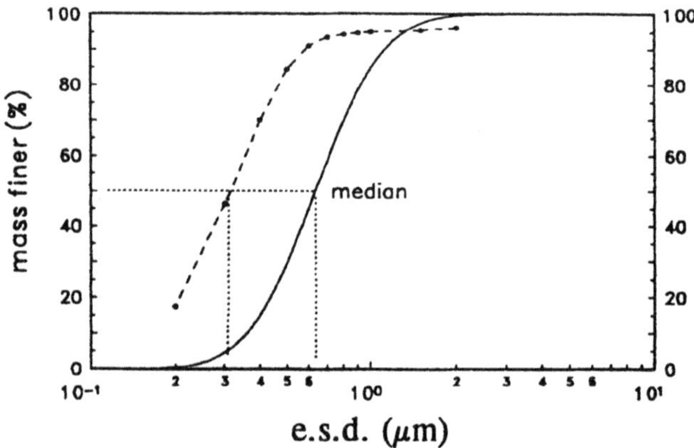

"Figure 5" - Comparison of sedimentation (broken curve) and rotary diffusion (full curve) distributions in the equivalent spherical diameter for a sample of kaolinite.

5. Deflocculation Monitoring

Light scattering intensity measurements have long been used as an indicator of aggregating systems. In general, larger particles scatter more intensely. Scattering from flocs is therefore greater and deflocculation can be indicated by a decrease in the scattered intensity I.

The electrically induced change in the scattered intensity (ΔI) is however more sensitive to the particle anisotropy than to the basis size. A large spherically shaped particle gives rise to little change in ΔI upon rotation in an electric field whilst a highly non-spherical particle, even if small, orientates and gives rise to significant changes in ΔI. Theories have been developed for ΔI for rods[16], discs[17] and polymer chains [18] under a variety of conditions. Electrically induced scattering changes afford an extremely sensitive means of indicating when somewhat symmetrical flocs disperse into individual non-spherical, albeit much smaller, particles. In this case, full theoretical interpretations are not necessary as relative responses are sufficient.

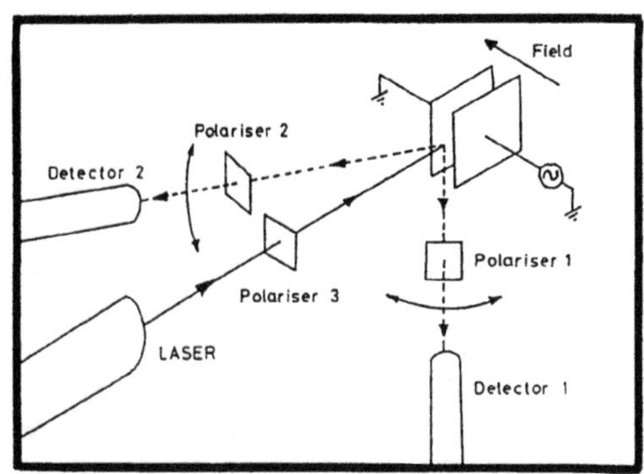

"*Figure 6*" - Transient back-scattering from concentrated suspensions. The system is conveniently arranged using fibre-optic guides. Two detectors are shown as depolarisation data are also interesting, though not discussed here.

The principle of a suitable apparatus is shown in figure 6, where considerations have been given to highly concentrated optically dense media. For such a case, back scattering can be used and fibre optics employed with considerable benefit both for remote sensing and for ease of experimentation. The changes in ΔI in response to pulsed electric fields can be recorded.

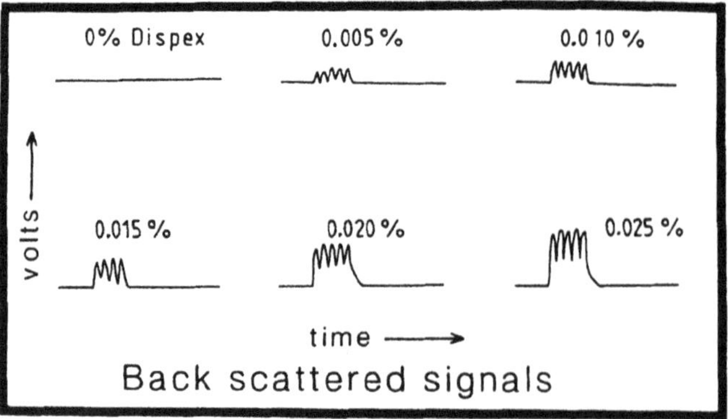

"Figure 7" - Transient scattering responses from a flocculated 30% solids kaolinite sol as a sodium polyacrylate (Dispex) deflocculating agent is increasingly added at pH 8.

An example is given in figure 7 for the addition of sodium polyacrylate to a 30% suspension of kaolinite clay mineral sol. This polymer is a dispersing agent. As the polyacrylate dose is increased, the amplitude of the electro-optic transient grows indicating the continuous evolution of highly anisotropic particles, which in this case are taken to be the individual, small, kaolin particles. Plotting the amplitude of the transient responses as a function of polyacrylate dose yields figure 8 which fully characterises the deflocculation process.

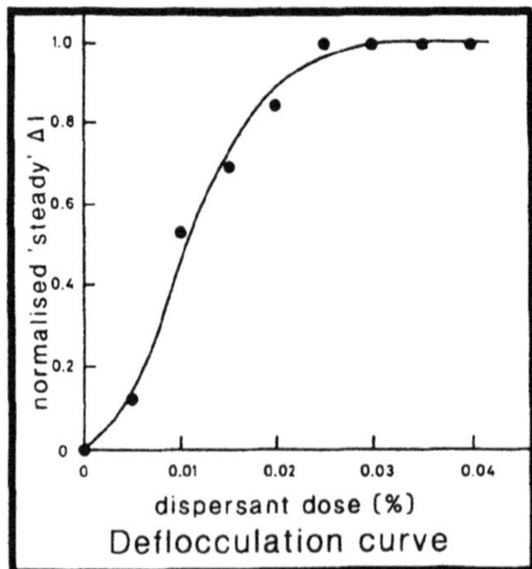

"Figure 8" - The transient amplitudes of figure 7 plotted as a function of polyacrylate dose. Large signals correspond to high particle anisotropy.

6. Conclusions

From the foregoing examples, it can be seen that transient electro-optic measurements of colloidal suspensions in general and mineral sols in particular are able to characterise highly anisotropic particles, to indicate non-sphericity, to measure particle size distribution in major, and in the minor dimension and to follow the changes in these parameters as colloidal media are subjected to flocculating and dispersing agents. When the sols are too dense for light transmission, fibre based reflection methods can be used.

7. References

1. Stoylov, S.P. (1991) *Colloid Electro-Optics*, Academic Press, London.
2. Barth, H.G. (Ed.) (1984) *Modern Methods of Particle Size Analysis*, Wiley Interscience, New York.
3. Allen, T. (1990) *Particle Size Measurement*, Chapman & Hall, London.
4. Peterlin, A. and Stuart, H.A. (1939) Über die bestimmung der grösse und form, sourie der elektrischen, optisdien und magnetischen anisotropie, *Z. Phys.* 112, 129-147.
5. O'Konski, C.T., Yoshioka, K., and Orttung, W.H. (1959) Determination of electric and optical parameters from saturation of electric birefringence in solutions, *J. Phys. Chem.* 63, 1558-1565.
6. Benoit, H. (1949) Theorie de l'effet Kerr d'und solution soumise à une impulsion electrique rectangulaire, *Compt. Rend.* 229, 30-32.
7. Jennings, B.R., Foweraker, A.R. and Morris, V.J. (1978) Particle size distributions from transient electric birefringence data, *Adv. Mol. Relxn. Proc.* 12, 211-220.
8. Fredericq, E. and Houssier, C. (1973), *Electric Dichroism and Electric Birefringence,* Oxford Univ. Press, London.
9. Oakley, D.M., Jennings, B.R., Waterman, D.R. and Fairey, R.C. (1982) An electro-optic birefringence fine-particle sizer, *J.Phys. E. (Sci. Instr.)* 15, 1077-1082.
10. Tsuji, K., Watanabe, H. and Yoshioka, K. (1976) A numerical method of obtaining the relaxation distribution function from decay curves, *Adv. in Mol. Relxn. Proc.* 8, 49-62.
11. Ingram, P., and Jerrard, H.G. (1963) Study by the Kerr effect of the action of enzymes on macromolecules, *Brit. J. Appl. Phys.* 14, 572-574.
12. Perrin, F. (1934) Mouvement Brownien d'un Ellipsoide *J. Phys. Rad.* Ser. vii, 497-516.
13. Watson, R.M.J. and Jennings, B.R. (1992) Polydisperse size data from single electric birefringence transients, *Powder Technology* 72, 63-69.
14. Hastings, N.A.J. and Peacock, J.B. (1975) *Statistical Distributions*, Butterworth & Co., London.
15. Parslow, K. and Jennings, B.R. (1988) Particle size measurement; the equivalent spherical diameter, *Proc. Royal Soc.* A419, 137-149.
16. Wippler, C. (1954) Elude théorique de la diffusion de la lumiere par des sols de batonnets orientés, *J.Chim. Phys* 51, 122-128.
17. Stoylov, S.P. (1967) Influence of electric fields on the light scattering by plate-like particles in an electric field, *Izv. Inst. Fizkhim* 6, 79-86.
18. Isihara, A., Koyama, R., Yamada, N. and Nishioka, A. (1955) Effect of electric field on light scattering of solutions of polar chain polymers, *J. Polymer Sci.* 17, 341-349.

POLYPHOSPHATE NANOPARTICLES AND GELS

From macroscopic monolyths to nanoparticles

F. GALEMBECK*, E.C.O. LIMA#, M.M. BEPPU, R.M. SASSAKI,
N.C.MASSON, V.A.R.MONTEIRO and E.F.SOUZA
Institute of Chemistry, Universidade Estadual de Campinas
CP6154, CEP13081-970, Campinas SP, Brazil
#Institute of Chemistry and Geosciences, Universidade Federal de Goiás, Goiânia GO, Brazil

Abstract

Amorphous metal (Al, Ca, Fe, Zr) polyphosphate particles and monolyths may be obtained in a range of chemical compositions, by admixture of aqueous solutions of sodium polyphosphate, metal salts and ammonium hydroxide.

In the Al-polyphosphate system, phosphate-rich solids (P/Al>1.5) are soft glasses, which contain sodium ions, too. Aluminum-rich particles are refractory and have a low electrical conductivity up to 800-1000°C, which is assigned to a restrained ion mobility.

Two different, macroscopically opposite behaviors are also observed in the preparation of these materials: in one case, monolythic gels are obtained, and these are easily dried into glasses. In the other case, nanoparticles are formed.

Most reports on nanoparticles preparation in the literature use kinetic control to limit particle size. This is the case of the well-known Szigmondy colloidal gold particles, as well as the recent reports on nanoparticle synthesis using aerosols and templates. In the present case, the nanoparticles are obtained under dynamic quasi-equilibrium dissolution-precipitation conditions, in the presence of excess reagents in the aqueous medium. For this reason, we propose that their small size is due to a low interfacial tension, as *e.g.* in microemulsions.

This assumption is strengthened by the observation of thermoreversible gel formation, from aluminum polyphosphate solutions: a thermoreversible gel is the result of the reversible formation of a highly solvated supramolecular network, in which interfacial tension between different domains is necessarily low and does not contribute a significant driving force for further phase seggregation.

Aluminum polyphosphate powders containing nanoparticles can be cold-pressed into translucent, low-porosity and mechanically resistant macroscopic solid discs and tubes. This attractive property is assigned to two factors: the particle nanodimensions and the large water content of the powders. Together, these may open a new path towards low temperature materials processing.

E. Pelizzetti (ed.), Fine Particles Science and Technology, 267–279.

1. Introduction

Phosphates are widely found in the nature, both as minerals and as inorganic components of biomaterials [1]. Most phosphates are non-toxic and can be used and disposed off without concern. The fabrication of phosphates largely follows a classical chemical process, which is energy-intensive and uses important amounts of sulfur, as a source of sulfuric acid. This process is widely used in the making of phosphoric acid and of phosphate fertilizers, from apatites. On the other hand, the major waste in this process is calcium sulfate, for which many applications have been developed, as a raw material. For these reasons, phosphate-based materials rely on an industry which operates on an environmentally acceptable basis, and may evolve towards a "zero-emissions" standard, in the near future.

There has been great interest in phosphates, as a source of new materials. For instance, the molecular tectonics [2] approach for the making of biomimetic materials focuses strongly on hidroxyapatite, not only because of the natural examples of materials like bone, dental enamel and coral reefs, but also because of the chemical and morphological versatility of this substance. Beyond that, phosphates have been used in the making of sorbents, special glasses and catalysts. The case of phosphate glasses [3] is interesting: they are attractive not only because of their high ultra-violet transparency but also because they have large thermal expansion coefficients, which makes easier glass-to-metal bonding.

Non-crystalline metal polyphosphates can be obtained in a broad range of chemical compositions, making a large family of solids with many interesting morphologies. We have previously described the formation of rigid foams [4] and of particles containing closed pores [5]. More recently, we found that these materials can be used to make nanoparticles, in large amounts and using a rather simple procedure. In this work, we describe the formation of aluminum polyphosphate nanoparticles as well as the formation and characteristics of aluminum and other phosphates and polyphosphates gels and macroscopic monolyths. Besides, we present some ideas on the factors leading to the formation of nano- or macrodomains of these materials.

Our approach to the making of phosphate materials has been based on energy unintensive, chemically intensive low-temperature processes using alkaline phosphates as the main raw materials. We have been following the routes described in the literature as the inorganic (hydrolysis) sol-gel [6] and the molecular tectonics [2] approaches. Beyond that, we have been relying on the low stability and great morphogenetic ability of non-crystalline solid particles.

The sol-gel approach has been quite successful in the making of silica and some other oxide glasses. There have been some attempts to make phosphate gels by hydrolysis of phosphate esters,[7] but these are perhaps too stable. On the other hand, there is an interesting possibility in the making of materials which is intermediate between room-temperature sol-gel and thermal processing, and which has not been fully exploited: this is the preparation of non-crystalline particles followed by their thermal processing at temperatures lower than these used for sintering crystalline particles.

A few years ago [8], we considered that glassy ionic solids should have a temperature threshold for physico-chemical transformations, akin to the glass transition temperature, T_g , observed in polymers and in silicate and other glasses. In one case studied (non-crystalline iron (III) hydroxoacetate), the solid is highly stable up to *ca.* 205°C, but at this temperature chemical reactions and crystal formation start and proceed steeply. There is one difference between this case and that of glass transition: the process is not fully reversible, because the products of the spontaneous transformations triggered by heating are highly stable. This is an interesting feature of these systems, for two reasons: first, it is an yes-or-no behavior, which depends on the respective temperature threshold. Second, the involved temperatures are much lower than Tammann temperatures for the corresponding solids. For this reason, non-crystalline ionic solids may undergo sintering at temperatures in the 200-500°C, which is much less energy-demanding than the sintering temperatures for *e.g.* transition metal crystalline oxides. Finally, non-crystalline nanosized solid particles should be further endowed with the attribute of superplasticity [9-11] adding another interesting possibility to the processing of materials. Our work has been directed at verifying these possibilities.

2. Preparation and Composition of Non-Crystalline Ortho- and Polyphosphates

Non-crystalline ortho- and polyphosphates can be easily prepared by admixture of aqueous solutions of a polyvalent metal ion (*e.g.* iron, aluminum [12], calcium, zirconium) salt (chloride, nitrate, sulfate), a sodium phosphate or polyphosphate and a pH-control reagent (sodium or ammonium hydroxide), at room temperature and under strong stirring. The product may form a monolythic gel or it may be found as small, swollen gel-like particles. This depends on their actual composition, both concerning phosphate-to-metal ion ratio and overall reagent concentration, as well as on the intensity and extent of stirring after mixing the solutions.

The chemical composition of the gels and particles obtained from a single polyvalent cation salt also changes depending on the actual reagent concentrations, as shown in Table 1. We did not find any correlation between the composition of the feed reagents and that of the precipitates or gels [13].

TABLE 1. Elemental composition in the feed reagents and in the particles obtained by admixture of aluminum nitrate, sodium phosphate and ammonium hydroxide, at room temperature.

Sample	Amounts of reagents in the feed (mols)			Relative amounts of reagents (mol P/mol Al)	
	$Al(NO_3)_3$	NaH_2PO_4	NH_4OH	in the feed	in the products
1	0.015	0.045	0.060	3.00	0.80
2	0.030	0.045	0.060	1.50	0.88
3	0.015	0.090	0.060	6.00	1.52
4	0.030	0.090	0.060	3.00	1.69
5	0.015	0.045	0.120	3.00	0.70
6	0.030	0.045	0.120	1.50	0.66
7	0.015	0.090	0.120	6.00	1.19
8	0.030	0.090	0.120	3.00	1.04

There is not a prevalent stoichiometry, for these compounds. This stoichiometric variability has important implications and its origin requires some discussion. The relevant facts are: i) these particles or gels do not contain crystalline domains, as shown by X-ray diffraction; ii) particle or gel formation are the results of a phase separation, which is akin to a liquid-liquid phase separation process, in which solute-rich and solute-poor domains are formed and coexist. Indeed, in the specific case of chromium (III) nitrate - sodium polyphosphate admixture, neither a gel or particles are formed, but two liquid phases separate: one is a deep green, viscous liquid and the other is a pale green, low viscosity solution; iii) mass transfer within the particles and throughout the gel is diffusion-limited. For this reason, when the reagents are mixed and the gel domains are formed, ions are trapped at various concentrations. Multivalent ions associate and form an ion-exchanger network, to which monovalent counter-ions are bound. For instance, sodium ions are obtained in the aluminum polyphosphate gels and particles, whenever larger P/Al ratios are used in the preparation. The network responsible for the formation of these gels and particles can be aassembled following a great number of combinations and ratios of the ionic building blocks. In a sense, it is as complex and variable as the networks formed *e.g.* by lignin and other condensation polymers which contain polyfunctional monomers.

In a recent paper, we described [3] in greater detail the conditions for the formation of gels, particles and solutions by admixture of aluminum nitrate and sodium polyphosphate aqueous solutions. This case is particularly interesting since the gels formed within some P/Al mol ratios are thermoreversible, this means, they are stable at room temperature but become liquid at higher temperatures and gel again, upon cooling. Gel stability is greatest at the P/Al ratios for which the sum of negative (polyphosphate) and positive (aluminum) charges is close to zero, in which case electroneutral aluminum polyphosphate networks are obtained [13].

The water content of the gel composition may also vary widely, as shown by data in Table 2. This refers to gels made out of sodium polyphosphate (n=6) and mixed calcium and iron (III) nitrate solutions.

TABLE 2. Gravimetric data on the gels formed by admixture of calcium nitrate, iron (III) nitrate (total 30 ml, 1M) and sodium polyphosphate (30 ml, 2M) solutions, at room temperature.

Gel weight (g) after each operation	Ca/(Ca+Fe) mol ratio						
	1	0.85	0.70	0.50	0.30	0.15	0
as prepared, following admixture of reagents (A)	8.86	18.49	67.29	67.15	68.31	68.45	68.97
(A), followed by 24 hours, syneresis, 25°C (B)	7.26	7.95	9.31	15.95	46.21	55.00	60.15
(B), dried at 120°C to constant weight (C)	4.83	5.45	6.29	8.19	10.44	11.12	12.58
(C), dried at 450°C to constant weight (D)	3.92	4.16	5.08	6.23	7.74	8.08	9.03

We observe that calcium polyphosphate gels contain much less water than iron (III) polyphosphate gels, neither of which display extensive syneresis. However, some calcium-iron mixed gels undergo contraction and expel most of the included liquid, within a 24-h period. This shows that plasticity may be built into these gels and the corresponding particles, by changing their chemical composition.

3. Thermal Properties of Aluminum Phosphate and Polyphosphate Particles

Thermalconductimetric analysis was done on the aluminum phosphates and polyphosphates powders previously dried at 120 °C, by measuring the AC electrical resistance of pressed discs as a function of temperature. Some representative curves are in Figure 1.

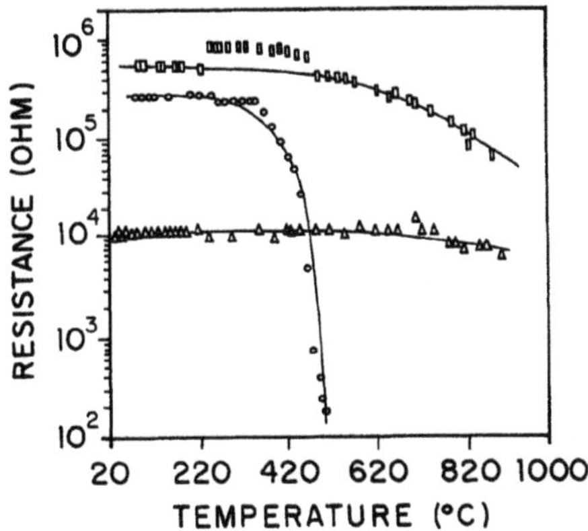

Figure 1. Electrical resistance *vs.* temperature of aluminum polyphosphate pressed powder pellets. P/Al ratios: (□), 1.0; (0), 1.7; (Δ), 0.5.

Curves recorded for various samples of eight different compositions followed one of these three major patterns of behavior:

1) the resistance of the disc is high at room temperature, and does not change appreciably up to ca. 1000 °C; this is observed in most samples containing a large Al/P ratio;
2) the resistance is high at room temperature, but it decreases by some orders of magnitude at some higher temperature. This is characteristic of the samples containing a low Al/P ratio, which do also contain significant amounts of sodium;
3) the resistance of the sample is lower at room temperature, and does not change considerably as the temperature increases. This is observed for samples in which the P/Al ratio is in the 0.4 to 0.6 range, which is the lower end of our experimental concentrations domain.

Thermogravimetric analysis revealed that the non-crystalline aluminum polyphosphates do retain water at temperatures as high as 500 °C. We have previously found that this is an important feature for the formation of hollow particles from these powders [5]. Some curves are given in Figure 2.

The sample having the lower P/Al ratio loses nearly one-third of its weight, when it is heated to 500 °C. This large degreee of hydration is probably a relevant factor determining the lower electrical resistance of this sample, at room-temperature.

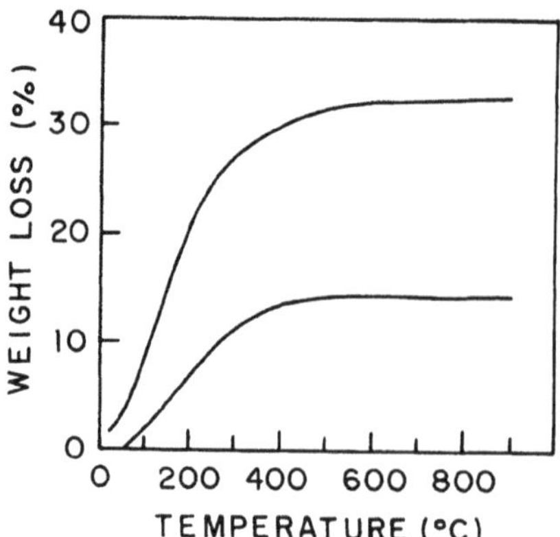

Figure 2 - TGA data from two aluminum polyphosphates. P/Al ratios: 0.5 (upper), 1.7 (lower).

The behavior of the sample yielding the lower electrical resistance at room temperature is also remarkable for another one peculiarity: the pressed powder discs made from these samples are translucent (see further comments on this matter, in Section 4). As we shall see in Section 3, these preparations contain large amounts of aluminum phosphate nanoparticles.

The temperature sensitivity of the P-rich particles can be understood considering that these are ionic glasses in which the multivalent cation has been partly replaced by a monovalent cation, Na^+. This leads to decreased electrostatic interactions among the ions in the glass, the same as in the well-known behavior of sodium and potassium-rich silicate glasses, which are softer than their calcium or aluminum-rich counterparts.

4. Preparation and Morphology of Aluminum Polyphosphate Nanoparticles

Many preparations of aluminum polyphosphate particles were made, by admixture of sodium polyphosphate, aluminum nitrate and ammonium hydroxide aqueous solutions, in the following range of final concentrations in the admixture: C_P= 0.8-1.5 M, C_{Al} = 2.0 M, C_{OH} = 3.6-4.0 M. Other preparations were also made, using concentrations as low as 1/10 of these, while C_P/C_{Al} ratio ranged from 1 to 2.

These preparations always yielded white precipitates at the higher concentrations, and milky liquids at lower total concentrations. The samples were then subjected to various treatments, before examination in the transmission electron microscope. They underwent dialysis for soluble salt removal and were dried by liophylization or by solvent exchange followed by evaporation or just by water evaporation, in the air. Prior to their application to the TEM microscope grids, the samples were redispersed in water or in various other, less-polar solvents. Representative images are presented in Figures 3-7.

Figure 3 was obtained from the supernatant of an aluminum polyphosphate dispersion, which was liophylized. The dry powder was then dispersed in isopropanol sonicated and placed on a microscope grid. In this case, large areas of the grid are covered by particles which sizes are in the 1-3 nanometer range. This shows that this technique can yield large amounts of very fine particles, easily.

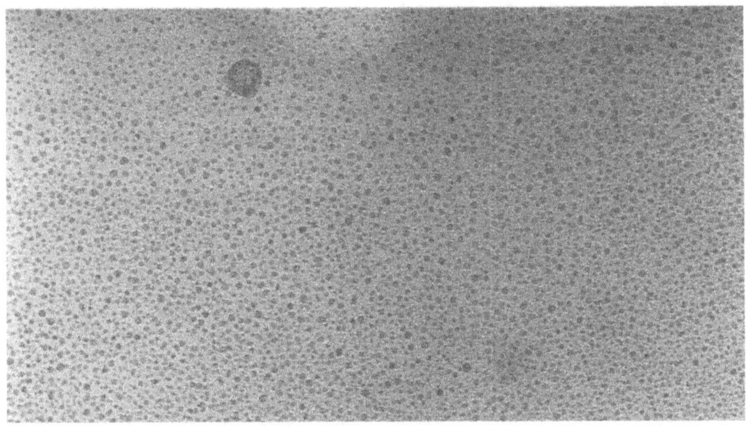

Figure 3. TEM micrograph from aluminum polyphosphate nanoparticles. P/Al = 0.4 mol ratio in the reaction admixture. Particles were obtained by liophylizing the supernatant of a dialysed aqueous sol, dispersing in isopropanol, applying on the microscope grid and drying over phosphorus pentoxide. Bar length = 50nm.

The observation of the nanoparticles is much more difficult if they are dispersed in water instead of isopropanol, prior to application in the microscope grid. In this case, the morphologies obtained are as presented in Figure 4.

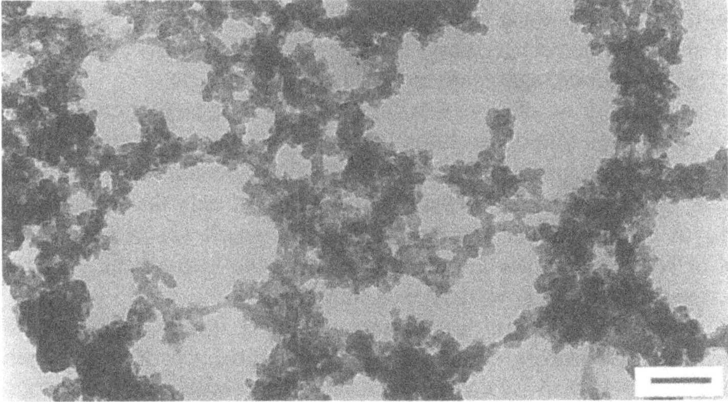

Figure 4 - TEM micrograph of aluminum polyphosphate particles. P/Al mol ratio = 0.4, in the reaction mixture. The particle dispersion was dispersed in isopropanol and applied on the microscope grid. Bar length = 100nm.

The formation of the aggregates shown in Figure 4 can be well understood if we assume that these are small, swollen and soft particles, which are pulled together during water drying, by capillary adhesion. The soft particles forced over one another do then coalesce, forming these larger aggregates.

Beyond that, there are also cases in which at least two different entities are observed, in the transmission microscope: round nanoparticles and larger, non-crystalline but fairly uniform pieces, as seen in Figure 5. It is not clear, from this picture only, if the large blocks are made by coalescence of nanoparticles, or not. At the rim of the larger blocks we observe what may be considered as deformed nanoparticles, which were perhaps joining the block when drying was being completed. However, we note also that some discrete particles are laying on top of the larger blocks, and they do not appear to melt away within these blocks.

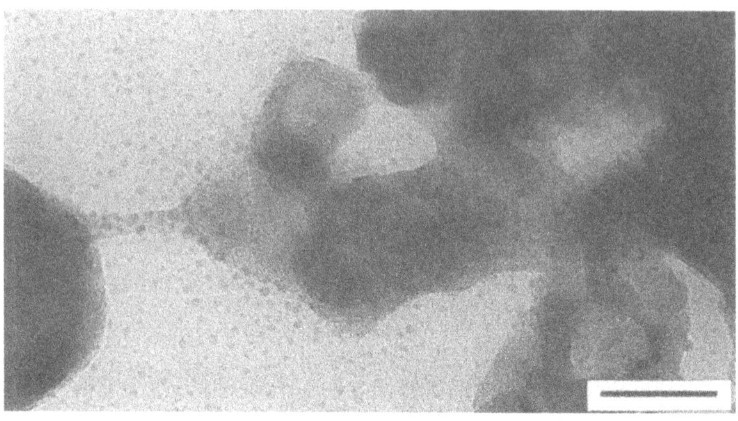

Figure 5 - TEM micrograph of aluminum phosphate particles P/Al mol ratio = 0.5 in the reaction admixture. The particle dispersion was dispersed in isopropanol and applied on the microscope grid. Bar length = 50 nm.

On the other hand, particle examination with the help of the EELS spectrometer built in the electron microscope shows that the well-defined particles are phosphate-rich, while the softer, more aggregate species are aluminum-rich.

Indeed, in Figure 6 we observe that the background is largely coated with a very thin, almost continuous film of an aluminum-rich solid, while the discrete particles in the nanosize range are phosphate-rich but aluminum-poorer than the background. Consequently, the nanoparticles preparation contains not only the particles which are clearly seen in the TEM, but also other particles which are perhaps more easily deformable or smaller than the former. These cover the hydrophobic carbon film in the microscope screeen, to a great degree of uniformity.

In the nano size domain we should expect a great degree of chemical heterogeneity among particles, due to the increased probability of larger fluctuations, which is expected in a small system in which the stoichiometric parameters may cover a broad range.

Two conclusions can now be drawn: i) nanoparticles preparation yields a dispersion of non-uniform particles, which chemical, morphological and coalescence properties are different; ii) large amounts of nanoparticles are obtained, which separation from the medium should not pose any major difficulty, if required.

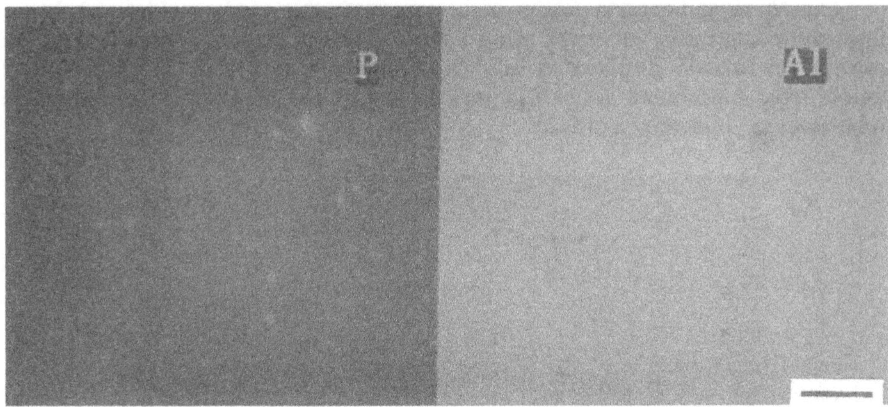

Figure 6 - P- and Al- selective energy-selective images (ESI) of a same area of a screen. Clear points are richer in the selected element than the surroundings. P/Al ratio = 2, in the admixture. Bar length = 350nm.

5. Fabrication of Objects with Aluminum Phosphate Nanoparticles by Cold-pressing.

A powder was prepared by collecting, washing and drying the precipitate from an admixture with the following analytical concentrations: c_P = 1.2 M; c_{Al} =2.0 M and c_{OH} = 3.6 M. The powder was dried at 120°C and uniaxially pressed at room temperature, within a steel mold. Figure 7 shows a picture of an aluminum polyphosphate cold-pressed disc, illuminated from the opposite side to the camera.

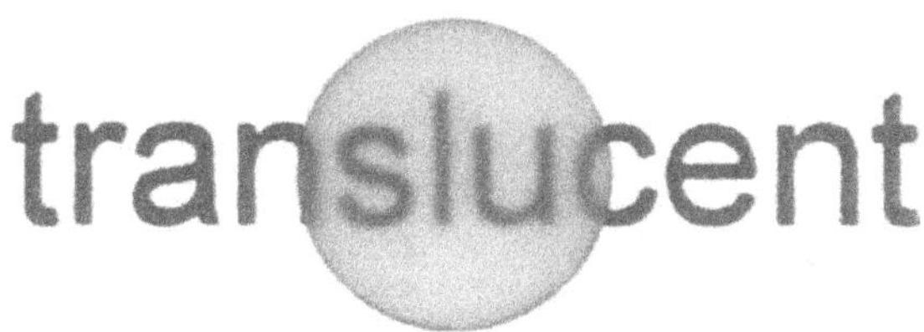

Figure 7 - A 1.1 mm-thick disc uniaxially pressed from aluminum polyphosphate powder under 500 bar, at room temperature. Disc diameter, 1.5 cm.

The disc is translucent, which shows that the pores are in a dimension range below the wavelength of light. This can only be achieved if there are many sub-micrometric particles within the powder, and/or the powder particles are highly plastic.

Tube-shaped objects (Figure 8) were also obtained by isostatic compression of the powder within a rubber sleeve, under 2000 bar at room temperature. These objects have smooth surfaces and resist handling.

This shows an important feature of these powders: they can be molded and shaped fast, by compression at room temperature. This is expected considering the superplastic behavior displayed by solid nanoparticles, as well as the large mobility of ions at room temperature, which is in turn due to both the large surface area and high water content, from these particles.

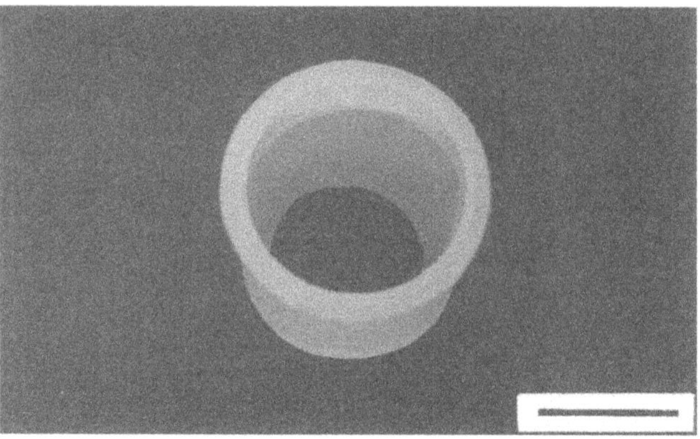

Figure 8 - An object made by isostatic compression of an aluminum polyphosphate powder. Bar length = 2 cm.

Figure 9 is a micrograph from a fracture surface of a cold-pressed aluminum polyphosphate object. This confirms the low porosity of these solids.

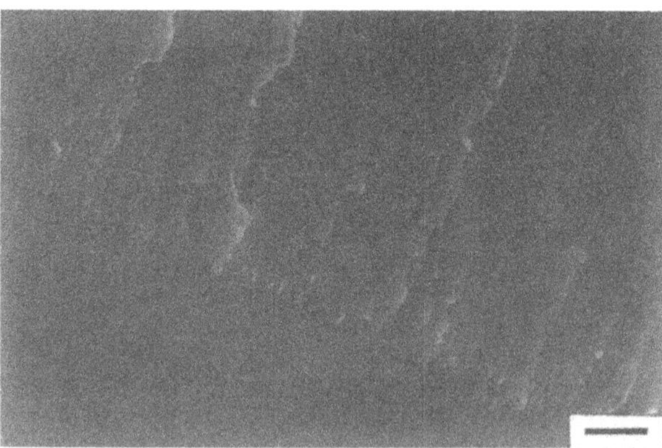

Figure 9- Scanning electron micrograph of a fracture surface from a disc as shown in Figure 8. Bar length = 10 μm.

Since the aluminum polyphosphate nanoparticles can be obtained in a very large scale, the property of their transformation and shaping at room temperature may well evolve into a new technology for materials fabrication.

6. Discussion

6.1 NANOPARTICLE FORMATION AND STABILITY

A few groups of nanoparticle systems are found in a thermodynamic equilibrium state, but most are only in a steady-state, or in a state of colloidal (or kinetic) stability.

Nanoparticle systems found within free-energy minima are:

i) solutions of macromolecular particles such as protein molecules or aggregates, in which particle size is determined by the covalent bonds responsible for joining the atoms within the macromolecule and by intermolecular bonds responsible for specific aggregation. All systems which have been considered as "reversible colloids" belong to this category, as well as fullerenes, cyclodextrins and other interesting substances [15];

ii) microemulsions [16], in which particle size is the result of a balance between enthalpic and entropic forces. In this case, dispersion of one liquid phase within the other is entropically driven and this is allowed by exceedingly low interfacial tensions, which in turn require the use of at least two different surfactants. In this case, we can distinguish among two types of entropic contributions: those arising from the increased mobility of molecules at the interfaces and those associated with nanodroplet dispersion in the continuous phase.

On the other hand, nanoparticles can be made of almost any substance by techniques relying in some sort of kinetic control. Most of these techniques belong to one of the two following categories:

i) single particle growth is limited by starvation, this means, by restricting the supply of precursor reagent. Examples are: the Faraday and Cassius gold sols [17], La Mer's sulphur sols [18] and the polystyrene latex dispersions of Ottewill and Shaw [19].

ii) single particle dimensions and shape are determined by some kind of template, which acts as a geometrical constraint. An example is given by the nanofibers synthesized within membrane pores [20].

Even if a strict control on particle formation is achieved, control of particle aggregation still has to be exercised, otherwise nanoparticles will be transformed into larger particles or particle aggregates. At this point, a further distinction can be made: upon touching each other, nanoparticles may become aggregates of individual particles or else, they may coalesce into a single aggregate. This will depend largely on the plasticity of particle surfaces. Soft surfaces will easily deform under capillary forces, allowing for particle coalescence into a single block, which can only be ruptured by mechanical action. The result is a kind of room-temperature sintering. On the other hand, hard and immobile surfaces can touch upon each other, but this will not lead to particle coalescence. At most, an aggregate will be formed within which the individual units can still be discerned. A similar behavior is displayed by particles which surface is strongly solvated. These may approach and associate reversibly, but in this case the solvation layer prevents particle coalescence.

6.2. THE STABILITY OF ALUMINUM POLYPHOSPHATE NANOPARTICLES

In the present case, nanoparticle formation takes place under quasi-equilibrium conditions: they are formed in the presence of excess reagents, so that starvation is not the factor controlling particle size; beyond that, particles are stable in dilute dispersion, and they do not show any of the consequences of Ostwald ripening, particle growth or aggregation. They are not stabilized by covalent bonds, and the rates of exchange of ionic neighbors, in this system, is in the order of magnitude of a few minutes.

For these reasons, we can better understand nanoparticle stability as a result of low interfacial tension, in the polyphosphate dispersions. Some arguments can be raised, to substantiate this assumption:

i) among the many different noncrystalline polyphosphates which we have already prepared, those forming nanoparticles have the highest water content; since the interfacial tension between hydrous particles and the mother solution should decrease as the compositions of both phases get closer, this justifies that the interfacial tension is lower for these than for any other aluminum polyphosphate particles;
ii) further evidence for the possibility of low interfacial tension in aluminum polyphosphate dispersions comes from the observation of thermoreversible gel formation, in this system. In these gels, there is a temperature in which a solution undergoes phase separation, but the two phases are still intimately intermingled, a behavior which would be impossible in the case of a large interfacial tension. Gels prepared with different contents of aluminum and polyphosphate ions undergo syneresis but at widely different rates - and lower rate and extent of syneresis point towards a low driving force for domain separation, i. e., a low interfacial tension.

This last argument indicates that there is a connection between seemingly opposite, macroscopic behaviors for a same system: the formation of small particles and of gels, depending on the actual experimental conditions. In both cases, low interfacial tension between the solvent-rich and solvent-poor domains is a positive factor contributing towards system stability. We may then hypothesize that the decisive factor for obtaining nanoparticles or gels, in system like this, is the particle concentration: if this is above the percolation threshold, a gel is formed. Below this threshold, nanoparticles and their aggregates are obtained. If this hypothesis proves correct, we may have a guideline for the preparation of nanoparticles under quasi-equilibrium conditions, in large amounts and from very many substances.

Acknowledgements. FG acknowledges the support of the Brazilian agencies, Fapesp and Finep. ECOL was a graduate fellow from Fapesp; MMB, RMS and NCM received graduate fellowships from CNPq. EFS received post-doctoral fellowships from Fapesp and CNPq.

7. References

1. Schrödter,K., Bettermann, G., Staffel, T. and Hofmann, T. (1991) Phosphoric acid and phosphates, in B. Barbara, S. Hawkins, G. Schulz (eds.), *Ullmann's Encyclopedia of Industrial Chemistry*, VHC Publishers, New York, pp. 465-503.
2. Mann, S. (1993) Molecular tectonics in biomineralization and biomimetic materials chemistry, *Nature* **365**, 499-505.
3. Lima, E.C.O. and Galembeck, F. (1994) Thermoreversible gel formation from aqueous aluminum polyphosphate solutions, *J. Colloid and Interface Science* **166**, 309-315.
4. Abreu Filho, P.P., Galembeck, F., Gandra, F.C.G., Baesso, M.L., Silva, E.C. and Vargas, H. (1990) Genesis of a solid foam: Iron (III) metaphosphate transformation in a sol-gel crystallisation process, *Langmuir* **6**, 1013-1016.
5. Lima, E.C.O. and Galembeck, F. (1993) Particle of aluminum metaphosphate containing closed pores. Preparation, characterization and optical properties, *Colloid and Surfaces A: Physicochemical and EngineeringAspects* **75**, 65-74.
6. Brinker, J.P. and Scherer, G.W. (1990) *Sol-Gel Science*, Academic Press, Boston.
7. Livage, J., Barboux, P., Vanderborre, M.T., Schmutz, C. and Taulelle, F. (1992) Sol-Gel synthesis of phosphates, *J. Non-Crystalline Solids* **147 & 148**, 18-23.
8. Pinheiro, E. A., Abreu Filho, P.P., Galembeck, F., Silva, E.C. and Vargas, H. (1987) Magnetite crystal formation from iron (III) hidroxide acetate. An ESR study, *Langmuir* **3**, 445-448.
9. Averback, R.S., Höfler, H.J. and Tao, R. (1993) Processing of nano-grained materials, *Materials Science and Engineering* **166**, 169-177.
10. Chokshi, A.H.(1993) Superplasticity in grained ceramics and ceramic composites: current understanding and future prospects, *Materials Science and Engineering* **166**, 119-134.
11. Langdon, T.G., Watanabe, T., Wadsworth, J., Mayo, M.J., Nutt, S.R., Kassner, M.E. (1993) Future research directions for interface engineering in high temperature plasticity, *Materials Science and Engineering* **166**, 237-241.
12. Montagne, L., Palavit, G. and Draoui, M. (1993) Mechanism of polyphosphate gel formation in the Na_2O-Al_2O_3-P_2O_5 system, *J. Non-Crystalline solids* **155**, 115-121.
13. Lima, E.C.O., Moita Neto, J.M., and Galembeck,F. (in press), *Aluminum polyphosphate thermoreversible gels: A study by* ^{31}P *and* ^{27}Al *NMR spectroscopy.*
14. Beppu, M.M., Lima, E.C.O. and Galembeck, F. (in press), *Aluminum Phosphate Particles Containing Closed Pores. Preparation, Characterization and Use as a White Pigment.*
15. Kruyt, H.R.(ed.) (1949) *Colloid Science* vol. II , Elsevier, Amsterdam.
16. Prince, L.M. (ed.) (1977) *Microemulsions*, Academic Press, New York.
17. Zsigmondy, R. (1926) *Traité de Chimie Colloidale*, Dunod, Paris, pp 187.
18. Zaiser, E.M. and La Mer, V.K. (1948) The kinetics of the formation and growth of monodispersed sulfur hydrosols, *J. Colloid Science* **3**, 571.
19. Ottewill, R.H. and Shaw, J.N. (1967) Stability of monodisperse polystirene latex dispersions of various sizes, *Kolloid* Zh. **215**, 161.
20. Martin, C.R. (1994) Nanomaterials: a membrane-based synthetic approach, *Science* **266** , 1961-1966.

LUMINESCENCE PROPERTIES OF SILVER BROMIDE: FROM NANOCRYSTALS TO MICROCRYSTALS

M. I. FREEDHOFF,[a,b] W. CHEN,[a,b] J. M. REHM,[a,b] C. MEYERS,[a] A. MARCHETTI,[a,c] and G. MCLENDON[d]
[a]*NSF Center for Photo-induced Charge Transfer, Rochester, NY 14627*
[b]*Department of Chemistry, University of Rochester, Rochester, NY 14627*
[c]*Image Research and Advanced Development, Eastman Kodak Company, Rochester, NY 14652-4708*
[d]*Department of Chemistry, Princeton University, Princeton, NJ, 08544*

1. Abstract

Using a modified reverse micelle protocol with acetyl bromide as the bromide source, AgBr nanocrystals can be grown over a size range of 30 to 1000 Å radius. These clusters display quantum confinement indicated by an increasing blue shift of the luminescence arising from the recombination of the free exciton as a function of size. As the cluster size decreases, the apparent radiative lifetime decreases as well. This is predicted by the expected breakdown of the $\underline{k}$ selection rules for small crystal sizes. However, the effects are quite modest, indicating that even for extremely small nanocrystals the indirect character of the transition is largely retained. In an attempt to overcome the problem of sample polydispersity and to resolve individual bound and phonon assisted excitonic transitions, a non-aqueous preparation of AgBr was developed, capable of providing more homogeneous size distributions of AgBr nanocrystals in the 100-400 Å diameter range. Low temperature luminescence studies show size restriction effects associated with these excitonic transitions. An examination of the dynamics of trapping of excitons at iodide impurities is undertaken, and the radiative lifetimes of some of the excitonic processes are measured in order to obtain a more global perspective on the photophysics of AgBr.

2. Introduction

Silver bromide is a material essential to the photographic process, and as such is of practical interest. The effect quantum confinement has on this material is of fundamental interest since AgBr is an indirect gap semiconductor. In order for an optical transition from the valence band maximum to the conduction band minimum to

E. Pelizzetti (ed.), Fine Particles Science and Technology, 281–293.

be allowed, absorption or emission of a momentum conserving phonon must also take place. The indirect transition itself is formally forbidden as per this **k** selection rule, but it has been postulated [1] that size restriction will result in the relaxation of these selection rules as the surface atoms' wavefunctions are incorporated into the band structure.

Quantum confinement studies on indirect gap semiconductors have been directed towards finding out what effect size has on this forbidden transition. When the size of the microcrystals becomes comparable to that of the exciton radius, a size dependent term is added to the Hamiltonian that results in three experimentally observed effects: 1) The intensity of the near bandgap exciton luminescence increases [2], 2) the energies of the free and shallowly bound exciton transitions are shifted to higher energies [3], and 3) the lifetimes of some of the transitions are changed from those of the bulk [4]. While some of these effects have been thought to be due to a breakdown in the momentum selection rules as the transition becomes less forbidden, [5] it is probable that impurity exclusion and surface assisted recombination are also factors. In order to examine the changes in the apparent radiative lifetime of the free exciton with cluster size, a synthesis was developed that was able to access a wider size regime than previous methods.

In addition to the free exciton and its phonon replicas [6], there exist a set of excitonic transitions in AgBr that lie close to the bandgap in energy [6,7,8] and these have been assigned to shallow traps. In fact, nine such excitons have been catalogued as "BX_N" and are attributed to trapping by ionized donors [7], impurities such as Cd^{2+} [7] or Pb^{2+} [7], interstitial silver ions [8], or vacancies and deformations of the lattice that occur during growth and fabrication [7]. Since these excitons' binding energy all lie within a few meV of the free exciton, their wavefunctions so closely resemble that of the FE that they undergo transitions utilizing the same momentum conserving phonons as the FE does. These excitonic transitions are absent or obscured in the smallest silver bromide nanocrystals due to impurity exclusion effects and sample polydispersity. In order to examine the effect size restriction has on these transitions, a non-aqueous synthesis of mid-size AgBr microcrystals was developed.

3. Experimental

3.1 MICELLAR PREPARATION OF AGBR:

Silver nitrate (Fisher) and Acetyl bromide (Aldrich) were used as received. Dioctyl-sulfosuccinate sodium salt (AOT, Aldrich) was purified by first stirring with activated charcoal for 24 hours, then filtering repeatedly, and finally drying under vacuum. The micellar solutions were made from a 0.2 M solution of AOT in HPLC grade n-heptane.

The AgBr preparations were made from solutions of $\omega = 2$ to 10, (where $\omega = [H_2O]/[AOT]$). For instance, an $\omega = 3$ reverse micelle solution was made by adding 0.27 ml of 0.1 M $AgNO_3$ solution to 25 ml of 0.2 M AOT in heptane. The solution was stirred in the dark until clear (about 30 minutes). A solution of 0.02 M CH_3COBr in heptane was freshly prepared.

To make the AgBr nanocrystals, 2 ml of the silver reverse micelle solution was mixed with 1 ml of the acetyl bromide solution and was vortexed for about 20 s. This solution was diluted to 8 ml with heptane and vortexed for another 30 s. The sample was then mounted and cooled to liquid helium temperatures within 4-5 minutes.

Sample sizes were determined by transmission electron microscopy (TEM). An $\omega = 10$ solution yielded nanocrystals with a mean particle diameter of 150 Å ± 70 Å. For an $\omega = 3$ solution, the mean particle diameter was 70 Å ± 15 Å (figure 1).

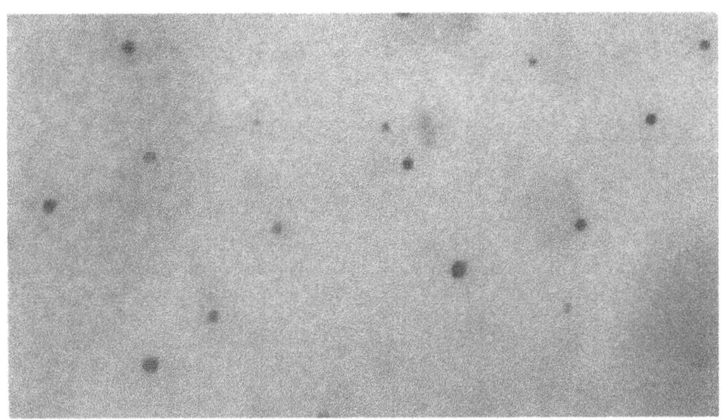

Figure 1: Transmission electron micrograph of an $\omega = 3$ micellar preparation of AgBr nanocrystals. The average particle diameter is 70 Å ± 15 Å.

3.2. NON-AQUEOUS PREPARATION OF AGBR

This synthesis was performed by pumping solutions of 1 M silver trifluoroacetate (in acetone) and lithium bromide (in acetone) at constant flow rates into a solution of 5% Butvar (in acetone) for 25 minutes. The silver ion concentration was monitored by an electrode system and kept constant by a feedback mechanism. Samples were withdrawn at various stages during the synthesis in order to collect a variety of sizes. Figure 2 shows the structure of the polymer Butvar which is thought to separate the microcrystals as they form, thereby preventing them from growing to larger sizes. The solubility of silver bromide in acetone is such that problems such as Ostwald ripening encountered in aqueous preparations [4] did not occur; The samples remained the same size for days when refrigerated.

Figure 3 shows a transmission electron micrograph of a sample collected 2 minutes into the synthesis, corresponding to a microcrystal size of 21 ± 4 nm diameter. A sample collected 25 minutes into the synthesis had an average diameter of 52 ± 14 nm. The increase in sample polydispersity with crystal size possible arises from renucleation occurring at some intermediate stage in the preparation.

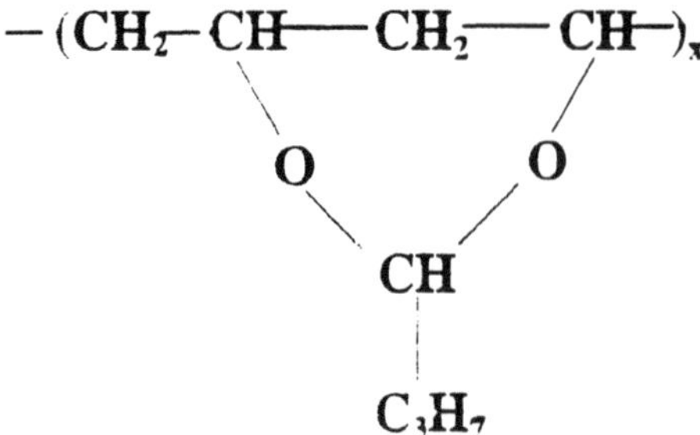

Figure 2: The structure of the polymer Butvar.

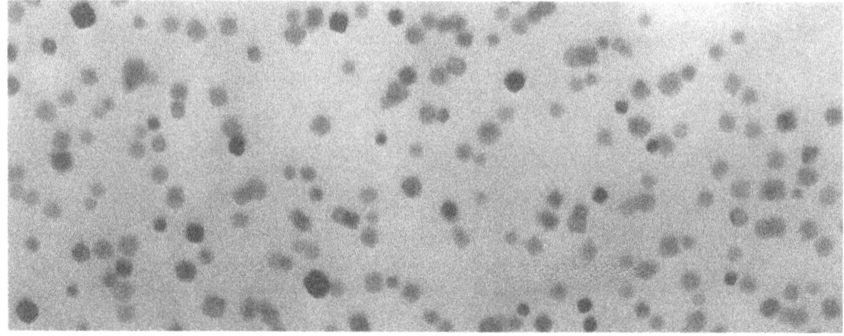

Figure 3: Transmission electron micrograph of a sample collected 2 minutes into the synthesis corresponding to microcrystals of 21 ± 4 nm diameter.

The low temperature luminescence apparatus has been described previously in [3].

4. Results and Discussion

4.1. MICELLAR SYNTHESIS

The preparation of semiconductor nanocrystals can be challenging since they are thermodynamically unstable relative to the bulk. By varying the kinetic parameters of the reaction (i.e. solvent, temperature and concentration of reactants) it can be possible to synthesize clusters of a variety of sizes. Inverse micelles have been used [3, 9-11] in order to systematically vary the size of the nanocrystals through the variation of the size of the water pool.

AgBr particles with a 50 Å radius have been synthesized by mixing micellar solutions containing Ag^+ and Br^-, but it proved impossible to create a range of sizes without further kinetic control of the reaction.

Acetyl bromide is a non-ionic bromide source that is stable in heptane but will hydrolyze on contact with the Ag^+ containing water pools at a controlled rate. The mechanism is a three step process:

1. $CH_3COBr + H_2O \rightarrow CH_3COOH + HBr$
2. $HBr + AgNO_3 \rightarrow AgBr + HNO_3$
3. $AgBr + nAgBr \rightarrow (AgBr)_n$

Step 1 provides the additional kinetic control required to produce AgBr clusters in a range of sizes. We believe that TEM tends to overestimate the size of the nanocrystals due to ripening that occurs while preparing the samples.

4.2. LOW TEMPERATURE STEADY-STATE LUMINESCENCE SPECTRA

4.2.1. Micellar Preparations:

Band edge luminescence has yielded information about the indirect exciton in AgBr [3]. For the micellar samples, this peak is broadened by the inhomogeneous sample sizes. Figure 4 shows the low temperature luminescence of three sizes of micellar AgBr. The peak of the indirect exciton shifts to higher energy as the volume of the water pool is decreased. A comparison of the observed blue shift in the indirect exciton peak position with an estimate based on Brus' particle in a box theory [12] has been given in [4].

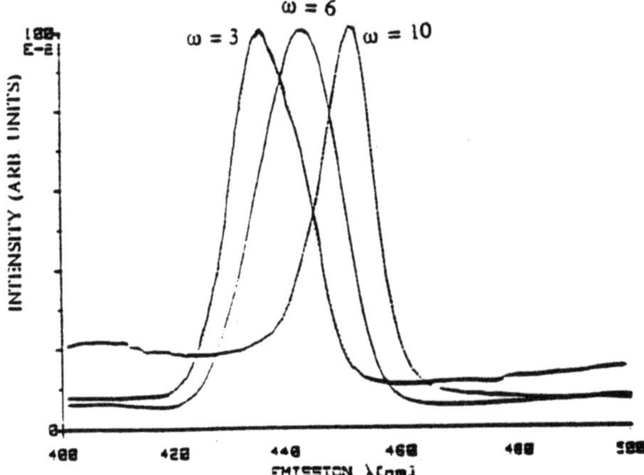

Figure 4: Low temperature luminescence spectra of three sizes of micellar AgBr taken at 6 K at an excitation wavelength of 325 nm. The ω = 3 sample corresponds to a crystal size (as determined by TEM) of 84 Å ± 19 Å diameter, ω = 6 to a size of 153 Å ± 71 Å diameter and the ω = 10 to a size of 146 Å ± 33 Å diameter.

4.2.2. Non-aqueous Preparations

While the micellar preparations serve to illustrate quantum confinement effects on the indirect exciton, it fails to address the analogous effects on the previously mentioned phonon assisted and shallowly bound excitonic transitions. To address these questions, low temperature luminescence studies of mid-size AgBr microcrystals prepared using a non-aqueous synthesis were undertaken. Figure 5 shows low temperature luminescence spectra of three different sizes of AgBr microcrystals prepared this way.

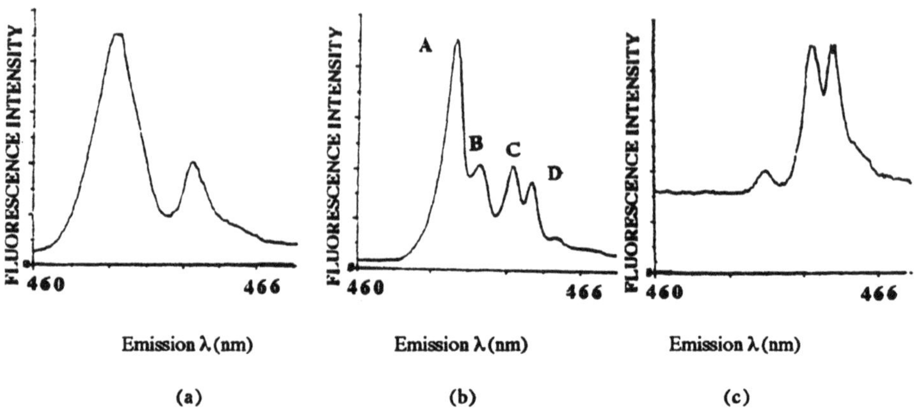

Figure 5: High resolution scans of the FEL region. (a) 1.5 minutes into the synthesis, (b) 4 minutes into the synthesis and (c) 25 minutes into the synthesis.

The smaller microcrystals exhibit a broad peak located at around 461.5 nm that becomes narrower, shifts to the red in energy and eventually resolves into two shifting peaks as the particles grow. By the end of the synthesis, these peaks have all but disappeared. The second feature of these spectra is the set of peaks at 464.16 nm and 464.64 nm that continue to grow in intensity throughout the synthesis but never display shifts in their peak positions.

As described previously, several shallowly bound excitons have been identified for AgBr. All of these undergo the same phonon assisted (i.e. momentum conserving) transitions as the free exciton does. Some have been shown to undergo strong 0,0 transitions while others possess intense TO phonon assisted transitions. This information was used to assign the peaks. Peak A (as labeled in figure 5b) was assigned to be the 0,0 transition of the BX8 exciton which is thought to be associated with trapping by ionized donors. The peak position of this transition converges to 2.6800 eV in the larger crystals, which corresponds to the accepted value for the bulk BX8 exciton. BX8 is also known to exhibit a strong 0,0 transition but only a weak TO phonon assisted transition. Since we only observe the 0,0 transition, this is also consistent with our assignment.

Peak B was more difficult to assign since it was observed only over a narrow size range. Energetically it corresponds to the TO phonon assisted transition of the free exciton which is known to be a strong transition.

Peaks C and D are more enigmatic. While they appear to be exciton peaks in the energy region where exciton peaks are expected, we observed no shift in energy with size, and the peaks do not correspond energetically to any previously observed bound excitons. The positions of the peaks do not change with excitation energy, nor do they have the same temperature dependence as the other excitonic transitions do. Measurements of the fluorescence lifetimes of these peaks do not vary with crystal size, and are orders of magnitude different from lifetime measurements made on the known excitonic transitions. Since they do not appear in any other measurements of AgBr microcrystals made using alternate synthetic techniques, the conclusion reached as to the identity of these peaks is that they are probably a synthetically related impurity or trap state, and that their location in the FEL region is purely coincidental.

4.3. TIME RESOLVED MEASUREMENTS AND EXCITON DYNAMICS:

4.3.1. Micellar Preparations

While steady-state measurements can provide information about the energetics of quantum confinement in AgBr, they do not directly probe how the selection rules for this indirect band gap material are modified by size restriction [13,14]. It is possible that confinement of the materials will result in a relaxation of the momentum selection rule, making the indirect transition more allowed. This would manifest itself as a decrease in the radiative lifetime as the material begins to resemble a direct gap semiconductor. However, several caveats should be noted. First, it is the radiative rate, k_r, which is related to the transition matrix element associated with the recombination of the indirect exciton. However, one does not directly measure the radiative rate but rather an observed rate $k_{obs} = k_r + k_{nr}$ where k_{nr} includes all non-radiative pathways such as trapping and thermal processes. In the simplest case, k_r can be obtained by combining measurements of k_{obs} and quantum yield: $\phi_r = \frac{k_r}{k_r + k_{nr}}$. Thus $k_{obs}/\phi = k_r$.

For AgBr emission, such an analysis is complicated by the quantum yield of exciton formation upon excitation with above bandgap light. This information is not precisely known and might be a size dependent parameter. However, earlier calculations suggest that free electrons and holes in these nanocrystals will rapidly collapse (due to multiple encounters resulting from the restricted space) to produce free excitons. This effect will minimize the effect of such branching. If significant branching occurs, then the apparent quantum yield ϕ_{obs} will be less than the actual radiative quantum yield ϕ_r. Thus the estimates of k_r reported here represent maximum values. We have therefore made the assumption that the quantum yield of exciton formation does not change over the range of cluster sizes we have examined. With these caveats, the measurements of time resolved exciton emission are discussed below.

In earlier work, quite rapid (ns) emission from the recombination of the indirect exciton was observed, suggesting a fundamental change in momentum selection rules for AgBr upon strong confinement. These results proved inconsistent with the absorption (or equivalently excitation hole burning) spectra which show no evidence of an exciton peak. With a new range of cluster sizes available, we have re-investigated the time-resolved luminescence. A fast (ns) component is present in all samples, but its lifetime does not vary with cluster size. More importantly, detailed single photon counting studies showed that over 95% of the emission had not decayed within the 100 nanosecond window of the photon counting apparatus. When we increased the time window of observation, we do observe a small amplitude nanosecond component, but the primary decay occurs with a much longer lifetime. Depending on the cluster size, this lifetime ranges from 200 μs (R = 35 Å) to 1000 μs (R ≥ 150 Å). A representative lifetime measurement is given in figure 6.

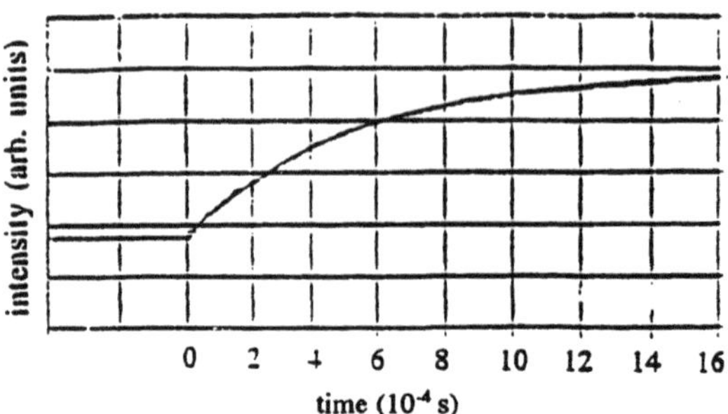

Figure 6: Lifetime of the emission of the recombination of the indirect exciton in 50 Å AgBr nanocrystals. The excitation wavelength was 337 nm and the emission was monitored at 444 nm.

It is apparent that the lifetimes, while somewhat dependent on cluster size, are still quite long, suggesting that within this size regime confinement of the AgBr exciton has not had a major effect on the momentum selection rules: Bandedge absorption and emission are still quite forbidden even in 35 Å nanocrystals. The noted uncertainties in quantum yield would only increase the radiative lifetime, which will not change this conclusion. Although the observed lifetime effects are not dramatic, the lifetimes do decrease with decreasing crystal size. This is in qualitative agreement with the supposition of the breakdown of the momentum selection rules that should occur for small samples. It is however clear that a change in the material's character from an indirect gap to direct gap semiconductor will probably not be observed in any size regime that can still be described using a solid state physics description.

4.3.2. Non-aqueous Preparation

Figure 7 shows a schematic representation of our overall understanding of the dynamics that occur upon excitation of AgBr. Since we are exciting with light with about 1 eV energy in excess of the bandgap, we are not creating excitons directly but rather free electrons and holes. These carriers then thermalize rapidly (ps-fs) [15] to the conduction and valence band edges, where the Coulombic attraction between the electrons and holes can result in the formation of excitons. Since the absorption of light must be a momentum conserving process, the heavier holes will possess less kinetic energy than the electrons upon excitation. In addition, the holes can couple to the lattice more effectively than the electrons can [16] and will therefore thermalize to the valence band edge before the electrons reach the conduction band edge. Since some carriers will become trapped before forming excitons, and since in contrast to the micellar samples these larger samples possess a significant number of impurity trapping sites, the quantum yield of exciton formation will certainly be less than unity.

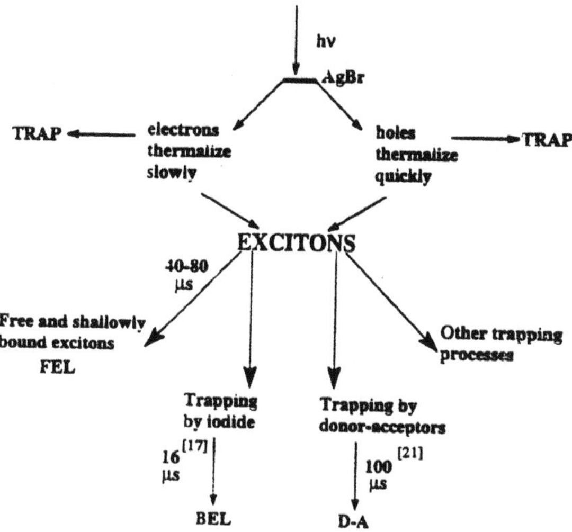

Figure 7: The pathways available to free carriers upon excitation with above band gap light.

Once an exciton is formed it can follow several different routes that result in its elimination.

Recombination of the Free or Shallowly Bound Exciton. The electron and hole can recombine radiatively either with or without the assistance of a momentum conserving phonon resulting in FEL. Shallow trapping of excitons at impurities or other lattice irregularities also result in recombination of this variety. Lifetime measurements of the 0,0 transition of the BX8 exciton (peak A-figure 5b) indicated that the measured fluorescence lifetime of 40-80 μs was in between that of the bulk (ns [6]) and that of the micellar samples (100-700 μs [4]). This can be explained by the fact that the very small micellar samples have fewer vacancies, interstitial ions, impurities and defects in their lattices and thus fewer relaxation pathways open to the exciton other than direct recombination. As the particles grow, other more favorable processes become available to the excitons, and thus the observed lifetimes become much shorter.

Recombination of carriers trapped at iodide impurities. A second process the exciton can undergo is trapping by an iodide impurity and subsequent BEL. The lifetime of this luminescence has been found to be 16 μs in the bulk [17] with a faster component that begins to contribute to the dynamics as the crystal size is decreased [16], but so far the trapping rate has not been modeled or measured.

In the simplest model, free excitons are able to move freely through a crystal with an RMS velocity determined by the crystal temperature,

$$v_{RMS} = (3kT/2M^*)^{1/2} \tag{1}$$

where k is Boltzmann's constant, T is the temperature and M^* is the exciton's translational mass. At 6 K, the RMS velocity is 1.40 x 10^4 m/s. The iodide trapping rate would be determined by comparing the volume swept out by a traveling exciton in its lifetime with the probability of finding an iodide impurity within that volume.

The volume swept out by the exciton is the cross sectional area (eqn. 2) times the root mean square velocity (eqn. 1):

$$\sigma = \pi R_{ex}^2 \tag{2}$$

Here R_{exc} is assumed to be the free exciton radius of 25 Å [18] (as calculated using a scaled hydrogen atom approximation). The cross sectional area is found to be 19.63 nm^2.

$$\sigma v_{RMS} = 2.75 \times 10^{14}\ nm^3/s = \text{Volume/time.} \tag{3}$$

If we postulate a unit trapping probability upon encounter, then the excitons in our largest (50 nm) crystals will have swept out a volume equivalent to that of the crystal volume in less than 1 ns. From this it can be assumed that even with a non-unit trapping probability, the timescales in question are so much faster than the radiative decay of the FEL in this crystal size regime that if the exciton encounters an iodide impurity, it will become trapped prior to direct recombination. It must be pointed out that this model not only assumes a unit trapping probability upon encounter, but also neglects other trapping processes.

The next step considers the probability of such an encounter. Moser and Lyu [19] outline a way to determine the iodide impurity concentration in melt grown bulk samples of AgBr. The ratio of the luminescence intensity of the iodide bound luminescence to that of the donor-acceptor luminescence gives a direct estimate of the concentration of iodide ions in a single crystal sample. This method was utilized to find that the largest samples prepared by this non-aqueous method contain an intrinsic iodide concentration of 0.37 ppm. Assuming that concentration to be constant throughout the growth process, Poisson statistics were employed to determine the probabilities of finding iodide impurities in the sample.

The distribution of iodide impurities in a group of microcrystals is governed by Poisson statistics:

$$P_i = (\lambda^i/i!)\, e^{-\lambda} \tag{4}$$

where P_i is the probability of finding i iodide impurities per microcrystal, and λ is the expected iodide concentration assuming a linear iodide incorporation rate.

Figure 8 shows such probabilities for finding zero and at least one iodide per crystal. As can be seen, even at the largest particle sizes of 50 nm, 40% of the sample should have no iodide impurities in it, and for the 20 nm crystals, that number increases to over 90%. It then becomes necessary to try to explain why the BEL is still such a dominant spectral feature even when 90% of the microcrystals in a sample contain no iodide ions. This can be rationalized by recalling that the iodide bound exciton transition is a more allowed transition than the free and shallowly bound exciton transitions are [20], and even though quantum size effects relax the momentum selection rules for the indirect transition somewhat, it still remains quite forbidden. Thus the oscillator strength for the iodide bound transition is so much stronger than that for the indirect transition that BEL dominates the spectrum even when the number of iodide impurities is very small.

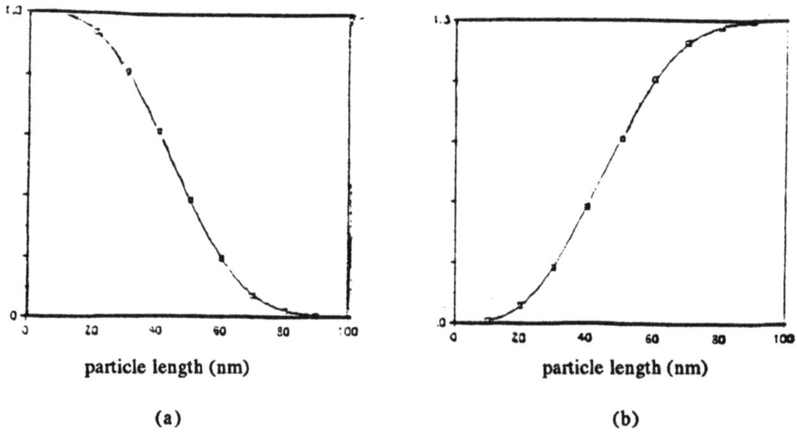

Figure 8. Probability (based on a Poisson distribution of iodide impurities) for finding (a) no and (b) one iodide impurities per crystal as a function of crystal size.

Measurements of the Iodide Trapping Time. Single photon counting experiments on several different samples were performed. The iodide bound luminescence was monitored at 500 nm. A measure of the rise time of this luminescence would give a direct estimate of the iodide trapping time. The results indicated that the trapping at iodide ions was taking place on a timescale limited by the picosecond laser pulsewidth, and no size dependence was observed. This is in clear disagreement with the results obtained via Poisson statistics.

There are several possible explanations for this. One is that the method utilized to calculate the iodide ion concentration was based on one used for bulk and melt grown samples; It may be that for this non-aqueous synthesis this is not a viable technique.

A more likely explanation draws on what has always been a generalized proposal for how bound exciton formation proceeds. Instead of exciton formation occurring prior to the trapping by iodide ions, perhaps the free carriers themselves are sequentially trapped. The hole would first become bound to the negative iodide ion, and then the electron would become bound to it. That process would occur on a picosecond timescale [15].

It is however an established fact that excitons that have already been created do get trapped at iodide ions. By exciting the samples right at the bandgap energy, excitons are being created exclusively, and iodide bound luminescence is still observed. Thus it is not sufficient to postulate that the discrepancy between theory and experiment is due solely to the measurement of a sequential trapping lifetime, as a slower component arising from the trapping of formed excitons is expected as well. A representative excitation spectrum is given in figure 9.

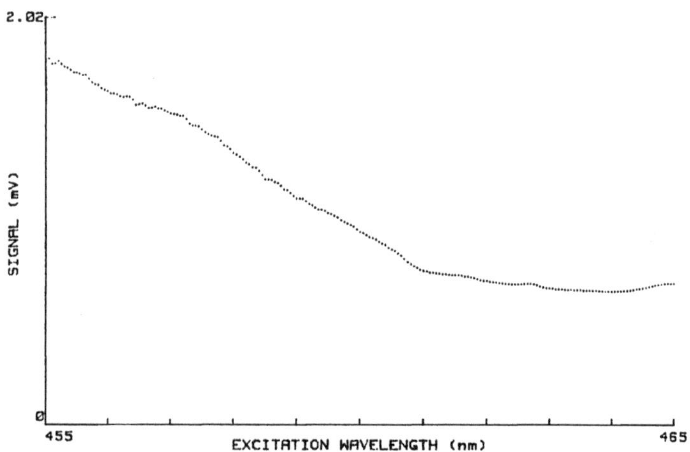

Figure 9: Excitation spectrum of 39 nm AgBr microcrystals.
The iodide bound luminescence was monitored at 500 nm.

A third and final possibility for the failure of the theory is that the trapping times quoted are in fact maximum trapping times; That is, the theoretical assumption made is that the excitons are formed as far away from the iodide impurity as possible, and that they then have to traverse the length of the microcrystal before even encountering the trapping species. Since radiative recombination from the iodide ions is a far more allowed process than indirect recombination of the free exciton, it stands to reason that absorption taking place close to the impurity is also a less forbidden process.

It is possible that the absorption of light is taking place preferentially close to the trapping species, and thus the excitons become trapped immediately rather than having to travel through the microcrystals at all.

This work was supported by NSF, through a grant to the Center for Photoinduced Charge Transfer.

5. References

1. Rossetti, R., Hull, R., Gibson, J. M. and Brus, L. E. (1985) Hybrid electronic properties between the molecular and solid state limits: Lead sulfide and silver halide crystals, *J. Chemical Physics*, **83**, 1406-1410.
2. Johannson, K. P., McLendon, G. L. and Marchetti, A. P. (1991) The effect of size restriction on silver bromide. A dramatic enhancement of free exciton luminescence, *Chemical Physics Letters*, **179**, 321-324.
3. Johannson, K.P., Marchetti, A.P. and McLendon, G. L. (1992) Effect of size restriction on the static and dynamic emission behavior of silver bromide, *J. Physical Chemistry*, **96**, 2873-2879.
4. Chen, W., McLendon, G., Marchetti, A., Rehm, J. M., Freedhoff, M. I. and Meyers, C. (1994) Size dependence of radiative rates in the indirect bandgap material AgBr, *J. American Chemical Society*, **116**, 1585-1586.
5. Kanzaki, H. and Tadakuma, Y. (1991), Indirect-Exciton Confinement and Impurity Isolation in Ultrafine particles of silver bromide, *Solid State Communications*, **80**, 33-36.
6. Von der Osten, W. and Stolz, H. (1990) Localized exciton states in silver halides, *J. Physics and Chemistry of Solids*, **51**, 765-791.
7. Von der Osten, W. (1987), Shallow bound excitons in silver halides, *Physica*, **146B**, 240-255.
8. Sliwczuk, U. and Von der Osten, W. (1988) Optical studies of weakly localized exciton states in silver halides, *J. Imaging Science*, **32**, 106-113.
9. Steigerwald, M. L., Alivasatos, A. P., Gibson, J. M., Harris, T.D., Kortan, R., Muller, A. J., Thayer, A. M., Duncan, T. M., Douglass D. C. and Brus, L. E. (1988) Surface derivatization and isolation of semiconductor cluster molecules, J. American Chemical Society, **110**, 3046-3050.
10. Chen, W., Rehm, J. M., Meyers, C., Freedhoff, M. I., Marchetti, A. and McLendon, G. (1994) Luminescence properties of indirect bandgap semiconductors: Nanocrystals of silver bromide, *Molecular Crystals and Liquid Crystals*, **252**, 79-86.
11. Kuczynski, J. and Thomas, J. K. (1982) Photochemistry at the surface of colloidal cadmium sulfide, *Chemical Physics Letters*, **88**, 445-447.
12. Brus, L. E. (1984) Electron-electron and electron-hole interactions in small semiconductor crystallites: The size dependence of the lowest excited state, *J. Chemical Physics*, **80**, 4403-4409.
13. Johannson, K., Cowdery, R., O'Neil, M., Rehm, J., McLendon, G., Marchetti, A., and Whitten, D.G. (1993) Comparative semiconductor photophysics of II-VI (CdS) and I-VII (AgBr) quantum dots, *Israel J. of Chemistry*, **33**, 67-70.
14. Takagahura, T. (1987) Excitonic optical nonlinearity and exciton dynamics in semiconductor quantum dots, *Physical Review B*, **36**, 9293-9296.
15. Seeger, K. (1989) *Semiconductor Physics*, Springer-Verlag Press, Berlin Heidelberg.
16. Marchetti, A. P., Johannson, K. P. and McLendon, G. L. (1993) AgBr photophysics from optical studies of quantum confined crystals, *Physical Review B*, **47**, 4268-4275.
17. Marchetti, A.P. and Burberry, M. S. (1983) Optical and optically detected magnetic resonance studies of $AgBr:I^-$, *Physical Review B*, **28**, 2130-2134.
18. Knox, R. S. (1963) *Theory of Excitons*, Academic Press Inc., New York.
19. Moser, F. and Lyu, S. (1971) Luminescence in pure and I-doped AgBr crystals, *J. Luminescence*, **3**, 447-458.
20. Czaja, W. and Baldereschi, A. (1979) The isoelectronic trap iodine in AgBr, *J. Physics C:Solid State Physics*, **12**, 405-424.
21. Marchetti, A. P., Burberry, M. S. and Spoonhower, J. P. (1991) Characterization of an intermediate case exciton in the 580-nm emission of Cd-doped and pure AgBr, *Physical Review B*, **43**, 2378-2383.

CARRIER DYNAMICS IN SILICON NANOCRYSTALS AND POROUS SILICON FILMS

LOUIS BRUS
AT&T Bell Labs
Murray Hill, NJ, 07974, USA

ABSTRACT In this short contribution I discuss recent work attempting to understand the photoluminescence of silicon nanocrystals and porous Si thin films. I also outline a simple model for transport and electroluminescence in porous Si liquid junction diodes.

1. Introduction

Crystalline silicon is an indirect gap semiconductor in which luminescence is forbidden by dipole selection rules. Thus there was great interest in the microelectronics industry when it was discovered that porous silicon (hereafter p-Si) photoluminesces with high quantum efficiency at visible wavelengths(1,2). Further work showed that liquid junction p-Si diodes show efficient electroluminescence, in fact better electroluminescence that dry p-Si diodes(3). Practical interest exists because an efficient electroluminescent device, that could be directly and cheaply grown on a Si VLSI chip, could be quite important in telecommunications and computing. In this short manuscript I discuss and review recent work aimed at understanding these phenomena(4). We studied the radiative and radiationless transitions of a single electron-hole pair in surface oxidized Si nanocrystals, and also built a simple model for carrier transport in both wet and dry forms of p-Si.

2. Synthesis and Optical Properties of Si Nanocrystals.

P-Si is composed of an open, random network of crystalline Si wires with undulating diameters of 1-10 nm size. This structure, which can be 75% porous, can be thought of as composed of H atom passivated Si nanocrystals that are partially fused to their neighbors. As a simple prototype system, we studied the properties of surface oxidized Si nanocrystals made in a high temperature aerosol apparatus(4). Extensive structural characterization showed the nanocrystals to have a 0.8 nm surface oxide layer over a crystalline core of Si whose lattice

E. Pelizzetti (ed.), Fine Particles Science and Technology, 295–301.

constant was unchanged from the bulk. Nanocrystals with a rather broad size distribution were collected as a colloid in ethylene glycol. Samples with more narrow distributions were made by size exclusion chromatography and by size selective precipitation. As the size of the crystalline core decreased into the 1-2 nm range, the photoluminescence shifted from the bulk band gap of Si in the infrared, to the visibly red 600-700 nm range. The luminescence quantum yields were as high as 5% at room temperature, but the lifetimes were very long, typically on the order of 50 microseconds. The Si nanocrystal spectroscopic data were very similar to the p-Si data(5), and served to confirm the assignment of a surface passivated Si nanocrystal as the emitting chromophore in p-Si.

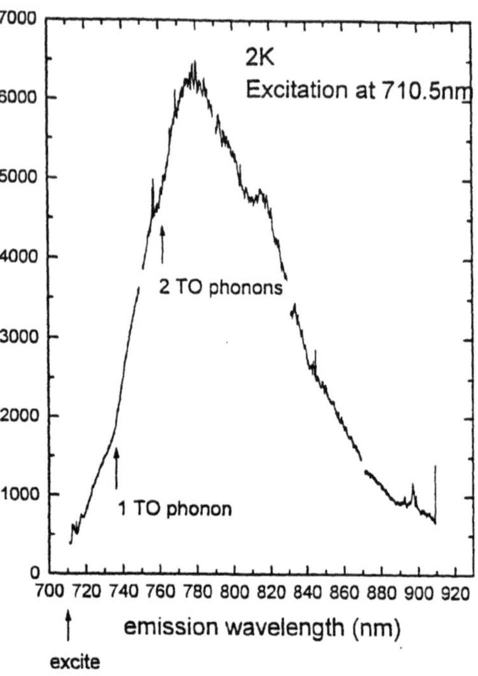

Figure 1. Nanocrystal luminescence spectra adapted from ref. 4e

In one experiment we excited photoluminescence with Ti-sapphire laser excitation at 710 nm, near the band gap of the larger nanocrystals in the size distribution of that particular sample, as shown in figure 1. Such excitation helps to remove part of the inhomogeneous broadening that obscures spectral structure in samples with a range of sizes. The luminescence observed was that of a vibronically induced transition that had a very weak origin (ie, band gap) transition. The emission was induced principally by TO phonons as marked in the figure. This result shows that Si nanocrystals remain indirect gap, even in the ca. 1.5 nm diameter particles emitting near 600 nm wavelength. This conclusion was confirmed by analysis of the luminescence excitation spectra, and the long radiative lifetimes.

3. Comparison of Bulk and Nanocrystal Si Optical Physics

Both nanocrystal Si and bulk Si are indirect gap semiconductors with TO phonon induced spectra, and long electron-hole pair radiative lifetimes (4,5). Nanocrystal Si luminescence is shifted to higher energy by as much as 1 eV due to the three dimensional quantum size effect. Bulk Si typically shows almost no luminescence, while nanocrystal Si shows efficient emission. This difference is caused by a pronounced decrease in non-radiative processes at room temperature as size decreases(4e). In bulk Si, three carrier Auger non-radiative recombination dominates carrier decay. In nanocrystals, the Auger process is decreased by orders of magnitude, as electron-hole pairs in separate nanocrystals are electrically isolated and do not undergo the three carrier Auger process. Also, if the Si contains rare deep trap defects that catalyze recombination, such non-radiative decay will become unimportant when size decreases to the point where there is less than one defect per nanocrystal.

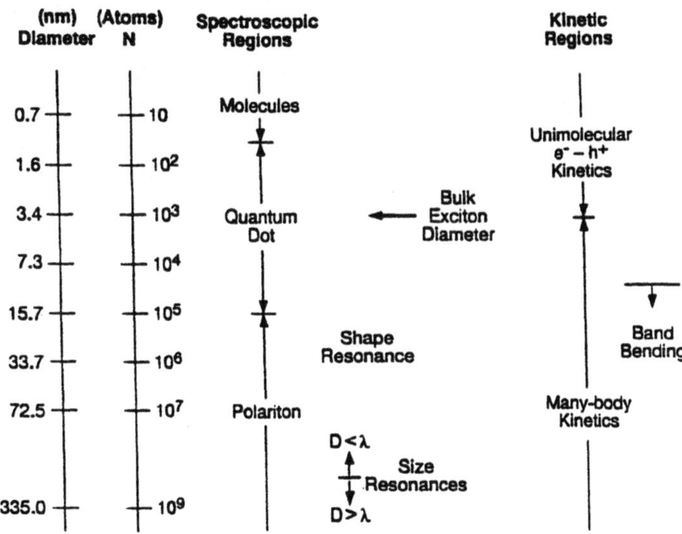

Figure 2. Schematic Size Regimes adapted from ref. 4e.

Si nanocrystals thus exemplify the change in carrier dynamics, from solid-state-like in bulk semiconductors, to molecule-like in nanocrystals. Molecule-like implies that a single electron-hole pair, interacting via Coulomb and exchange forces, and coupled to the lattice vibrations, decays unimolecularly. Figure 2 is a schematic diagram that outlines the various size regimes that are thought to occur.

4. Porous Si and Electroluminescence

Liquid junction p-Si electrochemical cells are remarkable(3). If electrons flow into the p-Si layer from the underlying crystalline Si substrate, and holes are injected into the layer from reducible molecular species in solution, then efficient electroluminescence occurs. The electroluminescence spectrum is relatively narrow, and can by tuned across the broader photoluminescence spectrum via change in the applied voltage. The broad photoluminescence spectrum is caused by the wide distribution of particle sizes present. The narrowed electroluminescence apparently implies that only that subset of nanocrystals at the Fermi level electroluminesce. The efficiency of such cells is significantly better than that of dry diodes that have an external second contact made by gold evaporation.

Because of the wide size distribution, porous Si is a disordered medium for carrier transport. Size dependent electrostatic energies are important in the relative stability of a hole or an electron on a nanocrystal(6). As the p-Si layer is charged with electrons, the Fermi level moves up from larger nanocrystals into smaller nanocrystals, and eventually the larger nanocrystals become doubly charged. Smaller nanocrystals have greater confinement energy, and less dielectric polarization energy, than larger nanocrystals, and hence higher potential is required to charge them with an extra electron.

In transport, when a hole or electron hops from nanocrystal to nanocrystal under an applied external field, the electronic levels will not be resonant unless the two nanocrystals are exactly the same size. In nonresonant electron transfer, energy must be absorbed from the environment if the acceptor nanocrystal is smaller, and deposited in the environment if it is larger. In dry p-Si, energy can only be exchanged with the environment via the relatively weak deformation potential coupling to Si acoustical phonons. If the p-Si pores are filled with water, then carriers can also couple to the motions of the polar fluid molecules via electrostatic forces(7). A quantitative treatment of the two types of "electron-phonon" coupling, using the Marcus model(8) originally developed for molecules and proteins in fluids, shows that the coupling to the polar fluid is typically an order of magnitude stronger than coupling to acoustical vibrations.

In electroluminescence, a critical step is injection of an electron on one nanocrystal into a second, touching nanocrystal containing a hole. In this process two charged nanocrystals convert into two neutral nanocrystals, and thus considerable electrostatic energy is released. Figure 3 is a logarithmic plot of the vibrational activation energy term in the rate of electron injection, for a 2nm donor nanocrystal injecting an electron into variable diameter nanocrystal containing a hole(7). In a water environment, the rate is activationless and fast for all sizes, as the water polarization readily accepts the injection exothermicity. In vacuum, direct transfer into the electron ground 1S state of the acceptor nanocrystal is very slow for all sizes. Injection into the 2S state is resonant for 2.1 nm diameter acceptors, and into the 3S state for 4.5 nm acceptors. Water greatly facilitates electron-hole injection in p-Si samples composed of a wide range of sizes.

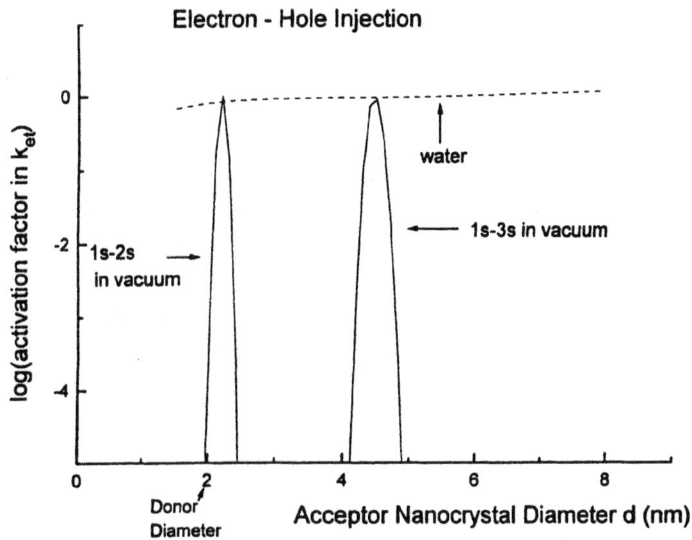

Fig 3. Log(base 10) plot of injection rates in water and in vacuum, after ref. 7.

In general, transport in p-Si changes from solid-state-like in vacuum, to protein-like in water. Solid-state-like implies that only electronically resonant process occur, because electron-phonon coupling is weak. Protein-like implies that exothermic transfer occurs without activation energy, because of coupling to polar molecular motions(9). The characteristic Coulomb blockade transport characteristic of a single nanocrystal(10) should not be present in a water environment.

References

1 (a) L. T. Canham, *Appl. Phys. Lett.* 57(1990) 1046; (b) N. Koshida, H. Koyama: *Jpn. J. Appl. Phys.* 30(1991) L1221. (c) V. Lehmann and U. Gosele, *Appl. Phys. Lett.* 58, 856 (1991).

2. Recent reviews and symposium proceedings: (a) Proceedings of the European Materials Research Society, Symposium F, in *Thin Solid Films* 255, Nos. 1-2 (1995) (b) Microcrystalline and Nanocrystalline Semiconductors, *Mat. Res. Soc. Sym. Proc. vol. 358,* R. W. Collins etal editor, 1995 (c) Porous Silicon Science and Technology, J. C. Vial and J. Derrien, eds., (Springer-Verlag, Berlin,) 1995 (d) Light Emission from Novel Silicon Materials, Y. Kanemitsu etal eds., *J. Phys. Soc. Jpn.* 63 (1994) Suppl. B.

3 (a) A. Halimaoui etal, *Appl. Phys. Lett.* 59, 304 (1991) (b) P. Bressers, J. Knapen, E. Meulenkamp and J. Kelly, *Appl. Phys. Lett.* 61, 108 (1992). (c) A. Bsiesy etal, *Phys. Rev. Lett.* 71, 637 (1993). (d) M. Ligeon etal, *J. Appl. Phys.* 74, 1265 (1993).(e) E. Kooij, R. Despo, and J. Kelly, *Appl. Phys. Lett.* 66, 2552 (1995).

4. (a) K. Littau, P. Szajowski, A. Muller, A. Kortan, L. Brus, *J. Phys. Chem.* 97, 1224 (1993). (b) W. Wilson, P. Szajowki, and L. Brus, *Science* 262, 1242 (1993) (c) S. Schuppler etal, *Phys. Rev. Lett.* 72, 2648 (1994). (d) S. Schuppler etal, *Phys.Rev.B* (to be published) (e) L. Brus, P. Szajowski, W. Wilson, T. Harris, S. Schuppler and P. Citrin, *J. Am. Chem. Soc.* 117, 2915 (1995).

5. (a)P. D. Calcott, K. J. Nash, L. T. Canham, M. J. Kane, D. J. Brumhead: J. Phys.: Condens. Matter 5(1993) L91. (b)P. D. Calcott, K. J. Nash, L. T. Canham, M. J. Kane, D. J.Brumhead: J. Lumin. 57(1993) 257.(c)T. Suemoto, K. Tanaka, A. Nakajima: J. Phys. Soc. Jpn. 63(1994) Suppl. B 190.

6. L. Brus, *J. Chem. Phys.* 79, 5566 (1983). (b) D. Babic, R. Tsu, and R. F. Greene, *Phys. Rev.* B45, 14150 (1992). (c) R. Tsu and D. Babic, *Appl. Phys. Lett.* 64, 1806 (1994) (d) C. Delerue, G. Allan and M. Lannoo, *Phys. Rev.* B48, 11024 (1993). (e) M. Lannoo, C. Delerue, and G. Allan, *Phys. Rev. Lett.* 74, 3415(1995). (f) J. Chazalviel, F. Ozanam, and V. Dubin, *J. Phys. I France* 4, 1325 (1994).

7. L. Brus, "Model for Carrier Dynamics and Photoluminescence Quenching in Wet and Dry Porous Silicon Thin Films", *Phys. Rev. B*, (submitted July 1995).

8. (a) R. Marcus, *J. Chem. Phys.* 24, 966 (1956). (b) R. Marcus, *Faraday Disc. Chem. Soc.* 29, 21 (1960). (c) V. Levich and R. Dogonadze, *Dokl. Akad. Nauk. SSR* 124, 123 (1959). (d) J. Jortner, *J. Chem. Phys.* 64, 4860 (1976). (e) for a review of chemical dynamics applications see M. Newton and N. Sutin, Annu. Rev. Phys. Chem. 35, 437 (1984).

9 For example, see M. Marchi, J. Gehlen, D. Chandler and M. Newton, *J. Am. Chem. Soc.*115, 4178 (1993).

10. M. Kastner, *Rev. Mod. Phys.* 64, 849 (1992).

Photoinduced Charge Transfer Processes in Semiconductor Heterostructures. Capped vs. Coupled Systems.

Prashant V. Kamat, Idriss Bedja
Radiation Laboratory, University of Notre Dame
Notre Dame, Indiana 46556, U.S.A.
and
K. Vinodgopal
Department of Chemistry
Indiana University Northwest
Gary, IN 46408

Abstract

By coupling or capping a metal oxide nanocrystallite with one or more semiconductors it is possible to design semiconductor heterostructures that are potentially useful in microelectronics or molecular devices. Photoinduced charge transfer processes in multicomponent semiconductor particulate systems can provide valuable information for improving the efficiency of charge separation. We have now synthesized a variety of semiconductor heterostructures with coupled and capped geometries. The charge transfer processes in metal oxide nanostructures of two different geometries have been investigated with nanosecond and picosecond laser flash photolysis. The geometry of these heterostructures has been found to control the interfacial charge transfer or charge collection efficiencies. The photoelectrochemical properties of capped and coupled semiconductor systems have also been made by preparing nanocrystalline thin films of these materials.

1. Introduction

The feasibility of coupling a metal oxide nanocrystallite with another semiconductor by chemical precipitation or electrochemical deposition opens up a wide array of possibilities in utilizing these materials for microelectronics or molecular devices. For example, in a recent study, we have shown that it is possible to rectify the flow of electrons in nanostructured TiO_2 film by coating a thin CdSe film [1]. Uekawa et al [2] have also succeeded in developing a p–n junction diode. The advantages of using coupled semiconductor films in photoelectrochemical cells are twofold: (i) to extend the photoresponse by coupling a large–bandgap

E. Pelizzetti (ed.), Fine Particles Science and Technology, 303–316.

semiconductor with a short–bandgap semiconductor and (ii) to retard the recombination of photogenerated charge carriers by injecting electrons into the lower lying conduction band of the large bandgap semiconductor.

In our continuing efforts to understand the charge transfer processes in multicomponent semiconductor particulate systems we have synthesized both coupled and capped type semiconductor systems. While the mechanism of charge separation in a capped semiconductor system is similar to that in a coupled semiconductor system, the interfacial charge transfer or charge collection at this multicomponent semiconductor system is significantly different. The principle of charge separation in coupled and capped semiconductor systems is illustrated in Figure 1. In a coupled semiconductor system two particles are in contact with each other and both holes and electrons are accessible for selective oxidation and reduction processes on different particle surfaces. On the other hand, capped semiconductors have a core–shell geometry. Only one of the charge carrier is accessible at the surface in a capped semiconductor system, thus making selective charge transfer possible at the semiconductor electrolyte interface. The other charge carrier (i.e., the electron in example (b) of Figure 1) gets trapped within the inner semiconductor particle and is not readily accessible.

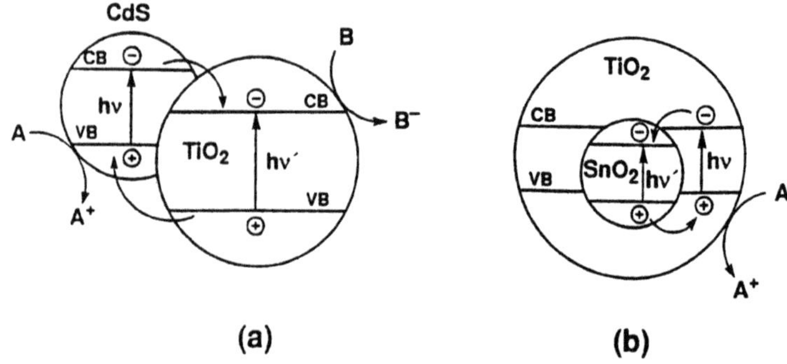

Figure 1. Principle of charge separation in semiconductor heterostructures: (a) Capped (or Core-Shell) geometry and (b) Coupled geometry. Acuumulation of electrons occurs at the conduction band (CB) of SnO_2 and holes at the TiO_2.

A variety of interesting properties of coupled and capped semiconductor systems have recently been reported by several researchers (see for example, [1, 3–6]). Such coupled semiconductor thin films not only extend the photoresponse of large bandgap semiconductors but also rectify the flow of photogenerated charge carriers [1] and improve the efficiency of dye sensitization [7]. Similarly, it has been possible to enhance the efficiency of SCN^- oxidation in a ZnSe capped ZnO system [8]. Recently, Eychmüller and his coworkers [9, 10] have demonstrated interesting fluorescence properties of CdS colloids capped with HgS. Such a capping process is not only convenient to build tandem semiconductor structures but also provides an economical way to deposit expensive semiconductor material on an inexpensive support. In view of these interesting features of these multicomponent semiconductor systems we have now investigated the charge transfer processes in metal oxide (SnO_2, TiO_2, and ZnO) nanostructures of two different geometries. The distinction between the capped and coupled semiconductor systems has been made by preparing colloids and nanocrystalline thin films of these materials.

2. Synthesis of Heterostructured Semiconductor Colloids and Films

Hydrolysis or chemical precipitation method is a convenient method for the synthesis of colloidal semiconductor suspensions. The method employed in the preparation of capped colloids involves preparation of the inner core semiconductor colloid of desired size and is followed by the controlled deposition of second semiconductor layer [11]. Titaniumisopropoxide (Aldrich) was added dropwise to a SnO_2 colloidal suspension(Alfa Chemicals). A precipitate settled down as the SnO_2 particles became coated with a layer of TiO_2. The total amount of titaniumiso-propoxide addition controls the thickness of the TiO_2 layer. Typically 0.1 mL of titaniumisopropoxide was added to the 1 mL solution of 18% colloidal SnO_2. This procedure yields approximately 25 Å thick TiO_2 layer. The particles were resuspended by the addition of small amounts of HCl and diluting with water. The final pH of the colloidal suspension was maintained between 1.6–2.0. Similar procedure was also employed for capping SiO_2 colloids with TiO_2. These capped colloids will be referred as **SnO_2@TiO_2** and **SiO_2@TiO_2** respectively in the following discussion. These colloids can be further coated with a layer of CdS by the addition of $Cd(ClO_4)_2$ solution and exposure to H_2S.

The colloids can spread on a conducting glass plate to prepare thin nanostructured films. The details on the preparation and characterization of TiO_2, SnO_2, ZnO and WO_3 colloids have been reported earlier [12–14]. It is also possible to carry out chemical precipitation of the semiconductor directly on the surface of a desired substrate. This technique is especially convenient to grow thin films of II–VI compound semiconductor nanocrystallites [4, 15]. For example, by successively dipping the ZnO film in Cd^{2+} and S^{2-} solutions one can cast a thin film of CdS nanocrystallites. The yellow coloration of the film indicates formation of CdS on the ZnO surface.

Coupled films of SnO_2/TiO_2 were prepared by mixing the two colloid suspensions at the desired ratio and applying it to the conducting surface of OTE. Coupled films of SnO_2/TiO_2 were prepared by mixing the two colloid suspensions at the desired ratio and applying it to the conducting surface of OTE. The composition of the two semiconductors was varied by varying the TiO_2 or SnO_2 content during mixing of two colloids. After air drying on a warm plate all these films were annealed at 673K. The uncovered area of the OTE plate was used to make the electrical contact. These electrodes with immobilized TiO_2 (referred to as OTE/TiO_2), SnO_2 (OTE/SnO_2), and coupled semiconductor (OTE/SnO_2/TiO_2) films were directly employed as a working electrode in the electrochemical cell.

3. Capped Semiconductors: SnO_2@TiO_2 System

When SnO_2@TiO_2 colloidal suspension was irradiated with UV–light, the solution turned blue. The absorption changes are shown in Figure 2. We attribute these absorption bands to the trapped holes and electrons respectively. No significant absorption changes were observed upon UV excitation of either TiO_2, SnO_2, or SiO_2@TiO_2 colloids. The UV absorption band at 335 nm is slightly blue shifted compared to the previously observed absorption band at 350 nm for the TiO_2 colloids [16, 17]. The natures of surface defect sites are likely to influence the characteristics of the absorption band. The formation of these absorption bands were also confirmed from the γ–radiolysis of SnO_2@TiO_2 suspension.

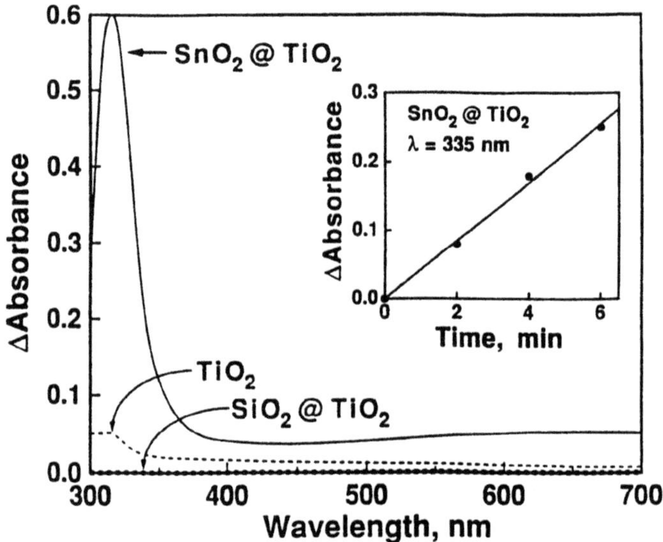

Figure 2. Difference absorption spectra recorded following the steady state UV photolysis of colloidal suspension in water: (a) TiO_2 (16 mM); (b) SiO_2@TiO_2 (29 mM SiO_2, 16 mM TiO_2) and (c) SnO_2@TiO_2 (29 mM SnO_2, 16 mM TiO_2). (Corresponding colloidal suspension before photolysis was used as a reference for recording of these spectra.) Duration of photolysis, 10 min. The insert shows the growth in the absorption at 335 nm arising from trapping of holes in SnO_2@TiO_2 colloids. (From Ref. 6)

The absorption peaks arising from trapped electrons and holes in $SnO_2@TiO_2$ colloids (Figure 2) indicate that the charge separation is quite efficient only in the capped semiconductor system. Since both SnO_2 and TiO_2 are large bandgap semiconductors with bandgap energy greater than 3.0 eV, they absorb strongly in the UV. Upon bandgap excitation, the initial charge separation occurs at both semiconductors. The photogenerated electrons and holes quickly diffuse in opposite directions to accumulate at SnO_2 and TiO_2 particles respectively (Reaction 1).

$$TiO_2 \longrightarrow TiO_2\ (e.....h) \qquad SnO_2 \longrightarrow SnO_2\ (e.....h)$$

$$\downarrow$$

$$SnO_2\ (e_t) + TiO_2\ (h_t) \tag{1}$$

The energetics of the conduction ($E_{CB(SnO_2)} = 0.0$ V and $E_{CB(TiO_2)} = -0.5$ V vs. NHE at pH 7) and valence bands ($E_{VB(SnO_2)} = 3.5$ V, $E_{VB(TiO_2)} = 2.7$ V) of these particles favor charge rectification within the capped semiconductor system.

Figure 3 shows transient absorption spectra recorded at 100 ps and 4 ns after 355 nm laser pulse excitation of an aqueous solution containing $SnO_2@TiO_2$ colloids. Because of the weak monitoring light intensity we couldn't probe the absorption changes at wavelengths below 400 nm. The tail absorption of the trapped holes in the 400 nm region could still be seen in the spectra recorded in Figure 3. The insert shows the absorption–time profile at 410 nm. The growth in the absorption which is completed within 100 ps shows the timescale of chemical changes associated with hole trapping in $SnO_2@TiO_2$ colloids. A fraction of the trapped holes undergo recombination as evidenced from the absorbance decay at 410 nm. The changes in the spectra recorded at longer times show little decay suggesting the stabilization of a small fraction of photogenerated electrons at SnO_2 and holes at TiO_2 layers.

Laser flash photolysis is also a convenient method to probe the interfacial charge transfer processes in colloidal semiconductor systems [18, 19]. The effectiveness of $SnO_2@TiO_2$ colloid as an oxidative photocatalyst was demonstrated by carrying out laser flash photolysis experiments. Figure 4 shows the transient absorption spectrum recorded immediately after 308 nm laser flash excitation of $SnO_2@TiO_2$ colloidal suspension containing I^-. As discussed in the previous section, the photogenerated electrons accumulate at the inner core of SnO_2 and holes at outer TiO_2 shell. Surface adsorbed I^- quickly scavenges away the holes at the interface and forms $I_2^{-\bullet}$ in the presence of excess I^- (reaction 2).

$$TiO_2(h_t) + I^- \rightarrow TiO_2 + I^\bullet \xrightarrow{I^-} I_2^{-\bullet} \tag{2}$$

The absorption bands observed at 380 and 700 nm confirm the formation of $I_2^{-\cdot}$.

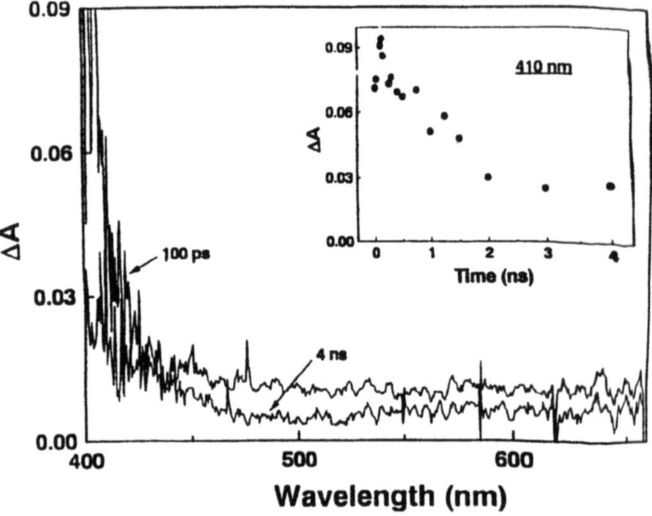

Figure 3. Transient absorption spectra recorded following 355 nm laser pulse excitation of $SnO_2@TiO_2$ colloidal suspension at delay times of 100 ps and 4 ns. The absorption-time profile at 410 nm is shown in the inset.

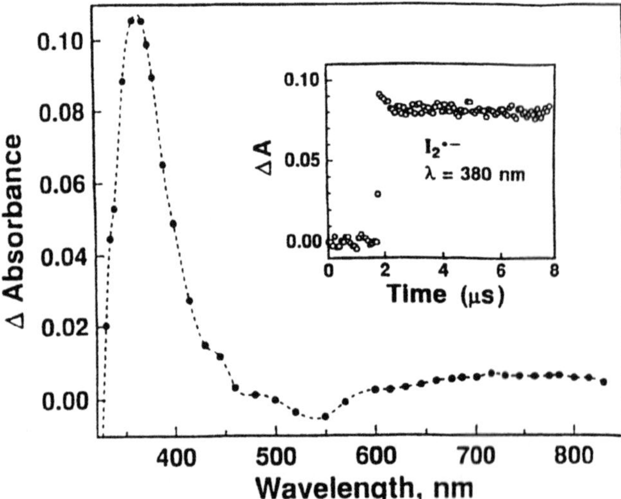

Figure 4. Oxidation of I^-: Transient absorption spectrum recorded immediately after 308 nm laser pulse (pulse width 10ns) excitation of $SnO_2@TiO_2$ (29 mM SnO_2, 8 mM TiO_2) colloidal suspension in water containing 0.05 M KI (deaerated with N_2). The insert shows the absorption-time profile of $I_2^{-\cdot}$ at 380 nm.

In order to ascertain the beneficial role of charge rectification in capped semiconductor systems, the oxidation of I^- was probed at varying concentration of TiO_2 (Figure 5). The concentrations of TiO_2 indicate the amount employed during its precipitation on SnO_2 surface. The concentration of SnO_2 or SiO_2 is kept constant throughout. With increasing TiO_2 concentration an increase in the I^- oxidation is seen. The relative quantum efficiency for I^- oxidation at SnO_2@TiO_2 colloid was two–to–three fold higher than SiO_2@TiO_2 or TiO_2 colloid alone. These results further demonstrate the usefulness of SnO_2@TiO_2 colloids in carrying out photocatalytic oxidation reactions.

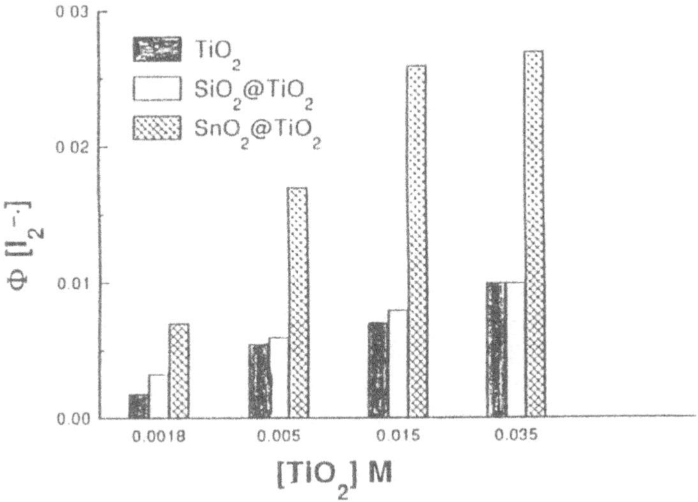

Figure 5. Dependence of $I_2^{-\cdot}$ yield on the concentration of TiO_2: The maximum absorbance at 380 nm was monitored following the 308 nm excitation of (a) SnO_2@TiO_2 (b) SiO_2@TiO_2 and (c) TiO_2 colloidal suspension containing 0.05 M KI (deaerated with N_2). The concentrations of SnO_2 (29 mM) and SiO_2 (15 mM) were kept constant in the corresponding set of experiment. (From Ref. 6)

4. Coupled Semiconductors

In our earlier studies we have shown that excitation of CdS colloids coupled with a metal oxide colloid such as TiO_2 [20] or ZnO [4] results in the injection of electrons into the lower lying conduction band of the metal oxide. The charge injection process in these systems was shown to occur within the pulse duration of 20 ps. Recent studies have indicated that such an inter–particle electron transfer occurs within 0.5–2 ps [21, 22]. Use of coupled semiconductor particles for improving the photocurrent generation in thin TiO_2/CdS, [3, 5, 15, 23] ZnO/CdS, [4] and TiO_2/CdSe [1, 24] nanocrystalline film has also been demonstrated. Efforts have been made by Serpone and his coworkers [25, 26] to employ coupled semiconductor configurations in slurry systems for photocatalytic degradation. Such coupled semiconductor thin films not only extend the photoresponse of large bandgap semiconductors but also rectify the flow of photogenerated charge carriers [1] and improve the efficiency of dye sensitization [7].

Recently, we have demonstrated that a significant enhancement in the photocatalytic degradation rate of Acid Orange 7 can be achieved by employing SnO_2 and TiO_2 nanoclusters in the coupled configuration [27]. By coupling these two large bandgap metal oxide semiconductors we were able to achieve efficient charge separation. Since the conduction band (CB) of SnO_2 (E_{CB} for $SnO_2 = 0$ V vs NHE at pH 7) is lower than that of the TiO_2 ($E_{CB} = -0.5$V vs. NHE at pH 7) the former acts as a sink for the photogenerated electrons. The holes move in the opposite direction from the electrons, and accumulate on the TiO_2 particle thereby making the charge separation more efficient. Our particular success in improving the efficiency of photocatalytic degradation can be attributed to the fact that we deposit these nanostructured films of TiO_2/SnO_2 on an optically transparent electrode (OTE) and transport the electrons efficiently to the external circuit with an externally applied electrochemical bias. The principle of charge separation in such a coupled semiconductor system deposited on an OTE is illustrated in Figure 1. Unlike in slurry systems, the photocatalytic degradation experiments can be carried out in the absence of electron scavengers such as O_2.

By employing a two compartment electrochemical cell we have now investigated the photocatalytic degradation of textile azo dyes, Acid Orange 7 (AO7) and Naphthol Blue Black (NBB). The principle and methodology of electrochemically assisted photocatalysis is reported in our earlier publications [28, 29]. The major photoelectrochemical reactions that initiate redox processes in the electrode compartments can be summarized as follows:

At photoanode: $SnO_2/TiO_2 + h\nu \rightarrow \{SnO_2(\mathbf{e}....\mathbf{h})\}\{TiO_2(\mathbf{e}.....\mathbf{h})\} \rightarrow SnO_2(\mathbf{e}) + TiO_2(\mathbf{h})$ (3)

$$TiO_2(\mathbf{h}) + OH^- \rightarrow TiO_2 + {}^\bullet OH \quad (4)$$

$$TiO_2(\mathbf{h}) \text{ or } {}^\bullet OH + \text{Azo Dye} \rightarrow \text{products} \quad (5)$$

At dark cathode: $Pt(\mathbf{e}) + O_2 \rightarrow O_2^-$ (6a)

$$O_2^- + e + 2H^+ \rightarrow H_2O_2 \quad (6b)$$

The absorption spectra of an aqueous solution of NBB recorded following the excitation of an $OTE/SnO_2/TiO_2$ are shown in Figure 6. The $OTE/SnO_2/TiO_2$ is maintained at a positive bias of 0.8 V vs. SCE to facilitate efficient charge separation in the semiconductor film. The oxidation potential of NBB is determined to be around 1.0 V at a glassy carbon electrode vs. SCE. Therefore the choice of 0.8 V as an applied bias potential in these electrochemically assisted photocatalytic experiments ensured that no direct oxidation of NBB occurred during these experiments. The solution was continuously bubbled with a stream of nitrogen so that electrochemically assisted photocatalysis could be carried out under anaerobic conditions.

As can be seen from Figure 6, the absorption peaks corresponding to the blue dye disappear completely following photolysis of the semiconductor film. Even more surprising is the relative speed with which the dye degradation can be achieved. An aqueous solution of 37 ppm. dye is 67% degraded following 30 minutes of irradiation of the semiconductor electrode indicating a pseudo first-order rate constant, k, of 3.6 x 10^{-2} min^{-1}. While the observed pseudo first-order degradation rate constant of the OTE/SnO_2 electrode was similar to that of OTE/TiO_2 electrode (k= 1.1 x 10^{-2} $min.^{-1}$), it is about three times slower than the OTE/SnO_2/TiO_2 system. Nearly ten times higher degradation rate was also reported by us for the degradation of another textile dye, Acid orange 7 [27]. These results highlight the role of coupled semiconductor films in enhancing the photocatalytic degradation rates of azo dyes.

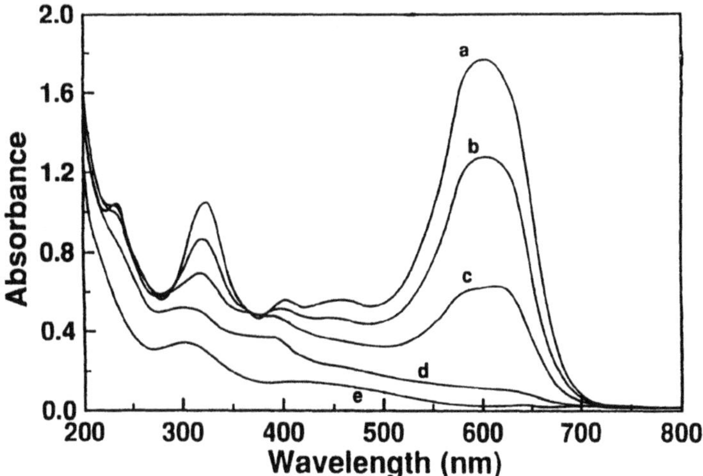

Figure 6. Absorption spectrum of a 37 ppm. (0.2mM) Naphthol Blue Black azo dye solution recorded at different time intervals following electrochemically assisted photocatalysis at an OTE/(SnO_2/TiO_2) electrode (0.30mg/cm^2 of SnO_2 and 0.06mg/cm^2 of TiO_2). The OTE/(SnO_2/TiO_2) electrode was maintained at an anodic bias of 0.8 V vs SCE and the solution was continuously bubbled with a slow stream of nitrogen. The absorption spectra were recorded at time intervals of (a) 0, (b) 10, (c) 30 and (d) 60 and (e) 90 min..

We also attempted to determine the optimum ratio of the two semiconductors necessary to obtain fastest degradation of the azo dye. An example of the electrochemically assisted photocatalytic degradation rates of Acridine Orange 7 (AO7) as a function of the total mass of the semiconductor when both TiO_2 and SnO_2 are present in equal amounts is shown as a bar diagram in Figure 7. For comparison, the rates obtained when either SnO_2 or TiO_2 is used separately (but at the same mass) is also included in the bar diagram. It is clear from Figure 7, that degradation is rapid with the coupled systems, but the fastest rates are obtained when the mass ratio of SnO_2 to TiO_2 is at least 2:1 or higher.

It should be noted that SnO_2 particles are smaller (particle diameter 3–5 nm) than TiO_2 (particle diameter 30 nm). Thus, at the optimum composition of the two semiconductor particles we would expect each TiO_2 particle to be surrounded by hundreds of SnO_2 particles. Such a configuration is likely to favor electron transfer from excited TiO_2 into SnO_2 particles and their transport towards the collecting surface of OTE. The holes that remain on the TiO_2 particles are efficiently utilized in the oxidation of the dye molecules.

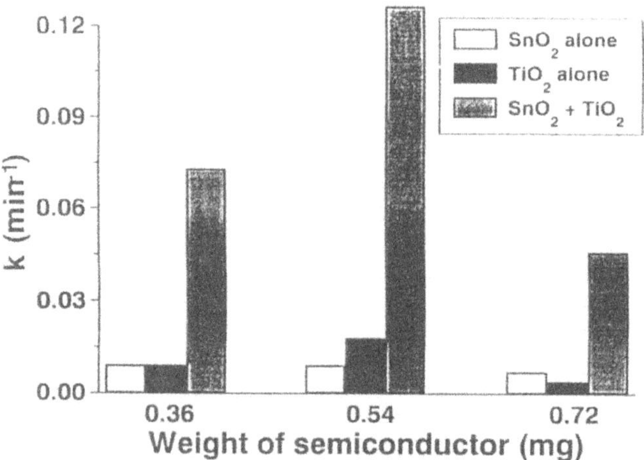

Figure 7. Comparison of photocatalytic degradation rates of AO7 with SnO_2, TiO_2 and SnO_2/TiO_2 particulate films coated on OTE electrodes. In each set of experiment, the total weight of the semiconductor catalyst was kept constant at the indicated value. The bias potential was 0.83 V vs. SCE and the electrolyte was 42 ppm AO7 in water at unbuffered pH ~6.

In order to make the coupled SnO_2/TiO_2 system useful for photocatalysis it is essential to carry out these experiments with an externally applied electrochemical bias. It provides necessary potential gradient within the nanostructured semiconductor film to drive away the accumulated electrons via the external circuit and thus promote oxidation at the semiconductor/electrolyte interface. The results presented here show how a simple approach of mixing two semiconductor particulate systems and packing them as an immobilized particulate film on a conducting glass surface leads to the increased rate of photocatalytic degradation. We are currently carrying out photoelectrochemical and transient absorption spectroscopy measurements to correlate the enhancement in the photocatalytic degradation rate with the degree of charge separation in coupled semiconductor films.

5. Photoelectrochemical Behavior of Capped and Coupled Films

Thin films of SnO_2 and TiO_2 were synthesized with capped and coupled geometries. The photocurrent generation at these two electrodes was recorded at various excitation wavelengths by constructing a photoelectrochemical cell with Pt counter electrode and 0.01 M NaOH electrolyte. The mechanism of photocurrent generation at the nanocrystalline semiconductor film is described elsewhere [28, 30, 31]. The photocurrent response as measured from the incident photon–to–photocurrent conversion efficiencies (IPCE) are compared in Figure 8. Only the OTE/SnO_2/TiO_2 electrode is effective in generating anodic photocurrent. Negligibly small photocurrents were observed with the semiconductor film prepared from capped semiconductor colloids.

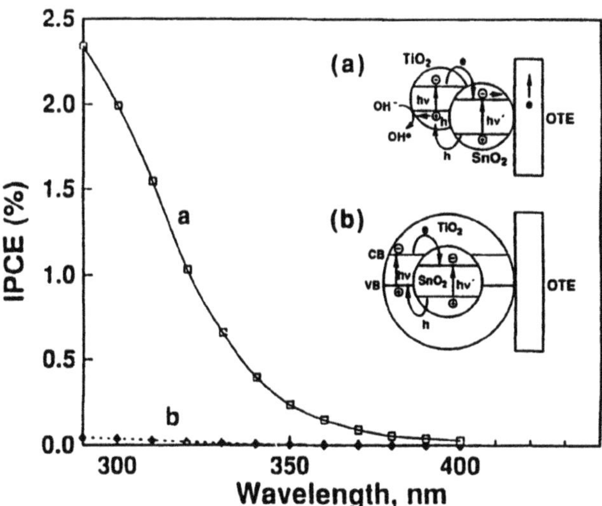

Figure 8. Photoelectrochemical response of nanocrystalline semiconductor films consisting of SnO2 and TiO_2 in a coupled and capped geometry: (a) OTE/SnO_2/TiO_2 and (b) OTE/SnO_2@TiO_2. (Electrolyte: 0.02 M NaOH. IPCE refers to incident photon-to-photocurrent conversion efficiency. See reference [13] for the definition of IPCE. From Ref. 6).

The inset in Figure 8 illustrates the mechanism of charge separation in these two types of electrode geometry. Although the initial charge separation in the capped and coupled systems is achieved by the accumulation of electrons in the SnO_2 layer and holes in the TiO_2, layer only the coupled semiconductor geometry facilitates electron transport towards the collecting surface of OTE. In the case of capped semiconductor films (OTE/SnO_2@TiO_2) the electrons trapped within SnO_2 nanocrystallites cannot cross the barrier of TiO_2 and reach the collecting surface of OTE. The capped semiconductor colloids obviously are not preferable for developing photoelectrochemically active electrode materials. Coupled semiconductor geometry on the other hand is quite efficient for charge transport and thus should be the choice for developing photosensitive electrodes. The results described here show how one can control the charge separation in a multicomponent semiconductor system by varying the semiconductor layout.

6. Acknowledgments

We would like to thank Surat Hotchandani, Di Liu and Chauhaid Nasr for their contribution in investigating the behavior of nanostructured semiconductor materials. The work described herein was supported by the Office of the Basic Energy Sciences of the US Department of Energy. This is Contribution No. 3827 from the Notre Dame Radiation Laboratory.

7. References

1. Liu, D. and Kamat, P.V. (1993) Photoelectrochemical behavior of thin CdSe and coupled TiO_2/CdSe semiconductor films, *J. Phys. Chem.* **97**, 10769–10773.
2. Uekawa, N., Suzuki, T., Ozeki, S. and Kaneko, K. (1992) Rectification effect by a p–n junctioned oxide film, *Langmuir* **8**, 1–3.
3. Gerischer, H. and Luebke, M. (1986) A particle size effect in the sensitization of TiO_2 electrodes by a CdS deposit, *J. Electroanal. Chem. Interfacial Electrochem.* **204**, 225–227.
4. Hotchandani, S. and Kamat, P.V. (1992) Charge–transfer processes in coupled semiconductor systems. Photochemistry and photoelectrochemistry of the colloidal CdS–ZnO system, *J. Phys. Chem.* **96**, 6834–6839.
5. Vogel, R., Hoyer, P. and Weller, H. (1994) Quantum–sized PbS, CdS, Ag_2S, Sb_2S_3 and Bi_2S_3 particles as sensitizers for various nanoporous wide–bandgap semiconductors, *J. Phys. Chem.* **98**, 3183–3188.
6. Bedja, I. and Kamat, P.V. (1995) Capped semiconductor colloids. Synthesis and Photoelectrochemical properties of TiO_2 capped SnO_2 surfaces, *J. Phys. Chem.* **99**, 9182–9188.

7. Hotchandani, S. and Kamat, P.V. (1992) Modification of electrode surface with semiconductor colloids and its sensitization with chlorophyll *a*, *Chem. Phys. Lett.* **191**, 320–326.

8. Kamat, P.V. and Patrick, B. (1992) Photophysics and photochemistry of quantized ZnO colloids, *J. Phys. Chem.* **96**, 6829–6834.

9. Haesselbarth, A., Eychmueller, A., Eichberger, R., Giersig, M., Mews, A. and Weller, H. (1993) Chemistry and photophysics of mixed CdS/HgS colloids, *J. Phys. Chem.* **97**, 5333–5340.

10. Eychmueller, A., Vobmeyer, T., Mews, A. and Weller, H. (1994) Transient photobleaching in the quantum dot quantum well CdS/HgS/CdS, *J. Lumin.* **58**, 223–226.

11. Ocana, M., Hsu, W.P. and Matijevic, E. (1991) Preparation and properties of uniform–coated colloidal particles. 6. Titatania on zinc oxide, *Langmuir* **7**, 2911–2916.

12. Hotchandani, S. and Kamat, P.V. (1992) Photoelectrochemistry of semiconductor ZnO particulate films, *J. Electrochem. Soc.* **139**, 1630–1634.

13. Bedja, I., Hotchandani, S. and Kamat, P.V. (1994) Preparation and characterization of thin SnO2 nanocrystalline semiconductor films and their sensitization with bis(2,2'–bipyridine)(2,2'–bipyridine–4–4'–dicarboxylic acid)ruthenium complex, *J. Phys. Chem.* **98**, 4133–4140.

14. Bedja, I., Hotchandani, S., Carpentier, R., Vinodgopal, K. and Kamat, P.V. (1994) Electrochromic and photoelectrochemical behavior of thin WO_3 films prepared from quantized colloidal particles, *Thin Solid Films* **247**, 195–200.

15. Vogel, R., Pohl, K. and Weller, H. (1990) Sensitization of highly porous, polycrystalline TiO_2 electrodes by quantum sized CdS, *Chem. Phys. Lett.* **174**, 241–246.

16. Lawless, D., Serpone, N. and Meisel, D. (1991) Role of $^{\bullet}$OH radicals and trapped holes in photocatalysis. A pulse radiolysis study, *J. Phys. Chem.* **95**, 5166–5170.

17. Kamat, P.V., Bedja, I. and Hotchandani, S. (1994) Photoinduced charge transfer between carbon and semiconductor clusters. One–electron reduction of C_{60} in colloidal TiO_2 Semiconductor suspensions, *J. Phys. Chem.* **98**, 9137–9142.

18. Kamat, P.V. (1994) Interfacial charge transfer processes in colloidal semiconductor systems, *Progr. React. Kinetics* **19**, 277–316.

19. Kamat, P.V. and Vinodgopal, K. The role of support material in the photodegradation of colored organic compounds. in Helz G.R., Zepp R.G., Crosby D.G., ed. *Aquatic and Surface Photochemistry*, Boca Raton, Fl: CRC Press, Inc., 1994: 443–450.

20. Gopidas, K.R., Bohorquez, M. and Kamat, P.V. (1990) Photoelectrochemistry in semiconductor particulate systems. 16. Photophysical and photochemical aspects of coupled semiconductors. Charge–transfer processes in colloidal CdS–TiO_2 and CdS–AgI systems, *J. Phys. Chem.* **94**, 6435–6440.

21. Kietzmann, R., Willig, F., Weller, H., Vogel, R., Nath, D.N., Eichberger, R., Liska, P. and Lehnert, J. (1991) Picosecond time resolved electron injection from excited cresyl violet monomers and Cd_3P_2 quantum dots into TiO_2, *Mol. Cryst. Liq. Cryst.* **194**, 169–180.

22. Evans, J.E., Springer, K.W. and Zhang, J.Z. (1994) Femtosecond studies of inter–particle electron transfer in a coupled CdS–TiO_2 colloidal system, *J. Chem. Phys.* **101**, 6222-6225.

23. Kohtani, S., Kudo, A. and Sakata, T. (1993) Spectral sensitization of a TiO_2 semiconductor electrode by CdS microcrystals and its photoelectrochemical properties, *Chem. Phys. Lett.* **206**, 166–170.
24. Liu D. and Kamat, P.V. (1993) Electrochemical rectification in CdSe + TiO_2 coupled semiconductor films, *J. Electroanal. Chem. Interfacial Electrochem* **347**, 451–456.
25. Serpone, N., Borgarello, E. and Graetzel, M. (1984) Visible light induced generation of hydrogen from H_2S in mixed semiconductor dispersions: Improved efficiency through inter–particle transfer, *J. Chem. Soc., Chem. Commun.* 342–343.
26. Hidaka, H., Asai, Y., Zhao, J., Nohara, K., Pelizzetti, E. and Serpone, N. (1995) Photoelectrochemical decomposition of surfactants on a TiO_2/TCO particulate film electrode assembly, *J. Phys. Chem.* **99**, 8244–8248.
27. Vinodgopal, K. and Kamat, P.V. (1995) Enhanced rates of photocatalytic degradation of an azo dye using SnO_2/TiO_2 coupled semiconductor thin films, *Environ. Sci. Technol.* **29**, 841–845.
28. Vinodgopal, K., Hotchandani, S. and Kamat, P.V. (1993) Electrochemically assisted photocatalysis. TiO_2 particulate film electrodes for photocatalytic degradation of 4–chlorophenol, *J. Phys. Chem.* **97**, 9040–9044.
29. Vinodgopal, K., Stafford, U., Gray, K.A., and Kamat, P.V. (1994) Electrochemically assisted photocatalysis. II. The role of oxygen and reaction intermediates in the degradation of 4–chlorophenol on immobilized TiO_2 particles, *J. Phys. Chem.* **98**, 6797–6803.
30. Hodes, G., Howell, I.D.J. and Peter, L.M. (1992) Nanocrystalline photoelectrochemical cells. A new concept in photovoltaic cells, *J. Electrochem. Soc.* **139**, 3136–3140.
31. Hagfeldt, A., Bjorksten, U. and Lindquist, S.E. (1992) Photoelectrochemical studies of colloidal TiO_2–films: The charge separation process studied by means of action spectra in the UV region, *Sol. Energy Mater. Sol. Cells* **27**, 293–304.

SYNTHESIS AND CHARACTERIZATION OF GaP, InP, AND $GaInP_2$ QUANTUM DOTS

O. I. MIĆIĆ, J. R. SPRAGUE, C. J. CURTIS, K. M. JONES, and A. J. NOZIK
National Renewable Energy Laboratory
1617 Cole Blvd., Golden, CO 80401

Introduction.

Although excellent progress has been made in the preparation and characterization of quantum dots (QDs) made from II-VI compounds, Si, oxides (Ti, In, Cu, W), and iodides (Hg, Pb, Bi); the preparation of high-quality QDs of III-V semiconductors has proven to be problematic.[1-8] Here, we report the synthesis and properties of excellent quality binary III-V InP and GaP colloidal QDs, and the ternary III-V QD, $GaInP_2$.[9,10]

It is interesting to compare the properties of InP and GaP QDs because the former is a direct gap semiconductor, while the latter is an indirect gap semiconductor; the effects of size quantization on the optical properties of the indirect semiconductor GaP are explored here. $GaInP_2$ is of interest because its bandgap depends on the degree of atomic ordering in the material, and the effect of quantization on this behavior has not been investigated.

The preparation of well-crystallized QDs of III-V compounds in solution requires high temperature, and the synthesis at high temperatures creates several problems. First, at these temperatures the relevant solution chemistry is not well-developed. Second, quantum dots also tend to aggregate at high temperature, and the resulting precipitate cannot be redissolved to form transparent, colloidal solutions. Also, the particles tend to grow sufficiently large to lose the quantum confinement.

Our results show that well-crystallized III-V semiconductor QDs can be produced with a relative narrow particle size spread (about 10%); an exciton peak can be resolved in their absorption spectra. We also report results measuring the relaxation dynamics of photogenerated carriers in InP QDs.

Experimental.

We used indium or gallium salts to react with trimethylsilylphosphine to yield InP, GaP and $GaInP_2$ quantum dots.[9,10] Upon mixing at room temperature, a precursor is formed first , which then thermally decomposes at 270-400°C in the presence of thecolloidal stabilizer, trioctylphosphine oxide (TOPO), to yield quantum dots. The growth of GaP nanocrystallites requires higher temperature (>350°C) than that for InP (250°C).[10]

OPTICAL CHARACTERIZATION AND TIME-RESOLVED SPECTROSCOPY

Optical absorption spectra were collected at room temperature using a Cary 5E UV-VIS-NIR spectrophotometer with 0.2-cm quartz cuvettes. Samples were prepared by dispersing washed QDs in toluene. Photoluminescence spectra were obtained at room temperature using a SPEX

E. Pelizzetti (ed.), Fine Particles Science and Technology, 317–330.

Fluorolog-2 spectrometer. Emission lifetimes were determined by time-correlated single photon counting. A cavity-dumped synchronously-pumped dye laser (Spectra-Physics 3500) operating at 585 nm provided pump pulses of 10 picoseconds. A Hamamatsu microchannel plate detector provided a typical instrument response function of 70 picoseconds. The lifetimes were estimated by fitting the decay using a sum of exponentials as an indicator of the decay time.

Femtosecond laser pump-probe (i.e., hole burning) experiments were performed at the Optical Sciences Center at the University of Arizona. The laser system is composed of a colliding-pulse mode-locked laser and a six-pass dye amplifier pumped by a tightly folded resonator (TFR) Nd:YLF laser operating at 1000 Hz and 527 nm. The output of this amplifier pumps an ethylene glycol jet to produce a chirped 1 ps continuum pulse from 450-700 nm. The continuum pulse is split with one part becoming the probe pulse while the other part is passed through an interference filter producing a spectrally narrow (~ 10 nm bandwidth) pulse that is amplified in a second six-pass amplifier pumped by another TFR Nd:YLF laser operating at 1000 Hz and 527 nm. The pump pulse is 115 fs and nominally 420 nanojoules focused to a 45μ diameter spot in the sample, producing a power density of 230 GW/cm^2; its center wavelength can be adjusted by tuning the interference filter. The pump pulse can be optically delayed over a 500 picosecond range and is typically 100 times more intense than the probe pulse.

Zero time delay refers to the peak of the probe pulse overlapping with the peak of the pump pulse. Time delay values in the figure captions are relative time delays between pulses for all the data in that figure. A time delay of -5ps means that the probe pulse passed through the sample 5ps ahead of the pump pulse. Since the repetition rate of the laser is 1000 Hz, the effective time delay in this case is 1 millisecond.

Results and Discussion

GaP QUANTUM DOTS

The reaction of $GaCl_3$ with $P(SiMe3_3)_3$ yields the well-defined pale yellow precursor, $[Cl_2GaP(SiMe_3)_2]_2$,[11] which dissolves in TOP/TOPO/toluene to form a transparent yellow solution. Chlorogallium oxalate also forms a precursor upon reaction with $P(SiMe3_3)_3$; both precursors begin to absorb light at about 450 nm.

Figure 1 shows the absorption spectra of the precursor $[Cl_2GaP(SiMe_3)_2]_2$ and GaP QDs produced from this precursor by heating at 270, 310, 370, and 400°C.

The work of Bawendi et al. on CdSe quantum dots demonstrated the remarkable ability of trioctylphosphine oxide to stabilize semiconductor QDs at high temperature.[12] We used this stabilizer for the preparation of our QDs.

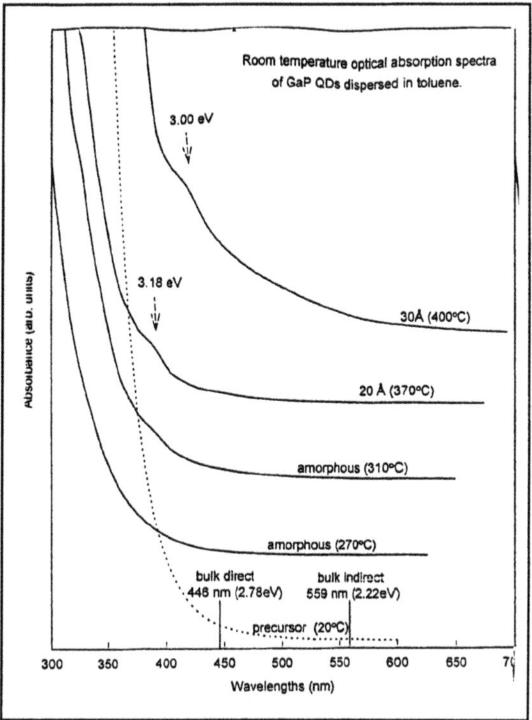

Figure 1. Optical absorption spectra at room temperature of GaP QDs heated at different temperatures and dissolved in toluene; also shown (dotted line) is the absorption spectrum of the GaP precursor $\{[Cl_2GaP(SiMe_3)_2]_2\}$. The direct and indirect band gaps of bulk GaP are indicated by the vertical line markers on the abscissa. (From Ref. 10)

Figure 2 shows the X-ray diffraction spectra for GaP QD samples heated and dried at different temperatures. The diffraction lines for bulk GaP with the zinc blend structure are presented in the figure. The samples heated at 400°C and 360°C show bulk crystal structures. The spectra of the QD particles show evidence of finite broading for all peaks and the line width sharpens as the particle size increases. The mean particle diameters are calculated from the Debye-Scherer formula. Diffraction patterns of the sample heated at < 320°C have two very broad diffraction lines and we attributed them to amorphous material.

The absorption spectrum of the 30 Å diameter GaP QD colloid (heated at 400°C) shown in Figure 1 exhibits a shoulder at 420 nm and a shallow tail that extends out to about 650 nm. For 20 Å diameter GaP QDs (heated at 370°C) the shoulder is at 390nm and the tail extends to about 550 nm.

Bulk GaP is an indirect semiconductor with an indirect bandgap of 2.22 eV (559 nm) and a direct bandgap of 2.78 eV (446 nm). Theoretical calculations by Rama Krishna and Friesner[13]

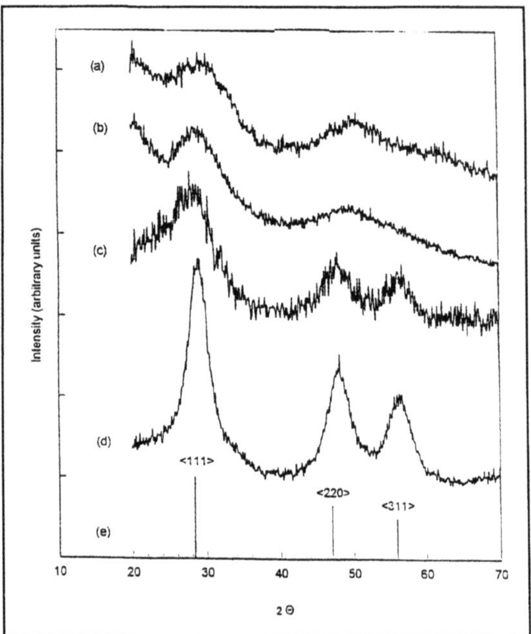

Figure 2. X-ray diffraction pattern for dried GaP QD colloids heated at (a) 270°C, (b) 320°C, (c) 370°C, and (d) 400°C, and compared with the peak positions of bulk zincblende GaP (e).
(From Ref. 10)

on GaP QDs show that the increase of the indirect bandgap with decreasing QD size is much less pronounced than that for the direct gap; for 30 Å diameter GaP QDs the direct and indirect bandgaps are predicted to be 3.35 eV and 2.4 eV, respectively. Below 30 Å the direct bandgap is predicted to <u>decrease</u> with decreasing size while the indirect bandgap continues to increase. As a result, GaP is expected to undergo a transition from an indirect semiconductor to a direct semiconductor below about 20 Å.

In Figure 1, we attribute the steep absorption and shoulder at 420 nm to a direct transition in the GaP QDs; the shallow tail region above 500 nm is attributed to the indirect transition. In Figure 1 the position of our assigned direct transition at about 3.0 eV is less than that predicted by theory (3.35 eV) for 30 Å GaP QDs[13]; the reason for this discrepancy is not currently understood. Also, it should be noted that the absorption tail extends below the indirect bandgap of bulk GaP. The origin of this subgap absorption is also not understood at the present time; it could be caused either by a high density of subgap states in the GaP QDs, by impurities created by the high decomposition temperature, or by Urbach type band-tailing produced by unintentional doping in the QDs[14]. We note that such subgap absorption below the bandgap was also observed in GaP nanocrystals that were prepared in zeolite cavities by the gas phase reaction of trimethylgallium and phosphine at temperatures above 225 C[6]. This latter result implies that the subgap absorption in GaP QDs is related to intrinsic behavior, and is not caused by synthetic byproducts.

InP QDs

Different particle sizes of InP QDs can be obtained by changing the concentration of the orange precursor in the TOPO/TOP solution or by changing the temperature at which the solution is heated. The duration of heating also improves the QD crystallinity. The precursor has a high decomposition temperature (>270° C); this is advantageous for the formation of InP quantum dots because the rate of QD formation is controlled by the rate of decomposition of the precursor. This slow process leads to InP QDs with a narrow size distribution.

Figure 3 shows the absorption spectra of InP QD colloids with different particle diameters; bulk InP is black with a band gap of 1.35 eV (918 nm). The size distrubution of the QDs is about 10%.

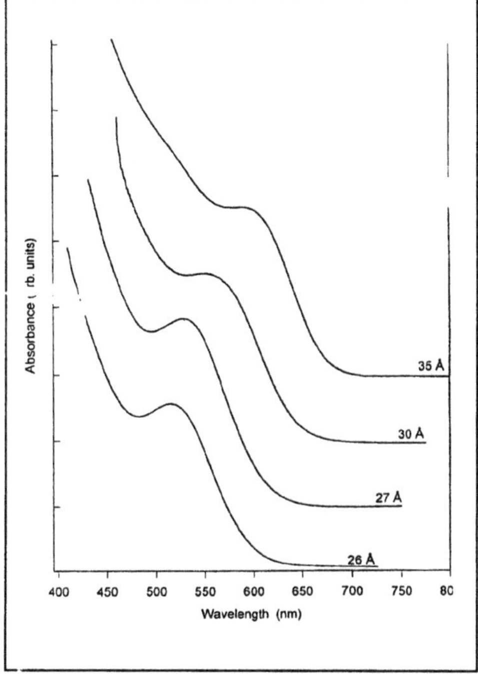

Figure 3. Absorption spectra for InP QD colloids with different diameters.

Figure 4 shows the x-ray diffraction patterns of a 25, 27 and 35 Å InP QD-particles formed into a film by drying the colloids. The peaks assigned to diffraction from the <111>, <220>, and <311> planes of crystalline zincblende bulk InP are seen at 2θ of 26.2 ± 0.2°, 46.3 ± 0.2° and 51.7 ± 0.2°, respectively. These diameters are in agreement with the values obtained from

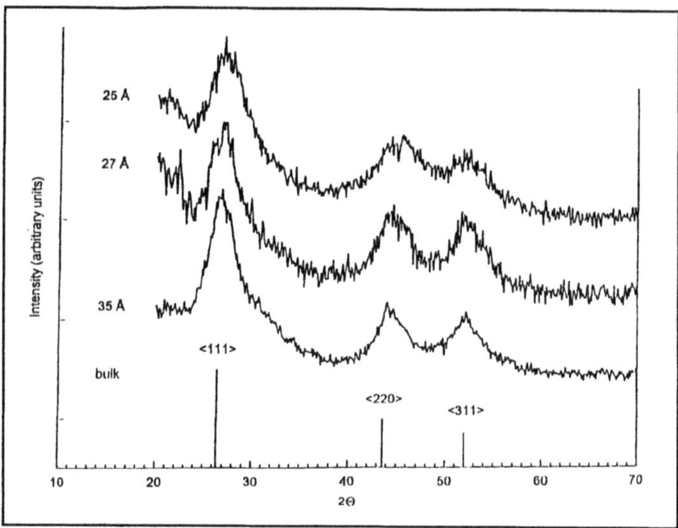

Figure 4. X-ray diffraction pattern for dried InP QDs colloids with different diameters compared with the peak positions of bulk zincblende InP.

transmission electron microscopy and from SAXS data. In the absence of the TOPO stabilizer, the particles grow large and the sharp peaks of bulk InP are obtained.

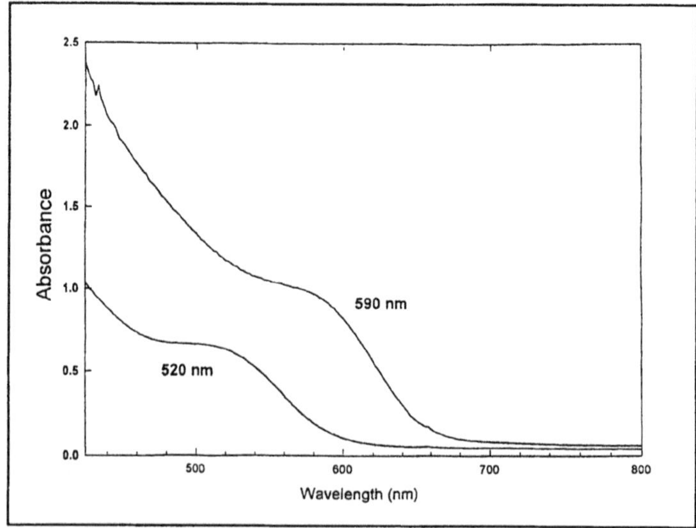

Figure 5. Spectra for two InP QD colloids used in pump-probe experiments, designated as 520 nm and 590 nm. (From Ref. 10)

The shape and size distribution of the InP QDs were determined by TEM. TEM pictures of InP preparations with TOPO that were only heated to 220° C for three days do not show the formation of either amorphous or crystalline InP. Upon heating to 240° C for three days the formation of zincblende nanocrystallites is evident, but the product is primarily amorphous. However, when the preparation is heated at 270° C for 3 days electron diffraction patterns show the <111>, <220>, and <311> planes of zincblende InP.
The photoluminescence for InP QD preparations showed two emission bands; one band in the visible at the band edge of the QD (50 nm Stokes shift), and a second band above 800 nm. The near-IR PL is attributed to deep traps, presumably phosphorus vacancies on the QD surface.

The absorption spectra of the two InP QD samples used in the pump-probe experiments are shown in Figure 5; they are labeled by the approximate wavelength of their excitonic peaks. Three types of time-resolved pump-probe experiments were recorded for both samples. First, short time delays with 50 fs steps were used to determine the dynamics of the onset of the transient bleach signal. Second, longer time delays from 2 to 200 ps were used to observe the decay of the transient bleach signal; and third, the pumping wavelength was changed to see if any shift in the transient spectrum occurred with pumping wavelength.

Figure 6 shows the fs transient absorption (bleaching) behavior for the 590 nm QD sample in 100 fs steps after being pumped at 560 nm. Clearly, the onset of the bleach (burned hole) is very fast reaching its maximum in 250-300 fs.

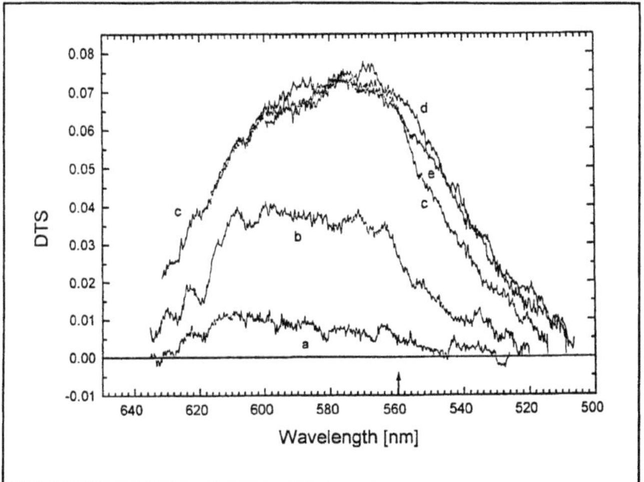

Figure 6. Pump-probe experiments for InP QDs showing the differential absorbance (-ΔA) spectra at room temperature taken every 100 fs after time zero for the 600 nm InP QD colloid pumped at 560 nm. Positive values on the figure correspond to bleaching, which becomes saturated after about 250-300 fs. Time zero is defined in the text in the Experimental Section; the curve labeled (a) is at time zero. (From Ref. 10)

Figure 7 shows the transient absorption signal for the 520 nm sample (after being pumped at 553 nm) using longer time delays from 2 to 200 ps. The signal appears to decay about 30% over this time range. With a time delay of 1 ms we could still observe that the transient signal had not completely decayed.

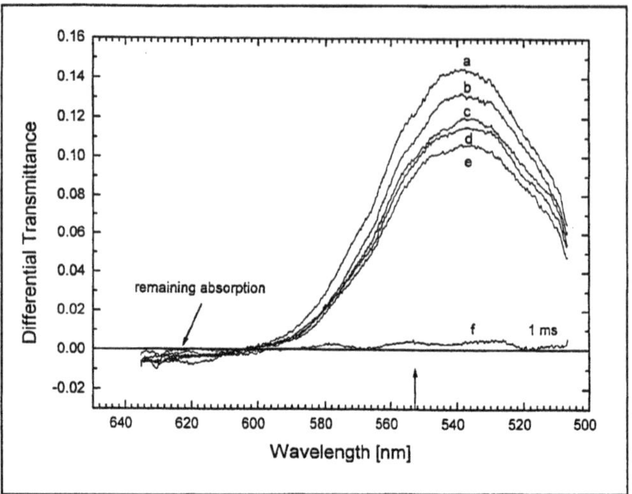

Figure 7. Pump-probe experiments for the 520 nm InP QD sample (pumped at 553 nm) with longer time delays (from 2 to 200 ps). Curve (a) is at 2 ps, (b) is at 10 ps; (c) is at 50 ps; (d) is at 100 ps, and (e) is at 200 ps. Curve (f) is at -5 ps (equivalent to +1 ms, see text).(From Ref. 10)

Finally, we changed the pump wavelength and monitored the transient absorption as a function of excitation energy.

In Figure 8(a) we show the transient bleaching signal for the 520 nm QD sample for three different pump wavelengths that are all to the red of the excitonic absorbance peak; the burned hole is always blue-shifted with respect to the pump energy and the peak moves slightly bluer with bluer excitation. The results for the 590 nm QD samples are shown in Figure 8(b). In Figure(b) we see that pumping relatively far to the blue of the excitonic peak for the 590 nm QD sample produces a bleached peak that is red-shifted with respect to the pump, but the peak of this burned hole does not move much when the wavelength of the pump is changed over a small range. If the 590 nm sample is pumped to the red of the excitonic peak, then the burned hole is blue-shifted and moves slightly with changing pump wavelength. The near instantaneous bleaching of the InP quantum dot absorption signal is consistent with QD state filling; this has also been observed in II-VI QD samples[15].

Figure 8. Pump-probe experiments showing differential absorbance (-ΔA) spectra as a function of the pump wavelength.

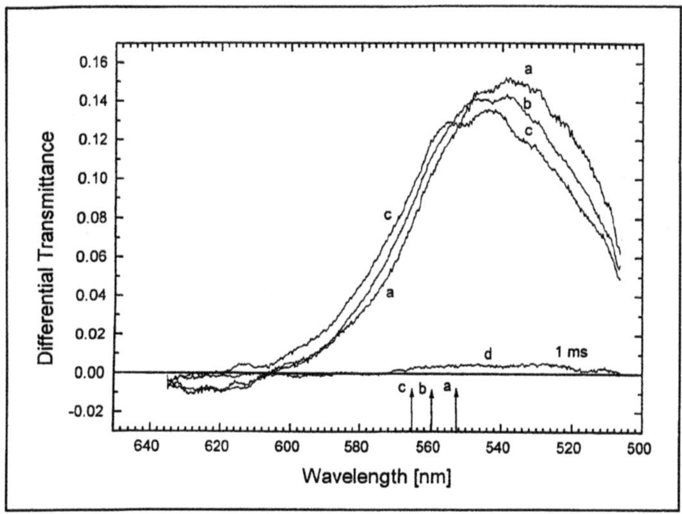

Figure 8(a): 520 nm InP QD colloid

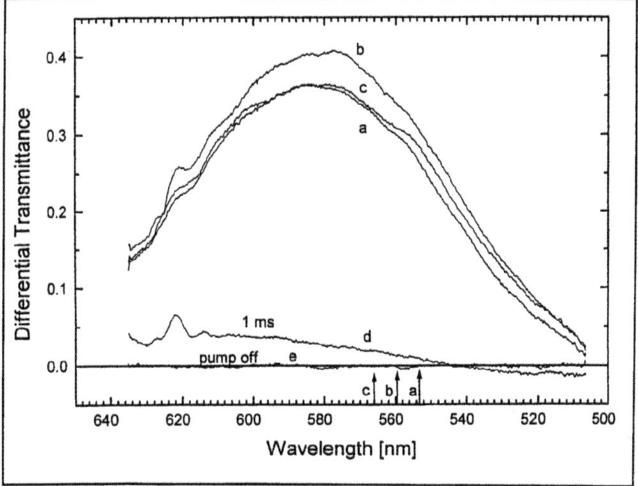

Figure 8(b): 590 nm InP QD sample. The pump wavelengths are indicated by the arrows along the wavelength axes. The spectra are shown after a time delay of 2 ps. (From Ref. 10)

The position of the burned hole as a function of the pump wavelength can yield information on whether the optical transitions in QDs are homogeneously or inhomogeneously broadened.[15] From Figures 5 and 6 we see that pumping to the blue (at 560 nm) of the exciton peak for the 590 nm QD results in a red shift of the burned hole (peak about 570 nm). From Figures 5 and 7 we see that pumping to the red (at 553 nm) of the exciton peak for the 520 nm QD results in a blue shift of the burned hole (peak about 530 nm). The same effects are seen in Figure 8. These results indicate that the transitions in the 520 nm and 590 nm samples are homogeneously broadened since only such a system could show burned holes that are blue-shifted when pumping to the red of the excitonic peak and red-shifted holes when pumping to the blue of the excitonic peak. The independence of the burned-hole peak position with pump wavelength that is shown in Figure 8(b) is also consistent with homogeneous broadening.

The small movement of the red-shifted peaks of the burned holes with pump wavelengths (Figure 8(a)) suggests some degree of inhomogeneous line broadening is also present in this sample. We conclude, therefore, that both homogeneous and inhomogeneous broadening affect the experimental linewidths, but that the latter has a smaller influence.

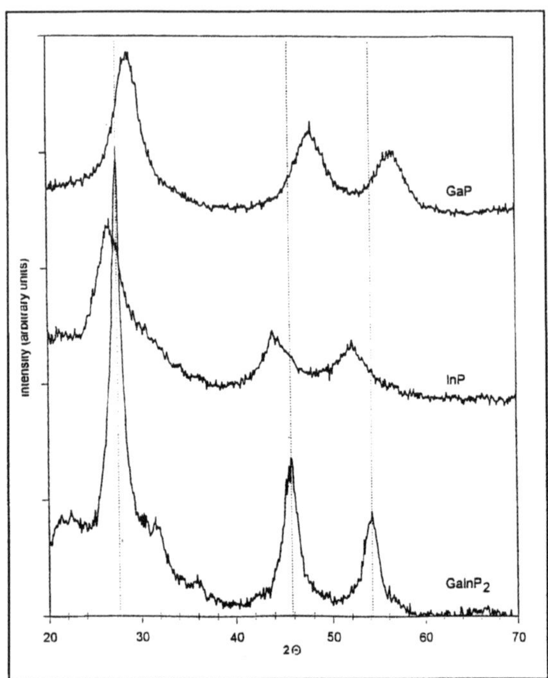

Figure 9. X-ray diffraction patterns for QD colloids of $GaInP_2$ (65 Å), GaP (30 Å), and InP (35 Å). (From Ref. 10)

The very long time for the bleaching recovery for InP QDs at room temperature is intriguing; this effect could be caused by very slow hot electron relaxation in QDs[3,16]. However, further work is necessary to distinguish between long decay times due to slowed carrier cooling and those due to other effects, such as traps.

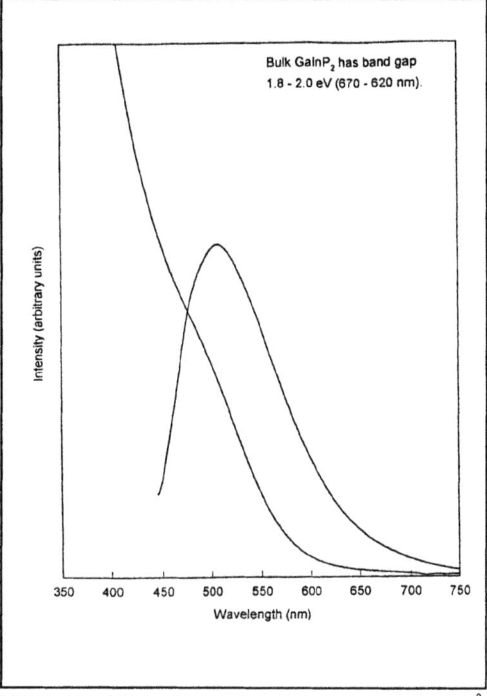

Figure 10. Absorption and emission spectra for 25 Å $GaInP_2$ colloids.

$GaInP_2$ QUANTUM DOTS

Colloidal $GaInP_2$ QDs were synthesized that had mean particle diameters of 25 Å and 65 Å; the x-ray diffraction pattern for the 65 Å sample is shown in Figure 9 along with that of the GaP and InP QDs. As expected for $GaInP_2$, the lattice spacings of these QDs is approximately the average of that for GaP and InP.

The ternary Ga-In-P system forms solid solutions which can exhibit direct bandgaps ranging from 1.7 to 2.2 eV, depending upon composition and growth temperature[17-19] At the composition $Ga_{0.5}In_{0.5}P$, the structure can be either atomically ordered or disordered (random alloy)[17-19]; the bandgap is direct, but it can range from about 1.8 to 2.0 eV, depending upon the degree of atomic ordering. Although one issue of interest is how size quantization will affect the dependence of bandgap on atomic ordering, we cannot at the present time specify the degree of atomic order in our $GaInP_2$ Qds. The absorption and emission spectra of the 25 Å $GaInP_2$ QDs is shown in Figure 10. An estimate of the direct bandgap of the $GaInP_2$ QDs from a plot of the square of the absorbance times photon energy vs. photon energy indicates a value of about 2.7 eV; this value is blue-shifted from the bulk value of 1.8-2.0, as expected for QDs. No excitonic structure is observed in Figure 11; this may be caused by a sufficiently wide size distribution which could easily mask excitonic peaks in QDs. The emission peak is nearly coincident with the absorption peak, indicating that surgace traps are minimal in these QDs.

Acknowledgement

This work was supported by the U.S. Department of Energy, Office of Basic Energy Sciences, Division of Chemical Sciences.

References

(1) Byrne, E. K.; Parkanyi, L.; Theopold, K. H. *Science* **1988**, *241*, 332.

(2) Olshavsky, M. A.; Goldstein, A. N.; Alivisatos, A. P. *J. Am. Chem. Soc.* **1990**, *112*, 9438.

(3) Uchida, H.; Curtis, C. J.; Nozik, A. J. *J. Phys. Chem.* **1991**, *95*, 5383.

(4) Uchida, H.; Curtis, C. J.; Kamat, P. V.; Jones, K. M.; Nozik, A. J. *J. Phys. Chem.* **1992**, *96*, 1156.

(5) Butler, L.; Redmond, G.; Fitzmaurice, D. *J. Phys. Chem.* **1993**, *97*, 10750.

(6) MacDougall, J. E.; Eckert, H.; Stucky, G. D.; Herron, N.; Wang, Y.; Moller, K.; Bein, T. and Cox, D. *J. Am. Chem. Soc.* **1989**, 111, 8006.

(7) Douglas, T.; Theopold, K. H. *Inorg. Chem.* **1991**, *30*, 594.

(8) Uchida, H.; Ogata, T.; Yoneyama, H. *Chem. Phys. Lett.* **1990**, *173*, 103.

(9) Mićić, O. I.; Sprague, J. R.; Curtis, C. J.; Jones, K. M.; and Nozik, A. J. *J. Phys.Chem.* **1994**, *98*, 4966.

(10) Mićić, O. I.; Sprague, J. R.; Curtis, C. J.; Jones, K. M.; Machol, J.L.; Nozik, A. J. *J. Phys.Chem.* **1995,** *99*, 7754.

(11) Aubuchon, S. R.; McPhail, A. T.; Wells, R. L.; Giambra, J. A.; Bowser, J. R. *Chem. Mater.* **1994**, *6*, 82; Wells, R. L.; Self, M. F.; McPhail, A. T.; Auuchon, S. R.; Wandenberg, R. C.; Jasinski, J. P. *Organometallics* **1993**, *12*, 2832.

(12) Murray, C. B.; Norris, D. J.; Bawendi, M. G. *J. Am. Chem. Soc.* **1993**, *115*, 8706.

(13) Rama Krishna, M. V.; Friesner, R. A. *J. Chem. Phys.* **1991**, *95*, 525.

(14) Pankove, J. I. "Optical Processes in Semiconductors," Dover, New York, **1971**, p. 43.

(15) Peyghambarian, N.; Fluegel, B.; Hulin, D.; Migus, A.; Joeffre, M.; Antonetti, A.; Koch, S. W.; Lindberg, M. IEEE J. Quant. Elec. **1989**, 25(12), 2516.

(16) Benisty, H.; Sotomayor-Torres, C. M.; Weisbuch, C. *Phys. Rev.* **1991**, *B44*, 10945.

(17) DeLong, M. C.; Ohlsen, W. D.; Viohl, I.; Taylor, P. C.; and Olson, J. M. *J. Appl. Phys.*, **1991**, *70*, 2780.

(18) Wei, S.-H.; Zunger, A. *Phys. Rev. B* **1989**, *39*, 3279; Wei, S.-H.; Ferreira, L. G.; Zunger, A. *Phys. Rev. B* **1990**, *41*, 8240; Froyen, S.; Zunger, A. *Phys. Rev. Lett.* **1991**, *66*, 2132.

(19) Mascarenhas, A.; Olson, J. M.; *Phys. Rev. B* **1990**, *41*, 9947; Horner, G. S.; Mascarenhas, A.; Froyen, S.; Alonso, R. G.; Bertness, K. A.; Olson, J. M. *Phys. Rev. B* **1993**, *47*, 4041; Olson, J. M.; Kurtz, S. R.; Kibbler, A. E.; Faine, P. *Appl. Phys. Lett.* **1990**, *56*, 623.

A COMPARISON OF PRESSURE-INDUCED STRUCTURAL TRANSFORMATIONS IN CDSE, INP, AND SI NANOCRYSTALS

A.B. HERHOLD, S.H. TOLBERT*, A.A.GUZELIAN AND
A.P. ALIVISATOS
Department of Chemistry, University of California at Berkeley and Materials Sciences Division, Lawrence Berkeley National Laboratory Berkeley, California 94720

ABSTRACT. Pressure-induced phase transitions in CdSe, InP and Si nanocrystals of 2 to 50 nm in diameter were studied with high pressure X-ray diffraction and optical absorption measurements. All samples were found to transform via a single nucleation event. In addition, all samples transformed at an elevated pressure with a large hysteresis compared to that of the bulk. These phenomena are explained by an overall change in shape of the crystallite after it goes through the phase transition. Evidence for this shape change is shown in the elevation of the phase transition pressure in all samples and in the high pressure shape of the Si nanocrystals. These experiments present an opportunity to investigate first order solid-solid phase transitions on a nanometer scale.

1. Introduction

As synthesized, semiconductor crystallites in the nanometer size regime have the bulk bonding geometry. However, the phase stability of nanocrystals differs from bulk away from ambient pressure and temperature. This is observed in the depression of the melting temperature in metallic [1,2,3] and semiconductor [4] particles and the elevation of the pressure-induced solid-solid phase transition pressure in semiconductor particles [5,6,7,8,9,10,11]. Understanding the effect of size on phase stability is important to many fields ranging from materials science to geophysics. Recent advances in the synthesis of semiconductor crystallites in the nanometer size regime [12,13,14] have made possible the study of finite size effects in several different semiconductors.

We have chosen to study size effects on first order solid-solid phase transformations in semiconductor nanocrystals. In this paper, we review the results of high pressure studies on CdSe [5,6,7,8] and Si [9] nanocrystals and introduce new high pressure data on InP nanocrystals. At atmospheric pressure, these materials all display tetrahedral bonding. However, they form different crystal structures because of their

E. Pelizzetti (ed.), Fine Particles Science and Technology, 331–342.

different ionicities. Si crystallizes in the diamond structure, which has a stacking order of planes along the (111) direction of ABCABC... and has cubic symmetry. InP has a large amount of covalent character and thus is stable in the zincblende structure, which has the same plane stacking as diamond and also has cubic symmetry. CdSe, however, is more ionic and therefore forms the wurtzite structure, which has a stacking order of ABABAB... in the (111) direction and hexagonal symmetry.

Under pressure, these semiconductors undergo first order phase transitions to more densely packed structures: the six-fold rock salt structure in CdSe and InP and the distorted six-fold β-tin structure in Si. As the degree of covalent character increases in this set of materials, the high pressure phase is more likely to be metallic. Thus, bulk CdSe transforms from wurtzite to rock salt at an upstroke pressure of 3 GPa [15,16,17], where the rock salt phase is a semiconductor [16], while bulk InP does not transform to the rock salt phase until 10.8 GPa [18,19,20]. In addition, rock salt InP is a metal [19,20]. Bulk Si transforms to the β-tin phase at approximately 11 GPa [21,22,23,24]. In general, Si has a very complex high pressure phase diagram. Of interest to this paper is the transition of bulk Si under pressure from diamond to two other structures and then to a primitive hexagonal phase at approximately 16 GPa [25,26,23,24]. All of the high pressure phases of Si are metals.

First order pressure-induced solid-solid phase transitions in tetrahedral bulk semiconductors are typically hysteretic [27] and result in a decrease of crystalline domain size [27,28]. This decrease in domain size arises from the mechanism of the phase transition. As the phase transition is approached, nuclei of the new phase grow inside the old phase. Multiple nucleations occur, forming many smaller crystalline domains out of the original crystal after the phase transition. The observed hysteresis is a result of strain energy created in the interface between the two phases. To overcome this strain energy, the pressure must be raised beyond the thermodynamic phase transition point.

When a small crystallite approaches the size of the fluctuations characteristic of a bulk phase transition, bulk nucleation theory may no longer apply. In addition, the increasing fraction of atoms on the surface must play a role in the phase stability. The decrease in melting point with decreasing size is explained by a surface effect [1,2,3]. Since the liquid has a lower surface energy than the solid, melting will occur at a lower temperature for a nanocrystal than for bulk. Similarly, the difference in surface energy between the high and low pressure phases affects the thermodynamics of solid-solid phase transitions in nanocrystals [7,8,9]. The surface may also have an influence on the kinetics of the phase transition. To understand these effects, we have performed high pressure X-ray diffraction on CdSe, InP, and Si nanocrystals and high pressure optical absorption measurements on CdSe nanocrystals. In this paper we discuss the results of these measurements to develop an understanding of the effect of size on the kinetics and thermodynamics of solid-solid phase transitions.

2. Experimental

CdSe nanocrystals were synthesized by a modification of the method of Murray and Bawendi [12,29]. InP nanocrystals were synthesized from organometallic precursors in solution at high temperature [30]. Si nanocrystals were synthesized by the gas pyrolysis method described elsewhere [14,31]. Samples were characterized by powder X-ray diffraction, TEM, and optical absorption. Si samples were also characterized by luminescence, and CdSe and InP samples were also characterized by XPS.

CdSe and InP nanocrystals were homogeneously dispersed in 4-ethyl pyridine for the high pressure experiments. 4-ethyl pyridine was found to be hydrostatic to pressures above 10 GPa. Si samples were dispersed in ethylene glycol, which was found to be a reasonable pressure medium at the pressures reached in the experiment [9].

For X-ray diffraction data, samples were loaded in either a Merrill-Bassett style diamond anvil cell with a beryllium support plate or a Mao-Bell style cell with a diffraction slit in one rocker. Data were collected on the Stanford Synchrotron Radiation Laboratory wiggler beam line 10-2. The focused beam was sent through a series of slits and collimated with either a 200, 150, or 100 micron pinhole. Diffraction images were collected in an angle-dispersive geometry with Fuji image plates read with 100 micron resolution. Analysis of the collected data was achieved using numerical integration of the digitized image. For optical absorption data, samples were loaded in a modified Mao-Bell style diamond cell. Absorption data were taken on a scanning Cary model 118 UV/visible spectrometer. In all cases, standard ruby fluorescence techniques were used to determine pressure.

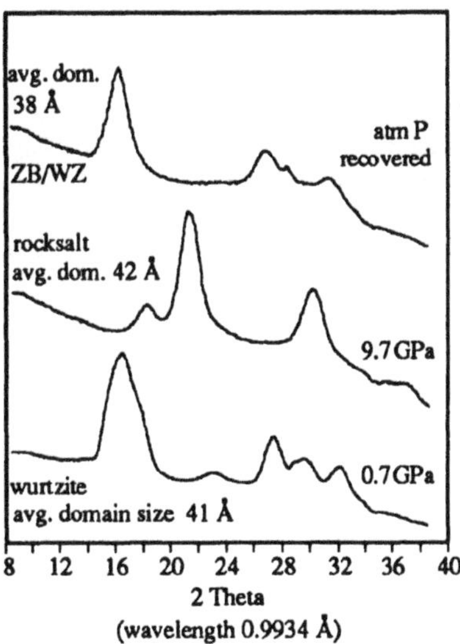

Figure 1. X-ray diffraction data on 4 nm CdSe nanocrystals with increasing pressure and recovered at atmospheric pressure. The crystalline domain size remained the same after each phase transition. For indexing, see references 6 and 7.

3. Discussion of Results

3.1 SINGLE NUCLEATION

High pressure X-ray diffraction on 4.2 nm CdSe nanocrystals is shown in Figure 1. In X-ray diffraction, the peak widths are inversely proportional to the crystalline domain size. The diffraction peaks of the nanocrystals are very broad due to their small size. As previously demonstrated, the wurtzite phase nanocrystals are stable to well above the bulk phase transition at 3 GPa [5,6,7,8]. In fact, the 4.2 nm CdSe nanocrystals do not transform to the rock salt phase until above 6 GPa. This large elevation in phase transition pressure will be discussed below. With release of pressure, the nanocrystals revert back to a mixture of wurtzite and zincblende structures.

Careful fitting of the diffraction patterns in Figure 1 reveals that after the phase transition, the peak width of the nanocrystals remains the same. The domain size

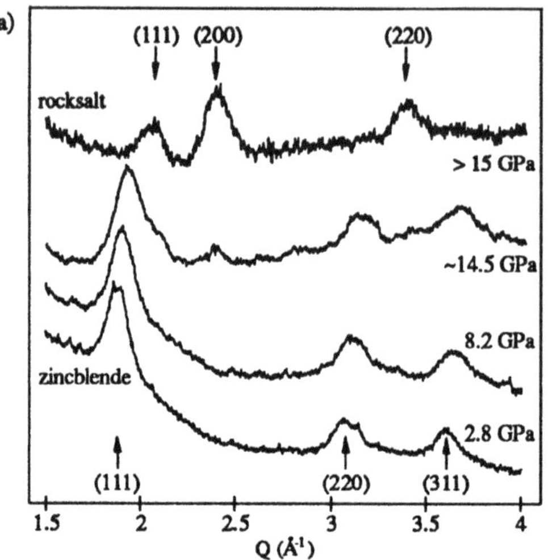

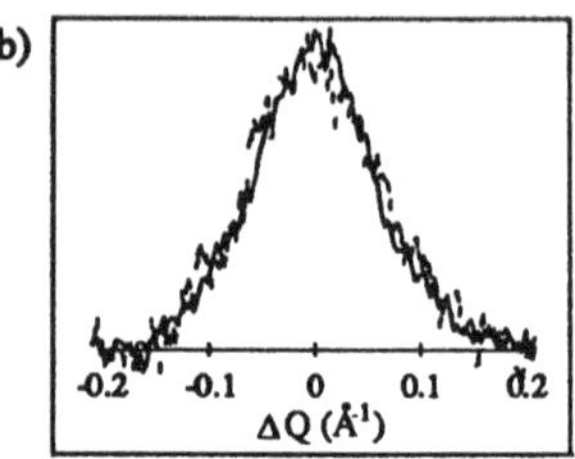

Figure 2. a) X-ray diffraction on 5.0 nm InP nanocrystals showing the phase transition from zincblende to rock salt with increasing pressure. The sample is stable above the bulk phase transition at 10.8 GPa. The (200) zincblende peak is very weak and not visible in these scans. The x-axis is in units of Q, where $Q=4\pi\sin\Theta/\lambda$. Displayed in this manner there is no angle dependence on the peak widths and data taken with different X-ray energies will have the same peak positions. b) The zincblende (111) peak (solid line) at ~ 0.5GPa and the rock salt (200) peak (dotted line) at > 15 GPa plotted together and scaled to the same height. The peak widths are the same in the two phases, indicating that the nanocrystals transformed via a single nucleation event.

is 4 nm in both phases, indicating that the nanocrystals transform coherently without domain fracture. This implies that the structural change proceeds in nanocrystals via a single nucleation event. However, it is difficult to make a clear visual comparison of the wurtzite and rock salt peak widths in the CdSe nanocrystals because the hexagonal symmetry of the wurtzite phase makes the diffraction pattern complicated. Although the individual lines are well-separated in bulk wurtzite CdSe, the broad widths of the nanocrystal diffraction makes it difficult to resolve the different peaks in small size. Instead, the peak widths are determined with a fitting program.

Since rock salt has cubic symmetry, the diffraction pattern has fewer lines and thus are clearly resolvable even in 4 nm diameter CdSe rock salt particles. In InP, both the low and high pressure phases have cubic symmetry, making InP an ideal system for visually comparing the domain sizes of the two different phases in nanocrystals. 5.0 nm InP nanocrystals are stable well above the bulk phase transition from zincblende to rock salt at 10.8 GPa. High pressure X-ray diffraction data on 5.0 nm InP particles are shown in Figure 2a. It can be seen that the nanocrystals do not transform until above 14.0 GPa.

Since both phases have cubic symmetry, a visual comparison of peak widths in the two phases of nanocrystals can be made. Figure 2b shows an overlay of the (111) wurtzite peak and the (200) rock salt peak from the InP nanocrystals. The domain size remained the same through the phase transition, giving further proof that the structural transformation in semiconductors of this domain size occurs with single nucleation. In addition, this implies that the critical size for multiple nucleation to occur in these materials is larger than 5.0 nm.

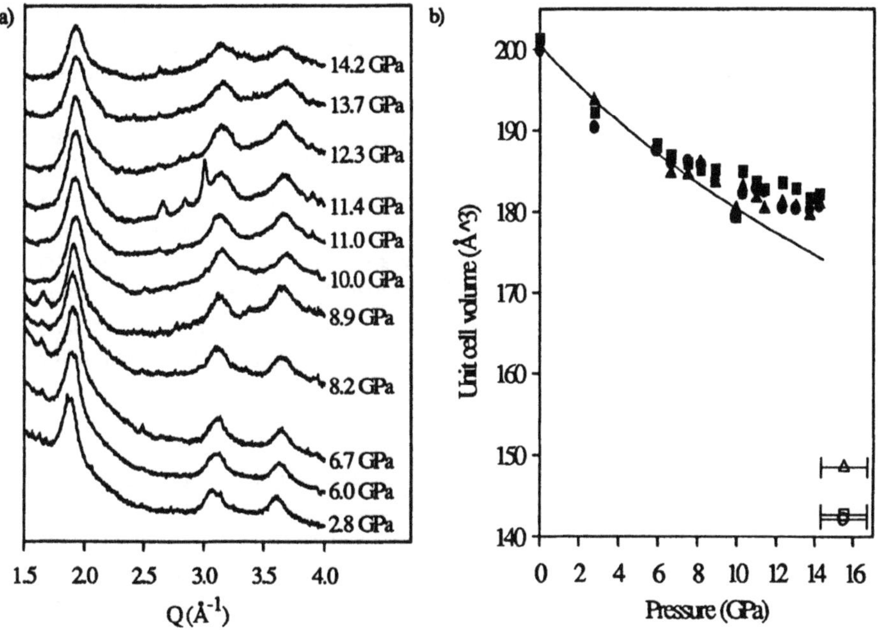

Figure 3. a) X-ray diffraction on 5.0 nm zincblende InP nanocrystals with increasing pressure. See Figure 2 for indexing of the peaks. The extra peaks at 11.4 GPa are due to gasket diffraction. b) Volume contraction of the unit cell with increasing pressure for the same sample. Unit cell volumes are calculated from the (111) (solid triangles), (220) (solid circles), (311) (solid squares) zincblende peaks and (111) (open triangle), (200) (open circle), and (220) (open square) rock salt peaks. The solid line is the Murnaghan equation of state fit with bulk InP values. The nanocrystals have a smaller compressibility at high pressures than predicted by the bulk values.

The X-ray diffraction data can be used to determine the compressibility of the nanocrystals. Several diffraction patterns on 5.0 nm zincblende InP nanocrystals with increasing pressure are shown in Figure 3a, and the corresponding change in unit cell volume with pressure is shown in Figure 3b. The Murnaghan equation of state [32] fit with the bulk values [20] for B_0 and B_0' is shown for comparison. The nanocrystals have a small lattice contraction of approximately 0.3% at atmospheric pressure. This lattice contraction was used to calculate the unit cell volume at atmospheric pressure in the Murnaghan equation. The bulk values appear to fit the data well at low pressures but fail at high pressures. Since the nanocrystals are stable past the bulk phase transition, they sample the anharmonic part of the compressibility, becoming increasingly harder to compress at high pressure. This causes the shift of fractional volume with pressure to flatten at high pressures. This result has also been observed in CdSe nanocrystals [6].

The stability of zincblende InP nanocrystals at high pressures allows an analysis of the compressibility at pressures where bulk zincblende InP cannot exist. In addition to structural transformations, other interesting properties to study at high pressures are optical characteristics. It has been proposed that zincblende InP has a direct to indirect gap crossover at high pressure [33]. There is experimental and theoretical controversy over whether this crossover occurs before the phase transition to rock salt or at a pressure just above the bulk phase transition pressure and thus cannot be observed [34,35,36,37]. It may be possible to address this issue by performing optical measurements at high pressure since InP nanocrystals are stable above the bulk phase transition. A comparison of direct and indirect materials in finite size has been performed previously in wurtzite and rock salt phase CdSe nanocrystals [38], where the features of the rock salt spectra were generalized to other indirect gap nanocrystals such as Si and AgBr. Studying the optical properties of InP nanocrystals under pressure may provide another example of indirect gap materials in finite size.

3.2 SURFACE EFFECTS

More insight into the elevated phase transition can be gained by studying the size dependence of the phase transition, which has been measured using optical absorption for CdSe [7,8]. Hysteresis curves for three different sizes of CdSe nanocrystals are shown in Figure 4. All three sizes exhibit a large hysteresis of about 6 GPa. In bulk CdSe, however, the hysteresis is only 2 GPa [15,16,17]. The nature of the large hysteresis in the nanocrystals will be discussed below. The dependence of the phase transition on size can be determined by taking the midpoint of the hysteresis curves as the thermodynamic phase transition point. From this, it can be seen that the phase transition pressure shows a strong dependence on radius. As the size of the nanocrystal decreases, the phase transition pressure increases [7,8]. To determine the cause of this elevation, we can look at the phase transition path.

In a solid-solid phase transition, there is often a specific path that the atoms follow when going from one phase to the next [39,40,41]. This path acts as the reaction coordinate for the phase transition. If all of the atoms are moving in a well-defined path, this must cause the crystal to change shape. This is usually not observable in macroscopic crystals because multiple nucleation will cause the crystal to break into many domains, thus obscuring any overall change in shape.

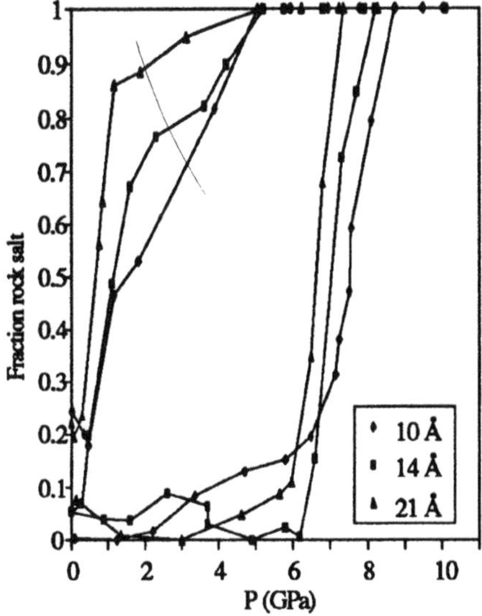

Figure 4. Hysteresis curves taken with optical absorption for 3 sizes of CdSe nanocrystals. The hysteresis for each size is ~3 times that in bulk CdSe. The phase transition pressure increases with decreasing nanocrystal size.

Since the shape change will change the position of the atoms on the surface as well as in the interior, it must change the surface energy of the crystal. In bulk, where most of the atoms are in the interior, the change in surface energy is negligible. However, in a nanocrystal, where a large fraction of the atoms are on the surface, this is not the case. Wurtzite CdSe nanocrystals as synthesized are approximately spherical with faceted, low index surfaces. A proposed transition path has been described previously [7,8] that is based on Burdett's model of the rock salt to wurtzite transition as a 3-dimensional Peierl's distortion [42]. When the atoms move along the proposed transition path, the nanocrystal changes shape to an ellipsoidal structure with a mixture of high and low index surfaces [7,8]. Rock salt structures in general can form surfaces with few dangling bonds and thus a very low surface energy. However, because of the path the atoms must follow in the phase transition, the resulting rock salt nanocrystal has an overall high surface energy.

Since the rock salt particles have a higher energy surface than the wurtzite particles, more work must be put into the system to cause the sample to change phase than in the bulk case. As the size of the crystallite is decreased, the fraction of the atoms on the surface increases, increasing the difference between the wurtzite and rock salt surface energies. This gives the observed dependence of phase transition pressure on size [7,8]. In addition, changing the ligands that bind to the surface of the nanocrystals can also change the phase transition pressure, as has been observed for 4 nm CdS nanocrystals where the effects of two different surface stabilizers were studied [10]. In that study, the phase transition pressure was elevated with respect to the bulk but by different amounts for the two different surfaces.

The large hysteresis of the structural transformation in CdSe nanocrystals shown in Figure 4 is observable in Si [9] and most likely InP nanocrystals as well. Hysteresis in a phase transition results from a barrier to the transition. In bulk, this barrier results from strain energy caused by nucleation. However, since the hysteresis is much larger in nanocrystals, there must be an additional kinetic barrier. One possible cause of the hysteresis is the need to break surface bonds. InP and CdSe nanocrystals are capped with organic ligands, and Si nanocrystals have a thick SiO_2 coating. In order for the nanocrystal to go through the phase transition, some of these bonds must be rearranged, requiring an increase in pressure past the nanocrystal thermodynamic phase transition point. Since the surface plays such a large role in the total energy of the nanocrystals, this can result in a large hysteresis.

By raising the temperature of the system, the sample can more easily go over the kinetic barrier. Heating has been observed to narrow the hysteresis curves in bulk systems [43]. The activation energy can be calculated by measuring the change of the width of the hysteresis curve with temperature [44,45]. The energetics of the barrier in nanocrystals would provide an interesting comparison to bulk. In addition, in nanocrystals, heating may have an additional effect; at high enough temperatures, the surfaces of the rock salt phase could be annealed, resulting in a rock salt structure with low energy surfaces. If the annealed rock salt sample was first cooled and then lowered back to atmospheric pressure, it would resist transformation back to wurtzite. It may then be possible to have rock salt nanocrystals stable at atmospheric pressure. This would be an example of using kinetics to stabilize new structures in finite size. Investigations into phase transitions at high temperature are currently in progress.

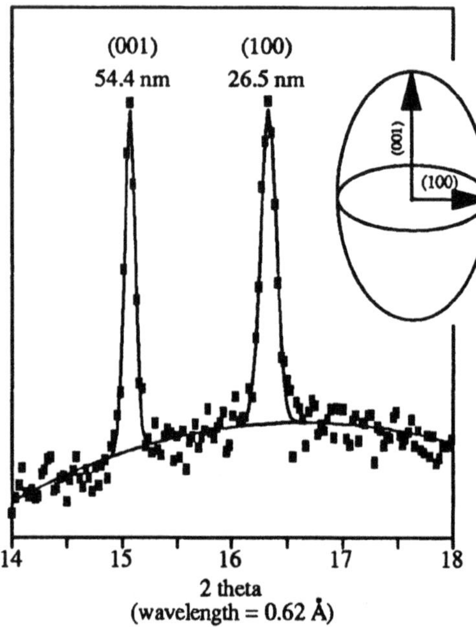

Figure 5. X-ray diffraction on 49.2 nm Si nanocrystals in the primitive hexagonal phase at 21 GPa. The squares are the data and the solid line is the fit to the data. The ratio of the domain sizes in the (001) and (100) directions give an ellipsoidal shape that agrees with the shape predicted by the most likely transition path.

3.3 SHAPE CHANGE

In both CdSe and InP, the high pressure phase has cubic symmetry. Since the crystallographic axes are degenerate in a cubic structure, each diffraction peak will be a convolution of the long and short dimensions of the ellipsoidal rock salt structure described earlier. Given the size and pressure inhomogeneity in our experiments, all of the peak widths in the rock salt phase are the same. Therefore, in CdSe and InP nanocrystals, the shape change cannot be observed experimentally by X-ray diffraction.

However, in the primitive hexagonal phase of Si, the crystallographic axes are no longer completely degenerate. This makes Si nanocrystals an ideal system for studying the path-induced shape change. High pressure X-ray diffraction data on 49.2 nm Si nanocrystals has been collected previously [9]. Despite their large size, these nanocrystals still show a significant elevation in phase transition pressure. This elevation can be explained again in terms of surface energies. The Si nanocrystals have a well-passivated SiO_2 surface, therefore, in order to change phase, the Si nanocrystals must break very stable surface bonds. Even in Si nanocrystals as large as 49.2 nm, the structural transformation does not occur until well above the bulk phase transition.

Using the most likely transition path, an approximately spherical diamond phase Si nanocrystal is transformed into a primitive hexagonal structure with a long axis in the (001) direction and shorter axes in the degenerate (100) and (010) directions [9]. Since the (001) direction is crystallographically distinct, comparing the widths of the (001) and (100) peaks will reveal any experimental shape change in the crystallite.

The (001) and (100) peaks of 49.2 nm Si nanocrystals in the primitive hexagonal phase are shown in Figure 5. The (100) peak is clearly broader than the (001) peak. After deconvoluting the data for pressure inhomogeneity and experimental resolution, the peak widths gave crystalline domain sizes of 26.7±1.0 nm in the (100) direction and 54.9±6.8 nm in the (001) direction. These different domain sizes fit an ellipsoidal shape as suggested by the transition path model with an aspect ratio of approximately 2.05, which agrees well with an aspect ratio of 1.9 predicted by the transition path model. In addition, since the long axis of the primitive hexagonal phase nanocrystal is longer than the initial domain size of the diamond phase nanocrystal, the nanocrystals did not fragment during the phase transition, indicating that each nanocrystal changed phase via a single nucleation event. The different primitive hexagonal peak widths and the evidence for single nucleation combined show that the Si nanocrystals systematically changed shape in a manner dictated by the transition path.

4. Conclusion

By examining the high pressure behavior observed previously in 2 to 4 nm CdSe and 50 nm Si nanocrystals and recently in 5.0 nm InP nanocrystals, we have found results that are general to all of these materials in finite size. X-ray diffraction shows that structural transformations in these samples proceeds via a single nucleation in each nanocrystal. Evidence for this is that the high and low pressure phase nanocrystal peak widths and thus crystalline domain sizes are the same as shown for InP. As a result, since the crystalline domain remains intact, when the atoms move along the transition path, the nanocrystal changes shape. This shape change produces high energy surfaces in the high pressure phase which leads to an elevation in phase transition pressure that is observed in all samples studied. Since the nanocrystals are stable above the bulk phase transition, they experience a different compressibility than the bulk at high pressures, as shown for 5.0 nm InP nanocrystals. In addition, the nanocrystals exhibit a large hysteresis which is possibly due to the requirement that surface bonds be broken so that the nanocrystal can change shape.

Crystallographic degeneracy prevents the observation of a shape change within experimental resolution in the high pressure phases of CdSe and InP nanocrystals. However, a change to an ellipsoidal shape is observed in Si nanocrystals, giving an aspect ratio of long to short axes that agrees with the shape predicted by the possible phase transition path. Since the materials studied have a wide range of high pressure behavior in the bulk and yet still show the same finite size effects, it appears that this nanometer scale picture of the phase transition is generalizable to other semiconductor materials.

5. Acknowledgments

The authors would like to thank Christopher Johnson and Dr. Louis Brus for their assistance with this research effort and Prof. Raymond Jeanloz for the use of his diffraction cell. This work was supported by the Department of Energy (DOE) under contract DE-AC0376SF00098. The X-ray experiments described here were performed using the facilities of the University of California-Lawrence Livermore National Lab PRT at the Stanford Synchrotron Radiation Laboratory, which is operated by the Department of Energy, Division of Chemical Science.

6. References

* Present address: Department of Chemistry, University of California at Santa Barbara, Santa Barbara, Ca 93106

1. Coombes, C.J. (1972) The melting of small particles of lead and indium, *J. Phys. F* **2**, 441-449.
2. Peppiatt, S.J. and Sambles, J.R. (1975) The melting of small particles. I. Lead, *Proc. R. Soc. London Ser. A.* **345**, 387-399.
3. Buffat, Ph. and Borel, J-P. (1976) Size effects on the melting temperature of gold particles, *Phys. Rev. A* **13**, 2287-2298.
4. Goldstein, A.N., Echer, C.M., and Alivisatos, A.P. (1992) Melting in semiconductor nanocrystals, *Science* **256**, 1425-1427.
5. Tolbert, S.H. and Alivisatos, A.P. (1993) Size dependence of the solid-solid phase transition in CdSe nanocrystals, *Z. Phys. D.* **26**, 56-58.
6. Tolbert, S.H. and Alivisatos, A.P. (1994) Se EXAFS study of the elevated wurtzite to rock salt structural phase transition in CdSe nanocrystals, in G.C. Hadjipanayis and R.W. Siegel (eds.), *NATO ASI Proceedings on Nanophase Materials: Synthesis-Properties-Applications*, Kluwer Academic Publishers, Dordrecht, pp. 471-482.
7. Tolbert, S.H. and Alivisatos, A.P. (1994) Size dependence of a first order solid-solid phase transition: the wurtzite to rock salt transformation in CdSe nanocrystals, *Science* **265**, 373-376.
8. Tolbert, S.H. and Alivisatos, A.P. (1995) The wurtzite to rock salt structural transformation in CdSe nanocrystals under high pressure, *J. Chem. Phys.* **102**, 1-15.
9. Tolbert, S.H., Herhold, A.B., Brus, L.E., and Alivisatos, A.P., Pressure induced structural transformations in Si nanocrystals: Surface and shape effects, submitted

10. Haase, M. and Alivisatos, A.P. (1992) Arrested solid-solid phase transition in 4-nm-diameter CdS nanocrystals, *J. Phys. Chem.* **96**, 6756-6762.
11. Tolbert, S.H. and Alivisatos, A.P. (1995) High pressure structural transformations in semiconductor nanocrystals, to be published in Ann. Rev. Phys. Chem.
12. Murray, C.B., Norris, D.J., and Bawendi, M.G. (1993) Synthesis and characterization of nearly monodisperse CdE (E=S, Se, Te) semiconductor nanocrystallites, *J. Am. Chem. Soc.* **115**, 8706-8715.
13. Micic, O.I., Sprague, J.R., Curtis, C.J., Jones, K.M., Machol, J.L., Nozik, A.J., Giessen, H., Fluegel, B., Mohs, G., and Peyghambarian, N. (1995) Synthesis and characterization of InP, GaP, and $GaInP_2$ quantum dots, *J. Phys. Chem.* **99**, 7754-7759.
14. Littau, K.A., Szajowski, P.J., Muller, A.J., Kortan, A.R., and Brus, L.E. (1993) A luminescent silicon nanocrystal colloid via a high-temperature aerosol reaction, *J. Phys. Chem.* **97**, 1224.
15. Edwards, A.L. and Drickamer, H.G. (1961) Effect of pressure on the absorption edges of some III-V, II-VI, and I-VII compounds, *Phys. Rev.* **122**, 1149-1157.
16. Onodera, A. (1969) High pressure transition in cadmium selenide, *Rev. Phys. Chem. Jpn.* **39**, 65-77.
17. Yu, W.C. and Gielisse, P.J. (1971) High pressure polymorphism in CdS, CdSe, and CdTe, *Mat. Res. Bull.* **6**, 621-638.
12. Minomura, S. and Drickamer, H.G. (1962) Pressure induced phase transitions in silicon, germanium and some III-V compounds, *J.Phys. Chem. Solids* **23**, 451-456.
19. Jamieson, J.C. (1963) Crystal structures at high pressures of metallic modifications of compounds of indium, gallium, and aluminum, *Science* **139**, 845-847.
20. Menoi, C.S. and Spain, I.L. (1987) Equation of state of InP to 19 GPa, *Phys. Rev. B* **35**, 7520-7525.
21. Jamieson, J.C. (1963) Crystal structures at high pressure of metallic modifications of silicon and germanium, *Science* **139**, 762-764.
22. Weinstein, B.A. and Piermarini, G.J. (1975) Raman scattering and phonon dispersion in Si and GaP at very high pressure, *Phys. Rev. B* **12**, 1172-1186.
23. Olijnyk, H., Sikka, S.K., and Holzapfel, W.B. (1984) Structural phase transitions in Si and Ge under pressures up to 50 GPa, *Phys. Lett.* **103A**, 137-140.
24. Hu, J.Z. and Spain, I.L. (1984) Phases of silicon at high pressure, *Sol. St. Comm.* **51**, 263-266.
25. McMahon, M.I. and Nelmes, R.J. (1993) New high-pressure phase of Si, *Phys. Rev. B* **47**, 8337-8340.
26. McMahon, M.I., Nelmes, R.J., Wright, N.G., and Allan, D.R. (1994) Pressure dependence of the *Imma* phase of silicon, *Phys. Rev. B* **50**, 739-743.
27. Venkateswaran, U.D., Cui, L.J., Weinstein, B.A., and Chambers, F.A. (1992) Forward and reverse high-pressure transitions in bulklike AlAs and GaAs epilayers, *Phys Rev. B* **45**, 9237-9247.
28. Hanneman, R.E., Banus, M.D., and Gatos, H.C. (1964) High pressure transition in InSb, *J. Phys. Chem. Solids* **25**, 293-302.
29. Katari, J.E. Bowen, Colvin, V.L., and Alivisatos, A.P. (1994) X-ray photoelectron spectroscopy of CdSe nanocrystals with applications to studies of the nanocrystal surface, *J. Phys. Chem.* **98**, 4109-4117.

30. Guzelian, A.A., Katari, J.E.B., Kadavanich, A.V., Banin, U., Hamad, K., Alivisatos, A.P., Wolters, R.H., Arnold, C.C., and Heath, J.R., Synthesis of size-selected, surface passivated InP nanocrystals, to be submitted.
31. Wilson, W.L., Szajowski, P.F., and Brus, L.E. (1993) Quantum confinement in size-selected, surface-oxidized silicon nanocrystals, *Science* **262**, 1242-1244.
32. Murnaghan, F.D., (1944) The compressibility of media under extreme pressures, *Proc. Natl. Acad. Sci. USA* **30**, 244-247.
33. Kobayashi, T., Takamasa, T., Kazunori, A., Yamamoto, K., Abe, K., (1981) Inversion of the Γ and X conduction bands in InP under high pressure, in J.S. Schilling and R.N. Shelton (eds.), *Physics of Solids Under High Pressure*, North-Holland Publishing Company, Amsterdam, pp. 141-147.
34. Müller, H. Trommer, R., and Cardona, M. (1980) Pressure dependence of the direct absorption edge of InP, *Phys. Rev. B* **21**, 4879-4883.
35. Menoi, C.S., Hochheimer, H.D., and Spain, I.L. (1986) High-pressure study of photoluminescence in indium phosphide at low temperatures, *Phys. Rev. B* **33**, 5896-5898.
36. Leroux, M. (1989) Photoluminescence study of the pressure-induced phase transition in InP at low temperature, *Semiconductor Science and Technology* **4**, 231-232.
37. Gorczyca, I., Christensen, N.E., and Alouani, M.(1989) Calculated optical and structural properties of InP under pressure, *Phys. Rev. B* **39**, 7705-7712.
38. Tolbert, S.H., Herhold, A.B., Johnson, C.S., and Alivisatos, A.P. (1994) Comparison of quantum confinement effects on the electronic absorption spectra of direct and indirect gap semiconductor nanocrystals, *Phys. Rev. Lett.* **73**, 3266-3269.
39. Chadi, D.J. and Martin, R.M. (1976) Calculation of lattice dynamical properties from electronic energies: application to C, Si, and Ge, *Sol. St. Comm.* **19**, 643-646.
40. Weinstein, B.A. (1979) High pressure phonon dispersion of the zinc chalcogenides and the metallic transition, in K.D. Timmerhaus and M.S. Barber (eds.), *High Pressure Science and Technology*, Plenum, New York, pp. 141-151.
41. Dougherty, T.P., Wiederrecht, G.P., Nelson, K.A., Garrett, M.H., Jensen, H.P., and Warde, C. (1992) Femtosecond resolution of soft mode dynamics in structural phase transitions, Science 258, 770-774.
42. Burdett, J.K. (1984) From bonds to bands and molecules to solids, *Prog. Solid St. Chem.* **15**, 173-255.
43. Jayaraman, A., Klement, W., and Kennedy, G.C. (1963) Melting and polymorphic transitions for some group II-VI compounds at high pressures, *Phys. Rev.* **130**, 2277-2283.
44. Jeanloz, R. (1987) Coexistence curves and equilibrium boundaries for high-pressure phase transformations, *J. Geophys. Res.* **92**, 10352-10362.
45. Li, X. and Jeanloz, R. (1987) Measurement of the B1-B2 transition pressure in NaCl at high temperatures, *Phys. Rev. B*, **36**, 474-479.

DIRECTED PHOTOCURRENTS IN NANOSTRUCTURED $SnO_2/TiO_2/Ru(II)L_2(CNS)_2$ HETEROJUNCTIONS

B. LEVY and W. LIU
Department of Chemistry, Boston University
Boston, MA 02215, U.S.A.

S. GILBERT
Institute of Physical Chemistry, Swiss Federal Institute of Technology
Ecublens, CH-1015 Lausanne, Switzerland

Abstract

An electrochemical photovoltaic cell utilizing a nanoporous dye sensitized TiO_2 electrode has been described displaying higher efficiency for current collection when exposure is through the SnO_2(F-doped) conducting glass support for the TiO_2 as compared to exposure from the reverse side [1]. Using the same electrode components, time resolved photocharge, TRPC, measurements are used to ascertain the vectorial directions of transient photocurrents, with and without dye sensitization of the nanoporous TiO_2 for coatings on glass and on transparent conducting glass.

The use of UV and visible laser light pulses, with and without CW UV-Visible illumination, make possible the evaluation of the presence and polarity of space charge fields. TiO_2 contact with a transparent conducting layer and an adsorbed sensitizing dye, upon equilibration of Fermi levels of the individual phases, spontaneously creates a field across the TiO_2 which aids in directing photoelectron currents toward the SnO_2(F-doped) electrode, regardless of sample orientation with respect to the incident light - explaining observed TRPC polarity reversals. When exposure is through the conducting glass, the proximity of the region of greatest light absorption to the electrode enhances photoelectron collection by the SnO_2(F-doped) coated glass.

Energy band structures are proposed for the various structures investigated, which are consistent with the TRPC results, and with other relevant information available in the literature. For the SnO_2(F-doped)-TiO_2(nanoporous)-$Ru(II)L_2(CNS)_2$ system, a combined *n-i-p/quantum well* band structure is proposed. The *n-i-p* segment is considered to consist of SnO_2(F-doped) in contact with bulk property TiO_2, to which dye is adsorbed on the available surface not in contact with the SnO_2. The *quantum wells* are comprised of nanoporous TiO_2 aggregates surrounded by adsorbed $Ru(II)L_2(CNS)_2$ aggregates.

E. Pelizzetti (ed.), Fine Particles Science and Technology, 343–370.

1. Introduction

Porous, nanocrystalline, semiconductor thin films are rapidly coming into the forefront as a new class of materials for catalysis and photocatalysis, due to their unusual optical and electronic properties. In this connection, the past decade has witnessed the development of several new types of electrochemical and chemico-electronic devices based on colloidal transition metal oxide films, such as the "Grätzel" liquid junction photovoltaic cell [1], electrochemical electrochromic devices [2], and solid state sensors [3]. The afore-mentioned examples are distinguished from their compact-film counterparts by their superior performance. The properties afforded these devices are derived primarily from the enormous surface area of the nanocrystalline films - up by a factor of ~1000 or more from compact thin films. The electronic properties of these materials are largely derived from the enhanced surface area, as the role of localized states in electronic conductivity is greatly exaggerated in comparison to compact films. In addition, because of the extremely small size of the individual particles that comprise the film structure (typically 1-50 nm), some size quantization effects are emphasized, particularly confinement of photogenerated charge carriers in the particles [4,5]. Dielectric confinement effects are also displayed under frequently encountered conditions, which control the regions of highest light absorption within individual particles or aggregates and concomitantly the vectorial directions of photo-induced charge carrier transport [6,7].

These materials can be described as a contiguous network of lightly sintered colloidal particles and aggregates, generally with primary particle size 1-10 nm in diameter, accompanied by a substantial void volume (50%), giving rise to the extremely high surface to volume ratio, and porosity. Concerning the latter, the pores are interconnected, with pore diameters on the order of several nanometers. The films are usually structurally sound, and are attached to conducting substrates. Generally, degenerately doped SnO_2 films on glass sheets are used if transparent electrodes are desired. In the case of transition metal oxide colloidal films, the materials are not intentionally doped, and exhibit very large resistivities at normal temperatures (~10^{12}W-cm) in air [8], although measured resistivities in vacuum of 10^9W-cm have been reported [9]. Through-thickness resistances for 10μm films are on the order of 10^9W for contact areas of 1 cm^2. Superficially, the films appear to be dielectrics, but in fact display unusual transient doping properties related to the high trap density that they typically possess.

Due to the complex structure of nanocrystalline oxide films, attempts to model underlying conduction mechanisms based on a detailed knowledge of the internal electric field structure have been impeded. It is not possible to use standard models based on compact materials. However, this difficulty has been circumvented to some extent by the advancement of simple models that explain experimentally observed results with relatively good success. Specifically, transient electrochemically derived photocurrents have been simulated using a trap-filling and -emptying model based on Shockley-Read-Hall kinetics [10]. It was shown in this model that deep traps can control the rise and decay time constants of the transients, and that a wide distribution of traps was needed in the simulations to adequately mimic the curves. Convincing experimental evidence indicating that traps are operative in colloidal films include measurements of slow non-exponential photocurrent decay which depends on the applied reverse bias [9]. Several other related studies of note are cited here [11-13].

The electric field distribution in colloidal oxide films is not known. However, calculations show that undoped individual particles are smaller than the Debye screening length known for the bulk material, thus precluding intra-particle band-bending [1]. In addition, the drift length of electrons in solid state diode devices under reverse bias have been estimated to be on the order of 100 nm (particle sizes in these films are typically 20 nm) [9]. Thus, it is apparently clusters of particles that support the electric fields for which the present investigation provides evidence. Interconnection of these clusters forms a microporous bulk structure that permits penetration of electric fields to the extent that they are not screened internally.

The nature of the SnO_2/TiO_2 junction has not been carefully investigated, particularly for the formation of contact potentials. In one report, it is claimed that the junction is ohmic [14]. In the present study, we present results which experimentally demonstrate the existence of a built-in contact potential, in the form of downward band-bending from the TiO_2 to the SnO_2, at the SnO_2/TiO_2 junction. When dye is added to this system, to form the SnO_2/TiO_2(nanoporous)/Dye device, it is considered to be fall within the *quantum well* class of devices discussed by Bube, if one includes a self-biasing feature common in photovoltaic cells [15]. Here, time resolved photocharge measurements, TRPC [16], are used for the first time to detect the presence of such a built in field, and shed some insight into the electric field distribution in nanoporous TiO_2 films.

2. Experimental

2.1. PREPARATION OF NANOPOROUS TiO_2 COATINGS

Thin films of colloidal TiO_2 (anatase) were prepared by spreading a suspension of nanosized-colloidal TiO_2 to a thickness of 10 microns on glass sheets, with and without a coating of fluorine doped SnO_2. The colloidal suspension was prepared following the procedure of O'Reagan and Grätzel [17], but autoclaved at 250°C to allow aggregates of ~100 nm diameter to form. This procedure permits film thickness greater than 5 μm. The wet films comprised of the 100 nm aggregates were then lightly sintered in air at 450°C for 15 minutes, forming films of ~8 μm thickness for the samples of the present investigation.

Several of the films were further coated with an efficient sensitizing dye used in the Grätzel solar cell, namely the cis-$(SCN)_2$bis(2,2'-bipyridyl-4,4'-dicarboxylate) Ru(II) complex, $Ru(II)L_2(SCN)_2$ [1]. To this end, the sintered films are soaked in a $3x10^{-4}$ M ethanolic solution of the complex for 1 hr. at 60°C. TRPC measurements were performed on dry samples from which solvent has been allowed to evaporate.

2.2. TIME RESOLVED PHOTOCHARGE (TRPC) MEASUREMENTS

The basis for the TRPC measurement, and an equivalent circuit analysis has been presented [16], including a description of the calibration of the apparatus for purposes of evaluation of the mean ambipolar charge carrier separation distance, Lm. A brief description is repeated here, along with a schematic diagram of the apparatus in Fig. 1.

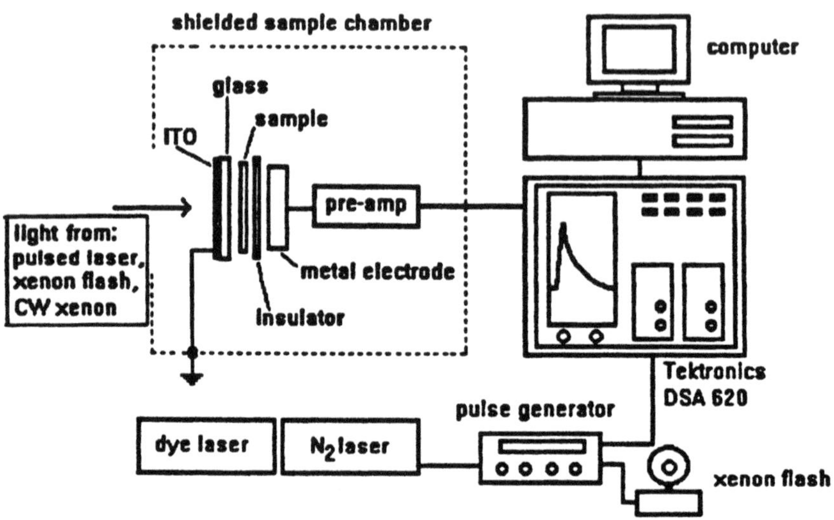

Figure 1. Schematic diagram the time resolved photocharge apparatus (TRPC).

For fixed light intensity and constant substrate thickness, both the equivalent circuit analysis , and the principle of measurement of charge carrier transport delineated by Tiedje for Time of Flight studies [18], lead to the conclusion that,

$$Lm = C\,|Vm|, \quad (1)$$

where |Vm| is the magnitude of the maximum of the TRPC signal amplitude. The constant, C, contains experimentally accessible terms, or terms that can be reasonably estimated, such as the sample dielectric constant, sample thickness and sample cell calibration parameters, and includes the number of absorbed light quanta. Thus |Vm| may be considered to be a photo-moment whose magnitude is dependent on the number of photo-induced charge carriers and their mean separation distance, Lm. From a pragmatic standpoint, the relative values of Lm or |Vm|, may be used to evaluate trends resulting from variations in material parameters and configuration. The mean velocity of charge carrier separation, v, is given by,

$$v = Lm/tm = C\,|Vm|/tm \qquad (2)$$

where tm is the time required to attain |Vm|. The sign of the TRPC response is the resultant of the vectorial photocurrents in the direction perpendicular to the blocking sample cell capacitor parallel plate electrodes. In the absence of space charge fields, the signal depends on inhomogeneously absorbed light and on a difference in charge carrier velocity for the photoelectrons and photoholes. Dielectric confinement effects can influence the location of the region of highest light absorption and thereby the sign of the TRPC response [6, 7].

Depletion layers, inversion layers and surface contact potentials also influence the magnitude and direction of the photo-induced currents [19, 20]. For isolated particles of sufficiently small diameter, surface band-bending is not expected [19, 20]. However, if the small particles are aggregated, or sintered as in the present investigation, the effective particle size may be in a size regime where band-bending can influence charge carrier transport. Experiments indicate that continuous high intensity exposure collapses surface space charge layers, if present, making it possible to obtain the intrinsic transport parameters characteristic of the photoconducting sample [20]. This is demonstrated in Fig. 2 for a 200 μm diameter TiO_2 sample - which displays upward band-bending; in Fig. 3 for a 20 nm diameter sample of Degussa P25 TiO_2 - which has little or no band-bending; and in Fig. 4 for an aggregated P25 sample - which displays upward band-bending.

It is also possible to study interfacial effects due to adsorbed or epitaxial deposits, and electrode contacts, by performing experiments with and without CW high intensity exposure, where the orientation of the sample is reversed with respect to the surface facing the light source [22]. The principle of such measurements is provided in Figs. 5-8 for the situations of strong and weak light absorption and for monolayer and multi-layer dye deposits. These techniques are applied extensively in the present investigation.

Thin film photovoltaic devices often utilize *p-i-n* heterojunction structures. The field formed across the intrinsic i-layer as a result of band gap mismatch and Fermi level equilibration, aids in charge carrier separation and minimizes recombination, as seen in Fig. 9 [23]. Similar considerations have been applied to Dye/Ag_2S/AgBr/Dye photographic transistor systems, as seen in Fig. 10 [15, 21a], and to Pt/TiO_2/Dye photocatalytic transistors, as seen in Fig. 11 [21b,c,d]. In the present investigation, TRPC measurements are used to test the premise that the Dye-nanoporous TiO_2-SnO_2 (F-doped) electrode is also analogous to a *p-i-n* electronic device, but also having *quantum well* device features.

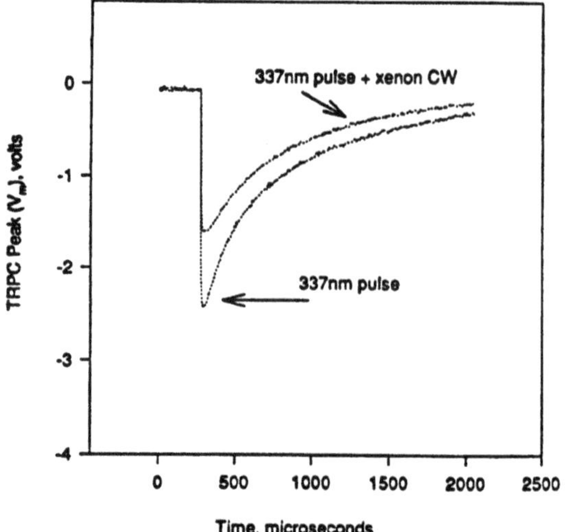

Figure 2. Determination of direction of band bending in TiO_2 pellet prepared with 200 nm mean particle size rutile. TRPC waveforms on the same sample are performed with: a) 337 nm laser probe pulse and b) 337 nm probe pulse + CW 200 watt xenon illumination. CW excitation decreases Vm - suggesting upward band bending, in agreement with interpretation of photocatalytic results by M. Anpo *et. al* for this sample of TiO_2. (see (1991) *Bull. Chem. Soc. Jpn.* **64**, 543.)

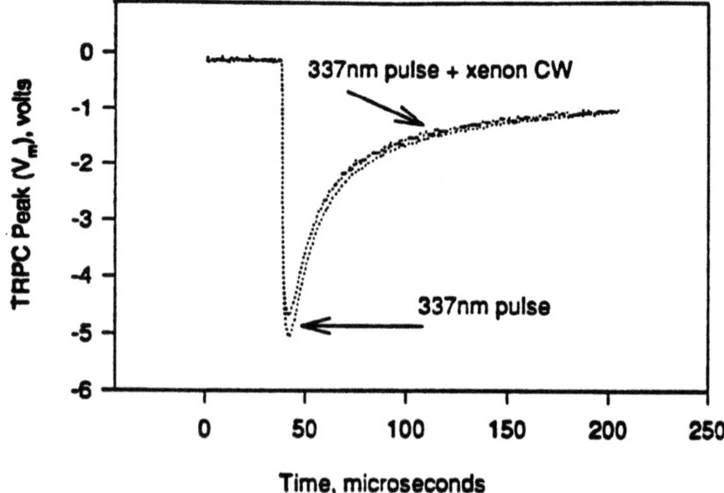

Figure 3. Determination of direction of band bending in TiO_2 pellet prepared with Degussa P25 - mean particle size 20 nm. The closely similar TRPC waveforms, with and without CW xenon illumination, indicate that the 20 nm particle size regime is too small to support substantial band bending.

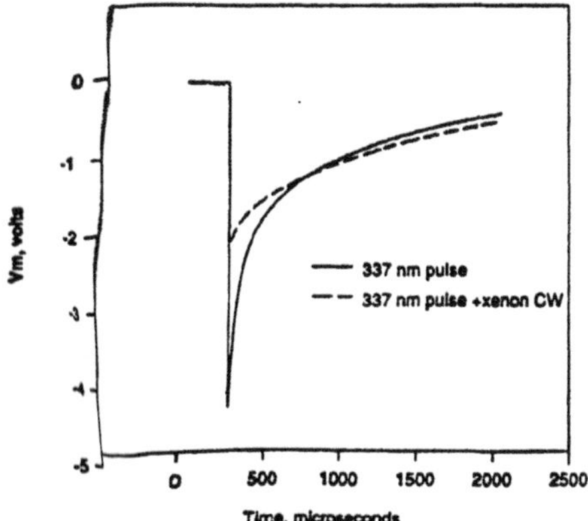

Figure 4. Determination of direction of band bending in TiO_2, Degussa P25, pre-treated with i-C_3H_7OH, filtered and dryed at 60°C for 3 hours, prior to TRPC measurements. Particle aggregation due to treatment with i-propanol leads to larger light absorbing unit which can support band bending.

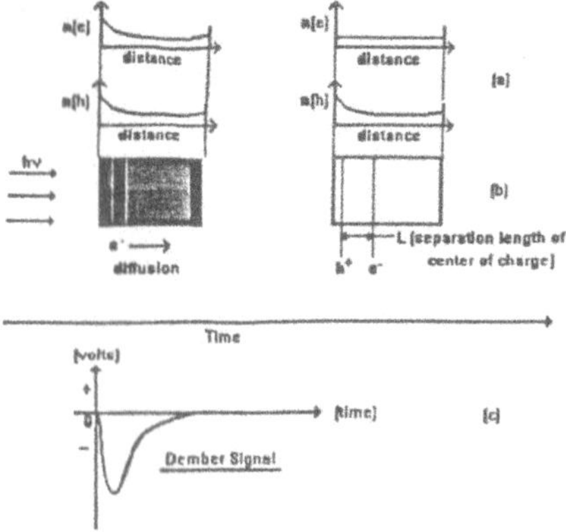

Figure 5. Front surface light absorption by photoconducting solid in the absence of surface space charge. Photoelectron diffusion velocity >> diffusion velocity of photoholes. a) concentration of photoelectrons, e^-, and photoholes, h^+ *vs* distance from the front surface, immediately after light absorption, t_0, and after time, t, necessary for the concentration gradient of photoelectrons to equalize by diffusion, b) similar to (a) with the concentration of e^- indicated by gray scale density. Separation length of the center of charge of the photoelectrons from that of the photoholes, L, referring to time, t. c) time resolved Dember signal for strongly absorbed light, using apparatus described in Fig. 1. Under these conditions, when the net initial transport of charge from the front to back surface is negative, the polarity of the waveform is negative, as shown (see [6, 22]).

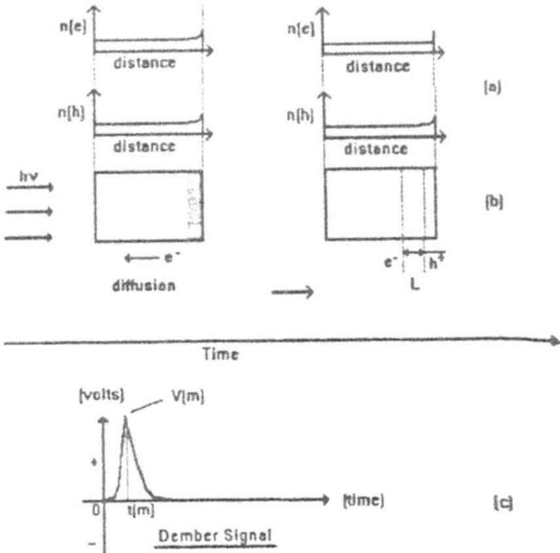

Figure 6. Photo-induced charge carrier distribution in a thin film or a small particle, for a weakly absorbed light pulse: a) concentration *vs.* distance; b) separation length, L. Light is uniformly absorbed, except at the back surface, where the incident and reflected light are reinforced due to the dielectric confinement effect; c) Dember signal is of positive polarity - opposite to that for strongly absorbed light of *Figure* 6. (see [6, 22])

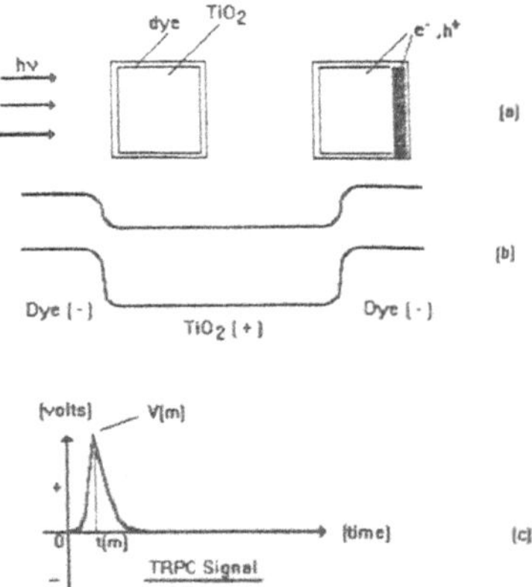

Figure 7. Photo-induced charge carrier distribution in a spectrally sensitized small particle for a weakly absorbed light pulse: a) concentration *vs.* distance; b) energy band diagram of dye sensitized AgX after Fermi level equilibration in the dark, showing a space charge at the dye/AgX interfaces; c) TRPC waveform for a spectrally sensitized AgX grain. Polarity is positive. (see [22])

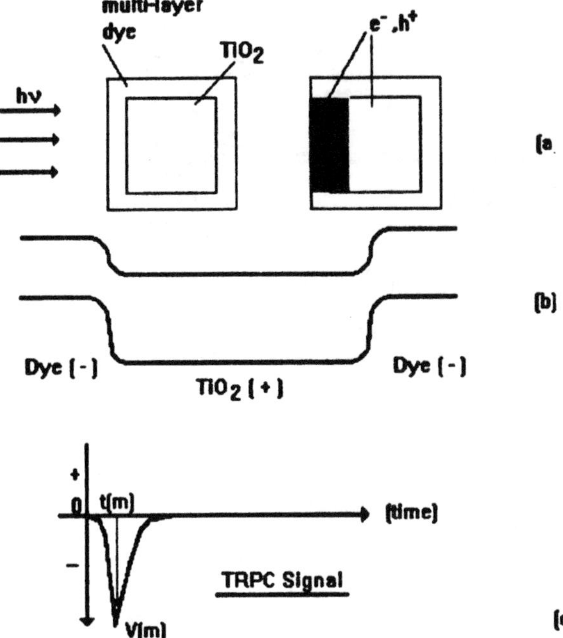

Figure 8. Same as *Figure 7.*, except that we now have multi-layer dye adsorption on TiO_2 and light is strongly absorbed at the front surface of the aggregates. This leads to a negative polarity TRPC signal.

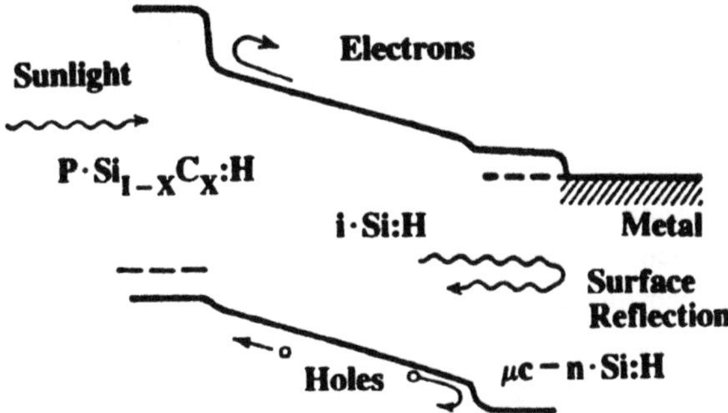

Figure 9. Energy band diagram for a thin film *p-i-n* amorphous silicon solar cell. Mismatch in band gaps and Fermi levels of each component creates a field across the i-Si:H layer, which directs photoelectrons to the μc-n-Si:H and holes to the carbon doped Si:H.(see ref. 23)

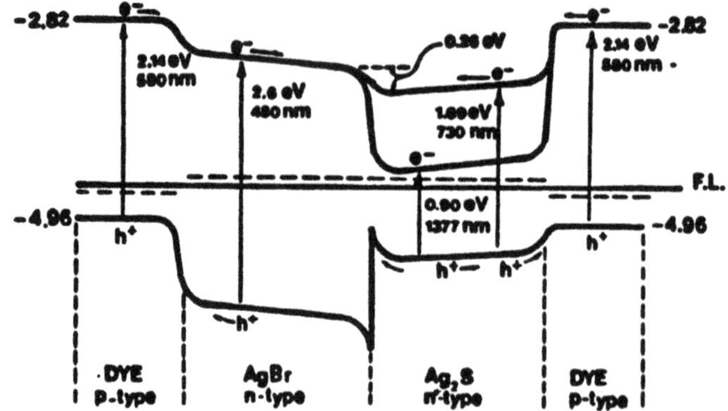

Figure 10. Energy band diagram of AgBr-Ag_2S-dye photograpic transistor. Basis for construction of the diagram is given in reference [21].

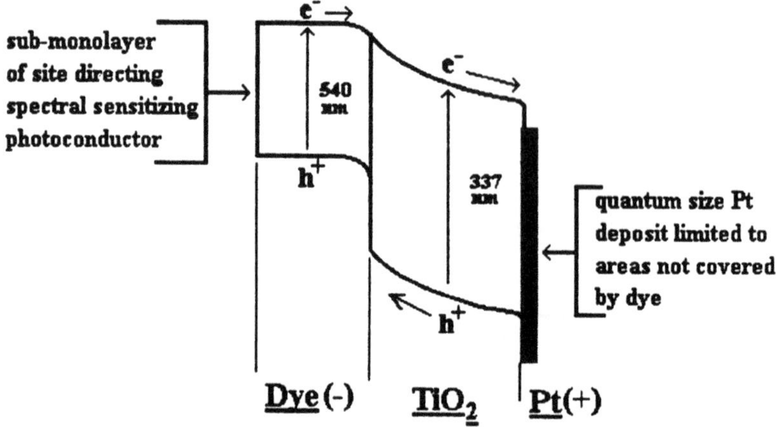

Spontaneously formed field across the TiO2 directs photoelectrons to the quantum sized Pt deposits and photoholes to the Dye surface, where reduction and oxidative reactions with adsorbed reactants are initiated, respectively.

Figure 11. Energy band diagram of a Dye/TiO_2/Pt photocatalytic heterojunction, as described in [21].

3. Results

TRPC transient waveforms are presented in Figs. 12-19, for samples consisting of nanoporous TiO_2 in various coating configurations. Single 10 ns pulse duration laser excitation at 337 nm and 540 nm were employed using a Molectron Model UV-24 Nitrogen Laser to pump a Molectron DL-II Tunable Dye Laser, with and without concurrent CW xenon illumination, and as a function of sample orientation in respect to the surface facing the incident light. TiO_2 does not absorb 540 nm radiation, while both the dye used, $Ru(II)L_2(SCN)_2$, and the TiO_2 absorb 337 nm light. The effect of sample orientation was investigated for each coating structure and exposure condition.

Table 1. presents a summary of the maximum TRPC signal amplitudes, Vm, corresponding to the transient waveforms of Figs. 12-19. Of special note in Figs. 12-19 and in Table 1. is the dependence of signal polarity on sample orientation with respect to the direction of the laser probe pulse and the CW illumination. ***Sign reversals are observed with samples coated on the transparent SnO_2(F-doped) conducting electrode, which are not observed when the corresponding materials are coated on glass.*** Information derived from these transient waveforms are used in formulating the proposed energy band diagrams, Figs. 21, 22, 24, and 25, discussed below. Since the TiO_2 is porous, the dye is adsorbed on the outer surface and also within the interior pore structure, but is considered to be partially shielded from adsorption directly on the SnO_2(F-doped) due to the increased TiO_2 density at the SnO_2 (F - doped)/TiO_2 interface, as observed in electron micrograpghs [1]. In each of these proposed energy band diagrams, the dashed lines represent shifts in energy band levels resulting from collapse of space charge layers during CW xenon illumination.

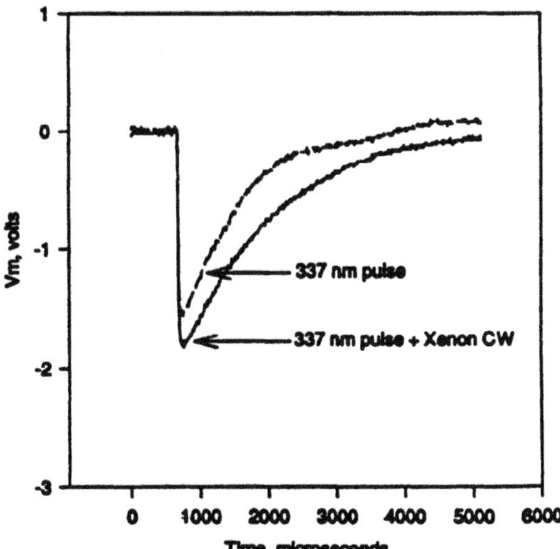

Figure 12. TRPC waveforms for nanoporous TiO_2 coated on glass, *i.e.*- TiO_2/glass. Excitation with: a) 337 nm laser pulse; b) 337 nm laser pulse + CW xenon lamp. TiO_2 surface is facing incident light.

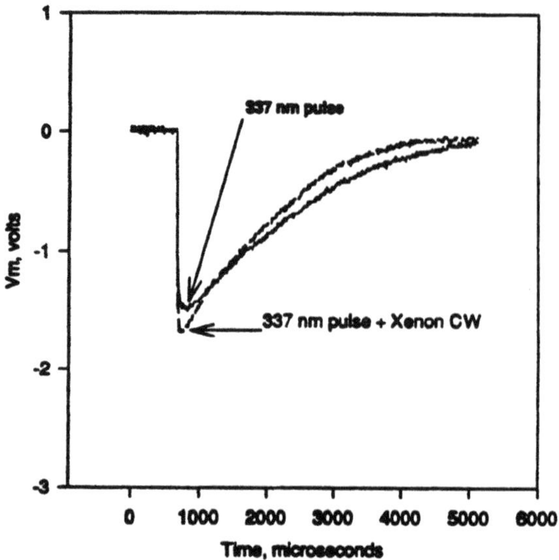

Figure 13. TRPC waveforms for nanoporous TiO_2 coated on transparent conducting glass, *i.e.*- TiO_2/SiO_2(F-doped)/Glass. Excitation with: a) 337 nm laser pulse; b) 337 nm laser pulse + CW xenon lamp. TiO_2 surface is facing incident light.

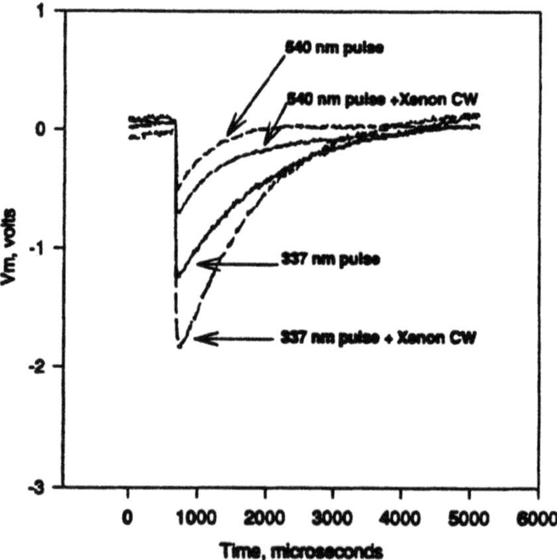

Figure 14. TRPC waveforms for dye sensitized nanoporous TiO_2 coated on glass, *i.e.*- $[\{Ru(II)(L)_2(CNS)_2\}_y/\{TiO_2\}_x]_n$/Glass. Excitation with: a) 337 nm laser pulse; b) 337 nm laser pulse + CW xenon lamp. Dye surface is facing incident light.

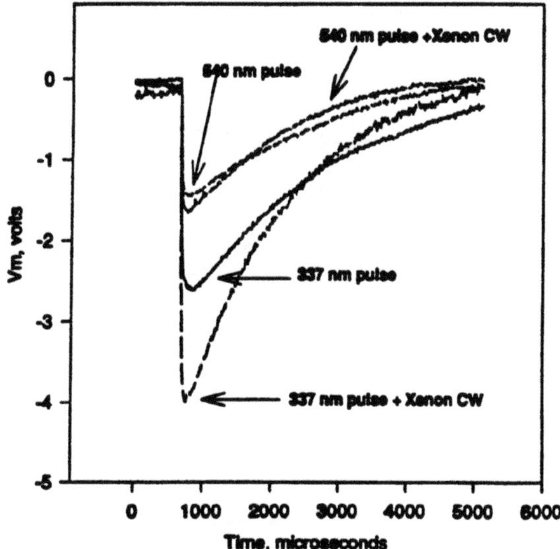

Figure 15. TRPC waveforms for dye sensitized nanoporous TiO_2 coated on conducting glass, *i.e.*-$[\{Ru(II)(L)_2(CNS)_2\}_y/\{TiO_2\}_x]_n/SiO_2$(F-doped)/Glass. Excitation with: a) 337 nm laser pulse; b) 337 nm laser pulse + CW xenon lamp. Dye surface is facing incident light.

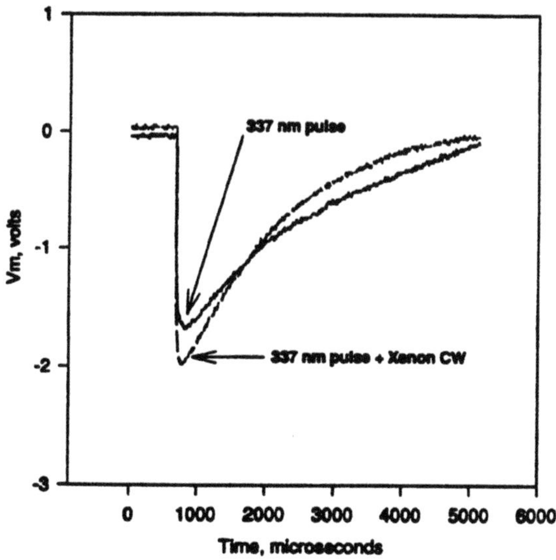

Figure 16. TRPC waveforms for nanoporous TiO_2 coated on glass, *i.e.*- Glass/TiO_2. Excitation with: a) 337 nm laser pulse; b) 337 nm laser pulse + CW xenon lamp. Glass surface is facing incident light.

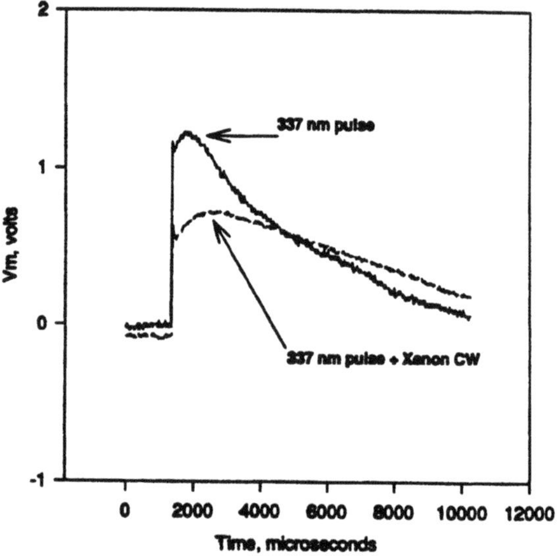

Figure 17. TRPC waveforms for nanoporous TiO_2 coated on transparent conducting glass, *i.e.*- Glass/SnO_2 (F-doped)/TiO_2. Excitation with: a) 337 nm laser pulse; b) 337 nm laser pulse + CW xenon lamp. Glass surface is facing incident light.

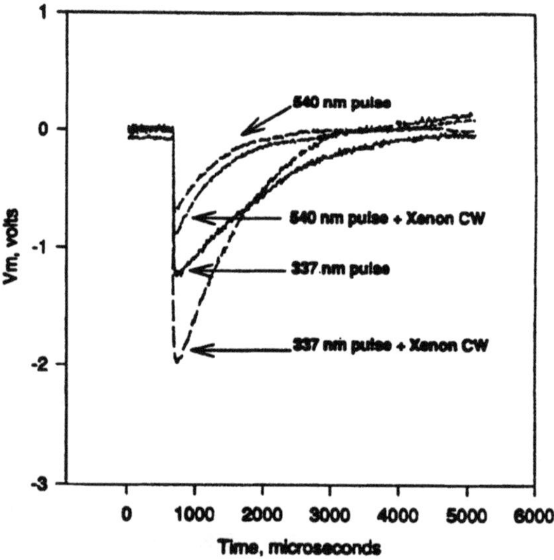

Figure 18. TRPC waveforms for dye sensitized nanoporous TiO_2 coated on glass, *i.e.* Glass/$[\{TiO_2\}_x/\{Ru(II)(L)_2(CNS)_2\}]_n$ Excitation with: a) 337 nm laser pulse; b) 337 nm laser pulse + CW xenon lamp. Glass surface is facing incident light.

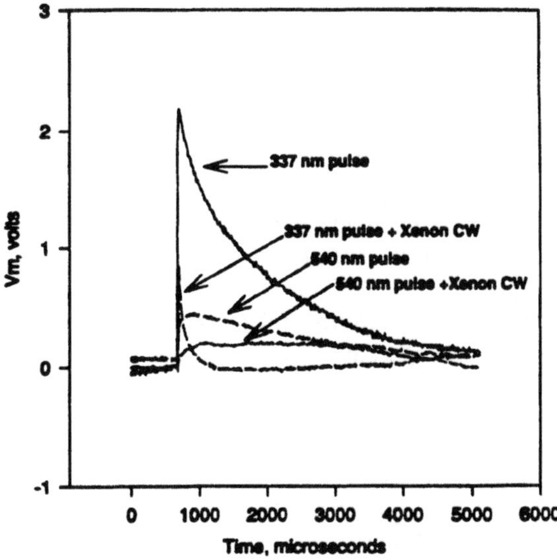

Figure 19. TRPC waveforms for dye sensitized nanoporous TiO_2 coated on transparent conducting glass, *i.e.*- Glass/SiO_2(F-doped)/[{Ru(II)$(L)_2(CNS)_2$}$_y$/{TiO_2}$_x$]$_n$. Excitation with: a) 337 nm laser pulse; b) 337 nm laser pulse + CW xenon lamp. Glass surface is facing incident light.

TABLE 1. Time resolved photocharge amplitudes, Vm, using 337 nm and 540 nm pulse excitation, with and without CW illumination, and with different sample orientation with respect to the direction of the incident light

[b]SAMPLE/ LIGHT SOURCE (exposure - left to right)	[a]Vm, mv 337 nm	337 nm + CW	540 nm	540 nm + CW
#1 TiO_2/Glass	-1.64	-1.85	0	0
#2 TiO_2/SnO_2(F)	-1.45	-1.59	0	0
#3 Ru(II)(L)2$(CNS)_2/TiO_2$/Glass	-1.38	-1.84	-0.6	-0.784
#4 Ru(II)(L)2$(CNS)_2/TiO_2$ /SnO_2(F)	-2.64	-3.92	-1.4	-1.6
#5 Glass/TiO_2	-1.78	-2.09	0	0
#6 SnO_2/TiO_2	1.43	0.96	0	0
#7 Glass/TiO_2/ Ru(II)(L)2$(CNS)_2$	-1.28	-2.02	-0.648	-0.92
#8 SnO_2(F)/TiO_2/ Ru(II)(L)2$(CNS)_2$	2.24	0.824	0.408	0.26

[a]Signal magnitudes before amplification of 1000X.
[b]The sample groups #(1-4) and #(5-8) are the same four samples exposed from opposing sides.

4. Discussion

4.1. NANOPOROUS ELECTROCHEMICAL SYSTEMS

The currently accepted paradigm concerning photoinjected charge transport in nanocrystalline colloidal TiO_2 films *in electrochemical systems* is that photoelectrons generated far from the transparent conducting contact are transported primarily by diffusion within the bulk of the colloidal network. This is because electrolyte contact is assumed to be so extensive that virtually all particles are depleted, and no potential gradient exists [12, 13].

Supporting this view are electron micrographs indicating less dense packing toward the top of the film, suggesting that the greatest amount of exposed surface is found here [1]. It appears that electrolyte can contact virtually all of the small aggregates and particles in the upper reaches of the film, thus depleting the top of the film. But less contact would be made toward the bottom, so depletion due to electrolyte or adsorbate contact could be less effective, permitting a one-dimensional depletion layer to form at the TiO_2/SnO_2 junction. The porosity of the film and verification of electrolyte contact at the SnO_2 layer has been established by comparison of voltammograms for the reduction of I^{3-} on a sputtered SnO_2 film electrode (used as substrate) and on nanoporous TiO_2 coated on SnO_2. The results are very similar, the only difference being that the TiO_2 acts as a blocking film, reducing the current, but surprisingly not by much, perhaps a factor of 2 [24]. The higher TiO_2 packing density at the SnO_2/TiO_2 interface would tend to act more like bulk material, considered capable of supporting a mesoscopic depletion layer.

4.2. NANOPOROUS SOLID STATE SYSTEMS

In solid state devices based on colloidal films, as might apply to the samples of our present investigation, such massive depletion would not necessarily take place, and it is possible that a potential gradient could exist throughout the nanoporous TiO_2 layer. In earlier solid state investigations, photoelectrons were collected at the conducting substrate by positively biasing this contact, always assuming that the SnO_2/TiO_2 junction forms an ohmic contact. Higher collection yields were attained concomitant with higher positive biases [9,11].

4.2.1. *Non-Dye Sensitized Nanoporous TiO_2*

337nm Pulse Excitation. Since the vectorial direction of the photoelectron current is in the same direction as the incident light for the TRPC results presented in Figs. 12, 13 and 16- and Table 1., #1, #2 and #5, in the absence of electrolyte and without an externally applied field, it is evident that the 337 nm laser pulse is most strongly absorbed at the TiO_2 surface facing the incoming direction of the incident light. Compelling evidence points to the presence of a built-in contact potential at the SnO_2/TiO_2 junction, engendering downward band-bending from the TiO_2 to the SnO_2. This conclusion is based on the TRPC polarity reversal when exposure is through the transparent SnO_2 coating, as shown in Figs. 13 and 18 and Table 1., for samples #2 and #6, using a 337 nm pulse excitation. Thus regardless of sample orientation, the photoelectron current is always directed toward the SnO_2 coating. In #1 and #5 from Table 1., where the TiO_2 is in contact with glass, the polarity of Vm is negative regardless of the orientation of the sample with respect to the direction of incident light.. This type of response is expected if one considers a simple energy band diagram for a junction between two wide band gap semiconductors, one degenerately doped, and the other virtually intrinsic, as shown in Fig. 20 for bulk materials. In fact, a downward band-bending scheme at the SnO_2/TiO_2 junction of a photoelectrochemical cell was already proposed to explain the vectorial charge flow in colloidal TiO_2photoelectrodes, where it was proposed to act as a sink for diffusing photoelectrons, thus sustaining the concentration gradient [12]. However, it was assumed that this field was produced by positive bias at the SnO_2 contact.

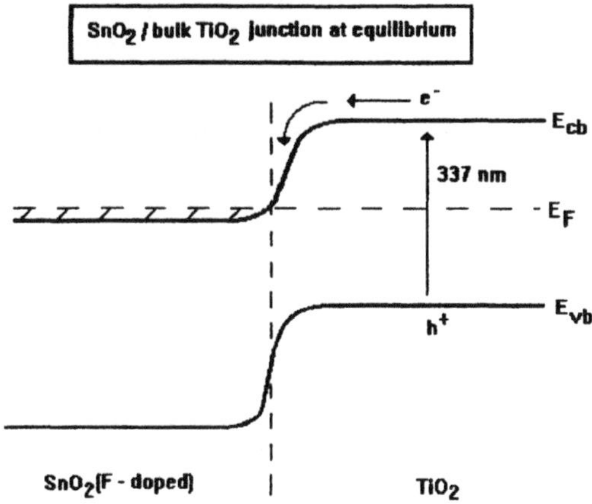

Figure 20. Energy band diagram for bulk TiO_2 coated on transparent tin oxide degenerately doped with fluorine, SnO_2 (F-doped).

The spontaneously formed field at the SnO_2/TiO_2 junction certainly does not appear to aid the separation of electron-holes pairs when exposure is directed on the nanoporous TiO_2 - not through the Glass/SnO_2 interface first, since the magnitudes of the photocharge responses, both in the presence and absence of SnO_2, are about the same, as seen in by comparison of Vm for #1 and #2 in Table 1. When exposure is through the SnO_2 transparent electrode, the situation is very different, as seen in Table 1., #6. Here, the proximity of the highest concentration of photo-induced charge carriers in the TiO_2 to the positive contact potential induced in the SnO_2(F-doped) layer, results in the net vectorial flow of photoelectrons in a direction which opposes the diffusion current. Diffusion dominates the transient photoelectron current of sample #5, consisting of TiO_2 coated on glass, where the drift component is absent. The field exerted by the SnO_2(F-doped) electrode appears to be effective over relatively short distances, i.e. << 8 μm - the thickness of the TiO_2 layer, as seen by comparison of #2 and #6 of Table 1., for two phase systems, *i.e.* TiO_2 and SnO_2.

337 nm Pulse Excitation + CW Illumination. The effect of CW illumination shown by comparison of samples #1, #2 and #3 of Table 1., is to increase Vm. This is interpreted as indicative of upward band-bending at the interfaces between TiO_2 particle aggregates, as is typical for n-type interparticle energy barriers [25]. For #6, the situation is different - CW illumination decreases Vm, implying downward band-bending from the TiO_2 to the SnO_2. Energy band diagrams believed to be consistent with these observations for nanoporous TiO_2 coated on glass and on SnO_2, are presented in Figs. 21 and 22, respectively.

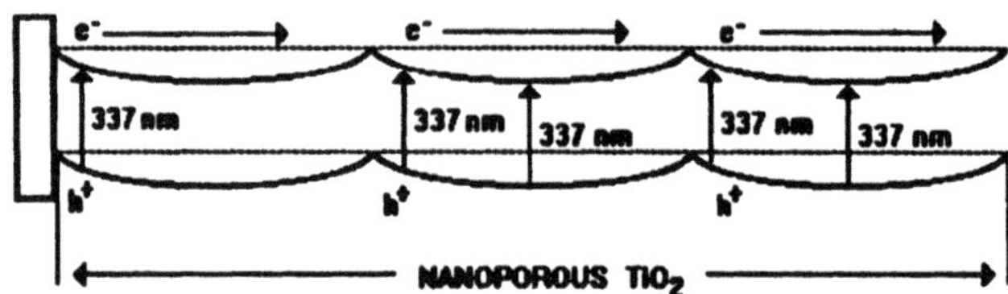

Figure 21. Proposed energy band diagram for nanoporous TiO_2 coated on glass.

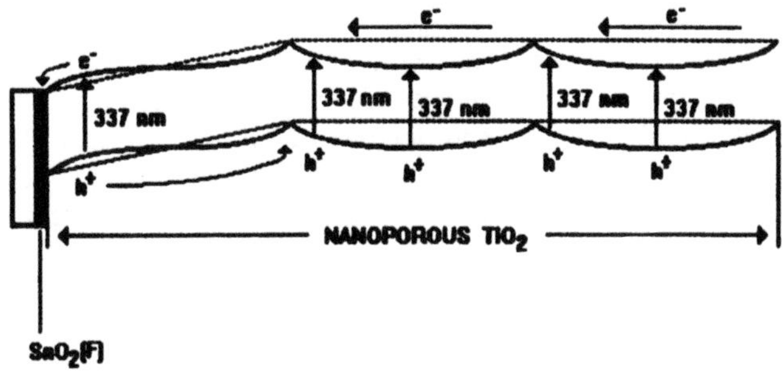

Figure 22. Proposed energy band diagram for nanoporous TiO_2 coated on transparent conducting glass.

4.2.2. *Dye Sensitized Nanoporous TiO_2 Systems*

337 nm Pulse Excitation. The presence of dye greatly enhances the 337 nm photocharge signals obtained with nanoporous TiO_2 coated on the SnO_2(F-doped) conducting glass substrate, but decreases the 337 nm signal when the substrate is ordinary glass, as seen in Table 1. by comparison of #2 with #4, #6 with #8, #1 with #3 and #5 with #7. It is known that the dye can react rapidly with holes, perhaps acting as a recombination center. On the other hand, dye can act as a hole trap, permitting electrons to migrate away from these trap centers, toward the SnO_2 layer, under the influence of the electric field. Hence the stronger signals with the SnO_2-TiO_2-Dye system compared to non-spectrally sensitized systems and the diminished signals when dye sensitized TiO_2 is in contact with insulating glass.

In addition, the "amplified" photoelectron current is always toward the SnO_2, regardless of sample orientation with respect to the incident light, implying a long range attractive force found only with the dye sensitized sample coated on SnO_2(F-doped). The 8 μm thick TiO_2 coating appears to absorb a constant number of the incident 337 nm quanta, independent of the presence of dye, which also absorbs at 337 nm [1]. This is supported by the very closely matched values of Vm for dye sensitized and unsensitized samples coated on ordinary glass, using 337 nm pulse excitation and 337 nm pulse + CW illumination, as shown in Table 1.. Remembering also that the magnitude of the TRPC signal is proportional to the number of charge carriers and the mean separation distance of photoelectrons from photoholes, the increased signal with the dye sensitized TiO_2 coated on SnO_2, is strong evidence of an increase in mean charge carrier separation distance, Lm, due to a field across segments of the TiO_2 layer that are bounded by SnO_2(F-doped) on one side and the adsorbed dye layer on the other side, as represented in the energy band diagram of Fig. 23, for bulk TiO_2. However, with the nanoporous TiO_2, there may be segments within the coating thickness, at some distance from the SnO_2 layer, where TiO_2 aggregates are more or less symmetrically surrounded by adsorbed dye.. Since the TiO_2 is sintered, with dye adsorbed over most or all of the available surface, the field across the TiO_2 may extend for a considerable distance in the TiO_2 layer, as appears to be the case when one considers the large enhancement of Vm when dye and SnO_2 are employed - compare #2 with #4 and #6 with #8 of Table 1.. A proposed energy band diagram is given in Fig. 24, combining features of a thin film *p-i-n* device [23] with *quantum well devices* discussed by Bube [15].

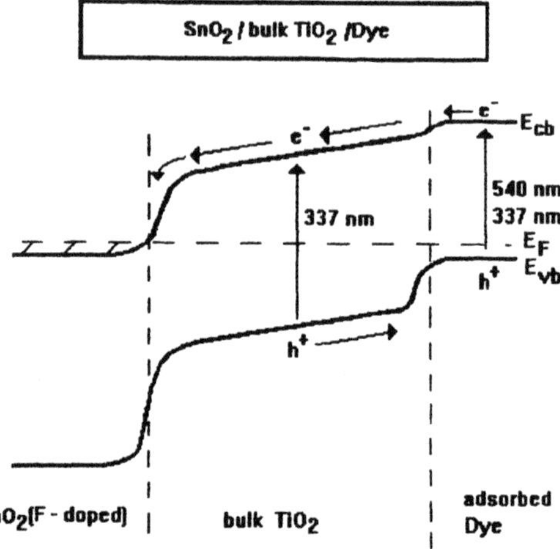

Figure 23. Energy band diagram for dye sensitized bulk TiO_2 coated on SnO_2 (F-doped).

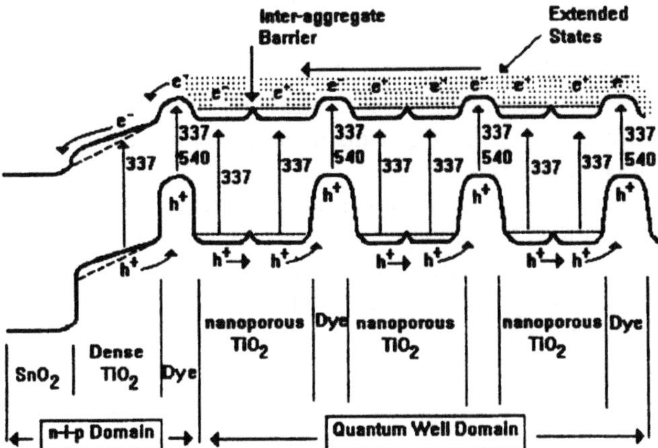

Figure 24. Proposed energy band diagram for dye sensitized nanoporous TiO_2 coated on transparent conducting glass.

337 nm Pulse + CW Illumination. With reference to the energy band diagram of Fig. 24, in the region where dye aggregates are symmetrically adsorbed on the TiO_2 aggregates, there is also contact with adjoining sintered TiO_2 aggregates in all directions - except where a pore is encountered. If the interparticle barriers are lowered sufficiently by CW illumination and if traps are already filled and recombination is not overly pervasive, this will allow for photoelectrons generated within the TiO_2 and for those injected by the dye into the TiO_2, to be transported toward the SnO_2 via the TiO_2 conduction band. However, the pore volume has been estimated to be about 50% and dye aggregates could fill these voids forming the equivalent of a *quantum well* structure. If during CW illumination, electrons are unable to escape from some or all of the TiO_2 *quantum wells*, they will be forced, under favorable conditions, to populate *extended states* above the barriers at the TiO_2/(adsorbed dye aggregate) interfaces. By virtue of having injected electrons into the TiO_2 *quantum wells*, the dye aggregates acquire a positive charge. The net effect of this confinement is the population of extended energy states in the TiO_2 and the depopulation of excited levels within the dye aggregate manifold. When degenerately doped SnO_2 is included in the structure, the TRPC measurements indicate that a field is exerted which extends into the *quantum well* domain of the structure. Electrons may be excited by > band gap energy directly from the TiO_2 valence band into these *extended states* or from heavily populated *quantum wells* using lower energies as well. Quantum size effects aid in this process, *i.e.* band gap broadening in the TiO_2 and band gap narrowing in the adsorbed dye aggregates, by creating a favorable condition for electron migration in *extended states* above the barriers [15]. The photoholes need not be transported over distances as large as the photoelectrons in the photoelectrochemical cell, since one of the roles of the electrolyte is to diffuse to the hole originating in the dye or to a trapped hole, where it is oxidized and the hole neutralized.

When dye sensitized TiO_2 is coated on glass, the energy band diagrams of Figs. 25 and 26 are considered, with Fig. 26 representative of densely sintered TiO_2 and Fig. 25 of nanoporous TiO_2. That the energy band structure of Fig. 26 does not apply to the situation of nanoporous TiO_2, is seen immediately from the negative polarity obtained with #3 and # 7 of Table 1 for 540 nm excitation as discussed below.

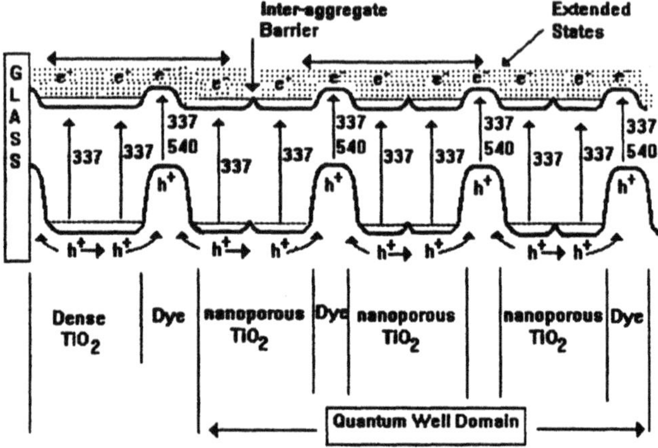

Figure 25. Proposed energy band diagram for dye sensitized nanoporous TiO_2 coated on glass.

337 nm Pulse Excitation + CW Illumination. As is seen in Table 1., excitation with 337 nm pulse + CW illumination of the dye sensitized sample coated on the glass substrate gives negative signals, regardless of sample orientation. Vm becomes more negative compared to signals obtained with 337 nm pulse excitation. This suggests that barriers created by upward band-bending are diminished by the CW illumination. These barriers may be the inter-particle or inter-aggregate TiO_2 barriers or *quantum wells* created by surrounding dye aggregates.

When CW illumination is used in combination with 337 nm pulsed excitation of the dye sensitized TiO_2 coating on SnO_2(F-doped), signal amplitudes are seen to be dependent on sample orientation with respect to the incident light direction. For exposure through the conducting glass the signal is positive, but the amplitude is much smaller than that obtained with 337 nm excitation alone. This implies that the field across the TiO_2, close to the interface with SnO_2 , has been substantially decreased due to the CW illumination, suggesting that band-bending from the TiO_2 to the SnO_2 is downward.

For 337 nm pulse + CW exposure of dye sensitized TiO_2 on SnO_2, with light incident on the dye, as in sample #4 of Table 1, the negative signal shows a marked increase over that obtained with a 337 nm pulse alone. This implies that the field spontaneously formed across the TiO_2, in the SnO_2(F-doped)-TiO_2/Dye system is effective over considerable distance once interparticle TiO_2 energy barriers are lowered by CW, and conduction in *extended states* above the *quantum wells* illumination is made possible. The energy band diagrams of Figs. 24 and 25 are believed to be applicable to these configurations and exposure conditions.

Evidence for Multi-Layer Dye Adsorption. Since the nanopores in the TiO_2 coating have been observed to be as small as only several nanometers in diameter [1], they could easily be partially bridged by nano-sized dye aggregates. Fig. 27 provides experimental evidence for ***quantum narrowing of the dye aggregate "band gap" with increase in aggregate size***, for a cyanine sensitizing dye in solution and also adsorbed on AgBr grains. This terminology was not in use at the time this work was performed [26], but it seems appropriate now.

Fig. 28, shows TRPC signal amplitudes, obtained with 540 nm laser excitation, for dye sensitized TiO_2 (P25) and for dye sensitized P25 with photolytically deposited Pt nano-clusters, as a function of equilibrium dye concentration used for adsorption of dye. Langmuir isotherms were used to determine the equilibrium dye concentration for monolayer coverage, which occurs within the domain where Vm is positive for each of the curves shown. Vm is negative at higher dye concentrations, where multi-layer adsorption occurs [21].

540 nm Pulse Excitation. Excitation with a laser pulse at 540 nm, in the spectral region where only the sensitizing dye absorbs light, gives negative values for Vm in samples #3, 4, and 7. These negative polarity signals suggest that dye is present as a multi layer coating. The basis for this assignment is discussed above and in Figs. 27 and 28.

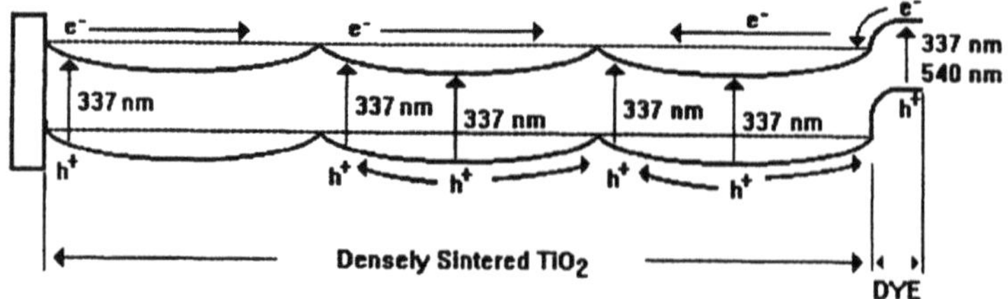

Figure 26. Proposed energy band diagram for dye sensitized densely sintered TiO_2 on glass.

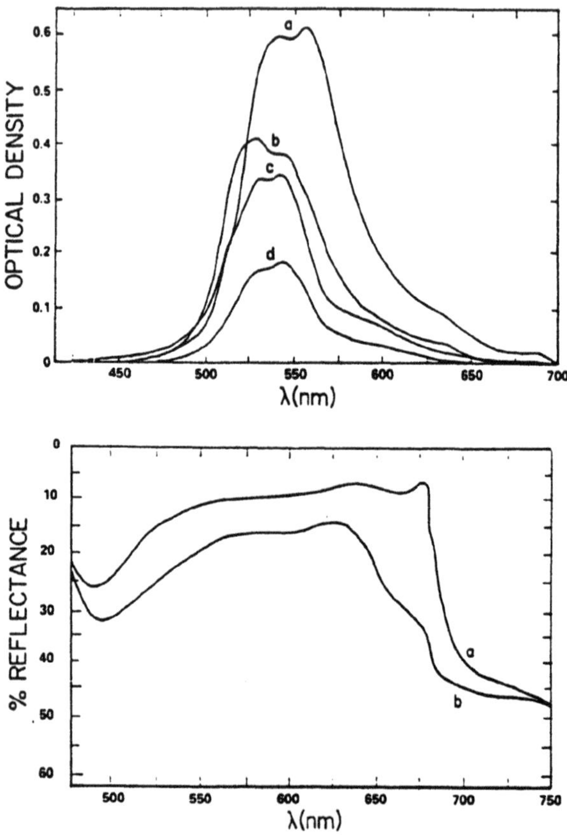

Figure 27. Spectral evidence for "band gap" narrowing with increase in number of dye monomer units in cyanine dye aggregates: : a) absorption spectra of aqueous dye solutions at various concentrations - 10^{-4} 5 X 10^{-5} M, 10^{-5} M, and 10^{-6} M and b) reflectance spectra of dye adsorbed on AgBr at known surface coverages, 1.14 X 10^{-6} M/m^2 and 5.45 X 10^{-7} M/m^2, as determined from Langmuir isotherms (see [26]).

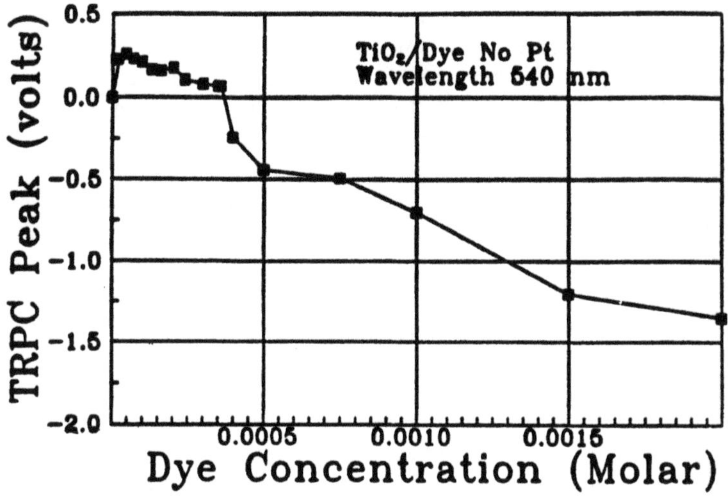

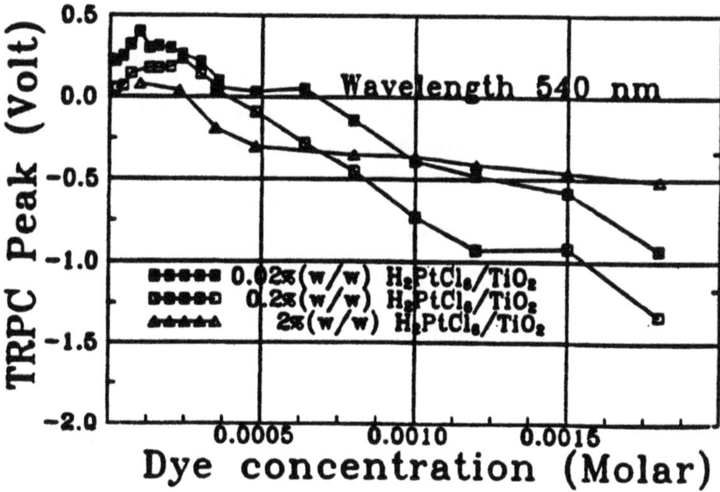

Figure 28. TRPC determinations of Vm as a function of equilibrium concentration of rhodamine 640 solutions after dye adsorption on TiO_2 P25. Measurements were performed on the solids, following filtration and drying. The positive signals were obtained at low dye coverage, with the positive maximums roughly corresponding to half monolayer coverage, as determined from Langmuir isotherms. The negative signals correspond to multi-layer adsorption (see [21]),

Positive values of Vm are obtained only for sample #8, where the SnO_2/TiO_2/Dye sample is exposed through the conducting glass. Here, the field across the TiO_2 assists in the transport of photoelectrons, generated by light absorbed in the dye, toward the SnO_2 via the TiO_2 conduction band. The energy band diagrams of Figs. 24 and 25 are believed applicable for these configurations.

540 nm Pulse Excitation + CW Illumination. For samples #3, 4, and 7, 540 nm pulse excitation + CW Illumination is seen to increase the negative values of Vm, while in sample #8, the positive value of Vm decreases, as compared to 540 nm pulse excitation alone. Qualitatively, the reasons offered as similar to that for 337 nm pulsed excitation of these samples. The energy band diagrams of Figs. 24 and 25 explain the situation.

5. Conclusions

TRPC measurements with nanoporous TiO_2 coatings on glass and on degenerately doped tin oxide, with and without sensitization by $Ru(II)(L)_2(CNS)_2$, reveal a number of interesting basic electronic properties of these structures having relevance to the operation of devices using these or similar materials in their fabrication. Among these properties are:

i. SnO_2, degeneratively doped with fluorine, in contact with nanoporous TiO_2, creates a diode structure, with photoelectrons attracted to the strongly *n-type* SnO_2, independent of sample surface facing the incident light;

ii. TRPC measurements using pulsed probe light excitation and pulsed probe + CW illumination, reveal the presence and direction of band-bending. The observed band-bending confirms that the fundamental light absorbing unit in the nanostructured coating is considerably larger than that of the unaggregated colloid.

iii. Band-bending at the SnO_2/TiO_2 interface is downward from TiO_2 to SnO_2. CW illumination lowers the field at this interface.

iv. There are TiO_2 interparticle and/or aggregate upward bending energy barriers, that are lowered by CW illumination.

v. Dye sensitization of the SnO_2/TiO_2 diode with $Ru(II)(L)_2(CNS)_2$, amplifies the TRPC response, suggesting of a *p-i-n* device behavior.

vi. The negative polarity of the 540 nm excitation TRPC signal for dye sensitized nanoporous TiO_2 coated on non-conducting glass, demonstrates that the dye is mainly present as a multi-layer coating, which is considered capable of bridging many of the nanopores in the TiO_2 structure.

viii. Supporting this concept is experimental evidence showing that cyanine dye aggregates, in solution and adsorbed to AgBr, undergo a quantum band gap narrowing effect with increase in number of dye monomers in the dye aggregates.

ix. The three component system, SnO_2-TiO_2-Dye, has properties analogous, in some respects, to thin film *p-i-n* photovoltaic devices, and in other respects to *quantum well* devices. These features are combined to explain their varied TRPC responses.

In terms of device behavior, for exposure with CW light through the SnO_2 conducting glass, the decreases in band-bending at the SnO_2/TiO_2 interface in the *p-i-n* domain is not considered desirable, since it decreases the TRPC signal magnitude, and might respond similarly in device applications. However, CW illumination also reduces barriers introduced by interparticle band-bending and makes possible the population of *extended states* in the dye/TiO_2/dye *quantum wells* - both of which are believed to promote charge carrier separation. Taken together, these opposing trends suggest the possibility of an optimum CW intensity for best efficiency in light utilization.

Analysis of the photocharge decay kinetics shows a substantial second order contribution, possibly due to recombination, for the SnO_2/TiO_2/Dye sample, exposed through the conducting glass, using a 337 nm probe pulse, with and without CW illumination. All other exposure and sample orientation conditions described above, produced TRPC signals that decayed mainly according to first order kinetics. The other samples decayed substantially by first order processes for all measurement conditions. Blocking contacts were used in all of the TRPC measurements reported. As yet untested is the effect on the photocharge decay kinetics of grounding the samples through the SnO_2 conducting layer.

Although the relevent experiments have not as yet been performed, we speculate that, intermittent exposure, at a relatively rapid on-off cycle, might improve efficiency, by allowing sufficient time for band-bending at the SnO_2/TiO_2 interface to become re-established during the off cycle. Intermittent exposure of single phase TiO_2 particulates, used in photocatalytic studies, have already been demonstrated to provide higher efficiencies than CW excitation, perhaps for similar reasons [27].

6. References

1. Nazeeruddin, M.K., Kay, A.,.Rodicio, I., Humphrey-Baker, R., Muller, E., Liska, P., Vlachopoulos, N. and Grätzel, M. (1993) Conversion of light to electricity by cis-X_2Bis(2,2'-bipyridyl-4,4'-dicarboxyate)ruthenium(II) charge-transfer sensitizers (X = Cl^-, Br^-, I^-, CN^-, and SCN^-) on Nanocrystalline TiO_2 Electrodes, *J. Am. Chem. Soc.* **115**, 6382-6390.
2. Hagfeldt, A., Vlachopoulos, N. and Grätzel, M *J. Electrochem. Soc.* in press.
3 Lin, HM, Hsu, CM, Yang, HY , Leeb, PY and Yang, CC (1994) *Sensors and Actuators B-Chemical* **22**, 63.
4. Hoyer, P., Eichberger, R. and Weller, H. (1993) Spectrochemical investigations of nanocrystalline ZnO films, *Ber. Bunsenges. Phys. Chem.* **97**, 630-635 (and refs. therein).
5. Kavan, L. , Stoto, T., Grätzel, M., Fitzmaurice, D.. and Shklover, V. (1993) Quantum size effects in nanocrystalline semiconducting TiO_2 layers prepared by anodic oxidative hydrolysis of $TiCl_3$, *J. Phys. Chem* **97**, 9493-9498.
6. Dickson, C.R. and Levy B. (1974) Wavelength and Angular dependence of the Dember Effect - an explanation for "anomolus" sign reversals, *J. Photogr. Sci. Engr.***18**, 524.
7. Wang, Y and Heron, N. (1991) Nanometer-sized semiconductor clusters: materials sysnthesis, quantum size effects, and photophysical properties, *J. Phys. Chem.* **95**, 525-532.
8. Gilbert, S., unpublished results.

9. Könenkamp, R., Henninger, R., and Hoyer, P. (1993) Photocarrier transport in colloidal TiO_2 films, *J. Phys. Chem*, **97**, 7328-7330.
10. Schwarzburg K., and Willig, F. (1991) Influence of trap filling on photocurrent transients in polycrytalline TiO_2, *Applied Phys. Lett.* **58**, 2520-2522.
11. Könenkamp, R., Henninger, R. (1994) Recombination in nanophase TiO_2 films, *Applied Physics* **A58**, 87-90.
12. Hagfeldt, A., Bjorksten, U., Lindquist, (1992) Photoelectrochemical studies of colloidal TiO_2 films: the charge separation process studied by means of action spectra in the UV region, S.-E. *Sol. Energy Mat., Sol. Cells* **27**, 293-304.
13. Sodergren, S., Hagfeldt, A., Olsson, J., Lindquist, S.-E. (1994) Theoretical models for the action spectrum and the current-voltage characteristics of microporous semiconductor films in photoelectrochemical cells, *J. Phys. Chem.* **98**, 5552-5556.
14. Wahi, A, Engelhardt, R., Hoyer, P., Könenkamp, R. *Proc. 9th European Conf. Photovoltaic Science and Technology*, Montreux, Oct. 1992 (Harwood Acad. Publishers, Switzerland 1993); p. 714.
15 Bube, R. (1992) *Photoelectronic Properties of Semiconductors*, Cambridge University Press, Cambridge, Great Britain, Ch. 12, pp.280-305.
16. Levy, B. (1991) Photocatalytic and photographic heterojunctions, in *Photochem. Conv. Storage of Solar Energy*; E. Pelizzetti and M. Schiavello, eds. (Kluwer Acad. Pub. 1991); pp. 357-392.
17. O'Regan, B. and Grätzel, M. (1991) A low-cost, high-efficiency solar cell based on dye-sensitized colloidal TiO_2 films, *Nature* **353**, 737-740.
18. Tiedje, T. (1984) Information about Band-Tail States from Time-of-Flight Experiments, in *Semiconductors and Semimetals*, **Vol.21, part** C, ed. J. I. Pankove, Academic Press, Inc., NY, pp. 207-238.
19 Memming, R., (1990) Photoelectrochemical utilization of solar energy, in *Photochemistry and Photophysics*, **vol.II**, ed J. F. Rabek, CRC Press, Boca Raton, FL, pp. 143-189.; Memming, R., (1988) Photoelectrochemical solar energy conversion, in *Topics in Current Chemistry*, Vol. 143, ed., E. Steckhan, Springer-Verlag, Berlin, Heidelberg,
20. Grätzel, M. (1989) *Heterogeneous Photochemical Electron Transfer*, CRC Press, Boca Raton, FL.
21. Levy, B., (1991) Heterojunction models for chemical, spectral and supersensitization of AgX by S, Au and sensitizing dye, *J. Soc. Photogr. Sci. Technol. Japan*, **54**, 477; Levy, B., (1991) Heterojunctions in imaging and photoconducting catalytic systems, in *Proceedings of the Symposium on Electronic and Ionic Properties of Silver Halides*, May 12-17, 1991, eds. B.Levy, L.Slifkin, J.Deaton, A.Muenter, T,Tani, I.Leubner and P.Kamat; Society for Imaging Science and Technology, Springfield, VA, pp. 189- 197; Levy B., and Zhang, T-G, (1994) Field assisted charge transport in dye sensitized AgX and Pt-TiO_2-Dye particulate heterojunctions, in *Proceedings of the International Conference of Photographic Science (ICPS' 94)*, May 15-20, , Rochester, New York; Levy B., and Zhang, T-G, (1994) Charge transport in TiO_2 and in Pt-TiO_2 and Pt-TiO_2-Dye heterojunction photocatalysts, in *Proceedings of the Tenth International Conference on Photochemical Conversion and Storage of Solar Energy (IPS-10)*, July 24-29, 1994, Interlacken, Switzerland.
22. Levy, B. (1983) Photographic diode and transistor heterojunction imaging, *Photogr.Sci.Engr.* **27**, 204; Zhang, TG and Levy, B. (1993) Modulation of photocharge polarity by light intensity in optically formed oriented AgX/Ag heterojunction arrays, *Bulgarian Chemical Communications* **26(3/4)**, 412-427.
23. Takahasi K,.and Konagai, M. (1986) *Amorphous Silicon Solar Cells*, John Wiley and Sons, NY, 1986, pp.124-163.
24. Gilbert, S., unpublished results.
25. see ref. 15, p.38.
26. Levy, B. and Mattucci, N. (1970) The adsorption and aggregation of cyanine dyes on AgBr, *Photgr. Sci. Eng.*, **14**, 308-315.
27. Sczechowski, J.G., Koval, C.A. and Noble, R.D. (1993) Improved photoefficiencies for TiO_2 photocatalytic reactors through the use of controlled periodic illumination, in *Photocatalytic Purification and Treatment of Water and Air*, eds. D.F. Ollis and H. Al-Ekabi, Elsevier, Amsterdam,

COLLOIDAL DISPERSION OF BIMETALLIC NANOPARTICLES: PREPARATION, STRUCTURE AND CATALYSIS

Naoki TOSHIMA
Department of Applied Chemistry, The University of Tokyo
Hongo, Bunkyo-ku, Tokyo 113, Japan

Abstract: Colloidal dispersions of polymer-protected bimetallic cluster nanoparticles were prepared by refluxing the solutions of two kinds of noble metal ions in ethanol/water in the presence of poly(*N*-vinyl-2-pyrrolidine). The bimetallic clusters thus prepared had higher activity as catalysts than the corresponding monometallic clusters. Pt/Pd, Au/Pd, Au/Pd and Pt/Rh bimetallic clusters prepared by the present method were structurally analyzed by an EXAFS technique as well as UV-VIS absorption, etc.. They have a special core-shell structure. The order of the tendency to be a core is estimated to be Au>Pt>Pd>Rh. This order could be explained by the redox potential of the corresponding metal ion and the coordination ability of the metal to the polymer.

1. Introduction

Colloidal dispersions of metals have a long history. The first scientific report appeared in 1857 by M. Faraday, who prepared some colloidal metals by reduction of the corresponding metal ions. He had known that macromolecules like gelatin can stabilize the colloids.

Recently much attention has been paid to colloidal metals as nanoparticles because recent polymer-protected metal clusters have an enough small size (from 1 to 8 nm) and narrow size distribution. For example, the colloidal dispersion of polymer-protected noble-metal

E. Pelizzetti (ed.), Fine Particles Science and Technology, 371–383.

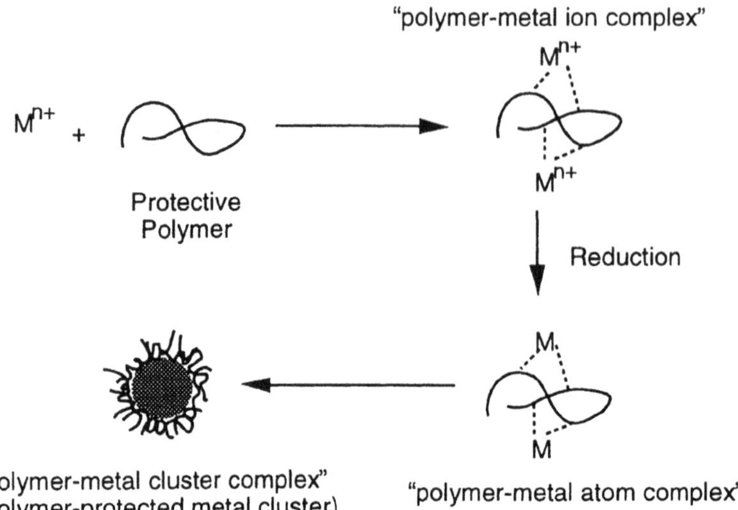

Figure 1. Schematic illustration of reduction process of metal ions in the presence of protective polymer.

clusters, of which we have much focused our attentions in recent years, can be prepared by refluxing the solution of the corresponding metal ions in water/alcohol in the presence of water-soluble polymers like poly(*N*-vinyl-2-pyrrolidine)[1]. The coordination of the polymer to the metal ions could result in mild reduction to produce small metal cluster particles as shown in Fig. 1.

The colloidal dispersions thus prepared are enough stable to work as active and selective catalysts for many organic reactions under mild conditions. Metal clusters can bridge between bulk metal and atomic or ionic metal, and have the special properties which are quite different from both bulk and atomic metals. This special properties have attention not only in catalysis but also in physical, biological, and other chemical field. Quantum size effect is one of the greatest problems for physicists.

On the other hand, the additive effect of the second metallic component in bimetallic catalysts is very important from the viewpoint not only of practical purposes in industries but also of scientific meanings in academia, and can be often understood as an ensemble effect and/or a ligand effect. From this viewpoint, the bimetallic clusters have been extensively investigated to improve their preparation procedures as well as the activity and selectivity as catalysts[2,3]. The gold/silver bimetallic colloids were prepared by γ-irradiation for the study of electrochemical reactions[4]. The same method was applied to preparation of trimetallic particles for investigation of optical properties[5]. Colloidal silver/platinum[6] and silver/palladium[9] alloys were obtained by the chemical reduction of the corresponding bimetallic double complexes. Bönnemann et al. recently reported the hydrogenation activity of the palladium/rhodium bimetallic colloids supported on charcoals, which exhibits a maximum activity for the crotonic acid at Pt/Rh=1/4[8].

In recent years, we have extensively investigated the preparation and characterization of colloidal dispersions of bimetallic clusters protected by polymers and their application to catalysis. The mixtures of two kinds of metal ions in solution were only simultaneously treated by heat to reflux in the presence of polymers. This simple procedure did give not the mixtures of two kinds of monometallic clusters but a homogenous dispersion of one kind of bimetallic clusters.

Here I would like to present general preparation procedure and aspects of colloidal dispersion of polymer-protected bimetallic cluster nanoparticles, and then show special structures and catalytic properties of some practical examples of bimetallic cluster nanoparticles.

2. General Preparation Procedure and Aspects

In our method, the colloidal dispersions of polymer-protected bimetallic clusters have been prepared by simultaneous reduction of two kinds of metal ions in the presence of poly(N-vinyl-2-pyrrolidine)(PVP) by using ethanol as a reducing agent.

The practical procedure is as follows: Ethanol solutions of palladium(II) chloride were prepared in advance by stirring dispersions of $PdCl_2$ powder in ethanol for 48 h. Solutions of H_2PtCl_6, $HAuCl_4$ and $RhCl_3$ were prepared by dissolving the corresponding crystals in water. Two kinds of the solutions and pure ethanol and/or water were mixed to produce a bimetallic ethanol/water(1/1) solution (50 cm^3) with a total metal concentration of 6.6×10^{-4} mol dm^{-3} containing 151 mg PVP(Tokyo Kasei Co., K30, MW=40,000). The mixed solution was refluxed for 1 h.

The colloidal dispersions of polymer-protected bimetallic clusters usually have a transparent dark brown color except otherwise mentioned and are stable for months at room temperature. The size and size distribution of the cluster nanoparticles, measured by transmission electron micrograph(TEM), are much small and narrow, respectively, on comparison with the conventional colloidal metals and even with the polymer-protected monometallic cluster particles prepared by the same alcohol reduction method.

3. Pd/Pt Bimetallic Clusters

3.1. PREPARATION AND CHARACTERIZATION

A solution of PVP, $PdCl_2$, and H_2PtCl_6 in ethanol/water(1/1,v/v) was heated to 90-95 °C for 1 h under ambient air, resulting in a clear, homogeneous, and stable dispersion of polymer-protected Pd/Pt bimetallic clusters.

The UV-VIS absorption spectra of the colloidal dispersions at various Pd/Pt ratios showed monotonous curves without exhibiting a maximum. However, the spectrum of the bimetallic colloidal dispersion differed from that of both the monometallic dispersion of palladium and platinum, and a mixture of the two, suggesting the formation of bimetallic cluster nanoparticles which showed involuement of both elements in each particle.

The similar suggestion were given by transmission electron micrograph(TEM) observations. Thus, the contrasts of the particles in the background were compared. The Pt monometallic particles have the highest contrast, while Pd monometallic particles have rather low contrast. The Pd/Pt bimetallic cluster particles did actually have the contrast just between the two. The size and size distribution also depend on the composition. The data of the average diameter and standard deviation are shown in Table 1, which suggests again that the bimetallic systems are different from either of the monometallic cluster particles. Especially, Pd/Pt=4/1 bimetallic particles are smallest in size among the bimetallic particles examined, which may suggest some specific meaning in this ratio.

TABLE 1. A comparison of average diameter and standard deviation of Pd/Pt bimetallic cluster nanoparticles at various Pd/Pt ratios

Pd/Pt ratio	Average diameter (nm)	Standard deviation (nm)
1/0	2.5	0.9
4/1	1.5	0.4
3/2	1.6	0.3
2/3	1.7	0.3
1/4	1.8	0.5
0/1	1.9	0.6

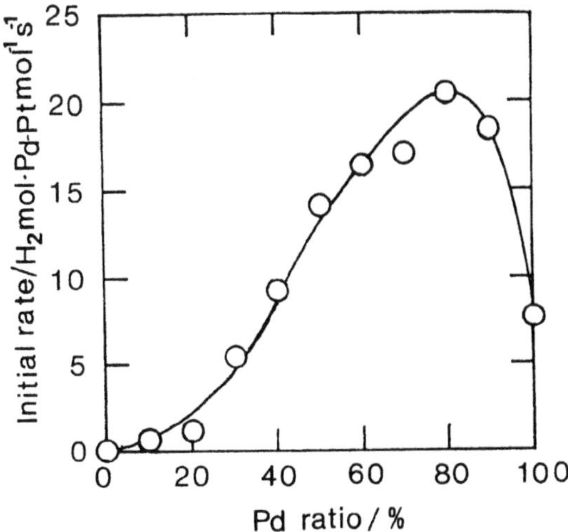

Figure 2. Dependence of the catalytic activity of Pd/Pt bimetallic clusters on the metal composition for partial hydrogenation of 1,3-cyclooctadiene.

3.2. CATALYSIS FOR PARTIAL HYDROGENATION

A colloidal dispersion of polymer-protected palladium clusters acts as an effective catalyst for the partial hydrogenation of 1.3-cyclooctadiene to cyclooctene, with high selectivity under mild reaction conditions[9]. Thus, the colloidal dispersions of PVP-protected Pd/Pt bimetallic clusters, prepared here under ambient air conditions, were applied to the catalysis for the partial hydrogenation of 1,3-cyclooctadiene[10]. The results are illustrated in Fig. 2. The Pd/Pt bimetallic clusters with palladium contents from 40 to 90 mol% show the higher catalytic activity than Pd monometallic clusters. The selectivity for cyclooctene formation by the bimetallic clusters is also very high as much as that by the Pd monometallic clusters[11].

This high catalytic activity and selectivity of the Pd/Pt bimetallic cluster particles are attributed to the special core-shell structure of the nanoparticles, described later. In fact. the catalytic activity, normalized by the palladium atoms on the surface of the bimetallic cluster particles, keeps the constant high value in the region from 50 to 90 % of the palladium content. This means that the surface palladium atom adjacent to platinum atom(s) has higher catalytic activity than that adjacent to only palladium atoms.

The Pd/Pt bimetallic clusters prepared under nitrogen instead of air have the characteristic physical properties are quite similar to those prepared under air. The catalytic activity of the bimetallic clusters prepared under nitrogen is a little lower than those prepared under air, but the behavior are quite similar each other.

3.3. STRUCTURAL ANALYSIS

As mentioned in section 3.1., The results by UV-VIS and TEM measurements suggest the formation of the Pd/Pt bimetallic cluster, which contains both Pd and Pt elements in each particle. However, these are not a direct evidence for the formation of the bimetallic clusters.

Extended X-ray absorption fine structure(EXAFS) is a powerful tool for investigation of local structure of the metal cluster particles. EXAFS measurements can give information on the environment about a particular atom in the metal clusters, i.e., the coordination number, the kind of neighboring atoms, and their distances, from the X-ray absorbing atom without requiring long-range ordered structure. The EXAFS technique was used to characterize some bimetallic clusters supported on inorganic supports[12], but there are little reports on EXAFS analysis of colloidal dispersions of bimetallic clusters.

Thus, EXAFS was applied to the determination of the structure of colloidal dispersions of the PVP-protected Pd/Pt bimetallic clusters[13]. In this EXAFS analysis, the reliable fitting parameters were empirically obtained from model reference compounds such an Pd foil, Pt foil, Pd/Pt(1/9) alloy foil, and Pd/Pt(9/1) alloy foil.

Coordination numbers and interatomic distances were determined for the Pd/Pt=1/1 and Pd/Pt=4/1 bimetallic clusters, the latter of which is the most active catalyst for selective partial hydrogenation of 1,3-cyclooctadiene as mentioned before. The selected data are collected in Table 2 with coordination numbers of these model structures. The data do clearly show that the observed coordination numbers are well fit for the Pt-core model in both cases of Pd/Pt=1/1 and 4/1. The Pt-core models are shown in Fig. 3, where Pt atoms are preferentially located in the centeral part of the nanoparticles.

The present Pt-core models are well fit not only for the EXAFS data but also for the TEM observation. In the present Pt-core model for the Pd/Pt (4/1) bimetallic clusters, 13 Pt atoms form a core and 42 Pd atoms surround the core forming a shell with one atomic layer. With decreasing Pd/Pt ratio, the Pd atoms on the surface of the cluster particle are substituted for the Pt atoms. Then, in the model of Pd/Pt (1/1) bimetallic cluster, 28 Pt atoms are located both in the core and on the surface connecting directly each other, and 27 Pt atoms form three islands on the surface of the cluster particle.

Taking the results on the dependence of both catalytic activity and surface structure upon the Pd/Pt ratio into account, the high activity of Pd atoms on the surface adjacent to Pt atoms can be caused not by an emsemble effect but by the ligand effect of Pt.

The PVP-protected Pd/Pt bimetallic clusters prepared under air, which are the more active catalyst than those prepared under nitrogen, were submitted for EXAFS analyses as well. it is concluded from EXAFS analyses that the bimetallic clusters prepared under air have Pd-O bond(s) but not Pt-O bond(s)[14]. This result suggests that only the surface of Pd atoms can be oxidized by treatment with oxygen, which can provide more active site(s) on the surface of the bimetallic clusters when treated by hydrogen for hydrogenation of 1,3-cyclooctadiene.

TABLE 2. Coordination numbers around the Pd and Pt atoms of the Pd/Pt bimetallic clusters determined by an EXAFS technique

Pd/Pt ratio	Absorbing metal	Scattering metal	Coordination number		
			Observed	Pt-core	Random
1/1	Pd	Pd	2.321.0	3.1	3.9
〃	Pd	Pt	3.021.1	3.0	3.9
〃	Pt	Pt	4.821.7	6.6	3.9
〃	Pt	Pd	1.321.0	2.9	3.9
4/1	Pd	Pd	4.421.0	4.6	6.0
〃	Pd	Pt	2.321.3	2.0	1.9
〃	Pt	Pt	5.521.7	5.5	1.9
〃	Pt	Pd	3.521.5	6.5	6.0

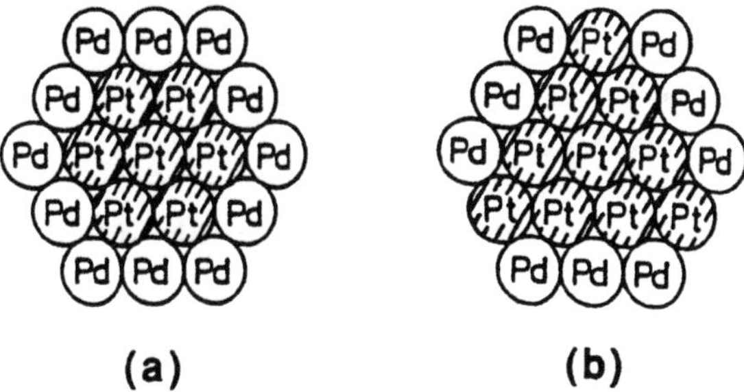

Figure 3. Cross section of the models for (a) Pd/Pt(4/1) and (b) Pd/Pt(1/1) bimetallic clusters.

4. Au/Pd Bimetallic Clusters

4.1. PREPARATION AND CATALYSIS

Colloidal dispersions of PVP-protected Au/Pd bimetallic clusters were prepared by the same method as the case of Pd/Pt bimetallic clusters from $HAuCl_4$ and $PdCl_2$[15]. In this case, the formation of Au/Pd bimetallic clusters is more clearly demonstrated by UV-VIS absorption spectra and TEM observations, because gold clusters and bimetallic clusters having gold aggregates on the surface have a peak, so-called plasma absorption at ca. 550 nm, and pure gold clusters have exceptionally large size in TEM photographs.

Although gold itself has no catalytic activity for usual chemical reactions, Au/Pd bimetallic clusters have higher activity than palladium monometallic clusters for the selective partial hydrogenation of 1,3-cyclooctadiene to cyclooctene. Again the Au/Pd(1/4) cluster has the highest activity.

4.2. STRUCTURAL ANALYSIS

The average diameter of PVP-protected Au/Pd(1/4) bimetallic clusters was determined to be 1.6 nm on the basis of TEM observation. Thus, the Au/Pd(1/4) bimetallic cluster particle is a little larger than the Pd/Pt(4/1) bimetallic particle in average.

EXAFS analyses for Au/Pd(1/4 and 1/1) bimetallic clusters indicate that the surface of the particle entirely consists of palladium atoms, and the core of gold atoms, i.e., an Au-core Pd-shell structure. The model structure for the Au/Pd(1/4) cluster is shown in Fig. 4.

5. Au/Pt Bimetallic Clusters

5.1. PREPARATION AND CATALYSIS

Colloidal dispersions of polymer-protected Au/Pt bimetallic clusters were prepared by the same alcohol reduction method from $HAuCl_4$ and H_2PtCl_6[16]. Although polymer-protected gold clusters have large average diameter(5.42 nm) and wide size distribution, polymer-protected

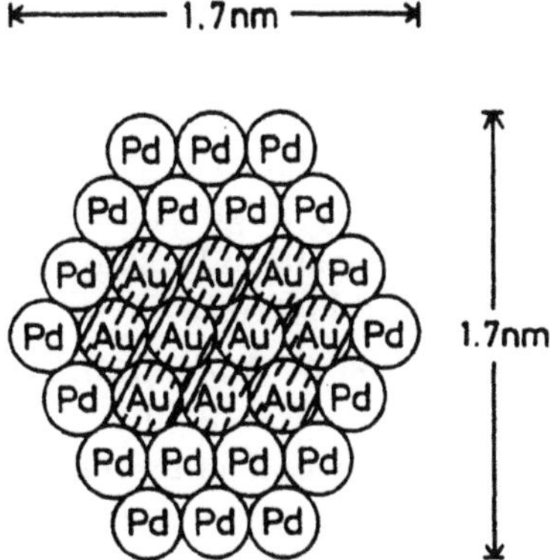

Figure 4. Cross section of the core-shell model for the Au/Pd(1/4) bimetallic cluster.

Au/Pt bimetallic clusters have smaller average diameter(e.g., 2.3 nm for Au/Pt(2/3) cluster) and narrower size distribution than gold clusters.

UV-VIS absorption spectra have gold plasma absorption at ca. 530 nm for gold cluster and Au/Pt(4/1) bimetallic clusters, but do not have it for Au/Pt(3/2, 2/3, and 1/4) bimetallic clusters. This result suggests that gold aggregates are not located on the surface of the bimetallic clusters if the Au/Pt ratio is less than 3/2. In other words, platinum atoms could cover the gold core in these bimetallic clusters.

The Au/Pt bimetallic clusters were used as the catalyst for electron transfer from methyl viologen radical cation to proton in water to produce hydrogen molecules on visible-light irradiation in the system of EDTA/$Ru(bpy)_3^{2+}$/MV^{2+}/bimetallic cluster catalyst. The Au/Pt bimetallic clusters were found to be more active than platinum or gold monometallic clusters.

5.2. FORMATION PROCESS AND STRUCTURE

In the case of Au/Pt bimetallic clusters, the EXAFS technique cannot be applied to the structural analysis because the X-ray absorption edges of gold and platinum are so close each other that a large error in calculations could be accompanied. Thus, the in-situ UV-VIS absorption spectra were used for consideration of the structure of Au/Pt bimetallic clusters.

The in-situ UV-VIS spectra during the simultaneous reduction of gold and platinum ions by ethanol can provide an evidence for the order of the reduction of both ions. Thus, the absorption peaks at ca. 320 nm attributed to gold ions disappears at first and then the peak at ca. 265 nm attributed to platinum ions disappears. Further heat treatment gives new peak at ca. 530 nm, which can be attributed to gold plasma absorption. This means gold clusters are produced at this stage. However, this gold plasma absorption is covered by large tailing absorption in whole area after additional heating, which strongly suggests that gold clusters are surrounded by platinum atomic layer(s).

This process of formation of Au/Pt bimetallic clusters is illustrated in Fig. 5. This process can answer to the question why and how the bimetallic clusters have the special core-shell structure. Thus, redox potential of metal ion and coordination ability of metal atom to polymer can control the special structure.

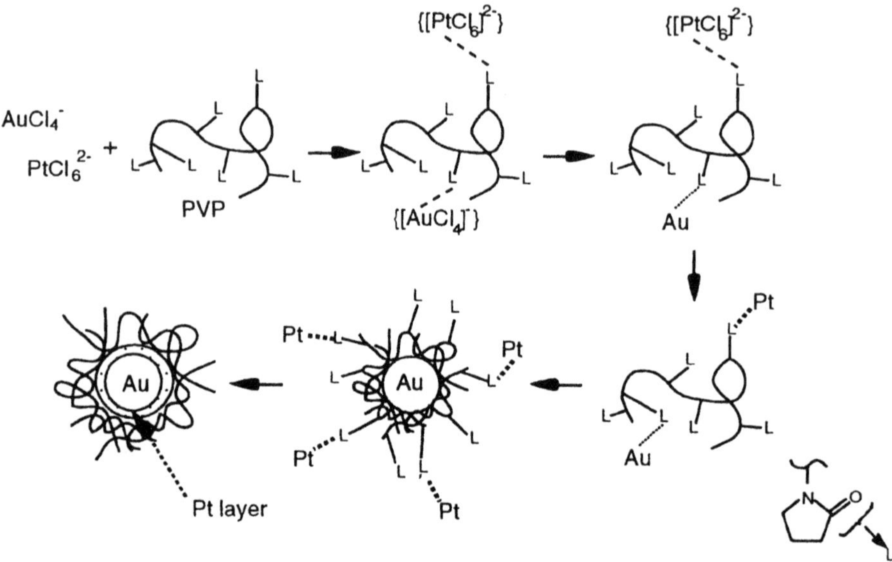

Figure 5. Schematic illustration of formation process of the polymer-protected Au/Pt bimetallic cluster with an Au-core Pt-shell structure.

6. Conclusion

Polymer-protected bimetallic cluster nanoparticles were prepared by reduction of the corresponding noble metal ions with an alcohol-reduction method in the presence of poly(*N*-vinyl-2-pyrrolidone). The bimetallic clusters are more active catalysts than the corresponding monometallic clusters for the reactions like partial hydrogenation of diene to monoene and visible-light-induced hydrogen generation.

The structures of the bimetallic cluster nanoparticles were analyzed by an EXAFS techniques in-situ UV-VIS spectra, HRTEM, XRD, and so on. The special core-shell structure has been proposed for Pt/Pd, Au/Pd, Au/Pt, and Pt/Rh[17] bimetallic clusters. The tendency to be a core in the core-shell structure is considered to be Au>Pt>Pd>Rh among these four atoms. This order will be due to both redox potential of metal ion and coordination ability of metal atom to polymer. Thus, the polymer plays an important role during the formation of metal clusters in controlling the structure as well as after the formation of the clusters in stabilizing the nanoparticles as the colloidal dispersions.

The higher catalytic activity of the bimetallic clusters than the corresponding monometallic clusters will probably be caused by a ligand effect of the core metal and not by an ensemble effect, because, for example, the Pd/Pt(4/1) cluster has only Pd atoms on the surface and no Pt atoms are located on the surface of the cluster.

Recently we have also succeeded to prepare Cu/Pd and Cu/Pt bimetallic clusters by the similar method[18].

Acknowledgment. The author gratefully acknowledges his coworkers, especially Drs. M. Harada and T. Yonezawa for their great contribution to the present work. This work was supported by a Special Grant by the Asahi Glass Foundation, a Grant-in-Aid for Scientific Research(B) (No. 05453141), and a Grant-in-Aid for Scientific Research in Priority Areas of "Macromolecular Complexes", "Non-equilibrium Processes in Solutions", "Reactive Organometallics", and "Photoreaction Dynamics" from the Ministry of Education, Science and Culture, Japan.

References

1. Hirai, H. and Toshima, N. (1986) Polymer-attached catalysts, *Tailored Metal Catalysts*, Ed. Iwasawa, Y., D. Raidel Pub., Dordrecht, pp. 87-148.
2. Toshima, N. (1990) Polymer-protected bimetallic clusters. Preparation and application to catalysis, *J. Macromol. Sci.-Chem.*, **A27**, 1225-1238.

3. Coq. B., Kumbhar, P.S., Moreau, C., Moreau, P., and Figueras, F. (1994) Zirconia-supported monometallic Ru and bimetallic Ru-Sn, Ru-Fe catalysts: Role of metal support interaction in the hydrogenation of cinnamaldehyde, *J. Phys. Chem.*, **98**, 10180-10188.
4. Mulvaney, P., Giersig, M., and Henglein, A. (1993) Electrochemistry of multilayer colloids: Preparation and absorption spectrum of gold-coated silver particles, *J. Phys. Chem.*, **97**, 7061-7064.
5. Mulvaney, P., Giersig, M., and Henglein, A. (1992) Surface chemistry of colloidal gold: Deposition of lead and accompanying optical effects, *J. Phys. Chem.*, **96**, 10419-10424.
6. Torigoe, K., Nakajima, Y., and Esumi, K. (1993) Preparation and characterization of colloidal silver-platinum alloys, *J. Phys. Chem.*, **97**, 8304-8309.
7. Torigoe, K. and Esumi, K. (1993) Preparation of bimetallic Ag-Pd colloids from silver(I)bis(oxalato)palladate(II), *Langmuir*, **9**, 1664-1667.
8. Bönnemann, H., Brijoux, W., Brinkmann, R., Fretzen, R., Joussen, T., Köppler, R., Korall, B., Neiteler, P., and Richter, J. (1994) Preparation, characterization, and application of fine metal particles and metal colloids using hydrotriorganoborates, *J. Mol. Catal.*, **86**, 129-177.
9. Hirai, H., Chawanya, H., and Toshima, N. (1985) Selective hydrogenation of cyclooctadiene using colloidal palladium in poly(N-vinyl-2-pyrrolidone), *Bull. Chem. Soc. Jpn.*, **58**, 682-687.
10. Toshima, N., Kushihashi, K., Yonezawa, T., and Hirai, H. (1989) Colloidal dispersions of palladium-platinum bimetallic clusters protected by polymers. Preparation and application to catalysis, *Chem. Lett.*, **1989**, 1769-1772.
11. Toshima, N., Yonezawa, T., and Kushihashi, K. (1993) Polymer-protected palladium-platinum bimetallic clusters: Preparation, catalytic properties and structural considerations, *J. Chem. Soc., Faraday Trans.*, **89**, 2537-2543.
12. Sinfelt, J.H. (1987) Structure of bimetallic clusters, *Acc. Chem. Res.*, **20**, 134-139.
13. Toshima, N., Harada, M., Yonezawa, T., Kushihashi, K., and Asakura, K. (1991) Structure analysis of polymer-protected Pd/Pt bimetallic clusters as dispersed catalysts by using extended X-ray absorption fine structure spectroscopy, *J. Phys. Chem.*, **95**, 7448-7453.
14. Harada, M., Asakura, K., Ueki, Y., and Toshima, N. (1992) Structure of polymer protected palladium/platinum bimetallic clusters at the

oxidized state. Extended X-ray absorption fine structure analysis, *J. Phys. Chem.*, **96**, 9730-9738.

15. Toshima, N., Harada, M., Yamazaki, Y., and Asakura, K. (1992) Catalytic activity and structural analysis of polymer-protected Au-Pd bimetallic clusters prepared by the simultaneous reduction of $HAuCl_4$ and $PdCl_2$, *J. Phys. Chem.*, **96**, 9927-9933.
16. Yonezawa, T. and Toshima, N. (1993) Polymer- and micelle-protected gold/platinum bimetallic systems. Preparation, application to catalyst for visible-light induced hydrogen evolution, and analysis of formation process with optical methods, *J. Mol. Catal.*, **83**, 167-181.
17. Harada, M., Asakura, K., and Toshima, N. (1994) Structural analysis of polymer-protected platinum/rhodium bimetallic clusters using extended X-ray absorption fine structure spectroscopy. Importance of microclusters for the formation of bimetallic clusters, *J. Phys. Chem.*, **98**, 2653-2662.
18. Toshima, N. and Wang, Y. (1994) Preparation and catalysis of novel colloidal dispersions of copper/noble metal bimetallic clusters, *Langmuir*, **10**, 4574-4580.

Electrochemical Studies on Nanosized Sols of Titanium Dioxide and Cadmium Sulfide

R. Andriamanampisoa, C. Karakus and R.A. Mackay*
Center for Advanced Materials Processing
Clarkson University
Potsdam, NY 13699-5665

* To whom correspondence should be addressed.

The electrochemical behavior of nanosized hydrosols of titanium dioxide and cadmium sulfide was examined at a mercury electrode by means of polarography, cyclic voltammetry and differential pulse voltammetry. No evidence was obtained for direct reduction of the particle. The solubility of the titanium dioxide sol in perchloric acid over the pH range of 1-3 was determined, as was the adsorption of cadmium and sulfide ions by the CdS sol.

1. Introduction

There has been increasing interest in recent years in ultra-small particles, ranging from metal clusters of less than 1 nm in size to nano particles up to 10-20 nm (1). These studies have generally focused on colloidal metal and semiconductor particles, although other inorganic materials (e.g. - Fe_2O_3) have been examined. Much of this interest was stimulated by the use of these dispersions as catalysts, particularly photocatalysts. However, it is also clear that a study of these particles as precursors to their growth or agglomeration to larger colloids is also of great importance (2).

Since many studies involve electron transfer between a (usually semiconductor) particle and a species in solution, the focus of this study was to determine if direct electron transfer between a nanosized hydrosol containing a reducible cation and an electrode could be observed. In addition, the nanoparticles themselves can frequently be stabilized with respect to agglomeration by a variety of methods (3) and materials such as polyphosphate (4), polyvinyl alcohol (5), and others (6-9). Therefore, if the direct electrochemical reactions of these sols yields sufficiently useful results, they may be utilized for the applications indicated above, and perhaps many others.

The possibility of such electron transfer has already been demonstrated by a few studies dealing with anodic processes involving the transfer of conduction band electrons in semi-conductor colloids (10-12). The direct oxidation of conduction band electrons in n-type semiconductor titania

E. Pelizzetti (ed.), Fine Particles Science and Technology, 385–402.

and cadmium sulfide hydrosols in the 10 nm size range, at tin oxide, glassy carbon and platinum electrodes was examined. It was shown that, at least for these cases, the same electrochemical equations are obeyed as for simple dissolved species. The average number of electrons transferred per particle (per event) ranged from about four to over four hundred, depending on the doping level (density of electrons in the conduction band). Although a number of studies on the use of submicron-sized semiconductor particle slurries in photoelectrochemical cells have been conducted (13-16), little has been done with smaller sols. One report (17) deals with the use of nanosized dispersions of titanium dioxide in a photogalvanic cell. In this case, however, mediators are employed and direct electron transfer from the colloid to the electrode does not seem to be involved.

On larger particles, excess electrons on TiO_2 were able to be collected at an electrode (13, 14, 18). However, in nanosized colloidal TiO_2 particles, there is evidence from pulse radiolysis studies that the accumulation of only one electron will lead to the reduction of Ti(IV) to a Ti(III) blue color center at a site close to or at the surface (19). Some of this reduction product may subsequently leave the particle to form an aqueous Ti(III) species. In similar studies on CdS particles, the first added electron was stable for minutes, but a second electron resulted in the reduction of Cd(II) to Cd^0 which upon continued reduction could produce a metal deposit on the particle (20). However, to the best of our knowledge, a study of the direct electrochemical reactions of ultra-small particles which contain oxidizable or reducible components does not seem to have been carried out.

In light of the many reported studies carried out on titanium dioxide and cadmium sulfide sols as indicated above, their corresponding methods of preparation (7, 10, 19, 21-24), and the fact that they both contain a cation which is reducible in the range of a mercury electrode, they were chosen for this investigation.

2. Experimental

2.1 REAGENTS

All of the chemicals were analytical grade and used without further purification: 99.99% titanium tetra-isopropoxide, 99.9% titanium tetrachloride and $Cd(ClO_4)_2$ (Aldrich); 70% perchloric acid and concentrated (36.5-38%) hydrochloric acid (Fisher Scientific); polybrene (hexadimethrine bromide) and 99% sodium dodecyl sulfate (Sigma); sodium sulfide, 9-hydrate and isopropyl alcohol (AR, Baker).

2.2 PREPARATION AND MEASUREMENTS OF COLLOIDAL TiO_2

The nanosized sols of TiO_2 were prepared principally by hydrolysis of $TiCl_4$ (21, 22), but some lower concentrations ($\leq$ 1.6 mM) were prepared by hydrolysis of Ti $(i\text{-}PrO)_4$ (10, 19, 23).

2.2.1 *From Ti(i-Pro)$_4$*
In a typical preparation, 1.6 ml of titanium tetra-isoproxide was dissolved in 30 ml of isopropanol and was slowly added dropwise to 300 ml aqueous 0.05M $HClO_4$ at 0 ± 0.2°C under nitrogen. The solution was stirred overnight, then dialyzed against an aqueous solution of pH 3.0 $HClO_4$ (replaced several times). The final suspension is transparent and stable for several weeks in a refrigerator. A particle size distribution obtained using a JEOL model 1200EX transmission electron miscopy (TEM) operating at 120 kV is shown in Figure 1a. The number average diameter is 71±28Å.

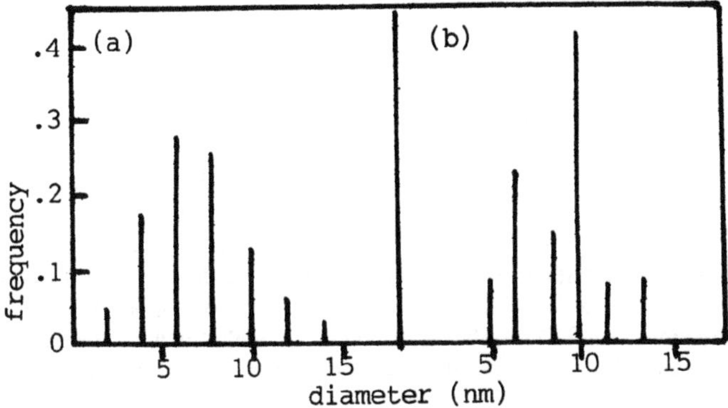

Figure 1. TEM particle size histograms for: (a) TiO_2; (b) CdS.

2.2.2 *From TiCl$_4$*
Colloidal suspensions of TiO_2 were prepared by slowly adding 6 ml of $TiCl_4$ to 200 ml of deionized water at 0°C under a nitrogen atmosphere and vigorously mixing using a magnetic stirrer. Total adding time was approximately 20-30 minutes. The resulting fluid was not clear. A clear suspension was obtained after 3-4 hours of mixing at room temperature. A 100 ml portion was dialyzed against 1 liter of 0.1 M $HClO_4$ for 3 hours, the fluid remaining transparent. A precise determination of the total TiO_2 concentration was made by dissolving 10 ml of the dialyzed solution of TiO_2 colloid in 20 ml of concentrated HCl. A Ti(IV) peroxo complex was formed by addition of 3 drops of 30% hydrogen peroxide (25). The resulting colored solution (maximum

absorption at 415 nm, with molar extinction coefficient of 640 $m^{-1}cm^{-1}$) was analyzed using a Milton Roy Spectronic 3000 diode array UV-Vis spectrophotometer. The total TiO_2 concentration in the suspension after dialysis was around 80 mM. A given sol preparation was stored in the refrigerator, and was stable for many days. However, all measurements on a given preparation were carried out within 48 hours.

2.2.3 *Physical Measurement*

Although most of the studies were performed in 0.1M $HClO_4$, some measurements were carried out at pH3 and a few at other values of the pH. The pH3 suspensions were obtained by repeated dialysis against $HClO_4$ (pH3) until the dialysate reached pH3. In other cases, $HClO_4$ was added to either the pH3 or 0.1M dialyzed systems to give the final pH. It was difficult to obtain good TEM pictures in 0.1M $HClO_4$ due to the formation of large crystals, probably consisting of a titanium perchlorate or oxyperchlorate salt. However, from the TEM and quasi-elastic light scattering (QLS) measurements using a Lexel Corporation Model 85 Argon-ion laser operating at 514 nm and a Brookhaven model BI-2030 AT digital autocorrelator with CONTIN data analysis, a particle diameter on the order of 100-200 Å was obtained.

2.3 PREPARATION OF COLLOIDAL CdS

These sols were prepared in sodium dodecyl sulfate (SDS) essentially by the method of Pileni, et.al. (24). Cadmium dodecyl sulfate, $Cd(DS)_2$, was prepared as previously reported (26). Two equal volumes of 1.0 mM Na_2S in 10 mM SDS and 1 mM $Cd(DS)_2$ in 10 mM SDS were then mixed. The size of the CdS particles was determined using a Leeds & Northrup Microtrac UPA Fiber Optic QLS (FOQLS) particle size analyzer operating at 633 nm, as well as by TEM. A particle size distribution from TEM is shown in Figure 1b, with a number average diameter of 90±22Å. In some cases, the CdS suspension was dialyzed against 10 mM SDS, as previously described for the TiO_2 sol. The dialysis membrane employed was a Spectra/Pro from Spectrum Medical Industries, with a nominal molecular weight cut off (MWCO) of 1000 daltons. Electrophoresis measurements were performed on a Delsa-440 instrument from Coulter Electronics.

2.4 ELECTROCHEMICAL MEASUREMENTS

The normal and sampled (tast) d.c. polarographic measurements were carried out with a Princeton Applied Research (PAR) model 174A polarographic analyzer and PAR model 303A static mercury drop electrode, using a saturated Ag/AgCl reference electrode and platinum counter electrode. The drop times, drop masses and scan rates employed were 2-5s, 1.4-5.7 mg, and 2-10 mV/s, respectively. The cyclic voltammetry (CV) measurements employed a hanging Hg drop electrode and a PAR model 173 potentiostat/galvanostat coupled with a PAR model 175 universal programmer at scan rates of 10-500 mV/s. For the differential pulse voltammetry (DPV), the pulse width was 5 mV. In all cases, the solutions were degassed by bubbling N_2 gas through the cell for 2 minutes, and then keeping a blanket of N_2 over the cell. All measurements were taken at

$22\pm1^{\circ}C$ in a Faraday cage. Little light reached the cell in the cage which was closed on all sides but the front, which had a fine copper mesh. Experiments in which this was also covered yielded identical results.

3. Results and Discussion

Since the results for TiO_2 and CdS are quite different due to the relative solubility of the two colloids in the media in which they are dispersed, they will be considered separately. It is known, although frequently not discussed, that titanium dioxide has some solubility in aqueous acid. On the other hand, the cadmium sulfide, particularly in neutral and non-complexing aqueous media, is effectively totally insoluble with a Ksp of 8.0×10^{-27} in water (27).

3.1 TITANIUM DIOXIDE

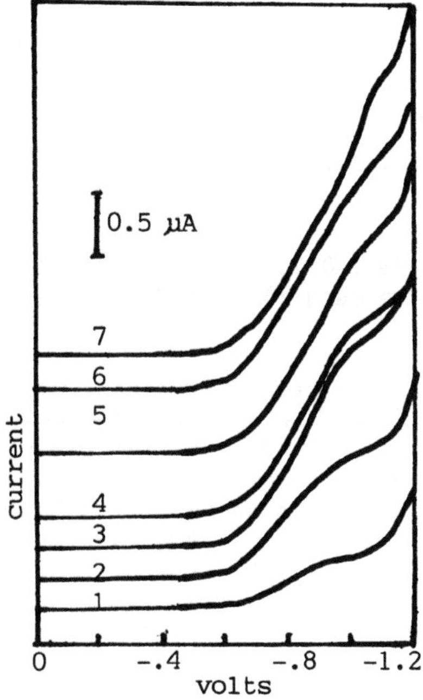

Figure 2. Tast polarograms (smoothed) of TiO_2 in 0.1 M $HClO_4$, drop time 2s, drop mass 2.8 mg, scan rate 5 mV/s at various TiO_2 concentrations, mM. (1) 0.16; (2) 0.32; (3) 0.64; (4) 0.96; (5) 1.28; (6) 1.6; (7) 3.2.

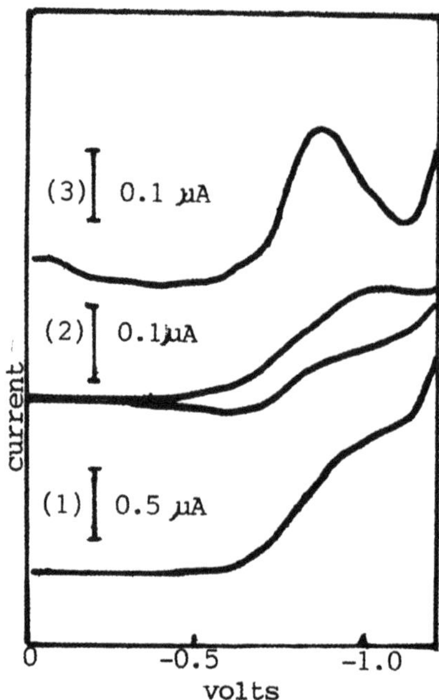

Figure 3. Current-voltage curves for 0.32 mM TiO_2 in 0.1 M $HClO_4$, drop mass 2.8 mg. (1) Tast polarogram, drop time 2s, scan rate 5 mV/s; (2) CV, scan rate 100 mV/s; (3) DPV, same conditions as (1).

The polarograms of titania sol at various concentrations of TiO_2 are shown in Figure 2. Although it may not be readily apparent from visual inspection of these half-waves, there is essentially a single wave at the lower concentrations, and a second wave appearing at higher concentrations. This is illustrated in Figure 3 which compares the polarogram, CV, and DPV of 0.32 mM TiO_2 in 0.1 M $HClO_4$. It is clear that there is essentially a single, irreversible wave with a half-wave potential ($E_{1/2}$) of -0.85 V vs Ag/AgCl (-0.89 V vs SCE). A $E_{1/2}$ value of -0.79 to -0.84 V vs SCE has been reported for Ti(IV) in 0.1 to 1.0 M $HClO_4$ (28). The reduction is irreversible at low scan rates, but becomes quasi-reversible at higher scan rates (Figure 4). This indicates that the electron transfer, presumed to be a one electron reduction of Ti(IV) to Ti(III), is followed by a slower chemical reaction of the product to a form which is electrochemically inactive in the potential range examined (0 to -1.2 V). This is curious, since the species $TiOH^{2+}$, purported to be formed by hydrolysis of aqueous Ti(III), is reported to be the electrochemically (and chemically) active oxidant (29). At higher concentrations, a second wave appears at about -1.1 V which is irreversible at all scan rates examined as shown in Figure 5. Lastly, it may be noted that the limiting current of the first half-wave increases up to about 0.5 mM TiO_2 in 0.1 M $HClO_4$, and then levels off. The value of $n_a\alpha$, where n_a is the number of electrons involved in the rate determining step and α is the transfer coefficient, is on the order of 0.3 - 0.4 for the first wave. If, as expected, $n_a = 1$ corresponding to reduction of Ti(IV) to Ti(III), then $\alpha = 0.3 - 0.4$. For the second wave, $\alpha \geq 0.5$.

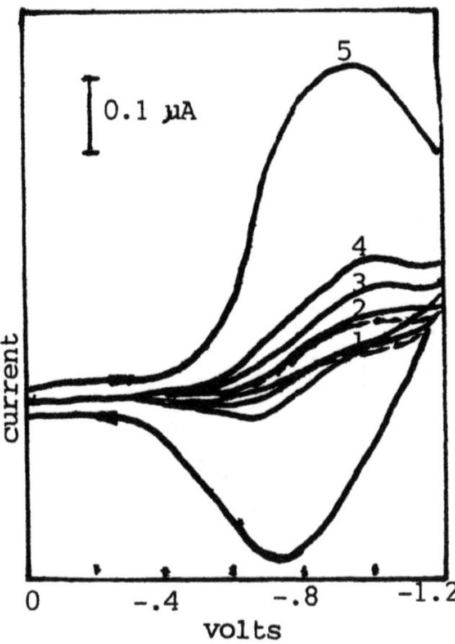

Figure 4. Cylic voltammograms of 0.32 mM TiO_2 in 0.1 M $HClO_4$ at various scan rates (mV/s). Drop mass is 2.8 mg. (1) 10; (2) 20; (3) 100; (4) 200; (5) 500. The switching potentials are at 0 and -1.2 V.

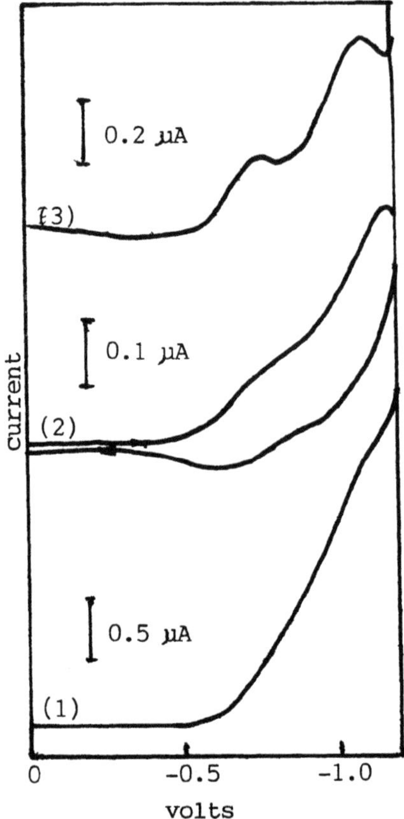

Figure 5. Current-voltage curves of 3.2 mM TiO_2in 0.1 M $HClO_4$, drop mass 2.8 mg. (1) Tast polarogram, drop time 2s, scan rate 5 mV/s; (2) CV, scan rate 100 mV/s; (3) DPV, same conditions as (1).

3.1.2 *Titanium Dioxide Solubility*
The limiting current of the first wave as a function of total TiO_2 concentration is shown in Figure 6. It is clear that the polarographic wave is due to soluble Ti(IV) species, not to the TiO_2 colloid. First, the current levels off (saturates), which should not occur if the current was due to the sol. Second, light scattering indicates that there are no particles for curve (2) in Figure 6(a) below the "break point" at about 0.5 mM. There are still some particles present in the case of curve (1), but this results from a slow rate of dissolution of the TiO_2. For example, for 0.6 mM TiO_2 in 0.1 M $HClO_4$ made by dilution of an 80 mM dialyzed suspension, it takes approximately four hours for the current to slowly increase to a limiting value, increasing by about 42%. An expanded plot of curve (2) is shown in Figure 6(b). The linear portion of the curve corresponds to a diffusion coefficient (D) of the soluble, electroactive Ti(IV) species of 1.9×10^{-5} cm^2/s. A value of $D = 7 \times 10^{-6}$ cm^2/s may be obtained from data for the reduction of Ti(IV) in 0.1 M H_2SO_4 (28). This large value of D could only correspond to soluble Ti(IV), most likely a monomeric species.

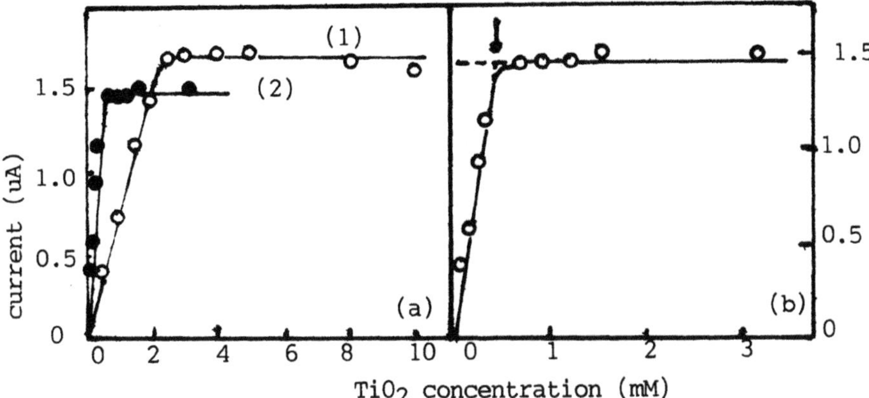

Figure 6. Limiting tast polarographic current of first half-wave ($E_{1/2} \sim -0.85$ V) as a function of total TiO_2 concentration in 0.1 M $HClO_4$. Drop time is 2s, drop mass 2.8 mg., and scan rate is 5 mV/s. (a) Samples prepared by dilution of an 80 mM dialyzed suspension. Curve (1), dilutions made directly into polarographic cell and run immediately. Curve (2), samples made up separately and run after sufficient time for equilibration; (b) Curve (2) from (a) on an expanded scale.

Therefore, the solubility of the TiO_2 sol as prepared in this study is taken as the intersection of the linear rising portion with the saturation (plateau), designated by the arrow in Figure 6(b). This corresponds to a solubility in 0.1 M $HClO_4$ of 0.48 mM. As a point of reference, it has been reported that TiO_2 colloids of about 24 Å diameter readily dissolve at low pH, and that such a 0.4

mM sol dissolved completely in about 2 hours in 0.1 M HCl (30). The resulting solution was purported to contain Ti(IV) oligomers. It was also noted that TiO_2 powder (Degussa P-25) at a concentration of 0.25 mM does not dissolve at pH1, even under reflux for 3 hours.

Values of the plateau current were also measured at other $HClO_4$ concentrations over the pH range 0.70 - 3.05. As expected, this current increases with decreasing pH, indicating a greater TiO_2 solubility with increasing acidity. If it is assumed that the diffusion coefficient of the electroactive Ti(IV) species is the same over this pH range, then the solubilities (S) can be calculated for each pH based on the value in 0.1 M $HClO_4$. These values are plotted in Figure 7. The line through the points is a least squares fit given by equation (1), where S has the units of mM.

$$\log S = -0.46\ pH + 0.21 \quad (1)$$

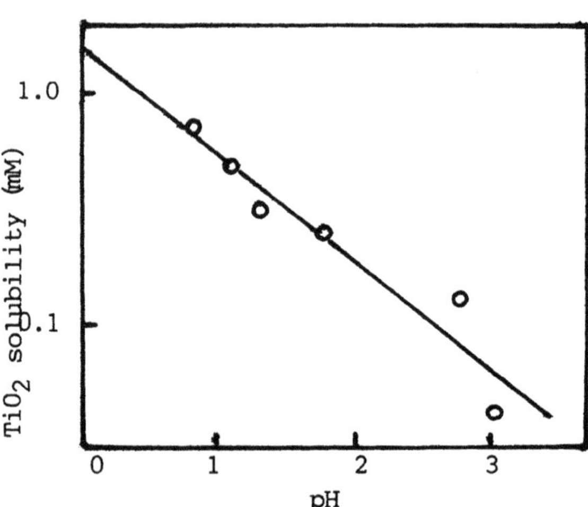

Figure 7. Solubility of the TiO_2 colloid in $HClO_4$ as a function of pH. The line is a least squares fit with a correlation coefficient of 0.956.

If the dissolution of the TiO_2 colloid is represented by a general process such as equation (2), then

$$TiO_2\ (\text{hydrated, } H^+) + mH^+ = Ti(IV)aq \quad (2)$$

the value of m is essentially ½. The presence of oxy-bridged Ti(IV) species, in addition to monomeric $TiO^{2+}_{(aq)}$, has been reported in concentrated (≥ 0.05 M) Ti(IV) solutions at hydrogen ion concentrations of 1-2 M (31). In particular, two different trimers and a tetramer were reported to be the principal oligomeric species present. The fraction of such species were expected to increase with increasing pH. Since the value of D would be expected to be greater for an oligomer as compared with the monomer, the actual values of S at higher pH would be lower than those in Figure 7. For example, an increase in D by a factor of four will result in a decrease in the calculated value of S by a factor of two. If anything, this will decrease the slope m even further. This solubility behavior of the TiO_2 sol as expressed by equation (1) is very puzzling. Dissolution of TiO_2 in acid to give TiO^{2+} would yield a value of m = 2. If the dissolved species was predominately any oligomer(s), a value of m = 1 is expected. A possible explanation is that TiO_2 sol dissolves to yield two monomeric species, such as illustrated in equation (3).

$$2\ TiO_2\ (sol) + H^+ = TiO^{2+} + TiO_3H^- \qquad (3)$$

Naturally these are aqueous species, and the TiO^{2+} species can be in equilibrium with $Ti(OH)_2^{2+}$. Also, the $TiO^{2+}_{(aq)}$ species may not be dominant in the 1-3 pH range, as illustrated in equation (4).

$$2\ TiO_2\ (sol) + H^+ \xrightarrow{2\ H2O} Ti(OH)_3^+ + TiO(OH)_2 \qquad (4)$$

These species may not be unreasonable, since hydroxy species form even at acid concentrations greater than 1 M. Both of these monomeric species would be expected to be in rapid equilibrium based on the reported exchange rates of oxo, hydroxo and aqua ligands (31). For equation (3) or (4), the $K_{sp} = 6.4 \times 10^{-7}$ M. In any event, the solubility behavior of TiO_2 nanosols in aqueous media deserves additional attention.

It is well known that, due to the operation of the Kelvin effect, the solubility of small particles should be greater than that of large particles. For example, using a typical solid surface energy of 300 erg/cm^2, a 200 Å particle would have a solubility approximately three times that of the bulk, while that of a 100 Å particle would be seven times greater. The average particle size estimated from TEM for 8 mM TiO_2 in 0.1 M $HClO_4$ is smaller than that for 2 mM. Both were made by diluting a dialyzed 80 mM suspension. Since the solubility is about 0.5 mM, only about 6% of the TiO_2 will dissolve in the 8 mM suspension, while 33% will dissolve in the 2 mM one. Thus, the original size distribution in the 80 mM suspension will be shifted to a slighter higher average value at 8 mM, but to a significantly higher value at 2 mM, as observed.

3.1.3 *Second TiO_2 Half Wave*

As may be seen from Figure 8, the peak current of the second wave increases approximately linearly with increasing TiO_2 concentration. This, coupled with the fact that the second wave is not observed at TiO_2 concentrations below the solubility limit, would tend to indicate that this reduction is related to the presence of the sol. In contrast to the first wave associated with the soluble Ti(IV), which is irreversible at low scan rates but becomes quasi-reversible at higher scan rates (see 3.1), the second wave is irreversible at all scan rates examined (e.g., see Figure 5, curve 2). Since the peak current should be proportional to the square root of the diffusion coefficient, a comparison of the slopes of the peak current *vs* TiO_2 concentration (inset, Figure 8) for the first

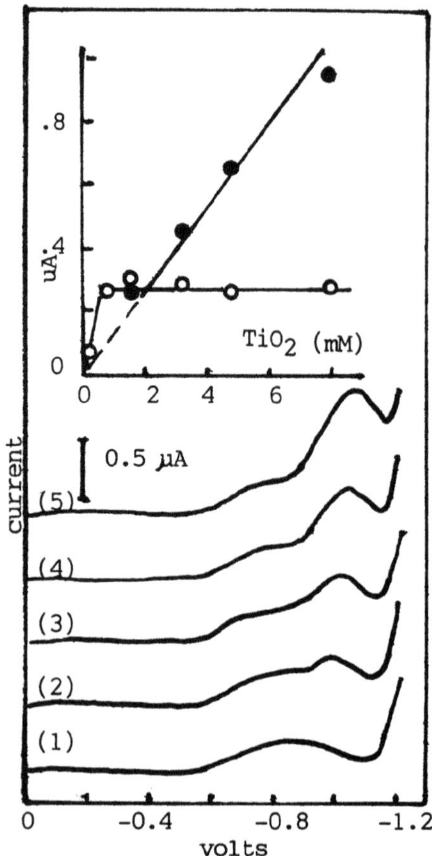

Figure 8. DPV of TiO_2 (mM) at various concentrations in 0.1 M $HClO_4$. Drop time 2s, drop mass 2.8 mg., scan rate 2 mV/s with low pass filter. (1) 0.8; (2) 1.6; (3) 3.2; (4) 4.8; (5) 8.0. The inset at the top shows the peak current for the first (open circles) and second (closed circles) waves. Curve (1), at the appropriate voltage, was used to correct the second peak current values for curves (2) - (5).

wave below the solubility limit (plateau) and the second wave yields a value for the latter of D = 5 x 10^{-7} cm^2/s. This corresponds to a particle of 100 Å diameter. However, this assumes that one electron is transferred per formula unit ("molecule") of TiO_2. Therefore, if this wave is in fact ascribed to a direct reduction of TiO_2 nanoparticles at the electrode, essentially the entire particle would have to be reduced. Although an attempt to observe direct electron transfer to the TiO_2 particle with reduction of some Ti(IV) was the motivation for this investigation, reduction of the entire particle, which would involve about 16,000 electrons, seems highly unlikely. The possibility that the aqueous Ti(III) resulting from the reduction of the soluble Ti(IV) chemically reacted with the TiO_2 particle to produce a particle containing Ti(III) with the release of Ti(IV) was considered. However, such a catalytic process would be expected to still produce only a single wave, albeit with a shifting half wave potential. At present, the nature of this second process remains unknown.

3.2 CADMIUM SULFIDE

The cadmium sulfide sols stabilized by SDS were generally on the order of 100 Å in diameter. For many of the experiments, the sols were prepared with a slight excess of either Na_2S or $Cd(DS)_2$. Dialysis of the sols reduced, but never eliminated, this excess. In one set of experiments, to be described below, cadmium perchlorate was mixed with Na_2S in the presence of polybrene (hexadimethrine bromide), a cationic polyelectrolyte, as stabilizer.

3.2.1 *CdS Electrochemistry*

The reduction of Cd(II) on mercury occurs around -0.6 V, even in the form of $Cd(DS)_2$ micelles (26). The flat band potential (E_{fb}) of CdS as a polycrystalline or single crystal electrode varies from about -0.7 V to -1.4 V *vs* SCE, depending upon the electrolyte and pH (32). It has also been reported that E_{fb} for powder and colloidal systems can be shifted to more positive values by about 0.7 V due to intergap states (33). The electrochemical behavior on mercury of the CdS sols in this study were carefully examined over the voltage range from 0 to -1.5 V at nanoampere sensitivity. No evidence could be found for any reduction process ascribed to the sol. A similar result was obtained on glassy carbon electrodes.

3.2.2. *Ion Adsorption*

As mentioned above, it was never possible to produce or obtain by dralysis a sol free of some small excess of ions in solution. Therefore, some experiments were carried out by starting with an excess of cadmium or sulfide ions, and then titrating with the other ion. A corresponding set of cyclic voltammograms are shown in Figure 9. In this case, a 0.5 mM CdS sol in 10 mM SDS is produced with a nominal Na_2S excess of about 0.1 mM, and a CV is obtained after each successive addition of $Cd(DS)_2$. A Cd(II)/10 mM SDS stock solution of sufficient concentration was used such that there was negligable dilution of the suspension in the electrochemical cell.

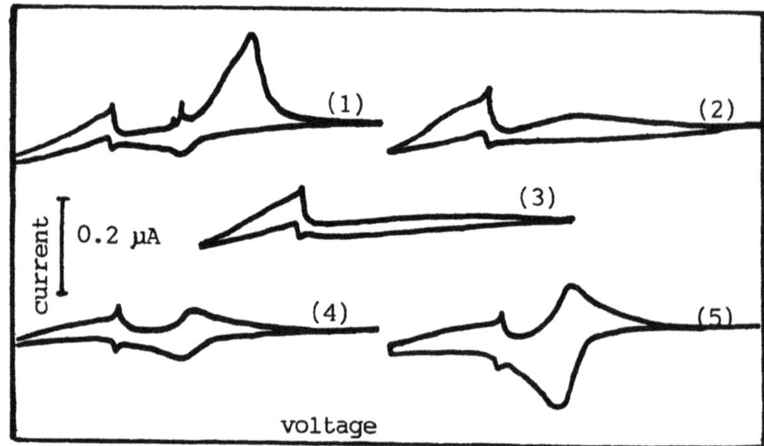

Figure 9. CV for 0.5 mM CdS sol in 10 mM SDS containing a 0.1 mM excess of Na_2S, with added mM concentrations of $Cd(DS)_2$. (1) 0.020; (2) 0.039; (3) 0.057; (4) 0.091; (5) 0.153. The drop mass was 1.4 mg., and the scan 50 mV/s. All scans run from -0.2 to -1.0 V.

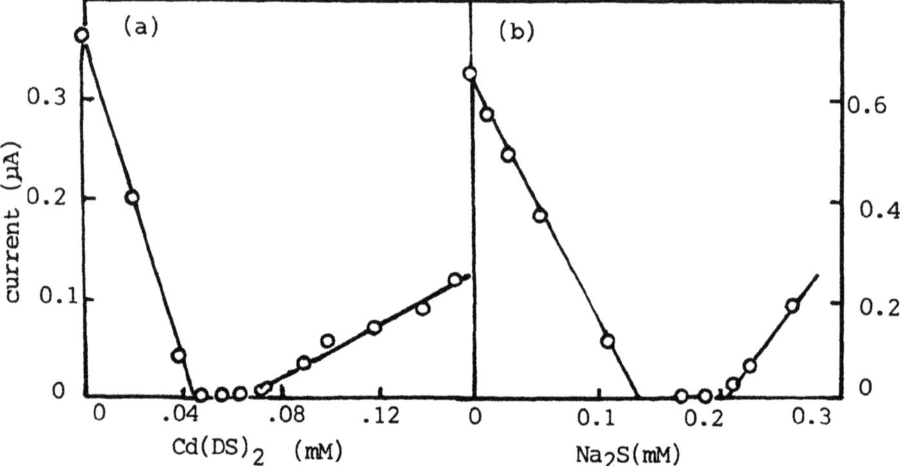

Figure 10. Cathodic peak currents from CV of 0.5 mM CdS sols containing an excess of Na_2S or Cd(II), as a function of added Cd(II) or Na_2S, respectively. (a) 0.1 mM excess Na_2S, 10 mM SDS; (b) 0.1 mM excess $Cd(ClO_4)_2$, 0.1% (w/w) polybrene.

By comparison with solutions containing 10 mM SDS and either only Na_2S or $Cd(DS)_2$, the voltammograms shown in Figure 9 by curves (1) and (2) are due to sulfide, and those shown by curves (4) and (5) are due to Cd(II). Curve (3) is identical to a background trace containing only 10 mM SDS. The sharp peaks at -0.43 V in all of the CV scans are capacitance waves due to SDS adsorbed on the mercury electrode. It was considered that these voltammograms involving either sulfide ion or cadmium ion might be due to ions adsorbed on the CdS particles. However, if this were the case, the ions would diffuse with a value of D corresponding to that of the particle, resulting in much lower currents. These voltammograms are therefore due to the ions in solution. A plot of the cathodic peak currents (i_{pc}) for the example discussed above, as well as for a system stabilized by polybrene (a cationic polyelectrolyte) and initially containing an excess of $Cd(ClO_4)_2$ titrated with Na_2S, is shown in Figure 10. The values of i_{pc} vary linearily with concentration, as expected. The slopes of the sulfide portion in both Figure 10 (a) and (b) are equal, and only slightly greater than the slope for aqueous Cd(II) in Figure 10(b). The slope for Cd(II) in Figure 10(a) is considerably less, since the diffusing species is a $Cd(DS)_2$/SDS micelle.

However, the two straight line portions do not intersect at a point, but result in a zero-current concentration "gap". The most likely explanation is that in this concentration range the ions are in fact adsorbed by CdS particle, and are not reduced. For the data in Figure 9(a), the ratio is 0.044 moles of adsorbed ions per mole (formula weight) of CdS. In order to cast this in terms of ions per particle, a particle size is needed. From FOQLS, the average diameter is about 100 Å, which combined with the density of CdS yields a particle concentration of 5×10^{-8} M, or 10^4 moles of CdS per mole of particles. This corresponds to 440 ions per particle, or about 75 Å^2 of particle surface area occupied by each adsorbed ion. This is less than monolayer coverage, possibly due to coordination and/or interference by the adsorbed surfactant (SDS). In fact, electrophoresis measurements show that the CdS particle is always negatively charged, even in the presence of excess $Cd(DS)_2$. It therefore seems likely that all of the adsorbed ions in this case are Cd(II).

For CdS stabilized by cationic electrolyte in Figure 10(b), the concentration range is larger (0.14 moles per mole of CdS). The average particle diameter in this preparation, as determined from the onset of the UV-*vis* adsorption spectrum of the sol (1, 7, 34), is 46 Å. This corresponds to a particle concentration of 4×10^{-7} M, and an area of 49 Å^2 per absorbed ion. Thus, the increase in moles of adsorbed ions per mole of CdS is due not only to a smaller particle size, and therefore an increased surface area, but also to an increase in the density of adsorbed ions on the particle surface. It may well be in this case that all of the adsorbed ions are sulfide, the Cd(II) being coulombically excluded.

4. Conclusion

No evidence was obtained in support of direct reductive electron transfer from a mercury electrode to nanosized hydrosols of TiO_2 and CdS. The TiO_2 sol particles exhibited appreciable solubility in perchloric acid, about 0.5 mM in 0.1 M $HClO_4$. In the case of TiO_2, in addition to the reduction of soluble Ti(IV), a second electrochemical process was observed above the saturation concentration with a current proportional to the total TiO_2 concentration. The nature of this process remains to be elucidated. The (log) solubility of the TiO_2 colloidal particles over the pH range of 1-3 increases linearly with the square root of the acid concentration. It is difficult to explain this dependence with conventional equilibria. The CdS sols can be titrated with cadmium or sulfide ions to yield suspension containing no (detectable) free ions of either species. Adsorbed Cd(II) or sulfide ions are not electroactive, and the extent of adsorption is less than "monolayer" coverage.

5. Acknowledgements

This work was supported by the New York State Science and Technology Foundation through the Clarkson University Center for Advanced Materials Processing. The authors wish to thank Professor M.P. Pileni for assistance with the preparation of the cadmium dodecyl sulfate and the cadmium sulfide sols, well as for valuable discussions. The support of a NATO Collaborative Research Grant CRG941221 for this purpose is gratefully acknowledged.

6. References

1. Henglein, A., (1989) *Chem. Rev.*, **89**, 1861.

2. Fischer, C.H., and Giersig, M., (1992) *Langmuir,* **8**, 1475.

3. Fischer, C.H., Weller, H., Katsikas, L., and Henglein, A., (1989) *Langmiur,* **5**, 429.

4. Henglein, A., (1989) *Chem. Phys. Lett,* **154**, 473.

5. Duonghong, D., Ramsden, J., and Gratzel, M., (1982) *J. Am. Chem. Society,* **104**, 2977.

6. Tricot, Y.M., and Manassen, J., (1988) *J. Phys. Chem,* **92**, 5239.

7. Petit, C., Lixon, P., and Pileni, M.P., (1990) *ibid,* **94**, 1598.

8. Willner, I., and Eichen, Y., (1987) *J. Am. Chem. Society,* **109**, 6862.

9. Imae, T., Muto, K., and Ikeda, S., (1991) *Colloid Polym. Sci.*, **268**, 43.

10. Albery, W. J., Bartlett, P. W., and Porter, J. D. (1984) *J. Electrochem. Soc.*, **131**, 2896.

11. Despic, A. R. and Pavlovic, M. G., (1984) *Electroanal. Chem.*, **180**, 31.

12. Riefkohl, J., Rodriguez, L., Romero, L., Sampoll, G., and Souto, F., (1987) *J. Electrochem. Soc.,* **134**, 1679.

13. Dunn, W. W.., Aikawa, Y., and Bard, A. J., (1981) *ibid,* **128**, 222.

14. Ward, M. D., White, J. R., and Bard, A.J., (1983) *J. Am. Chem. Soc.,* **105**, 27.

15. White, J. R., and Bard, A. J., (1985) *J. Phys. Chem.,* **89**, 1947.

16. Leland, J.K., and Bard, A.J., (1987) *ibid* **91**, 5083.

17. Kamat. P. V., (1985) *J. Chem. Soc. Faraday Trans.,* **81**, 509.

18. Dunn, W. W., Aikawa, Y., and Bard, A.J., (1981) *J. Am. Chem. Soc.,* **103**, 3456.

19. Henglein, A., (1982) *Ber. Bunsenges, Phys. Chem.,* **86**, 241.

20. Henglein, A., Kumar, A., Janata, E., and Weller, H., (1986) *Chem. Phys. Lett.,* **132**, 133.

21. Lawless, D., Serpone, N., and Meisel, D., (1983) *J. Phys. Chem.,* **95**, 5166.

22. Moser, J. and Gratzel, M., (1983) *J. Am. Chem. Soc.,* **105**, 6547.

23. Duonghong, D., Bargarello, E., and Gratzel, M., (1981) *J. Am. Chem. Soc.,* **103**, 4685.

24. Pileni, M.P., Motte, L., Jain, T.K., Petit, C., and Billoudet, F., (1992) in *Electrochemistry in Colloids and Dispersions*, Mackay, R.A., and Texter, J., Eds., VCH Publishers, Inc., New York, Ch. 28.

25. Moser, J., and Gratzel, M., (1982) *Helv. Chim. Acta.,* **65**, 1436.

26. Andriamanampisoa, R., and Mackay, R.A., (1994) *Langmuir,* **10**, 4339.

27. Dean, John A. (ed.), *Lange's Handbook of Chemistry*, 12th edition, McGraw Hill Book Co., New York, 1979.

28. Mate, L., Zuman, P., Rupp, E., Fenner, T. and Naraguanan, N., *CRC Handbook Series in Inorganic Electrochemistry*, Vol. VIII, CRC Press, Boca Raton, FL, 1986.

29. Kavan, L., O'Regan, B., Kay, A., and Gratzel, M., (1993) *J. Electroanal. Chem.,* **346**, 291, and references therein.

30. Kormann, C., Bahnemann, D.W., and Hoffmann, M.R., (1988) *J. Phys. Chem.*, **92**, 5196.

31. Comba, P., and Merbach, A., (1987) *Inorg. Chem.*, **26**, 1315.

32. Meissner, D., Memming, R., and Kastening, B., (1988) *J. Phys. Chem.*, **92**, 3476.

33. Finlayson, M.F., Wheeler, B.L., Kakuta, N., Park, K.-L., Bard, A.J., Campion, A., Fox, M.A., Webber, S.E., and White, J.M., (1985) *J. Phys. Chem.*, **89**, 5676.

34. Henglein, A., (1987) *J. Chim. Phys.*, **84**, 1044.

PHOTOELECTROCHEMISTRY WITH TiO_2 AEROGELS

MICHA TOMKIEWICZ AND SEAN KELLY
Dept. Of Physics, Brooklyn College of CUNY, Brooklyn, NY 11210

1. Introduction

One way of understanding light-induced charge separation processes on semiconducting substrates is to arrive at a parametrization that correlates material properties with output parameters. The Gärtner equation and its descendants [1] are classic examples in which we try to correlate output parameters of photovoltaic devices,

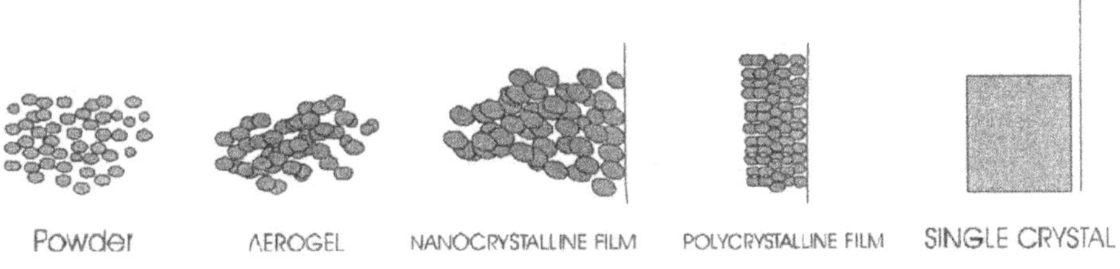

Figure 1. Different modes of aggregation for particulate matter

such as short circuit current and open circuit voltage, with material parameters such as minority carriers' diffusion length and absorption coefficient These four parameters, and many similar ones, are parameters of the collective system. In the general language of complex systems, they are emerging properties of the aggregates - in our case, in its simplest presentation - of the single crystal substrates. Photoelectrochemistry on semiconducting substrates has now reached the point where we need to develop parametrization that spans different aggregation modes. Figure 1 demonstrates the central issue; we show there five different modes of aggregation. The first mode is a powder of isolated nanocrystalline grains such as Degussa P25. The particles are well defined but the interparticle space is constantly changing. Interparticle transport can take place only through the surrounding electrolyte. The second mode is represented by TiO_2 aerogels in which the fundamental units remain the nanocrystallites but they now aggregate to establish connectivity. In this mode the surface area and most of the bulk properties are similar to ones obtained with the powders but the interparticle pore space is well defined and interparticle transport can take place not only through the electrolyte but also between particles. The next mode takes the form of nanocrystalline electrodes

E. Pelizzetti (ed.), Fine Particles Science and Technology, 403–411.

in which the morphology is similar to the aerogels but is deposited on a substrate. In this mode one can measure collective macroscopic charge-distributions such as current and voltage. The forth mode represents polycrystalline electrodes in which much of the nanocrystalline grain structure remains, although at a much higher packing density, and the void structure is replaced by grain boundaries. The final aggregation mode in this procession takes place when the individual grains and their associated grain boundaries disappear and the solid crystallizes to form a single crystal. The specific surface area across these aggregation modes varies between 1000 m^2/gr for some of the aerogels [2] to $2.5x10^{-5}$ m^2/gr for the single crystal. There is now a vast and growing literature on photoelectrochemistry with almost every one of these aggregation modes using the same semiconducting material - TiO_2. The output parameters are different - photocatalytic products with the non-electrode configurations and photovoltaic performance parameters with the electrode-style configurations. One can also find considerable overlap - photocatalytic activity is being investigated throughout the series and photovoltaic performance starting with the nanocrystalline electrodes. Subject to their intrinsic limitations (Bandgap and band location) almost all the systems are among the most efficient in their respective categories. This brings into focus the need to parametrize these systems in terms that can be applied across the different aggregation modes.

Any attempt at correlating photoelectrochemical properties across the different aggregation modes critically depends on our ability to parametrize, measure and control the morphologies of these materials and on our ability to measure in a reproducible way the quantum efficiency of output parameters that correlates with the efficiency of the charge separation process. The conversion efficiencies are output parameters that are the end result of a complex chain of events that include absorption of the light, efficiency of carrier separation within a particle, adsorption properties of the electrolyte, charge transfer to the electrolyte, turnover number of the electrolyte, diffusion of reactants and products, etc. In addition to the intrinsic complexity, an important contribution to the balkanized efforts to understand these systems rests on disciplinary segregation imposed by training constraints.

In this paper we will describe parametrization of the morphology and measurements of quantum efficiency of photodecomposition of salicylic acid by TiO_2 aerogels. Photoelectrochemistry on aerogels is perhaps the least described among the aggregation modes that werementioned above. Aerogels are highly porous materials that are produced via sol-gel processing and supercritical drying. The very high surface area and pore volume of aerogels makes them attractive candidates for catalytic applications. The morphology of the TiO_2 aerogels has been characterized in terms of two length scales. Typical morphology constitutes of 5 nm diameter, rough, crystalline nanoparticles of anatase closely packed into mesoaggregates near 50 nm in size. The mesoaggregates are, in turn, packed to a loosely linked structure with an overall porosity of 80%. The total surface area of these aerogels is attributed to the sum of the

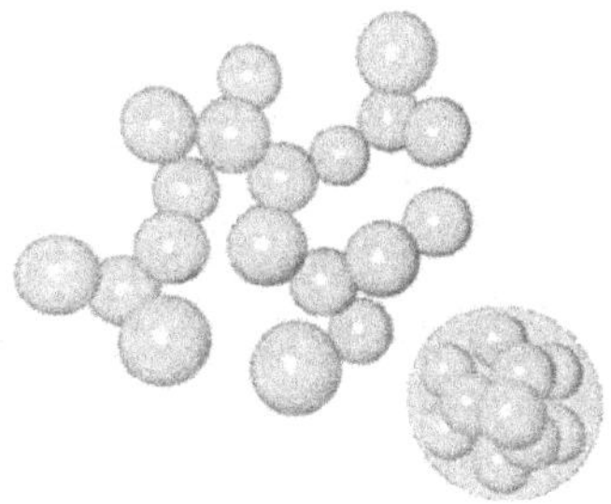

Figure 2. Schematic representation of aerogel

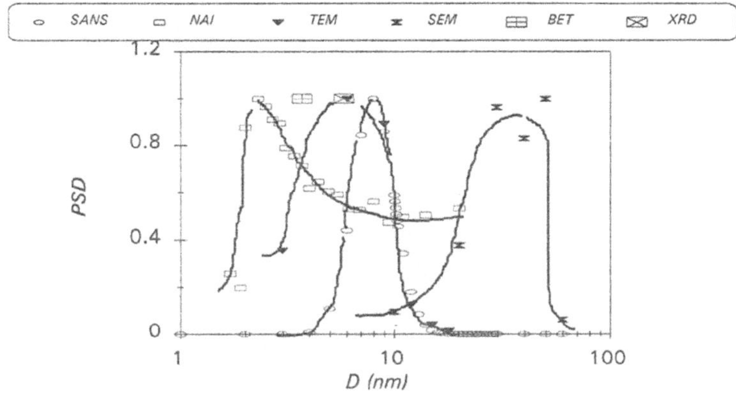

Figure 3. Particle Size distribution in a typical TiO_2 aerogel as determined by various techniques [5]. The techniques include SANS(Small Angle Neutron Scattering), NAI(Nitrogen adsorption isotherm [measures the pore distribution], TEM (Transmission Electron Microscopy), SEM (Scanning Electron Microscopy), BET (Brunauer-Emmett-Teller isotherm) and XRD (X-ray diffraction)

surfaces of the nanoparticles [4-6]. Schematic representation of this structure is given in figure 2. In figure 3 we show the particle and pore size distributions as determined by various experimental techniques [5,6]. Aside from the data obtained from the nitrogen adsorption isotherms at high partial pressure which measure pore size distribution, the rest of the data describe the particle size distribution. The bimodality of the distribution is evident, with SEM the only technique that is sensitive to the mesoparticles. In the original publications [4-6] we emphasized the agreement between the various techniques, taking into account that some of the techniques (SANS) are model sensitive, some are not, and some of the techniques involve averaging from macroscopic samples (XRD, BET) while others are local (TEM, SEM). Within these differences and considering the available dynamic range, a factor of two difference in particle size seems to support the notion that the system is homogeneous and that the particle distribution is approximately uniform.

The function-morphology correlation is an issue that needs to be addressed on a number of different scales. Previously we argued [3] that the simultaneous need to optimize mass transport and adsorption of reactants requires morphology of pore structure that is most likely fractal. We have also shown that preliminary comparative photocatalytic activity in photodecomposition of salicylic acid with various aerogels, xerogels and powders of TiO_2 approximately scales with the total BET surface area [2]. These experiments were run under conditions not limited by mass transport. Qualitatively, the arguments stated in [3] apply globally to catalytic reactions. The catalyst acts as an extended surface on which the catalytic reaction takes place. The quality and nature of the bulk of the catalyst makes very little difference. Semiconducting photocatalysts are different in the requirement that the light absorption process takes place in the bulk. The light absorption induces charge separation and the separate charges need to diffuse to the solid-liquid interface to participate in the catalytic process. Thus a key additional set of parameters that plays an important role with these materials are the bulk properties of the individual crystallites.

The morphology of the TiO_2 aerogels is complex enough for an attempt at a holistic approach at the photocatalytic process that will take into account apparently conflicting requirements. The essence of such an approach is to be able to make meaningful comparisons between different photocatalysts. Such comparisons require sophisticated parametrization that is not easily available. It is not required that such a parametrization be unique, but it is required that it will be consistent. For meaningful comparisons, it is necessary not only to be able to measure the appropriate parameters, but also develop synthetic techniques for a selective control of these parameters.

For the TiO_2 aerogels, the control of the morphological parameters can come at almost every stage of the preparation procedure: At the onset - during the colloidal stages of the sol - appropriate control parameters are water contents, temperature, acidity, ionic strength, etc. During the aging process, one can affect the strength of the connectivity between the individual mesoparticles by adjusting the chemistry of the gelation environment. Finally - at the aerogel stage - annealing changes the crystallite size, and at around 700 ^{0}C, one can induce a phase change from Anatase to Rutile. Many of the synthesis - morphology correlations seem to be independent of the oxide and thus the work on photoactive oxide can benefit from the much larger parallel effort in the SiO_2 aerogels.

2. Experimental

The detailed preparation procedure and the catalytic activity of TiO_2 aerogels in the photodecomposition of salicylic acid were described in [2]. Surface area (S_{BET}) of the aerogels was determined by the BET method, in the partial pressure range of $0.05 < P/P_0 < 0.3$. The X-ray measurements were carried out using a Rigaku's diffractometer with horizontal goniometer D/MAX-B, Cu sealed tube as X-ray radiation source and a scintillation counter detector. The power settings were 40 kV and 20 mA. The average crystallites sizes were calculated from the peak width using the Scherrer equation .

2.1 QUANTUM EFFICIENCIES:

The difficulty in measurements of the quantum efficiencies for photodecomposition of substrates on such aggregates has been discussed in various forums. The spread of quantum efficiencies for oxidation of the same substrate by

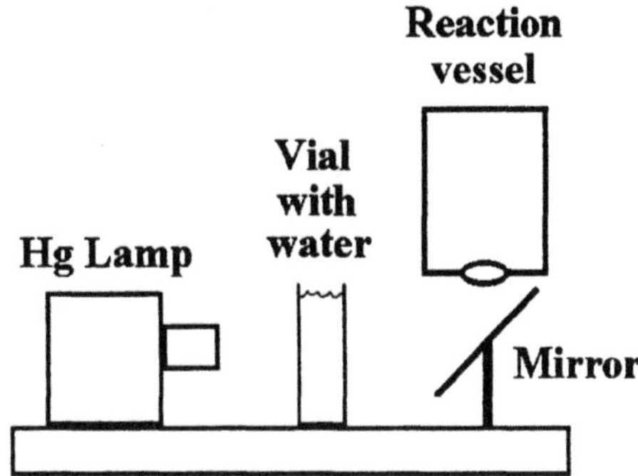

Figure 4. Experimental setup for photodecomposition measurement

different commercial sources of TiO_2 [7], combined with other difficulties, was used as an argument to abolish the attempts to measure this quantity and substitute instead a concept of "relative photon efficiency" [8] which normalizes the results to the degradation efficiency of a standard compound such as phenol under a given set of experimental conditions. There is no doubt that such a procedure can increase reproducibility and should serve as an important intermediate step in reporting efficiencies. However it masks the expected variation with the solid state properties of the catalysts, which is the subject of interest here.

Figure 4 shows the experimental setup for quantum efficiency measurements that we presently use. It differs from our previous setup [2] in two major aspects: The light enters the sample from the bottom of the reactor and the experimental conditions are adjusted to insure total absorption. This setup is based on the premise that quantum efficiencies can be accurately measured, irrespective of the morphology of the absorber, as long as one is assured of total light absorption of the incident photons. Under these conditions the amount of TiO_2 will be irrelevant. The bottom-up configuration was designed to insure that measurements take place after the samples are settled at the bottom. Although our primary, practical, objective in using aerogels was to be able to adjust their density to be below water so that they can float, in the absence of a hydrophobic coat, the pores will fill-up with the solvent and the aerogels will sink. We have measured the light intensity at the entrance and the exit (top and bottom) of the reactor. The reactor was constructed from Teflon such that a minimum amount of light will be absorbed by the walls. The light source was an 200W Hg lamp. Irradiation intensity in the spectral range between 300-400 nm, as measured with a radiometer (Eppley PSP) and cut off filters, was 8 W/m^2. All the quantum efficiencies that we quote here are for reactant disappearance and not for product formation. The quantum efficiencies are not optimized. Oxygen was not deliberately added to the reaction vessel. Thus, oxygen used in the reaction is atmospheric oxygen dissolved in the solution. Photo-degradation was quenched by filtration of the samples.

Table 1 shows the properties of the seven aerogels that we discuss here, together with a Degussa sample. The morphology and crystallinity of the first three samples were controlled by adjusting the water contents of reactants while those of the last four samples were modified by heating the dry aerogels at the specified conditions. The crystal structure in the first five aerogels is anatase, the crystal structure of the last two aerogels is rutile and the crystal structure of degussa is mixed anatase/rutile.

TABLE I

BET surface area, crystallite size and the method of adjustment of samples discussed in this paper.

Sample	BET Surface Area (m^2/g)	Crystallite Size (nm)	Adjustment	Calculated Surface Area (m^2/g)
A	735	1.7	$H_2O:Ti(OP^i_r)_4$	464
B	472	2.0	$H_2O:Ti(OP^i_r)_4$	395
C	449	2.2	$H_2O:Ti(OP^i_r)_4$	359
D	362	4.3	200^0C - 2Hrs.	184
E	134	9.7	500^0C - 2Hrs.	81
F	72	13.8, rutile	700^0C - 6Hrs.	56
G	48	20.3, rutile	750^0C - 14Hrs.	38
Degussa	54	>20		39

3. Results and Discussion

Figure 5 shows the dependence of the quantum efficiencies on the BET surface area and figure 6 shows the variation of the surface area normalized quantum efficiencies with the crystal size. Figure 5 shows a typical monotonic increase of the conversion efficiency with the total surface area that approaches saturation at high surface area. Figure 6 shows a linear increase in efficiency with crystal size.

It is obvious that the two figures are contradictory in terms of a simple geometrical interpretation. If the only structural features were the nanocrystallites, then a simple geometrical consideration would require that the surface area/unit weight be inversely proportional to the crystallite size. In the last column in table 1, we show the surface area as calculated from the crystallite size as measured by X-ray diffraction. For all the samples listed, this surface area is smaller than the BET area. In some cases (samples B and C) the two values are close and in other (sample D) the difference is by close to a factor of two. The others are somewhere in between. The obvious explanation is that X-ray diffraction is biased in favor of the large crystallites, while BET surface area, at least in principle, is more sensitive to the small crystallites. This bias can be seen in the particle size distributions shown in figure 3. Even BET measurements depend on the penetration of the adsorbing gas. Close-packed small crystallites imply

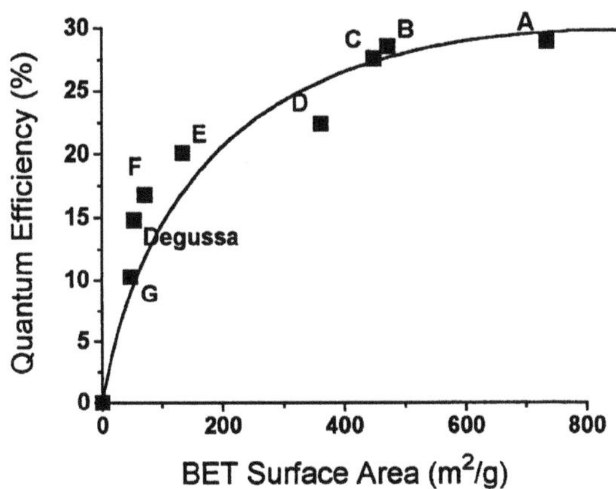

Figure 5. Monotonic increase in quantum efficiency with increasing surface area. Refer to Table 1 to identify individual points

very small pores. We have already demonstrated, in samples that showed a much better agreement between the X-ray crystallinity and the BET surface area, that the skeletal density obtained with Helium pyknometry indicates that the Helium does not penetrate into the mesoparticles [4,5]. This implies that the BET is an underestimate and that thedifference between the "real" surface area and the one calculated from the X-ray

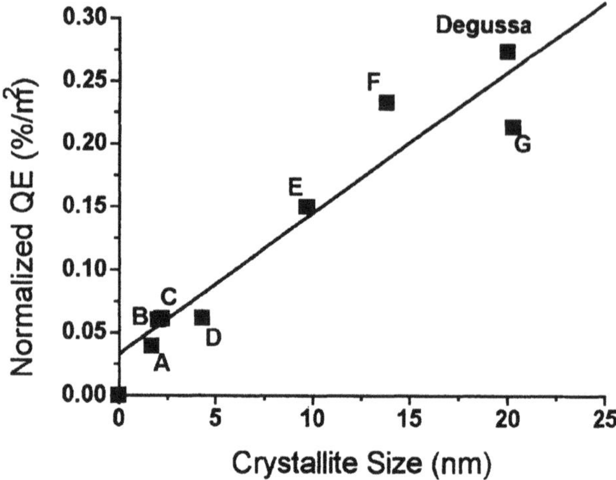

Figure 6. Normalized quantum efficiency vs. crystallite size

diffraction radius is even larger. This is obviously not a problem for the Degussa sample where the particles are separated. One possible consequence of such an analysis is that the apparent saturation in the efficiency vs. surface area behavior is not only due to saturation in adsorption but also due to the presence of very small crystallites (< 1.5 nm) that contribute to the surface area but very little to the charge separation process.

An interesting conclusion that emerges out of the results shown in figures 5 and 6 is that the quantum efficiencies seem to be independent of the crystal structure of the crystallites.

It is, perhaps, better at this stage to regard the X-ray measured crystalline radius and the BET surface area as two independent parameters. The crystalline radius will help to quantify the bulk properties such as light absorption and charge separation efficiencies, while the surface area characterizes the surface properties, adsorption, interaction between the charge carriers and substrates etc. Better correlations and cross correlations will come from better control over the morphology and crystallinity and in particular synthesis of a more uniform particle size and a larger data base from which one might draw the correlations.

ACKNOWLEDGMENTS

This work was supported by the US Department of Energy and the New York State Energy Research and Development Authority

4. References

1. See M.A. Butler, J. Appl. Phys., 48, 1914 (1977).
2. G. Dagan and M. Tomkiewicz; J. Phys. Chem, 97, 12651 (1993).
3. M. Tomkiewicz, G. Dagan and Z. Zhu; Research on Chemical Intermediates, 20, 701 (1994.)
4. Z. Zhu and M. Tomkiewicz; Proc. Mater. Res. Soc., 346, 751 (1994).
5. Z. Zhu , L.Y. Tsung and M. Tomkiewicz; J. Phys. Chem (1995).
6. Z. Zhu, M. Lin , G. Dagan and M.Tomkiewicz; J. Phys. Chem (1995).
5. V. Augugliaro, A. Franco, F. Inglese, L. Palmisano and M. Schiavello in Proc. Of the 9th. Int. Conf. On Photochem. Conv. And Stor. Of Solar Energy; E. Pelizzetti and M. Schiavello edts. 274 (1992).
6. N. Serpone, G. Sauve, E. Pelizzeti, P. Pichat and M.A. Fox in Proc. Of the 10th. Int. Conf. On Photochem. Conv. And Stor. Of Solar Energy; G. Calzaferri edt. B1 (1994).

CONTROL OF THE SIZE AND SHAPE OF NANOPARTICLES

M. P. Pileni[1,2*]; A.Hammouda[1,2]; I.Lisiecki[1,2]; L. Motte[1,2]; N.Moumen[1]; and J.Tanori[1,2]

1- Université P. et M. Curie, Laboratoire S. R. S. I. , BP 52, Bat 74, 4 place Jussieu, 75005 Paris, France

2- C.E.N. Saclay, DRECAM-SCM, Bat 522, 91191 Gif sur Yvette, France

1. Introduction

The use of dispersed media to synthetize microparticles in "situ" has made considerable progress in the last few years: Reverse micelles[1], Langmuir-Blodgett films[2], zeolites[3], vesicles[4], glass matrices and sol-gel methods[5]. Semiconductor nanocrystals present quantum confinement effects, called Q-dots[6]. To control the size and the shape of nanoparticles reverse and normal micelles, bicontinuous and lamellar solutions have been used.

In the present paper it is shown a control in the particle size by colloidal assemblies[7-11]. In reverse micellar solution the size is controlled by the size of the droplet whereas in oil in water droplet, the size is controlled by the concentration of the reactants. This has been observed in reverse micelles, with copper, silver and cobalt metallic particles and silver sulfide semiconductors. Furthermore, by using reverse micelles latexes having a diameter close to 2nm is obtained. In oil in water micelles the control of the size is shown with copper and cobalt ferrite particles. The shape of the nanosized particles is controlled either by using bicontinuous structure or lamellar phase as the microreactor or

E. Pelizzetti (ed.), Fine Particles Science and Technology, 413–429.

by using oil in water micellar solutions.

2. Experimental section

2.1. PRODUCTS

Sodium di(2-ethylhexyl) sulfosuccinate usually called Na(AOT), sodium sulfide Na_2S and dodecanethiol were obtained from Sigma, Janssen and Merck respectively. The solvents such as isooctane, cyclohexane, heptane and ethanol were purchased by Fluka. Silver and copper di(2-ethylhexyl) sulfosuccinate, Ag(AOT) and $Cu(AOT)_2$, were prepared as described previously[12].

2.2. SYNTHESES

2.2.1. *Synthesis of silver sulfide particles:*

Aqueous solution containing sodium sulfide is mixed to reverse micelles formed by Na(AOT) and Ag(AOT). The water content is fixed by the amount of aqueous solution containing sodium sulfide added. Dodecanethiol is added to reverse micellar system containing Ag_2S nanocrystallites. The nanocrystallites coated by dodecanethiol are dispersed in heptane forming an optically clear solution.

2.2.2. *Synthesis of copper metallic particles*

i) In reverse micelles: Hydrazine, N_2H_4, used as a reducing agent is added to micellar solution formed either by mixed Na(AOT)-$Cu(AOT)_2$ or $Cu(AOT)_2$ reverse micelles. A change in the water content of the micelle, w, is obtained by adding H_2O to the micellar solution before the reaction takes place. The total AOT concentration is kept equal to 0.1M.

ii) In oil in water micelles: Same procedure as above has been used by replacing $Cu(AOT)_2$ by copper dodecyl sulfate, $Cu(DS)_2$.

2.2.3 *Synthesis of cobalt ferrite particles*

A synthesis is performed in oil in water micellar solution by mixing $Co(DS)_2$, and $Fe(DS)_2$ micellar solution to methylamine, CH_3NH_3OH. The change in the particle' size is obtained by changing $Fe(DS)_2$ concentration keeping the ratio

$[Co(DS)_2]/[Fe(DS)_2]$, and $[Fe(DS)_2]/[CH_3NH_3OH]$ constant.

2.2.4. *Synthesis of latexes*

Didecyldimethylammonium methacrylate, $C_{10}MA$, reverse micellar solution is UV irradiated for 7 hours at various water contents.

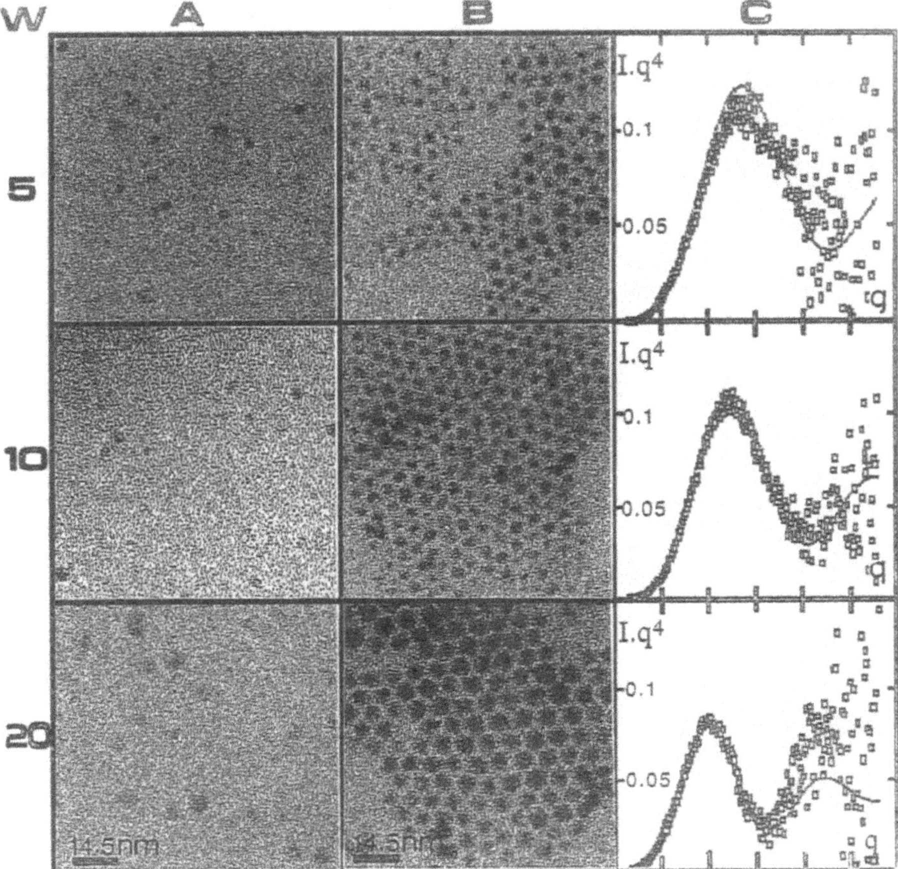

Figure 1: A: Electron microscopy patterns of silver sulfide particles in reverse micelles at w=5; w=10 and w=20. B: Coated silver particles obtained by extraction from reverse micelles having water content equal to w=5, 10 and 20. C: Porod plot obtained by S.A.X.S. of coated particles dispersed in heptane.

3. Results and discussions:

1. Monolayer made of silver sulfide particles[13]:

The size of the Ag_2S particles synthetizes, observed by T.E.M. (fig.1A) increases linearly from 2 to 10nm with the water content,w. This indicates a control of the crystallite size by the water content. At low water content (below w=10), the size distribution (30%) is large compared to what it is observed at higher water content (13%).

The silver sulfide nanosize particles synthesized in reverse micelles are coated by dodecanthiol and dispersed in heptane (fig. 1B). The extraction has been possible for particles having a diameter lower or equal to 6nm. The absorption spectra of various sizes coated Ag_2S particles are similar to those obtained in reverse micelles. The sizes of coated particles are determined by T.E.M. (fig. 1B) and compared to those observed by S.A.X.S.. Figure 1C shows the simulated and experimental data obtained on the Porod plot representation of particles made in reverse micelles having various water contents, w. It is observed a well-defined maximum and minimum indicating a low polydispersity in size (fig. 1C). The particle diameter is deduced from the maximum and minimum of the Porod plot representation. A good agreement between the average size determined by S.A.X.S. and T.E.M. is obtained.

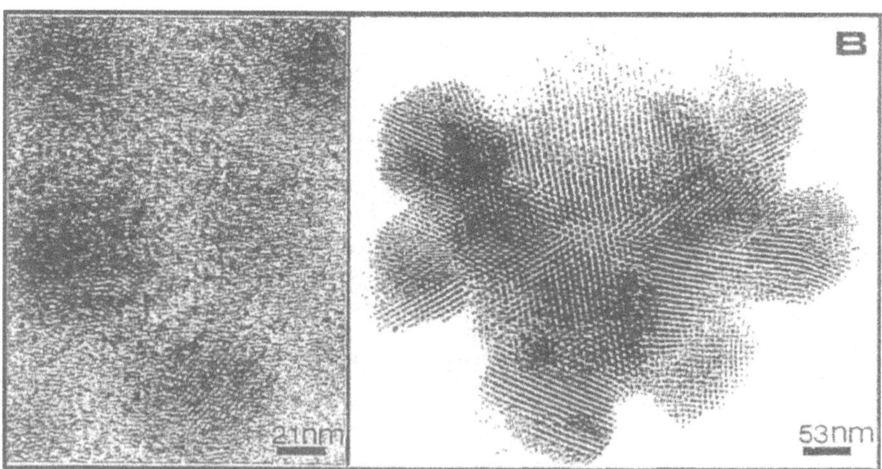

Figure 2: High resolution of Ag_2S particles (**A**), large domain of observation by T.E.M. of particles (**B**)

The silver sulfide particles are well crystallized. As matter of fact, Figure 2A shows the intereticular distance of the lattices obtained from high electron microscopic resolution. The intereticular distance is found equal to 0.26 nm and is attributed to (-1 2 1) lattices' planes.

Figure 1B shows a monolayer made of nanosized particles organized in a hexagonal network. This process takes place for various particle sizes. For a given size of particles figure 2B shows formation of multilayers in a very large domain keeping its hexagonal network. From the differences in the electron microscopic contrast and from preliminary results obtained by X rays diffraction superposition of monolayer takes place with an organization in which the particles' seat under another particle and not in a compact network.

The interparticle spacing is unchanged with the particle size. It is closed to the length of hydrocarbon tail of dodecanethiol derivative. As matter of fact, from T.E.M., the average distance between particles is found close to 1.8nm. The length of dodecanethiol tail, l, is evaluated from the empirical equation given by Bain and al[14]: $l = 0.25 + 0.127.n$ (nm) where n is the number of CH_2 groups. The value calculated is found equal to 1.77nm which is very close to the average distance determined by T.E.M. (fig. 2A).

When the particles are coated a powder made of nanosize particles is obtained. The X rays diffragram has be performed with particles powder having 6nm as an average size. The X rays diffraction pattern is compared to simulated diffractogram of the two phases which exist in the bulk phase. As matter of fact, bulk silver sulfide exhibits a reversible crystallographic phase transformation at T_c= 180°C. The stable phase below T_c is a monoclinic structure and is called βAg_2S. Above T_c, a cubic phase appears which is called αAg_2S. A good agreement between the experimental and simulated spectra is obtained for a monoclinic structure, β Ag_2S.

2. Change in the size and shape of copper metallic particles

From absorption spectroscopy it has been demonstrated that colloïdal copper particles are characterized by a plasmon peak. The intensity and the position of this peak depend on the particles size[15] and on the shape[17]. The particles are observed by T.E.M. High-resolution electron microscopy shows an interreticular distance of 0.2nm which is in good agreement the lattice of metallic copper.

Electron diffraction showed concentric circle characteristic of a face-centered-cubic phase with a lattice dimension equal to 0.361nm. The presence of oxide is detected by electron diffraction.

3.1. OIL USED AS BULK PHASE

3.1.1 *Change in the particle sizes*[15]

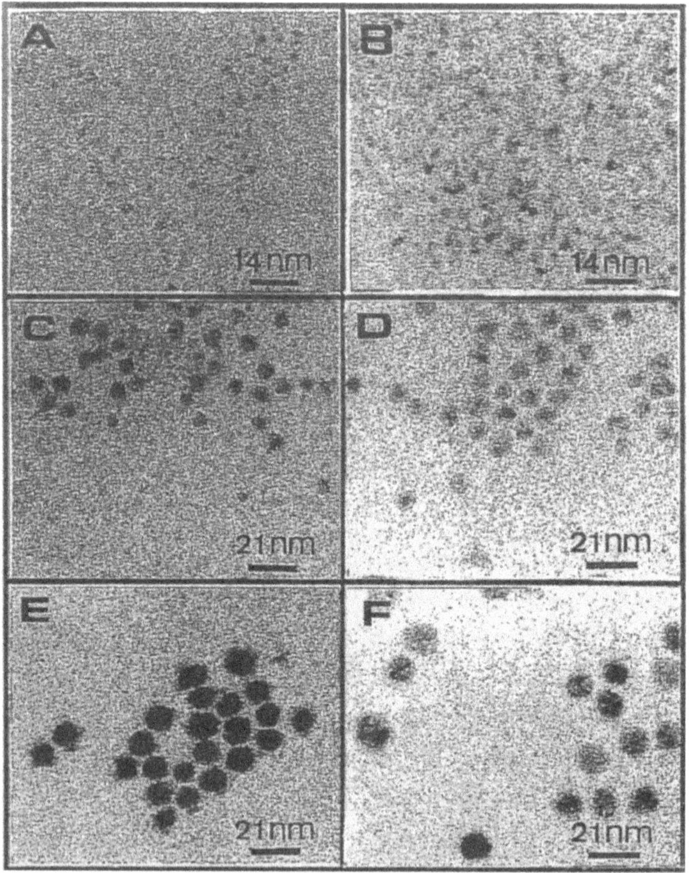

Figure 3: Electron microscopy patterns of copper metallic particles obtained at various w values: **(A)** w=1; **(B)** w=2; **(C)** w=3; **(D)** w=4; **(E)** w=5; **(F)** w=10.

In reverse micelles, the size of copper metallic particles is controlled by the water content. As matter of fact, figure 3 shows the data obtained by T.E.M. at various water contents. It is clearly observed, below w= 10, an increase in the particle size from 2 to 12nm with the water content. At water content up to 10, the size of the particles remains unchanged with an increase in the polydispersity.

3.1.2. *Change in the particle shape* [16]

Previous structural study of $Cu(AOT)_2$-water-isooctane reverse micelles[12], demonstrates a change in the shape of the water droplets with the water content, w=[H_2O]/[$Cu(AOT)_2$]. Below w=3, spherical droplets are formed whereas between w=3 to 5, the droplets are elongated with a cylindrical structure. Syntheses performed in cylindrical droplets are shown on figure 4A. It is observed the formation of spherical copper metallic particles with an average radius equal to 10nm. However it is observed the appearance of small amount of cylindrical particles.

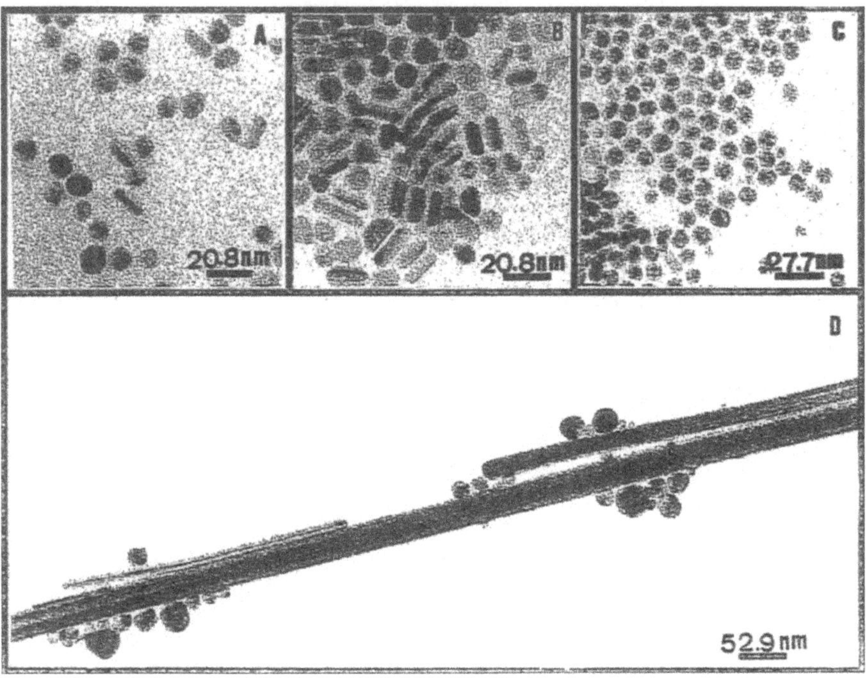

Figure 4: Electron micrographs of metallic copper particles prepared in a homogeneous cylindrical reverse micellar solution (**A**) in the L_2(I) (**B**), L_2(III) (**C**) phases and in lamellar phases L_α [**D**]

At w=5.5, the water in oil droplets (L_2 phase) is destabilized with the appearance of a liquid-gas phase transition. The upper phase is pure isooctane and the surfactant is located in the lower phase. Synthesis performed at w=5.6 shows the formation of 35% of cylindrical copper metallic particles with the presence of above 65 % of spherical particles. The average cylinders are 19nm long with a diameter of 7nm, whereas the average diameter of the spheres remains equal to 10nm (fig. 4B).

At w=9.8, a third turbid and birefringent phase appears. Synthesis performed at w=9.8, shows the formation of rods metallic copper particles (figure 4D). The diameter of the rods varies from 10 to 30nm and the length from 300 to 1500nm. Similar average size of copper metallic particle is obtained at w=14. High resolution electron microscopy shown on figure 2B indicates a high crystalinity with a very low amount of defects. The formation of such large rods having a high crystalinity indicates that the phase structure governs a slow nucleation.

Above w=16, a new isotropic phase appears in coexistence with $L(\alpha)$ and isooctane phases. As above, the increase in the water content induces a decrease in the $L(\alpha)$ and isooctane phases and an increase of the isotropic phase called $L_2(II)$ phase. Copper metallic particles made at w=20 are spherical nanosized crystallites with an average diameter equal to 10nm (figure 4C). Again a some amount of cylindrical particles is obtained.

3.2. WATER USED AS BULK PHASE

At the C.M.C., a well-defined plasmon peak centered at 570nm is observed without absorption at 800nm (fig. 5). This indicates the formation of pure copper metallic particles. By T.E.M., it is observed isolated aggregate forming an interconnected network.

By a progressive increase of $Cu(DS)_2$ concentration (above C.M.C.), the intensity of the plasmon peak decreases keeping a low absorption at 800nm. By T.E.M. it is observed a progressive fragmentation of the network with formation of isolated particles. Furthermore, it is observed a change in the shape of the particles from elongated to spherical one.

Spherical metallic copper particles characterized by a plasmon peak centered at 558nm. The maximum of the absorption spectrum obtained for elongated particles

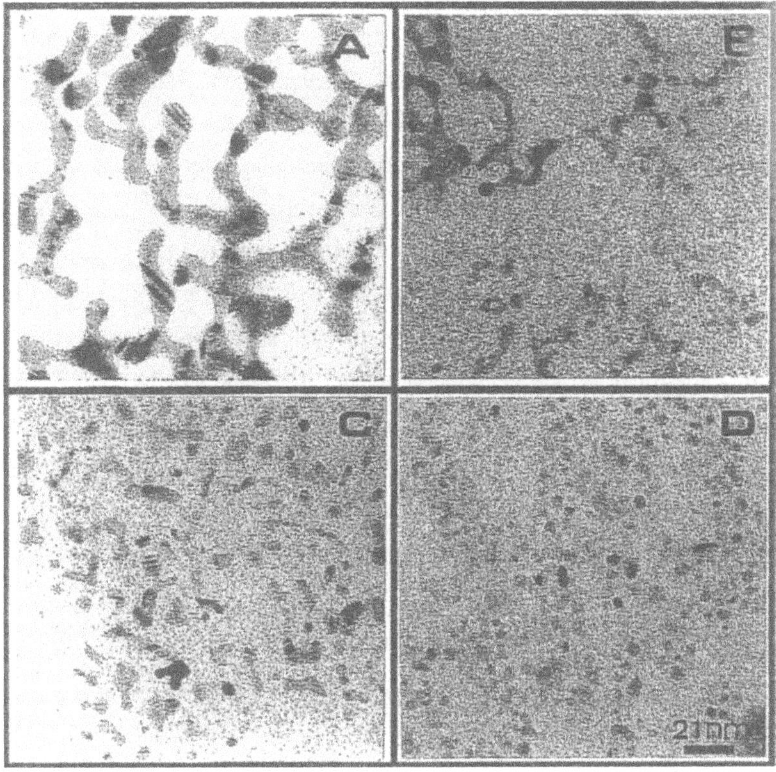

Figure 5: T.E.M. patterns and absorption spectra of copper metallic particles synthetized at various $Cu(DS)_2$ concentrations. **(A)** $[Cu(DS)_2]=1.2.10^{-3}M$; **(B)**; $[Cu(DS)_2]=2.10^{-3}M$; **(C)** $[Cu(DS)_2]=3.10^{-3}M$; **(D)** $[Cu(DS)_2]=.10^{-2}M$;

(corresponding to $[Cu(DS)_2]= 2.10^{-3}M$), is characterized by a maximum centered at 564nm whereas for interconnected network of copper metallic particles, (corresponding to $[Cu(DS)_2]= 1.2\ 10^{-3}M$), the maximum is centered at 570nm. Figure 6 shows a red shift in the plasmon peak when the shape of the copper metallic particles changes from spheres to rods. Such changes in the absorption spectra of copper metallic particles can be related to those predicted from simulation[17] at various r values, where r is defined as the ratio of the length over the diameter of a cylinder. From simulation, the red shift in the plasmon peak from 558 to 620nm corresponds to a change in the r value from 1 (sphere) to 3.5 (cylinder). Figure 6 shows formation of rods with a plasmon peak centered to 570nm that corresponds to an r value equal to 2.5. From the imagery analysis of the skeleton of the interconnected network corresponding to a plasmon peak centered at 570nm, the average length of linear strands is found equal to 22nm. The average minor diameter is equal to 6.5nm. This corresponds to an r value equal to 3.5. From the theoretical predictions such r value (r=3.5) corresponds to a higher shift in the plasmon peak (620nm) than that obtained (570nm). These differences in the correlation between the maximum of the plasmon peak and the r value could be explained by the fact that the simulation is related to one size of cylindrical particles. Furthermore, the polydispersity has to be taken into account. Hence a qualitative shift in the maximum of the plasmon peak with the particle shape is in good agreement to those predicted. Hence the shift and the ripening in the absorption band, compared to spherical and elongated particles indicate that the interconnected network exist in solution with a change in the effective mean free path of conduction electron.

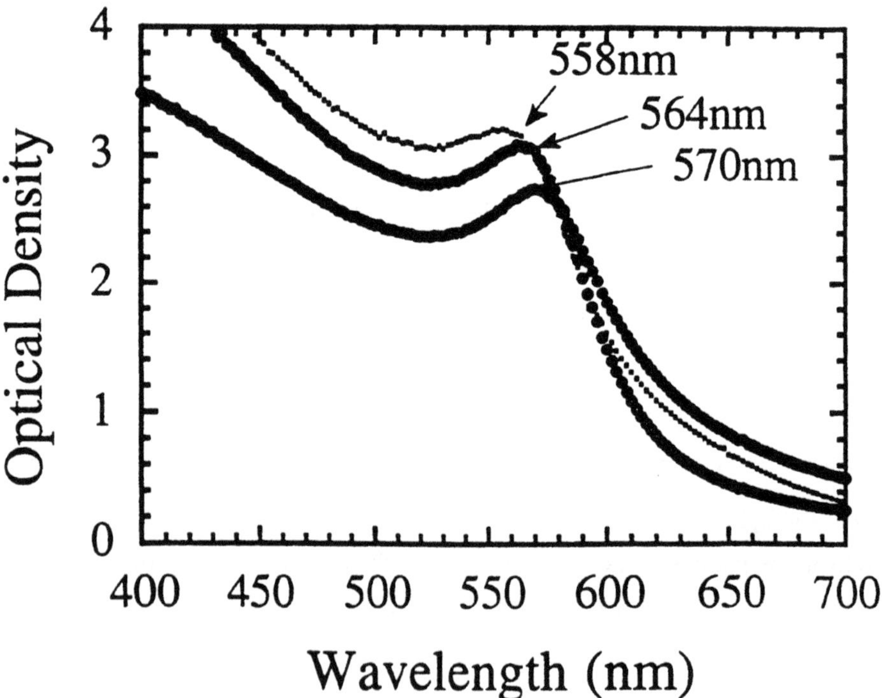

Figure 6: Absorption spectra of copper metallic particles differing by their shape.

4. Control of the size of Cobalt ferrite fluids[18]

By changing the $Fe(DS)_2$ concentration the size of cobalt ferrite changes from an average diameter equal to 5nm to 2nm (fig. 7). The size distribution remains unchanged with the average diameter and is found close to 30%. The electron diffractogram pattern indicates the formation of particles having a spinel crystalline structure as in the bulk phase. The average diameter of the particles determined by S.A.X.S. confirms that obtained by T.E.M.

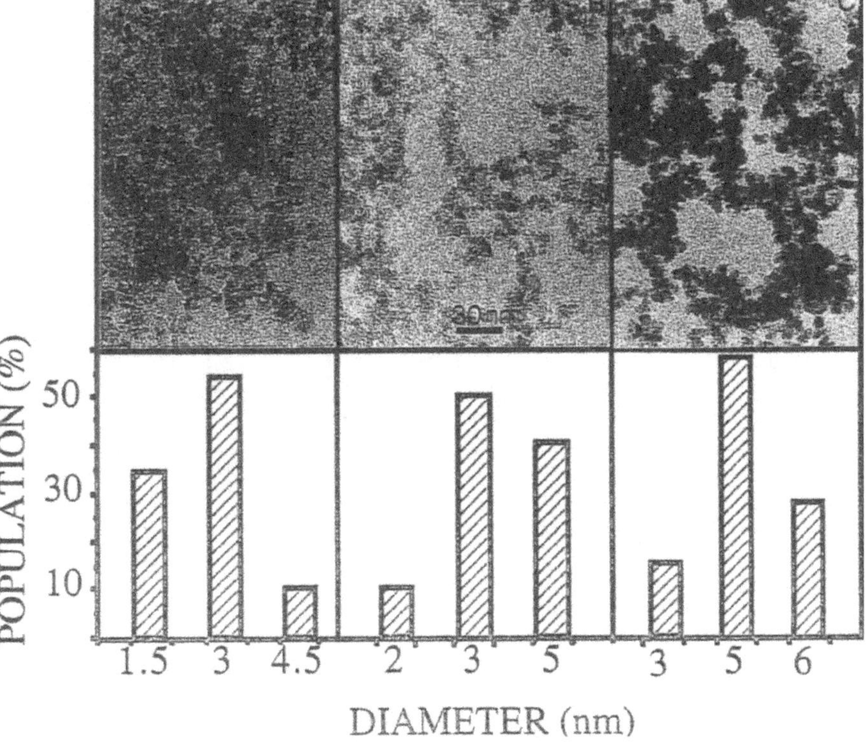

Figure 7: Electron microscopy pattern and histogram of magnetic fluid made at various surfactant concentrations keeping $[Co(DS)_2]/[Fe(DS)_2]$ = 0.325, $[Co(DS)_2]/[NH_2CH_3]$ = $1.3.10^{-2}$. $[Fe(DS)_2]$= 6.5 10^{-3}M **(A)**, $[Fe(DS)_2]$= 1.3 10^{-2}M **(B)**, $[Fe(DS)_2]$=2.6 10^{-2}M **(C)**.

Magnetic particles are dispersed in 50% of ethylene glycol in water. For particles having 5nm as an average diameter the variation of magnetization, M, with applied field, H, is given, at room temperature, on figure 8A. As expected, no hysteresis is observed indicating that the particles are in a superparamagnetic regime. The saturation magnetization is found equal to 35 $emu.g^{-1}$ whereas for bulk phase it is equal to 78 $emu.g^{-1}$ (at 300K)[19]. By decreasing the diameter of the particles to 2nm, the saturation magnetization is not be reached for an applied field equal to 30kOe. This decrease in the saturation magnetization with the particle size can be attributed to a surface effect with formation of inert layer(s).

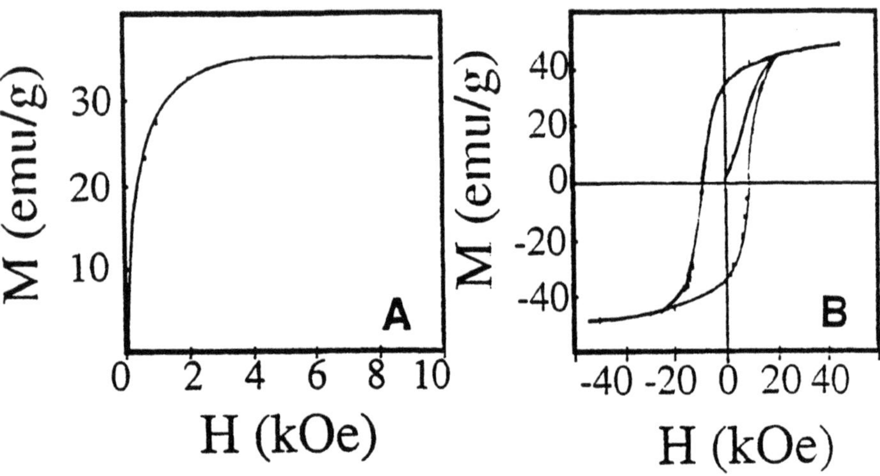

Figure 8: Magnetization as function of applied field at room temperature (A) and at 20K (B) for particles having an 5nm average diameter.

The field needed to reach the saturation magnetization of bulk $CoFe_2O_4$ depends on the crystallographic direction. As matter of fact, for 100 and 111 atomic density orientations, the saturation magnetization is equal to 1 and 9kOe respectively[20]. For particles having an average diameter equal to 5nm, the saturation magnetization is reached at 4kOe. This value is in good agreement to the data obtained in bulk phase.

The magnetic particles having an average diameter equal to 5nm are frozen in zero field at 20K. Figure 8B shows the presence of an hysteresis. For random distribution of easy axes of particles with cubic magnetocrystalline anisotropy, the ratio of the remanence to saturation magnetizations is expected to be equal to 0.83 at 0K. At 20K it is found equal to 0.74. If the anisotropy had been primarily due to shape, then the ratio would have been 0.5 at 0K. The coercivity at 20K is found 8.8kOe for particle having a diameter equal to 5nm. By decreasing the diameter of particle from 5 to 2nm, the coercitivity decreases from 8.8kOe to 4.8kOe. The large remanence and coercivity values and the remance to saturation magnetization of 0.83 indicate that the system consists of randomly oriented equiaxial particles with cubic magnetocrystalline anisotropy[21].

Figure 8 shows an increase in the saturation magnetization with decreasing temperature from 300K (figure 6A) to 20K (figure 8B). This is similar to the data obtained in bulk phase. As matter of fact, the ratio of saturation magnetizations at 300K and at 20K is found equal to 0.83 and 0.88 for particles having 5nm diameter and for the bulk phase respectively.

5. Control of the size of latexes[22]

The polymerization conversion is obtained from proton N.M.R. spectra by measuring the relative decrease of the ethylenic protons in comparison with the ammonium ones. It depends on the initial conditions and finds about 50-70 %. It has been impossible to isolate the polymer from the residual monomer by precipitation techniques because of the low weight of the polymer or due to monomer-polymer interactions preventing the separation of the polymer from the residual surfactant. Size exclusion chromatography, S.E.C., experiments have been performed to measure the polymerized micelles hydrodynamic volume. Under the conditions described elsewhere[14], it has been possible to isolate the polymer from the residual monomer. The radii determined by such technique are equal to 1.6 and 2nm when the water contents are 5 and 10 respectively.

Electron micrographs (fig. 9) obtained after the polymerization of C_{10}MA reverse micelles, show the presence of small particles. At low w value (w = 5), main size of the particles is equal to 1.9nm. By increasing the water content, the size increases. At w up to 20, the average size of latexes remains unchanged (2.8nm), but the polydispersity decreases (fig. 9A,B,C). This is explained by the fact that the percentage of initiator is the same with respect to the monomer, so it decreases with respect to one micelle.

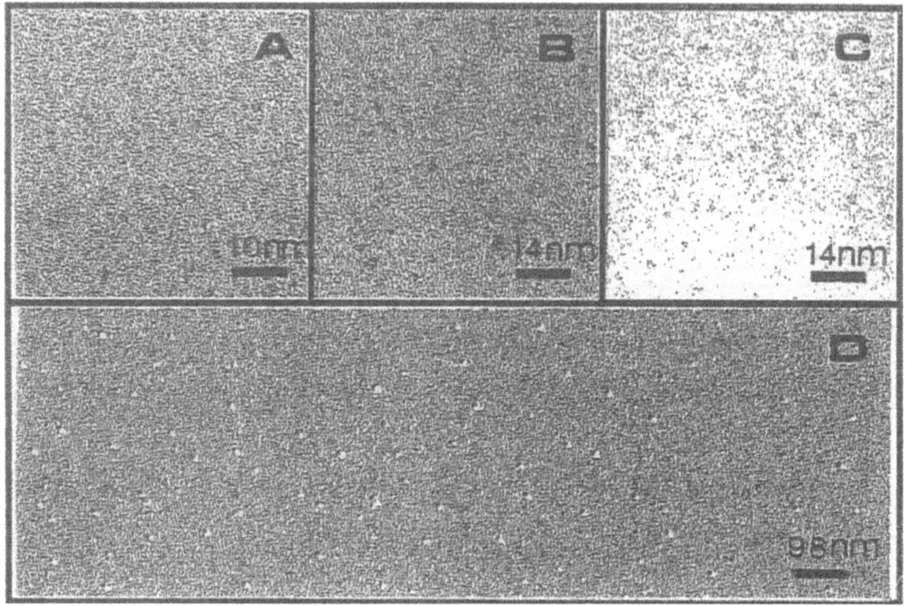

Figure 9:
[A]: Electromicrographs of latexes; w = 5, $[C_{10}MA]_o$ = 0.05 M. 1 mm = 10Å
[B]: Electromicrographs of latexes; w = 10, $[C_{10}MA]_o$ = 0.05 M. 1 mm = 14Å
[C]: Electromicrographs of latexes; w = 15, $[C_{10}MA]_o$ = 0.05 M. 1 mm = 14Å
[D]: Electromicrograph of platinated latexes,.w = 10, $[C_{10}MA]$ = 0.1 M 1 mm = 98 Å

To confirm the electron microscopy data presented in figure 9D, T.E.M. experiments have also been performed on platinum shadowed samples. Figure 9D shows a micrograph obtained for latexes synthesized at w value equal to 10. The latexes are spherical and the dispersion is quite homogeneous. The main size of the particles is equal to 7nm and the particles of larger diameter are due to aggregation of several latexes. The size difference between unshadowed (3nm) and shadowed (7nm) particles is probably related to the fact that the direct observation reveals mainly the core of the polymerized $C_{10}MA$ reverse micelles, while the shadowing allows to visualize the whole micelle, including the ammonium moiety, surrounding the polymer (≈ 2 x 1 nm). The remaining difference in size (about 2 nm) may be attributed to some shrinkage of the air-dried, unfixed polymer, when observed directly in the microscope and to some enlargement of the shadowed particle due to the metal deposit.

References:

[1] Pileni, M.P., J.Phys.Chem.; **1993**, 97, 6961.

[2] Zhao, X.K.; Yang, J.;.McCormick, L.D.;.Fendler, J.H., J.Phys.Chem. **1992**, 96, 9933.

[3] Herron,N.;Wang,Y.;Eddy,M.M.;Stucky,G.D.;Cox,D.E.;Moller,K.;Bein,T. J.Amer.Chem.Soc.; **1989**, 111, 530.

[4] Youn, H.C.; Baral, S.; Fendler, J.H., J.Phys.Chem ; **1988**, 92, 6320.

[5] Osakada,K.; Taniguchi, A.; Kubota, E. and all Chem.Mater **1992**, 4, 562.

[6] Henglein, A.; Chem. Rev. **1989**, 89, 1861 and the references included.

[7] Structure and reactivity in reverse micelles; Pileni, M.P. ed; pub. elsevier **1989**.

[8] Pileni, M.P.; Zemb, T. and Petit, C.; Chem. Phys. Lett.; **1985**, 118, 414.

[9] Satake, I.; Iwamatsu, S.; Hosokawa, R.; Matuura, Bull. Chem. Soc. Japan Vol.; **1962**, 36, 204.

[10] Hayter, J.B.; Penfold, J. Colloid & Polymer Sci. ; **1993**, 261, 1022,

[11] Moroi, Y.; Motomura, K.; Matuura, R. J. Colloid Interface Sci.; **1974** , 46, 111.

[12] Petit, C; Lixon,P.and Pileni, M.P.; J.Phys.Chem.; **1993**, 97, 1274.

[13] Motte,L.; Billoudet,F.; and Pileni, M.P.; Submitted for publication.

[14] Bain, C.D.; Evall, J.; Whitesides, G.M., J.Am.Chem.Soc.; **1989**, 111, 7155.

[15] Lisiecki I., and Pileni M.P; J. Am. Chem. Soc.; **1993**, 115, 3887.
Lisiecki I., and Pileni M.P.; J.Phys.Chem.; **1995**, in press.
Lisiecki I.,Borjling.M.; Motte,L.; Ninham;B.; and Pileni M.P.; Langmuir, **1995**, in press.

[16] Tanori,J.; Pileni,M.P.; Advanced in Material, **1995**, (in press).
Tanori,J.; Duxin, J.; Petit,C.; Lisiecki,I.; Veillet, P.; and Pileni, M.P. Progress Colloid Int.Sciences.; **1995**, in press.

[17] Lisiecki, I.; Billoudet, F.; and Pileni, M.P.; submitted to publication.

[18] Moumen, N.; Veillet, P.; Pileni, M.P.; J.Mag.Mag.Mat.; **1995**, in press.

[19] Ferromagnetic materials ed. Wohlfarth, E.P; Publ.Northholland Publishing Company Amsterdam, New York, Orford **1982**.

[20] Landolt-Bornstein Numerical data and functional relationships in Science and Technology vol. 12. Ed..Hellwege, K.H and Hellwege, A.M. Publ. Springer-Verlag Berlin , Heidelberg, New York **1980**.

[21] "Magnetism and Mettallurgy Berkowtiz, A.E and Kneller, E editors, Academic Press Publ., **1969**, vol 1 chap. 8.

[22] Hammouda, A.; Gulik,T.; and Pileni, M.P.; Langmuir **1995**, in press.

SYNTHESIS OF MONODISPERSE GeO_2 PARTICLES IN REVERSED MICELLAR SYSTEMS

K.KON-NO

Department of Industrial Chemistry, Faculty of Engineering, Science University of Tokyo, Institute of Colloid and Surface Chemistry,1-3, Kagurazaka, Shinjuku-ku, Tokyo 162

Abstract

In order to examine the relationship between formation of particles and reaction field, the solubilized states of water in sodium 1,2-bis(2-ethylhexyloxycarbonyl)ethanesulfonate (AOT) in cyclohexane were studied by using some spectroscopic techniques at 30 ℃. From the changes of these spectra features with Rw(=[H_2O]/[AOT]), it is found that three typed water exist with varying proportions in the interior of micelles, i.e., at below Rw=4 water is bound directly to the polar groups of AOT, and in the range of Rw=4 to 11 the succeeded water is held together with the hydrated polar groups by the hydrogen bonds and at above Rw=11 bulk-like water is built up. These regions corresponding to each types of water were named as reversed micelles, swollen micelles and W/O micro-emulsions, respectively. On the basis of the phase diagram of water obtained, the synthesis of GeO_2 particles by the hydrolysis of $Ge(OC_2H_5)_4$ were carried out at varing Rw in each region. Monodisperse polyhedron particles were produced in both the reversed and swollen micelle regions, whereas in W/O microemulsion ones cubic particles were formed. The size of these particles increased at Rw which change from reversed micelles to swollen ones, whereas that produced in W/O microemulsions rather decreased with Rw. All the particles obtained here were identified as α-quartz typed crystal, but that obtained at Rw=3 transformed to rutile-typed one at 1000 ℃. These facts were recognized by X-ray diffraction analysis or FT-IR measurements.

1. Introduction

Utilizing water microdroplets solubilized by reversed micelles, the synthesis of uniform and small particles has been demonstrated for a variety materials. In the preparation of particles in reversed micelles, it is very important to examine the relation of the

E. Pelizzetti (ed.), Fine Particles Science and Technology, 431–442.

formation of particles with the states of water droplets in reversed micelles as a reaction field. However, attention has not always been paid to the solubilized states of water for the formation of particles, though the effect of the size of water droplets on the formation of Ag or $CaCO_3$ particles has been investigated [1,2]. The results concerning the solubilized states of water in ionic reversed micelles have widely differed by measuring techniques even though identical surfactant/aplar solvent systems [3-5].
we report here for the solubilized states of water in sodium 1,2-bis(2-ethylhexyloxy-carbonyl)ethanesulfonate (AOT) micelle in cyclohexane measured by varying measuring techniques at 30 ℃. We focus on the size or morphology of GeO_2 particles synthesized by the hydrolysis of germenium tetraethoxide in different solubilized states of water obtained and discussed on the transformation of their crystal forms by heat treatment.

2. Experimental

2.1 MATERIALS

AOT was purified according to Magid's method [6]. Germanium tetraethoxide (GET) of purity of 99.999%) was used.

2.2 MEASUREMENTS OF SOLUBILIZED STATES OF WATER

The limiting amounts of water solubilized by AOT in cyclohexane were measured at 30 ℃ by Karl-Fischer titration according to a previous paper [7], i.e., water was added into the AOT solution until very small amounts water or precipitates were separated from the solubilization systems and they were setteled for one week in temperature controlled water bath. The water content dissolved in the obtained transparent solutions was determined by the Karl-Fischer method. Here, the molar ratio of water to surfactant is abbreviated as Rw.
The chemical shifts of solubilized water (δ_H) were measured using a JEOL FX-900Q spectrometer. TMS was used as an internal standard.
The rotational correlation time (τ_c) and hyperfine splitting constant (a_N) of 2,2,6,6-tetramethyl-4-(phosphonooxy)piperidine used as an ESR probe in solubilized water was measured using a JEOL-FEIX spectrometer.
The internal relaxation time (ρ) of sodium 3,4,9,10-peryenetetracarboxylate used as a fluorescence probe in solubilized water was measured using a Horiba NAES-1100 time resolved fluorescence spectrometer. The sample solutions were degassed under vacuum by repeating three times.
The near-infrared spectra of mixtures (90% D_2O+10%H_2O) solubilized by AOT were measured with a Hitachi model 330 spectrometer. D_2O was used to obviate the coupling of OH strretches within one water molecule.
The intensities of scattered light and refractive indices of solutions were measured using a

Marvern 4700 Submicron particle analyzer equipped with an argon ion- laser and a Union Giken RM-102 double beam differential refractometer, respectively. All the measurements were carried out at 90° ,since there was no angular dependence. The average micellar weight was determined from Debye plots. The hydrodynamic diameter of the micelles was calculated from the Einstein-Stokes equation.

2.3 SYNTHESIS AND IDENTIFICATION OF PARTICLES

The synthesis of particles was carried out as follows, i.e., the amounts of water corresponding to a desirable Rw was dissolved in cyclohexane solutions of 0.1 mol kg^{-1} AOT and same concentration of AOT solutions containing GET was sequentially added to the solubilized solutions of water, with stirring at 25℃. The GET concentration was 0.02 mol kg^{-1}.

The size and morphology of the particles were observed with a Hitachi S-5000n scanning electron microscope (SEM) and a Nippon Denshi JEM-100B transmittion electron microscopes (TEM).

The particles were identified by a Rigaku RAD X-ray diffractometer. From the diffraction pattarns showed in Fig. 1, all the particles obtained in any systems were found to be α-quartz typed GeO_2 by referring to a JCPDS CARD 4-497.

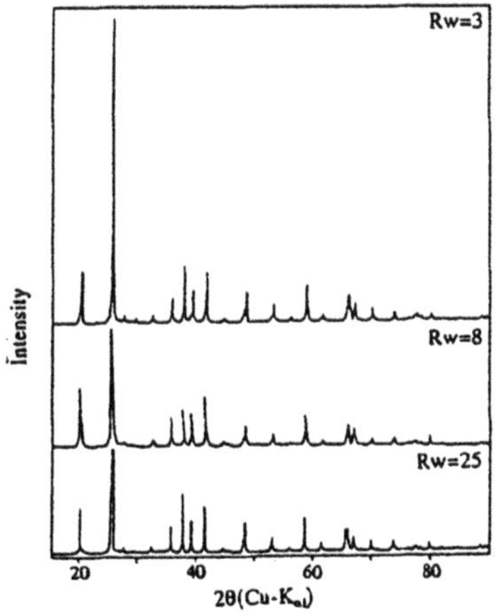

Fig. 1 X-ray diffraction pattarns of GeO_2 particles produced at various Rw

Differential thermal analysis-thermal gravimetric analysis (DTA-TG) of the particles were carried out by a Regaku TG-DTA 8087 G2DTA analyzer.

FT-IR spectra of the particles were measured with Nicolet 510 spectrometer.

3. Results and Discussion

3.1 SOLIBILIZED STATES OF WATER

In order to examine the solubilization capacity of water by AOT, the limiting amounts of solubilized water was measured as a function of the concentration of AOT at 30 ℃. The results obtained are shown in Fig. 2.

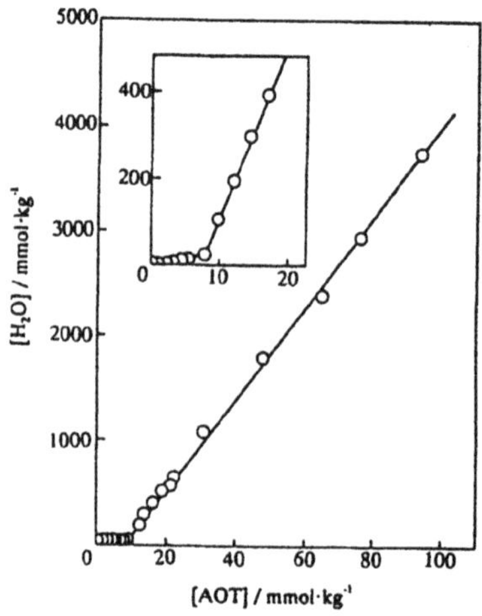

Fig.2 The change of limiting amounts of water with concentration of AOT in cyclohexane at 30 ℃

The limiting amounts increased abruptly at 8 mmol kg^{-1}, followed by a linear increase. This abrupt increase in the limiting amounts is due to AOT micelles having a constant micellar weight which form at a concentration above about 8 mmol kg^{-1} in cyclohexane [8]. Therefore, the slope of the straight lines obtained above 8 mmol kg^{-1}, indicating the solubilization power of AOT micelles, was 42.2 as Rw.

Then the solubilized states of water in the region broadened at concentration greater than 8 mmol kg^{-1} were examined as a function of Rw at various concentration of AOT by using ^{1}H-NMR, ESR, Fluorescence and near-infrared spectroscopic techniques. The change of the a_N, τ_c, ρ, and δ_H values with Rw are shown in Fig. 3. The a_N values, indicating a measure of the micropolarity of water around the probe, increased with increasing Rw; at Rw=5, however, it clearly inflected and closely approached that in bulk water in the region above Rw=11. Such an inflection at Rw=5 and Rw=11 was also observed in the δ_H - Rw profile at 50 mmol kg^{-1} AOT. On the other hand, both the values of τ_c and ρ, which indicate a measure of the microviscosity of water around the probes, decreased monotonously up to Rw=11, and followed by the values in bulk water. Similar profiles were also observed at other concentration.

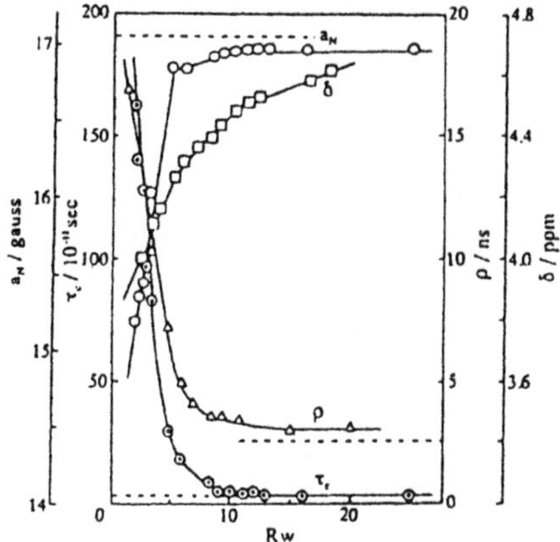

Fig. 3 The change of a_N, τ_c, ρ, and δ_H with Rwin cyclohexane
○: a_N, ⊙: τ_c, △: ρ, □: δ_H at 50 mmol kg^{-1} AOT. Dashed lines indicate the observed values in bulk water

Such an appearance of inflection points suggests that three types of water exist in varying proportions in the interior of AOT reversed micelles. This suggestion might be clarified from the near-infrared spectra obtained in the region of water OH overtone frequency ($2\upsilon_{OH}$), where "free" OH and hydrogen-bonded streching bands. Figure 4 shows the spectral change in the $2\upsilon_{H_2O}$ region with Rw at 100 mmol kg^{-1} AOT solution. The band at 1420 nm as assigned to the "free" OH, while those at 1450 and 1650 nm were due to energetically unfavored hydrogen-bonded and hydrogen-bonded species, respectively. The growth of the 1420 nm band increased with increasing Rw, whereas those of the 1450 and 1650 nm bands were seen above Rw=5.7. These spectral features clearly reveal that three states of water exist in varying proportions as a function of Rw in the interior of reversed micelles.

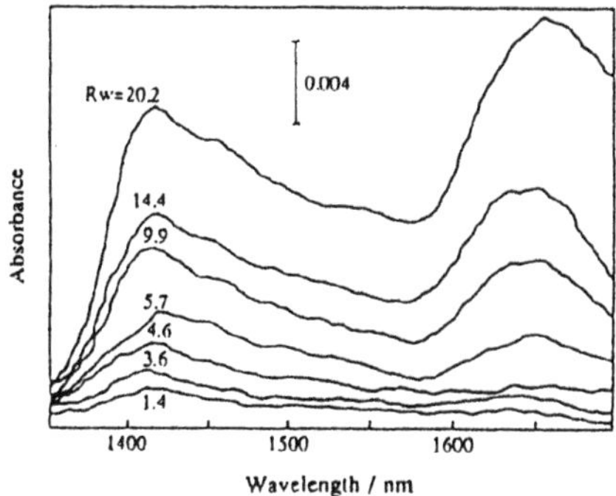

Fig. 4 The change of FT-IR spectra with Rw in 0.1 mol kg^{-1} AOT

The first is apparent only below Rw=5 and probably water solvating ionic head groups, -SO_3^- Na^+ of AOT, since neither the bands of 1450 or 1650 nm appear. The second is present in the range of Rw=5-11 and corresponds to the water interacting with the ester and the hydrated ionic groups of AOT. This fact is clear from the appearance of 1450 and 1650 nm bands. However, the water existing in this region is not bulk-like water, since a_N, τ_c, and ρ did not approach those in bulk water, as seen in Fig. 3. In the region above Rw=11, a remarkable increase in the bond at 1650 nm reflects the build-up of bulk-like water, so-called W/O microemulsions could also be recognized from the approach of the observed values of a_N, τ_c, and ρ to those in bulk water, as seen in Fig. 3

3.2 RELATIONSHIP BETWEEN THE SOLUBILIZED STATES OF WATER AND MICELLE FORMATION

When the Rw values corresponding to the inflection points of the a_N, or δ_H values at various concentrations of AOT were plotted within the solubilization region above 8 mmol kg^{-1} AOT, the region was divided into three realms (I, II, and III) by two straight lines (dashed and solid-drawn) passing through the inflection at 8 mmol kg^{-1} AOT, as seen in Fig. 5. In order to examine the relationship between the solubilized states of water and micelle formation, the change of hydrodynamic diameter (d_h) and aggregation number (n) with Rw were measured. No inflection in the d_h - Rw profile was found, whereas in n profile n values incresed with increasing Rw and at Rw=5 and 11 two inflections were found.

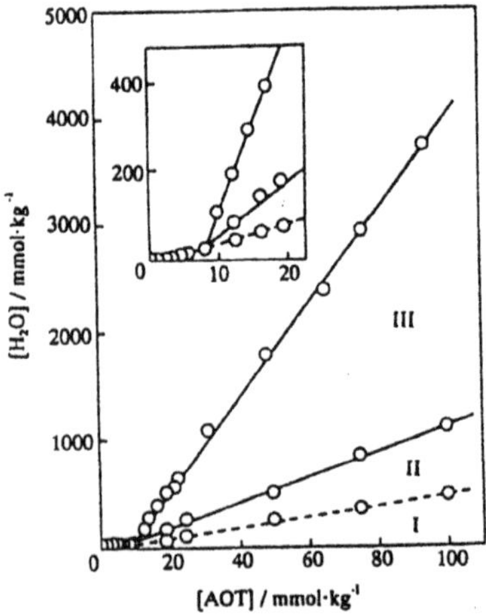

Fig. 5 The phase diagram of solubilized water in cyclohexane systems 30 ℃

The Rw values corresponding to two inflections obtained here were found to agree with

those of the inflections in the a_N and d_h values. This fact indicates that each realm can be applied into two reversed micelles and W/O microemulsions, i.e., each realm I and II below and above the dashed line corresponds to the reversed micelles formed by the water bound to the ionic head groups of AOT and by the water interacting with the hydrated head groups. Realm III was taken, of course, as W/O microemulsions. Here, we call the reversed micelles formed in realm II as being swollen micelles, since the solubilized water is still bound to AOT molecules. Therefore the phase diagram of solubilized states of water can be also designated as the micellar phase diagram.
From these results, the dashed and solid lines drawn within the solubilization region shown in Fig. 5 could be denoted as the minimum amounts of water required for the formation of swollen micelles and W/O microemulsions, respectively. These amounts for the former and the latter could be estimated as Rw=4.9 and 12.0 from the slopes of these straight lines. Accordingly, the surfactant concentration corresponding to the inflection point in the limiting amounts of solubilization, i.e., 8 mmol kg^{-1} should be regarded as "critical concentration of formation of swollen micelles or W/O microemulsions".

3.3 SIZE AND MORPHOLOGY OF PARTICLES

Then GeO_2 particles were synthesized as a function of Rw in each AOT micelle regions including different solubilized states of water. Figure 6 shows SEM photographs of GeO_2 particles obtained at Rw=3, 8, and 35, as a representative example. In both the reversed and swollen micelle systems of Rw=3 and 8, uniform polyhedral particles were formed, but their size were fairly larger in the latter than in the former.

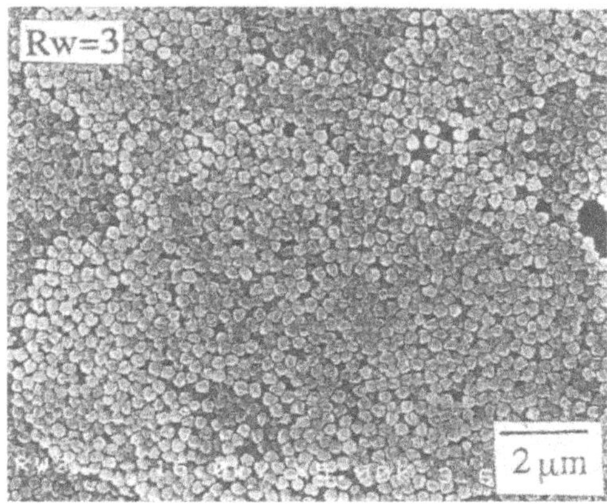

Fig. 6 SEM photographs of GeO_2 particles produced at Rw=3, 8 and 35

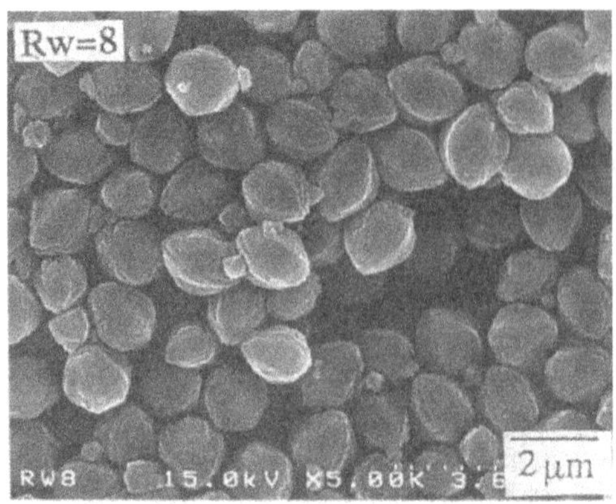

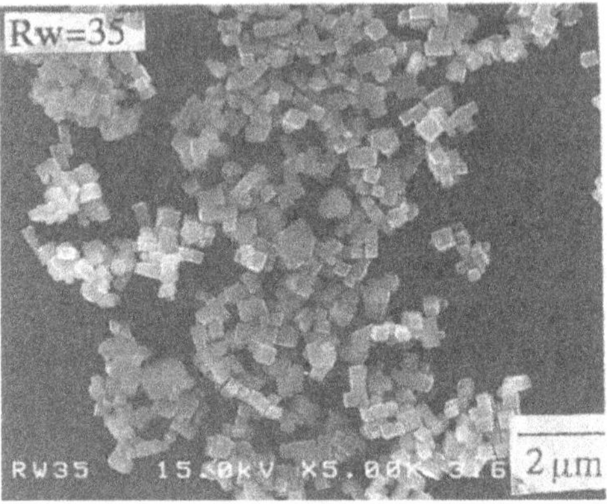

Fig. 6 SEM photographs of GeO_2 particles produced at Rw=3, 8 and 35 (continued)

These particles were found to be a coagulates of a fine primary particles from the SEM photographs of high magnifications. In W/O microemulsion systems of Rw=35, however, relatively uniform cubic particles were formed. Similar shape of particles were also observed at other Rw within each micelle and W/O microemulsion systems, but in the systems of Rw=4 and 5 which change from reversed micelles to swollen micelles, the produced particles were a mixtures of particles formed in each reversed and swollen micelle systems. Then about 200 particles were selected from the photographs and the lengths of apsis for the polyhedral particles and of a side for the cubic particles were measured as a measure of the particle size. The average lengths and standard deviations of the particles are summarized in Table 1 together with the existing ratio of the particles and each micellar state.

Table 1 The micellar states and the size parameters of GeO_2 particles

Rw	Micellar state	Polyhedron/nm small	Polyhedron/nm large	Existing/% small	Existing/% large	Cubic/nm
2	RMs	80±9	-	-	-	-
3	RMs	272±21	-	-	-	-
4	RMs	623±130	972±96	43	57	-
5	SMs	643±140	1003±140	41	59	-
6	SMs	-	1128±73	-	-	-
8	SMs	-	1650±110	-	-	-
10	SMs	-	1750±110	-	-	-
13	MEs	-	-	-	-	993±127
15	MEs	-	-	-	-	1021±84
25	MEs	-	-	-	-	479±53
35	MEs	-	-	-	-	265±39

RMs: Reversed Micelles, SMs: Swollen Micelles, MEs: W/O microemulsions

As seen in Table 1, the shapes of the particles were clearly differed between reversed or swollen micelle systems and W/O microemulsion ones. These facts indicate that the polyhedral particles were formed in two types of water bound to the ionic head groups of AOT in reversed micelles or water interacting with the hydrated head groups in swollen micelles, and in bulk-like water of W/O microemulsions cubic particles were produced. On the other hand, the size of particles were larger in swollen micelle systems in comparison with that in reversed micelle ones, but these particles size increased with increasing the extent of water droplets, characterized as Rw and in W/O microemulsion systems they were inverse. The size of particles in any systemswere exceedingly large in comparison with each micellar size determined by light scattering. This means that the micellar size dose not a determining factor of the particle size.
Here, the reason why diffrent shape of particles were formed in different solubilized states of water droplets could not be clarified in this study. However, the reason why large polyhedral particles were formed in swollen micelle systems in comparison with reversed micelle system may be explained from the difference in the dispersability of primary GeO_2 particles which construct the particles in these particles. That is, in contrast to reversed micelle systems, TiO_2 particles of 40 to 90 nm in diameter are due to the coagulation by the hydrogen bond between the water molecules interacting with the hydrated head groups of AOT in swollen micelle systems [9]. If this suggestion is true, polyhedral particles appear to form by collision among reversed or swollen micelles including primary GeO_2 particles formed at initial stage of reaction. Nevertheless, it is

notable that different size and shape of particles were clearly formed in different solubilized states of water droplets in AOT micelle systems.

3.4 CRYSTAL FORMS OF PARTICLES

In order to examine the crystal transformation of particles by heat treatment, at first, the DTA-TG analysis of each particles were undertaken with the heating rate of 10 ℃ min^{-1}.

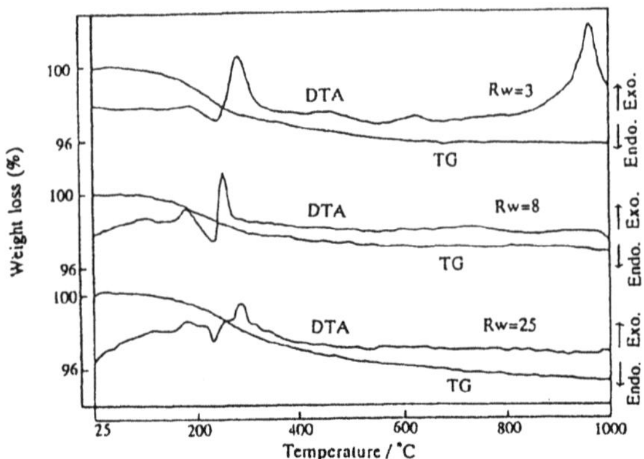

Fig. 7 DTA-TG analysis of GeO_2 particles produced at Rw=3,8 and 25

Figure 7 shows the results of the analysis of the particles obtained at Rw=3, 8, and 25. The endothermic peak around 250 to 270 ℃ and the exothermic peak around 300 ℃, accompanied with about 3-4 % of weight loss, were observed for any particles. However, at vicinity of 970 ℃ relatively large exothermic peak, unaccompanied with weight loss, appeared for the particles formed Rw=3. The identification of these peaks have tried by comparison with FT-IR spectra or X-ray diffraction pattams of particles obtained after heat treatment for 2 hr at each temperature shown in Figs. 8 and 9, respectively. As seen in Fig. 8, the intensities of the absorption bands of 3450 and 1630 cm^{-1} assigned to the O-H stretch and deformation vibrations decreased gradually with heat treatment until 600 ℃, and that of a trace bands of 2920 and 2960 cm^{-1} assigned to the C-H stretch vibrations of $-CH_2-$ and CH_3- groups disapeared by the treatment at 300 ℃, respectively. These results appear to indicate that the endothermic peaks around 250 to 270 ℃ is due to the elimination of water and the exothermic peak around 300 ℃ corresponds to the combustion of organic residues adsorbed onto the particles. This is also clear from TG curves showing about 3-4 % of weight loss until 600 ℃. On the other hand, the exothermic peak around 970 ℃ is due to the transformation from α-quartz typed crystals to rutile typed ones. This is also clarified from the appearance of diffrent absorption bands of 600 and 690 cm^{-1} in FT-IR spectra and the X-ray diffraction pattams of the particles treated at 1000 ℃ shown in Fig. 9.

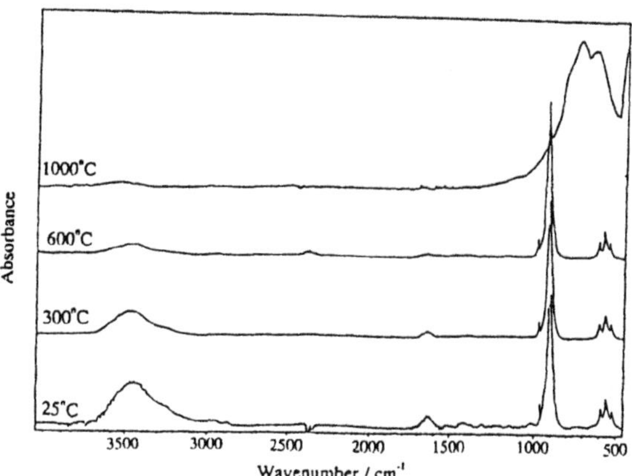

Fig. 8 FT-IR spectra of GeO_2 particles obtained Rw=3 at varying temperatur

Such crystal transformation was also observed between 1000 and 1100 ℃ for the particles of Rw=35, while the particles of Rw=8 was fused above 1100 ℃ without transformation. These results appear to indicate that different transition temperature of GeO_2 crystals is induced from a different size of particles, irrespective of the morphology of particles.

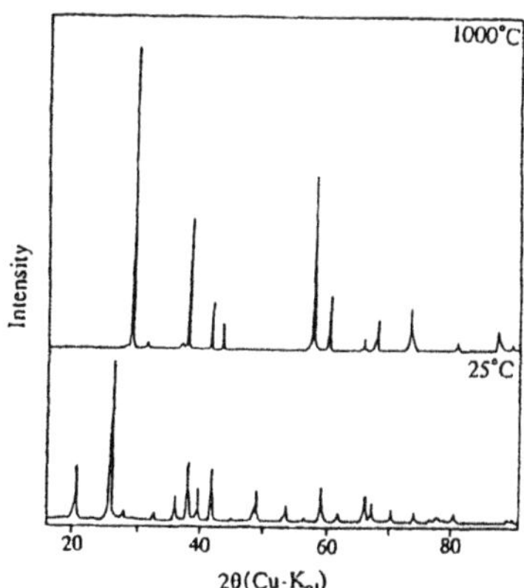

Fig. 9 X-ray diffraction pattarns of the particles at Rw=3 treated at 1000 ℃

4. References

1. Petit,C., Lixon,P. and Pileni,M.P. (1993) In Situ Synthesis Silver Nanocluster in AOT Reverse Micelles, J.Phys.Chem., **97**,12974-12983.
2. Kandori,K., Kon-No,K. and Kitahara,A. (1988) Formation of ionic water/oil micro-emulsions and their application in the preparation of $CaCO_3$ particles, J.Colloid Interface Sc., **122**,78-82.
3. Kotlarchyk,M., Huang,J.S. and Chen,S. (1985) Structure of AOT reversed micelles determined by small-angle neutron scattering, J.Phys.Chem.,**89**,4382-4386.
4. Haering,G., Luisi,P.L and Hauser,H. (1988) Characterization by electron spin resonance of reversed micelles consisting of the ternary system AOT-isooctane-water, **92**,3574-3581.
5. Casillas,N., Puig,J.E., Olayo,R., Hart,T.J. and Franses,E.I. (1989) States of water and surfactant in lyotropic liquid crystals, Langmuir,**5**,384-389.
6. Magid.L.J. and Martin,C.A. (1984) ^{13}C NMR studies of molecular conformations and interactions in the curved surfactant monolayers of Aerosol OT water-in-oil micro-emulsions, in P.L.Luisi and B.E.Strub (eds), Reversed Micelles, Plenum Press, New York and London, pp.181-193.
7. Kawai,T., Shindo,N. and Kon-No,K. (1992) Solubilization behavior of water in hexaethylene glycol dodecyl ether reverse micelles in apolar solvents, J.Surface Sci. Technol.,**8**,145-153.
8. Kon-No,K., Asano,H. and Kitahara,A. (1983) Study on the size of reversed micelles of anionic and cationic surfactants, Prog. Colloid Polym.Sci.,**68**,20-24.
9. Kon-No,K., Nakayama,T, and Kawai,T., Unpublished.

PREPARATION OF CdS AND ZnS PARTICLES IN NANOPHASE REACTORS PROVIDED BY BINARY LIQUIDS ADSORBED AT LAYERED CLAY AND SPHERICAL SILICA PARTICLE SURFACES

IMRE DÉKÁNY
Department of Colloid Chemistry, Attila József University,
H-6720, Szeged Aradi V.t.1, Hungary

JANOS H. FENDLER
Department of Chemistry, Syracuse University,
Syracuse, New York 13244-4100, U.S.A.

1. Abstract

The polar component of binary (liquid alcohol-cyclohexane) liquids has been shown to adsorb selectively at the surfaces of clay mineral (montmorillonite) and silica (aerosol) particle surfaces and, thus, to provide a unique nanophase reactor.

Determination of the adsorption excess isotherms from alcohol-cyclohexane liquid mixtures on this solid particles provided information on the composition and the volumes of the adsorption layer as nanophase reactor. CdS and Zns particles were generated in the nanophase reactor by introducing the reactants into the surface of clay in the interlamellar space and the adsorption layer of silica particles. The nanoparticles in the nanophase reactor have been elucidated by absorption spectrophotometry in order to determine the absorption threshold and the band gap energy. Small-angle X-ray scattering were performed in order to determine the particle size of semiconductors(d=3.5-10nm).Growing semiconductor nanoparticles on the silica or on clay mineral sheets showed a very sensitive increase of the rheological parameters in semiconductor-silica organic suspensions. The heat of wetting was also measured to detect the surface modification of silica by CdS and ZnS nanoparticles in methanol.

E. Pelizzetti (ed.), Fine Particles Science and Technology, 443–455.

2. Introduction

Nanoparticles and nanoparticulate films are increasingly used as advanced materials [1]. Their large surface area to volume ratios are fruitfully exploited in developing novel nanoparticle based catalysts [2]. Electron and hole confinements in size quantized semiconductor nanoparticles have led to unique optical, electrical and electro-optical properties and, hence, to the construction of advanced devices and sensors [3]. Furthermore, a systematic change of particle diameters permitted the fine tuning of the band-gaps, as well as of the oxidation and the reduction potentials of a variety of semiconductors [4].

Nanoparticles and nanoparticulate films are customarily prepared in the solid state, usually in ultra high vacuum. More recently, alternative "wet" colloid chemical techniques have been developed for the generation of nanoparticles [5]. Metallic, semiconducting and magnetic particles have been *in situ* generated in aqueous and reversed micelles, and surfactant vesicles, as well as under monolayers and between Langmuir-Blodgett films [5].

In a preliminary communication we have reported the use of surfactant modified montmorillonites, layered silicates, often referred to as clay organo-complexes, as templates for CdS nanoparticle formation [6]. Generalization of the colloid chemical and liquid sorption principles involved in the *in situ* generation of subcolloidal particles in nanophase reactors provided by binary liquids adsorbed at solid surfaces is the subject of the present work.

Manipulation of the nature and composition of the adsorption layer, composed of binary liquid mixtures is expected to influence profoundly the formation and growth of semiconductor nanoparticles. Experimental examination of the conditions required for the optimization of the adsorption layer at the solid interface (ie., in the nanophase reactor) for controlling the growth of CdS and ZnS nanoparticles is the subject of the present report.

3. Theoretical Section

Advantage has been taken in the present work of the adsorption phase at swelling silicate layer interfaces to provide a nanophase reactor within which the growth of semiconductor particles could be confined. Importantly, the thickness of the adsorption layer, ie., the volume of the nanophase reactor has been controlled by the judicious selection of binary liquid mixtures. Established principles of adsorption equilibrium govern the formation of an adsorption layer on a surface [7]. These principles will be summarized for the general case of adsorption on any smooth or porous surfaces, and then extended to our special case of layered hydrophobic clays. Further, the equations derived will permit the assessment of the reactant concentrations in the adsorption layer whose adjustment provides an additional control of the sizes of the incipient semiconductor particles.

3.1. VOLUME OF THE ADSORPTION LAYER (THE NANOPHASE REACTOR) ON SMOOTH SOLID SURFACES.

The binary liquid mixture selected should have the following properties for an efficient control of the semiconductor nanoparticle formation. First, one component of the liquid mixture (x_1) should be preferentially adsorbed at the solid interface (indicated by the superscript s), hence its concentration in the bulk (indicated by the superscript b) should be negligible (ie., $x_1^s >> x_1^b$). Second, the other component of the liquid mixture (x_2) should not effectively adsorbed on the solid (ie., $x_2^s << x_2^b$). Third, the semiconductor precursors should be highly soluble in the liquid which is preferentially adsorbed on the solid surface and they should be highly insoluble in the liquid which is not adsorbed on the solid particles. These conditions would effectively limit nucleation and growth of the semiconductor nanoparticles to the nanophase reactor provided by the adsorption layer at the solid holder. Appreciable solubility of the reaction partners in the liquid preferentially present in the bulk phase would lead, of course, to the formation of large irregular micron sized semiconductor particles in the bulk phase.

The relationship between the molar composition of the adsorption layer and the reduced adsorption excess, $n_1^{\sigma(n)}$, is given by the Ostwald-de Izaguirre Equation [7]:

$$n_1^{\sigma(n)} = n_1^s - (n_1^s + n_2^s)x_1 = n^s(x_1^s - x_1) \tag{1}$$

where the material amount of the adsorption layer $n_1^s + n_2^s = n^s$ can be evaluated by the graphical extrapolation method after Schay and Nagy determined [7-9].

The adsorption volume on a solid surface (indicated by the subscript s), V^s, is determined by the adsorption capacities n_i^s according to:

$$V^s = a^s t = n_1^s V_{m,1} + n_2^s V_{m,2} + n_3^s V_{m,3} + \ldots (i = 1,2,3) \tag{2}$$

where a^s is the surface area of the adsorption layer, t is the thickness of the adsorption layer, n_i^s is the amount of i-th component, and $V_{m,i}$ is the molar volume of the i-th component at the interface. For binary liquids (i=1,2) the composition of the adsorption layer is:

$$x_1^s = n_1^s/(n_1^s + n_2^s) = n_1^s/n^s \quad \text{or} \quad \phi_1^s = rx_1^s/(x_1^s + rx_2^s) = n_1^s V_{m,1}/V^s \tag{3}$$

where x_1^s is the molar, ϕ_1^s the volume fraction of the adsorption layer and $r=V_{m,2}/V_{m,1}$.

If component 1 is adsorbed preferentially at the interface, ie., $x_1^s >> x_1$, then the volume of the adsorption layer (ie. the volume of the "nanophase reactor") is given by: $V^s = n_1^s V_{m,1}$.

Thus, knowing the surface area, allows the assessment of the thickness of the adsorption layer, hence the reaction volume of the "nanophase reactor", from Equation 2 (see Figure 1a).

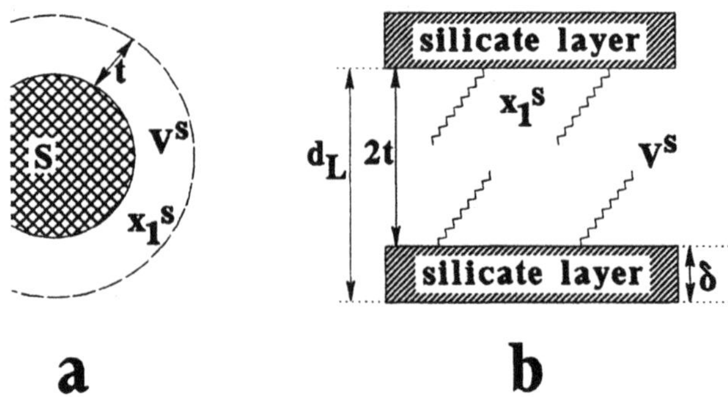

Figure 1. Schematic picture of the adsorption layers (a)on smooth spherical, (b)on layered hydrophobic silicate (organoclay). V^s=adsorption layer volume, x_1^s=adsorption layer composition, t=layer thickness, d_L=basal spacing of the silicate lamella.

3.2. VOLUME OF THE ADSORPTION LAYER (THE NANOPHASE REACTOR) IN LAYERED SILICATES AND CLAY ORGANOCOMPLEXES

Volumes within interlamellar spaces are assessed by the interlamellar space-filling model (see Figure 1b). An important advantage of layered systems is that interlamellar distances can be accurately determined by X-ray diffraction measurements. The interlamellar free volume, V_{int}, is given by $V_{int} = (a \times b)(d_L - \delta)$, where $a \times b$ is the surface area of a layer silicate unit cell (its area, $a \times b$, is taken to be 0.232 nm^2), δ is the thickness of the unit cell (0,96 nm), and d_L is the basal distance, available from X-ray diffraction measurements (from Bragg-reflections). Interlamellar distances in hydrophobic clay has been shown to increase upon the introduction of an appropriate organic liquid. This process is the swelling of the organoclay particles. Using a binary organic liquid mixtures, the interlamellar free volume, V_f, is given by:

$$V_f = V_{int} - V_{alk} = 0.232(d_L\text{-}0.96) - 0.205[(0.127n_c+0.28)+V_{pol}]\xi \quad (4)$$

where V_{alk} is the volume occupied of the surfactant alkyl-chains, containing n_c number of carbon atoms, in the interlamellar space, V_{pol} is the volume of the surfactant head groups in the interlamellar space and ξ is surface charge density/unit cell. Previous experiments have demonstrated good agreements between values of V_f calculated by Equation 4 and V^s assessed by Equation 2.

3.3. ASSESSMENT OF REACTANT CONCENTRATIONS IN THE ADSORPTION LAYER AND GROWTH CONTROL OF NANOPARTICLES

Selection of a binary liquid in an appropriate composition permits, as described above, the preferential adsorption of one component liquid at the solid interface (V^s) or in the interlamellar free volume ($V_f = V_{int} - V_{alk}$). Further, the thickness of the adsorption layer, and hence the volume of the nanophase reactor, is also governed by the selection of the binary liquid mixtures. It is also important to select the second liquid component such that the reactants are insoluble therein. Under these optimal conditions, Equation 2 becomes:

$$V^s = a^s t = n_1{}^s V_{m,1} + \sum n_r{}^s V_{m,r} \quad (5)$$

indicating the presence of $n_r{}^s$ molar reaction partners in $V_{m,r}$ volume. Thus Equation 5 permits the calculation of reactant concentrations in the adsorbed layer. Sizes of the incipient semiconductor nanoparticles are determined by the volume of the adsorption layer and by the concentration of the precursors introduced into this nanophase reactor. Thus, selection of a binary liquid with nanometer thick adsorption layer will lead to nanometer sized semiconductor particles. Further reduction in the size of the incipient semiconductors is accomplished by decreasing the concentrations of their precursors below their solubility in nanophase reactor.

4. Experimental Section

4.1 MATERIALS AND PREPARATION OF CDS AND ZNS NANOPARTICLES

Two colloidal silica (SiO_2) particles, hydrophilic A-200 and hydrophobic R-972, (treated by $(CH_3)_3SiCl$) were obtained from Degussa AG (Hanau, Germany).The specific surface areas of the A-200 and R-972 silica particles were determined to be

a^s=192m^2/g and a^s=121m^2/g by N_2 adsorption at 77.4 K after the BET-Equation. The mean diameter of this particles are 10nm.

The preparation, purification and characterization of the hydrophobic clay mineral (hexadecylpyridinium-montmorillonite, HDP-montm.), have been described [9]. Hydrogen sulfide was prepared in a Kipp-apparatus, purified by washing by water and dried over calcium carbonate. Appropriate amounts were withdrawn from the vessel by a microliter syringe. Reagent grade cadmium acetate and zinc acetate (Reanal, Hungary) were used as received. Ethanol, methanol and cyclohexane (Reanal, Hungary) were dried over 0.4 nm molecular sieve(Merck, Germany).

Semiconductor nanoparticles were prepared by adding alcoholic solutions of cadmium (or zinc) acetate in amounts appropriate to those calculated for 0.5 - 1.0 of the SiO_2 or HDP-montm. particles. Composition of the desired binary liquids were then adjusted by adding appropriate amounts to cyclohexane (to give ethanol, x_1, = 0.05 and cyclohexane, x_2, = 0.95; and methanol, x_1, = 0.01 and cyclohexane, x_2, = 0.99) to the suspensions. These suspensions remained stable for at least a month (as evidenced by absorption spectrophotometric measurement.

4.2 CHARACTERIZATION OF THE NANOPHASE REACTORS

Adsorption excess isotherms $n_1^{\sigma(n0} = f(x_1)$ in methanol - cyclohexane and ethanol - cyclohexane mixtures were determined on A-200, R-972 silicas and HDP-montm. organoclay at 298±0.1 K. The composition of the supernatant was determined by a Zeiss liquid-interferometer. The excess isotherms were calculated by: $n_1^{\sigma(n} = n^o(x_{1,0} - x_1) = n^o\Delta x_1$,where $x_{1,0}$ and x_1 are the molar fractions of alcohol/cyclohexane prior and subsequent to adsorption and n^o is the total amount of the liquid mixture in mmol/g adsorbent [7-9].

Small angle X-ray diffraction (SAXS) measurements were performed on a Philips PW 1830 generator (CuK_α λ=0.154nm radiation) by using the compact Kratky Camera (Model KCEC/3 1129, Anton Paar Co. Austria) with 80 μm and 100 μm entrance and detector slits. Samples were placed in a 1.00 mm diameter 2.0 cm long capillary. Absolute intensity of the X-ray beam was determined by the moving slit method. Scattered X-ray intensity differences (ΔI) were taken of the sample and the blank as the function of the scattering vector, h. The date were reported in terms of logΔI *vs.* log h as scattering functions. The lnΔI *vs.* h^2 Guinier function were used to asses information on the semiconductor particle sizes (R_G)in the Gunier region [10].

Rheological measurements were taken by a Haake Type RV20/CV-100 (Germany) rotational viscosimeter equipped with a ME-15 head at 298±0.1K.

Absorption spectra of the CdS and ZnS nanoparticles on the holders in the organic suspensions were taken on a two-beam UVIKON 930(Italy) differential spectrophotometer.

The immersion enthalpies of wetting ($\Delta_w H$-values) were determined by isothermal sorption microcalorimeter LKB 2107 (Sweden) at 298$\pm$0.01 K.

5. Results and Discussion

The adsorption excess isotherms for hydrophilic A-200 and hydrophobic R-972 SiO_2 in methanol-cyclohexane mixtures are illustrated in Figure 2. The partial miscibility of methanol in cyclohexane limited the available range of alcohol composition in the bulk to a low mole fraction of methanol(x_1=0.1), but the composition in the adsorption layer was x_1^s=0.9-0.99 because the preferential adsorption of methanol on the hydrophobic silica surface. In binary liquids such as ethanol-cyclohexane, the adsorption of ethanol is positive, but there is a sign change for hydrophobic R-972 silica (Figure 2b). In this case, the composition of the surface layer and of the bulk are x_1^s=0.65 and x_1=0.05.

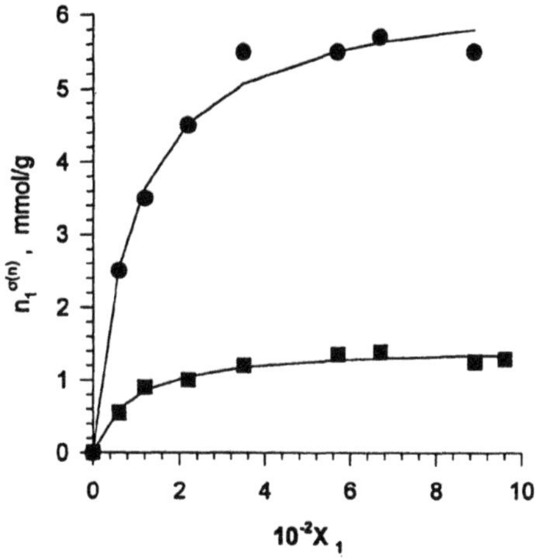

Figure 2a. Adsorption excess isotherms in methanol(1)-cyclohexane(2)mixtures on A-200(●) and R-972(■).

Excess isotherms of hydrophobic clay minerals on such hexadecylpyridinium-montmorillonite (HDP-montm.) were determined similarly. The adsorption volumes are greater due to swelling of the clay platelets. Knowing the basal distances from X-ray diffraction measurements, the V_{int}-V_{alk} volumes can be also calculated (see Eq.4.). The results of the liquid sorption experiments are summarized in Table 1.

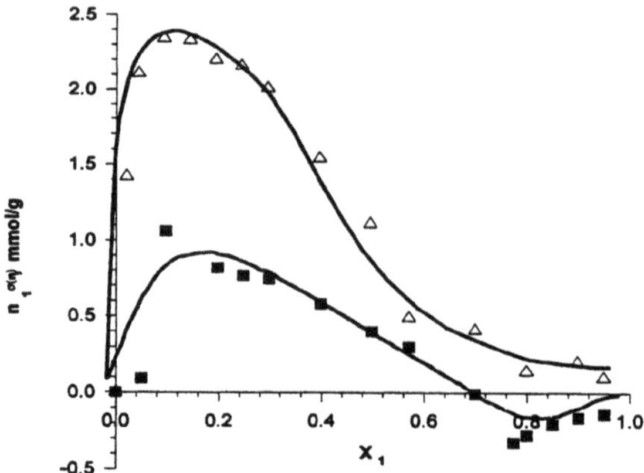

Figure 2.b. Adsorption excess isotherms in ethanol(1)cyclohexane(2) mixtures on A-200 (●) and R-972 (■).

TABLE 1. Calculation of the volume of the nanophase reactor from adsorption excess isotherms and X-ray diffraction data

Solid adsorbent	Liquid mixture					
	methanol(1)-cyclohexane(2)			ethanol(1)-cyclo-hexane(2)		
	n_1^s mmol/g	V^s cm^3/g	V_f cm^3/g	n_1^s mmol/g	V^s cm^3/g	V_f cm^3/g
A-200 SiO_2	5.45	0.218	—	4.32	0.266	—
R-972 SiO_2 treated by $(CH_3)_3SiCl$	1.40	0.056	—	2.78	0.171	—
HDP-montmorillonite	4.25	0.170	0.231	11.25	0.692	0.746

V_f=V_{int}-V_{alk} according to Equation (4).

Formation of CdS and ZnS particles in the Cd^{2+} and Zn^{2+}-containing, alcohol-rich adsorption layer was readily observable by the development color upon the introduction of H_2S. Typical absorption spectra are illustrated in Figure 3 and the absorption edges and the optical bandgaps of the nanoparticles are collected in Table 2.

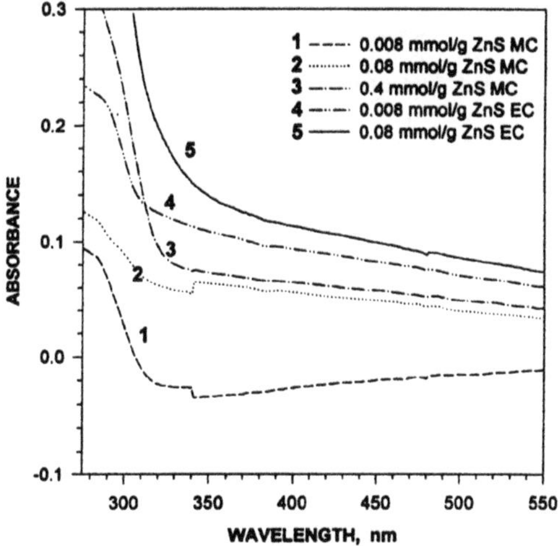

Figure 3. Absorption spectra of ZnS nanoparticles on A-200 SiO2 at various surface density in methanol(1)-cyclohexane(2) $x_1 = 0.01$ (MC) and ethanol(1)-cyclohexane(2) x1 = 0.05 (EC) mixtures.

TABLE 2. CdS and ZnS nanoparticles on hydrophilic A-200 silica generated in ethanol(1)-cyclohexane(2) x_1=0.05 mixture composition

CdS(mmol/gSiO$_2$)	8.10^{-4}	8 10^{-3}	8 10^{-2}
λ_g (nm)	389	433	450
E_g (eV)	3.19	2.86	2.76
R_G (nm)	—	5.96	6.63
ZnS(mmol/gSiO$_2$)	8 10^{-4}	8 10^{-3}	8 10^{-2}
λ_g (nm)	302	307	314
E_g (eV)	4.10	4.04	3.55
R_G(nm)	4.88	5.31	6.90

If the Cd^{23} and Zn^{2+} ions are mixed in the nanophase reactor we will have nearly the same absorption spectra as with the 1:1 amount mixture of CdS and ZnS on the silica surface (see Figure 4).

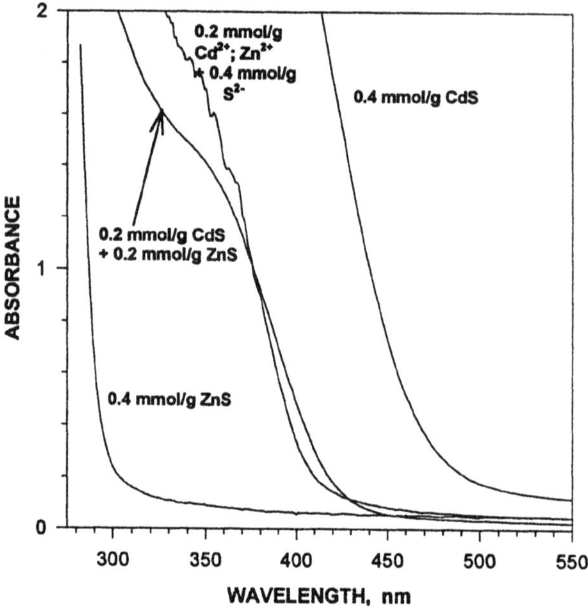

Figure 4. Absorption spectra of CdS/ZnS nanoparticles mixed 1:1 on the surface of R-972 SiO_2 in methanol(1)-cyclohexane(2)x_1=0.01 mixtures.

In our earlier experiments, we established that HDP-montmorillonite behaves similarly [6].

Formation of size quantized semiconductor particles also affected the rheological properties of their silica hosts. Plots of shear stress (τ) versus shear rates (D) for silica dispersions in ethanol - cyclohexane ($x_1 = 0.05$) in the absence and in the presence of different amounts of ZnS are shown in Figure 5. Increasing the concentration of the semiconductors are seen to increase the shear stress as expected for aggregated particles through "nanosize-bridges" in the silica suspension.

The heat of wetting in methanol on the hydrophilic A-200 powder was determined to be 41.5 J/g in the absence and in the presence of semiconductors (Figure 6).

SAXS measurements also provided valuable insight into the structural changes which accompanied the formation of ZnS in the nanophase reactor provided by the SiO_2 surface. Increasing concentrations of ZnS led to higher scattering intensities, as indicated by the Porod-plots on Figure 7. Treatment of data in terms of the Guinier Equation, permitted the evaluation the radius of gyration(R_G) of the semiconductor particle (see Table 2).

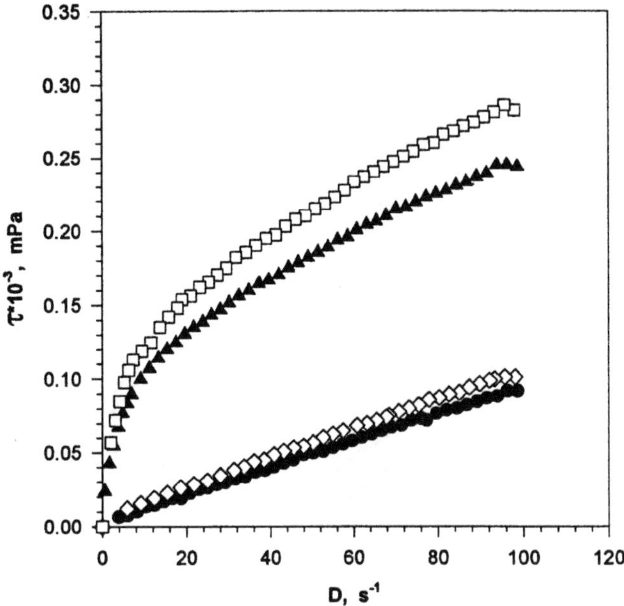

Figure 5. Rheological flow curves of ZnS nanoparticles on A-200 silica holder at various surface concentration in ethanol(1)cyclohexane(2) $x_1 = 0.05$) mixture suspension ($c_{susp.}$=0.82g ZnS/SiO_2/100ml mixture).

6. Acknowledgments

Support of this research by grants from the U.S.-Hungarian Joint Fund (Hungary), the National Science Foundation (U.S.A.) and OTKA I.5 (Hungary) is gratefully acknowledged.

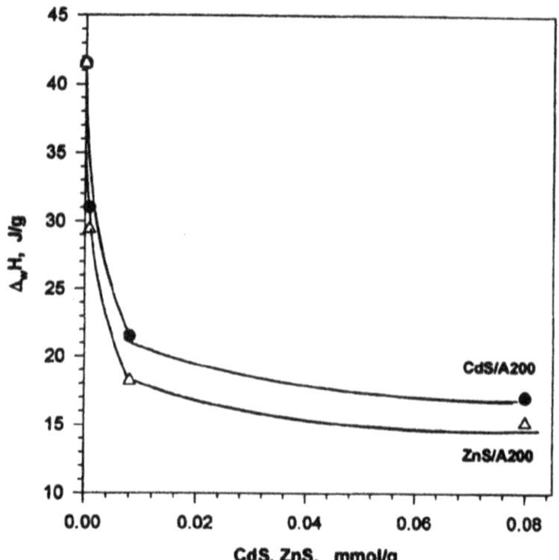

Figure 6. Plots of immersion enthalpies (ΔH_W) *vs.* the concentration of CdS or ZnS, generated on hydrophilic A-200 SiO_2 particles (expressed in terms of their precursor ion concentrations) in methanol.

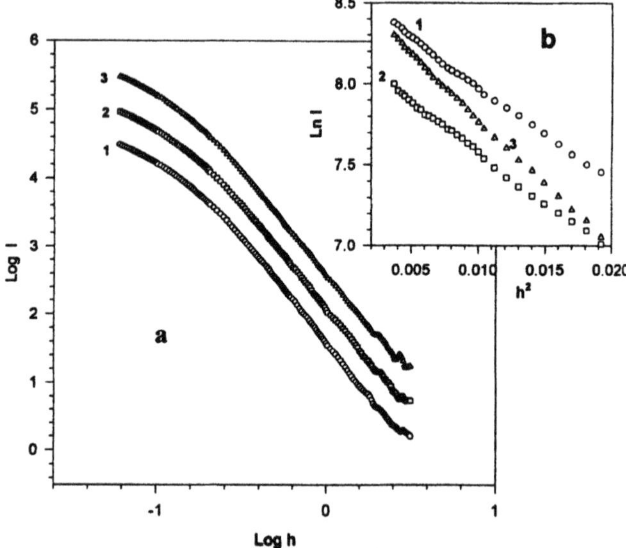

Figure 7. SAXS curves (a) for ZnS nanoparticles on R-972 silica at 1:0.004, 2:0.04, 3:0.4 mmol $ZnS/gSiO_2$ surface concentrations; (b) the Guinier plot of the same systems.

7. References

1. Ozin, G. A. (1992) Nanochemistry: synthesis in diminishing dimensions *Adv. Materials* **4**, 612-649.
2. Siegel, R. W. (1993) Mechanical properties of nanophase materials, in G. C. Hadjipanayis and R. W. Siegel (eds.), *Nanophase Materials*, Kluwer Academic Publishers, The Netherlands, pp. 233-261.
3. Henglein, A. (1993) Physicochemical properties of small metal particles in solution: "microelectrode" reactions, chemisorptin, composite metal particles, and the atom-to-metal transition, *J. Phys. Chem.* **97**, 5457-5471.
4. Kamat, P. V. (1993) Photochemistry on nonreactive and reactive (semi-conductor) surfaces, *Chem. Rev.* **93**, 267-300.
5. Fendler, J. H. (1994) *Membrane-Mimetic Approach to Advanced Materials*, Advances in Polymer Sciences Series, Vol. 113, Springer-Verlag, Berlin.
6. Kipling, J. J. (1965) *Adsorption from Solution of Non-Electrolytes*, Academic Press, London.
7. Dékány, I., Nagy, L. G., Schay, G. (1978) Effect of surface modification on solid-liquid interfacial adsorption of mixtures, *J. Colloid Interface Sci.* **66**, 197-200.
8. Dékány, I., Szántó, F., Weiss, A., Lagaly, G. (1985) Interlamellar liquid adsorption on hydrophobic silicates, *Ber. Bunsenges. Phys. Chem.* **89**, 62-67.
9. Kotov, N. A., Putyera, K., Fendler, J. H., Tombácz, E., Dékány, I. (1993) Cadmium sulfide particles in organomontmorillonite complexes, *Colloids and Surfaces A: Phys. Chem .Eng. Asp.* **71**, 317-326.
10. Guinier, A., Fournet, G. (1955) *Small-Angle Scattering of X-rays*, John Wiley, New York.

EXCESS CHARGE IN SMALL SEMICONDUCTOR PARTICLES

C. LUANGDILOK, D. LAWLESS AND D. MEISEL
Chemistry Division
Argonne National Laboratory
Argonne, IL 60439 USA

Abstract: We address the question of the effect of excess charge on the spectroscopy of small semiconductor particles. A variety of reducing species was used to inject charge into small CdS particles. A blue-shift or broadening of the spectrum is often observed at the low energy edge of the absorption. The effect of the excess charges on the absorption spectra of the particles is independent of the source of the electrons or the pH of the solution. The spectral shifts are essentially independent of the loading of electrons into the particle. Time resolved conductivity measurements show that no protons are released to the bulk of the solution upon injection of electrons from H atoms below the point of zero charge. We infer that a dipole is created across the particle in that case.

To demonstrate the effect of excess charge at the particle surface we added electrons to capping molecules attached to the surface. The surface of the particles was modified to include various electron acceptors of known redox potentials. Injection of electrons into these acceptors at the surface generates the radical anion of the corresponding molecule and its absorption spectrum could easily be monitored. However, the same spectral effects on the absorption edge of the particles could be observed in this case as in the direct injection of the electrons into the particle. It is concluded that the effects result from the electric field associated with the excess charge.

1. Introduction

The effect of excess charge on the absorption spectra of CdS particles was first reported by Albery et al (1). Using flash photolysis they were able to show that the absorption near the threshold shifts to higher energies. Direct injection of electrons from the hydrated electron using pulse radiolysis was demonstrated by Henglein and coworkers to induce a similar effect. Many other materials in the form of small colloidal particles have since been reported to exhibit the same phenomenon. The interest in these observations stems from their possible utilization as nonlinear optical or electrical effects in switching or imaging devices (3). The explanation for the effect, however, remains unclear. A band filling model had been originally proposed but was challenged soon thereafter (4). Further experimental and theoretical evidence seemed to verify the original proposition, presumably by thermal repopulation from shallow traps (5-7).

E. Pelizzetti (ed.), Fine Particles Science and Technology, 457–465.

We report here the results from two types of experiments in which excess charge was generated either by directly injecting electrons into CdS particles or by injection of electrons into acceptors that were attached to the particles surface. In the latter case the redox potential of the acceptor is known and in all cases it is positioned well within the bandgap. Nonetheless, we observe similar effects in both cases. Since in this case Boltzmann excitation into the conduction band is very inefficient, we conclude that it is the presence of charge at the surface that causes the effect. Thus, we suggest that the effects result from the generation of an electric field, induced by the injection of electrons, much like the effects induced by an external field (8).

2. Experimental

Colloidal CdS solutions in the concentration range of $(0.1\text{-}2)\times10^{-3}$ M (in terms of total CdS molecules) were prepared by injecting H_2S saturated water into degassed solutions containing $CdClO_4$, and hexametaphosphate. To control the size of the particles, capping with thiolate was used as previously described (9-12). In the present study, the thiolates mercaptopropionic acid (MPA) or 3-mercapto-1,2-propanediol (RSH) were used. The thiols were added to the solution prior to the addition of H_2S; as noticed earlier, the order of addition of the thiol and H_2S is a major factor in determining the size of the resultant particles. All solutions were deaerated by argon bubbling using the syringe technique. Excess of free thiols (or any other molecular species) was removed from the solution by ultrafiltration or dialysis using 8000 or 15000 molecular weight cutoff filters. For irradiation, electron pulses of 2-40 ns in width from the Argonne electron linear accelerator were used to produce 2.8×10^{-6} to 3.0×10^{-5} M e_{aq}^-.

3. Results and Discussion

3.1 RADIOLYTIC CONSIDERATIONS

Deposition of the energy in aqueous solutions generates the products shown in eq 1.

$$H_2O \rightsquigarrow OH,\ e_{aq}^-,\ H,\ H_2,\ H_2O_2 \qquad (1)$$

The yields of OH and e_{aq}^- are identical, 2.8 molecules per 100 eV, while that of H atoms is much smaller (0.6 molecules per 100 eV). All solutions contained 0.1 M tert-butanol to scavenge the oxidizing OH radicals generated in reaction 1 by the radiolysis of water. The radical produced from this alcohol has been shown to be inert to CdS in this system and to disappear via radical-radical recombination. In acid solutions most of the solvated electrons are scavenged by protons to yield H atoms, reaction 2:

$$H^+ + e^-_{aq} \rightarrow H \qquad k_2 = 2.3\times10^{10}\ M^{-1}s^{-1} \qquad (2)$$

The presence of trace amounts of free thiols that may remain in the solution can complicate the experiments quite significantly. In the presence of the free capping molecules, unless the thiol is rigorously dialyzed out, the β-hydroxyl radical from t-butanol and the H atoms react with the thiol to generate a sulfur centered radical. The latter in turn may react with the thiolate parent anion to produce disulfide radical ions(13). Therefore, any excess of free thiols was rigorously removed in these experiments. On the other hand, the presence of free thiols in the bulk of the solution is easy to detect by the characteristic absorption of the disulfide radical ion at ~ 450 nm upon irradiation (13). None was observed in the experiments described below.

3.2. CHARGE INJECTION INTO THE PARTICLE

Figure 1 shows the spectral changes observed upon the injection of electrons into CdS particles of r=23 and 34 Å. The spectra were recorded a few μs after the pulse when the injection reaction was complete. Both spectra of Fig. 1 were measured at pH 10. At this pH the rates of the disappearance of e^-_{aq}, measured at 600 nm, and the appearance of bleaching are identical. Under these conditions the reaction is clearly injection of electrons directly from e^-_{aq} into the particle, as shown in reaction 3.

$$(CdS)_p + e^-_{aq} \rightarrow (CdS)_p^- \quad (3)$$

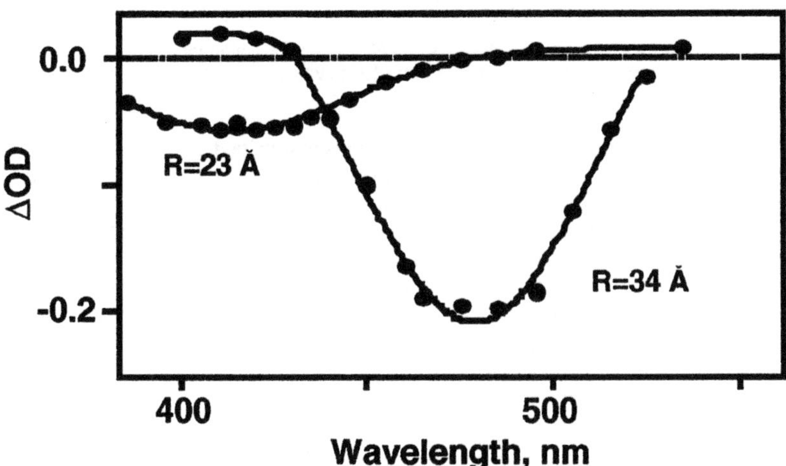

Figure 1. Difference spectra obtained upon injection of electrons into CdS particles of r = 23 and 34 Å. All solutions were at pH 10; [CdS] = 1 mM, $[e^-_{aq}] = (2.5\pm0.3)\times10^{-5}$ M.

The difference spectra shown in Fig. 1 translate into the spectra shown in Figure 2. Also shown in that figure are the spectra of the sol prior to the injection of electrons for both sizes . In both cases the shift amounts to approximately 70 meV to higher energies. Not so pronounced but visible in both cases is weak enhancement of the absorption on both sides of the strong bleaching. These results are in qualitative agreement with earlier observations (3). The change in the wavelength of maximum bleaching in the two cases of Fig. 1 follows the shift of the band gap to higher energies with decreasing the size. It is also clear from Fig. 1 that the change in absorbance is much more pronounced for the larger particles than the smaller ones. The dose in both experiments was nearly the same and total concentration of CdS is similar also. Therefore, the average charge density in both is essentially the same. Yet the effect is more pronounced for the larger particles.

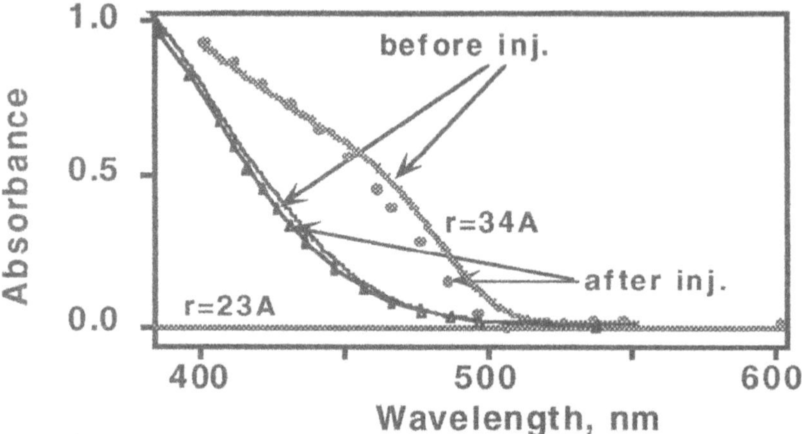

Figure 2. The spectra of the CdS particles shown in Fig. 1 before and after the injection of electrons. Solid spectra were recorded before injection and marker points are after the injection. Circles and triangles are for r=34 Å and 23 Å respectively.

Similar difference spectra were obtain from solutions maintained at pH 4. One therefore concludes that the same product, excess electrons in the CdS particle, is generated in that pH as well. However, in this case reaction 3 cannot be the reaction that generates the excess charge in the particles. From the known rate constants and concentrations of the species, it can be recognized that the solvated electrons will protonate via reaction 2 to generate H atoms. Nonetheless, the bleaching does occur, but as expected, at a slower rate than the disappearance of e^-_{aq}. In the acidic range the sequence of reactions that generates excess charge in the particles is reaction 2 followed by reaction 4.

$$H + (CdS)_p \rightarrow (CdS)_p^- + H^+ \qquad (4)$$

From the dependence of the rate of appearance of the bleaching on the particles concentration we estimate a rather high rate constant for reaction 4, $k_4 = (7.0 \pm 1.0) \times 10^{10}$ M^{-1} s^{-1}. However, attempts to detect the protons that are released in reaction 4, using the transient conductivity technique, failed. Hardly any conductivity changes were observed during the same time regime at which the bleach can be detected at pH 4. From the known specific conductivity of protons and from the known dose we estimate that less than 1% of the protons released in reaction 4 are released into the bulk of the solution. Therefore, we suggest these protons remain at the particle surface or deep within the diffuse double layer surrounding the particles where their mobility is much hindered. Thus, while the electron is transferred to the particle and its presence in the solid is observable for long periods of time (>1 ms), the electrolyte adjusts instantaneously to the newly created electric field at the interface. Using similar techniques, we have previously shown that when a solvated electron is injected into

TiO_2 (or metallic Pt) particles a proton is immediately taken from the water to adjust for the imbalance in charges (14).
Figure 3 shows the dependence of the bleaching on loading of electrons into the particles. These results were collected over a wide range of doses ($[e^-_{aq}] = (3\text{-}30)\times10^{-6}$ M) and particles concentration ($[CdS] = 0.5\text{-}2.0)\times10^{-3}$ M). The striking observation is

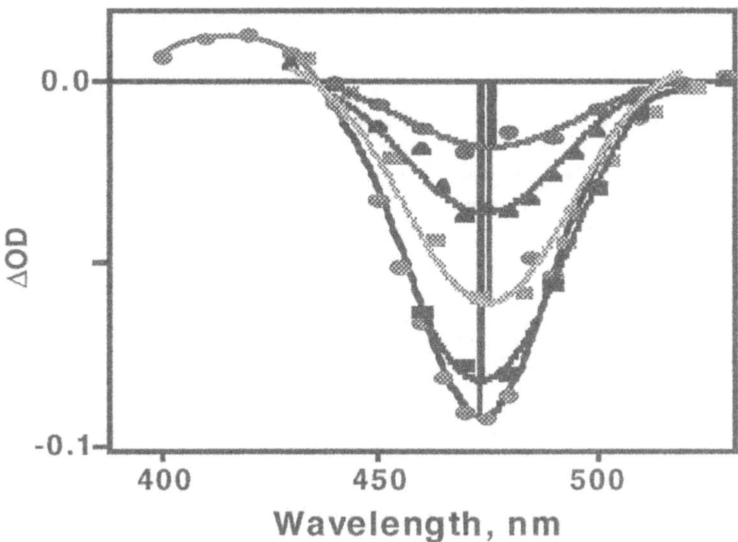

Figure 3. The difference spectra of the r=34 Å particles shown in Fig. 1 as a function of increasing loading of the particles with electrons. [CdS] = 1 mM for all spectra. The concentration of e^-_{aq} increases from 1.5 to 28 μM on going to stronger bleaching. Vertical lines from the maximum amplitude are drawn to show the very small shift in energy upon increased loading.

the rather small shift in the energy of the bleaching as the loading increases. In the set of experiments shown in Fig. 3 the loading increases from <1 to nearly 20 electrons per particle. In the same range of loading the band filling model of Liu and Bard (5,6) would predict a shift of ~70 nm in that band.
The efficiency of bleaching is much higher for the bleaching in the larger particles (e.g. compare ΔOD for the large and small particles in Fig. 1 where the dose and [CdS] are the same). Each electron causes a stronger bleaching in the larger particles than in the small ones. Furthermore, our studies of the effect of loading on the efficiency show that the effect saturates at lower charge densities in the smaller particles than in the larger one. The higher efficiency of bleaching in the larger particles than in the smaller ones is in stark contrast to the effect that one would expect from the band filling model; because of the lower density of states in the smaller particles a more pronounced effect would then be expected.
The size of the particles in both preparations is smaller than the Debye length within the solid. In both cases more than one electron per particle is required before the effect of the charge saturates. Thus, we suggest that overlap and cancellation of fields, which

result from different electrons, occur in both particle sizes. This cancellation of fields occurs at lower charge densities in the small particles than in the large ones. This is probably the reason for the higher efficiency of bleaching in the larger particles. Similarly, the earlier saturation, that is observed in the smaller particles than in the larger ones is attributed to the more extensive overlap and cancellation of fields in the former.

3.3 CHARGE INJECTION INTO SURFACE STATES

In the experiments described above, charge was injected directly into the particle. However, its spatial location and energetic level cannot be confidently identified. Presumably, the injected electrons in these cases are all trapped somewhere within the band gap close to the interface. The experiments described below were designed to remove this ambiguity. Particles were synthesized as before, but reactive functional moieties were attached to their surface. This was achieved either via covalent binding of capping molecules (nitrothiophenols) or by adsorption to the surface (quinones). The hydrated electron was then allowed to react with the molecule at the surface generating the corresponding radical ion; reaction 5, where Q is the capping moiety.

$$(CdS)_p - Q + e^-_{aq} \rightarrow (CdS)_p - Q^- \qquad (5)$$

For the molecules chosen, the redox potential of the Q/Q^- couple is well established and is known to be at energies deep within the band gap. Furthermore, the spectrum of the radical ions, Q^-, is well documented and reasonably well separated from that of the particles. Thus, localization of the excess charge can be shown to be on the acceptor molecule at the particle surface. Its energy level is at the redox potential of the Q/Q^- couple deep in the gap.

The histograms of Figure 4 are the results from such an experiment. To CdS particles (r = 13 Å) we added anthraquinone sulfonate [AQ] which strongly adsorb at the surface. Quenching of the emission from the particles, as well as the absence of an effect of AQ concentration on the rate of decay of e^-_{aq}, were used as indicators for the adsorption on the particle. The top histograms were recorded at $\lambda = 500$ nm, where the absorbance is dominated by the formation of the radical anion of the quinone. The fast decay at the first μs after the pulse is the decay of e^-_{aq} to produce the semiquinone. It is clear that in both histograms, with and without the CdS particles, essentially the same absorbance is obtained. One concludes, therefore, that in both cases the electron is localized on AQ and that the same amount of AQ^- is obtained in the two experiments.

In spite of the similarity of the results at 500 nm, very different histograms were obtained at 380 nm, near the edge of the absorption of the CdS particles (bottom Fig. 4). While for the free AQ^- the tailing end of its absorption band can be observed, bleaching appears for the same semiquinone when it is adsorbed on the particles. Note again that from the upper part of Fig. 4 we concluded that the electrons, essentially all of them, are on the quinone. Yet their effect on the particle is still the same as when the electrons were injected directly into the particles.

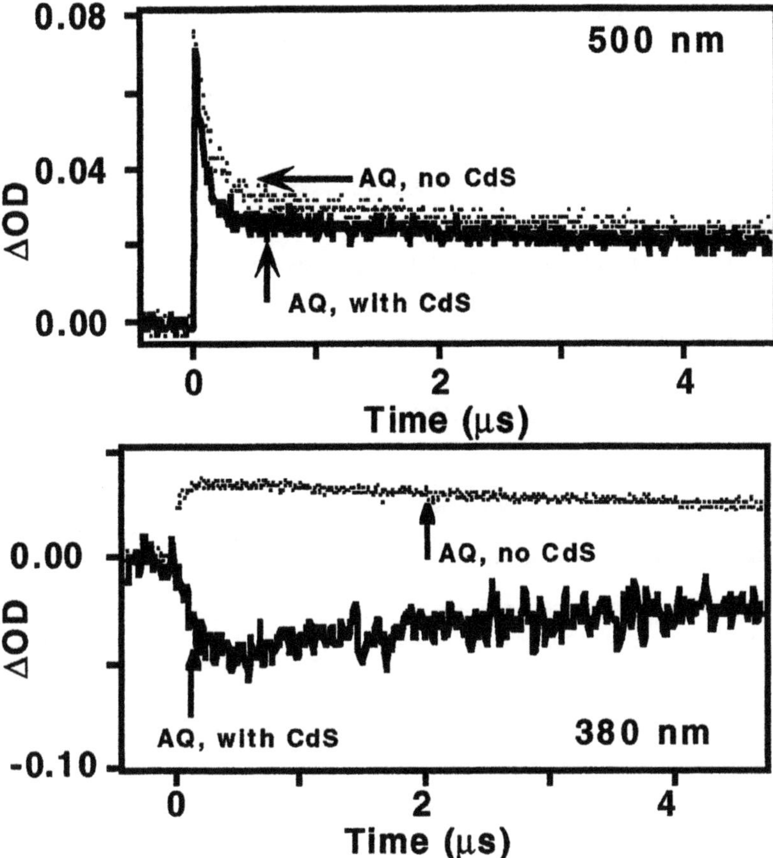

Figure 4. Histograms showing the change in absorbance following the attachment of e^-_{aq} to anthraquinone sulfonate (AQ). Top: at λ = 500 nm (dominated by AQ^-). Bottom: λ = 380 nm (edge of CdS). Dotted curve is for free AQ; solid curve: at the surface of CdS particles of r = 13 Å. Experimental conditions: [CdS] = 2×10^{-3} M, [Anthraquinone]=2×10^{-5} M; 10 ns electron pulses in all cases.

The complete difference spectrum of the system described above is shown in Figure 5 in comparison to the spectrum of the free radical anion. The combination of the two parts of Fig. 4 and Fig. 5, leads one to conclude that when the excess charge is injected into the surface of the particle, even when these electrons are energetically deep within the band gap (E^0 = -0.38 V (15)), the effect on the spectrum of the particle is still observable. This redox level is approximately 0.9 V more positive than the potential of the conduction band and ~1.2 V more negative than the valence band, nearly in the middle of the gap. Similar results were obtained with other acceptors at the surface with redox potentials of up to 0.1 V.

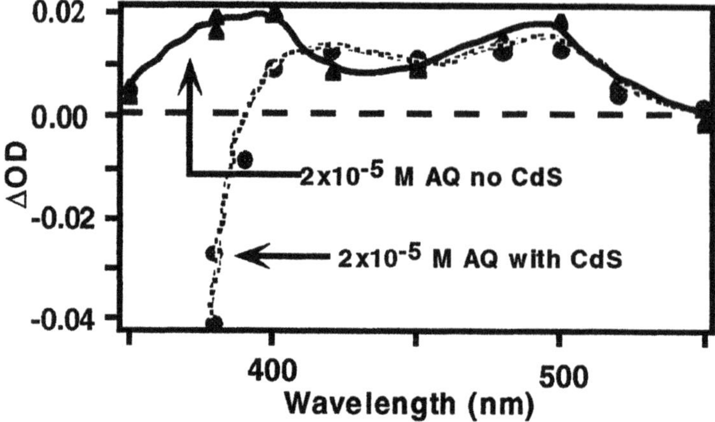

Figure 5. Difference spectra following the attachment of e^-_{aq} to anthraquinone sulfonate (AQ) free in solution (triangles; solid curve) and at the surface of CdS particles of r = 13 Å (circles; dotted curve). Experimental conditions: [CdS] = $2x10^{-3}$ M, [Anthraquinone]=$2x10^{-5}$ M; Spectra taken 10 µs after a 4 ns electron pulse.

We also compared the intensity of bleaching in the same preparations using the same doses in the presence and absence of the quinone (i.e., $(CdS)_p$-Q^- configuration vs. direct injection). The intensity of bleaching in both was very similar. Thus, one concludes that either the electrons localize at the surface in both cases (i.e. in the direct injection also) or that the effect is independent of the exact location of the trapped electrons.

4. Conclusions

Several observations in this study indicate that the effect of excess charge results from the electric field generated in the particle. The higher efficiency of bleaching that is obtained in larger particles than in smaller ones, its saturation at high loading and the lack of dependence of the band edge shift on charge density all concur with an electric field effect. They are also in contrast to an effect that may result from increasing the energy of the HOMO in the particles. In practical terms, if one can inject the same number of electrons to a large particle as to a small one, the former will lead to a larger effect than the latter as long as the diameter of the particle is smaller than the Debye length. We have also shown that hydrogen atoms can inject electrons into the CdS particles and generate essentially the same effect. The interface in this case responds immediately to the change of charge at the interface and the proton remains near the particle.

The effect of external field on small CdSe particles has been reported and was utilized to conclude that the first excited state in the particles is non-spherical (8). The shape of the spectral changes in that case was analyzed as the 2nd derivative of the absorption band. In the present study the changes could be described neither as the 1st nor the 2nd derivative of the exciton band. Close examination of many of the reported cases where excess charge was generated in small particles leads to the same conclusion. However, in many cases, enhanced absorption could be discerned on both sides of the bleaching signal. Whereas this is reminiscent of a second derivative signal, it clearly cannot be quantitatively described by a pure Stark effect. Whether this results from overlapping bands, electric field effects other than a Stark effect, the response of the double layer to the appearance of the charge or another combination of these effects is not clear at present.

5. Acknowledgments

Work performed under the auspices of the Office of Basic Energy Sciences, Division of Chemical Science, US-DOE under contract number W-31-109-ENG-38.

6. References

1. Albery, J., Brown, G. T.., Darwent, J.R., Saievar-Iranizad, E. (1986) *J. Chem. Soc. Faraday Trans. I* **81**, 1999.
2. Henglein, A., Kumar, A., Janata, E., Weller, H. (1986) *Chem. Phys. Lett.* **132**, 133.
3. Henglein, A. (1989) *Chem. Rev.* **89**, 1861.
4. Hilinski, E. F., Luccas, P.A., Wang, Y. (1988) *J. Chem. Phys.* **89**, 3435.
5. Liu, C., Bard, A. J. (1989) *J. Chem. Phys.* **93**, 3232.
6. Liu, C., Bard, A. J. (1989) *J. Chem. Phys.* **93**, 7047.
7. Kamat, P. V., Dimitrijevic, N. M., Nozik, A. J. (1989) *J. Phys. Chem.* **93**, 2873.
8. Colvin, V. L., Alivisatos, A. P. (1992) *J. Chem. Phys.* **97**, 730.
9. Nosaka, Y., Yamaguchi, K., Miyama, H., Hayashi, H. (1988) *Chem. Lett.* 605.
10. Nosaka, Y., Ohta, N., Fukuyama, T., Fujii, N. (1993) *J. Coll. Interface Sci.* **155**, 23.
11. Hayes, D., Micic, O. I., Nenadovic, M. T., Swayambunathan, V., Meisel, D. (1989) *J. Phys. Chem.* **93**, 4603.
12. Swayambunathan, V., Hayes, D., Schmidt, K.H., Liao, Y. X., Meisel, D. (1990) *J. Am. Chem. Soc.* **112**, 3831.
13. Asmus, K. (1990) *Sulfur Centered Reactive Intermediates in Chem. and Bio.* Plenum Press, New York and London, Vol. 197.
14. a. Borgarello, E., Pelizzetti, E., Mulac, W.A., Meisel, D. (1985) *J. Chem. Soc., Faraday Trans. 1* **81**, 143. b. Matheson, M. S., Lee, P. C., Meisel, D., Pelizzetti, E. (1983) *J. Phys. Chem.* **87**, 394.
15. Meisel, D., Neta, P. (1975) *J. Am. Chem. Soc.* **97**, 5198.

RADIATION INDUCED REACTIONS AT "SMALL" SURFACES AND ON SMALL PARTICLES[1]

GUOHONG ZHANG, XINSHENG LIU, YUN MAO, J. K. THOMAS
Department of Chemistry and Biochemistry
University of Notre Dame
Notre Dame, IN 46556, USA

This work describes the formation of hydrated electrons in the supercages of hydrated zeolites X and Y and the chemistry of organic adsorbates on silica surfaces by fast electron beam irradiation of these heterogeneous systems. Electron trapping by water clusters compartmentalized in zeolite supercages was observed to overwhelm the trapping by Na^{+} clusters in the hydrated forms of zeolites NaX and NaY, while only Na^{+} cluster trapped electrons were observed in both hydrated and dehydrated zeolites NaA and NaZK-4. The formation of the hydrated electron was examined with respect to the size of the water clusters confined in the supercages of the zeolites, and six "free" water molecules were found to be critical for the hydrated electron formation. Studies on the subsequent reactions of the hydrated electron demonstrate the restricted nature of its reactivity in the zeolite cage structure. Hydrogen atoms and hydrogen gas are produced with significant yields on hydroxylated silica surfaces upon high energy irradiation. Trapping of free charge carriers leads to the observation of surface trapped electrons in 150°C pretreated silicas and trapped holes in 600°C dried silicas. Formation of the 1-hydro-pyrenyl radical (PyH·), the pyrene cation radical and the pyrene anion radical was observed simultaneously with the reduction of hydrogen gas yield on 150°C dried silica sufaces loaded with pyrene. The hydrogen atoms produced on silica surfaces are shown to be the precursor of the molecular hydrogen and the pyrene free radical PyH·. G value measurements suggest that the production of hydrogen atoms is the major radiolytic process occurring on 150°C dried silica surfaces. Decreased yield of the H adduct and increased yields of ion radicals of pyrene were observed on silica surfaces pretreated at 600°C. The mechanism of production of the reactive intermediates of organic adsorbates on silica surfaces is discussed and correlated with the small particle based structures of the high surface area silicas used in the studies. The surface hydroxyls were found to be important in directing the chemistry occurring on silica surfaces through modulation of the energy transfer from inside the primary SiO_2 particles. A comparison between zeolites and silicas demonstrates the unique features of radiation induced chemistry at interfaces of heterogeneous systems.

E. Pelizzetti (ed.), Fine Particles Science and Technology, 467–497.

1. Introduction

The production of small particles via colloids is a historic event, gold colloids being produced in the last century. Recently interest has increased when it was realized that these colloids and suspensions mimic systems that store light energy.[2-17] In fact, photochemistry at interfaces on small particles has been of major interest in chemistry, as witnessed by this meeting.

This paper takes a slightly different view of this field of research where reactions in constricted media, i.e., pores of silica gel or zeolites, are quite reminiscent of their "cousins" in the colloid field. These systems also constrain liquids in small dimensions thus altering liquid properties.[18,19] Studies will be described of small pools of water and the influence of size on electron trapping in these systems.

Another feature is the generation of excited species in small particles of SiO_2 and the subsequent chemistry that is created when the energy migrates to the surface thus creating hydrogen atoms and charged species. This is reminiscent of semiconductor excitation but now the small particle is an insulator. This latter work is particularly important in catalysis, as many of the systems act as supports or as catalysts, and also in developing radiation resistant ceramics for high energy radiation sources.

2. Experimental Section

Chemicals and Sample Preparation. Zeolites NaA, NaX and NaY with Si/Al ratios of 1.0, 1.4 and 2.5 respectively were obtained from Aldrich. ZK-4 was synthesized following the literature procedures[20] using tetramethylammonium (TMA^+) as a template. The sample was examined with X-ray diffraction and chemically analyzed to give framework Si/Al ratio, 1.4. The as-synthesized ZK-4 was calcinated at 500°C for 8 h and followed by ion exchange with a 1M NaCl aqueous solution for 24 h at room temperature. The sample was then washed with distilled-deionized water and dried at 60°C. Methyl viologen dichloride hydrate ($MVCl_2$ $6H_2O$, 98%) was purchased from Aldrich and used as received. The MV^{2+}-exchanged zeolites were prepared by ion exchange with measured amounts of MV^{2+} in the solutions (solid : solution = 1:100). The supernatants were examined with spectrophotometer to ensure no residuals of MV^{2+} in the liquid phase.

Granular silica gel (Davisil, 60Å, 100~200 mesh) was supplied by the Aldrich Chemical and used as a model system of porous silica. The average surface area for Davisil-60 by the BET method is ~460m^2/g. Silica gels with different pore sizes such as Davisil-40 and Davisil-150 (from Aldrich) were also used for comparison. Cab-O-Sil fumed silica obtained from Cabot with high surface area brand CAB HS-5 (surface area S_A=325±25 m^2/g) and low surface area brand CAB L-90 (S_A=100±15 m^2/g) were used as model systems for non-porous silica. Pyrene (Aldrich, 99%) was purified by multiple chromatographic separation and recrystallization from cyclohexane solution. Cyclohexane is HPLC grade from Aldrich and used as received.

Pulse radiolysis experiments on zeolites were performed using disc samples. Discs were made by pressing zeolite powder under a pressure of ~$5x10^3$ kg/cm^2, with disc thickness typically around 20 μm. Silica powders were pretreated either in a 150°C oven over night to remove the physisorbed water or in a 600°C oven for 24 hours to remove ~80% of the surface hydroxyls.[21] Sample transfer and loading were carried out in cyclohexane solutions. Pyrene adsorption isotherms were measured for silica samples pretreated at 150°C and 600°C. The 1 mm radiation cell is made of quartz glass sealed with a 40 μm thick polyethylene teraphthalate film as the beam window, which can sustain vacuum up to 10^{-2} torr and provide a transient absorption window above 350 nm. Silica samples covered with cyclohexane solution were vacuum pumped in the radiation cell for 3 hours to completely remove the solvent. Silica/cyclohexane slurry systems were also used for index matched spectroscopic measurements. In some of the experiments, disc samples were prepared by pressing the vacuum dried silica powder and quickly transferred into a home made irradiation chamber connected with the vacuum line. The silica disc samples were checked against powders, and no significant effect of rehydration was observed.

Time Resolved Measurements. Pulse radiolysis experiments were performed on a Field Emission Febetron (Model 706) which delivers a ~2 ns pulse of fast electrons with the output energy of 0.4~0.6 MeV and a dose of ~200 krad per pulse. Transient species produced by the high energy radiation in solid samples were monitored using conventional transient absorption spectroscopy in the diffuse transmittance mode. A pulsed Xe lamp (450 W, Oriel) generating a short intense white flash (width ~500 μs) covering the wavelength region 300~800 nm was used as the analyzing light. The time response for microsecond measurements is ~20 ns, and it is ~7 ns for nanosecond measurements. Transient absorption spectra were collected by assembling the measurements at different wavelengths with a bandwidth of 3~4 nm. Spectra were corrected by monitoring the fluctuation of beam output using a separate transient absorption line. Irradiation of disc samples were carried out in a home made steel chamber with a sealed titanium thin film as the electron beam window. The whole chamber was evacuated to a pressure of about 10^{-2} torr under which 15~20% of the total content of water was removed from the fully hydrated zeolite samples. In scavenging experiments, gases and vapors were introduced into the irradiation cell or chamber via a vacuum line and pressures read out by a digital manometer.

G values of the produced transient species (number of transient species generated per 100 eV energy loss) were measured relative to that of the hydrated electron in water G=3.0.[22] The dose per pulse was determined relatively using the transient absorption of the hydrated electron at 720 nm measured at the end of the pulse (t~7 ns) in an deaerated aqueous solution with 0.1M ethanol and 0.01M NaOH in a 3 mm cell. The G values of transients were obtained using the following relation.

$$G(Tr) = \frac{A(Tr;\lambda)}{A(e_{aq}^-;720nm)} \cdot \frac{\varepsilon(e_{aq}^-;720nm)}{\eta \cdot \varepsilon(Tr;\lambda)} \cdot G(e_{aq}^-) \quad (1)$$

where, G(Tr): G value of the transient to be measured;

$G(e_{aq}^-)$: G value of the hydrated electron in water;

$\varepsilon(e_{aq}^-$; 720 nm): the extinction coefficient of the hydrated electron at 720 nm, ε=18,500 $M^{-1}cm^{-1}$ was taken from literature;[22]

ε(Tr; λ): the extinction coefficient of the transient at wavelength λ, which was taken as the same value measured in solutions;

$A(e_{aq}^-$; 720 nm): the absorbance of the hydrated electron at 720 nm measured at the end of the pulse;

A(Tr; λ): the absorbance of the transient species at wavelength λ measured at the specified time;

η: the fraction of the dose absorbed by solid samples, which is 1 for silica powder irradiated in the 3 mm cell, and ~0.3 for zeolite discs.

Experiments at low temperatures were performed in a home made vacuum dewar where disc samples were cooled down on a copper finger fitted into a liquid nitrogen reservoir. The sample temperature was controlled by the amount of liquid nitrogen added and the time of cooling, and was read out with a digital thermocouple.

Steady State Measurements. Steady state γ-ray irradiation of silica samples was carried out on a ^{60}Co-source with a dose rate of ~24 rad/s. The dosimetry was measured by the Fricke dosimeter, and corrected by the density and the mass absorption coefficient of silica powder. Silica samples were irradiated for 2 hours at room temperature for hydrogen gas measurements, and for 30 minutes at 77K for spectroscopic analysis. The yield of hydrogen gas evolved from high energy irradiation of Davisil-60 samples with different pyrene loading was measured on a Carle gas chromatography, equipped with a packed column (80% PPN and 20% PPΦ) and a thermoconductivity detector. Diffuse reflectance spectra of silica samples were taken at 77K in a home made low temperature cell on a Varian UV-visible spectrophotometer equipped with an integrating sphere. EPR spectra were recorded on an Varian EPR spectrometer (E-line century series) equipped with an X-band klystron and a cavity mode TE_{102}, operating at ~9.5 GHz and with 100 MHz magnetic modulation. G values of the hydrogen gas were determined for silica samples with different pyrene loading by measuring the absolute amount of hydrogen gas produced from irradiated solid samples against the absorbed dose.

3. Results and Discussion

3.1. Zeolites

The Structures of The Zeolites. The zeolites used in this study are NaA, NaZK-4, NaX and NaY. Both zeolites X and Y have similar structures but different framework Si/Al ratios (>1.5 for zeolite Y and <1.5 for zeolite X), and are isostructural to the natural mineral, faujasite.[23] The structure of faujasite is cubic and built from sodalite cages connected via the double-six-membered rings (D6R) of SiO_4 and AlO_4^- in a manner of tetrahedron. The structure contains three types of cages: D6R, sodalite cage, and supercage. The Na^+ cations, which compensate the negative charges of the framework due to isomorphous substitution of Si by Al, are distributed among at least four kinds of sites, site I in the center of the D6R cage, site I' inside the sodalite cage, site II on the wall of the supercage, and site III above the four-membered ring in the supercage.[24,25] Zeolite A is a synthetic material with no natural counterpart.[26] The structure is also composed of sodalite cages similar to faujasite but connected through double four-membered rings (D4R) of SiO_4 and AlO_4^-. By this connection, three cages are present: D4R, sodalite cage and α-cage. The structure of zeolite A is also cubic, having a Si/Al ratio of 1.0. The cation locations are found to be on the site A in the center of the six-membered-ring of sodalite cage and the site E on the window of the eight-membered-ring. ZK-4 is also a synthetic zeolite formed in the presence of tetramethylammonium ion and isostructural to zeolite A but of higher framework Si/Al ratio.[20] Scheme I shows a schematic drawing of the framework structures together with the possible cation sites. The occupation of the cation sites depends on the nature of the cations and on the conditions under which the samples are treated. On hydration, water molecules fill the spaces of the cages, generally on average a total of ~4 water molecules in the sodalite cage, ~24 in the α-cage of NaA, 28~32 in the supercage of NaY and NaX.[27] The water molecules are randomly located in the supercages of NaX and NaY, but structured (pentagonal dodecahedron) in the α-cage of NaA.[28]

The Radiolytic Processes in Zeolites. The following processes occur on subjection of zeolites to high energy radiation.

$$Z \xrightarrow{\beta} h^+ + e^- \quad \text{(framework ionization)} \quad (2)$$

$$h^+ + e^- \longrightarrow Z^* \quad \text{(exciton formation)} \quad (3)$$

$$h^+ \longrightarrow Z^+_{trap} \quad \text{(hole trapping in zeolite)} \quad (4)$$

$$e^- \longrightarrow e^-_{trap} \quad \text{(electron trapping)} \quad (5)$$

$$e^-_{trap} + Q \longrightarrow Q^- \quad \text{(trapped electron quenching)} \quad (7)$$

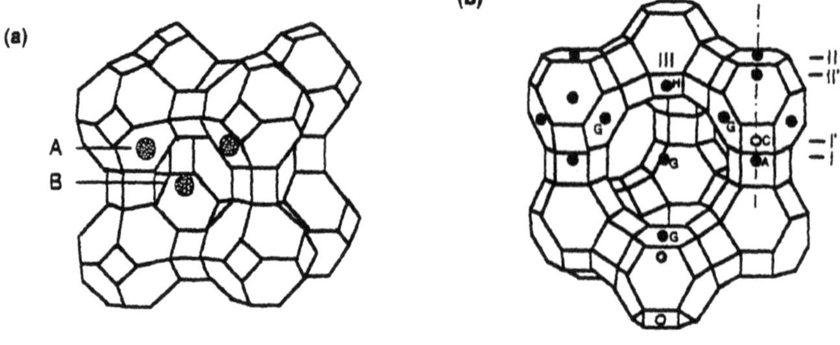

Scheme 1. Framework structures of zeolites used in this study and their possible cation locations. (a) Structure of zeolites A and ZK-4; (b) Structure of zeoites X and Y.

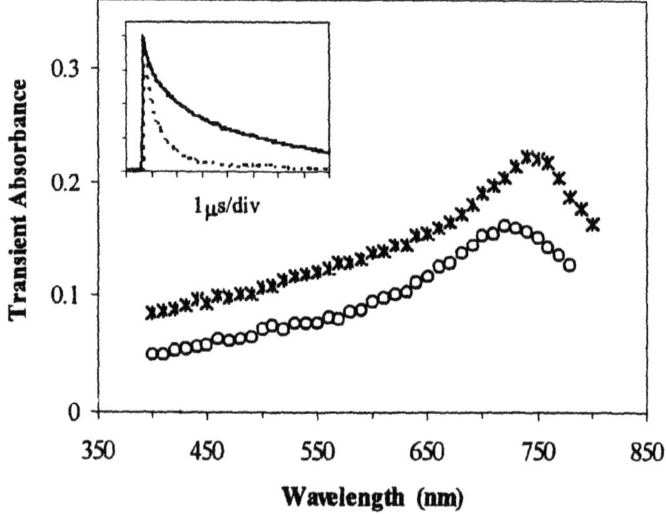

Figure 1. Transient absorption spectra of hydrated NaX (stars) and NaY (circles) taken at 20 ns after the pulse. The spectrum of NaY is offset by +0.04 absorbance unit. Insert: Decay curve monitored at 700 nm for NaX (dash line) and NaY (solid line).

The electron trapping process (5) can be further detailed for different trapping sites.

$$e^- + Na_n^{n+} \longrightarrow Na_n^{(n-1)+} \quad \text{(e}^-\text{ trapping by cation clusters)} \quad (8)$$

$$e^- + (H_2O)_n \longrightarrow e_{aq}^- \quad \text{(hydrated electron formation)} \quad (9)$$

$$e^- + Q \longrightarrow Q^- \quad \text{(electron scavenging)} \quad (10)$$

where Z denotes zeolite framework and Z_{trap}^+ the trapped hole in the zeolite framework, e^- is the electron in the conduction band, Na_n^{n+} the cation cluster capable of trapping an electron, and $(H_2O)_n$ the liquid-like water cluster in the zeolite cavities. Q is the electron quencher or scavenger including foreign molecules, metal ions introduced by ion exchange and impurities and defect sites in the zeolites.

Hydrated Electrons in Hydrated Zeolites. On irradiation by 0.4 MeV electrons, the fully-hydrated NaX and NaY (evacuated to ~10^{-2} torr at room temperature) show transient absorption spectra characteristic of the hydrated electron observed in aqueous solutions (Figure 1).[22] The broad absorption covers the entire spectral range examined, with a maximum centered at ~740 nm for NaY and ~720 nm for NaX. The assignment is confirmed by a spectral match with the spectrum of the hydrated electron reported in solution,[22] by the short lifetimes of the species compared to those of the cation cluster trapped electrons, and by quenching studies using electron scavengers (see below). The results therefore suggest that water clusters in the cages trap electrons instead of the cation clusters. The switch of the electron trapping center from the cation clusters under dry conditions to the water clusters upon hydration of the zeolites reflects the change of the systems on adsorption of water. The cation clusters may no longer be present in the hydrated samples or may not be strong enough to compete with the water clusters for electron trapping. Due to the presence of two kinds of cages, supercage and sodalite cage, therefore, there are two possible locations for the cation clusters. The destruction of the cation clusters is more likely to occur in the supercages. A competition of Na^+ cation clusters located in the sodalite cages with water for trapping electrons is observed, which gives a long-lived and O_2-unquenched transient signal of Na_4^{3+} around 550 nm (Figure 2). The O_2-quenched nature of the hydrated electrons, however, indicates that the hydrated electrons are exclusively located in the supercages. Except for the Na_4^{3+} trapped electrons, no other types of cation cluster trapped electrons are present, as confirmed by examinations of signals at other wavelengths. From the relative intensity of the cation cluster signal (about 10% of the total intensity of the 550 nm absorption), it is clear that the content of Na_4^{3+} trapped electrons in the hydrated sample is not high. The low yield of Na_4^{3+} trapped

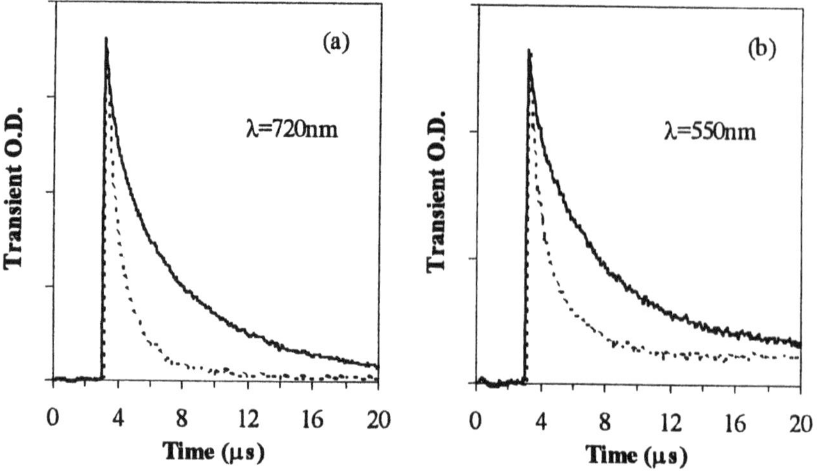

Figure 2. Time resolved absorption of hydrated NaY irradiated in vacuum (solid line) and under 20 mbar oxygen (dash line) monitored at (a) 720 nm, and (b) 550 nm.

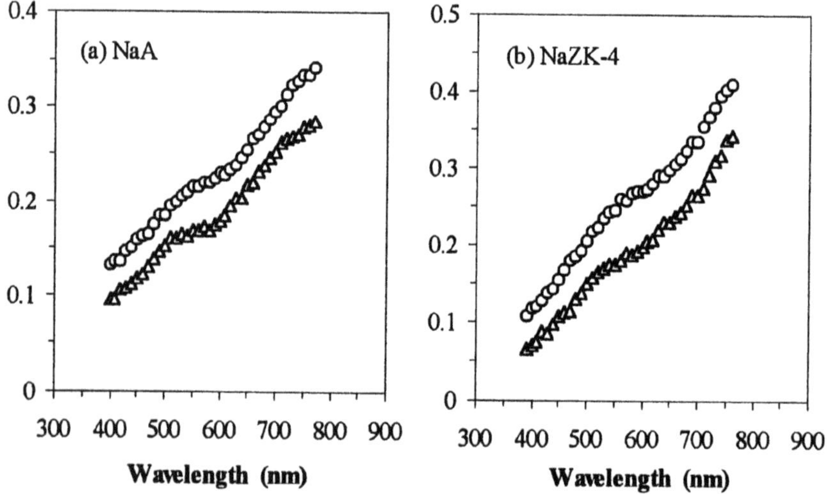

Figure 3. Transient absorption spectra of hydrated and dehydrated zeolites (in O.D. unit) taken at 20 ns after the pulse. (a) NaA: hydrated (circles, offset by +0.04) and dehydrated (triangles); (b) NaZK-4: hydrated (circles, offset by +0.04) and dehydrated (triangles).

electrons is either due to an inefficient competition of the Na_4^{4+} sites with water for trapping electrons and/or to a decrease of the number of Na_4^{4+} clusters by changing the cation location on hydration (if any).

In contrast to NaX and NaY, the water molecules present in the α-cage of the hydrated NaA are not capable of trapping electrons. Figure 3a shows the transient absorption spectra of the hydrated and dehydrated NaA recorded 20 ns after the electron pulse. The spectrum of the hydrated NaA is very similar to that of the dehydrated NaA, with only the long-lived Na^+ cluster trapped electrons. Compared to NaX and NaY, NaA has two different structural features: a slightly smaller α-cage (~11Å in diameter) and a structured arrangement (pentagonal dodecahedron) of water molecules (random in NaX and NaY). In order to clarify whether these are the reasons for the lack of electron trapping by water clusters, NaZK-4, which is structurally similar to zeolite A but of a higher Si/Al ratio (1.4), was examined. The sample was prepared in the presence of tetramethylammonium ions, followed by subsequent calcination at 500°C to remove the organic molecules and ion-exchange with Na^+. These attempts ensure the random distribution of water molecules in the α-cage. Figure 3b shows the spectra before and after dehydration. The spectra are very similar to those of NaA and are not affected by dehydration. The results therefore indicate that no matter how the water molecules are arranged, structurally or randomly in the α-cage, no hydrated electrons are generated. We therefore conclude that, when the cage of the zeolite becomes smaller or equal to 11Å, the water molecules in the cage can no longer compete with the cation clusters for trapping electrons.

Using electron spin echo technique to determine the structure of the hydrated electron, Kevan[29] has found that six water molecules are coordinated to the electron. Theoretical studies of the solvation structure of a hydrated electron in water by Rossky et. al.[30] have shown that the electron wavefunction spreads mainly over the six neighboring water molecules. A study of hydrated electrons in the water pools of inverted micelles in heptane by Wong et. al.[31] has shown that no hydrated electrons can be observed when the ratio of the number of water molecules to the number of sodium diisooctyl sulfosuccinate molecules reaches 4. In other words, the water molecules which participate in the solvation of Na^+ ions are not available to solvate electrons. In comparison with these literature reported results, we suggest that at least ~6 water molecules are "free" or liquid-like in each supercage of NaX and NaY for the formation of a hydrated electron and that there are non or insufficient "free" water molecules in the α-cage of NaA and ZK-4. The water in a zeolite cavity is found to be present in several forms as cation solvation water, similar to that seen in the inverted micelle systems, as hydrogen-bonded water to the framework atoms, and as bridging water in between pairs of cations or between cations and other water or framework oxygens. The number of water molecules directly coordinated to each cation depends on the nature of the cation, the location of the cation in the cavities, and on the coordination of the cation to the framework oxygen. In general, the smaller the cage,

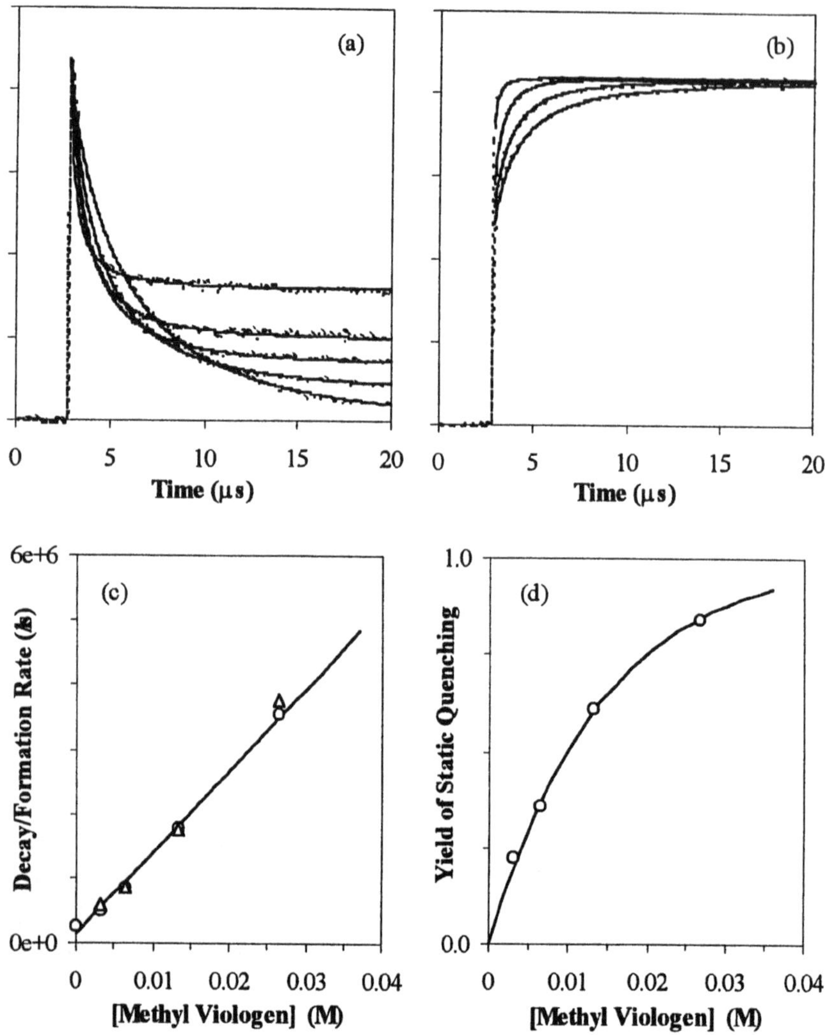

Figure 4. (a) Quenching of the hydrated electron in hydrated NaY by MV^{2+} monitored at 700 nm (signal normalized); (b) Formation of $MV^{+\cdot}$ monitored at 390 nm with different concentration of MV^{2+} (signal normalized); (c) Variation of decay rates of the hydrated electron (circles) and formation rates of $MV^{+\cdot}$ (triangles) in hydrated NaY with the concentration of MV^{2+}. The solid line is the fitting with a linear function; (d) The yield of the static formation of $MV^{+\cdot}$ at different concentrations of MV^{2+}. The growth pattern is fitted to the Perrin model (solid line).

the fewer water molecules that are coordinated to the cations. In NaA, for instance, only four water molecules are found in each sodalite cage.[28] If we assume that there are 4 Na^+ ions in each sodalite cage, then the number of Na^+ in each α-cage would become 8, and the ratio of the number of water molecules to the number of Na^+ in the α-cage would be $n(H_2O)/n(Na^+) = 24/8 = 3$, which is conspicuously smaller than the value found in inverted micelle systems.[31] For NaZK-4, even though its Si/Al ratio is 1.4, the $n(H_2O)/n(Na^+)$ ratio estimated is only ~4 if a similar situation in the sodalite cage is assumed to that in NaA. On the basis of these estimations and a comparison with the inverted micelle systems, one would not expect to observe hydrated electrons in NaA and NaZK-4, which is in agreement with the experimental observations. The same estimation may be performed on NaX and NaY using literature data.[32,33] If we assume that each Na^+ ion on the site II is coordinated only by 2 water molecules from the supercage due to its coordination to the framework oxygens and that there are 4 additional Na^+ ions in each supercage of NaX and 1 in each supercage of NaY, the $n(H_2O)/n(Na^+)$ ratio in the supercage is ~6 for NaX, and it is ~20 for NaY. Consequently the formation of hydrated electrons are expected in these zeolites, which is in accord with the experimental observations. If we also assume that the Na^+ ions in the supercages are similar to those in the water pool of the inverted micelle and each is coordinated by 4 water molecules,[31] then the number of "free" water molecules estimated is 4x6-4x4 = 8 for NaX and 1x20-1x4 = 16 for NaY. Even though 15~20% of the total water can be removed by evacuation, the number of "free" water molecules in a supercage of these zeolites is still 6.4 for NaX and 12.8 for NaY. From the above comparative study of electron trapping in hydrated zeolites with different cage sizes and Si/Al ratios, it appears that a minimum of 6 water molecules are required for the formation of hydrated electrons. As more "free" water molecules are present in the supercage of NaY, the hydrated electron is more confined by its environment in NaX than in NaY. The slight blue shift of the absorption band of the hydrated electrons in NaX compared to that in NaY (720 nm versus 740 nm) may be a consequence of this effect. Similar phenomena have been observed in inverted micelle systems[31] where the spectra of the hydrated electrons becomes narrower and blue shifted when the water pools become smaller and rather tightly confined by their environment.

Figure 1 also gives the decays of the hydrated electrons in NaX and NaY (see insert in Figure 1). From the decay curves, the average lifetime of the hydrated electrons were measured to be 0.46 μs in NaX and 3.61 μs in NaY. The shorter lifetime of the hydrated electrons in NaX compared to in NaY reflects the difference in environment experienced by the hydrated electrons in these two zeolites, which is parallel to that seen for the cation cluster trapped electrons.

Quenching studies of the hydrated electrons using electron scavengers give information about the mobility of hydrated electrons in the zeolite systems. Figures 4a and 4b show the decay of the hydrated electron and the concomitant formation of the reduced methyl viologen ($MV^{+\cdot}$) in zeolite Y, monitored at 700 nm and 390 nm respectively, as a function of the concentration of methyl viologen (MV^{2+}). The reactions involve

$$MV^{2+} + e^- \longrightarrow MV^{+\cdot} \tag{11}$$

$$MV^{2+} + e^-_{aq} \xrightarrow{k_{12}} MV^{+\cdot} \tag{12}$$

The first equation is the reaction of MV^{2+} with the conduction band electrons, which contributes to the instantaneous rise of the $MV^{+\cdot}$ in Figure 4b, and the second is the reaction of MV^{2+} with the hydrated electron over microseconds to form $MV^{+\cdot}$, which contributes to the slow growth of $MV^{+\cdot}$ seen in Figure 4b and the dynamic quenching of the hydrated electrons seen in Figure 4a. The presence of the slow formation of the $MV^{+\cdot}$ signal in Figure 4b unambiguously demonstrates that the second reaction (equation 12) indeed takes place. Computer fitting of the decay curves of the hydrated electrons and the corresponding slow growth of the $MV^{+\cdot}$ using the Gaussian distribution model[34] gives a quenching rate constant, k_{12}, for hydrated electrons reacting with MV^{2+} in NaY to be $1.26 \times 10^8\ M^{-1}\ s^{-1}$ (Figure 4c). The exact match of the decay rate of the hydrated electrons and the formation rate of $MV^{+\cdot}$ confirms the mechanism for the dynamic quenching of the hydrated electrons that occurs solely between MV^{2+} and hydrated electrons (see Figure 4c). Due to its size and charges, the mobility of MV^{2+} in the cavities could be neglected, and the diffusion coefficient of the hydrated electrons (D_e) in NaY is estimated using the Smoluchowski's relation[35]

$$k = 4\pi R D_e N \tag{13}$$

where N is the Avogadro constant and R is the effective reaction radius which includes the electrostatic interaction between MV^{2+} and hydrated electron and was in this case taken roughly as 16Å on the basis of the rate constant measured in aqueous solution,[36] $k = 7.5 \times 10^{10}\ M^{-1}\ s^{-1}$. The D_e value estimated then is $1.2\times10^{-7}\ cm^2\ s^{-1}$ which is about two orders of magnitude smaller than that of hydrated electrons in pure water, $4.9\times10^{-5}\ cm^2 s^{-1}$.[21] The slow mobility of the hydrated electrons in the cavities of zeolites reflects confinement of the hydrated electrons by the zeolite structures.

The instantaneous formation of the $MV^{+\cdot}$ in Figure 4b, which corresponds to the static quenching of the hydrated electron signals, originates from the reaction of MV^{2+} with e^- (equation 11). The change in the static yield of the $MV^{+\cdot}$ with increasing MV^{2+} concentration is plotted in Figure 4d. Computer curve fitting using the Perrin model[37] gives the radius of the "active" sphere, R_c, of MV^{2+} for its reaction with the precursor electrons in the zeolite.

$$\frac{Y}{Y_\infty} = 1 - \exp\left(-a\,[MV^{2+}]\right) \tag{14}$$

$$a = \frac{4}{3}\pi R_c^3 N \tag{15}$$

where Y_∞ is the yield of $MV^{+\cdot}$ at the highest MV^{2+} concentration, and N is the Avogadro constant. The R_c obtained from this equation is 30.3Å for reaction (11) in hydrated NaY.

Figure 5a gives the decay curves of oxygen quenching of hydrated electrons in the fully hydrated zeolites and Figure 5b shows the plots of the corresponding rate data using the Gaussian distribution model.[34] The quenching feature of the hydrated electrons by oxygen indicates that the cavities of NaX are not completely blocked by the adsorbed water molecules. This is consistent with the removal of 15~20% of the total amount of water from the cavities. The oxygen molecules can penetrate through the free volume into the cavities to quench the hydrated electrons.

$$e^-_{aq} + O_2 \xrightarrow{k_{16}} O_2^- \tag{16}$$

Loading large molecules such as tris(2,2'-bipyridine)ruthenium(II) ions, $Ru(bipy)_3^{2+}$, onto the external surface of the zeolite particles does not lead to observable quenching of e^-_{aq}, confirming the penetration of oxygen into the supercages.

The apparent O_2 quenching rate constants for the hydrated electrons in NaX and NaY, k_{16}, are 5.50×10^4 and 6.98×10^4 $mbar^{-1}s^{-1}$, respectively (see Figure 5b). The difference in the O_2 quenching rate may just reflect the different degree of blocking in these two zeolites due to their different surface polarities or hydrophilicities for water molecules. The quenching involves two mobile species, the hydrated electrons and the O_2 gas molecules. To understand the movement of O_2 in the zeolites, the O_2 quenching of pyrene singlet in the hydrated NaX was examined under identical conditions to those used in the e^-_{aq} studies. Such a comparison was justified by the fact that both pyrene and e^-_{aq} are located in the supercage. Due to the relative immobility of pyrene molecule in this system, only O_2 molecule moves to quench the excited pyrene singlet via a diffusion-controlled quenching mechanism. The O_2 quenching rate obtained for the hydrated Py-NaX is 5.34×10^4 $mbar^{-1}$ s^{-1} which is very close to the value given above for the hydrated electrons. This result therefore indicates that O_2 quenching of the hydrated electrons is mainly due to diffusion of O_2 through the free space.

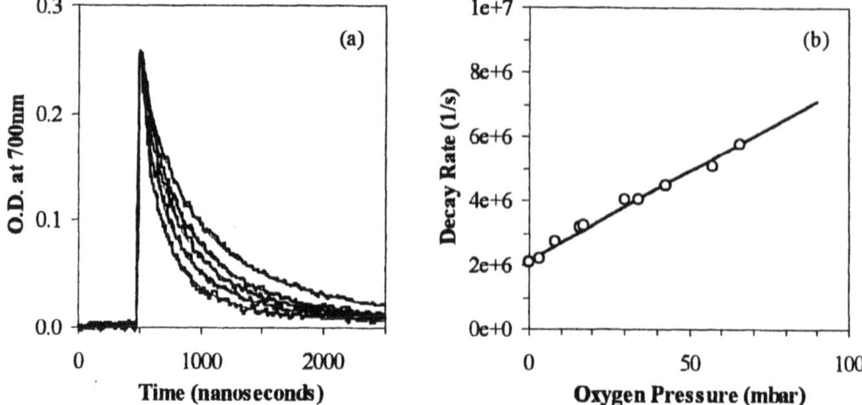

Figure 5. Oxygen quenching of the hydrated electron in NaX. (a) Decay curves monitored at 700 nm at different oxygen pressures 0, 8, 16, 33, and 66 mbar from top to bottom. Smooth lines are the Gaussian fitting. (b) Linear regression of the decay rates of the hydrated electrons against the oxygen pressure.

However, the movement of O_2 in the fully hydrated zeolites is obstructed severely by the water molecules compared with the dehydrated sample. This can be seen by a comparison of the O_2 quenching rate of the excited pyrene singlet in the dehydrated NaX. The O_2 quenching rate for $^1Py^*$ measured in the dehydrated case is 1.2×10^7 mbar^{-1} s^{-1}, which is about 200 times larger than that in the hydrated case. Under the condition of complete water blocking of the cavities (equilibrated with 15 mbar water vapor), the O_2 quenching rate becomes very small, 3.56×10^2 mbar^{-1} s^{-1}, which corresponds to O_2 diffusion through the water in zeolites.

The G value of the hydrated electrons in NaY was measured in electron pulse radiolysis to be 8.3±0.2, nearly 3 times that of the hydrated electron in water. This indicates that charge separation efficiency in the zeolites is high.

3.2. Silicas

The high surface area silicas used in the following studies have two different structures: porous for Davisils and non-porous for Cab-O-Sils. The common feature of these two types of silica powders lies in that they consist of ultimate SiO_2 particles of nanometer size. For Cab-O-Sils, these primary particles spontaneously adhere together in loose aggregations. The particle sizes are ~60Å in diameter for HS-5, and ~300Å for L-90. The primary particles in Davisils, however, are continuously linked into a rigid three dimensional network interpenetrated by a multiconnected pore structure. The particle diameters are ~75Å for D60, and ~210Å for D150. Adsorption of foreign molecules including aromatics occurs via hydrogen bonding interaction with the silanol groups on silica surfaces. Physisorbed water is removed before loading other organics by heating silica samples at 150°C overnight, while pretreatment at 600°C eliminates most of the hydrogen bonded silanol groups leaving isolated hydroxyls intact.[21] The concentration of hydroxyls is ~5 OH/nm^2 in 150°C dried silicas, and it is ~1.2 OH/nm^2 on 600°C dried surfaces.[21] Pyrene, used as the surface probe in this work, is adsorbed preferentially on those hydroxyl rich sites which can offer maximum hydrogen bonding.[38] As we will see in the following, the selective chemistry of pyrene on silica surfaces is favored by this preferential adsorption.

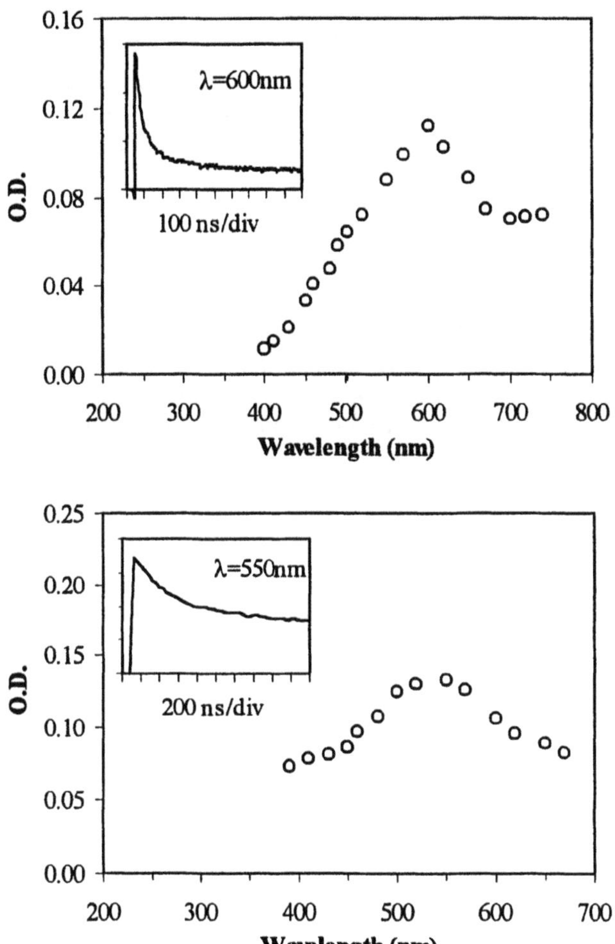

Figure 6. Transient absorption spectra of irradiated Davisil-60 taken at ~7 ns after the pulse. (a) Pretreated at 150°C. Insert shows the decay trace monitored at 600 nm. (b) Pretreated at 600°C. Insert gives the decay trace at 550 nm.

Irradiation of Blank Silicas. γ irradiation of D60 powder dried at 150°C produces copious amounts of hydrogen gas at room temperature and H atoms stabilized on silica surfaces at 77K. The results will be presented later in this paper. Time resolved experiments using 2 ns electron pulse revealed a short lived transient which exhibits a broad absorption spectrum in the visible region with its maximum around 600 nm (Figure 6a). The decay of this transient is nonexponential with a half lifetime of ~60 ns (Figure 6a insert). Introduction of oxygen and nitrous oxide quenches the transient band dynamically, indicating that (1) it is a surface species, and (2) it is due to electron trapping. Measurements were also made in a Davisil-60/cyclohexane slurry where the same absorption band was observed with a half lifetime of ~30 ns. Addition of chloroform in the slurry statically quenches the transient signal, again indicative of a surface trapped electron. Modification of the silica surface induces a variation in its lifetime. Longer lifetimes were observed in Na^+ exchanged D60 and D60 with triethyl amine adsorbed on the surface, while a shorter lifetime was observed on D60 with physisorbed water. The same surface trapped electron was also observed in other Davisils and Cab-O-Sils. Dehydroxylation of D60 at 600°C removes the surface trapped electron and gives rise to a new broad absorption band after pulsed irradiation (Figure 6b). The new band is relatively long lived (Figure 6b insert), and it is not affected by the typical electron scavengers such as oxygen, N_2O, and chloroform. However, adsorption of cyclohexane statically decreases the yield of the new transient, and it is completely removed in a D60/cyclohexane slurry. At this stage, it is tentatively assigned to a trapped hole on the dehydroxylated surface, and its structure is not clear. The removal of the surface trapped electron by dehydroxylation suggests that the electron might be trapped by the hydrogen bonded surface silanol clusters, but not by the isolated hydroxyls. The same surface trapped electron was also observed in two photon ionization of pyrene on the surface of Davisil-60 where electrons are ejected and trapped.

Irradiation of Silicas Loaded with Pyrene. Based on its well established chemical reactions and spectroscopic data,[39-41] pyrene was used as a probe molecule to examine the radiation chemistry on silica surfaces. Irradiation of a 1.8×10^{-6} mol/g pyrene/D60 sample produces a transient spectrum different from that of the blank silica. The surface trapped electron is little affected at this level of pyrene loading, and thus contributes to the short lived spectral component around 600 nm. The spectrum taken at ~0.5 μs after the decay of the surface trapped electron clearly shows the production of pyrene transient species (Figure 7a). The absorption peaks around 450 nm and 490 nm were identified as the pyrene cation radical and the pyrene anion radical, respectively. The absorption band around 400 nm has the highest intensity, and is assigned to the 1-hydro-pyrenyl radical (PyH·) based on the spectral shape and the mechanistic studies described in the following part of this paper. The transient absorption spectrum of PyH· produced via hydrogen addition reaction in acidic methanol was also shown in Figure 7a for comparison.[41] G values of these transients were measured for D60 samples with different pyrene loading (Figure 7b). The yield of

PyH· increases almost linearly with the increasing pyrene loading, and shows some saturation at high pyrene loading. The G value of PyH· production at the highest loading [Py]=2.45×10^{-5} mol/g is 3.6, while the yield of pyrene cation is G=1.2 and that of pyrene anion is G=0.3. The low yield of pyrene anion is consistent with the minimal effect of pyrene loading on the surface trapped electron. The inefficient electron scavenging by adsorbed pyrene, biphenyl, and even methyl viologen exchanged on surface might result from the efficient electron trapping by surface silanol groups which exist in high surface concentration (~5 OH/nm^2) on the 150°C dried silica.

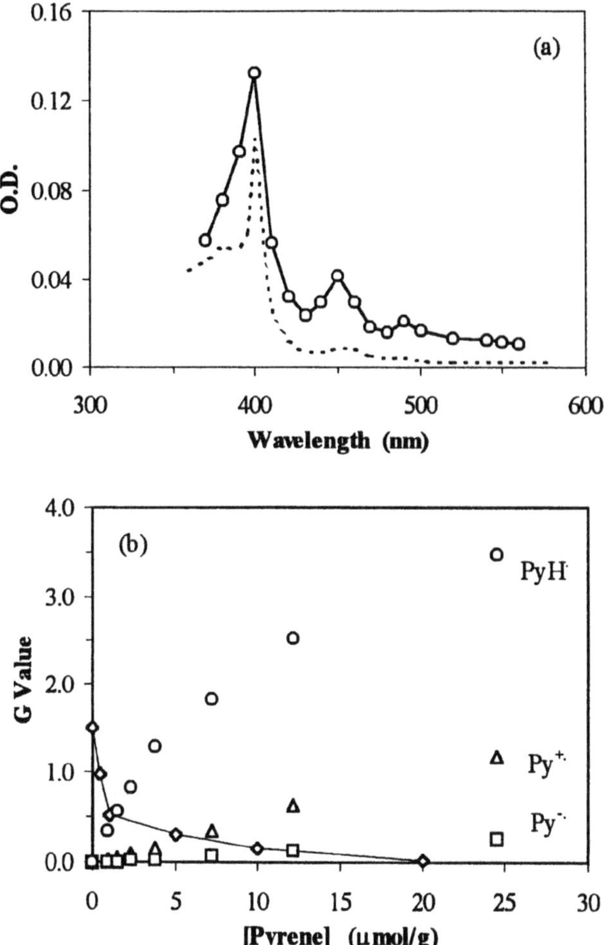

Figure 7. (a) Transient absorption spectrum of irradiated 1.8×10^{-6} mol/g pyrene/D60 powder (T_a=150°C) taken at 0.5 μs after the pulse. The dash line shows the spectrum of PyH· produced in acidic methanol; (b) G values of the pyrene transient species produced on the surface of D60 (T_a=150°C).

Irradiation of 600°C dried D60 with the pyrene loading of $2.5x10^{-6}$ mol/g also produces PyH·, $Py^{+\cdot}$, and $Py^{-\cdot}$, concomitant with a static quenching of the trapped hole. But the yields of the pyrene transients are different from those measured for the 150°C dried samples. A comparison was made between two samples with the same pyrene loading, one pretreated at 150°C and one dried at 600°C (Figure 8). Relative to the 150°C sample, the increased production of pyrene ion radicals was observed with a decrease of PyH· yield, indicative of a shift of energy transfer from the radical form to the ionic form. However, considering the dramatic decrease of the surface hydroxyls from ~5 OH/nm^2 at 150°C to ~1.2 OH/nm^2 at 600°C, the small drop of the PyH· yield is indicative of the selective formation of the PyH· radical which might be attributed to the energy transfer to those surface sites with hydroxyl groups where pyrene molecules are adsorbed.

Formation of The Pyrene Radical. The above measurements of the transient species produced on irradiated silica surfaces shows that the major process on irradiated silica surfaces is the formation of the PyH· radical. Previous work in liquid solutions has suggested that there are two possible mechanisms for the PyH· radical formation: hydrogen addition to pyrene[41] and protonation of the pyrene anion radical.[42]

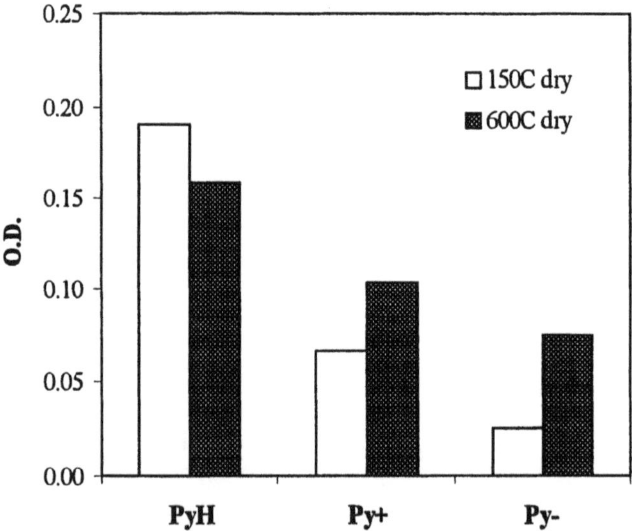

Figure 8. Comparison between the pyrene transient yields on irradiated Davisil-60 pretreated at 150°C and 600°C. The same pyrene loading was used, $[Py]=1.8x10^{-6}$ mol/g.

$$H + Py \longrightarrow PyH\cdot \tag{17}$$

$$H^{+} + Py^{-}\cdot \longrightarrow PyH\cdot \tag{18}$$

Both mechanisms seem possible on silica surfaces. For reaction (17), excitation energy moves to the surface hydroxyls ($\equiv SiOH$), splits the $O-H$ bond, and produces hydrogen atoms;[43,44] while for reaction (18), charge separation of an electron-hole pair leads to electron scavenging by pyrene, and hole trapping at the non-bridging oxygen hole centers (NBOHC) present on the surface and release of proton H^{+},[45] subsequent ion recombination on surface leads to PyH· formation. These two different processes show one common feature, i.e., they are surface hydroxyl controlled reactions.

Attempts were made to examine reaction (18) by monitoring the pyrene anion radical signal at 490 nm on both 150°C dried and 600°C dried D60 samples. For a 150°C treated D60 with pyrene loading of 1.8×10^{-6} mol/g, the yield of the pyrene anion, after the deduction of the trapped electron signal, is too small (G=0.02) to account for the large yield of PyH· (G=0.72), and the decay of $Py^{-}\cdot$ could not be correlated with the PyH· radical formation. The same mismatch was observed for a 2.5×10^{-6} mol/g pyrene loaded D60 pretreated at 600°C. Therefore, protonation reaction (18) to produce H atoms via ion recombination either does not occur or is an insignificant process on silica surfaces.

Time resolved measurements of PyH· formation kinetics at 400 nm on D60 samples with different pyrene loading observed a distinct two component growth pattern with a fast part which rises with the excitation pulse and a slow part extending over many microseconds (Figure 9a). The slow growth component exhibits the largest amplitude in between the low and high pyrene loadings. However, Cab-O-Sil samples only show the fast formation of PyH· without the slow component. Nanosecond measurements on semi-transparent Cab-O-Sil disc samples with different pyrene loading found that (1) the PyH· formation kinetics is highly nonexponential, with ~60% of the radical formed within the system response of ~7 ns and the other 40% formed over ~100 ns (Figure 9b); (2) no slow growth was observed after 400 ns; (3) the formation kinetics are independent of pyrene loading. The effect of oxygen on the PyH· radical formation kinetics is also shown in Figure 9b. The PyH· growth which is slower than ~5 ns (after the deduction of the instrument response) is removed by 50 mbar oxygen, indicating that oxygen reacts with the precursor of PyH·.

Scavenging experiments were carried out to identify the nature of the precursor which leads to the formation of the PyH· radical. Figure 10 shows the effect of foreign gases on the yield of the PyH· radical produced in 150°C dried D60. The slow formation of PyH· over microseconds is removed by oxygen with pressure as low as 1 mbar, leaving the fast part unaffected. With the increasing oxygen pressure the fast

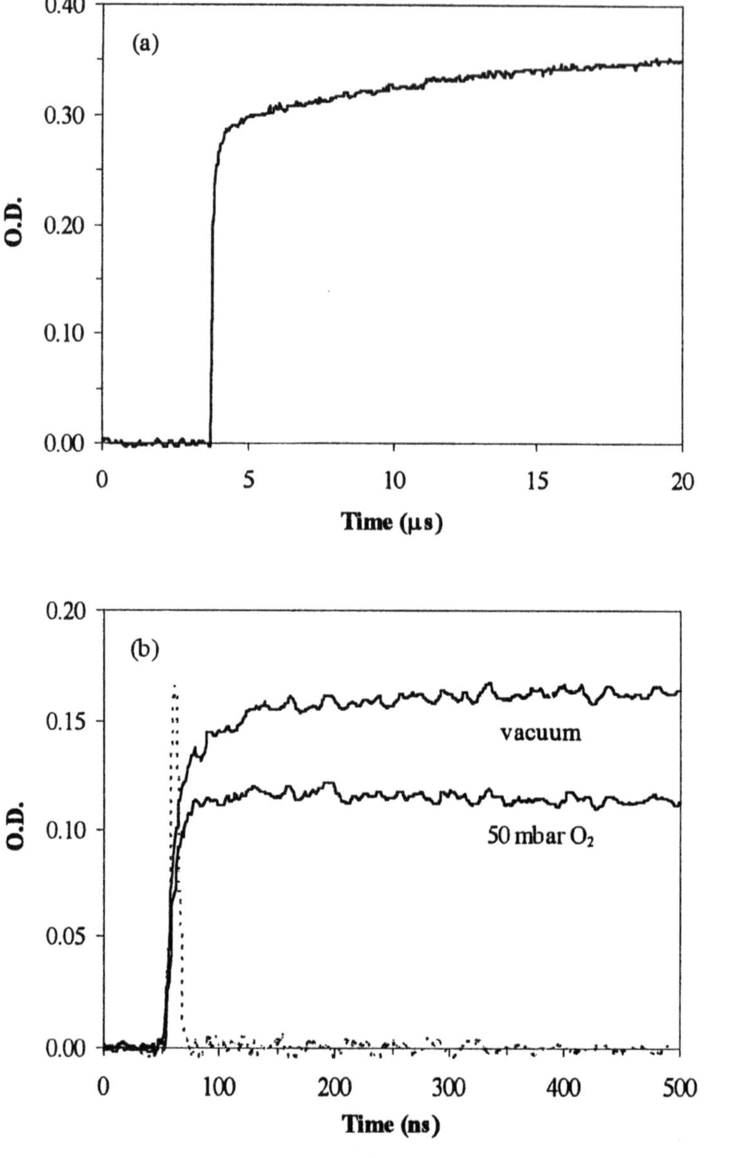

Figure 9. Formation kinetics of the PyH· radical: (a) measured on Davisil-60 powder sample with [Py]=5.0×10^{-6} mol/g; (b) measured on a Cab-O-Sil (HS-5) disc sample with [Py]=1.0×10^{-5} mol/g. The system response is obtained from measurement of the Cerenkov emission (dash line).

component is statically quenched, and the pressure at which 50% static quenching was observed is around 160 mbar. In contrast to oxygen quenching, nitrous oxide gradually removes the slow growth of PyH·, and shows only minimal quenching of the fast component. The comparable oxidation reactivities of oxygen and nitrous oxide (as we have observed for the surface trapped electrons) and the dramatic difference in static quenching of the PyH· radical clearly indicate that they are not reacting with any negatively charged precursor, neither the surface trapped electron nor the pyrene anion radical. The above experiments seem to suggest that the PyH· radical is formed on hydroxylated silica surface via reaction (17). Actually, O_2 and N_2O react with hydrogen atoms quite differently, which results in the observed static quenching patterns. The reaction between O_2 and H is predominated by a three-body process in the liquid phase and also on silica surfaces, while the one between N_2O and H is simply bimolecular.

$$H + O_2 + M \longrightarrow HOO\cdot \qquad (19)$$

$$H + N_2O \longrightarrow N_2 + OH\cdot \qquad (20)$$

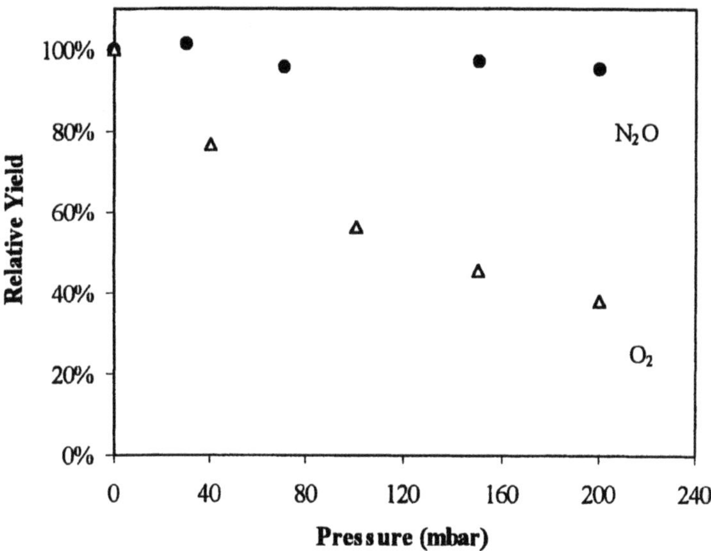

Figure 10. Effect of foreign gases on the yield of PyH· (λ=400 nm) measured at 100 ns after the pulse. [Py]=5.0×10^{-6} mol/g on D60 (Ta=150°C).

In the gas phase, reaction (19) is collision controlled with a negative activation energy,[46] while reaction (20) has an activation energy as high as 16.6 kcal/mol and a rate constant of 2.1×10^6 $M^{-1}s^{-1}$ even at 773K.[47] In aqueous solution, the rate constant for reaction (19) is ~3.0×10^{10} $M^{-1}s^{-1}$, which is much larger than the rate constant of 2.1×10^6 $M^{-1}s^{-1}$ for reaction N_2O + H.[36] The efficient static quenching of PyH· by oxygen in silica systems is due to reaction (19) where silica surfaces serve as the third body M. From the above nanosecond time resolved measurements on Cab-O-Sil (Figure 9b), the rate constant of reaction (19) was estimated to be ~4.0×10^6 $mbar^{-1}s^{-1}$ or 5.6×10^6 $torr^{-1}s^{-1}$. Therefore, at oxygen pressure of 160 mbar, the lifetime of H atoms is ~1.6 ns, which implies from Figure 10 that 50% of the PyH· radicals are formed within ~1.6 ns. The much slower rate of reaction (20) results in little quenching of the fast formation of the H-pyrene adduct over the pressure range used in the experiments. We attribute the fast part of PyH· formation to the H + Py reaction at the same site. The slow formation in porous silica gel over microseconds (Figure 9a) might be due to the H addition reaction limited by H atom movement across the multiconnected network of pores to find pyrene before it escapes the silica granule into vacuum. This inter-pore reaction is absent in non-porous Cab-O-Sil samples. Since it is difficult to completely capture the signal of the slow component due to the decay of PyH·, the yields of the PyH· radical in Davisil samples were only measured for the fast component with increasing pyrene loading (Figure 7b).

Vapors of cyclohexane and chloroform were introduced to 150°C dried D60 samples and the adsorption isotherms were measured for use in scavenging studies. Figure 11 shows the different quenching patterns by cyclohexane and chloroform at different surface coverage. Cyclohexane is known to react with hydrogen atoms with a rate constant of 7.8×10^7 $M^{-1}s^{-1}$ in the liquid phase, and a similar abstraction reaction exists for $CHCl_3$ but with a slower rate of 1.1×10^7 $M^{-1}s^{-1}$.[36] However, $CHCl_3$ exhibits much more efficient static quenching of PyH· than cyclohexane, indicating that they are not reacting with the same precursor, i.e., the hydrogen atom. The concomitant increase of the pyrene cation radical yield with the static quenching of PyH· by $CHCl_3$ suggests that chloroform might react with the electron in the conduction band. Such an electron scavenging reaction prevents the primary electron-hole pair from exciton formation, and therefore reduces the production of hydrogen atoms on the silica surface (see Mechanism).

$$H + c\text{-}C_6H_{12} \longrightarrow H_2 + c\text{-}C_6H_{11}\cdot \quad (21)$$

$$e^- + CHCl_3 \longrightarrow CHCl_2\cdot + Cl^- \quad (22)$$

Nanosecond measurements were also performed for 150°C dried and pyrene loaded D60 samples in a cyclohexane index matched slurry. The rise of the PyH· signal is within the system response (~7 ns), and the yield of PyH· is ~25% of that of a dry powder sample with the same surface loading of pyrene. Considering the very short lifetime (~1 ns) of H atoms due to the scavenging reaction (21) in the silica pores filled with liquid cyclohexane, less than 1% conversion of the hydrogen atoms to PyH· is estimated from the well established rate constants,[36,41] if a Stern-Volmer type of homogeneous kinetics is assumed. However, a significant fraction of the PyH· radicals are produced within ~1 ns due to the inhomogeneous nature of H + Py on silica surfaces, which is in agreement with the measurements on Cab-O-Sil discs and the oxygen quenching studies.

The formation of hydrogen gas bubbles were readily observed in both 150°C and 600°C dried D60 slurry systems with cyclohexane and ethanol as solvents but not with benzene and toluene, due to hydrogen abstraction and addition reactions, respectively. Gas chromatography was used to measure the hydrogen gas production in γ-ray irradiated dry silica samples where recombination of H atoms leads to the formation of molecular hydrogen.

$$H + H \longrightarrow H_2 \tag{23}$$

Figure 7b shows that the H_2 gas yield decreases with the increasing pyrene loading, clearly demonstrating the competition between reactions (17) and (23). Comparison between the steady state measurements of H_2 gas quenching and the time resolved measurements of the growth of PyH· yield found that the decrease of the H_2 yield is larger than the corresponding increase of the PyH· yield. This discrepancy might be settled considering that the PyH· yield was measured in the time resolved experiments without including the contribution of the slow growth component. The difference in dose rates between pulse and steady state radiolysis is another factor that affects the recombination rate of H atoms. The low dose rate of γ irradiation favors the H addition reaction over dimerization in the steady state measurements.

Temperature Effect. The yield of PyH· on the 150°C dried D60 surface has a strong dependence on temperature as shown in Figure 12. The slow component of PyH· formation is absent at lower temperatures and the amplitude of the fast component is gradually reduced with the decreasing temperature. The decrease of PyH· production was attributed to the activated nature of reaction (17) as is found for the other H addition reactions with aromatic compounds.[48] Assume the fast component is due to the reaction between H atom and pyrene at the same site, the escape rate for H atom k_s is a weak function of temperature, and the rate of the addition reaction is expressed in the Arrhenius form $k_r=k_0 \exp(-E_a/RT)$, the yield of PyH· can be simply related to temperature by the following equation.

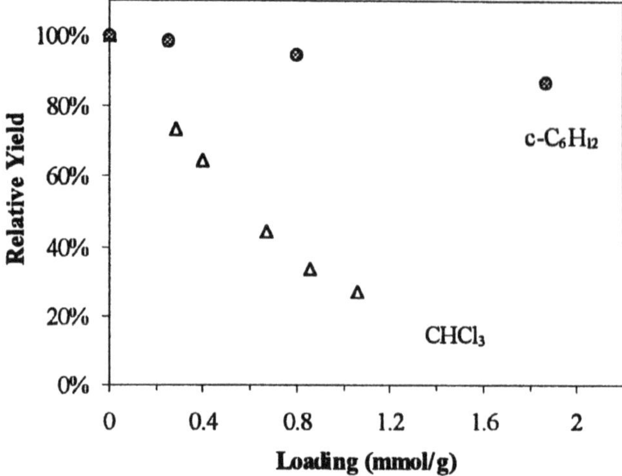

Figure 11. Effect of scavenger vapors on the yield of PyH· measured at 100 ns after the pulse. [Py]=5.0x10^{-6} mol/g on D60 (Ta=150°C).

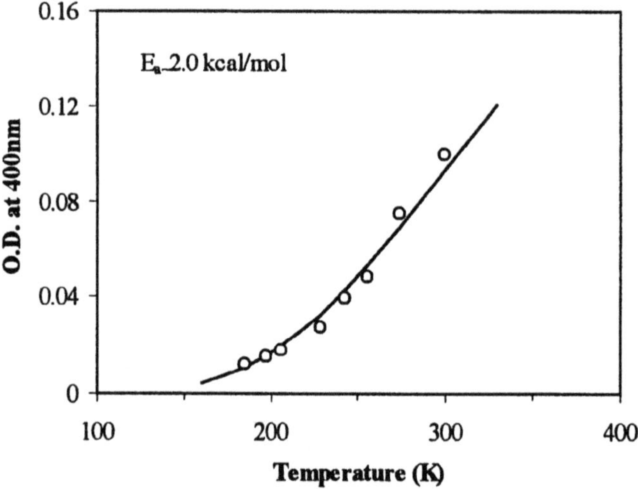

Figure 12. Temperature effect on the production of PyH· on the surface of Davisil-60 (Ta=150°C). Pyrene loading is [Py]=5.0x10^{-6} mol/g. The solid line shows the Arrhenius fit to equation (24).

$$\frac{Y_\infty}{Y} - 1 = \frac{k_s}{k_0} \exp\left\{\frac{E_a}{RT}\right\} \tag{24}$$

where, Y and Y_∞ are the yields of PyH· measured at temperature T and extrapolated to the high temperature limit, respectively. Fitting of the data in Figure 12 gives the activation energy E_a=2.0±0.3 kcal/mol, which compares well with those of benzene and toluene.[48]

As a result of the activation controlled addition reaction, more hydrogen atoms escape from the H + Py reaction at low temperatures. Steady state EPR measurements of γ irradiated silica samples at 77K clearly give a H atom signal with and without pyrene loading. The signal intensity is little affected by pyrene loading. Figure 13a shows the double line EPR spectrum of the hydrogen atom with the characteristic hyperfine splitting (508 G)[49] produced on a 150°C dried D60 with 4.0×10^{-6} mol/g pyrene loading at 77K. Diffuse reflectance absorption spectrum of the same sample shows the presence of the PyH· absorption at 400 nm along with a strong $Py^{+\cdot}$ band around 450 nm and a minimal $Py^{-\cdot}$ at 490 nm (Figure 13b). Using the cation band as the internal reference, the intensity of the PyH· radical band is less than ~10% of that observed in time resolved measurements at room temperature. The temperature effect on radiation products again confirms our argument that the precursor for the PyH· radical production is the hydrogen atom.

Mechanism of Energy Transfer. Reactions on silica and other oxide surfaces induced by photon and high energy radiation are of particular interest from the perspective of both material science and catalysis. The decomposition of organic adsorbates on silicas and minerals has been clearly observed in the early studies of radiation chemistry in such heterogeneous systems.[50-52] Polymerization was also reported for various vinyl monomers adsorbed on silica surfaces upon high energy radiation.[53,54] The chemistry occurring at the interfaces suggests that excitation initially produced in the solid particle absorbents moves out and induces reactions of adsorbates on the particle surfaces.[52,55] The exact mechanism of energy transfer is not clear, although charge transfer was proposed to be one of the pathways on the basis of observation of organic ion radical formation on silica surfaces.[52,55] The results reported here on high surface area silicas might be used as the model systems for the other oxide insulators.

The relatively lower yield of charge generation and the efficient free radical production on 150°C dried silica surfaces, both porous and non-porous, suggest that only a small fraction of the primary electron-hole pairs generated inside SiO_2 particles by the high energy radiation escape the exciton formation and recombination, and give rise to the trapped charges, such as the surface trapped electrons. The major proportion of the electron-hole pairs form excitons which are localized at the defect and impurity energy levels within the band gap, mainly the silanol groups on the silica surfaces.

(a)

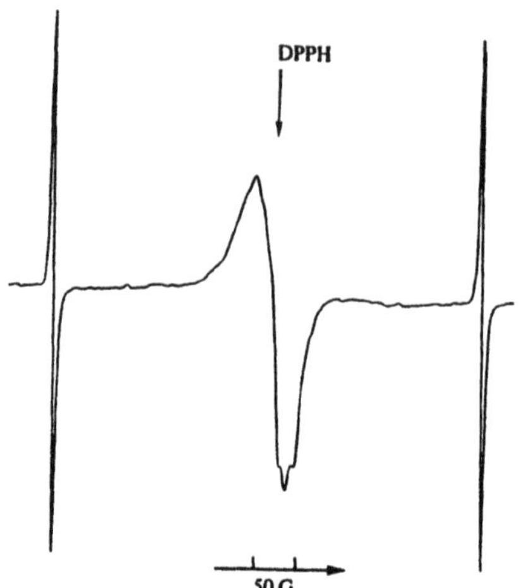

(b)

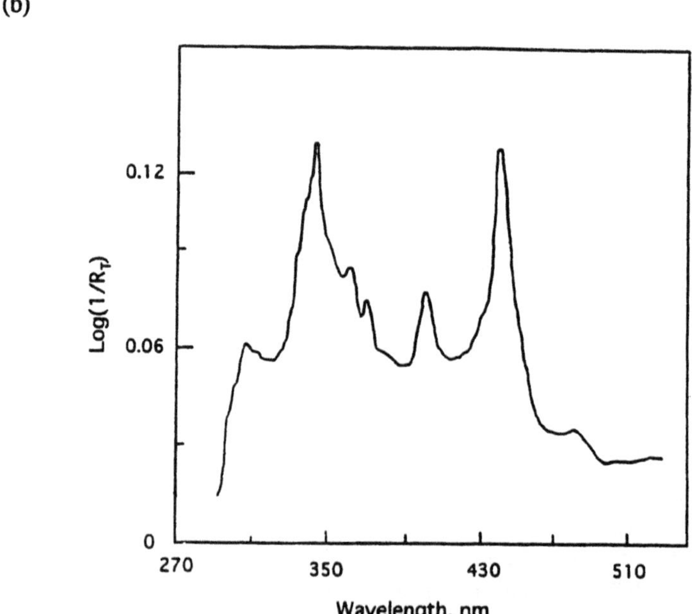

Figure 13. Steady state measurements of D60 (Ta=150°C) with [Py]=4.0×10^{-6} mol/g irradiated at 77K. (a) EPR spectrum; and (b) Diffuse reflectance absorption spectrum.

Like the OH group in H_2O and alcohols, the excited states of hydroxyls on silica surfaces is dissociative. The localization of unrelaxed excitons with energy above the band gap of amorphous SiO_2 (~9 eV) at the surface hydroxyls results in breaking of the O-H bond (bond energy ~6 eV) and production of H atoms.

$$\text{exciton} + \equiv \text{SiO-H} \longrightarrow \equiv \text{SiO}\cdot + \text{H} \quad (25)$$

Hydrogen atom production was also observed in high purity silica glass with a hydroxyl concentration as low as ~1200 ppm by EPR measurements at low temperatures,[56,57] which results from the above bond breaking localization of excitons generated in bulk SiO_2. Thermal bleaching experiments show that recombination with NBOHC and dimerization are the major reaction leading to the disappearance of H atoms in bulk silica.[57] The optical spectrum of amorphous silica often shows a subband gap transition due to the hydroxyl impurities.[45] These hydroxyl impurity levels can be even directly populated by the ArF excimer laser at 193 nm.[58] The high concentration of hydroxyls present on the surfaces of the small SiO_2 particles, which is much higher than those in silica glasses, efficiently withdraw the energy deposited inside the particles via exciton localization, leading to hydrogen production on the silica surfaces. Our experiments clearly show that only a small portion of electrons in the conduction band are trapped by surface hydroxyls.

$$e^- + (\equiv \text{SiOH})_n \longrightarrow e^-_{trap} \quad (26)$$

The adsorption of pyrene preferentially at the hydroxyl rich sites on silica surfaces facilitates the H addition reaction (17), and also leads to the pyrene cation radical formation via interaction with positive holes in the valence band of SiO_2 particles.

$$h^+ (SiO_2) + Py \longrightarrow Py^{+}\cdot \quad (27)$$

Due to the efficient trapping of electrons which escape the exciton formation by the associated surface silanols as described in reaction (26), electron scavenging by adsorbed pyrene is suppressed on fully hydroxylated silica surfaces, leading to the low yield of the pyrene anion radicals.

$$e^- + Py \longrightarrow Py^{-}\cdot \quad (28)$$

Pretreatment of silica surfaces at 600°C destroys the hydroxyl clustering sites for electron trapping and gives rise to the increased electron scavenging by pyrene via reaction (28). As the result of dehydroxylation, less H atoms are produced with more electron-hole pairs subject to charge scavenging, which leads to a shift of energy transfer to the ion radical channel and therefore ionic chemistry on silica surfaces.

The energy transfer processes in high surface area silicas are illustrated in Figure 14. The primary form of energy exists as unrelaxed electron-hole pairs in the SiO_2 particles. The splitting of the electron-hole pairs by the intrinsic charge trapping centers and the charge scavenging molecules adsorbed on surface competes with the formation of exciton whose localization at surface hydroxyls leads to instantaneous H atom production. The variation of the surface chemistry originates from the modulation of the hydroxyl content present as impurities within the band gap. Surface hydroxyls withdraw the energy via exciton trapping and direct the chemistry of adsorbates into free radical processes, while dehydroxylation switches the form of energy transfer from radical to ionic in nature.

The particle based structure of silica is different from the zeolite framework which is inter-penetrated by cage structures on the atomic level and therefore completely exposed to external charge balance ions and small molecules. The presence of Na^+ clusters as intrinsic electron trapping sites and water clusters as externally introduced electron traps in zeolite cavities leads to a very efficient charge separation upon high energy excitation of the zeolite framework.

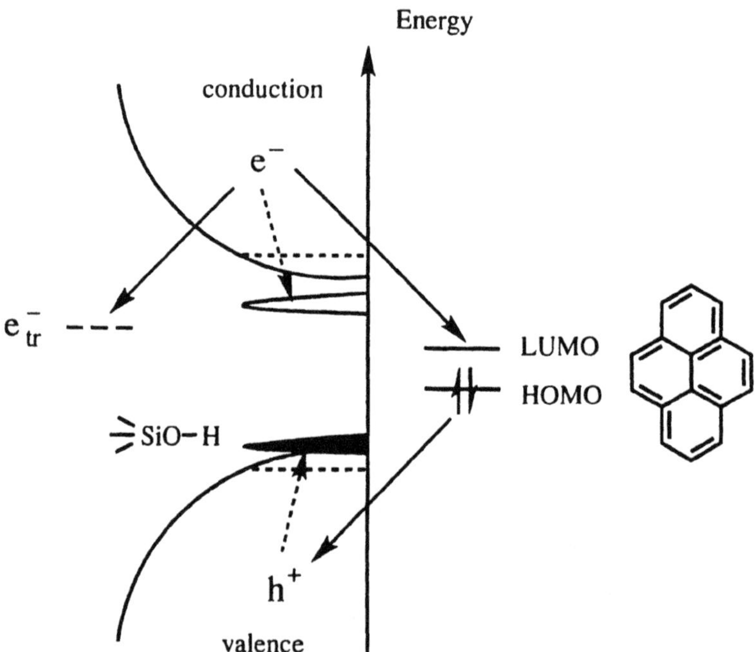

Figure 14. Schematic drawing of the electronic structure of silicas with high content of OH impurities and the energy transfer processes induced by high energy radiation.

Summary

As illustrated by the above studies on zeolites and silicas, the specific forms of chemistry are generated in different heterogeneous systems despite the nature of non-selective excitation of solid substrates by high energy radiation. Charge separation and trapping in zeolites are the major chemical processes which results in formation of Na_4^{3+} and e_{aq}^- in the zeolite cages. Cage structure stabilizes these charged species and prevent reaction with other species by limiting the accessibility. However, energy deposited in silicas induces radical processes on the surfaces of nanosized SiO_2 particles, which are initiated by hydrogen atoms produced in large yields from efficient exciton trapping. Dehydroxylation of the surfaces shifts the energy transfer from radical production to charge generation. The effect of particle size on the extent of such energy transfer to surfaces is presently under investigation.

References

1. Work supported by the National Science Foundation.
2. Matijevic, E.; Scheiner, D. (1978) *J. Coll. Interface Sci.* **63**, 509.
3. (a) Fojtik, A.; Weller, H.; Koch, V.; Henglein, A. (1984) *Ber. Bunsen-Ges. Phys. Chem.* **88**, 969. (b) Weller, H.; Fojtik, A.; Henglein, A. (1985) *Chem. Phys. Lett.* **117**, 484. (c) Baral, S.; Fojtik, A.; Weller, H.; Koch, V.; Henglein, A. *J. Am. Chem. Soc.* **108**, 375. (d) Spanhel, L.; Hasse, M.; Weller, H.; Henglein, A. (1987) *J. Am. Chem. Soc.* **109**, 5649. (e) Henglein, A. (1984) *Pure and Appl. Chem.* **56**, 1215.
4. Ramsden, J. J.; Weber, S. E.; Grazel, M. (1985) *J. Phys. Chem.* **89**, 1740. (b) Ramsden, J. J.; Grazel, M. (1984) *J. Chem. Soc., Faraday Trans. 1* **80**, 919. (c) Duong, H. D.; Ramsden, J. J.; Grazel, M. (1982) *J. Am. Chem. Soc.* **104**, 2977. (d) Kalyanasundaram, K.; Borgarello, E.; Duongrong, D.; Grazel, M. (1981) *Angew Chem. Intern. Ed. Engl.* **20**, 987. (e) Dimitrijevic, N. M.; Li, S.; Grazel, M. (1984) *J. Am. Chem. Soc.* **106**, 6565. (f) Grazel, M.(1985) in *Photoelectrochemistry, Photocatalysis and Photoreactions*, Schiavello, M., Ed.; Reidel, Holland, p217.
5. Chandrasekaran, K.; Thomas, J. K. (1985) *J. Coll. Interface Sci.* **106**, 532; (1984) *J. Coll. Interface Sci.* **100**, 116; (1983) *Chem. Phys. Lett.* **99**, 7; (1983) *Chem. Phys. Lett.* **97**, 357; (1983) *J. Am. Chem. Soc.* **105**, 6383; (1984) *J. Chem. Soc., Faraday Trans. 1* **80**,1163.
6. Stramel, R.; Thomas, J. K. (1986) *J. Coll. Interface Sci.* **110**, 121.
7. Korman, C.; Bahnemann, D. W.; Hoffmann, M. R. (1989) *J. Photochem. & Photobio. A. Chemistry* **48**, 161.
8. (a) Thomas, J. K. (1987) *J. Phys. Chem.* **91**, 267. (b) Kuczynski, J.; Thomas, J. K. (1985) *Langmuir* **1**, 158. (c) Kuczynski, J.; Thomas, J. K. (1983) *J. Phys. Chem.* **87**, 5498. (d) Kuczynski, J.; Milosavljevic, B. H.; Thomas, J. K. (1984) *J. Phys. Chem.* **88**, 980. (e) Lianos, P.; Thomas, J. K. (1986) *Chem. Phys. Lett.* **125**, 299. (f) Lianos, P.; Thomas, J. K. (1987) *J. Coll. Interface Sci.* **117**, 505. (g) Kuczynski, J.; Thomas, J. K. (1985) *J. Phys. Chem.* **89**, 2720.

9. Darwent, J. R.; Porter, G. (1981) *J. Chem. Soc., Chem. Commun.* **10**, 145.
10. (a) Kraeutler, B. K.; Bard, A. J. (1978) *J. Am. Chem. Soc.* **100**, 2239, 4317, 5985. (b) Bard, A. J. (1980) *Science* **207**, 139.
11. Dannhauser, T.; O'Neil, M.; Johanson, K.; Whitten, D.; McLendon, G. (1988) *J. Phys. Chem.* **90**, 6074.
12. (a) Wang, Z.; Herron, N. (1987) *J. Phys. Chem.* **91**, 257. (b) Herron, N.; Wang, Y.; Eddy, M. M.; Stucky, G. D.; Cox, D. E.; Moller, K.; Blin, T. (1989) *J. Am. Chem. Soc.* **111**, 530.
13. (a) Stramel, R. D.; Nakamura, T.; Thomas, J. K. (1986) *Chem. Phys. Lett.* **130**, 423. (b) Stramel, R. D.; Nakamura, T.; Thomas, J. K. (1988) *J. Chem. Soc. Faraday I* **84**, 1287. (c) Liu, X.; Thomas, J. K. (1989) *J. Coll. Interface Sci.* **129**, 476. (d) Liu, X.; Thomas, J. K. (1989) *Langmuir* **5**, 58.
14. Micic, O. I.; Meisel, D. (1986) in *Homogeneous and Heterogeneous Photocatalysis*, Pelizzetti, E.; Serpone, N., Eds.; Reidel, Holland, p227.
15. (a) Fox, M. A.; Chen, C. C. (1982) *J. Am. Chem. Soc.* **103**, 6757. (b) Fox, M. A.; Lindg, B.; Chen, C. C. (1982) *J. Am. Chem. Soc.* **104**, 28. (c) Fox, M. A.; Chen, C. C. (1983) *Tetrahedron Lett.* **24**, 547. (d) Fox, M. A.; Chen, C. C.; Younathan, J. N. (1984) *J. Org. Chem.* **49**, 9. (e) Fox, M. A.; Chen, M. J. (1983) *J. Am. Chem. Soc.* **105**, 4497. (f) Kamat, P. V.; Fox, M. A. *J. Phys. Chem.* **87**, 59. (g) Kamat, P. V.; Fox, M. A. (1983) *Chem. Phys. Lett.* **104**, 104. (h) Fox, M. A. (1988) in *Homogeneous and Heterogeneous Photocatalysis*, Pelizzetti, E.; Serpone, N., Eds.; Reidel, Boston, p363.
16. (a) Nozik, A. J.; Williams, F.; Nenadovic, M. T.; Rajh, T.; Micic, O. I. (1985) *J. Phys. Chem.* **89**, 397. (b) Nedeljkovic, J. M.; Memadovoc, M. T.; Micic, O. I.; Nozik, A. J. (1986) *J. Phys. Chem.* **90**, 12.
17. (a) Brus, L. E. (1987) *New J. Chem.* **11**, 123. (b) Rosetti, R; Nakahara, S.; Brus, L. E. (1983) *J. Chem. Phys.* **79**, 1086. (c) Brus, L.E. (1984) *J. Chem. Phys.* **80**, 4403. (d) Rosetti, R.; Ellison, J. L.; Gibson, J. M.; Brus, L. L. (1984) *J. Chem Phys.* **80**, 4464.
18. Kavanaugh, R. J.; Thomas, J. K. unpublished data.
19. Liu, X.; Zhang, G.; Thomas, J. K. (1995) *J. Phys. Chem.* **99**, 10024.
20. Kerr, G. T. (1966) *Inorg. Chem.* **5**, 1537.
21. Iler, R. K. (1979) *The Chemistry of Silica: Solubility, Polymerization, Collod and Surface Properties, and Biochemistry*, John Wiley & Sons, New York.
22. (a) Hart, E. J.; Anber, M. (1970) *The Hydrated Electron*, John Wiley & Sons, New York. (b) Hunt, J. W.; Wolff, R. K.; Bronskill, M. J.; Jonah, C. D.; Hart, E. J.; Matheson, M. S. (1973) J. Phys. Chem. 77, 425.
23. Breck, D. W. (1974) *Zeolite Molecular Seives: Structure, Chemistry and Use*, John Wiley & Sons, New York, p92 and references therein.
24. Eulenberger, G. H.; Shoemaker, D. P.; Keil, J. G. (1967) *J. Phys. Chem.* **71**, 1812.
25. Breck, D. W. (1964) *J. Chem. Ed.* **41**, 678.
26. Breck, D. W.; Eversole, W. G.; Milton, R. M.; Reed, T. B.; Thomas, T. L. (1956) *J. Am. Chem. Soc.* **78**, 5963.
27. See ref 23, p176.
28. Gramlich, V.; Meier, W. M. (1971) *Z. Kristallogr.* **133**, 134.

29. Kevan, L. (1981) *J. Phys. Chem.* **85**, 1628.
30. Rossky, P.; Schnitker, J. (1988) *J. Phys. Chem.* **92**, 4277.
31. Wong, M.; Grazel, M.; Thomas, J. K. (1975) *Chem. Phys. Lett.* **30**, 329.
32. Broussard, L.; Shoemaker, D. P. (1960) *J. Am. Chem. Soc.* **82**, 1041.
33. Hseu, T. (1972) Ph. D. Thesis, University of Washington, University Microfilms, No. 73-13835, Ann Arbor, Michigan.
34. Albery, W. J.; Bartlett, P. N.; Wilde, C. P.; Darwent, J. R. (1985) *J. Am. Chem. Soc.* **107**, 1854.
35. Rice, S. A. (1985) *Comprehensive Chemical Kinetics*, Vol. 25: Diffusion Limited Reactions, Bamford, C. H.; Tipper, C. F. H.; Compton, R. G., Eds.; Elsevier, Amsterdam, Chapter 2.
36. Buxton, G. V.; Greenstock, C. L.; Helman, W. P.; Ross, A. B. (1988) *J. Phys. Chem. Ref. Data* **17**, 513.
37. Birk, J. B. (1970) *Photophysics of Aromatic Molecules*, John Wiley & Sons, New York, p 441.
38. Kransnansky, R. (1990) *Photophysical and Photochemiscal Probing of the Silca Gel Surface*, Ph.D. Dissertation, University of Notre Dame, Notre Dame, Indiana.
39. Richard, J. T.; West, G.; Thomas, J. K. (1970) *J. Phys. Chem.* **74**, 4137.
40. Shida, T.; Iwata, S. (1973) *J. Am. Chem. Soc.* **95**, 3473.
41. Zhang, G.; Thomas, J. K. (1994) *J. Phys. Chem.* **98**, 11714.
42. Okada, T.; Karaki, I.; Mataga, N. (1982) *J. Am. Chem. Soc.* **104**, 7191.
43. Griscom, D. L.; Brinker, C. J.; Ashley, C. S. (1987) *J. Non-Cryst. Solids* **92**, 295.
44. Silin, A. R.; Skuja, L. N. (1980) *J. Mol. Struct.* **61**, 145.
45. Elliot, S. R. (1990) *Physics of Amorphous Materials*, 2nd Ed.; Longman Scientific & Technical, England.
46. Hsu, K.-J.; Anderson, S. M.; Durant, J. L.; Kaufman, F. (1989) *J. Phys. Chem.* **93**, 1018.
47. Dixon-Lewis, G.; Williams, D. J. (1977) in *Comprehensive Chemical Kinetics*, Vol. 17: Gas-Phase Combustion, as ref 35, Chapter 1.
48. Sauer, M. C.; Ward, B. (1967) *J. Phys. Chem.* **71**, 3971.
49. Wertz, J. E.; Bolton, J. R. (1986) *Electron Spin Resonance: Elementary Theory and Practical Applications*, Capman Hall, New York.
50. (a) Sutherland, J. W.; Allen, A. O. (1961) *J. Am. Chem. Soc.* **83**, 1040. (b) Abrams, L.; Allen, A. O. (1969) *J. Phys. Chem.* **73**, 2741.
51. Sagert, N. H.; Dyne, P. J. (1967) *Can. J. Chem.* **45**, 615.
52. (a) Rabe, J. G.; Rabe, B.; Allen, A. O. (1966) *J. Phys. Chem.* **70**,1098. (b) Waong, P. K.; Allen, A. O. (1970) *J. Phys. Chem.* **74**, 774.
53. Shimada, M.; Nakamura, Y.; Kusama, Y.; Udagawa, A.; Takehisa, M. (1982) *J. Appl. Polym. Sci.* **27**, 1259.
54. Bruk, M.A.; Pavlov, G.G. Isaeva; Yunitskaya, E.Y. (1986) *Eur. Polym. J.* **22**, 169.
55. (a) Wong, P. K.; Willard, J. E. (1968) *J. Phys. Chem.* **72**, 2623. (b) Joppien, G. R.; Willard, J. E. (1972) *J. Phys. Chem.* **76**, 3158.
56. Griscom, D. L.; Stapelbroek, M.; Friegele, E. J. (1983) *J. Chem. Phys.* **78**, 1638.
57. Griscom, D. L. (1984) *J. Non-Cryst. Solids* **68**, 301.
58. Tsai, T. E.; Griscom, D. L.; Friebele, E. J. (1989) *Phys. Rev.* **B 40**, 6374.

MIE-PLASMON SPECTROSCOPY: A TOOL OF SURFACE SCIENCE

U. KREIBIG, M. GARTZ, A. HILGER, H. HÖVEL
I. Physikalisches Institut der RWTH Aachen/Germany

1. Introduction

Clusters are regarded as tiny entieties consisting of few atoms or molecules only (of the order of 10^1 to 10^5 atoms/cluster); alternatively, they can be regarded as atomic systems with extremely high specific surface (i.e. surface to volume ratio). These are, of course, two sides of the same medal.

For one century physicists mainly focused upon the first picture, and search for cluster size effects - in particular quantum size effects - was their aim. Chemists - on the other hand - were mainly interested in the surfaces, in view of chemical effects at the cluster-surrounding interface in embedded cluster systems, in particular, of catalytic processes (e.g. [16]).

Surface physics is mainly the domain of planar surface geometries and free surfaces were treated as well as thin films or interfaces between two media (e.g. [1]). However, all these systems are as well met in the spherical geometry of clusters (Fig. 1).

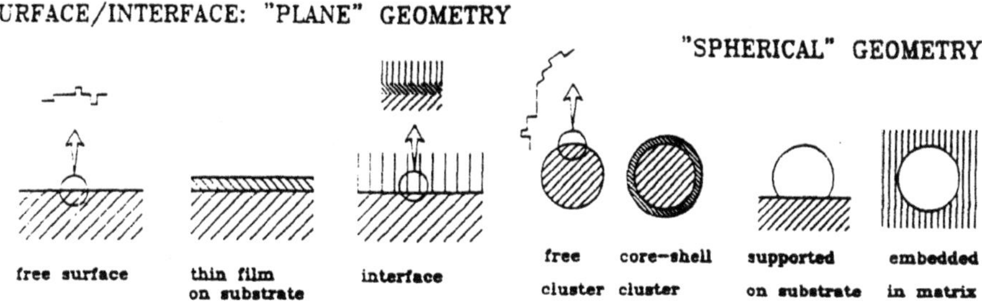

Figure 1. Surface and interfaces in plane and spherical geometrie

E. Pelizzetti (ed.), Fine Particles Science and Technology, 499–515.

These latter systems even are advantageous in having higher degrees of surface, and, hence, it is not necessary to develop special experiments with enhanced sensitivity for the surface region. Instead, conventional experiments, - i.e. not only surface selective ones - can be performed to investigate surface properties, like ESR, NMR, Mößbauer, optical etc. Moreover the cluster size acts as an additional parameter to increase and differentiate obtainable data.

In the following, we demonstrate, as an example, the high sensitivity of optical cluster properties against chemical manipulations of the cluster surface, which thus can be used as tools of surface analysis. The cluster surface is well defined only above a certain cluster size: while small clusters of the order of 10^1 units act as molecular entieties, a monoatomar surface layer surrounding an extended inner volume only appears at $N \geq 10^2$. Only then cluster properties can be separated into volume and surface effects. Strictly, they are not independent of each other: changes of the volume atomic ordering will induce surface reconstruction, etc. The stable geometric configurations follow from minimizing, both, surface and volume energy including their coupling. In larger clusters, volume ordering forces are favored against surface optimization, the bulk atomic ordering appears and the surface exhibits facets similar to the planar case. This may develop by a "phase transition" at a critical size, which may be different if the cluster surface is covered by some foreign material. Then, parts of the surrounding are included in addition to the cluster, into the energy minimization. Above this "critical size" cluster enlargement results merely in increasing surface area. It is well known that, e.g. gold clusters can exhibit shapes with [111] planes forming the surface, only. Edges and corners resemble step structure of the plane surface. So, planar and cluster surfaces are comparable in their topology and chemical reactivity and the cluster surface can be described by extrapolating models of the planar surface.

The charge distribution at the free metal surface, as calculated in self-consistent jellium approximation with the electron density of Ag [2], is shown in Fig. 2. It only slightly differs for spherical cluster symmetry [3]. The electrons develop Friedel oscillations at the steep ion profile and an almost exponential decay outside, the "spill-out". This position dependent electron density can easily be influenced by ions covering the surface, when the clusters are deposited upon a substrate or embedded in a liquid or solid embedding medium.

The electric field of an optical experiment causes changes $\delta\rho$ of the electron density as also shown in Fig. 2 and it is obvious from this picture that their localization close to the surface area renders optical (or electrical) experiments sensitive for probing the charge distribution at the cluster interface. This may be changed by foreign atoms and ions arranged clo-

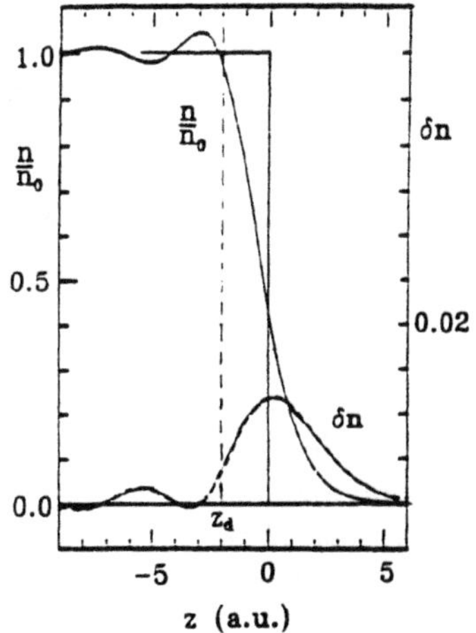

Figure 2. Electron density n/n_0 around the ion edge ($z = 0$) of a jellium cluster [3,4] with "spill-out". δ n: additional electron density by optional excitation (e.g. Mie resonances)

seby and may then impact optical spectra. Quantitatively more important, by far, beyond these static charge effects are the influences of surrounding foreign atoms and ions onto the electron dynamics, i.e. scattering and tunneling effects at the cluster interface. This will be shown in the following.

2. Mie's Theory

Optical properties of metal clusters belong to the most spectacular known cluster effects, yielding, for many cluster materials, characteristic and unique colors which are well known and applied since centuries.

In the frame of Maxwell's theory they are described by the theory of Mie, which is the analogue of Fresnel's formulae for spherical symmetry. It yields expressions for the absorption, the elastic scattering and the extinction of light, i.e. the linear response, of non-interacting spherical clusters of uniform size, where the sizes can be varied arbitrarily. The original paper of Mie [4] takes more than 70 pages; in the following, a short compilation of this theory will be given. The Mie theory is based upon the phenomenological electrodynamics. It treats one spherical cluster by solving the Helmholtz equation for spherical geometry:

$$(\nabla^2 + k^2)\, \Pi(r, \Theta, \varphi) = 0 \tag{1}$$

where Π stands for the electric and the magnetic potentials Π_j, the solutions of which are:

$$r \cdot \Pi_j(r,\Theta,\varphi) = \frac{1}{k^2} \sum_{l=1}^{\infty} \sum_{m=-l}^{+l} x_{l,m}^{(j)} \{\Psi_l^{(j)}(k \cdot r)\} \cdot Y_{l,m}(\Theta,\varphi) \tag{2}$$

where the Ψ and Y are Bessel functions and spherical harmonics, respectively. The according fields are derived from these potentials and, applying the Poynting law, the intensities in front of, in, and behind the clusters are obtained in the form of series expansions in terms of the spherical multipole order l. l describes the spherical symmetries: $l = 1,2$ means spherical dipole and spherical quadrupole, respectively. These two multipoles, exhibiting distinct resonance behavior in clusters of many different materials, are the only ones which contribute if the clusters are essentially smaller than the light wavelength and only $l = 1$ is important for clusters as small as treated in this paper. The boundary conditions for the fields yield the relative amplitudes i.e. the "Mie-coefficients". As will be shown later, realistic details of the physical interface properties can enter via these coefficients to be obtain an extended theory.
The differential intensity losses for extinction and scattering are obtained, in Mie's original paper, by:

$$\frac{dI_{ext}}{I_0} = \frac{\lambda}{2\pi\varepsilon_m} \cdot \mathrm{Im}\left\{\sum_{l=1}^{\infty}(-1)^{l-1} \cdot (a_l - p_l)\right\} \tag{3}$$

$$\frac{dI_{sca}}{I_0} = \frac{\lambda^2}{2\pi\varepsilon_m} \cdot \sum_{l=1}^{\infty}(|a_l|^2 + |p_l|^2) \tag{4}$$

with the Mie-coefficients:

$$a_l = (-1)^l \cdot \tfrac{l+1}{l} \tfrac{(2\pi)^{2l+1} \cdot (R/\lambda)^{2l+1}}{1^2 \cdot 3^2 \cdot \ldots \cdot (2l-1)^2} \cdot u_l \cdot \frac{\varepsilon(\lambda) - v_l \cdot \varepsilon_m}{\varepsilon(\lambda) + (l+1)/l \cdot \varepsilon_m \cdot w_l} \tag{5}$$

$$p_l = \ldots \cdot u_l \cdot \frac{1 - v_l}{1 + (l+1)/l \cdot w_l} \tag{6}$$

$\varepsilon(\lambda), \varepsilon_m, R$ are the dielectric functions of the particle material of the surrounding medium and the cluster radius, respectively. u_l, v_l, w_l are complex valued power series in (R/λ) and ε, all beginning with 1.
Computer aids to solve Mie's formulae are numerous on the market, e.g. in [5]. Various alternative derivations have been developed since the appearance of Mie's theory in 1908 and various slight physical extensions were performed since then [5]. Fig. 3 gives examples of Mie extinction spectra of several less common materials which are free or embedded in matrices with the indicated magnitudes of ε_m. (More examples are compiled in [5]).

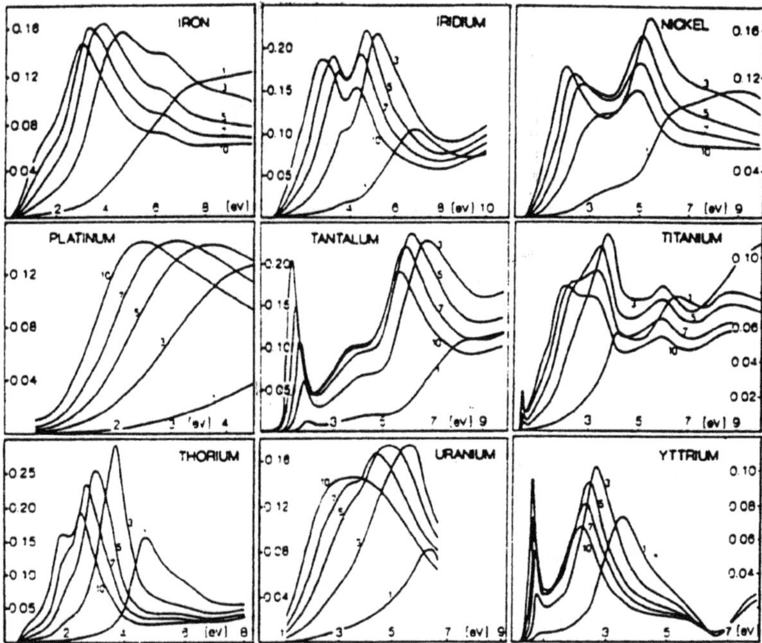

Figure 3. Calculated Mie extinction spectra of less common cluster materials. The cluster size is always $2R = 10nm$. Bulk values of $\varepsilon(\omega)$, taken from literature. The spectra for each material differ by the assumed embedding media, the dielectric constants ε_m of which are noted to amount between 1 and 10.

The resulting absorption and extinction spectra are, for many materials including the noble metals, governed by the Mie plasmon resonances which are due to the - purely collective - excitations of the "conduction electrons". Among all cluster materials, this resonance is strongest in the case of silver and, hence, the investigations reported here, were restricted to this latter material.
Mie's theory is as exact as are the Fresnel formulae, being directly derived from Maxwell's equations. Hence, it is a classical theory. These kinds of phenomenological theories treat the expansion of electromagnetic waves only, but do not give any information about the cluster properties, which enter the theory via the dielectric functions treated as unspecified, open parameters, to be taken from elsewhere. In the frame of the original Mie theory, the dielectric functions of bulk materials were intended to be used. Now, however, it is well established, both, from theoretical and experimental results that this assumption is wrong: in fact, particular size and surface dependent functions, describing the polarizability of the cluster as a whole have to be determined and used in eq. (5).

As a consequence, one important property of the classical Mie absorption spectra is lost: In the frame of the original theory, these spectra should become independent of cluster size, if $R \ll \lambda$. Instead, all investigati-

ons indicate strong size dependences which follow from size dependent ε-functions. It is the aim of the present paper to show that the experimental optical spectra also reflect chemical interface effects which are to be expressed by proper surface/interface contributions to these "effective" ε-functions. It is this ε (or, alternatively, the cluster polarizability) which is topic of modern cluster science, since the cluster size dependent electronic and phononic properties, the single electron and collective excitations, the volume and surface/interface influences may all be included in this quantity.
Interactions of a cluster with some surrounding material, which may consist of adsorbed gases forming interface shells, of a substrate or of an embedding liquid or solid matrix, have, in principle, two different implications on the optical properties.

First, the electrodynamic response is changed, since different ε_m enter the Maxwellian boundary conditions ("dielectric effect"). The prediction of this effect from Mie calculations is complicated, since the underlying assumption of a sharp step in the normal components of the electric and magnetic fields is only roughly approximative in the dimensions of nm, i.e. the dimensions of atomic surface layers, as Fig. 2 clearly demonstrates.

Secondly, as we will show in this paper, in addition, the dielectric function ε of the cluster material itself may be drastically influenced, thus rendering the Mie-spectra sensitive tools for the experimental investigation of chemical interface effects.

3. Experiments and Results

Cluster size dependences of the dielectric function have been established since more than 30 years [5]. First indications of strong additional effects due to the chemical surroundings of the clusters were found by the Schulze group [7] with gas-aggregation produced Ag-clusters embedded in different solidified gases.

Fig. 4 shows that there is not a constant contribution but it is the amount of size dependence of the Mie plasmon halfwidth Γ that varies when the embedding matrix of the clusters is changed. Chemical reactions at the interface of Ag-clusters in colloidal hydrosols could also be visualized in the Mie absorption spectra (Fig. 5).

These experiments suffered from the lack of spectra of the totally uncontaminated clusters which are inevitable for quantitative separation of the chemical effects from cluster size and structure effects.
Hence, we constructed a supersonic Ar-gas expansion source to produce a free uncontaminated Ag-cluster beam, which is dense enough to allow opti-

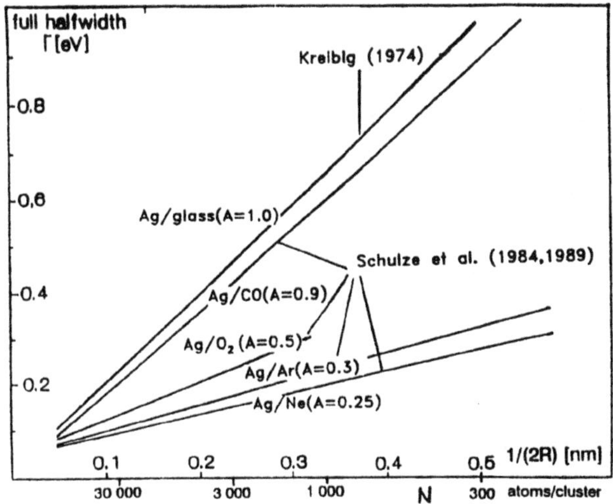

Figure 4. Experimental results of Γ versus $1/R$ for Ag clusters in various solid embedding matrices [7,8]. The according values of the A-parameter are added.

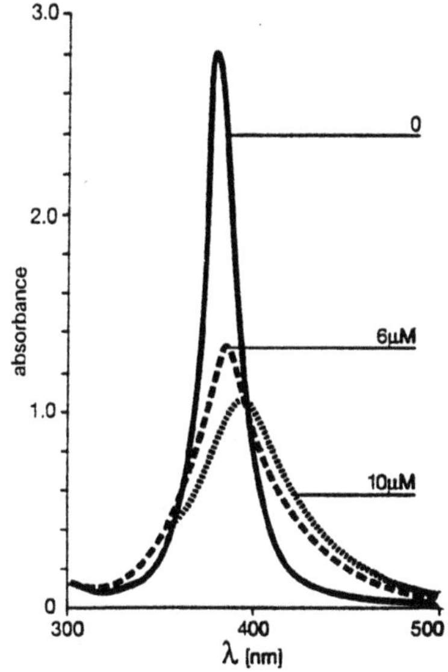

Figure 5. Measured absorbance of Ag clusters produced in hydrosol by radiolysis [14] after adding various amounts of KJ.

cal absorption spectroscopy at the free beam. Subsequently, these clusters can be manipulated, either by adding adsorbing gases, by depositing the clusters on some substrate and by embedding them in some co-evaporated embedding matrices, and their optical spectra can be recorded after each of these manipulations. The cluster source with attached experiment chamber

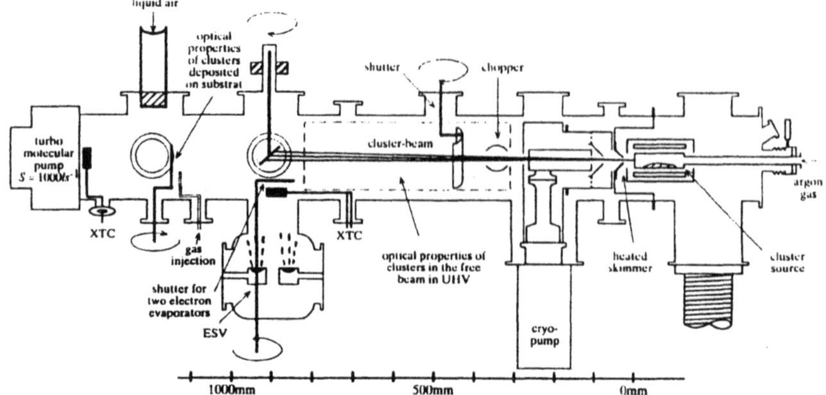

Figure 6. Cluster source (right) and experiment chamber (left) for free Ag cluster jets with $2\overline{R} = 2.0nm$ [9].

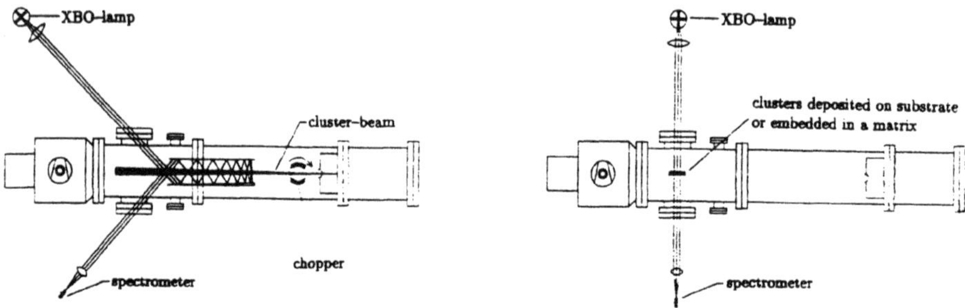

Figure 7. Different optical equipments of the apparatus of Fig. 6: multireflection arrangement for experiments at the free cluster beam (left) and transmission setup for deposited or embedded clusters [9].

is shown in Fig. 6, it was described recently [9].

The mean particle size, determined by TEM, is $2R \approx 2.0nm$ with a standard deviation of about $\pm 0.3nm$, and is almost independent of particular values of the source parameters. The filling factors of the free beam amount up to $f \sim 10^{-11}$ and a multiple reflection equipment, yielding effective optical path lengths in the beam of about one meter, allowed to record, in situ, the absorption spectra. The transmission of the substrate samples and the matrix samples could, because of markedly higher f's, be measured directly (Fig. 7).

Fig. 8 shows absorption spectra of the same Ag-clusters in these above described three different kinds of samples. The Mie plasmon shows drastic differences: From the free to the deposited to the embedded sample the peak is strongly red-shifted ($\sim 0.55eV$) and the width increases for a factor of four.

These results were highly surprising, since they do not follow from Mie's

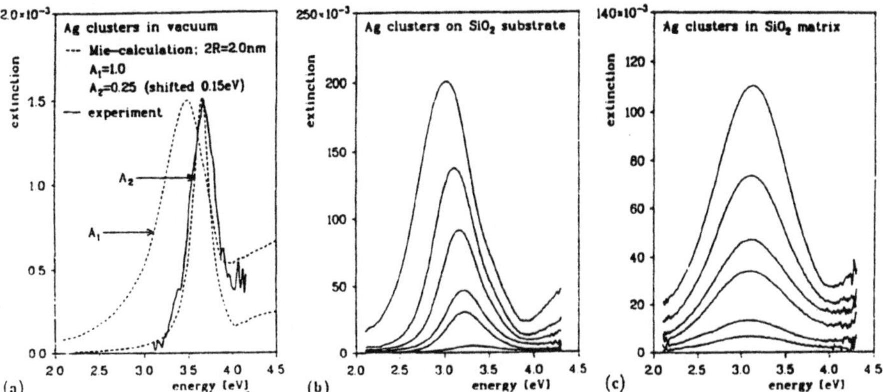

Figure 8. Measured absorption spectra of Ag-clusters produced with the apparatus of Fig. 6 [9]. a): measured in the free beam; b): clusters deposited on a SiO_2 substrate; c): embedded in SiO_2. Different curves in b) and c): different filling factors increasing from bottom to top.

theory. In the next section we will show that the differences can be explained by chemical interface effects to be described by a properly modelled dielectric function (or, alternatively, cluster polarization). Before we proceed with this interpretation we will present more experimental results.

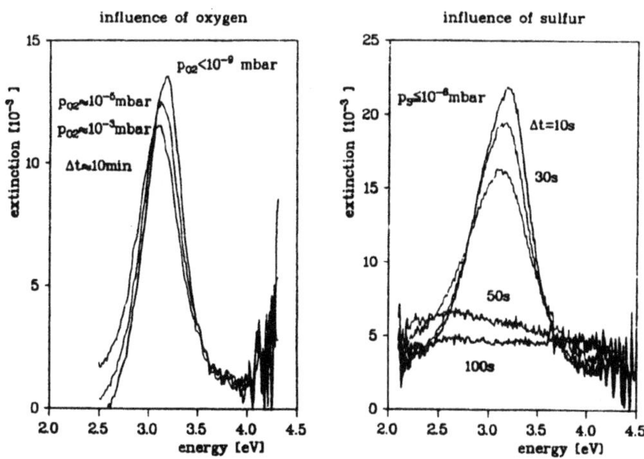

Figure 9. Optical absorption of Ag clusters ($2\overline{R} = 2nm$) deposited on SiO_2 substrates, exposed to oxygen and gaseous sulphur.

Fig. 9 a and b show how Ag clusters deposited on silica change their optical spectra by surface reaction with oxygen and sulphur, respectively. While, in the first case, less than one monolayer of oxide is expected to develop at the cluster surface (at planar Ag surfaces, oxidation takes place under comparable conditions on surface inhomogenities like atomic steps [10]), the reaction with sulphur only ends after the whole cluster has been transformed into Ag S, and the Mie peak vanishes completely. The strongly varying intermediate spectra indicate that such chemical surface reactions can be followed, by optical inspection of the Mie resonance, in detail.

Fig. 10 gives a compilation of spectra of Ag clusters embedded in various dielectric solid oxides. Both, strong red-shifts and drastic broadening are obvious. In these samples chemical surface reactions are less probable, however, the presence of some material surrounding the clusters influences the optical properties of the clusters. These changes differ clearly for the different matrix materials and this points to the high sensitivity of the Mie resonance on the cluster-matrix interface properties, which renders the Mie plasmon spectroscopy a promising tool to investigate surface effects by chemically differing embedding media.

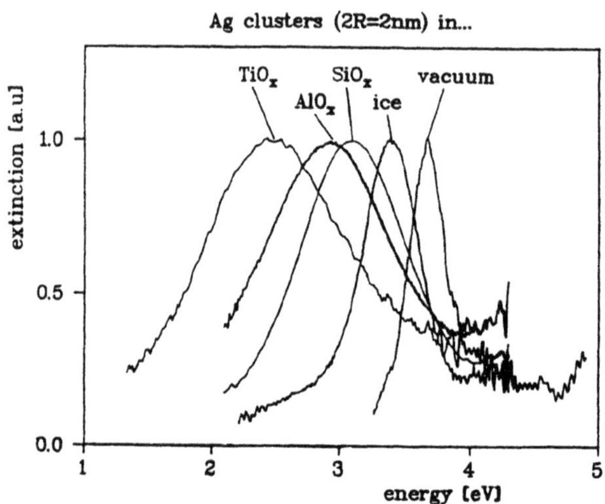

Figure 10. Optical absorption of Ag clusters ($2\overline{R} = 2nm$) produced in a free beam and embedded, alternatively, in different dielectric matrix materials. (The compositions denoted by x are close to SiO_2, Al_2O_3 and TiO_2, respectively).

To emphazise: these effects are not predicted by the classical Mie theory. The red shifts exceed those predicted by the above mentioned "dielectric effect", and the broadening is not predicted at all. They afford additional modelling of the appropriate cluster polarizability. This will be performed in the following section.

4. Modelling of the Cluster Interface Effects

In order to simplify the complex optical properties of metals, theoreticians prefer to take refuge with the strongly simplifying jellium-model, which, however, often does not yield numerical results that can be quantitatively compared to experimental findings. In this approximation cluster quantum size effects [e.g. 3] have as well been theoretically treated as chemical surface effects [11].

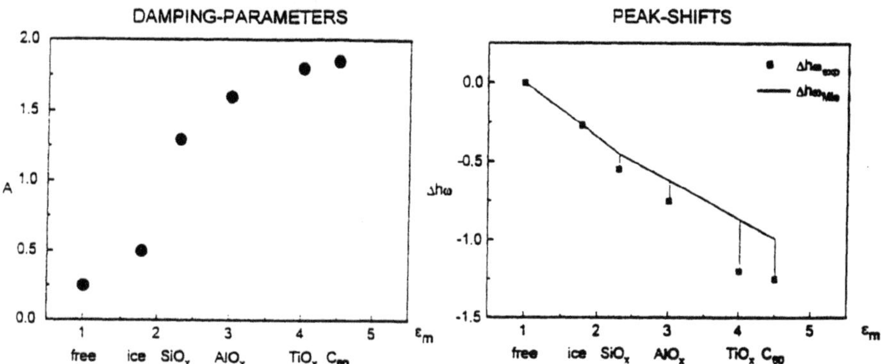

Figure 11. Compilation of A-parameters of eq. (left) and of Mie peak shifts (right) determined from the measured absorption spectra like in Fig. 10. An arbitrary classification parameter the ε_m of the matrix material was chosen for the abszissa). From Mie theory follow $A \equiv 0$ and $\Delta\,\hbar\,\omega_{Mie}$.

These investigations were performed in terms of the polarizability of the cluster. This quantity is directly related to the "effective" dielectric function of the cluster material, introduced before. This latter quantity has the meaning of having to have the very numerical values which, inserted in the Mie theory (eq. 5), yield optical Mie spectra quantitatively reproducing the recorded experimental spectra. Hence $\varepsilon(\omega)$ is an average over the cluster volume and the interface, which, as we will show, includes the adjucent atoms of surrounding foreign atoms, too. In fact, we use $\varepsilon(\omega)$ to model the samples more realistically, than done in Mie's original theory, and, thus, this quantity differs from the one defined for bulk material. We will, however, see that the main lack of the jellium model, the neglect of optical interband excitations, can be circumvented by applying $\varepsilon(\omega)$. Extrapolating from the case of bulk metals, we write:

$$\hat{\varepsilon}(\omega) = \varepsilon_1 + i\,\varepsilon_2 = 1 + \frac{N \cdot e^2}{\varepsilon_0 \cdot m_{eff}} \cdot \frac{1}{\omega^2 + \gamma^2}\left[-1 + i\frac{\gamma}{\omega}\right] + \hat{\chi}^{ib} \qquad (7)$$

with $N, m_{eff}, \gamma_0, \varepsilon_0, \hat{\chi}^{ib}$ the conduction electron density, effective mass, relaxation frequency, the electrical field constant and the susceptibility of the interband transition excitations, respectively.

In the special case of Ag clusters, the Mie resonance resembles a Lorentz profile and peak position $\hbar\omega_s$ and the FWHM width Γ are well suited for characterization. Both can approximatively expressed in terms of $\hat{\varepsilon}$ [5]:

$$\hbar\omega_s \approx \frac{1}{\sqrt{\varepsilon_m + 2}} \cdot \hbar\omega_p; \quad \omega_p = \left(\frac{Ne^2}{\varepsilon_0 m_{eff}}\right)^{1/2} \tag{8}$$

ε_m is the dielectric constant of the surrounding medium, which is assumed here to be non-absorbing.

$$\Gamma \approx \frac{\varepsilon_2(\omega_s)}{\partial\varepsilon_1/\partial\omega}; \quad (\omega \approx \omega_s) \tag{9}$$

In Fig. 3 the experimental data of Γ increase linearly with $1/R$, the slope $A^\star$ being different for different embedding media:

$$\Gamma(R) = \Gamma_0 + A^\star \cdot \frac{V_F}{R} \tag{10}$$

with V_F the Fermi velocity.

This can be traced back to the assumption of an analogue dependence of the conduction electron relaxation time in the Drude expression of eq.

$$\gamma(R) = \gamma_{bulk} + A \cdot \frac{V_F}{R} \tag{11}$$

Additional possible size dependences of ω_p and $\hat{\chi}^{ib}$ will be disregarded here.

The $(1/R)$-relation has been theoretically interpreted by classical considerations ("free path effect") as well as by the quantum theoretical confinement ("quantum size effect") [5]. These theories all disregarded the sensitivity of A to the chemical nature of the embedding material, as following from Figs. 4,5,9 and 10. The values of A and of $\hbar\,\omega_s$ owing to the spectra of Fig. 10 are compiled in Fig. 11 for the particular mean cluster size of $2R = 2nm$, which corresponds to a number of 250 atoms per cluster. The data were plotted against the dielectric constant of the matrix materials, ε_m, to characterize these materials somehow. In Table 1 all A-values presently known are compiled. It should be reminded that in the frame of the classical Mie theory A should be zero and $\hbar\omega_s$ should, due to the "dielectric

TABLE 1. Values of the A-parameter of Ag clusters and Au clusters surrounded by various materials:

Surroundings	A according to [11]	A experimental	
Ag in solid Ne	~ 0	0.25	
solid Ar	0.1	0.3	
solid O_2	0.3	0.5	[7]
solid CO	0.8	0.9	
$Na - SiO_2$ glass	1.0	1.0	[see 5]
ice		0.5	
$SiO_x(\sim SiO_2)$		1.3	
$AlO_x(\sim AlO_2O_3)$		1.7	present
$TiO_x(\sim TiO_2)$		1.8	results
solid CO_{60}		1.9	
ITO		0.7 - 1.2	
SiO_2 substrate		0.7	
$+O_2)$		0.5	
$+S$		1.2 - 1.7	
Au in $Na - SiO_2$ glass		1.5	[see 5]
Au_{55} in triphenylphosphine		> 2	

effect", follow the plotted line in Fig. 10. So, we have two new parameters selective to the properties of the cluster surface.

In the next sections we give theoretical models interpreting these results.

5. Chemical Interface Damping

Persson [11] developed a model to explain the dependence of the Mie band width Γ on the surrounding material. According to this model, electrons, involved in the excited dipolar Mie plasmon, can change into electronic states of adsorbate atoms at the interface by tunneling. These levels are assumed to be broadened compared to the free atom (see Fig. 12). After some residence time which suffices for transfer of excitation energy, or excitation momentum, the electron may tunnel back into the cluster.

If, as done by the author, it is assumed that this electron carries the whole Mie plasmon energy $\hbar\,\omega_s$, then resonance tunneling occurs if the adsorbate levels lie about $\hbar\,\omega_s$ higher than the Fermi level in the metal cluster. If, alternatively, we assume the electron to carry only the $1/n$-th part of $\hbar\,\omega_s$, where n is the total number of conduction electrons involved in the collective plasmon excitation, then lower lying adsorbate states are

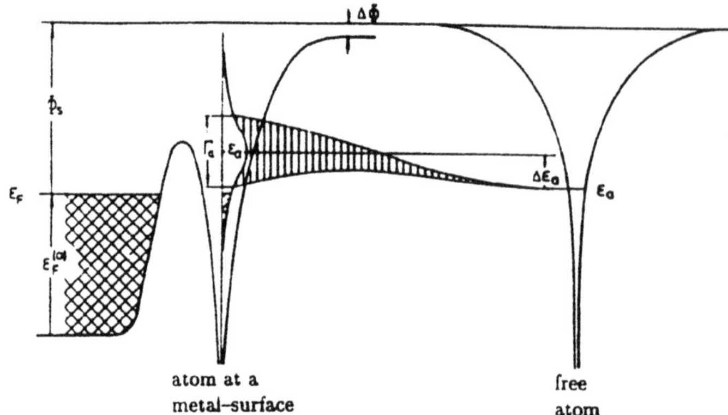

Figure 12. Engergy level-scheme of the adsorption at a metal surface (from [15]). The atomic level is shifted by $Delta\,\varepsilon_a$ and broadened to Γ_a. Left: the conduction band of the metal with ε_F the Fermi energy.

important. In this tunneling picture, the electron looses the phase of the collective motion in the cluster and, hence, after tunneling back into the cluster does no longer contribute to the correlated , collective plasmon excitation even, if no dissipation of energy or momentum takes place. Then, the lifetime of the Mie plasmon is additionally reduced by this "dephasing" effect if many such tunneling transitions occur in proper time.

Persson introduced a complex self energy into the polarizability α of the cluster, assumed to consist of jellium:

$$\alpha(\omega) = \frac{1-\varepsilon_m}{1+2\varepsilon_m} R^3 + \frac{3\varepsilon_m}{1+2\varepsilon_m} R^3 \frac{\Omega^2}{\Omega^2 - \omega^2 - \Sigma(\omega)}$$

with $\Omega = \omega_p(2\varepsilon_m + 1)^{-1/2}$ and $\Sigma(\omega) = Re\,\Sigma + i\,\mathrm{Im}\Sigma$ the "self energy". It can be separated, $\Sigma = \Sigma_{||} + \Sigma_{\perp}$; since it is due to interface currents $j = j_{||} + j_{\perp}$ parallel and normal to the cluster surface. The resulting changes of the widths $\Gamma_{||}$ and $\Gamma_{\perp}$ and the corresponding contributions to the A^*-parameter of eq. (10) are:

$$\begin{aligned} (1) \quad \Delta\Gamma_{||} &\approx Im\,\Sigma_{||}(\Omega)/\Omega = \frac{N_A v_F}{R} \sigma_0\, J(\omega) \\ \hookrightarrow A_{||} &= \text{const.} \cdot J(\omega, \rho_A, E_A, E_F, \Gamma^*, \omega_p) \end{aligned}$$

The physical effect is diffuse electron scattering into the Lorentzian adsorbate state characterized by the density $\rho_{A'}$ the maximum energy E_A and its width Γ^a (see Fig. 12).

$$(2) \quad \Delta\Gamma_\perp \approx Im\,\Sigma_\perp(\Omega)/\Omega = \frac{N_A}{R}\cdot\sigma_1\cdot \mathrm{Im}\,\alpha_\perp(\Omega)$$
$$\hookrightarrow A_\perp = \mathrm{const.}\cdot d^2\cdot \mathrm{Im}\,\alpha_\perp(\omega,\rho_A,E_a,E_F,\Gamma^*,\omega_p)$$

where $\alpha_\perp$ is the polarizability of an adsorbate d is the Feibelman parameter (distance of adsorbate center to image plane).

The resulting contribution to A is then given by $A = A_{\|} + A_\perp$. The problem remains that mostly the adsorbate states are unknown. Vice versa, we expect that it will be possible in future to derive information about these states from the optical Mie resonance spectra.

6. Plasmon band shifts

As is obvious from Fig. 11 the peak shifts $\Delta\hbar\omega_s$ which are induced by the existence of some surrounding medium, is larger than the "dielectric effect". The differences exceed the slight shifts induced by increased damping of the resonances, discussed before. Formally, they can be described by variation of the slope of $\varepsilon_1(\omega)$, which is, according to eq. (7) determined by (N/m_{eff}).

RESONANCE SHIFTS : Possible mechanisms

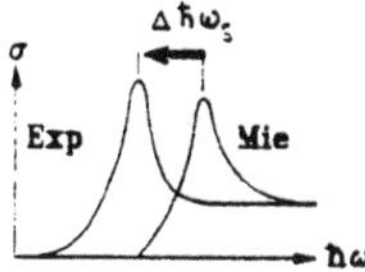

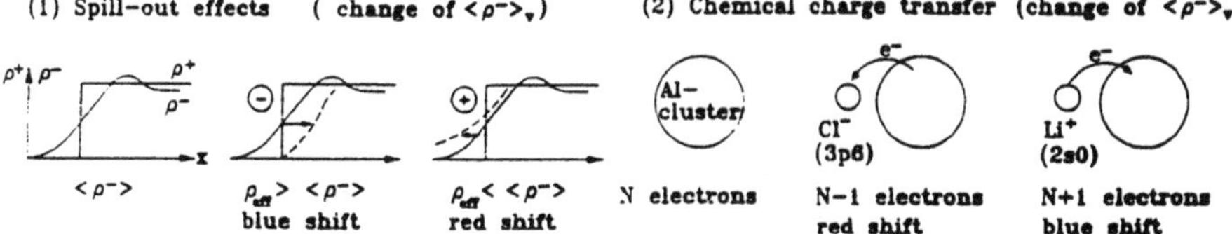

Figure 13. Schematic mechanisms of Mie resonance shifts. Left: spill-out is changed by foreign ions at the interface. The following change of the average electron density $<\rho>= N$ causes shifts according to eq. Right: Charge transfer due to formation of chemical interface bonds are shown for reactions with Cl^- and Li^+ ions, respectively. Again the average electron density is changed.

Fig. 13 gives an illustration of two different interface effects which influence the average electron density N in the cluster: 1) changes of the spatial extension of the "spill out" of Fig. 1 by Coulomb forces from adjacent foreign ions or polarizable atoms/molecules, and 2) charge transfer involved in the formation of chemical bonds between cluster surface atoms and the foreign surrounding material. As shown schematically, both effects may, alternatively, reduce or increase N, causing red or blue shifts of the plasmon peak. In the system of Ag clusters embedded in solid C_{60}, this shift was recently identified by Gartz et al. [12] and a net charge transfer of about one electron per adjacent C_{60} molecule from the silver is in accordance with recent ESCA experiments of Wertheim et al. [13] on planar Ag/C_{60} systems.

7. Summary

A novel method of spectroscopy of chemical surface effects is proposed which is based upon the recently discovered sensitivity of the Mie plasmon resonance of metal clusters upon the cluster surroundings, i.e. interface properties, instead of the direct spectroscopy of the interface atoms.

The deviations of the optical properties of metal clusters from the predictions of Mie's theory (fed with bulk dielectric functions) yields additional information about electronic size effects and about details of the cluster/surrounding interface and about chemical interface reactions.

As a general result, the electronic and optical properties of clusters in contact with foreign surrounding material cannot be modelled without including, at least, one monoatomic layer of the surrounding, which, thus, belongs to the "cluster".

For quantitative evaluations, calibration with the optical properties of free, uncontaminated clusters is required. Thus optical absorption was recorded for the same silver clusters in the free beam, deposited on a substrate and embedded in foreign surrounding materials. The chemical interface damping strongly affects the width of the Mie plasmon resonance. Charge transfer involved in the formation of chemical bonds at the interface causes the peak position to shift. It is hoped that quantitative evaluations of this novel tool of surface science will be successful.

References

[1] Lüth, H., (1994) *Surface and Interfaces of Solids*. Springer Series in Surface Science 15

[2] Lang, N.D., Kohn, W., (1970) *Phys. Rev.* **B 1**, 4577; Lang, N.D., Kohn, W., (1971) *Phys. Rev.* **B 3**, 1215

[3] Ekardt, W., (1984) *Phys. Rev.* **B 29**, 1558; Ekardt, W., (1985) *Phys. Rev.* **B 31**, 6360; Brack, M., (1989) *Phys. Rev.* **B 39**, 3533; Liebsch, A., (1993) *Phys. Rev.* **B 48**, 11317

[4] Mie, G., (1908) *Ann. Physik* **B 25**, 377

[5] Kreibig, U., Vollmer, M., (1995) *Optical Properties of Metal Clusters*. Springer Series in Materials Science 25

[6] Jackson, J.D., (1975) *Classical Electrodynamics*. Wiley, New York

[7] Charlé, K.P., Frank, F., Schulze, W. (1984) *Berichte Bunsenges. Phys. Chem.* **88**, 350; Charlé, K.P., Schulze, W., Winter, B., (1989) *Z. Physik* **D 12**, 471

[8] Kreibig, U., (1974) *J.Physics* **F 4**, 999

[9] Hövel, H., Fritz, S., Hilger, A., Kreibig, U., Vollmer, M., (1993) *Phys. Rev.* **B 48**, 18178

[10] Mrozek, I., Otto, J., (1990) *J. Electr. Spectr. Rel. Phen.* **54/55**, 985

[11] Persson, B.N.J., (1989) *Phys. Rev.* **B 39**, 8220; Persson, B.N.J., (1991) *Phys. Rev.* **B 44**, 3277; Persson, B.J.N., (1993) *Surf. Sci.* **281**, 153

[12] Gartz, M., (1994) *unpublished results*

[13] Wertheim, G.K., Buchanan, D., (1994) *Phys. Rev.* **B 50**, 11070

[14] Gutierrez, M., Henglein, A., (1993) *J. Phys. Chem.* **97**, 11368

[15] Hölzl, J., Schulte, F., Wagner, H., (1979) *Solid Surface Physics* Springer Tracts in Modern Physics 85

[16] Haberland, H. (ed.) (1994) *Clusters of Atoms and Molecules I,II* Springer Series in Chemical Physics 52, 56

LARGE METAL ION CLUSTERS - SMALL MAGNETIC PARTICLES

Dante Gatteschi and Roberta Sessoli

Department of Chemistry, University of Florence, Florence, Italy

Abstract. Large metal ion clusters can provide examples of mesoscopic materials in which novel magnetic properties can be observed. In particular the coexistence of quantum and classical properties can be observed. We provide here two examples, one comprising eight manganese(III) and four manganese(IV) ions and one comprising eight iron(III) ions both with a ground S= 10 state which at low temperature show very slow relaxation of the magnetization and a behaviour similar to that observed in superparamagnets. The current understanding of these new phenomena will be reviewed.

1. Introduction

The most important market based on magnetics technology is recording [1]. The last years have seen a remarkable growth in performance which can be measured by the number of bits stored per unit area. Currently the densities are around 100 megabits per square inch, and the aim is that of reaching 10 gigabits per square inch in the near future. The size of the magnetic particles will have to be 1 micron wide and only 70 nm long. It might be imagined to increase further the magnetic density by decreasing the size of the particles. However there are some limits for the process of miniaturization, because approaching the nanometre scale quantum phenomena become more and more important, introducing fluctuations which change dramatically the properties of the materials. Much research is currently performed in order to understand how much new physics can be found in nanometric size magnetic particles.

One of the challenges in producing very small magnetic particles is that of controlling their dimensions and the variance of size from particle to particle. This problem is being addressed to with both a top-down and a

E. Pelizzetti (ed.), Fine Particles Science and Technology, 517–531.

bottom-up approach. The former means to make small particles starting from bulk materials, with methods akin to lithography used for electronics, while the latter ideally considers to make them by assembling the magnetic building blocks one at a time. This can be done with chemical methods [3-4], with scanning tunneling microscopy techniques [4] or following the natural biomineralization processes [5,6].

The chemical approach uses several different techniques of controlled hydrolysis, blocking the growth of the particles with protecting polymers, inverse micelles, etc. A variant of this strategy consists in the isolation of large molecular clusters, which can be crystallized with the usual techniques of molecular chemistry, and this is the one we want to advocate here. The system isolated so far are generally closer to the sub-nano rather than to the nano size. The individual spins in the clusters can be provided by transition metal or rare earth ions, or even by organic radicals. Several different classes of cluster of transition metal ions have been reported so far ranging from oxo bridged systems [7-9], which correspond to small particles of oxides or hydroxides, to sulfur bridged systems which are models for metal sulfides [10]. The largest systems comprising magnetic centres have up to nineteen metal ions [11].

The synthesis of clusters of increasing size in principle provides the possibility of learning which is the minimum size of magnetic particles which can give rise to bulk behaviour. Most of them show only a complex paramagnetic behaviour, without any clear indication of incipient bulk behaviour, but at least a few cases show superparamagnetic like behaviour,

In the following we will first make a short resume of the properties of large magnetic clusters and then focus on a couple of examples which we feel show particularly exciting perspectives.

2. Theoretical Background

Finite size clusters can be considered to behave as complex paramagnets, in which there are in general strong to moderate interactions between the spins present in the cluster, and weak interactions between the clusters. We will assume negligible intercluster interactions. In order to intepret the magnetic properties of the clusters it is customary to express a spin hamiltonian in which the pairwise interactions between the different spins, i and j, are paramaterized as:

$$H = \Sigma_{i>j} J_{ij} \mathbf{S_i} \cdot \mathbf{S_j} \qquad (1)$$

In the case of pairs the levels can be expressed as a function of the total spin S

$$|S_i - S_j| \leq S \leq S_i + S_j \qquad (2)$$

with the energies

$$E(S) = J/2\,[S(S+1) - S_i(S_i+1) - S_j(S_j+1)] \qquad (3)$$

When $J<0$ (ferromagnetic coupling) the ground state has the maximum spin multiplicity $S = S_i + S_j$, while when $J>0$ (antiferromagnetic coupling) the ground state has $S = |S_i - S_j|$.

The different spin levels will be thermally populated, yielding a temperature dependent magnetic susceptibility. The use of this hamiltonian for dimers yields the well known Bleaney-Bowers equation [12] for the magnetic susceptibility.

Extention of (1) to large clusters is in principle easy, but in the practice it rapidly becomes extremely difficult. In fact for a cluster of N spin characterized by individual quantum numbers S_i, the total number of states which form the hamiltonian matrix are $(2S_i+1)^N$. For instance, for $S_i = 5/2$, as appropriate for iron(III), when N= 8 the number of states is already 1,679,616. Therefore the calculation of the spin levels, and of the magnetic susceptibility becomes a formidable task. The best approach so far reported is that of using irreducible tensor operators, ITO's, which exploit at best the symmetry associated with the total spin [13,14]. However even in this case it is necessary to calculate 21 matrices (corresponding to the total spin values S ranging from 0 to 20), the largest of which, S= 5 is 16,576x16,576.

A further reduction in the size of the matrices can be obtained using the point group symmetry of the cluster. For D_2 symmetry there are 81 matrices to be calculated, the largest of which is 4,170x4,170. At this level it is possible to perform the calculations on a work station [15].

As a result of the great number of spin levels these will tend to merge into a continuum: when this occurs the system must cross-over from a complex paramagnetic to bulk magnetic behaviour. This transition must be expected to occur gradually. From a theoretical point of view we may imagine to build large clusters by assembling them one at a time.

In this ideal process at some stage the interacting spins of the cluster will be blocked in a preferred orientation, corresponding to a ferro-, ferri-, or antiferromagnet, and the system becomes ordered. Associated with the ordering is the development of magnetic anisotropy, which will dictate the spatial orientation of the magnetic moment. The origin of the anisotropy can be associated to the shape of the particle, to the contribution of the anisotropy of the individual ions, or to a contribution associated with the magnetic interaction between the individual ions. For relatively large particles the anisotropy A can be assumed to be proportional to the volume of the particle [16]:

$$A = KV \tag{4}$$

A represents the barrier that a particle must overcome in order to reorient its magnetization to the energetically equivalent position, in the absence of an external field, shown in Figure 1.

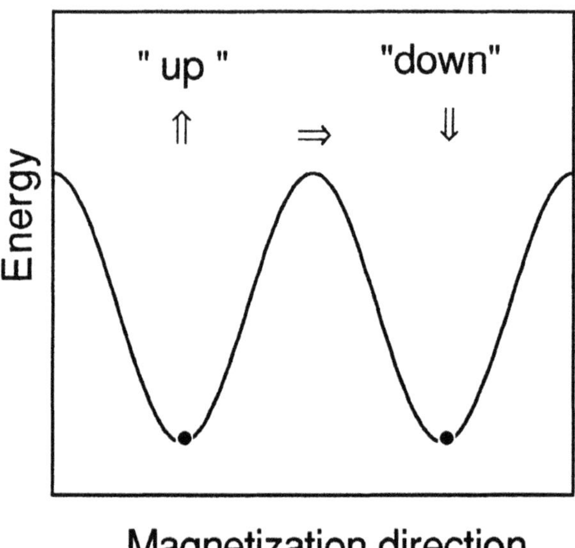

Figure 1. Anisotropy barrier for the reorientation of the magnetization in a small particle.

When A is large compared to thermal energy the magnetization will be blocked in one of the two minima, but when the barrier is small the magnetization will freely flip over from one minimum to the other. As a consequence the ordered system has no longer a permanent magnetization and it will behave as a paramagnet, but one characterized by a large magnetic moment. A particle which behaves like this is called a superparamagnet.

The relaxation of the magnetization τ follows a thermally excited process, described by the Arrhenius law:

$$\tau = \tau_0 \exp(KV/kT) \qquad (5)$$

τ_0 is typically 10^{-9}s for ferromagnets like iron.

A particle behaves like a bulk magnet when τ is larger than the characteristic time τ_c of the investigation technique used to monitor it. τ_c is 10^{-3}-10^0 s for ac susceptibility measurements and 10^{-8} s for Mössbauer spectroscopy. The temperature at which $\tau \approx \tau_c$ is called blocking temperature, T_B. Above T_B the magnetization freely fluctuates, while below T_B it is blocked.

If we consider the cluster as a small magnetic particle at low temperature it will be in the ground state, corresponding to an ordered state. Decreasing further the temperature its magnetization may eventually become blocked. Therefore we may assume that in the clusters the first indication of bulk behaviour comes from the slow relaxation of the magnetization observed at low temperature.

A further reason of interest for the investigation of the dynamics of the magnetization of the clusters at low temperature is that, since their size is intermediate between that of bulk objects, to which classical mechanics apply, and microscopic objects, to which quantum mechanics applies, the inversion of the magnetization occurs by a tunneling process. Indeed this is one of the reasons of the large interest of solid state physics in this new kind of magnetic materials [17-19]. In fact the macroscopic quantum tunneling effects of magnetization can prove to be useful testing grounds for quantum theory itself.

3 The Mn_{12} Clusters

The first unambiguous example of large metal ion cluster exhibiting superparamagnetic like behaviour is provided by $[Mn_{12}O_{12}(O_2CCH_3)_{16}(H_2O)_4)]$, $Mn_{12}Ac$, which was originally reported in the early eighties, without much characterization [20]. Other similar compounds were reported afterwards, which differed in the nature of the carboxylic acid [21]. The structures of the clusters are all very similar to each other, comprising eight manganese(III), S= 2, ions arranged in an external ring, as shown in Figure 2, and four internal manganese(IV), S= 3/2, ions.

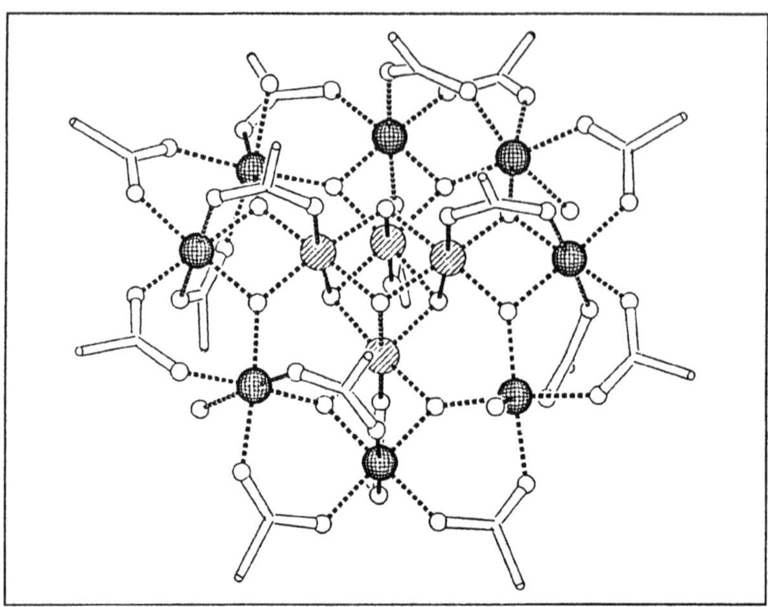

Figure 2. View of the Mn_{12} cluster where Mn(IV) ions are reported as hatched circles,Mn(III) as cross-hatched ones, and oxygen and carbon atoms by empty circles of dcreasing size

The ions are bridged by both oxo and carboxylato groups. After some controversy it was clearly seen that the ground state of the clusters is S= 10, which can be justified by the spin arrangement shown in Figure 3 [22,23].

Attempts were made also to fit the low temperature magnetic susceptibility with a simplified model which assumes that the exchange interaction between the manganese(III) and manganese(IV) ions connected by two μ-oxo bridges is strong and antiferromagnetic. In practice it was assumed that the four manganese(III)-manganese(IV) pairs are in the ground S= 1/2 state at low temperature, contributing a total spin multiplicity of $16= 2^4$, which multiplied by $625= 5^4$ (4 Mn(III) with S= 2), gives a total spin multiplicity of 10,000. The total spin values range from 0 to 10, the largest matrix being 248x248, corresponding to S= 3, making the problem tractable using ITO's. A reasonable agreement was found with

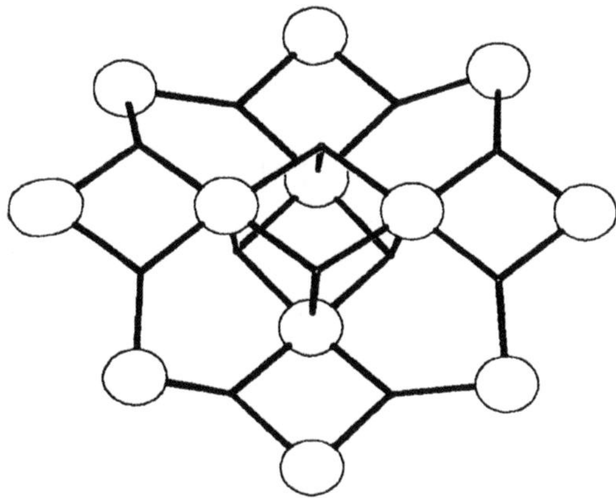

Figure 3. Preferred spin alignment within the $Mn_{12}Ac$ cluster.

experiment, by using $J_{Mn(III)Mn(III)} = 0\ cm^{-1}$, $J_{Mn(IV)Mn(IV)} = 140\ cm^{-1}$, and the manganese(III)-manganese(IV) exchange interaction transmitted through the single oxo bridge, $J_{Mn(III)Mn(IV)} = 180\ cm^{-1}$, The ground state was confirmed to be S= 10, with two degenerate S= 9 states ca. 25 cm^{-1} higher in energy. The benzoate and the propionate derivative were reported to have a ground S= 9 state [27]. Interestingly it was observed that the clusters are stable in solution, and they can undergo redox reactions, which afforded the synthesis of $[Mn_{12}O_{12}(O_2CEt)_{16}(H_2O)_4]^-$, which comprises seven manganese(III), four manganese(IV) and one manganese(II), S= 5/2. The ground state of this cluster was reported to be S= 19/2. Successfull attempts were also made to substitute some manganese(III) ions with iron(III) ions [24].

The Mn_{12} acetate cluster was shown to have a large zero field splitting in the ground state [25]. The first evidence was found through high frequency EPR, HF-EPR, spectroscopy which showed one transition between the ground Ms=-10 and the first excited Ms=-9 state which went to zero field when the quantum of radiation used in the experiment was ca. 10 cm^{-1}. These data provided the value and the sign of D=-0.5 cm^{-1}. At low temperature all the clusters will be in the $M_S = \pm 10$ states. In a weak field parallel to the symmetry axis of the clusters the two levels will split as $\pm g\mu_B M_S H$, while when the field is orthogonal to the axis no splitting will be observed. Therefore the system becomes extremely anisotropic, as confirmed by the magnetic anisotropy data reported in Figure 4. The magnetization data at 6.0 K showed that parallel to the tetragonal axis of the cluster the saturation of 20 μB is reached in an external field of 2 T, while perpendicular to the axis 10 T are needed.

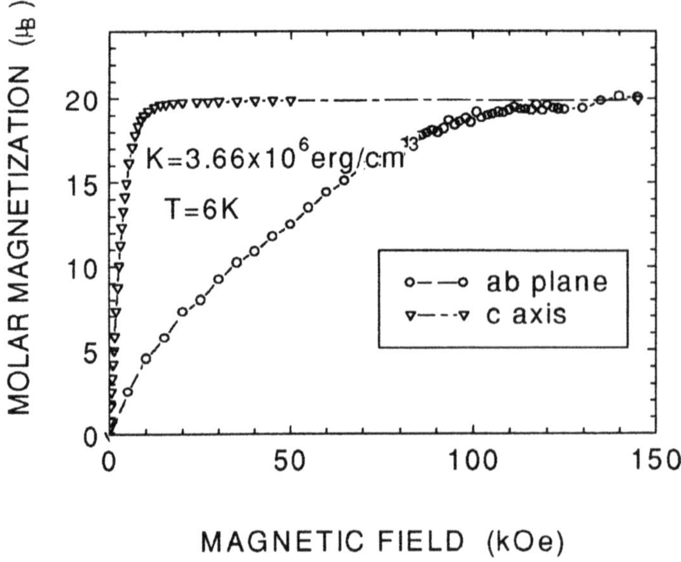

Figure 4. Magnetic anisotropy of $Mn_{12}Ac$. The anisotropy constant K has been evaluated as the area between the magnetization curves obtained with the field parallel and perpendicular to the easy axis.

The consequence of such a large ground state coupled with the large Ising type anisotropy is that the relaxation time of the magnetization becomes extremely slow at low temperature. The first experimental evidence for that was obtained through ac susceptibility measurements, which yielded frequency dependent out of phase components. From these the relaxation time was obtained, which was found to follow an exponential law of the type of eq. (3), with $\tau_0= 2\text{x}10^{-7}$ s, and A=61 K. The relaxation time becomes so long at low temperature that it can be directly measured by saturating the magnetization with an external field and then monitoring its decay with a SQUID, after removing the field. It is extremely interesting that essentially the same results are obtained in frozen solution, thus ruling out any possibility that the slow relaxation is determined by intermolecular interactions. Some typical time dependences of the magnetization from 2.12 K to 2.75 K are shown in Figure 5.

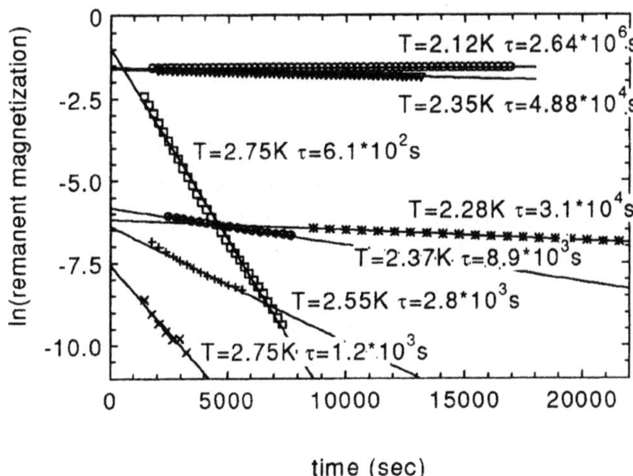

Figure 5 Time dependence of the remanent magnetization of $Mn_{12}Ac$ measured on single crystals (upper part) and on a frozen solution (lower part).

At 2 K the relaxation time is of the order of one month, and below that limit it becomes practically independent of temperature [26], as shown in Figure 6. A temperature independent relaxation time is typical for quantum tunneling of the magnetization. Therefore it seems that Mn_{12} Ac is indeed a system in which quantum and classical behaviour coexist.

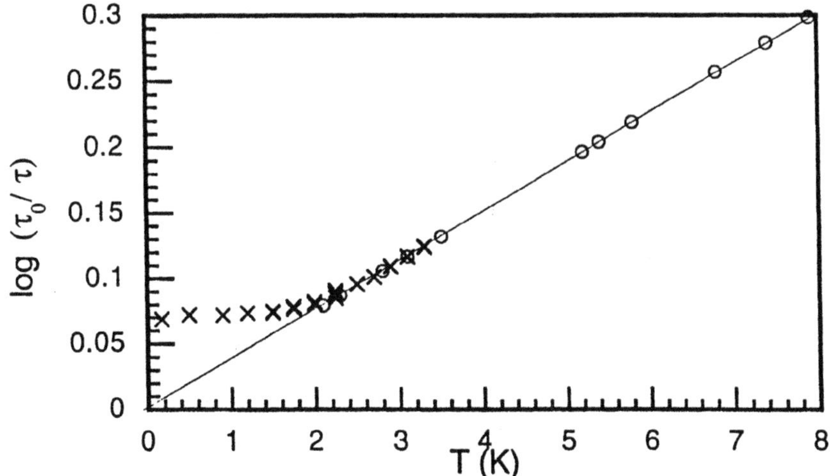

Figure 6. Temperature dependence of the relaxation time of $Mn_{12}Ac$.

Another consequence of the long relaxation time of the magnetization of $Mn_{12}Ac$ is that below 3 K it shows magnetic hysteresis effects for measuring times larger than 10^2 s. A typical hysteretic plot is shown in Figure 7. In general magnetic hysteresis effects are observed in bulk magnets due to the irreversibility in the movement of the domain walls. In the present case there are no domains present in the material, as shown by the paramagnetic behaviour at high temperature and by the absence of any long range order effects, as monitored by specific heat measurements. A further confirmation to the molecular origin of the hysteresis comes from the relaxation measurements in dilute frozen solutions mentioned above. The origin of magnetic hysteresis is just the slow relaxation of the magnetization and it has a molecular origin, so in principle it opens the possibility to store information at the molecular level, at least when the size of the molecules is nanometric.

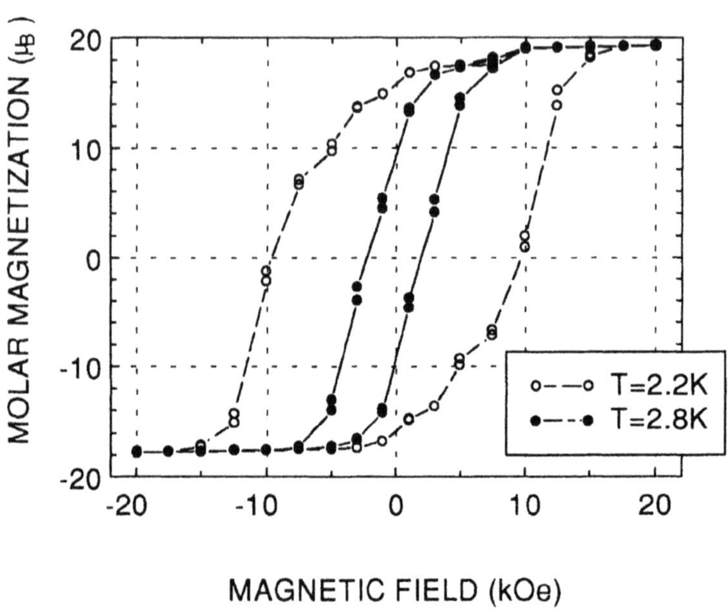

Figure 7. Magnetic hysteresis observed for $Mn_{12}Ac$ at two temperatures.

Attempts were made also in order to justify the slow relaxation of the magnetization in the thermally activated region [27]. The theoretical basis was the extention of the Orbach model [28] for the relaxation of the magnetization in simple paramagnets. If a given spin is excited in the M_S= +10 state at low temperature it cannot revert to the ground M_S= -10 by a direct decay, but it must climb up the ladder of levels up to M_S= 0, and then decay down, as shown in Figure 8. The relaxation rate at low temperature is calculated to be:

$$\tau^{-1} \approx C\,(\Delta/S^2)^3 \exp(-\beta\Delta)/[1-\exp(-\beta\Delta/S^2)] \qquad (6)$$

where Δ is the energy gap between the lowest and the highest spin component ($M_S = \pm S$, and $M_S = 0$ or $\pm½$, for system with even or odd numbers of unpaired electrons, respectively), and $\beta = k_BT$. C is a constant which depends on the spin-phonon coupling.

The relaxation rate is proportional to Δ^3, as already reported for S= 1 systems, and inversely proportional to S^6. Clearly the effect of a large spin is a very important one in giving rise to slow relaxation.

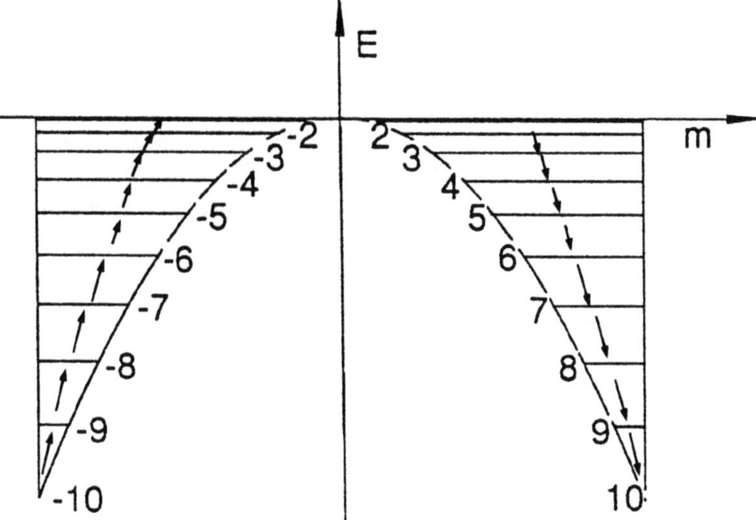

Figure 8. Schematic picture of the mechanism of reversal of the magnetization for an S=10 spin system splitted in zero field due to magneto-crystalline anisotropy.

The origin of the anisotropy can be associated with the single ion contribution of the manganese(III) ions, which are Jahn-Teller distorted. In fact, in this assumption, using spin projection techniques [29] it is possible to calculate the following relation between the individual zero field splitting, D_{Mn}, of the manganese(III) ions and the zero field splitting of the cluster, D_{10}, in the ground S= 10 state:

$$D_{10} = 0.126\, D_{Mn} \qquad (7)$$

From the experimental $D_{10} = -0.5\ cm^{-1}$ we calculate $|D_{Mn}| \sim 4\ cm^{-1}$, which sounds quite reasonable for a manganese(III) ion. It seems therefore that at least in this case it is not necessary to invoke shape effects [29] in order to justify the high anisotropy barrier observed for the reorientation of the magnetization.

4 Iron Clusters

Compared to the manganese clusters described in the previous section the iron clusters have the advantage that they can be investigated also by using Mössbauer spectroscopy, which provides easily the blocking temperatures of superparamagnets Indeed much work has been performed on iron(III) oxo clusters, and in at least three compounds of formula $[Fe_{11}O_6(OH)_6(O_2CPh)_{15}]$.6THF and $[Fe_{16}MO_{10}(OH)_{10}(O_2CPh)_{20}$, with M= manganese(II) or cobalt (II) it was found [30] that slow relaxation sets in at temperatures close to 4.2 K.

The common features of the structures of the clusters in which "superparamagnetic" behaviour was observed with Mössbauer spectroscopy consist in the fact that they are all pseudo three-dimensional clusters, while no anomalous slow relaxation was observed in clusters which have a pseudo one- or two-dimensional structure. It was suggested that indeed the observation of slow relaxation is easier for systems which are intrinsically three-dimensional rather than in systems which are lower dimensional, and the anisotropy barrier required for slow relaxation is due to shape, surface, and strain anisotropy contributions.

We have recently investigated [15] the magnetic properties of $[(tacn)_6Fe_8(\mu_3\text{-}O)_2\mu_3\text{-}OH)_{12}]^{7-}$, Fe_8, whose structure is shown in Figure 9. The cluster is flat, with the iron(III) ions lying approximately on a plane. Therefore Fe_8 is a model of two-dimensional magnetic material.

The temperature dependence of the susceptibility gives a clear indication of a ferrimangnetic arrangement of the spins, with a ground state which can be either S= 9 or S= 10. A quantitative fit was attempted, using ITO's, even if the size of the matrices is very large as reported in Section 2.

HF-EPR was of extreme importance even in this case [31], because it clearly showed that the ground state is S= 10, with a negative zero field splitting $D = -0.2\ cm^{-1}$. It is apparent that there is a similarity between Fe_8 and $Mn_{12}Ac$, the main difference being in the fact that the former has a much smaller zero field splitting. than the latter.

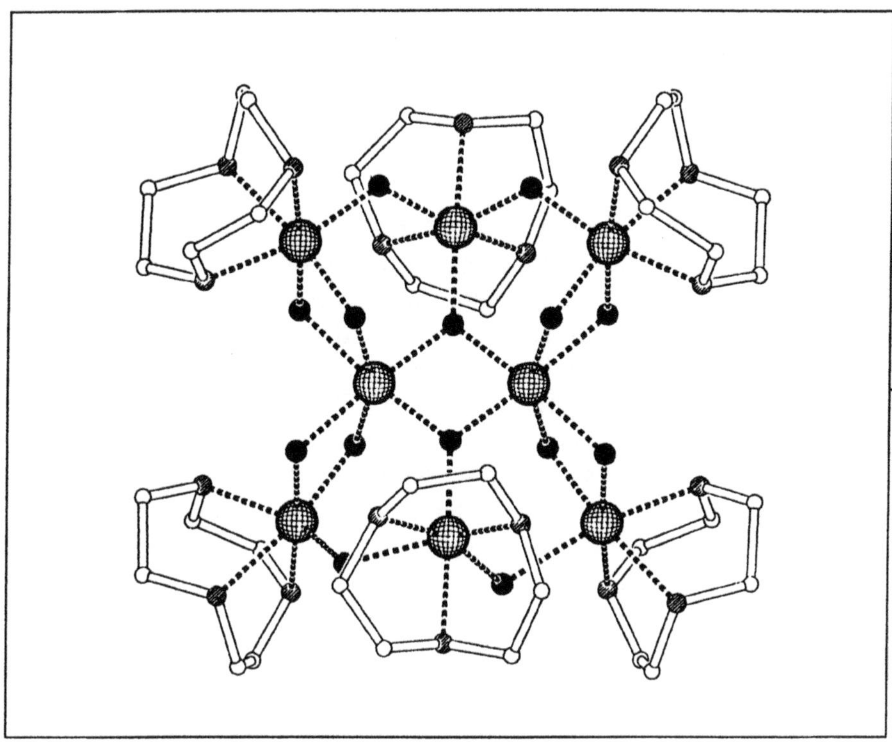

Figure 9. View of the structure of Fe_8 cluster. The cross-hatched circles represent the iron atoms, while full, hatched and empty circles stay for oxygen, nitrogen and carbon atoms respectively.

However, also for Fe_8 a barrier must be expected to the reorientation of the magnetization, of about 27 K. Therefore slow relaxation of the magnetization must be anticipated at low temperature, and indeed we could observe an exponential dependence on T below 3 K. Also in this case the preexponential factor, τ_0, is much smaller than usually observed in ferromagnets.

Mössbauer spectroscopy also showed slow relaxation, below 25 K. This is the first example of a molecular system showing superparamagnetic like behaviour both in ac susceptibility measurements and in Mössbauer spectroscopy.

The structure of Fe_8 is of the pseudo two-dimensional type, and nevertheless it is the smallest cluster so far reported with superparamagnetic properties. Therefore it must be concluded that the key role for observing such behaviour is not associated to the shape of the clusters, but rather to the anisotropy, which is presumably of single ion origin.

5 Conclusions

There is a a large potentiality of interest for large metal ion clusters, for the development of new kind of magnetic materials of nanometric size, for clarifying the mechanism of biomineraliztion of iron oxides, for checking fundamental theories like quantum theory. In order to be able to do this it is necessary to accept some fundamental challenge, the first of which is the development of suitable synthetic techniques. The largest magnetic cluster investigated so far has nineteen iron ions, but it is clear that this number must be dramatically increased if the similarity with the bulk must be better established. The recent report by Müller et al. of the crystal structure of a gigantic cluster comprising 154 molybdenum ions certainly opens new exciting perspectives [32]. The cluster is only slightly magnetic, because most of the ions are non-magnetic molybdenum(VI) ions, but it certainly shows that very large clusters which can be crystallized can eventually be produced.

References

1. Simonds, J. L., (1995), Phys. Today, , 26.
2. Awschalom, D. D.; Di Vincenzo, D. P., (1995), Phys. Today, , 40.
3. Gatteschi, D.; Caneschi, A.; Pardi, L.; Sessoli, R., (1994), Science, **265**, 1054.
4. Awschalom, D.D.; Di Vincenzo, D.P., Smyth, J.F. (1992) *Science*, **258**, 414.
5. Mann, S., Frankel, R.B. (1989), in Mann, S., Webb, J., Williams, R.J.P. (eds.), *Biomineralization Chemical and Biochemical Pespectives*, VCH, Weinheim, 389.
6. Meldrum, F.C.; Heywood, B.R.; Mann, S., (1992), *Science,* **257**, 522.
7. Lippard, S.J (1988). *Angew. Chem. Int. Ed. Engl.*, **27**, 344.
8. Wieghardt, K. (1989) *Angew. Chem. Int. Ed. Engl.*, **28**, 1153.
9. Hagen, K.S.(1992) *Angew. Chem. Int. Ed. Engl.*, **31**, 1010.
10. Holm, R.H.; Lee, S.C., (1990), *Angew. Chem. Int. Ed. Engl.*, **29**, 840
11. Heath, S.L., Powell, A.K. (1992) *Angew. Chem. Int. Ed. Engl.*, **31**, 191.

12. Bleaney, B.; Bowers, K.D., (1952), *Proc. Roy. Soc. (London), Ser. A.*, **214**, 451.
13. Gatteschi, D.; Pardi, L., (1993), Gazz. Chim. It., **123**, 231.
14. Gatteschi, D.; Pardi, L., in *Magnetochemistry*, O'Connor, C.J., Ed., (1993), World Scientific, Singapore
15. Delfs, C.; Gatteschi, D.; Pardi, L.; Sessoli, R.; Wieghardt, K.; Hanke, D., (1993), *Inorg. Chem.*, **32**, 3099.
16. Monish, A.H., *The Physical Principles of Magnetism*, (1966), J. Wiley & Sons, New York.
17. Leggett, A.J. (1984) *Phys. Rev. B.*, **30**, 1208.
18. Tejada, J., Zhang, X.X., Balcelles, L. (1993) *J. Appl. Phys.*, **73**, 6709.
19. Gider, S.; Awschalom, D.D.; Douglas, T.; Mann, S.; Chaparala, M., (1995), *Science*,
20. Lis, T. (1980) *Acta Crystallogr. Sect. B.*, **36**, 2042.
21. Eppley, H.J.; Tsai, H.-L.; de Vries, N.; Folting, K.; Christou, G.; Hendrickson, D.N., (1995), *J. Am. Chem. Soc.*, **117**, 301.
22. Caneschi, A.; Gatteschi, D.; Sessoli, R.; Barra, A.L.; Brunel, L.C.; Guillot, M. (1991), *J. Am. Chem. Soc.*, **113**, 5873.
23. Sessoli, R.; Tsai, H.-L.; Schake, A.R.; Wang, S.; Vincent, J.B.; Folting, K.; Gatteschi, D.; Christou, G.; Hendrickson, D.N., (1993), *J. Am. Chem. Soc.*, **115**, 1804.
24. Schake, A.R.; Tsai, H.L.; de Vries, N.; Webb, R.J.; Folting, K.; Hendrickson, D.N.; Christou, J., (1992), *J. Chem. Soc. Chem. Commun.*, 181.
25. Sessoli, R.; Gatteschi, D., Caneschi, A., Novak, M.A. (1993) *Nature*, **365**, 141.
26.Paulsen, C.; Park, J.-G.; Barbara, B.; Sessoli, R.; Caneschi, A., (1995), *J. Magn. Magn. Mater.*, **140-144**, 379
27. Villain, J.; Hartman-Boutron, F.; Sessoli, R.; Rettori, R., (1994), *Europhys. Letters*, **27**, 159.
28. Orbach, R., (1961), *Proc. Roy. Soc. (London), Ser. A.*, **264**, 458..
29. Bencini, A.; Gatteschi, D., (1990), *EPR of Exchange Coupled Systems*, Springer Verlag, Berlin.
30. Papaefthymiou, G. (1992) *Phys. Rev. B.*, **46**, 10366.
31. Barra, A.L.; Debrunner, P.; Gatteschi, D.; Sessoli, R., manuscript in preparation
32. Muller, A., personal communication

SYNTHESIS AND CONSOLIDATION OF ZnS/GaP POWDERS

Mufit Akinc and Yong Han
Department of Materials Science & Engineering
Iowa State University
Ames, IA 50011

1. Abstract

The zinc sulfide/gallium phosphide system was chosen as a model for studying nano composite synthesis and processing. ZnS/GaP composite powders were synthesized by impregnating nano porous zinc sulfide powders with an organometallic GaP precursor. Zinc sulfide was synthesized by homogeneous precipitation from aqueous solutions whereas gallium phosphide from organometallic routes. Nano-porous ZnS particles were impregnated with phosphinogallane liquid precursor followed by flash pyrolysis to precipitate GaP in the pore network. ZnS/GaP powders were characterized with respect to chemistry and structure. Compacts were hot isostatically pressed to dense parts. Compacts maintained nanocrystallinity even after HIPping at 900°C exhibiting crystallite size less than 85 nm. Nano composites exhibited hardness values that were double that of ZnS.

2. Introduction

Infrared transmitting windows materials used in high speed transport systems are expected to perform under harsh environments. The primary function of these components is to protect the thermal imaging devices. To serve this purpose, the window material is required to be transparent to the IR region of the spectrum, strong, tough and have high thermal shock resistance. A variety of oxide ceramics such as magnesium aluminate spinel ($MgAl_2O_4$), sapphire (Al_2O_3), yttria (Y_2O_3), are being used in the 3-5 μm region. However only a limited number of materials are known to show a tolerable combination of thermomechanical properties and transparency in the 8-12 mm region. Despite its

E. Pelizzetti (ed.), Fine Particles Science and Technology, 533–542.

poor mechanical properties, zinc sulfide (ZnS) is currently being used in 8-12 μm range. Therefore, some attempts have been made to improve the properties of ZnS by applying a protective coating(1), by solid solution strengthening(2) with Ga_2S_3, or by forming a composite with diamond(3).

Recent studies(4) indicated that the nanostructured materials show improved mechanical properties. Traditional composite processing is mainly based on either mechanical mixing or precipitation of a secondary phase from a melt. It is, however, difficult to control the size and the distribution of the secondary phase within the matrix. Uniform and ultrafine microstructures in composite materials may be achieved by controlling the chemistry and morphology of the constituents. In this study, we demonstrate that the scale of the microstructure can be reduced well below 100 nm using chemically designed materials. The basic concept is based on synthesis of nano-scal e powders of matrix phase followed by introduction of second phase within the pores of the nano scale particles. For this investigation, 0.2-0.3 μm spherical, agglomerates of nano-crystalline sphalerite particles were used as the matrix phase. For impregnation of the dispersed phase an organometallic precursor , namely tetra t-butyl phsophino gallane ($(t\text{-}Bu)_2P\text{-}Ga(t\text{-}Bu)_2$), was used. The liquid precursor can then be converted to solid GaP phase by thermal treatment to produce nano-scale intimate mixture of ZnS and GaP , i.e. , ZnS/GaP nano composite powder. Typical properties relevant to objectives of the present investigation is listed below.

Table 1. Selected physical and mechanical properties of ZnS and GaP

Property	ZnS	GaP
Crystal Structure	Zinc Blende	Zinc Blende
Melting Point, C	1830	1467
Density, g/cm^3	4.08	4.13
Thermal Conductivity, W/cm K	0.17	0.97
Thermal Expansion Coef. 10^{-6} K^{-1}	6.8	5.3
Hardness, GPa	1.6	8.3
Young's Modulus, GPa	74.5	102.6
Strength, MPa	96.6	102.6
Toughness, MPa $m^{1/2}$	0.5	0.8

The concept of synthesis and consolidation of ZnS/GaP composite is shown schematically in Figure 1. It is worth to note that the model described here is applicable to any system where a nano porous matrix and a liquid precursor of dispersed phase are available. In the process depicted in Figure 1, the critical step is pyrolysis of organo metallic precursor to solid phase cleanly and with

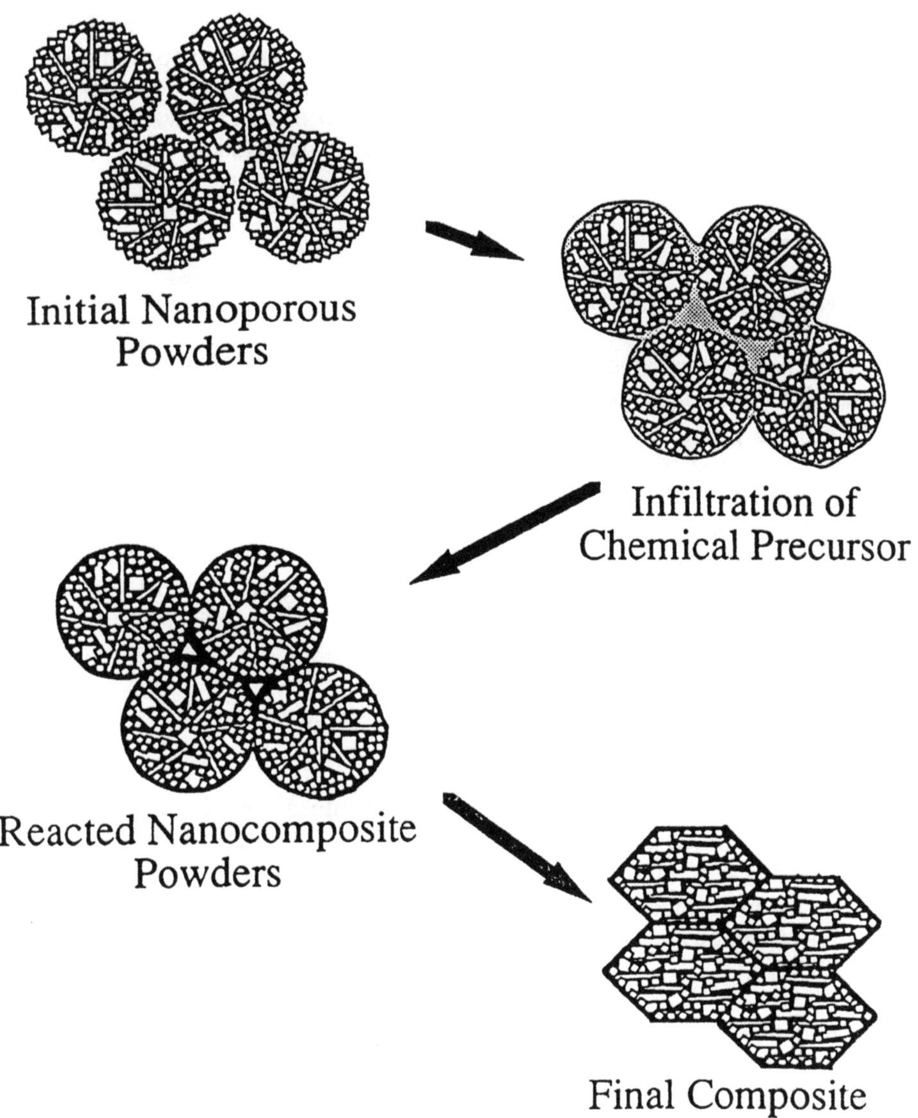

Figure 1. Schematic illustration of nanocomposite synthesis

quantitative yield. Therefore, it is essential that the precursor can be flash pyrolyzed without any considerable organic residue and without significant evolution of GaP. in order to accomplish the formation of nano composite it is also necessary that no significant chemical reaction occurs between the matrix and the precursor liquid during impregnation or flash pyrolysis.

3. Experimental

3.1 ZINC SULFIDE POWDER SYNTHESIS

Spherical nanoporous zinc sulfide powder was homogeneously precipitated by thermal decomposition of thioacetamide(5). A 0.05 M solution of zinc nitrate was acidified to a pH of 2.0 with nitric acid and heated to 70°C in a water bath. Thioacetamide was added to the solution to give a concentration of 0.8 M. The precipitation began in 4-6 minutes. After aging for 440 minutes, the solution was quenched in an ice bath and the precipitate was collected by centrifugation. The precipitate was washed with dilute ammonium hydroxide solution (pH=10.0) followed by centrifugation. Washing step was repeated until the pH of supernatant reached 10. Final wash was carried out with acetone, after which the powders were dried at 120°C.

3.2 PHOSPHINOGALLANE SYNTHESIS

Detailed procedure for the phosphinogallane synthesis is given elsewhere(6). Monomeric tert-butyl phosphinogallane, $(t\text{-}Bu)_2Ga\text{-}P(t\text{-}Bu)_2$, was synthesized by a stoichiometric reaction between $LiP(t\text{-}Bu)_2$ and $ClGa(t\text{-}Bu)_2$. Reaction was carried out in a double manifold vacuum line following the standard inert atmosphere and Schlenk techniques(7).

3.3 ZnS/GaP COMPOSITE POWDER SYNTHESIS

Synthesis of ZnS/GaP composite powders(8) was carried out in the aforementioned vacuum line. Zinc sulfide powder (20 g) was placed in a pyrex tube and outgassed at 400°C until the pressure stabilized at 0.02 Torr. While cooling the tube to room temperature, phosphinogallane stored in a glove box was weighed out to provide various GaP contents (0.5-10.0 weight % in the ZnS/GaP composite), then it was transferred into a 50 mL flask with a side arm. A septum was inserted into the flask joint and the flask was taken out of the glove box. Diethyl ether (10-15 mL), which was dried over sodium metal and outgassed just before use at liquid nitrogen temperature to remove oxygen, was added to the flask through the septum to dissolve phosphinogallane. The solution was injected to the Pyrex tube which contained ZnS powder and the solvent was vacuum evaporated at -15°C. To convert the phosphinogallane to GaP, the mixture was flash pyrolyzed at 600°C for 30 minutes in a flowing argon atmosphere.

3.4 DENSIFICATION OF ZNS/GAP COMPOSITE POWDERS

The powders were sieved through a 200-mesh nylon screen to break large agglomerates and compacted into pellets by dry pressing uniaxially with a 22 MPa pressure. The pellets were then cold isostatically pressed at 207 MPa. After spray coated with boron nitride to a minimum thickness of 0.1 mm, the pellets were encapsulated in Pyrex glass ampoules under vacuum. The pellets were hot isostatically pressed at 900°C and 207-276 MPa for 30-60 minutes.

3.5 CHARACTERIZATION

Structural and morphological changes brought about by heat treatment of ZnS and ZnS/GaP composite powders were studied. Powders were sealed in quartz ampoules at 0.05 Torr and heat treated for 24 hours at temperatures ranging from 500° to 900°C with 100°C intervals. The powders were characterized with XRD, SEM, and TEM. Using x-ray diffraction line broadening, average crystallite size was calculated. From the TEM image of the densified pellets, the average grain size was obtained using the intercept method to compare with that from x-ray line broadening. Young's moduli of the dense composites were measured from ultrasonic technique using longitudinal and shear velocities. Room temperature hardness and fracture toughness were evaluated by indentation method using 1.0 kg and 0.5 kg loads for 15 seconds.

4. Results And Discussion

4.1 ZnS and ZnS/GaP COMPOSITE POWDERS

SEM micrographs and XRD traces of the as-prepared and heat treated powders are shown in Figure 2 and 3, respectively. As-precipitated ZnS powders were spherical agglomerates of nano-crystallites. From the x-ray line broadening, the average crystallite sizes were estimated to be 8 nm for the as-precipitated ZnS and 12 nm for the as-prepared composite powders. Coarsening of the ZnS powders was observed at temperatures as low as 500°C. At 700°C, the initial powder morphology was lost and grains as large as few micrometers were observed. The average crystallite sizes were estimated to be 24 nm at 500°C and

41 nm at 600°C. Composite powders of ZnS/GaP, however, did not undergo any noticeable changes up to 600°C. Average sizes were 17, 29, and 86 nm at 700°, 800°, and 900°C, respectively. The effect of GaP impregnation on the grain coarsening behavior is quite significant. *In-situ* heating experiments using TEM equipped with a hot stage suggested that grain growth of nanocrystalline ZnS powder is due to material transport via fluid phase(9). Since ZnS crystallites are covered with GaP in the composite powders, material transport via vapor phase in composite is less significant than it is in pure ZnS powders.

Following hot isostatic pressing (HIPping), all samples showed relative densities greater than 95% with an average value of 96.3 % as determined by the liquid immersion method. Figure 4 is a TEM micrograph of HIPped composite with 3.0 wt.% GaP, where the nano-sized microstructure is evident. The average grain size was estimated to be 34 nm by the interception method(10), while grain size using x-ray line broadening was 20 nm. For the remainder of the work reported here, grain sizes obtained from XRD line broadening are used.

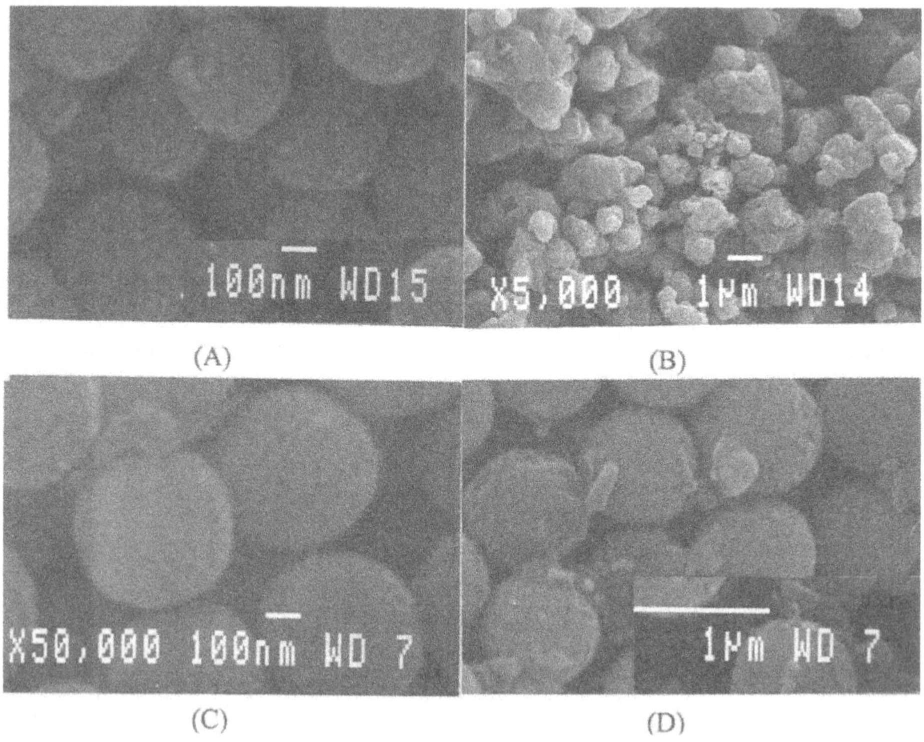

Figure 2. SEM micrographs of (A) ZnS (B) ZnS at 700°C (C) ZnS/GaP (D) ZnS/GaP at 700°C

The room-temperature values for hardness, Young's modulus, and indentation fracture toughness along with average grain size are tabulated in Table 2. The average grain size gradually decreased with an increasing GaP content. It can be seen that as GaP content increases, the Vicker's hardness value increases, which is accompanied by a corresponding decrease in grain size, while fracture toughness remains fairly constant. At ≈ 3 wt.% GaP, the hardness values are almost twice as high as that of

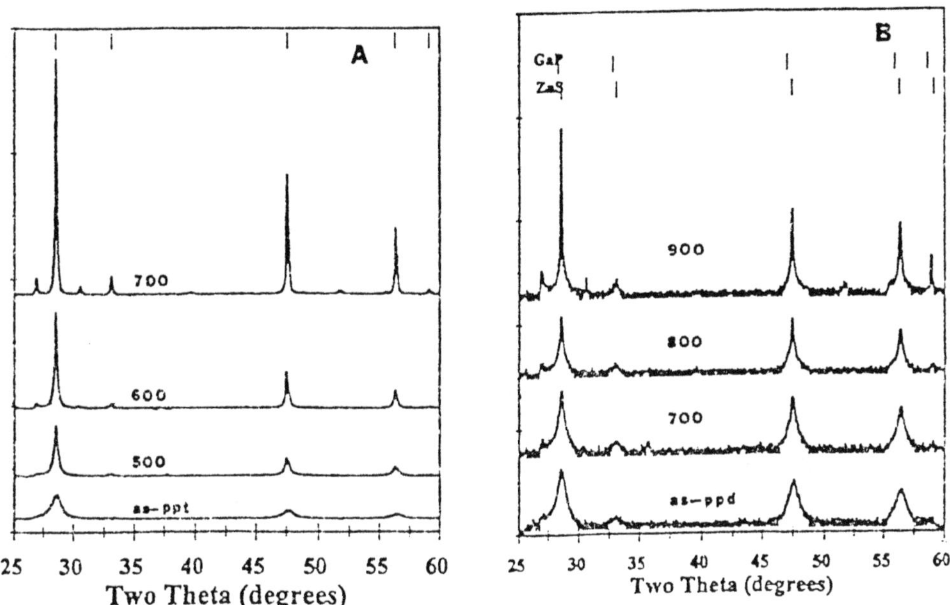

Figure 3. XRD traces of (A) ZnS (B) ZnS/GaP powders

pure ZnS. Plastic deformation in ceramics at ambient temperature mainly occurs via dislocation generation and its movement. As grain size decreases, the chance for dislocation pileup or pinning at the grain boundaries increases, thus increasing the hardness. This is the basis for the well known Hall-Petch relationship. However, Siegel(11) argued that as the grain size further decreases into the nano size regime, conventional dislocation generation and its movement become very difficult since the grain size is smaller not only than the critical bowing lengths necessary for a grain to serve as a dislocation source, but also than the typical spacing between dislocations in a pileup. Therefore, it appears that the hardness increase in nanophase material may be due to lack of an easier medium for plastic deformation, i.e. dislocation.

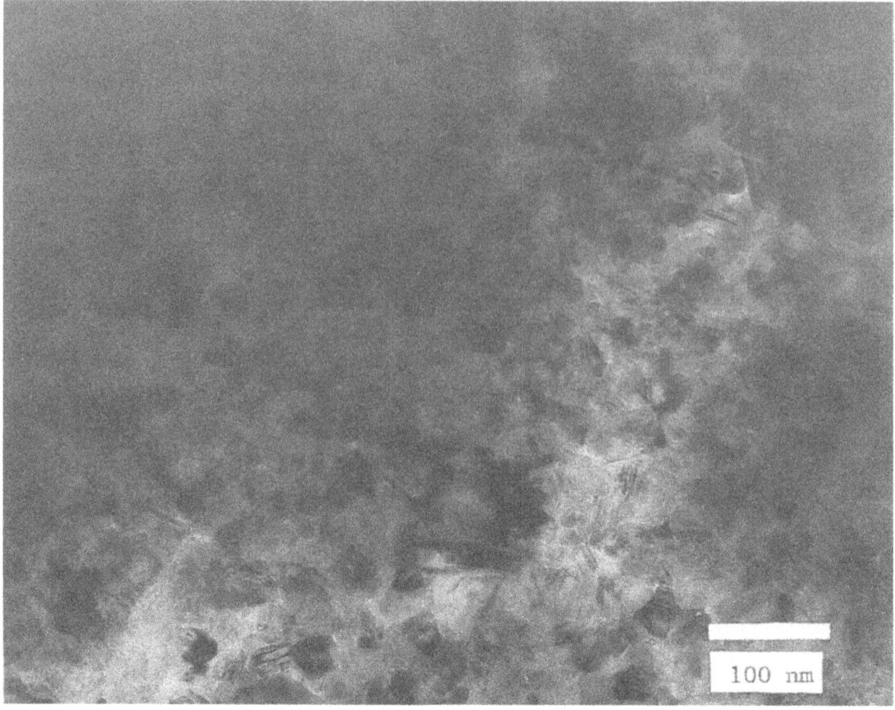

Figure 4. TEM bright field image of ZnS/GaP nanocomposite with 3.0 wt.% GaP HIPped at 900°C and 276 MPa for 30 minutes.

5. Summary

The preparation, characterization, and hot isostatic pressing of zinc sulfide/gallium phosphide composite powders wre investigated. Monodisperse, spherical, submicron, nanoporous ZnS powder was used as a matrix for the nano composite synthesis. GaP was introduced into the porosity utilizing phosphinogallane solution. Composite powders showed significant grain growth retardation compared to pure ZnS. After HIPping, dense, ultrafine microstructures (as low as 16 nm) were obtained. Hardness and fracture toughness were measured by an indentation technique. While fracture toughness remained comparable to ZnS, hardness was almost doubled by addition of 3.2 wt.% of GaP. Although the exact nature of hardening mechanism is not understood, the lack of an easy medium for plastic deformation, i.e. dislocation, appears to be responsible for the hardness increase.

Table 2. Room-temperature mechanical properties of ZnS and ZnS/GaP nanocomposites.

GaP Content (wt.%)	density (%)	E (GPa)	d (nm)	H (GPa)	K_{IC} $MPa \cdot m^{1/2}$
CVD ZnS[A]	> 99	79	5-60 μm	1.38 ±0.49	-
HP ZnS[B]	99.5	78	>500	2.41±0.23	0.57±0.09
0.51	94.8	67	30±3	3.40±0.04	0.39±0.09
0.64	95.1	70	31±2	3.32±0.04	0.48±0.05
1.20	95.3	70	24±2	3.63±0.07	0.57±0.11
3.00	96.8	73	20±1	4.23±0.05	0.51±0.10
3.20	96.9	74	16±2	4.31±0.08	0.46±0.04
GaP[C]	-	103	single xtal.	8.28	0.80

A. ClearTran: a commercial CVD grown ZnS disc (Wilmad Glass, Buena, NJ)
B. Hot pressed ZnS using homogeneously precipitated powder, 900°C 105 MPa.
C. Data taken in the (100) plane(12).

6. Acknowledgment

This work was supported in part by the Office of Naval Research and Iowa State University.

7. References

1. Saunders K.J. Wong, T.Y., Harnett,T.M., Tustison, R.W., and Gentilman, R.L., (1986) "Current and Future Development of Calcium lanthanum Sulfide," pp. 72-78 in *Infrared and Optical Transmitting Materials, Proceedings of SPIE* Vol. 683 (San Diego, CA, August, 1986). Edited by R.W. Schwartz. SPIE-The International Society for Optical Engineering, Bellingham, WA.

2. Zhang, J. , Chen, W.W.., and Ardell, A.J. (1990) "Solid Solution Strengthening of ZnS," pp. 93-98 in *Windows and Dome Technologies and Materials II, Proceedings of SPIE* Vol. 1326 (San Diego, CA, July 1990). Edited by P. Klocek. SPIE-The International Society for Optical Engineering, Bellingham, WA.

3. Xue, L.A. and Raj, R. (1991) "Grain Growth in Superplastically Deformed ZincSulfide/Diamond Composites," *J. Am. Ceram. Soc.*, **74**[7], 1729-32.

4. Niihara, K., Nakahira, A., and Sekino,T. (1993) "New Nanocomposite Structural Ceramics," pp. 405-12 in *Nanophase and*

Nanocomposite Materials, MRS Symposium Proceedings, Vol.286 (Boston, MA, December 1992). Edited by S. Komarneni, J.C. Parker, and G.J. Thomas. MRS, Pittsburgh, PA.

5. Celikkaya, A., and Akinc, M. (1990) "Preparation and Mechanism of Formation of Submicrometer Zinc Sulfide Powders," *J. Am. Ceram. Soc.*, **73**[8], 2360-65.

6. Naiini, A.A., Han, Y., Akinc, M., and Verkade, J.G., (1993) "Synthesis and Pyrolysis Studies of [(t-Bu)2GaPR2]x," *Inorg. Chem.*, **32**[23], 5394-95.

7. Shriver, D.F. and Drezdon, M.A. (1986). *The Manipulation of Air Sensitive Compounds*, pp.7-167. Wiley and Sons, N.Y. , N.Y.

8. Han, Y., Voiles, E.T. , Naiini, A.A., Chumbley, L.S. , Verkade, J.G. , and Akinc, M., (1994) "Synthesis and Characterization of Zinc Sulfide/Gallium Phosphide Nano-composite Powders," J. Am. Ceram. Soc., **77**[12], 3153-60.

9. Voiles, E.T. , Chumbley, L.S., Han, Y., and Akinc, M., (1994) "TEM Examination of Heat-Treated Nanocrystalline ZnS," *J. Am. Ceram. Soc.*, **77**[12], 3173-76.

10. ASTM Standard Test Methods for Determining Average Grain Size (E112), (1989), *Annual Book of ASTM Standards,* Vol.03.011, American Society for testing and Materials, Philadelphia, PA.

11. Klocek, P. , and Weirauch, D.F., (1990) "Pressure Sintering of III-V Compounds," pp. 105-18, in *Windows and Dome Technologies and Materials, Proceedings of SPIE* Vol. 1112 (Orlando, FL, March 1989), Edited by P.Klocek, SPIE-The International Society for Optical Engineering, Bellingham, WA.

12. Siegel, R.W. , (1993) "Synthesis and Properties of Nanophase Materials," pp.189-97 in *Materials under Extreme Conditions and Nanophase Materials, Proceedings of Symposium C of E-MRS* (Strasbourg, France, November 1992), edited by G. Kostorz, I. Gorynin, V. Trefilov, H.D. Kunze, B.Gunther, B. Skorokhod, and V. Troitsky, North-Holland, N.Y. , N.Y.

MAGNETIC AND TRANSPORT PROPERTIES OF NANOPARTICLES CONTAINING FERROCOLLOIDS

E. BLUMS
Institute of Physics, Latvian Academy of Sciences
Salaspils-1, LV-2169, Latvia

Abstract

The main concepts of magnetic properties of ferrofluids containing polydisperse particles and some new problems of nanoparticle transport properties in colloids in the presence of an external magnetic field are discussed in the paper. The following problems are considered: i) the Langevin-type magnetization of an assemble of polydisperse particles in one-particle approximation; ii) the particle magnetic interaction effects and problems of magnetogranulometry analysis; iii) the nanoparticle effective translation diffusion coefficient in the presence of uniform magnetic field; iv) thermomagnetophoretic particle transfer in nonisothermic ferrocolloids; v) nanoparticle separation in thermodiffusion column in the presence of high temperature gradients.

1. Introduction

The success made in last two decades in preparing stable colloidal dispersions of magnetic nanoparticles in various carrier liquids has evoked great interest of scientists dealing with the physico-chemical, physical and engineering problems of magnetic fluids. In the early stage of investigation theoretical models of magnetic fluids were based on a single-particle concept according to which the colloid has been macroscopically considered as a system containing monodisperse non-interacting magnetic dipoles. This concept enabled to interpret the qualitative level of main physical properties of magnetic fluids. Starting with the early 1980s many experimental results have been published which are not only in a quantitative but also in a principal qualitative disagreement with the single-particle concept. These results stimulated theoretical investigations of

E. Pelizzetti (ed.), Fine Particles Science and Technology, 543–556.

particle magneto-dipole interactions as well as these of phase transition into a magnetically ordered state of magnetic colloids. The magnetic relaxation and ordering processes determine a wide range of hydrodynamic phenomena which depend on both the internal structure of ferrocolloid and the relaxation kinetics of magnetization.

Real magnetic fluids are not well defined systems. There exists a significant dependence of the stability and properties of ferrofluids on their preparation method and stabilization procedure. The chemical nonhomogeneity of nanoparticles, properties of surfactant adsorbtion layer, characteristics of particle polydispersedness as well as other uncontrollable parameters may give a significant difference in the ferrofluid properties from sample to sample. The most popular method of preparation of ferrofluids is the chemical coprecipitation method which allow to produce ferrite nanoparticles of various sizes and different compositions. Temperature-sensitive magnetic fluids containing complex ferrite nanoparticles of low Curie temperatures for some new applications. are of special interest.

In the present paper a review of some principal problems of physical properties of polydisperse magnetic colloids is given. Besides the magnetic properties, special attention is paid to problems of mass transfer which play an important rule to the long-term stability of devices employing magnetic fluids.

2. Magnetic Properties of Ferrocolloids

The diameter of particles in ferrofluids usually does not exceed 10-12 nm. Thus, the colloidal particles of conventional ferro-and ferrimagnetic materials (Fe, Co, Ni, Fe_3O_4) are single-domain with a magnetic moment m proportional to the saturation magnetization of bulk material M_s and to the particle volume V: $m=M_sV$. According to single-particle approximation, the magnetization curve $M(H)$ of ferrocolloids regardless on the magnetic relaxation mechanism (Brownian or Neel type) obeys the Langevin function

$$M(H) = NmL(\xi), \quad L(\xi) = \coth\xi - \frac{1}{\xi}, \quad \xi = \frac{\mu_0 mH}{kT} \tag{1}$$

with H being the magnetic field strength, N the particle concentration, k the Boltzman constant and T the absolute temperature. For a polydisperse system this relations transforms to the integral

$$M(H) = \varphi M_s \int_0^\infty f(m) L(\frac{\mu_0 mH}{kT}) dm\, , \quad \int_0^\infty f(m) dm = 1 \tag{2}$$

where $\varphi=NV$ is the volume fraction of magnetic phase and $f(m)$ characterizes the distribution of particle magnetic moments proportional to the particle size.

2.1. MAGNETOGRANULOMETRY OF POLYDISPERSE COLLOIDS

The above relationship is a theoretical basis for the granulometric analysis of polydisperse ferrocolloids. The function $f(m)$ may be calculated from measurements of the integral magnetization curve by using a variation method of regularization (see, for example, Ref. [1]). To perform real calculations, the magnetization curve, instead of integral (2), is approximated by sum of Langevin functions with different discrete values of magnetic moments m_i:

$$M(H) = \sum_i N_i m_i L(\frac{\mu_0 m_i H}{kT}). \quad (3)$$

For diluted ferrocolloids of well-stabilized ultrafine magnetite or meghemite particles prepared by chemical coprecipitation method, the above analysis results usually for a log-normal form of function $f(m)$ which agrees satisfactory with the particle size distribution curve determined from direct electron microscope measurements. Here, a definite disagreement between the mean values of "magnetic" d_m and physical d diameters of particles has been observed frequently. Some authors interpret the result $d_m<d$ as a decrease of M_s for ultrafine particles [2], but the idea of particle nonmagnetic boundary layer formation due to a chemosorbtion of surfactant molecules on the particle surface [3] seems to be more likely.

Magnetic fluids of high particle concentration, in particular those with large particles, do not obey the Langevin-type magnetization curve (1). The particle interactions in colloids are often described by use of so called mean field approximation (see, for example, [4]). According to the Weiss model, the particle is subjected to an effective magnetic field $H_e=H_0+\lambda M$ where the constant λ usually is expected to be equal to $\lambda=4\pi/3$ (the local Lorentz field model). This model predicts the Curie-Weiss law of initial magnetic susceptibility κ.. For small ξ when $L(\xi)=\xi/3$ from (1) it follows that

$$\frac{1}{\kappa} = \frac{3kT}{\mu_0 N m^2} - \lambda, \quad (4)$$

i.e., an ordering temperature $T_C{}^*.=\lambda\mu_0 N m^2/3k$ is expected to be observed. Many experiments, (see for example [5]) in an interval near the room temperature confirm the relation (4) and give the $T_C{}^*$ values near to the theoretically predicted ones. Nevertheless, there have been published some results of magnetization measurements showing a remarkable deviation from the classical Curie-Weiss behavior [6]. Moreover, the abrupt increase of magnetic susceptibility near the region of phase transition into a magnetically ordered state at ordering temperature $T_C{}^*$ has not been confirmed experimentally. Therefore, in the low temperature region, the Onsager model of magnetic interactions, according to which the initial susceptibility remains finite at all temperatures and particle concentrations, seems to be more suitable. In ref. [7] it is shown that magnetic properties of highly concentrated magnetite ferrofluids may be described with a very high accuracy by using a mean-spherical model which was firstly developed to interpret the properties of liquid dielectrics.

In the hydrodynamics of magnetic fluids, an important rule plays not only the value of stationary fluid magnetization but also its relaxation time. There are two mechanisms of magnetic relaxation: the orientation of a magnetically "hard" particles in the presence of magnetizing field (the Brownian mechanism) and the supperparamagnetic relaxation of particle magnetic moments (the Neel mechanism). The Brownian relaxation time τ_B depends on particle volume V and on the viscosity of surrounding liquid η:

$$\tau_B = \frac{3\eta V}{kT} , \tag{5}$$

whereas the Neel relaxation is determined by V and by the magnetic anisotropy constant of particle material K:

$$\tau_N = \tau_0 \exp(\frac{KV}{kT}) . \tag{6}$$

When $\tau_B << \tau_N$, the Brownian mechanism of magnetic relaxation dominates whereas for colloids with $\tau_N << \tau_B$ the Neel type relaxation prevails. Due to the polydispersity of real colloids there exists a wide range of relaxation times. Spectrum analysis performed by use of Cole-Cole diagrams show that in cobalt ferrite colloids, when due to high value of K the Brownian relaxation mechanism dominates, the experimentally detected dependence of relaxation time on magnetic field strength agrees very well with the theoretical one which corresponds to the mean field approximation of particle interactions [8].

2.2. TEMPERATURE SENSITIVE MAGNETIC COLLOIDS

Usually, magnetic fluids for commercial application are prepared by using magnetite nanoparticles. The saturation magnetization of Fe_3O_4 is less that that of ferromagnetic metals Fe and Co, but magnetite is chemically stable in different carrier liquids, and nanoparticles may be produced by using the relatively simple chemical coprecipitation method. In last years, there is evoked a practical interest to magnetic colloids containing other complex ferrites. Some mixed ferrites are proposed to be used to prepare so called temperature sensitive magnetic fluids having high value of pyromagnetic coefficient $\beta_m = M^{1} \partial M / \partial T$.

It should be noted that β_m depends on both, on the Curie-Weiss behavior of nanoparticle assemble as well as on the pyromagnetic coefficient of particle material $\beta_p = m^{-1}\partial m/\partial T$. In low magnetic fields $H << \lambda Nm$ when according to (1) and (4)

$$\beta_m^0 = -\frac{1}{\kappa_0}\frac{\partial \kappa_0}{\partial T} = \frac{1}{(1-\lambda\mu_0 Nm^2/3kT)}\left(\frac{1}{T}-\frac{2}{m}\frac{\partial m}{\partial T}\right), \qquad (7)$$

the dependence of fluid magnetization on temperature, especially for fluids of high particle concentration N, is determined mostly by the ordering temperature T_C*, whereas the effect of pyromagnetic coefficient of the particle material on β_m is relatively small. On the contrary, in the presence of high magnetic field $H >> Nm^2$ when the colloid magnetization is close to the saturation regime $M = Nm(1-\xi^{1})$ instead of (7) we obtain

$$\beta_m^\infty = \left(\frac{1}{\xi T}-\frac{1}{m}\frac{\partial m}{\partial T}\right)\frac{\xi}{1-\xi} \approx \beta_p, \qquad (8)$$

So, the pyromagnetic coefficients of magnetically saturated colloids ($\xi \to \alpha$) is equal to that of particle material β_p.

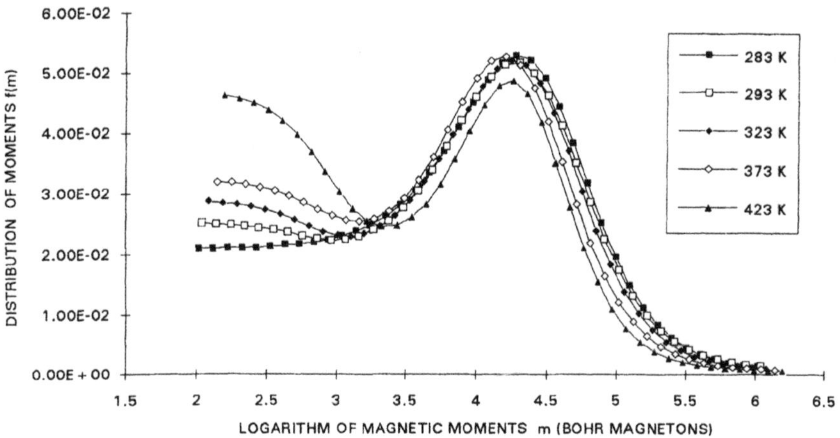

Figure 1. Distribution of magnetic moments of polydisperse Mn-Zn ferrite colloids, x=0.5 [12]

The most popular material for temperature-sensitive ferrofluids is Mn-Zn ferrite $Mn_xZn_{1-x}Fe_2O_4$ having low Curie temperature (T_C varies with x in the interval between 70 C and 300 C). It is interesting to note that that there exist some difference in thermomagnetic behavior of ferrofluid samples and that of bulk ferrite material. Magnetization measurements [9;10] do not indicate any exactly defined Curie temperature T_C of ferrofluid samples even at x values for which according to [11] T_C is expected to be near the 100 C. An asymptotic character of the magnetization temperature dependence curve close to the exponential one $M=M_0\exp(-k_T(T-T_0))$ was observed. The increase of k_T values with phase ratio x agrees qualitatively very well with the dependence of Curie temperature on x which has been found for the bulk ferrite material [11].Numerical values of k_T are significantly higher than the values of the corresponding pyromagnetic coefficient of magnetite $\beta_m^{\infty}=0.2\times10^{-3}$. For example, for magnetic fluid sample containing $Mn_{0.3}Zn_{0.7}Fe_2O_4$ nanoparticles it was found that the coefficient k_T of exponential decrease of magnetization in increasing temperatures is equal to $k_T=6\times10^{-3}$ [9].

Magnetogranulometry measurements show that, contrarily to the magnetite based ferrofluids, for colloids containing Mn-Zn ferrite nanoparticles, a peculiar bimodal curves of magnetic moment distribution usually are observed (see Fig. 1). Moreover, only the spectra region of low magnetic moments (paramagnetic particles) is sensitive to the temperature changes, whereas the effect of temperature on the second spectra maximum in high m region is relatively small.

In ref. [9] an attempt is made to investigate magnetic properties of both spectra regions separately. Centrifugal separation of ferrofluid sample gives the possibility to analyze different fractions of ferrofluid particles. Figure 2 represents the magnetic moment distribution curve of ferrofluid containing $Mn_{0.5}Zn_{0.5}Fe_2O_4$ particles as well as the corresponding spectra of heavy and light fractions separated for 3 hours in an ultracentrifuge at 50000 g. Sum of log-normal magnetic moment distribution curves of both fractions agrees very well with the bimodal spectra of initial sample.

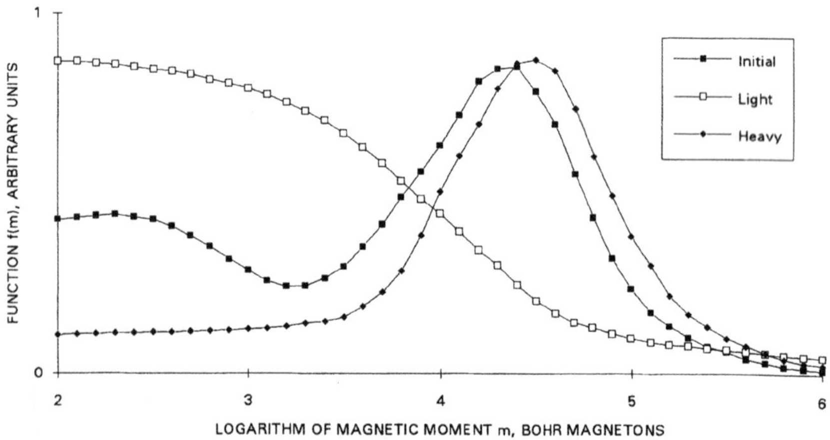

Figure 2. **Magnetogranulometry spectra of ferrofluid containing $Mn_{0.5}Zn_{0.5}Fe_2O_4$ particles before and after treatment in centrifuge (3 hours at 50000 g) [9].**

From X-ray diffraction spectra measurements it was found that the mean diameter of particles of the light fraction is d_l=3 nm whereas that of the heavy one is d_h= 13 nm. The corresponding particle mean volume V ratio $(d_h/d_l)^3$ value 81 is approximately 3 times less than the value of mean magnetic moment ratio m_h/m_l=240 (see Fig. 2). Calculations of saturation magnetization M_s made wit the respect of relation $m=M_sV$ give different values of the magnetic moment per molecule for both fractions: n_l=0.53 μ_B (Bohr magnetons) and n_h=1.63 μ_B. Both of them are essentially smaller than n=6.9 μ_B which corresponds to the bulk Mn-Zn ferrite material at x=0.5 [11]. Similar results for magnetite and some other ferrites have been reported in [2]. Most probable causes of decrease of the magnetization of small particles are the magnetic degradation of the particle surface or the changes in properties of surface layer due to the physico-chemical action of surfactants [2;13] Measurements of thermomagnetic properties of both fractions show the higher value of pyromagnetic coefficient for the light fraction comparing with that for the heavy one [9]. It is suspected that this is caused by a difference between Langevin parameters for both fractions. At H=20 kOe at which the pyromagnetic properties are measured, the magnetic saturation of light fraction is not reached and ordering processes (see (7)) play a significant rule. Contrarily, the heavy fraction is magnetically saturated and its pyromagnetic coefficient according to the relation (8) is determined mostly by properties of the particle material.

3. Mass Transfer in Ferrocoilloids Under the Effect of Magnetic Field

3.1. THE MAGNETOPHORESIS OF NANOPARTICLES

The stability of magnetic fluids is closely linked with problems of mass transfer of colloidal particles. It is assumed that the gravitation sedimentation of nanoparticles is negligible and the main processes of particle transfer are their Brownian diffusivity and magnetophoresis. In the presence of strong temperature gradients thermophoresis also may play a significant rule. For Brownian particles the diffusion coefficient obeys the relation $D=kT/f_v$, where f_v is the coefficient of hydrodynamic drag for spherical particles of radius a equal to $f_v=6\pi\eta a$ (η is the viscosity of surrounding liquid). Under such approximation for colloids of low particle mass concentration $n=Nm_g$, the mass flux usually is written in the form of

$$j=-D\nabla n+\frac{m_g}{f_v}\mu_0(\overline{M}-\overline{M}_l)n\nabla H-nD\frac{\alpha_T}{T}\nabla T \qquad (9)$$

with m_g being the particle mass, $\bar{M}$ and $\bar{M}_l$ the specific magnetization of particle and that of surrounding liquid, α_T the thermodiffusion ratio. The second term in (9) describe the magnetophoretic velocity of particles $\mathbf{u}_m$ which may be calculated also from the macroscopical viewpoint [3]. The classical Stokes approximation gives the velocity of spherical particle equal to

$$\mathbf{u}_m = \frac{2a^2}{9\eta}\mu_0(M - M_l)\nabla H \tag{10}$$

which agrees with the corresponding term in (9). Classical Stokes' problem assuming the potentiality of particle driving force is correct only for homogeneous and isothermic magnetic fluids. If temperature and particle concentration gradients ∇T and ∇n in fluid are present, it is necessary to take into account in the Stokes problem the nonpotentiality of magnetic force caused by both, the temperature and the concentration nonhomogenities of colloid.

Firstly, such analysis is performed in Ref. [14] considering the thermomagnetic particle transfer in the presence of an uniform external magnetic field $\mathbf{B}=\mu_0(\mathbf{M}+\mathbf{H})$ oriented parallel to ∇T. For $\mathbf{B}=const$ it is found [3;14]:

$$\mathbf{u}_m = -\left(\frac{8K_\mu}{15} - \frac{K_\mu^2}{20}\right)\frac{\mu_0 H^2 a^2}{\eta}\frac{\partial M}{\partial T}\nabla T. \tag{11}$$

Here K_μ is the particle magnetic saturation coefficient equal to $K_\mu=(\mu_p-\mu)/(\mu_p+2\mu)$ with μ_p being the magnetic permeability of particle and μ that of a surrounding liquid. In further calculations [15] it is shown that in the presence of arbitrarily orientated magnetic field the particles are transferred along the ∇T as well as along the $\mathbf{H}$. Recently, a more detailed analysis of particle magnetophoresis was been performed additionally taking into consideration both, the nonuniformity of an external magnetic field and the temperature perturbations caused by difference between thermal conductivities of particle λ_p and of surrounding liquid λ [16]. The magnetophoretic velocity of spherical particle is found to be the following:

$$\mathbf{u}_m = \frac{2\mu_0 a^2}{9\eta}\{\mu(\mathbf{H}\nabla)\mathbf{H}3K_\mu - \mathbf{H}(\mathbf{H}\nabla)\mu[\frac{3}{5}K_\mu + \frac{3}{80}K_\mu^2 - K_\lambda(\frac{9}{40}K_\mu + \frac{1}{35}K_\mu^2)] +$$

$$H^2\nabla\mu[\frac{6}{5}K_\mu + \frac{21}{80}K_\mu^2 - K_\lambda(\frac{3}{40}K_\mu + \frac{2}{35}K_\mu^2)]\}$$

where $K_l=(\lambda_p-\lambda)/(\lambda_p+2\lambda)$. (12)

If particle transfer in an uniform external magnetic field $\mathbf{B}$=const is considered, from (12) it follows that

a) when $\mathbf{B}$ is parallel to the magnetic permeability gradient ∇m:

$$\mathbf{u}_m = -\frac{2\mu_0 a^2 H^2}{9\eta}\nabla\mu \frac{12}{5}K_\mu f_1(K_\mu, K_\lambda),$$
$$f_1(K_\mu, K_\lambda) = 1 - \frac{3}{32}K_\mu - \frac{K_\lambda}{16}(1 - \frac{4}{21}K_\mu); \tag{13}$$

b) when **B** is orientated perpendicularly to $\nabla\mu$:

$$\mathbf{u}_m = +\frac{2\mu_0 a^2 H^2}{9\eta}\nabla\mu \frac{6}{5}K_\mu f_2(K_\mu, K_\lambda),$$
$$f_2(K_\mu, K_\lambda) = 1 + \frac{7}{32}K_\mu - \frac{K_\lambda}{16}(1 - \frac{16}{21}K_\mu). \tag{14}$$

3.2. TRANSLATION DIFFUSION AND THERMOMAGNETOPHORESIS

Nonhomogenity of magnetic field in a colloid is caused by temperature as well as by concentration gradients. Let us consider for simplicity the one-dimensional gradient $\nabla = d/dx$. For **B**=*const* from Maxwell equation div**B**=0 it follows [3]:

$$H^2\frac{d\mu}{dx} = \frac{\mu H}{1 + \partial M/\partial H}(\frac{\partial M}{\partial n}\frac{dn}{dx} + \frac{\partial M}{\partial T}\frac{dT}{dx}) \ . \tag{15}$$

Assuming that the magnetization of a fluid may be approximated by Langevin formula (1) and taking into consideration the relations (13), (14) and (15) one can see that the mass flux (9) may be written in the form

$$\mathbf{j} = -D_m\frac{dn}{dx} + D\frac{\alpha_{Tm}}{T}\frac{dT}{dx}n \tag{16}$$

where D_m and α_{Tm} now are the effective translation diffusion coefficient and the effective thermodiffusion ratio. Both of them are effected by a magnetic field. Besides, the effect depends also on the mutual orientation between **H** and ∇n. When **B** is oriented parallel to ∇n, theory predicts an increase of the effective diffusion coefficient of magnetic nanoparticles [17]:

$$D_m^p = D(1 + \frac{12}{5}K_\mu A f_1(K_\mu, K_\lambda)) \ , \tag{17}$$

(Coefficient A depends on particle concentration as well as on magnetic field H: $A = \mu\varphi\xi L(\xi)/(1 + \varphi\xi M_s L'(\xi)/H))$. On the contrary, magnetic field oriented perpendicularly to ∇n causes a decrease of D_m^n:

$$D_m^n = D(1 - \frac{6}{5}K_\mu A f_2(K_\mu, K_\lambda)) \ , \tag{18}$$

Thermodiffusion ratio α_{Tm} also depends on the mutual orientation of **B** and ∇T:

a) **B** parallel to ∇T:

$$\alpha_{Tm}^p = \alpha_T - \frac{12}{5}K_\mu A f_1(K_\mu, K_\lambda)\frac{1}{M}\frac{\partial M}{\partial T} \ ; \tag{19}$$

b) **B** perpendicular to ∇T:

$$\alpha_{Tm}^{n} = \alpha_T + \frac{6}{5} K_\mu A f_2(K_\mu, K_\lambda) \frac{1}{M} \frac{\partial M}{\partial T} \quad (20)$$

It is necessary to note that coefficient K_μ depends on n and H. Accounting for a superparamagnetic law (1) we obtain

$$K_\mu = \frac{(1-\varphi) M_s L(\xi)}{3H + (1+2\varphi) M_s L(\xi)} \quad (21)$$

By taking into account that above analysis in the one particle approximation is valid only for low values of particle concentration $\varphi \ll 1$, we can see from (21) that K_μ varies from $K_\mu = 1$ at low H to $K_\mu \approx M_s/3H_0 < 1$ for high magnetic fields $H_0 > M_s$. The magnetophoretic summand in equation (19) for $K_\lambda = 0$ agrees with the results reported in Refs. [3] and [14]. Dependence of coefficients f_1 and f_2 on the coefficient K_λ, which varies from $K_\lambda = 1$ to $K_\lambda = -1/2$, is relatively small.

Dependence of effective diffusion coefficient on magnetic field and on its orientation is observed in experiments on forced Rayleigh scattering in ferrocolloids [18]. Relations (17) and (18) are in rather good agreement with these experiments and with the authors thermodynamic theory accounting for particle interactions.

Numerical estimates of α_{Tm} lead to the conclusion that the thermophoretic velocity of micron sized particles in suspensions may exceed their sedimentation velocity. For nanoparticles in colloids it is expected that conventional thermophoresis could prevail in comparison with the thermomagnetophoretic transfer which in the low particle size region is calculated to be very small. Nevertheless, the predicted values of α_{Tm} are greater than those of conventional thermodiffusion in molecular gases and solutions.

In Ref. [19] the magnetic nanoparticle diffusion coefficient is analyzed by using the Batchelor model with the account for interparticle interactions. Besides the steric and the hydrodynamic forces a significant rule in magnetic fluids play also the magnetic dipole interaction. The last one causes an anisotropy of transport properties. Diffusion coefficients in the presence of a longitudinal or a transverse magnetic field are the following:

$$D_m^p = \frac{kT}{6\pi a \eta}[1 + 1.45 n V_p - \frac{4\pi}{3}\lambda \frac{\mu_0 n m^2 H_0}{3kT} + 8.92\pi \frac{\mu_0 n m^2 H_0}{3kT} L^2(\xi)], \quad (22)$$

$$D_m^n = \frac{kT}{6\pi a \eta}[1 + 1.45 n V_p - \frac{4\pi}{3}\lambda \frac{\mu_0 n m^2 H_0}{3kT} - 8.92\pi \frac{\mu_0 n m^2 H_0}{3kT} L^2(\xi)]. \quad (23)$$

Second terms in (22) and (23) are the conventional result of hard sphere dispersion whereas the last two terms represent the effect of magnetic interactions of particles. If the magnetization nonhomogenity in fluid is caused by a gradient of temperature, it may be shown that similar expressions describe also the effect of magnetic field on thermodiffusion ratio α_T^m. The last terms in (22) and (23) are in a qualitatively good agreement with the result of macroscopic analysis performed by using the single-particle approximation (17) and (18).

In Ref. [20] the effect of particle polydispersivity on mass transfer processes is analyzed. Numerical simulation of kinetics of gravitation sedimentation in a vertical column shows an intensive deformation of dispersion curve of particle sizes. The sedimentation of a relatively small fraction of large magnetic particles is responsible for a significant magnetic susceptibility difference between the top and the bottom of the magnetic fluid volume. It is in a good agreement with measurements of the electromagnetic inductivity changes in magnetofluid inclinometer [21].

4. Experiments on a Thermodiffusion of Colloidal Particles

Transport properties of magnetic nanoparticles play a significant rule in the problem of long- term stability of magnetic fluids. Usually, in research the attention has paid only to the gravitation sedimentation and to the magnetophoretic separation of ferroparticles. Nevertheless, in many magnetofluid devices, for example, in high speed rotary seals or in loudspeaker cooling systems when high temperature gradients are present, the long- term stability of magnetic fluid may be effected also by a thermophoretic nanoparticle transfer. It is well known that thermophoretic velocity of particles in aerosols and in gaseous dispersions may reach a value significantly higher than that in molecular gases and liquids. Some experiments, for example, these of particle grating in optical interference bands (the forced Rayleigh scattering experiment) [18], indirectly indicate a high thermophoretic mobility of particles in colloidal dispersions also.

In our knowledge, the first attempt to perform direct thermodiffusion coefficient measurements is reported in [22]. Experiments are performed by using a vertical thermodiffusion column. It includes a flat channel formed by heated and cooled side walls and upper and lower separation chambers. Due to the free convection and the thermodiffusive particle transfer across the channel a particle concentration difference in both chambers Δn develops. According to the column theory [23] the particle separation rate $d(\Delta n)/dt$ in the initial stage of separation is constant. If volumes of upper V_u and lower V_l containers are equal, $V_u=V_l=V$, from theory [24] follows a relation $\Delta n=2utS/V$ with S being the cross section of the vertical column of width d and u - the generalized particle separation velocity. which includes both the gravitation sedimentation velocity u_g and the thermodiffusive convective particle transfer velocity u_c. The later, according to the column theory [17] modified for colloids in which the particle separation is determined by mixed (thermal and concentration) convection, depend on the thermal Gr_T as well as on the concentration Gr_n Grashoff numbers: $u_c=(D/d)k(Gr_T \ +kGr_n \ )Sc/6!$ Here D is the particle Brownian diffusion coefficient, Sc is the Schmidt number and k is the nondimensional thermodiffusion parameter proportional to the thermodiffusion ratio α_T : $k=\alpha_T\Delta T/T$ (ΔT is the temperature difference between column walls).

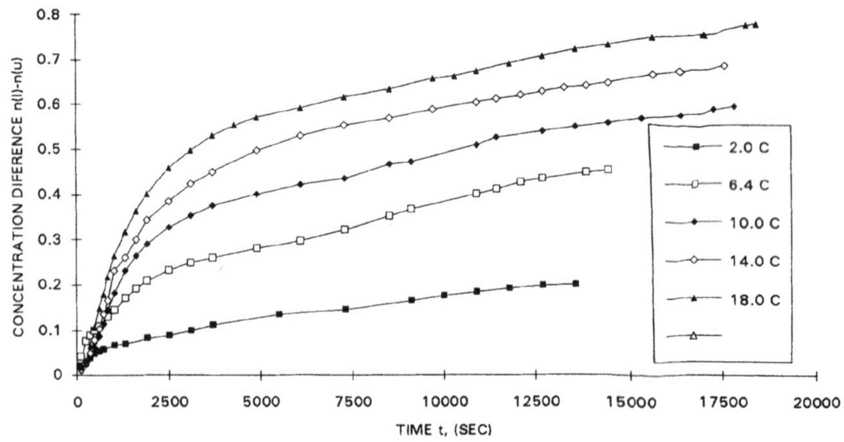

Figure 3. **Nonstationary particle separation in a thermodiffusion column at different temperature differences between column walls in tetradecane based ferrofluid containing Fe_3O_4 particles [22]**

Experiments are performed by using a column of height L=7.5 cm and of width d=0.5 mm, its volume is equal to the volume of upper and lower containers. Particle concentration during the separation experiment is determined by an inductive method using resonance frequency measurements. Induction coils are mounted inside the separation chambers. Tetradecane based ferrofluids containing magnetite particles is used in experiments. The saturation magnetization of fluid M_s = 15.2 G/g, the mean particle diameter determined by magnetogranulometry analysis $2a$ = 7.7 nm. Some results of the nonstationary separation measurements performed at different differences of temperature of column side walls are presented in Fig. 3. For ΔT=18 C a concentration difference Δn>0.8n_0 near the stationary regime of particle separation is found. Such separation rate correspond to the value of α_T approximately equal to -15. Magnetogranulometry measurements do not indicate any changes in magnetic moment distribution curves for ferrofluid portions in upper and lower containers during the separation process. It should be noted that the measured value of α_T is remarkable higher than those for molecular liquids and gases.

5. Conclusions

The polydispersivity of magnetic nanoparticles and the mass transfer in colloids play an important rule in the problem of magnetic fluid stability. Big particles in polydisperse colloids act as condensation centers which stimulate formation of aggregates or even a lose of colloid stability. Magnetogranulometry is an effective method which allow to characterize the particle polydispersivity parameters. One of the most important problem of magnetogranylometry analysis is the theoretical model of magnetization. Obviously, the mean-field approximation is valid for colloids containing very small particles of low and moderate concentration.

Bimodal magnetic moment spectra of complex ferrite colloids have to be investigated more deeply. In further investigations, it would be necessary to separate different subfractions and to analyze their properties separately. We hope that high gradient magnetic separation methods will help to perform additional measurements.

Thermodiffusive particle transfer in nonisothermic fluids is of great practical interest. In many technical devices the particle thermophoretic velocity evaluated by using the measured value of thermodiffusion ratio may reach the value approximately equal to magnetophoretic velocity caused by external magnetic field gradients. In further investigations we hope to measure the effect of magnetic field on particle diffusion and thermodiffusion coefficients.

This work is performed as part of the Magnetic fluid research program supported by Latvian Science Council, Grant No. 93.830. The experimental work on thermomagnetic properties of temperature sensitive ferrofluids is supported also by the International Science Foundation (Grant LBE000).

References

1. Maiorov, M. M. (1981) Magnetization curve of ferrofluid and distribution of particle magnetic moments, in *Proc. 10th Riga MHD Conference (Russ)* Institute of Physics, 1,191-192.
2. Sato, T.,Iisima, T., Seki, N. and Inagaki, N.(1987) Magnetic properties of ultra-fine ferrite particles, *J. Magn. and Magn. Materials*, **65**, 252-256.
3. Blums, E., Cebers, A. and Maiorov, M.,M. (1989) *Magnetic Fluids (Russ.)*, Zinatne, Riga.
4. Cebers, A. (1982) Thermodynamic stability of magnetic fluid, *Magnetohydrodynamics (Russ.)*, No. 2, 42.

5. Varlamov,Yu. D. (1987).Viscosity and magnetic susceptibility of medium-concentrated magnetic fluids, *PhD thesis*, Novosibirsk.
6. Holmes, M., O'Grady, K. and Popplewell, J. (1990) A study of Curie-Weiss behavior in ferrofluids, *J. Magn. and Magn. Materials*, **85**, 47-50.
7. Schliomis, M. I., Pshenichnikov, A. F. and Morozov, K. I. (1990) Magnetic properties of ferrocolloids, *J. Magn. and Magn. Materials*, **85**, 40-47.
8. Blums, E., Maiorov, M.M., Nikoaru, B. L. and Cebers, A. (1987) Experimental study of dispersion, depending on an external magnetic field, of magnetic susceptibility of hard magnetic particle colloid, *Magnetohydrodynamics (Russ.)*,No. 1, 53-57.
9. Blums, E., Maiorov, M.M. and Kronkalns, G. (1993) Thermomagnetic properties of ferrofluids containing chemically coprecipitated Mn-Zn ferrite particles, *IEEE Trans. Magn., MAG*-**29**, 3267-3269.
10. Maiorov, M.M. (1994) Magnetic properties of polydisperse complex ferrite magnetic colloids, in *Abstracts of the 7th Int. Conf. on Magnetic Fluids, Bhavnagar, India,* 92-93.
11. Guillaud,C. and Creveaux, H (1950) Proprietes ferromagnetiques de ferrite mixtes de cobalt et de zinc et de manganese et de zinc, *Compte Rendus*, **230**, 1458-1460.
12. Maiorov, M.M. and Kronkalns G. (1995), private communication.
13. Gangopadhyay S., Hadjipanayis, G. D., Sorensen, C. M. and Klabunde, K. J. (1993) Magnetism in ultrafine Fe and Co particles, in *Digests of the International Magnetism Conference INTREMAG-93*, Stockholm, EA-02.
14. Blums, E. (1979) Thermophoresis of particle in magnetizable suspension, *Magnetohydrodynamics (Russ.)*, No. 1, 23-29
15. Miroshnikov, V. A. (1980) Thermophoresis of ferrosuspensions in the presence of a magnetic field, in *Hydrodynamics and Heat Physics of Magnetic Fluids, Salaspils,* 177-182.
16. Naletova, V. A., Ţimotin, G. A. and Shkel, I. A. (1989) Force acting on the body immersed in non-homogeneously heated magnetic fluid, *Fluid Dynamics (English Trans.)*,**24**, 225-229.
17. Blums E. (1995) Some new problems of complex thermomagnetic and diffusion driven convection in magnetic fluids, *J. Magn. and Magn. Materials,* (to be published).
18. Bacri, J.-C., Cebers, A., Bourdon, A.,. Demouchy, D., Heegard, E. M. and Perzynski, R. (1994) Forced Rayleigh experiment in a magnetic fluid, *Phys. Rew. Letters* (to be published).
19. Morozov, K. I. (1993) The translation and rotational diffusion of colloidal ferroparticles, *J. Magn. and Magn. Materials*, **122**, 98-101.
20. Blums, E. and Chukhrov, Yu. A. (1993) Separation processes in polydisperse magnetic fluids, *J. Magn. and Magn. Materials*, **122**, 110-114.
21. Sablin, A. Yu., Chukhrov, A. Yu. and Shramek, V. B. (1991) Effect of gravitation sedimentation on the characteristics of magnetofluid inclinometer, in *Abstracts of the 6th USSR Magnetic Fluid Conference (Russ.)*, Moscow, **2**, 106-107.
22. Mezulis, A., Blums, E., Kronkalns, G. and Maiorov, M. M. (1995) Nonstationary ferrofluid particle separation in thermodiffusion column, in *Abstracts of the 14th Riga MHD Conference MAHYD-91*, Riga-Salaspils (to be published).
23. Rabinowich, G. D., Gurevich, R. J. and Bobrova, G. I. (1971) Thermodiffusion Separation of Liquid Dispersions, (Russ.), Nauka I Tehnika, Minsk.
24. Blums E., Kronkalns, G. and Ozols, R. (1983) The characteristic of mass transfer processes in magnetic fluids, *J. Magn. and Magn. Materials,* **39**, 142-146.

SELF-ASSEMBLY OF NANOSTRUCTURED SEMICONDUCTOR FILMS

JANOS H. FENDLER AND NICHOLAS A. KOTOV
Department of Chemistry, Syracuse University
Syracuse, New York 13244-4100, U.S.A.

IMRE DÉKÁNY
Department of Colloid Chemistry, Attila József University
Aradi Vt., t. 1., H-6720 Szeged, Hungary

1. Abstract

Spectroscopic (absorption and emission), microscopic (transmission electron and atomic force), electrochemical (cyclic voltammetry) and photoelectrochemical (photocurrent) measurements have provided evidence for the formation of stable ultrathin films with regular periodicities, by the layer-by-layer self-assembly of polycations, montmorillonite platelets, and cadmium sulfide, lead sulfide, and titanium dioxide nanoparticles.

2. Introduction

Self-assembly, the spontaneous attachment of molecules from a dilute solution to a substrate, has become an accepted alternative to the traditional monolayer formation in a Langmuir trough [1,2]. Particularly successful has been the covalent binding of long-chain organosilicon derivatives to polar substrates and the chemisorption of alkyl thiols, alkyl dithiols and thiol derivatives to coinage metal surfaces [1]. An appropriate balance of the interaction forces between the substrate surface and the headgroups and those between the hydrophobic regions of the molecules governs the formation of tightly packed and highly ordered monolayers. Introduction of a functionalized group at the terminal position of the organisilicon or thiol derivatives has permitted activation of the deposited monolayer by a suitable chemical

E. Pelizzetti (ed.), Fine Particles Science and Technology, 557–577.

reaction and the subsequent chemisorption of a second monolayer. Repeated activations and chemisorptions have, in turn, resulted in the step-wise construction of multilayers.

Successful demonstration of Langmuir film formation from poly-electrolytes, polystyrene microspheres, silylated glass beads, and surfactant-coated metallic, semiconductor, magnetic and ferroelectric nanoparticles has opened the door to the self-assembly of these supramolecular systems [3,4]. Thus, advantage has been taken of electrostatic interactions to self-assemble inorganic solids [5], charged colloidal particles [6], polyelectrolytes [7], and clay platelets [8] onto oppositely charged surfaces. The versatility and the ease of formation of relatively uniform three-dimensional structures consisting of monoparticulate layers have prompted the burgeoning of self-assembly.

Preliminary results, obtained in our laboratories, on the layer-by-layer self-assembly of ordered nanostructured films, composed of cationic poly(diallylmethylammonium chloride) and negatively charged solid particles, onto a variety of substrates are summarized here. Montmorillonite clay platelets (M), thiol or sodium hexamethaphosphate-stabilized cadmium sulfide (CdS), lead sulfide (PbS) and polyacrylic acid-stabilized titanium dioxide (TiO_2) were employed as anionic particles. The self-assembled films were characterized by cyclic voltammetry, absorption, emission and surface plasmon spectroscopy, transmission electron microscopy and diffraction measurements. The structures of the nanostructured films were shown to depend markedly on the potential that was applied during self-assembly and on the sequence in which the clay platelets, the CdS and the TiO_2 particles were layered.

3. Experimental Section

CdS and PbS nanoparticles were prepared in aqueous dispersions using a mixture of thiolactic acid (TLA) and ethylmercaptane (EM) as stabilizers. The optimum mixture was found by systematically varying the mole fraction of TLA to EM. Self-assembly of nanoparticles was found to be most efficient in a TLA:EM = 1:3 ratio. Typically, appropriate amounts (0.1-1.5 ml) of TLA and EM were dissolved in 100 ml of water. The pH of the solution was adjusted with 2.0 M NaOH to 9.5-10.2. The metal salt, $Cd(ClO_4)_2$ or $Pb(NO_3)_2$, was dissolved in 100 ml deionized water and added slowly to the aqueous solution of the stabilizers. The combined solution of precursors was purged with Ar for at least 20 min prior to the addition of Na_2S (0.053-0.24

g). A substantial excess (1.5-10 times) of metal ions over a stoichiometric amount of S^{2-} ions was maintained to ensure efficient binding of the stabilizer to the nanoparticles. Normally, a clear, slightly yellow (CdS) or black (PbS) dispersion was obtained in the synthesis. The absorption spectra of CdS dispersions revealed the presence of a pronounced peak and a sharp absorption edge. These results indicated a narrow size distribution of the nanoparticles [9]. Refluxing at 100 °C was found to stimulate the growth of particles while maintaining a narrow size distribution. This process was used to vary the diameter of the nanoparticles in the self-assembled films (SAF).

TiO_2 nanoparticles were synthesized by the slow injection of 1.5 ml of titanium isopropoxide (0.005 mole) into 300 ml of water in the presence of 0.1 g of polyacrylic acid (PAA; pH = 10.5). The immediate formation of a slightly opaque colloidal dispersion was observed. After aging for 24 hours, a portion of the stock solution was subjected to centrifugation and then used for self-assembly. The average particle size in the dispersion was estimated to be 25 Å by absorption spectroscopic measurements [10]. Electrochemical measurements were carried out on Polyscience TiO_2 samples (100-nm diameter), dispersed in deionized water at pH = 9.5 (1g of TiO_2 per 100 ml of water).

Na-montmorillonite was prepared by the standard procedure [11]. The stock dispersion was produced by sonication of a desired amount of dry Na-montmorillonite (0.2-1.0g) in 100-ml deionized water for 40 min (Braunsonic 2200 sonication bath). Then, the mixture was subjected to centrifugation at *ca.* 1500 rotations per minute for two hours; the solid precipitate was collected and the transparent supernatant was used for further experiments.

The preparation of SAFs started with the cleaning of quartz, glass, gold, platinum or Teflon substrates in Nocromix^{TM} solution in concentrated sulfuric acid for 0.5-3 hours. Then, the substrates were thoroughly rinsed with ample amounts of deionized water, dried and immersed into a 10% solution of polyelectrolyte (P) for 15-30 minutes. After that, they were rinsed and dried again and immersed into an aqueous 0.2 % dispersion of M for 15 minutes. The presence of a layer of montmorillonite platelets was found to promote uniformity of surface and to standardize the quality of the initial layer. Nanoparticulate films were found to form efficiently on quartz, glass and metal surfaces, even in the absence of a primer layer of M. However, priming was important when Teflon films were used as substrates. X-ray diffraction showed substantial improvement in the uniformity of films which were layered on substrates which had been primed by a layer of M.

Self-assembly of the second P/M or P/CdS sandwich unit was initiated by immersion into a 2% aqueous solution of P, kept at pH = 6, for 20 minutes. Subsequent to affluent rinsing and drying, the substrate was immersed into a dispersion of nanoparticles (CdS, PbS, TiO_2) for 24 hours or into a dispersion of clay platelets (M, montmorillonite) for 15 minutes. The immersion time was optimized by monitoring the absorption spectrum of a given self-assembly as a function of time. Immersion was considered to be complete when the absorption spectrum reached a level which no longer changed with further immersion. At this point, the substrate was removed and rinsed with an ample amount of water, dried and characterized spectrophotometrically (UV, fluorescence or FTIR).

Repeating n-times the sequence of immersion into the desired dispersion for the optimized time, rinsing and drying led to the self-assembly of n-layers of organic polyelectrolyte-inorganic particle sandwich units onto the selected substrate. The short-hand notation of $S\text{-}(P/M)_n$, $S\text{-}(P/CdS)_n$ or $S\text{-}(P/TiO_2)_n$ was adapted for the self-assembled systems investigated in this work. The schematics of the self-assembly of $S\text{-}(P/CdS)_n$ and $S\text{-}(P/M)_n$ films are illustrated in Figure 1.

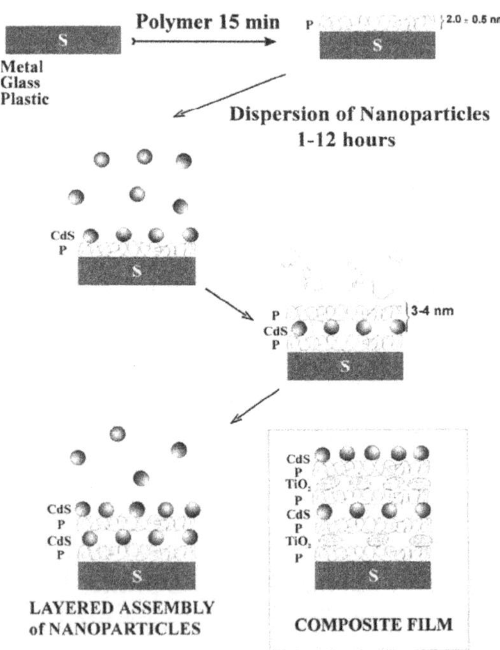

Figure 1a.

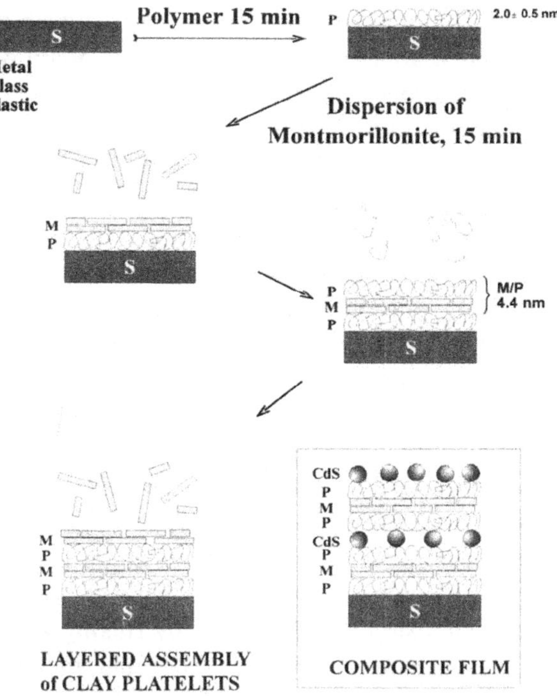

Figure 1b Schematics of the self-assembly of ultrathin films composed of alternating layers of polycations (P)-semiconductor nanoparticles (CdS) (a), polycations (P)-clay platelets (M) (b) and composite polycations (P)-clay platelets (M) and polycation (P)-semiconductor nanoparticles (CdS or TiO_2) (in the box) on metal, glass or Teflon substrates (S).

Water was purified by a Millipore Milli-Q system containing a 0.22-μm Millistack filter at the outlet.

X-ray diffraction measurements were performed by a Phillips PW 1830 diffractometer (CuK_α, λ =1.54 Å). The quartz, glass, ITO-coated glass and teflon plates, used as substrates, were found not to have peaks in the region of interest. Fluorescence spectra were taken on a Tracor-Northern 6500 rapid scan spectrofluorometer system [12].

Surface plasmon measurements were carried out by using a home-constructed system. Gold- or silver- (vacuum evaporated to 40-50 nm thickness) coated glass slides were used as the reflection element. The uncoated side of the slide was brought into optical contact with the base of a 90° glass prism (n = 1.52) by an index matching oil (n = 1.518±0.0005;

purchased from R. P. Cargille Laboratories, Inc., New Jersey, lot #081268). A p-polarized, 632.8-nm beam was directed to the base of the prism by a HeNe laser (Hughes, 3235H-PC, *ca.* 20 mW). The prism was mounted on a stepping motor driven rotator (Oriel) which was capable of synchronously varying the angle of incidence, θ, and the direction of a large area silicon detector (Newport, 818-SL) with an angle resolution of 0.01°. The angular sample interval was 0.05° within 2° of the resonance minimum and 0.2° elsewhere. Angular reflection scans required about 20 minutes. Each angular scan was fitted to a theoretical reflection curve, calculated by choosing appropriate one-, two- or three-layer models and refractive indices.

Transmission electron micrographs (TEM) were taken either on a JEOL 2000-FX electron microscope operating at 200 keV or on a JEOL 2000-EX electron microscope operating at 120 keV. Formvar-covered, carbon-coated, 200-mesh copper grids were allowed to float on the surface of a 2% polyelectrolyte solution for 12 hours. Care was taken to avoid contact of the uncoated grid face with the aqueous solution. After copious rinsing with water, the grids were put in the same manner on a dispersion of montmorillonite which was analogous to the one that was used for the priming of other substrates. This step was followed by rinsing and self-assembly of polyelectrolytes in an identical procedure. After that, the grids were ready for the self-assembly of semiconductor particles, which proceeded by floating on the surface of an aqueous dispersion of nanoparticles for 24 hours. Once again, care was taken to avoid deposition of self-assembled films on both sides of the grid.

AFM images were taken by a Topometrix Explorer 2000 scanning probe microscope using standard silicon nitride tips (force constant of 0.12 N/m) in the contact as well as in the non-contact mode. Freshly cleaved mica was used as the substrate for AFM measurements.

Cyclic voltammograms (CVs) were determined by using PAR 174A and EG&G PAR 273 electrochemical instruments interfaced with an IBM PC. The working electrode was prepared by the sequential self-assembly of the required number and types of sandwich units on a platinum substrate. Ag/AgCl was used as the reference electrode. CVs recorded in an aqueous 10-mM solution of $K_3Fe(CN)_6$ in 1M KCl immediately after deposition were shown to be identical after three minutes and after 24 hours of soaking in the solution. A slight rise (*ca.* 10%) in the cathodic peaks was observed between the first and the second voltammograms and then they remained constant. Potential-assisted self-assembly was performed without any additional supporting electrolyte by using the intrinsic conductivity of the dispersions upon application of a given potential for one (for CV) or five min. (for XRD).

Photocurrent kinetics were observed by illuminating the Pt electrode with a defocused, high-pressure, 100-W Hg lamp. Defocused light was used to avoid heating of the sample surface that could cause a substantial non-Faradaic current. The photocurrent was registered at potentiostatic conditions of $\Delta V = 0.6$ V (Ag/AgCl standard). A magnitude of the photocurrent reaching a constant level (i_{photo}) was used to characterize TiO_2, composite TiO_2/CdS and TiO_2/CdS/M films. The initial maximum of the photocurrent was found to depend on the frequency of dark/light periods, whereas i_{photo} was affected only by the structure of self-assembled semiconductor films on electrodes (number of layers, sequence of particles, etc.).

4. Results And Discussion

Demonstration of a versatile method of layer-by-layer self-assembly of polycations, clay platelets and semiconductor nanoparticles into stable, ultrathin films is the most significant accomplishment of the present work (Figure 1).

Cation poly(diallyldimethylammonium chloride) (P) was found to adhere strongly to a variety of surfaces. Indeed, immersion of a quartz slide, a platinum or gold electrode, an ITO-coated glass, a freshly cleaved sheet of mica and even a teflon film into a 1.0 % aqueous solution of P resulted in the adhesion of the polyelectrolyte onto these substrates. The thickness of P, self-assembled on a gold electrode, was determined to be 2.0±0.5 nm by surface plasmon spectroscopy (Figure 2). Furthermore, soaking in an aqueous solution for four hours did not remove P, as evidenced by the unaltered surface plasmon spectrum. AFM images of the polyelectrolyte film revealed featureless wave-like structures with height variations of about 0.2 nm (not shown). Apparently, multiple van der Waals forces maintain P on a variety of substrate surfaces, even in the absence of covalent bonding [13]. It should be pointed out that polyelectrolytes were not directly self-assembled onto substrates previously. Typically, the substrates were silanized or coated by functionalized monolayers [8,14] prior to immersion into the polyelectrolyte solution.

Self-assembly of the negatively charged surfactant-stabilized CdS, PbS and TiO_2 nanoparticles onto the positive surface of P was greatly facilitated by strong electrostatic interactions. Assuming the polycation to be

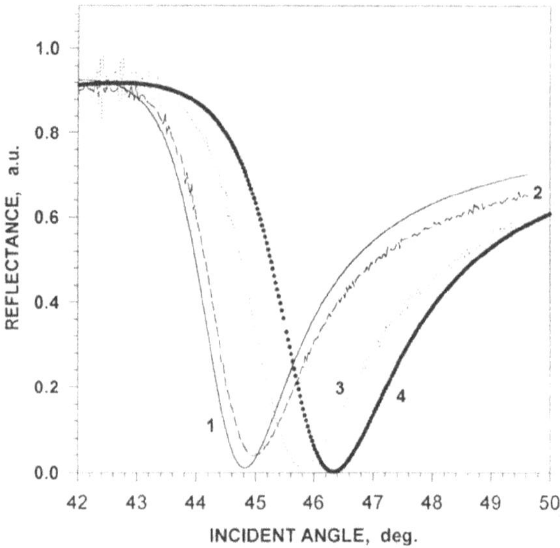

Figure 2. Surface plasmon spectroscopy of a gold substrate (1), a layer of polycation self-assembled on the gold substrate (2), a layer of polycation and a layer of clay platelets self-assembled on the gold substrate (3) and a layer of polycation and a layer of clay platelets self-assembled on the gold substrate under an applied potential of +1.3 V (4).

completely stretched on the surface of S, the mean area occupied by a positive charge of P can be estimated to be 10 Å^2. The corresponding mean distance between negative charges on the semiconductor nanoparticle surface is somewhat harder to estimate. We can assume a fairly dense packing of stabilizer molecules in a fashion similar to self-assembled thiol monolayers on gold. Recent NMR and XPS data substantiate this assumption [15]. It is important to mention that only one of the stabilizers used, TLA, carries a charge. Consequently, the area per negative charge will depend on the TLA:ET ratios. Nevertheless, the upper limit of the area occupied by a negative charge can be estimated to be about 50 Å^2. Taking only half of the particle surface area facing the plane of adsorption, the minimum number of interaction sites between the positive P and negative semiconductor particles is estimated to be to be about 20. This number can be much greater, of course. However, it is important to note that a highly charged surface does not necessarily ensure efficient self-assembly. The higher the charge of a nanoparticle, the higher its affinity toward water. Consequently, the Gibbs free energy of a nanoparticle does not change substantially upon adsorption on a polyelectrolyte film and, hence, it can also be easily desorbed by subsequent washing. Finding a proper balance between charge (to provide attraction to the polyelectrolyte layer) and hydrophobicity (to prevent lability of adsorbed species) is, therefore, of paramount importance for the optimization of self-assembly. This condition, for the present system, was found by using a 1:3 = TLA:ET ratio for coating the semiconductor nanoparticles.

The absorption spectra of CdS (Figure 3) and PbS (Figure 4) nanoparticle dispersions in aqueous solutions are retained in the self-assembled films. Furthermore, the absorption edges (*ca.* 430 nm for CdS and *ca.* 500 nm for PbS) correspond to mean particle diameters of 35 Å and 100 Å. That these particles are indeed in the nanometer region and are, thus, size quantized can also be seen in their transmission electron micrographs (Figure 5).

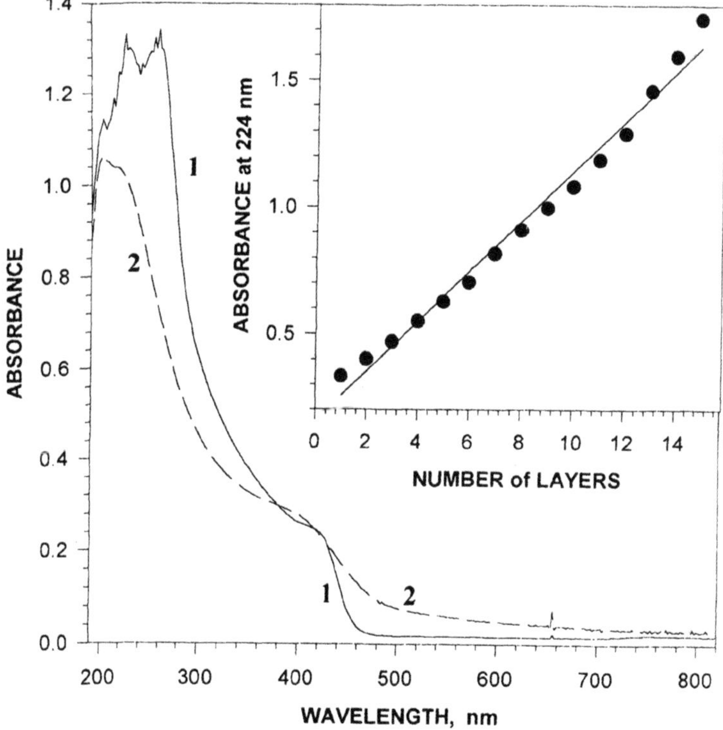

Figure 3. Absorption spectrum of a CdS nanoparticle dispersion in an aqueous solution (1) and that of one monoparticle layer of CdS self-assembled onto one layer of polycation-coated quartz substrate (2). The absorbances at 320 nm *vs.* the number of sandwich units (n) in $S\text{-}(P/PbS)_n$ are plotted in the insert.

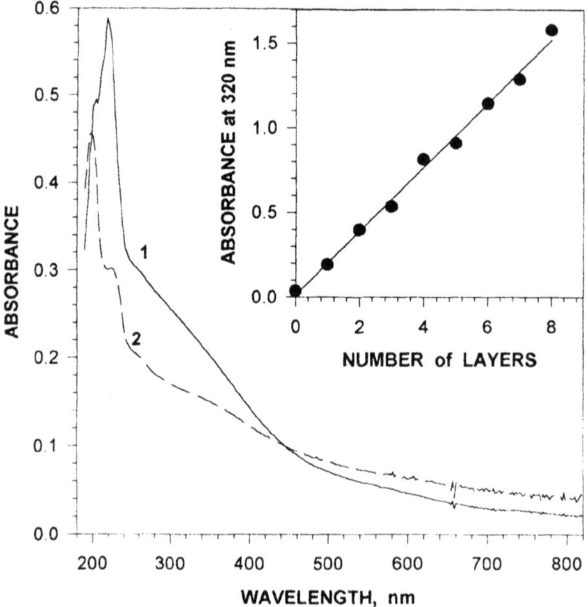

Figure 4. Absorption spectrum of a PbS nanoparticle dispersion in an aqueous solution (1) and that of one monoparticle layer of PbS self-assembled onto one layer of polycation-coated quartz substrate (2). The absorbances at 224 nm *vs.* the number of sandwich units (n) in $S\text{-}(P/CdS)_n$ are plotted in the insert.

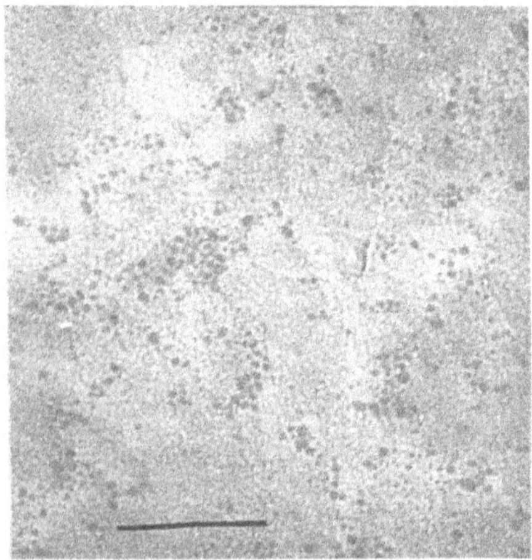

Figure 5. Transmission electron micrographs of one layer of PbS self-assembled onto one layer of polycation-coated substrate. The scale bar is 100 nm.

AFM images of self-assembled TiO_2 revealed the formation of a well-packed monoparticle layer (Figure 6). The tight packing of 100-nm-diameter TiO_2 particles precluded penetration of the AFM tip to the mica surface. The image of the film was quite similar in all of the acquisition modes that were used (topography, internal sensor, beam deflection, lateral force and z-force). Interestingly, lateral force microscopy provided the most detailed images. In the topography mode, the images which were obtained tended to have enlarged features. These observations indicate the presence of rather strong van der Waals forces between the AFM tip and the TiO_2 particles.

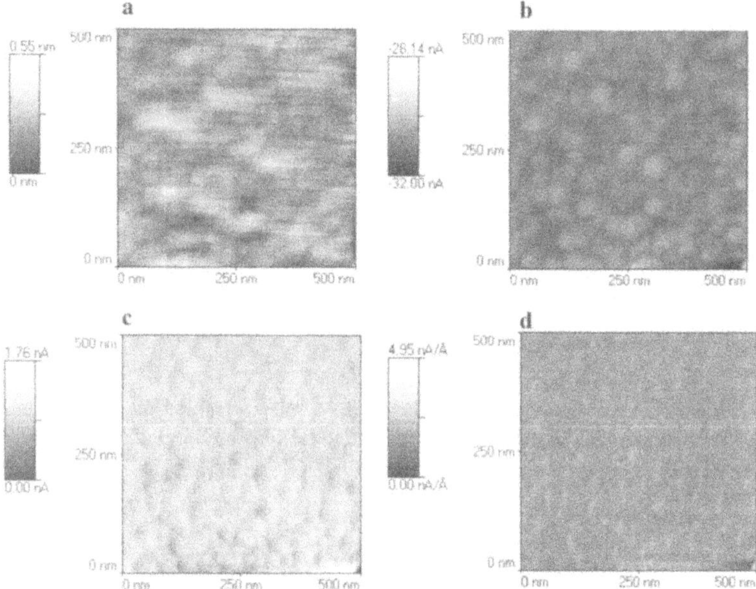

Figure 6. AFM images of 1 layer of TiO_2 nanoparticles self-assembled on mica. a = topography, b = internal sensor, c = lateral forces and d = hardness. Set point = -12 nA, scan rate = 100 nm/s, resolution = 200 lines.

Self assembly of clay platelets (M) was also observed by using the topography, internal sensor and lateral force modes of the AFM (Figure 7). The lateral force mode was found to be least informative due to the small height variation along the surface. The height variations did not exceed 5 nm in a 2700 nm interval as seen in the topographic images (see Figure 7a). The primary source of this height variation is the stacking of the clay platelets and adsorption of not-fully-exfoliated aggregates. Otherwise, the film is rather uniform and the planar orientation of the platelets is readily observable. The planar orientation of the montmorillonite platelets in self-assembled films is substantiated by their transmission electron micrograms (Figure 8).

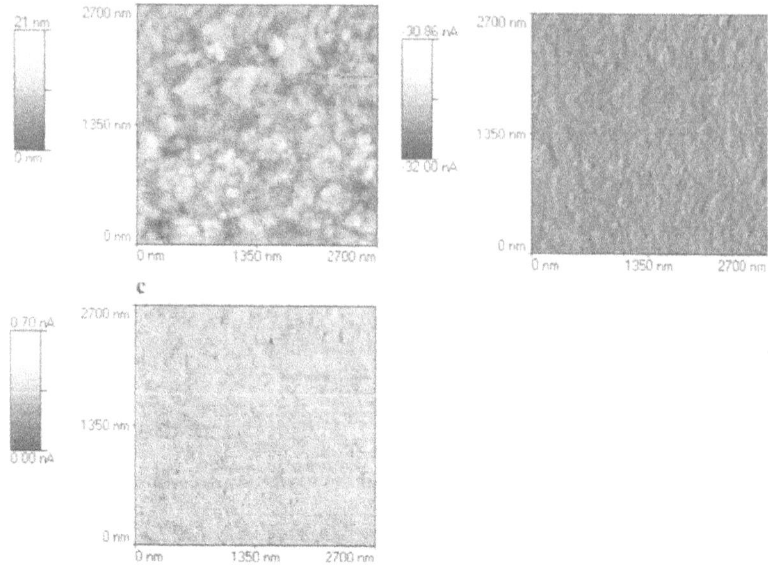

Figure 7. AFM images of one layer of M self-assembled on mica. a = topography, b = internal sensor and c = lateral forces. Set point = -10 nA, scan rate = 200 nm/s, resolution = 200 lines.

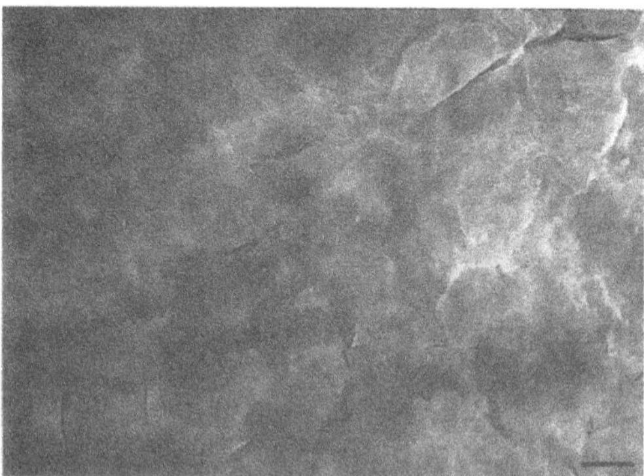

Figure 8. Transmission electron micrographs of one layer of M self-assembled onto one layer of polycation-coated substrate. The scale bar is 200 nm.

Successful self-assembly of a large number of repeating sandwich units of S-(P/CdS)$_n$, S-(P/PbS)$_n$, S-(P/TiO$_2$)$_n$ and S-(P/M)$_n$ were demonstrated by X-ray diffraction, absorption and emission spectroscopic, and cyclic voltammetric measurements. For example, the observed good linearities in the plots of absorbances *vs.* n in S-(P/CdS)$_n$ and S-(P/PbS)$_n$ indicates the uniformity of the sandwich units that were self-assembled (see inserts in Figures 3 and 4). Similarly, the observed linear plot of emission intensity *vs.* n in fluorescence-activated S-(P/CdS)$_n$ films indicates uniform self-assembly on a teflon film (Figure 9).

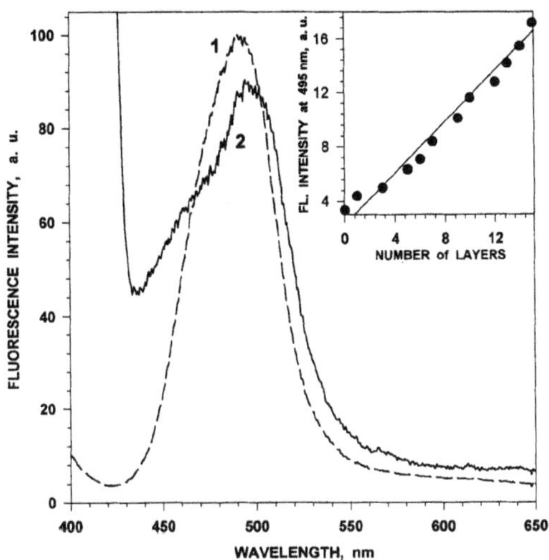

Figure 9. Emission spectrum of a CdS nanoparticle dispersion in an aqueous solution (1) and that of S-(P/CdS)$_{20}$ self-assembled onto a teflon film (2). λ_{ex} = 350 nm, front-face illumination. The fluorescence intensities at 495 nm *vs.* the number of sandwich units (n) in S-(P/CdS)$_n$ are plotted in the insert.

X-ray diffraction measurements showed sharp peaks for S-(P/M)$_{20}$ at 2θ = 2.1° (Figure 10), for S-(P/CdS) at 2θ = 1.70 (not shown) and for S-(P/TiO$_2$)$_{20}$ at 2θ = 2.2 (not shown), which corresponded to planar periodicities of 4.3±0.5 nm, 6.5±0.8 nm and 4.5±0.5 nm for the P/M, P/CdS and P/TiO$_2$ repeating units, respectively. Since the mean diameters of the CdS and TiO$_2$ (stabilized by PAA) used were determined be 4.0±0.5 nm and 2.5±0.5 nm, the difference between the assessed periodicity and the particle diameters (spacing) (6.5±0.8 nm - 4.0±0.5 nm = 2.5±0.5 and 4.5±0.5 nm -

2.5±0.5 nm = 2.0±0.5 nm) nicely substantiates the measured thickness of P (2.0±0.5 nm). The second broad peak in the X-ray diffraction of S-(P/M) films corresponds to the basal spacing between the clay platelets within one layer. Clays platelets are well known for efficient horizontal stacking [16]. This, in fact, occurs upon self-assembly, as evidenced by AFM and transmission electron microscopy (Figure 8). The basal spacing of 18±1.5 Å coincides very well with the distance between aluminosilicate sheets in swelled clay minerals. Thus, the value of the film thickness, obtained from the sharp peak at 2.1° (4.3 nm), and the thickness of the polyelectrolyte layer (20±0.5 Å) show that two layers of clay platelets (each 9 Å and *ca.* 9 Å gap) are being adsorbed during one deposition cycle.

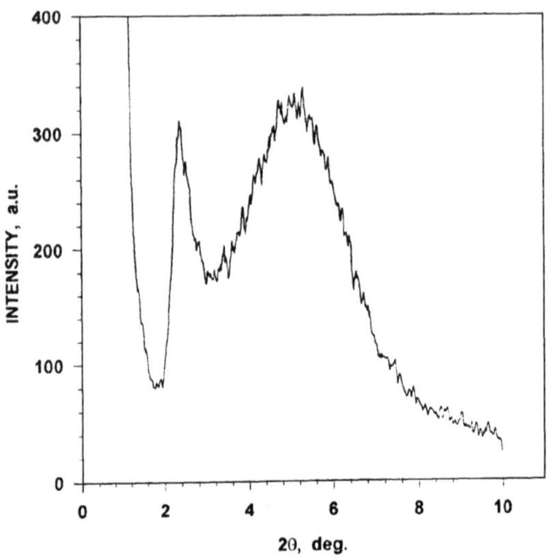

Figure 10. X-ray diffraction patterns of $S\text{-}(P/M)_{20}$.

Cyclic voltammograms of 10 mM $K_3Fe(CN)_6$ in aqueous solutions provided useful information on the properties of the self-assembled films (Figure 11). Oxidations and reductions occurred at the metal electrode interface as a result of the diffusion-controlled penetration of the $Fe(CN)_6^{3-}$ and $Fe(CN)_6^{4-}$ ions. Increasing current drops upon the deposition of successive layers of P/M are clearly seen in the cyclic voltammograms (Figure 11). The thickness of a given multilayer film was found to depend on the potential which was applied during the deposition(s) of M. Positive potentials increased the film thickness, while negative ones decreased it

(Figure 12). Not unexpectedly, negatively charged M is attracted to a positively charged electrode more efficiently than it is to a negatively charged one. Similar behavior has been noted in the electrophoretic deposition of metal particles onto conductive substrates [17].

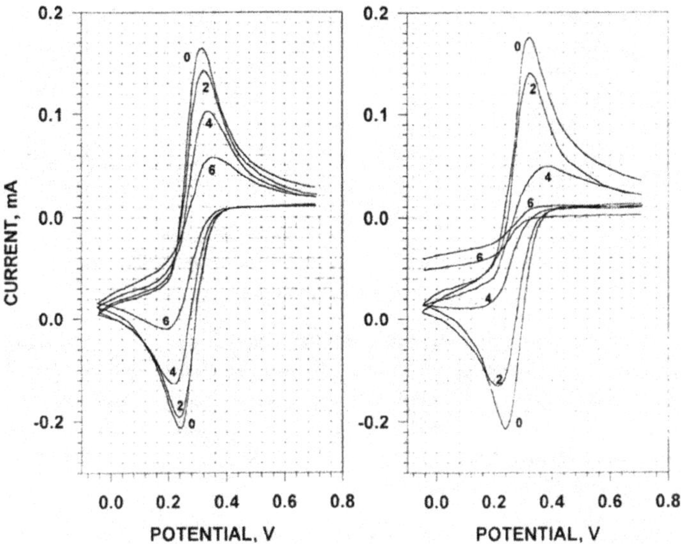

Figure 11. Cyclic voltammograms of 20 mM $K_3Fe(CN)_6$ in aqueous, 1.0 M KCl solutions. The working electrode was prepared by the sequential self-assembly of M/P/M/P... (see Figure 1b) in the absence (voltammograms in the left box) and in the presence of a +1.3 V potential (voltammograms in the right box) during the self-assembly of M. The number on the voltammograms indicates the number of deposition cycles (*ie.*, the number of M/P layers on the substrate).

The effect of potential on the thickness of $S\text{-}(P/M)_n$ can be understood by considering the interplay between self-assembly, adsorption and desorption. Self-assembly is a relatively slow process. Thus, the effect of applied potential is not obviated by the much faster adsorption desorption equilibria which occurs during the rinsing. Additionally, smectic clays have a great tendency toward stacking, in spite of the negative charges that the clay platelets carry. An applied positive potential diminishes this charge repulsion and results in more tightly stacked clay platelets. Conversely, an applied negative potential augments the electrostatic repulsions, which then brings about less tightly packed self-assembled films.

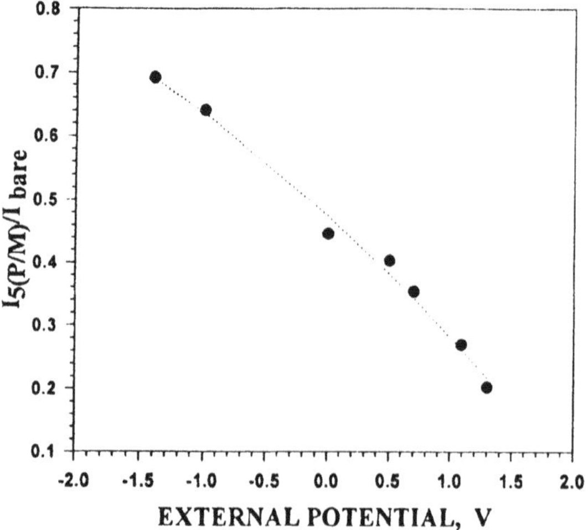

Figure 12. A plot of the ratios of cathodic peaks, $I_{(S\text{-}P/M)5}/I_{bare}$ (comparing the cathodic peaks obtained on using a working electrode covered by five layers of P/M with that obtained on using a bare platinum electrode), *vs.* the external potential applied during the self-assembly of M. Cyclic voltammograms were obtained for the $Fe(CN)_6^{3-}$-$Fe(CN)_6^{4-}$ oxidation-reduction pair (see Experimental Section for details).

Self-assembly allows the construction of composite films comprising different nanoparticles which can be layered in any desired order. In this work, we used photocurrent and photoelectron transfer measurements as the means to examine the consequences of interspacing layers of insulators (P/M) between layers of semiconductors (P/TiO_2 and P/CdS) in a variety of different arrangements (Table 1). Under irradiation, the S-(P/TiO_2)$_n$, and any combination of mixed the P/TiO_2 and P/CdS self-assembled films, behaved like a typical n-type semiconductor (Figure 13) whose properties were determined by the development of a depletion layer at the semiconductor-electrolyte interface [2]. It should be noted that the nanoparticles investigated here are too small to allow full development of the depletion layer whose thickness is supposed to be about 100 nm for TiO_2. It is conceivable that in our semiconductor films, the whole assembly, consisting of multiple layers, behaves as one electronic unit. The photoinduced current is caused by photo-generated holes (minority carriers). At positive electrode potentials, they accumulate and oxidize water molecules. The plateau value of the photo-current, i_{photo}, for a P/TiO_2 unit deposited up to a maximum at S-(P/TiO_2)$_{15}$ beyond it decreased somewhat (Figure 14).

TABLE 1. Photocurrents in differently organized composite semiconductor nanostructures

Structure of the working electrode[a]	i_{photo}, μA[b]
S-$(TiO_2/P)_{10}$	0.63
S-$(M/P)_3(TiO_2/P)_{10}$	0.10
S-$(TiO_2/P)_2(M/P)_3(TiO_2/P)_8$	0.23
S-$(TiO_2/P)_5(M/P)_3(TiO_2/P)_5$	0.40
S-$(TiO_2/P)_8(M/P)_3(TiO_2/P)_2$	0.47
S-$(TiO_2/P)_{10}(M/P)_3$	0.63
S-$(TiO_2/P)(CdS/P)(TiO_2/P)_2(CdS/P)(TiO_2/P)_2(CdS/P)$- -$(TiO_2/P)_2(CdS/P)(TiO_2/P)_2(CdS/P)(TiO_2/P)$	0.43
S-$(TiO_2/P)_{10}(CdS/P)_5$	0.23

[a]S = substrate; M = montmorillonite; TiO_2 = titanium dioxide; CdS = cadmium sulfide. The subscript number following the parenthesis indicates the number of repeating units which were self-assembled.
[b]i_{photo} = plateau value photocurrent.

The effects of organization of P/M, P/CdS and P/TiO_2 on the measured photocurrent (i_{photo}) are summarized in Table I. Introduction of one to three layers of insulating P/M between layers of P/TiO_2 had the expected consequence of photocurrent decrease. Aluminum silicates block the movement of photogenerated charge carriers within the film and, thus, diminish the photocurrent. A correlation between the position of P/M and photocurrent was observed. The further the P/M layer was placed from the electrode, the greater the photocurrent became. The transparent P/M film does not affect light absorption in the film, but it divides the composite film into active and non-active parts. The part which is adjacent to the electrode is responsible for the generation of the photocurrent. The closer the active part of the film to the electrode, the greater the photocurrent.

Close contact between P/TiO_2 and P/CdS units facilitates photosensitized electron transfer and, thus, increases i_{photo} values. In contrast, separated P/TiO_2 and P/CdS units have smaller i_{photo} values. Importantly, photocurrent measurements provide a convenient method for characterizing the interplay between the different semiconductor particles which comprise the nanostructured films.

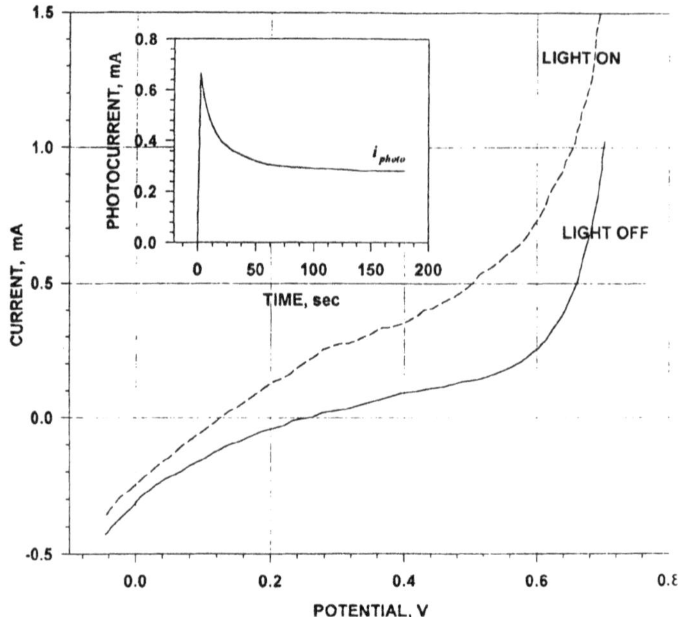

Figure 13. Current *vs.* potential plots, in the dark and under illumination, in a photoelectrochemical cell using a working electrode, prepared by self-assembling ten layers of TiO_2, $S\text{-}P/TiO_2)_{10}$. Scan rate = 0.5 mV/sec. Illumination onto the platinum electrode was provided by a defocused 100 W Hg-lamp. A typical photocurrent is illustrated in the insert. The current in the first couple of seconds originates from the oxidation of water and a small amount of adsorbed impurities. Hence, the peak photocurrent observed was found to depend on the sample history (extent of previous irradiation). However, the plateau value of the photocurrent (i_{photo}) was independent of the sample history and was reproducible (within 10%) for a given sample in the same electrolyte solution under the same extent of irradiation.

The spontaneous, layer-by-layer self-assembly of polyelectrolyte-semiconductor particle nanostructured films, reported here, exemplifies well our proposed membrane-mimetic approach to advanced materials [2]. While the dimensions, organization, and deceptive simplicity of spontaneous self-assembly are analogous to the corresponding properties of the biological membrane, the present system offers versatility and a relative ease of systematic physical and chemical manipulation. Judicious selection of components and their layer-by-layer self-assembly will result in nano-structured films with the desired mechanical, optical, electro-optical, electrical, magnetic and electromagnetic properties.

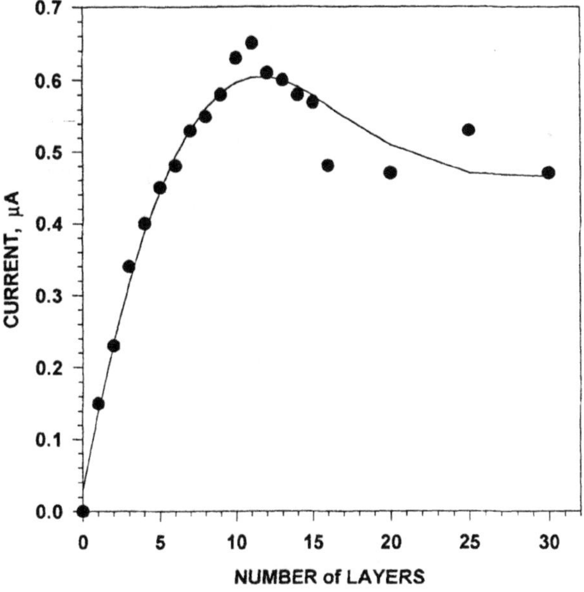

Figure 14. Photocurrent *vs.* the number of sandwich units (n) in S-$(P/TiO_2)_n$.

5. Acknowledgments

Support of this work by the National Science Foundation (U.S.A.) and the U.S.-Hungarian Joint Fund (Hungary) is gratefully acknowledged. Thanks are also due to Dr. Youxin Yuan for performing TEM measurements.

6. References

1. Ulman, A. (1991) *An Introduction to Ultrathin Organic Films from Langmuir Blodgett to Self-Assembly,* Academic Press Inc., Boston.
2. Fendler, J. H. (1994) *Membrane Mimetic Approach to Advanced Materials,* Springer-Verlag, Berlin.
3. Fendler, J. H. and Meldrum, F. C. (1995) Colloid-Chemical Approach to Nanostructured Materials, *Adv. Mater.*, in press.
4. Meldrum, F. C. and Fendler, J. H. (1995) Construction of organized particulate films by the Langmuir-Blodgett technique, in S. Mann (ed.), *Biomimetic Approaches in Materials Chemistry,* VCH Publishers, New York, in press.

5. Mallouk, T. E., Kim, H.-N., Ollivier, P. J., Kelter, S. W. (1995) Ultrathin films based on layered inorganic solids, personal communication.
6. Freeman, R. G., Grabar, K. C., Allison, K. J., Bright, R. M., Davis, J. A., Guthrie, A. P., Hommer, M. B., Jackson, M. A., Smith, P. C., Walter, D. G., and Natan, M. J. (1995) Self-assembled metal monolayers: an approach to SERS substrates, *Science* **267,** 1629-1632.
7 Lvov, Y., Essler, F., Decher, G. (1993) Combination of polycation/ polyanion self-assembly and Langmuir-Blodgett transfer for the construction of superlattice films, *J. Phys. Chem.* **97**, 13773-13777.
8. Kleinfield, E. R., Ferguson, G. S. (1994) Stepwise formation of multilayered nanostructured films from macromolecular precursors, *Science* **265**, 370-373, 1994.
9. Yoffe, A. D. (1993) Low-dimensional systems, *Adv. Phys.* **42**, 173-276.
10. Fox, M. A., Dulay, M. T. (1993) Heterogeneous photocatalysis, *Chem. Rev.* **93**, 341-357.
11. Dékány, I., Turi, L., Tombácz, E., and Fendler, J. H. (1995) Preparation of size-quantized CdS and ZnS particles in nanophase reactors provided by binary liquids adsorbed at layered silicates, *Langmuir*, in press.
12. Tian, Y., Qu, C., and Fendler, J. H. (1994) Fluorescence activation and surface state reactions of size-quantized cadmium sulfide particles in Langmuir-Blodgett films, *J. Phys. Chem.* **98**, 4913-4918.
13. Schonhorn, H. (1978) Surface modification of polymers for adhesive bending, in D. T. Clark and W. J. Feast (eds.), *Polymer Surfaces*, Wiley-Interscience, New York, pp. 345-406.
14. Decher, G., Lehr, B., Lowack, K., Lvov, Y., and Schmitt, J. (1994) New nanocomposite films for biosensors: layer-by-layer adsorbed films of polyelectrolytes, proteins or DNA, *Biosensors & Bioelectronics*, in press; Colvin, V. L., Goldstein, A. N., and Alivisatos, A. P. (1992) Semiconductor nanocrystals covalently bound to metal surfaces with self-assembled monolayers, *J. Am. Chem. Soc.* **114**, 5221-5230; Decher, G., Hong, J. D., and Schmitt, J. (1992) Buildup of ultrathin multilayer films by a self-assembly process: III. Consecutively alternating adsorption of anionic and cationic poly-electrolytes on charged surfaces, *Thin Solid Films* **210/211**, 831-835; Lvov, Y., Haas, H., Decher, G., and Möhwald, H. (1994) Successive

deposition of alternate layers of polyelectrolytes and a charged virus, *Langmuir* **10**, 4232-4236; Ferreira, M., Cheung, J. H., and Rubner, M. F. (1994) Molecular self-assembly of conjugated polyions: a new process for fabricating multilayer thin film heterostructures, *Thin Solid Films* **244**, 806-809; Cheung, J. H., Fou, A. F., and Rubner, M. F. (1994) Molecular self-assembly of conducting polymers, *Thin Solid Films* **244**, 985-989.

15. Sachleben, J. R., Wooten, E. W., Emsley, L., Pines, A., Colvin, V. L., and Alivisatos, P. (1992) NMR study of the surface structure and dynamics of semiconductor nanocrystals, *Chem. Phys. Lett.* **198**, 431-436; Becerra, L., Murray, C. B., Griffin, R. G., Bawendi, M. G. (1994) Investigation of the surface morphology of capped CdSe nanocrystallites by ^{31}P nuclear magnetic resonance, *J. Chem. Phys.* **100**, 3297-3300; Bowen Katari, J. E., Colvin, V. L., and Alivisatos, A. P. (1994) X-ray photoelectron spectroscopy of CdSe nanocrystals with applications to studies of the nanocrystal surface, *J. Phys. Chem.* **98**, 4109-4117.
16. Van Olphen, H. (1977) *An Introduction to Clay Colloid Chemistry*, 2nd Edition, Wiley-Interscience, New York.
17. Giersig, M. and Mulvaney, P. (1993) Preparation of ordered colloid monolayers by electrophoretic deposition, *Langmuir* **9**, 3408-3413.

OPTOELECTRONIC CHARACTERIZATION OF EPITAXIAL FILMS OF ELECTRODEPOSITED CDSE QUANTUM DOTS

B. ALPERSON, S. COHEN, Y. GOLAN, I. RUBINSTEIN AND G. HODES
Weizmann Institute of Science, Rehovot 76100, Israel

1. Abstract

Optical and electrical properties of films of CdSe quantum dots epitaxially electrodeposited onto gold have been measured by modulated optical spectroscopy, liquid junction photoresponse spectroscopy and local current-voltage spectroscopy using a metallized scanning probe tip. The blue shift in the bandgap resulting from quantum confinement in the quantum dots has been measured by all these techniques. Large differences in the charge collection efficiency between the first and subsequent layers of dots have been measured and explained by the positive role of grain boundaries in enhancing collection efficiency in these films. Single electron transfer through the dots has been observed and explained by the large charging energies of the very small (4-5nm) dots.

2. Background

Semiconductor nanocrystals possess the intriguing property of a continuous change in their electronic structure with decrease in crystal dimensions below a certain size, typically ca. 10nm. This change - the quantum size effect - is manifested as an increasing blue shift in the optical absorption with decrease in crystal size, together with breaking up of the bulk semiconductor bands into discrete energy levels [1,2]. Nanocrystals in this size regime are often referred to as quantum dots (QDs).

QDs can be formed by a variety of methods. Among chemists, the most popular and simple method is by arrested precipitation resulting in a colloid (usually dispersed in a liquid phase but sometimes in a glass) [1]. The most common characterization method used for these colloids is, not surprisingly, optical absorption spectroscopy, which shows the effective bandgap of the semiconductor. This parameter can then be correlated with crystal size measured by transmission electron microscopy (TEM) and/or X-ray diffraction (XRD). The same techniques are, in general, equally applicable to films of QDs which are made up of aggregated nanocrystals which adhere to a substrate, such as glass [3,4]. We have, however, been studying electrodeposited QDs, which are epitaxially oriented with the metal substrates [5,6], where normal optical absorption has been found to be insufficiently sensitive. The reason for this is due to the fact that the average thickness of the semiconductor "film" is ca. 5nm in many cases, coupled with the fact that the substrate is a thin, almost opaque metal film (usually Au, which additionally has a reflectance feature which masks the semiconductor absorption).

E. Pelizzetti (ed.), Fine Particles Science and Technology, 579–590.

For this reason, we have turned to other characterization techniques to measure the optical properties of these samples. The three methods we will describe here are: electrolyte electrotransmission (EET), current - voltage measurements using scanning probe microscopy (I-V/SPM) and photocurrent spectroscopy using the sample as a photoelectrode in a photoelectrochemical (PEC) configuration.

3. Experimental

CdSe nanocrystals were electrodeposited from a dimethylsulfoxide solution containing 50 mM $Cd(ClO_4)_2$ and saturated with elemental Se at a temperature of 90°C, onto evaporated {111} textured gold (30 nm thick) that had been annealed for 12 hours at 250°C [5]. The substrates were, in most cases, mica that had been cleaved prior to the gold evaporation, or glass. The first layer of crystallites were epitaxial with the Au and typically 4-5nm in diameter.

EET was performed in an electrolyte of aqueous sodium selenosulfate (0.2M Se and 0.4M Na_2SO_3). The modulating signal was a sine wave of 0.5V peak-to-peak.

The spectral response of the photocurrent was measured on films of various nominal thicknesses using the same electrolyte as for the EET measurements.

I-V/SPM spectroscopy was carried out at room temperature using an atomic force microscope with conducting tips (Pd or Au/Pd alloy coating on silicon nitride). Using the force between tip and surface as feedback control, I-V (and dI/dV) measurements were performed.

4. Results and Discussion

We first briefly describe the structural characteristics of the samples. ED CdSe on Au was found to be epitaxial with the Au, at least for the first monolayer of crystals, with a crystal size of ca. 4-5nm. The epitaxy decreased considerably in succeeding layers and crystallite size increased to ca. 2-3 times that size within a few crystal layers. The size distribution was fairly narrow [5]. Fig.1 shows a schematic diagram of the samples. Fig.2 shows a TEM micrograph of a sample with ca. 20-25% coverage by CdSe. The individual crystals of CdSe can be seen to be fairly evenly distributed on the Au substrate in this case. This distribution can be controlled by deposition temperature and current density to give a defined degree of aggregation of the nanocrystals. The selected area electron diffraction pattern (SAD) (inset in fig.2) of a single Au crystal with many CdSe nanocrystals shows a single crystal pattern for the CdSe oriented with the Au substrate, showing the epitaxial relationship between the CdSe and Au.

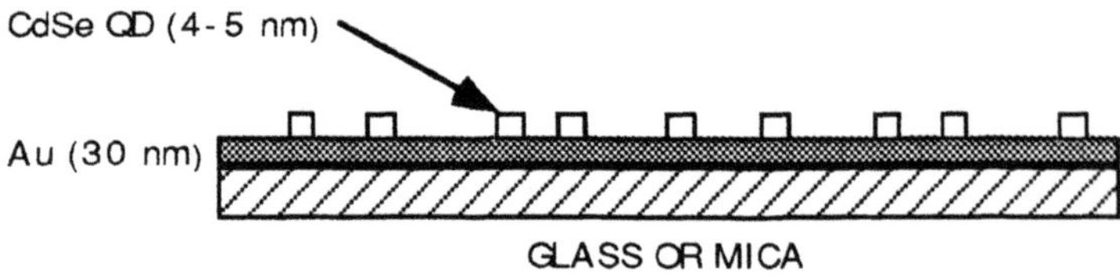

Figure 1. Schematic diagram of CdSe quantum dots electrodeposited on a gold film.

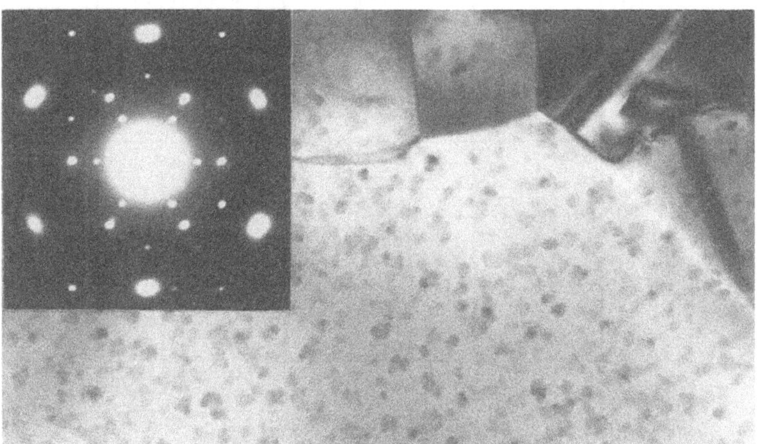

Figure 2. TEM micrograph of electrodeposited CdSe on Au (scale: 43 nm/cm). Inset: SAD pattern of Au + CdSe.

4.1 ELECTROLYTE ELECTROTRANSMISSION (EET)

EET belongs to the group of spectroscopies known as modulation spectroscopy. As the name suggests, modulation spectroscopy involves modulating some parameter (e.g., electric field, pressure, temperature) which affects the absorption coefficient of the semiconductor [7]. The measured optical property (reflection (R), transmission (T)) therefore varies at the modulation frequency and is detected by a lock-in amplifier set to that frequency. A normalized spectrum of dR(T)/R(T) vs. wavelength (or energy) is obtained. Since the absorption coefficient is only sensitive to the modulated parameter near an optical transition, the spectrum gives a signal near the optical transitiion but no signal elsewhere. Also, although the perturbation of the absorption coefficient by the modulated parameter is small, because of the lock-in technique used, the method is very sensitive. Applied to our system, the high sensitivity allows measurement of a very small amount of reflecting or absorbing material and, more importantly, the Au substrate does not severely interfere with the measurement (as it does in normal optical spectroscopy) since, being a metal, it does not give an interfering signal. We have used EET since the electric field modulation is easily applied by an electrolyte contact [7] and for the very thin films used here, dT is more easily detected than dR.

The EET spectrum of a ca. 10nm "thick" CdSe film is shown in fig.3. Two features can be clearly seen. The main one, at ca. 640nm (1.94eV), is the 'band gap' transition. A smaller one at ca. 510nm (2.43eV) can be attributed to the transition from the spin orbit split-off valence "band" to the first conduction level. The energy difference of ca. 0.49eV between the two features is similar to that between the (almost degenerate) light and heavy hole bands and the spin orbit split-off band in bulk CdSe. Since the hole effective masses in the light hole band and split-off band iare very similar in CdSe, we might expect both bands to behave similarly upon size quantization, ignoring complications such as mixing of levels. We have used modulation techniques to measure spectra of chemically deposited CdSe films exhibiting various degrees of size quantization, and do find that the energy separation between the two transitions is constant although the effective band gap varies considerably [unpublished].

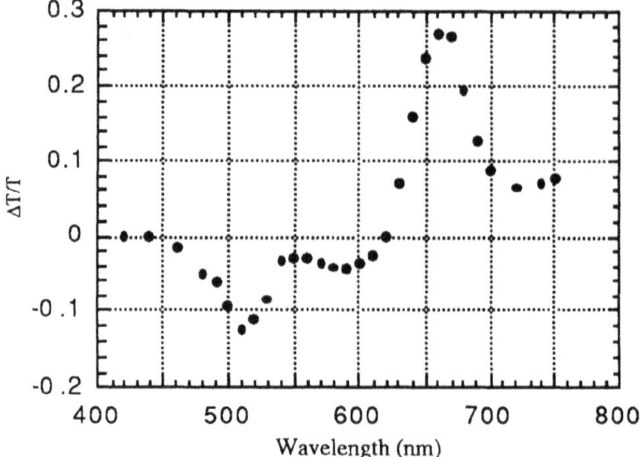

Figure 3. EET spectrum of 10nm (nominal; equivalent to 1.5-2 crystal layers) CdSe QD film.

A value of of 1.94eV for Eg is estimated from this spectrum based on the symmetry of the spectrum. We do not attempt an accurate curve fitting to this spectrum as is often done for modulated spectra. Since the film is made up of (almost) two layers of crystals - the first (adjacent to the Au) with a size of ca. 4-5nm and the second somewhat larger, the spectrum is a function of a moderate size distribution which would complicate any attempt at curve fitting. From the (room temperature) results of Murray et al. [8] this translates into an average size of 6nm which is somewhat larger than the first nanocrystal layer and smaller than the crystals in a considerably thicker layer (ca. 2-3 times that size).

4.2 PHOTOELECTROCHEMICAL (PEC) CHARACTERIZATION

PEC spectroscopy is a special type of photovoltaic photocurrent spectroscopy, where the sample is immersed in an electrolyte and the photocurrent flowing between the sample and a second (non-photoactive) electrode when the sample is illuminated is measured as a function of wavelength. As in normal optical absorption spectroscopy, there is a sharp increase in the signal with onset of light absorption at Eg. This forms the basis for the measurement of Eg using this method. This onset is often less well-defined than for the corresponding absorption spectrum since sub-Eg signals, not usually seen in absorption spectra, are quite common in PEC (and other photovoltaic) spectra and these signals blur the onset of photocurrent to some effect. However, the true supra-Eg signal is usually much more marked than any sub-Eg component and a value for Eg can usually be estimated with good reliability, if somewhat less than with optical spectroscopy. A plot of the square of the photocurrent vs. photon energy, extrapolated to zero photocurrent, is often used as a good approximation to Eg, although this analysis is often not applicable to our system due to the frequent presence of steps in the spectra.

PEC spectra of CdSe "films" are shown in fig.4 as a function of average film thickness. As noted above, the film thickness can be related directly to the number of layers of nanocrystals and to the crystlal size. There are several features of note in these spectra.

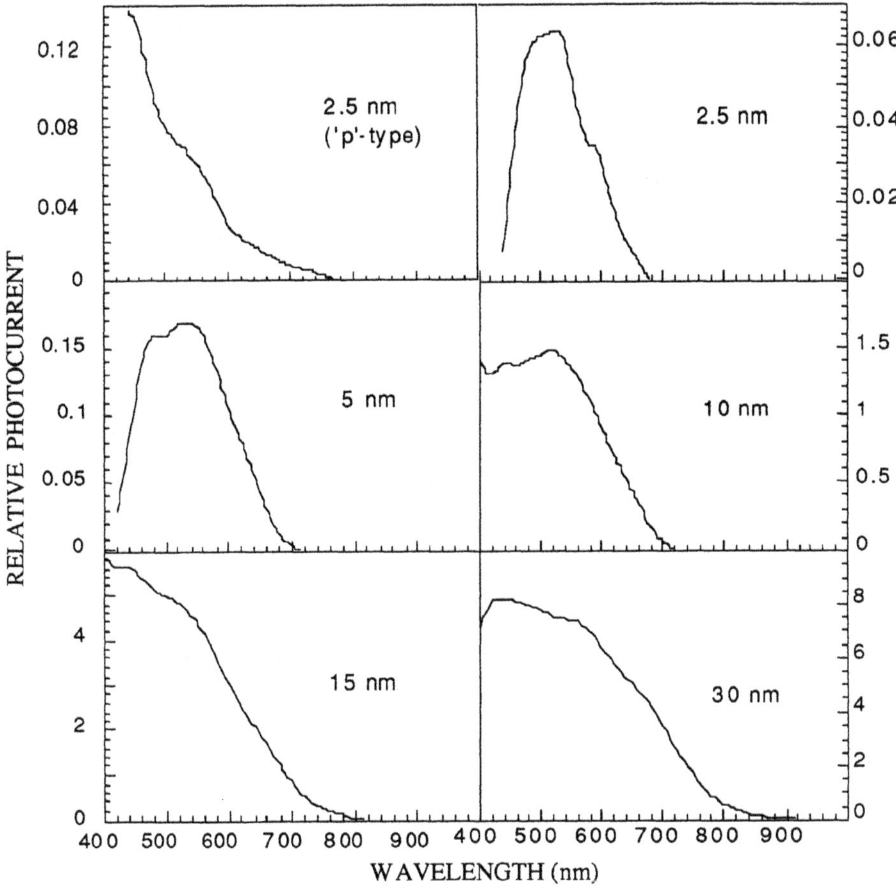

Figure 4. PEC photocurrent spectra of various thicknesses (given in the figures) of CdSe deposited on Au. 2.5nm is equivalent to half a momolayer of QDs; 5nm is a full monolayer; 15nm is a little more than two layers and 30nm at least three. The 'p'-type spectrum in the top left hand box is from a sample which behaved 'p'-type (in contrast to the usual 'n'-type behaviour). The values of the photoresponse are arbitrary but correctly shows the relative values of the different films.

The most obvious feature, and the one for which these experiments were originally carried out, is the "bandgap" cutoff. For the 2.5nm (ca. half a monolayer of crystals) and 5nm films (ca. a full monolayer of crystals), the cutoff values are similar and the bandgap can be estimated at ca. 1.95eV (the 5nm film has a slight tail extending to lower energies and which is probably due to a small concentration of larger crystals). This value is similar to that measured by EER for a somewhat thicker layer. The step in the 2.5nm film could be explained by two predominant crystal sizes - one corresponding to 1.95eV (ca. 6nm) and the other to ca. 2.2eV (<4nm). The 10nm film (estimated between 1.5 and 2 monolayers of crystals; less than two since the crystals in the second layer are already somewhat larger than the those in the first) exhibits a small red shift in Eg to an value estimated at ca. 1.9eV. The 15nm film (>2 layers of crystals) has a bandgap of ca. 1.8eV and a tail extending to 800nm (1.55eV), suggesting a sub-Eg component. This sub-Eg response is even more pronounced in the 30nm film (ca. at least 3 crystal layers containing some crystals which are no longer size-quantized [larger than 12nm] and should have a bandgap of 1.74eV); it extends considerably beyond this bulk bandgap. It is interesting also to note that the more highly quantized films do not show a sub-Eg response, particularly since the high surface-to-volume ratio of the crystals in these films might suggest such a response. The 30nm film also shows clear steps on the response onset which, like the 2.5nm film, might be due to the different spectral response of the different crystal sizes in this film, although they are not obvious in the 15nm film where they would also be expected to appear, if less pronounced.

A second feature is the relative photocurrent at the peak response of the different layer thicknesses. For such thin films, the photocurrent is expected to increase approximately proportional to the film thickness. As seen in fig.4, this is only true for the transitions from the 2.5nm to the 5nm film and from the 15nm to the 30nm film. The 10nm film gives an order of magnitude higher photocurrent than the 5nm film and the 15nm film is three times more active than the 10nm one. In order to understand this behavior, we must first consider the mechanism of loss in photocurrent for a PEC system. Fig.5 shows a schematic diagram of a single CdSe nanocrystal on the gold substrate (a,b and c) and of two crystals - one on top of the other - representing two layers of crystals (d).

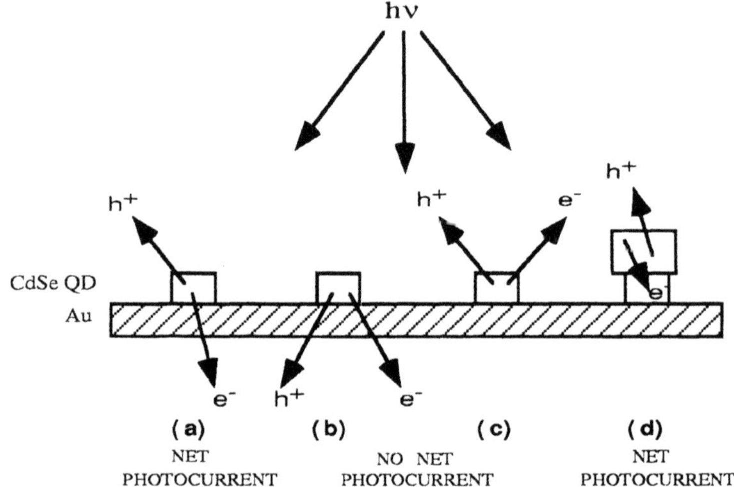

Figure 5. Schematic diagram of photogenerated charge transfer and loss pathways in PEC (see text).

In order to measure photocurrent, one charge must be transferred into the electrolyte and the counter charge into the Au (fig. 5a shows the situation for an "n-type" response; the direction would be reversed for a "p-type" response). Figs.5b and c represent electron-hole recombination through the Au substrate and the electrolyte respectively. In either of these cases, no photocurrent will be obtained. It is not surprising that the collection efficiency for the single layer (or half layer) is so low since, taking into consideration the fact that the electron-hole pair is generated so close to the Au, recombination through the Au substrate is intuitively expected to be an easier path than transfer of one of the charges to the electrolyte. When charge generation occurs in the second layer of crystals (fig. d), for recombination through the substrate to occur, *both* charges have to pass through an additional barrier - that between the two crystallites. In this case there is a better chance that one of the charges generated in the topmost crystal will be transferred into the electrolyte. The probability of recombination through the substrate decreases with the number of crystal boundaries that both charges must cross. However, since the major change in collection efficiency is between the first and second crystal layers and there is esentially no change after less than three layers, this suggests that the barrier for charge transfer between individual crystals is sufficiently high to allow one of the charges to react with an electrolyte species in preference to transfer from crystal to crystal. Acrually this is not surprising, since a similar assumption of a intercrystalline barrier to charge transport is necessary to explain the observance of the spectral shifts in aggregated nanocrystalline films. These shifts are characteristic of size quantization in isolated nanocrystals of the same size, where the good mechanical contact between individual crystals might be expected to preclude such a shift.

The above results present an interesting and counter-intuitive conclusion; grain boundaries *improve* the photocurrent collection efficiency rather than decrease it as is usually the case for photovoltaic cells in general. The reason for good collection efficiencies (0.6-0.7 peak efficiencies have been obtained in "thick" (ca. 100nm) nanocrystalline CdSe films) have been discussed previously [9]. Briefly, it can be explained by the fact that the electrolyte makes contact with each individual nanocrystal in the somewhat porous film. This means that the photogenerated charges are generated essentially at the crystal surface and one of them (usually the hole) can be readily transferred to the electrolyte. The other charge (usually the electron) has to pass through a fairly large number (on average) of crystal boundaries to get to the substrate; however, since the holes were asumed to be rapidly removed by the electrolyte (or possibly trapped at the crystal surface in sites of low recombination cross-section), and there are few, if any, non-photogenerated holes in the (intrinsic) CdSe, there is little opportunity for recombination unless the film thickness was much greater than 100nm (i.e., many more crystal boundarics).

The reason why one of the charges is preferentially transferred to the electrolyte is less obvious. We have recently shown, using surface photovoltage spectroscopy, that the charge which is preferentially injected into the electrolyte is determined predominantly *not* by the nature of the electrolyte, but rather by the trapping characteristics of the two charges at the nanocrystal surfaces; the more readily or more deeply trapped charge is preferentially injected into the electrolyte[10] (or conversely, the less-trapped charge more readily crosses the crystal boundaries).

Another feature can be seen in fig.4. The photoresponse of the thin films peaks at ca. 2.4-2.5eV, then decreases at higher energies. For the two thicker films (>2 crystal layers), no such peak occurs and the response is essentially flat at higher energies. This phenomenon was explained for thicker nanocrystalline films; the higher energy light is absorbed far from the substrate, resulting in a greater chance of recombination loss [9]. This explanation is not valid in the present case since few or even no grain boundaries are involved. We speculate that the phenomenon may reflect photocurrent arising from higher energy levels in these (quantized, and therefore discrete level) nanocrystals. The decrease in response occurs at an energy of 0.4-0.5eV above the first excited state. This could correspond, in theory, to either a transition from the first ($1S_{3/2}$) valence level to the second (1P) conduction level or from the spin-orbit split-off valence level ($1S_{1/2}$) to the first (1S) conduction level. While there is no experimental evidence for the occurrence of the first of these transitions, the second certainly can occur, as seen, e.g., by our EET results above. It therefore seems obvious to attempt to connect the decrease in the hole transfer to the electrolyte to the generation of higher energy (hot) holes. Based on our results outlined above concerning the role of charge trapping on the photocurrent, we speculate that the hot holes are more readily transferred through the crystal boundaries and to the substrate (or directly to the substrate for a single layer of crystals), resulting in increased recombination at the substrate. When more than one crystal boundary is involved (more than two layer of crystals), the hot hole will probably thermalise to the first valence level after passing through the first crystal boundary, and will then behave as a hole generated at that level with a good chance of being trapped and transferred into the electrolyte. This would explain why no decrease in response is seen in films thicker than two crystal layers.

While the above explanation is, admittedly, quite speculative, we do have further experimental evidence which strongly supports it. Fig.5a shows the PEC spectrum of a 2.5nm film which was, fortuitously, "p-type" as prepared (the films are usually "n-type" as deposited; however they occasionally show "p-type" behavior for reasons unknown to us at present). The striking feature of this spectrum is the *increase* in photocurrent at the same photon energy where the "n-type" samples exhibit a *decrease* in photocurrent. This behavior is consistent with our previous explanation. Hot holes are generated at ca. 2.5eV and more readily recombine at the Au with the large excess of free electrons there. Since in this case, this is the direction of charge flow required to produce photocurrent (in contrast to the "n-type" samples), the result is an increase in photocurrent due to the generation of hot holes.

4.3. I-V SPECTROSCOPY

We have developed this technique specifically to study our QDs. The technique, however, should be of much more general utility. It is based on the well-known method of I-V spectroscopy using a scanning tunneling microscope tip as one electrode and the sample as the other. This method has been used to obtain an approximation of Eg for semiconductor nanocrystals [11]. We have used the tip of an atomic force microscope (AFM) and evaporated a film ca. 40 nm thick of Pd or Pd/Au alloy onto the insulating tip. While the AFM tip is larger than that of the STM, due to the radius of curvature (typically 20 nm) and the relatively large average separation between individual QDs, only very few QDs, and possibly only a single one, will be contacted by the tip at any time.

Fig.6 shows I-V plots of two different samples (a, b) of CdSe on Au (nominal thickness of 1nm; equivalent to 20-25% coverage of the surface) together with their respective conductance plots (c, d). The two most notable features are a) the zero current region of ca. 1eV on both sides of zero bias and b) the steps in the regions where current flows, in particular in the tip negative bias region. This plot can also be given as differential conductance, dI/dV, as a function of bias (fig. b). Such a plot defines more accurately the gap between the regions of current flow. Even more noticeable is the sharpening of the steps in the regions of current flow into well-defined peaks.

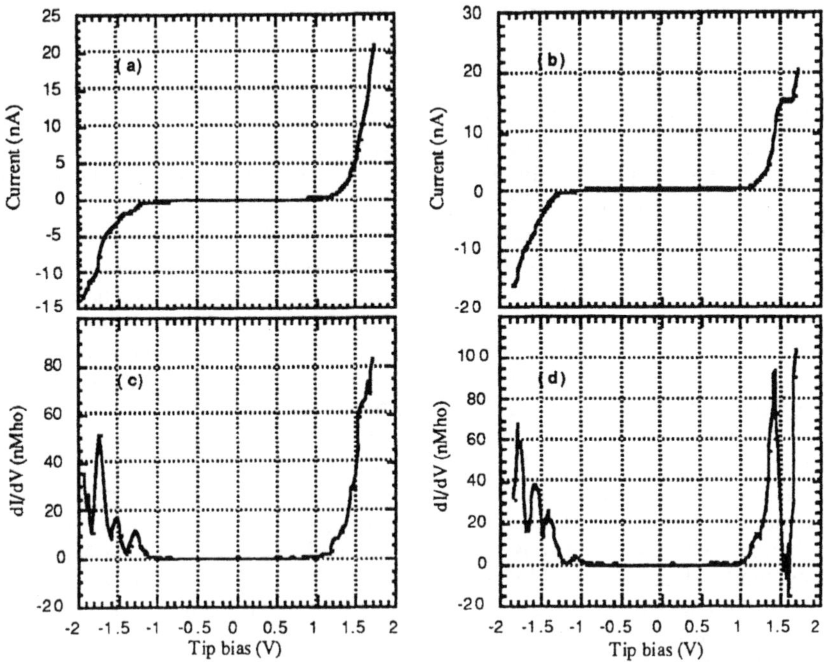

Figure 6. I-V spectra of a CdSe "film" (ca. 25% coverage with ca. 4-5nm QDs) for two different samples (a,b) and the corresponding conductivity spectra (c,d).

We first consider the steps in the current flow regions (the peaks or inflections in the differential conductance plots). The negative bias current flow region can be correlated with resonance between the tip and the empty conduction level(s), assuming almost all the bias voltage drops across a tip-sample "dipole" layer (see below). There are two possible obvious explanations for these conductance peaks. The first is that each peak represents resonance with a different level; the second that the peaks are due to a Coulomb charging effect.

The energy separation between the peaks is ca. 0.2eV (over many measurements, we find that it varies from 0.17-0.21eV for the separation between the first two peaks and somewhat less for that between the second and third). If this separation is a manifestation of higher energy, discrete levels, it should correlate with the known energy separation between the higher levels. There appears to be very little work carried out on higher energy levels in nanocrystalline quantum dots. Ekimov et al. have calculated some of the higher energy levels in CdSe and correlated these calculations with experimental measurements [13]. Their calculations appear to exaggerate the energy separation between the lowest excited states (the effective bandgap). Their values for the separation between higher levels appear to be more reasonable if corrected for this exaggeration in the bandgap and compared with the experimental results for the higher transitions in ref. 14 (the only other work we have seen which describes a higher lying ($1P_e$) unoccupied level). For a crystal size of 4nm, we estimate values of ca. 0.35 and 0.30eV for the energy spacing between the first ($1S_e$) and second ($1P_e$) empty levels and the second and third ($1D_e$) ones respectively. These values are considerably larger than the interpeak spacing measured by us, but not enough to disqualify this explanation in view of the uncertainties in the calculated values and since other factors, particularly interaction between electrons in the various levels, could modify the calculated values somewhat.

The second possible explanation for the conductance peaks is that they reflect Coulomb charging of the CdSe. The Coulomb charging effect is well known for metals, particularly in STM spectroscopy, where a tunneling barrier exists between STM tip and the sample. Briefly, it appears as a voltage gap in the I-V characteristics due to the discrete energy required to transfer a single electron into a very small junction [15]. This charging energy is given as $e^2/2C$, and becomes appreciable (i.e., greater than the thermal energy) for very small junctions where the capacitance is of the order of 10^{-18} Coulombs. A further characteristic of this phenomenon occurs if there is a barrier on the other side of the metal, and this barrier has a different resistance than the tunneling barrier between tip and sample. In this case, there will be a staircase-shaped I-V characteristic (the Coulomb staircase), with each step corresponding to an additional electron in the tiny metal structure.

For a semiconductor, and particularly for a QD, where there is a natural bandgap and discrete energy levels, Coulomb charging will introduce additional structure in the I-V characteristics which will depend on the relative energy spacings of the semiconductor energy levels and of the Coulomb charging energy [16-18]. Since the Coulomb charging energy is approximately additive to the semiconductor level energy, the bandgap measured by I-V spectroscopy will be greater than the actual gap by twice the charging energy, e^2/C, (once for the valence level and once for the conduction level).

We can make an approximate estimate of the charging energy for our system, based on a 4nm CdSe crystal and a parallel plate capacitor geometry with the electron density concentrated in the centre of the QD. We obtain a value for C of ca. 6.10^{-19} Coulombs. Using this estimation, we obtain a value for the charging energy of ca. 0.15eV. This value is reasonably consistent with, if a bit smaller than, the energy separation of ca. 0.2eV we find between the conductance peaks. Thus Coulomb charging appears to offer a reasonable explanation for the conductance peaks in the I-V characteristics. It should be noted that this interpretation requires that electron trapping occurs since the three peaks cannot represent three electrons in one level. In view of the rapid (sub-psec) trapping times characteristic of these nanocrystals and the maximum average residence time of electrons in the QDs (tens of psec estimated from the current at the inflection points in fig. 6 a and b), such trapping is expected to occur.

We now turn to the ca. 2eV zero current gap in the spectra. This gap is usually interpreted as the bandgap in corresponding STM studies of semiconductors and semiconductor nanocrystals [12,11]. Although not explicitly stated, this interpretation assumes that the bias voltage drops entirely across the tunneling gap and not across the semiconductor itself. If the latter occurred, applying bias would result in an increase in the semiconductor energy levels at the tip side relative to that at the sample side (i.e., a field across the semiconductor), and the barrier between the tip and the conduction level would (ideally) remain constant. Since the onset of current flow is assumed to occur when the tip Fermi level is in resonance with the conduction level, no current would flow until some other mechanism, probably tunneling through the field in the semiconductor, became operative. We assume the presence of a tunneling junction between tip and QD as a result of the oxide dipole layer that invariably exists on CdSe and related materials.

Based on the above argument that the intrapeak spacing (in the negative bias region) reflects the charging energy, we define the bandgap by the energy difference between the first negative bias peak and the first positive bias peak (inflection), *minus* twice the charging energy. From fig.6d , this energy is 2.55 - 0.42 = 2.13eV. This corresponds to a size of 4nm [8]. In addition, the essential symmetry of the current onset about the zero bias implies that the QDs are intrinsic (E_F is in the centre of the gap).

We have measured I-V plots of many QDs on different samples, and have also measured (by direct TEM imaging) the size distribution of the QDs. The sizes range from 3 - 6nm in diameter, with a sharp peak in the distribution at ca. 4nm for a sample deposited under the same conditions as those used for the I-V measurements. This distribution and peak corresponds to values for Eg of ca. 2.[illegible] - 1.95 and 2.15eV respectively [8]. These values agree quite well with the range of values of the bandgap measured by us on many different QDs (1.9 - 2.2eV).

We note that consideration of the charging energy argues more strongly against assigning the peaks to separate quantized levels. As we showed above, the peak spacing was somewhat too small to support such an argument. If we add the charging energy, the required level spacing would increase even further, resulting in an even greater discrepancy in the corellation between estimated and measured peak spacings. Unless our estimation of the charging energy is highly overexaggerated, the explanation whereby each peak reflects the addition of a single electron to the QD is therefore considerably more consistent with our measurements than that involving multiple levels.

We conclude with some speculation involving fig.6 (a,c). The spectrum in fig. 6c exhibits 4 negative bias peaks. However, the first peak is smaller than usual and is separated by nearly twice the other negative peak spacings from the second peak. The bandgap measured from this spectrum assuming this small peak is the first ($1S_e$) level is 1.75eV - an anomalously low value. If measured from the second peak, the value is 2.2eV, which is on the high side, but within, the bandgap distribution. We have observed such an anomalous peak in a number of measurements, and speculate that it represents a surface level close to the $1S_e$ level and that resonance occurs between the probe tip and this surface level.

Acknowledgment.

G. Hodes and I. Rubinstein acknowledge support of this work by the Office of Naval Research, grant No. N00014-1-1151.

5. References

1. Yoffe, A. D. (1993), Adv. Phys. **42**, 173.
2. Bányai, L. and Koch, S.W. (1993) *Semiconductor Quantum Dots*, World Scientific Series on Atomic, Molecular and Optical Physics - Vol.2.
3. Hodes, G., Albu-Yaron, A., Decker, F., and Motisuke, P. (1987), *Phys. Rev. B* **36**, 4215.
4. Hodes, G. (1993), *Isr. J. Chem.* **33**, 95.
5. Golan, Y., Margulis, L., Rubinstein, I., and Hodes, G. (1992), *Langmuir* **8**, 749.
6. Golan, Y., Margulis, L., Hodes, G., Rubinstein, I., and Hutchison, J.L. (1994), *Surface Science* **311**, L633.
7. Cardona, M., Shaklee, K.L., and Pollak, F.H. (1967), *Phys. Rev* **154**, 696.
8. Murray, C.B., Norris, D.J., and Bawendi, M.G. (1993), *J. Am. Chem. Soc.* **115**, 8706.

9. Hodes, G., Howell, I.D.J., and Peter, L.M. (1992), *J. Electrochem. Soc.* **139**, 3136.
10. Gorer, S., Hodes, G., Kronik, L., Fefer, E., Bachrach-Idan, N., Shapira, Y., and Leibovitch, M. (submitted for publication), Surface States and Photovoltaic Effects at CdSe Quantum Dot Films.
11. Zhao, X.K., McCormick, L.D., and Fendler, J.H. (1991), *Langmuir*, **7**, 1255.
12. Feenstra, R.M. and Mårtensson, P.(1988), *Phys. Rev. Lett.* **61**, 447.
13. Ekimov, A. I., Hache, F., Schanne-Klein, M.C., Ricard, D., Flytzanis, C., Kudryavtsev, I.A., Yazeva, T.V., Rodina, A.V., and Efros, Al.L. (1993), *J. Opt. Soc. Am. B* **10**, 100.
14. Norris, D.J., Sacra, A., Murray, C.B., and Bawendi, M.G. (1994), *Phys. Rev. Lett.* **72**, 2612.
15. Averin, D.V. and Likharev, K.K. (1991) *Mesoscopic Phenomena in Solids*, Elsevier, Amsterdam.
16. Beenakker, C.W.J. (1991), *Phys. Rev. B* **44**, 1646.
17. Averin, D.V., Korotkov, A.N., and Likharev, K.K. (1991), *Phys. Rev. B* **44**, 6199.
18. Johnson, A.T., Kouwenhoven, L.P., de Jong, W., van der Vaart, N.C., Harmans, C.J.P.M., and Foxon, C.T. (1992) *Phys. Rev. Lett.* **69**, 1592.

nm-SEMICONDUCTOR PARTICLES AND MOLECULAR AGGREGATES AS REDOX SPECIES

F. WILLIG, K. SCHWARZBURG, B. TRÖSKEN, J. MAHRT, L. MOTTE*, AND M.-P. PILENI*
Fritz-Haber-Institut der Max-Planck-Gesellschaft
Faradayweg 4-6, 14195 Berlin, Germany
*Universite P. et M. Curie (Paris VI)
4 Place Jussieu, 75231 Paris, France

1. Introduction

It is well established that the electronic levels in semiconductors shift when the latter are prepared in the form of nm-size particles [1]. Correspondingly the optical behavior of such small particles depends on their size and shape. One can expect that the redox behavior will also change with the size of nm-particles. Knowledge in the latter field is still in an infant state. In this paper experiments will be discussed that shed some light on the redox behavior of nm-semiconductor particles, firstly as donors in the excited electronic state and secondly as electron acceptors. Transport of excess charge carriers through a chain of nm- size semiconductor particles will be briefly discussed since this latter process is important for some of the suggested device applications for nm-particles. Presently available experimental results concerning electron transfer with semiconductor nm-particles suffer from the fact that the detailed structure and chemical nature of the interface is not really known for these systems. The experimental systems are complicated and the measurements can suffer from systematic faults. Experimental results obtained at closely related systems can serve as guidelines, e.g. electron transfer measurements on covalently linked molecular donor-spacer-acceptor systems and on quantum well-barrier-quantum well systems. Progress in this field appears highly desirable, since very interesting practical applications have been suggested for systems prepared from nm-size semiconductor particles involving electron transfer reactions and electron transport. Some of these systems have shown promising features. One prominent example is an electrode prepared from nm-size anatase TiO_2 colloidal particles. This electrode is prepared with nm-particles that retain to a large degree their individual properties but are glued together such that they can facilitate efficient transport of excess charge carriers. This electrode provides a greatly enhanced surface area, e.g. by about a factor of 500, and serves as electron acceptor and transport medium for excess electrons [2]. Injection of electrons into this electrode has been realized via suitable adsorbed dye molecules in the excited electronic state. Up to 10 percent efficiency was achieved in a two electrode electrochemical cell for the conversion of 1 AM sunlight into electrical energy. A similar type of electrode was also employed as photo-electrochromic device [3]. In other applications layers were prepared of nm-size TiO_2 or SnO_2 colloidal particles on solid substrates and a thin layer of metal was deposited on top. At the many tips formed on such a nm-structured gold or silver surface the electric field is strongly enhanced and facilitates the well-known surface-enhanced Raman effect (SERS) [4,5]. It has been suggested that such surfaces can useful in analytical chemistry applications. More recently, nm-size semiconductor particles were employed successfully as

E. Pelizzetti (ed.), Fine Particles Science and Technology, 591–608.

recombination centers in light emitting diodes [6,7]. The wavelength of the recombination light can be tuned with the size of the nm-particles. Already several years ago Weller and coworker [8] demonstrated that the process of dye sensitization, well known for the photographic emulsion and semiconductor photo-electrochemistry, can be performed also by nm-size semiconductor particles in the excited electronic state. Another interesting field is the application of nm-size oxide semiconductors as photo-catalytic agents [9]. For the improvement of any device application involving electron transfer and electron transport of nm-size semiconductor particles it is important to investigate and clarify the basic processes.

2. Electron transfer in well-defined related systems

It is helpful to consider first characteristic rate constants and electron transfer parameters for systems that are very well characterized with respect to the "interfacial" region through which the electron is transferred. These data are best understood with the help of some basic expressions describing electron transfer. In most experimental cases the rate constant of electron transfer can be described with an expression that is well known as Fermi's golden rule [10, 11, 12, 13].

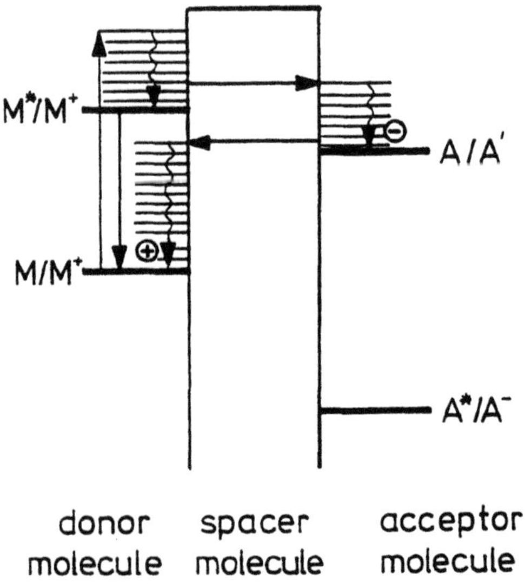

Fig. 1

When the reactant is a molecule or another type of localized state and electron transfer is coupled to intramolecular vibrations or to phonon modes of a surrounding medium the simplest expression is obtained for the harmonic oscillator approximation with the density of states 1/hω and the thermally weighted Franck-Condon factor that can be expressed as a well known analytic expression [11].

$$k = 2\pi/h \,|\, V^2 \,|\, 1/h\omega \; GFC \qquad (1)$$

In equ. 1 the important parameters are V = electron transfer integral describing the electronic coupling between reactant and product state and GFC = thermally averaged Franck-Condon factor describing the summation over the occupied excited vibrational states in the reactant state and their overlap with the isoenergetic nuclear wavefunctions in the product state. The GFC is controlled by two important parameter, the first is the reorganization energy that describes the change in the nuclear equilibrium coordinates when going from the reactant to the product state, and the second is the free energy difference, i.e. difference between the standard redox potentials, between the electronic level of the reactant and that of the product state. The high temperature version of the GFC, i.e. when $k_BT > h\omega$ for the phonons and vibrations that are coupled to electron transfer, gives an Arrhenius behavior where the activation energy has a quadratic dependence on the free energy difference of the standard redox potentials [14]. When there is only one electronic acceptor level the maxiumum attainable GFC is reached when the standard free energy difference is downhill from the donor to the acceptor state and equal to the reorganization energy. The value of the corresponding rate constant can give a good idea of the electron transfer integral V and thus of the electronic coupling when the standard free energy difference and the reorganization energy are known. A well defined experimental situation with respect to the electron transfer integral V is realized when a donor molecule is covalently linked via a rigid spacer molecule to an acceptor molecule. This experimental situation is illustrated in Fig. 1. A great number of time-resolved measurements of electron transfer reactions from the excited electronic state of a donor

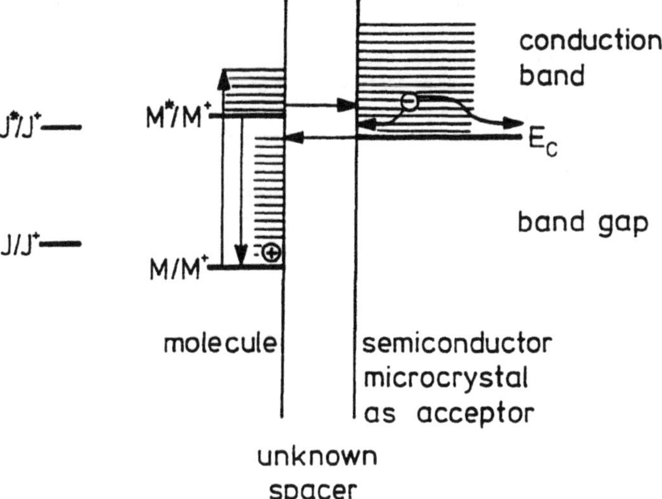

Fig. 2

rate constant of electron transfer from the excitonic state of a dye J-aggregate into the (111) surface of octahedral AgBr microcrystals of 0.9 μm length, the curve in the middle gives the rate constant for electron tranfer from the excited state of adsorbed dye monomers into the same AgBr microcrystals, and the lower curve illustrates the rate constant for a typical redox ion in solution. The parameter λ is the respective reorganization energy of the electron transfer process. In electrochemical nomenclature the data in Fig.3 show the temperature dependence of the fictitious photo-exchange currents in the respective experimental system. The experimental electron transfer integrals, i.e. 5-10 cm^{-1}, are of the same order of magnitude as those found for the molecular donor-spacer-acceptor systems shown in Fig.1, where the rigid molecular spacer is well defined. The corresponding detailed information on the structure and chemical nature of the interface is not ensured in the case of the wet chemistry prepared systems whose data are shown in Fig.3. However, there is indirect evidence that these thiocarbocyanine dyes, and in particular the J-aggregate, are directly adsorbed on the surface of AgBr, since the narrow absorption peak of this J-aggregate was found at an identical spectral position on several different substrate surfaces, but showed a distinct blue shift by by 5-10 nm on the (111) surface of AgBr [18]. The small reorganization energy in the case of the J-aggregates shows that electron transfer couples only to solid state phonons of low energy, i.e. the fit to the J-aggregate data gave 51.4 cm^{-1}, and not to the intramolecular vibrations of the individual dye molecules with energies in the range of 1000 cm^{-1}. This type of coupling in the electron transfer process is in accord with the coupling of the optical transition in the molecular J-aggregate to solid state like phonons but not to intramolecular vibrations [19].

In connection with the data shown in the two upper curves of Fig.3 it is important to take notice of some of the experimental difficulties in studying electron transfer in interfacial systems. There can be a serious difficulty in regenerating the true ground state of the system if the solid state part of the system is spatially extended and contains deep traps, or if there occurs an irreversible competing reaction step in either part of the system. In both cases the measured data can be severely distorted even at low photon exposure and can represent only inactive spots in a heterogeneous system. Fig.4 shows a clear example for such a behavior, i.e. the disappearance of the J-aggregate fluorescence peak on the (111) surface of AgBr upon continued exposure to a low photon dose. The fast almost "virgin" decay curve a in Fig.4 is converted into the remaining slow decay curve d in Fig.4 after exposure to a low dose of photons. Curve d originates only from dye molecules adsorbed on inactive spots on the surface. The progressive bleaching of the J-aggregate fluorescence peak that corresponds to the transition from decay curve a to the decay curve d is shown as inset in Fig.4. Data shown in the upper curves in Fig.3 were collected by minimizing the photon exposure to the "virgin" situation of curve a on a given spot and shifting the laser beam to another spot for each new measurement. This procedure had to be repeated for every different temperature reading down to 40 K. Since the reaction was slightly uphill, by about 50 cm^{-1}, the turn over due to electron transfer was negligible below 40 K, and the result of the bleaching process carried out at a higher temperature was safely recorded by taking the sample again down to 5 K and measuring the fluorescence spectrum, inset of Fig.4, before returning to a higher temperature for measuring the decay curve with the experimental procedure described already for curve a in Fig.4. We have presented this case in some detail here since it is very important to realize that many investigations of interfacial systems can have serious pitfalls that are perhaps not so well-known and have not been carefully identified for [illegible] published data in this field.

It is instructive to compare the molecular-molecular and the molecular-solid systems with a system wehre also the excited donor state is prepared in a solid layer, where upon optical excitation electrons are scattered from the electronic level in the solid donor layer to another electronic level in the adjacent solid acceptor layer via emission or absorption of a molecule were performed in the last years at such molecular systems with a well-defined molecular spacer. With rigid spacers corresponding to 5-7 saturated -C-C- bonds corresponding to an edge to edge distance between donor and acceptor molecule in the range of 7-9 Å the maximum rate constants were always found to fall into the range of a few reciprocal ps [15]. The corresponding electron transfer integrals are of the order of 10 cm^{-1} or a few meV.

The GFC can become larger than in the above experimental case and can approach its maximum value of 1 when there are many electronic acceptor levels with different energies serving as acceptors in the wide empty conduction band of a semiconductor [16]. This situation is illustrated in Fig.2. When the electronic level of the donor molecule is located sufficiently high above the edge of a wide conduction band the GFC can approach the value 1, and the rate constant reaches its maximum possible value that is only controlled by the level spacing of the main vibration coupled to electron transfer and by the electron transfer integral V. Considering Wannier acceptor functions centered at the adjacent atoms on the surface of the semiconductor [17] the corresponding electron transfer integral must be of the same order of magnitude as above when the same donor molecule is fixed at a similar distance, i.e. 7-9 Å, from the surface of a semiconductor and its excited donor level is located sufficiently high above the conduction band edge. Thus, the maximum rate constant of electron transfer in the latter system must be in the range of a few reciprocal ps or even faster. When the standard redox potential difference is zero the rate constant of electron transfer, and particularly its temperature dependence, shows the magnitude of the reorganization energy and of the electron transfer integral. Two experimental results are shown in Fig.3. The highest curve shows the

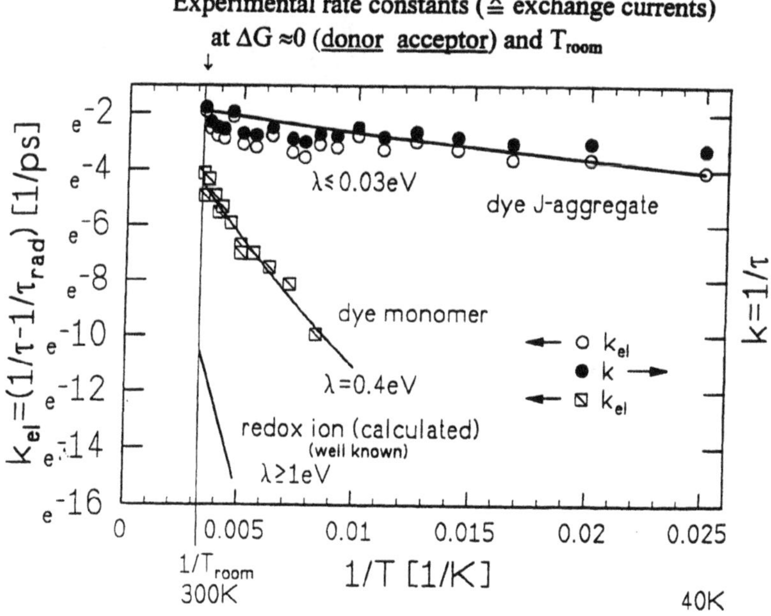

Fig.3

phonon. Again Fermi's golden rule expression can be applied but the electronic coupling is now of a non-adiabatic type corresponding to electron transfer connected with the absorption or emission of a phonon [13]. In contrast to equ.1 the molecular Franck-Condon factor expression GFC does not appear in this case and the summation is carried out over all the final electronic states that can be reached in the scattering process to the adjacent layer when the electron absorbs or emits a phonon. Corresponding experiments were carried out in Multiple-Quantum-Well structures by Feldmann et al. [20]. When the width of the well is sufficiently narrow the electronic wavefunction becomes sizable at the interface of the two layers and electron transfer times in the range of a few ps were measured. The temperature dependence of the rate constant is controlled by the phonon occupancy, and for the case of only phonon absorption, i.e. for a slightly uphill reaction, it is similar to the highest curve of Fig.3, becoming

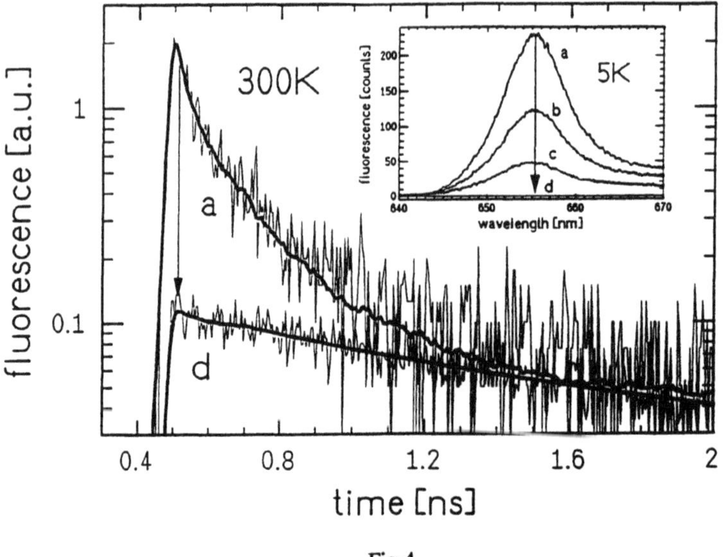

Fig.4

slightly steeper than this curve only around and above room temperature.

To summarize the main points, electron transfer times in the range of a few ps were measured at room temperature in a variety of different molecular-molecular systems, in combined interfacial molecular-solid state systems, and in interfacial solid-solid systems. To observe this ps time scale in a molecular system in solution, it is necessary that the downhill energy difference between reactant and product state compensates the large reorganisation energy. In most rigid solid-solid systems an isonergetic or even slightly uphill electron transfer reaction will show already this ps time scale since the reorganization energy is extremely small. Much slower electron transfer times, e.g in the range of ns, were and still are sometimes reported for interfacial systems prepared with wet chemistry methods. These slow rate constants can have a number of possible origins, e.g. an unknown spacer barrier to the electron transfer reaction, the reaction going strongly uphill, or a weakly downhill electron transfer reaction combined with a large molecular reorganisation energy as illustrated in the lower curve of Fig.3. Also the more trivial origin illustrated in Fig.4, i.e. incomplete regeneration of the system when exposed to a higher dosage of light, is a serious experimental difficulty, and its effect on the experimental data must be be carefully checked.

3. Optical and redox behavior of molecular aggregates and of nm-particles of inorganic semiconductors

In the upper part of Fig. 5 the change in the lowest excited electronic state e and in the lower optical transitions is illustrated when going from an isolated aromatic monomer molecule via a rigid dimer configuration, two cases are shown, to a strongly coupled J-aggregate [21]. Of

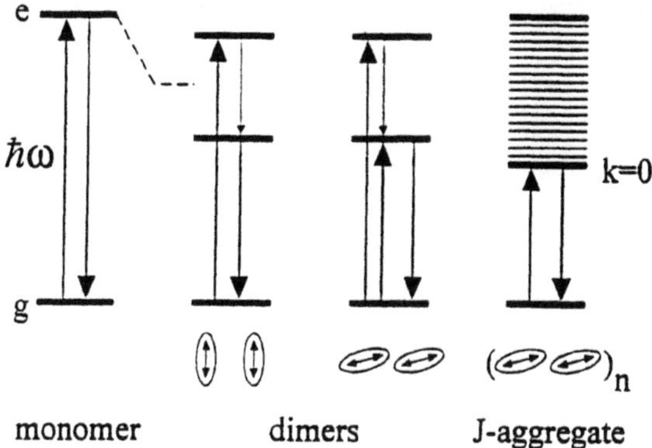

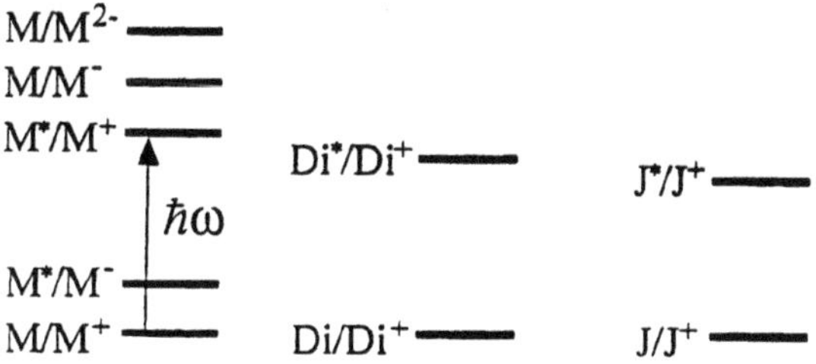

Fig.5

particular interest is here the case of the J-aggregate. In the $k = 0$ excited state at the bottom of the J-aggregate's excitonic band the oscillator strength of several monomer units are combined leading to a strong optical transition in absorption and luminescence with a correspondingly reduced lifetime. The excited state is a neutral excitonic state. Charge transfer excitons, i.e. those with a spatial separation between electron and hole, do not show up in the absorption spectrum, at least at lower photon energies. Unlike in the case of nm-semiconductor particle electron and hole quasiparticles with opposite k-vectors cannot describe the exciton. Only the transition to the $k = 0$ state in the excitonic band appears at low temperature in the absorption spectrum. The spectrum in Fig.6 shows the inhomogeneously broadened absorption and luminescence spectrum of a J-aggregate formed on a quartz substrate by the same thiocarbocyanine dye, whose electron transfer rate constants for the J-aggregate on AgBr was presented in the highest curve of Fig.3. The absorption spectrum of the dye monomer in solution is shown as dashed cuve in Fig.6 for comparison. The large red shift of the 0,0-level when going from the monomer dye to its J-aggregate is clearly demonstrated in Fig.6. Preparation of the J-aggregate on a solid substrate does not include all the dye species, and the absorption of some remaining dye monomers and smaller dye aggregates can be seen in the solid line in Fig.6 in addition to the absorption of the J-aggregates. The spatial extension of the excitonic state of the J-aggregate is usually much smaller than the geometrical extension of the J-aggregate. Scattering of the excitons at phonons and at stacking faults limits the spatial extension of the exciton. Until now a systematic increase of the J-aggregate's oscillator strength with the number of monomer units incorporated into the J-aggregate has not yet been demonstrated.

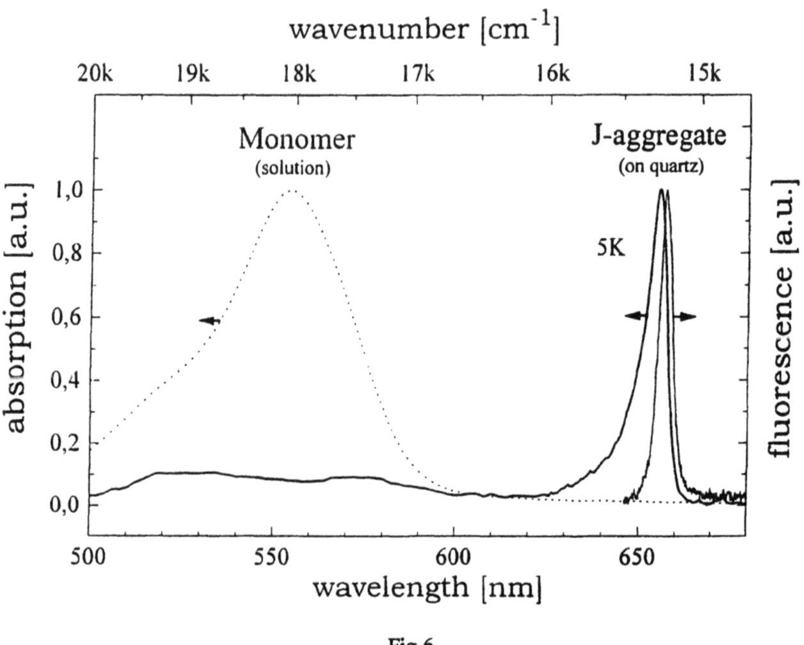

Fig.6

The lower part of Fig.5 illustrates redox energy levels in the dye monomer M, in the dimer Di, and in the J-aggregate J. Several different redox levels are shown for the monomer and only two levels, i.e. the oxidation potential in the ground and excited state, for the dimer and the J-aggregate. Excitation of the lowest excited singlet state of the monomer requires in general a much smaller photon energy than corresponds to the energetic separation between the oxidation and the reduction potential. This difference would disappear only in the case of complete screening of electron and hole. The addition of a second electron to the already reduced monomer requires a more reductive potential due to the repulsion energy of the already present excess electron and electron-electron correlation. Delocalization and screening of the hole in the dimer and in the J-aggregate lead to a decrease in the ionisation energy, or oxidation potential, with respect to the monomer. On platinum the oxidative shift of the reduced monomer requires a more reductive potential due to the repulsion energy of the already present excess electron and electron-electron correlation. Delocalization and screening of the hole in the dimer and in the J-aggregate lead to a decrease in the ionisation energy, or oxidation potential, with respect to the monomer. On platinum the oxidative shift of the oxidation potential was 0.15 eV when going from the monomer of the above thiocarbocyanine dye to the J-aggregate [18]. The oxidation potential for removal of a second electron from the same J-aggregate has not yet been identified. The potential difference between the first and second oxidation potential must depend strongly on the spatial extension of the J-aggregate and on the screening of the charges within the J-aggregate and via the electronic and nuclear response to the excess charge in the environment.

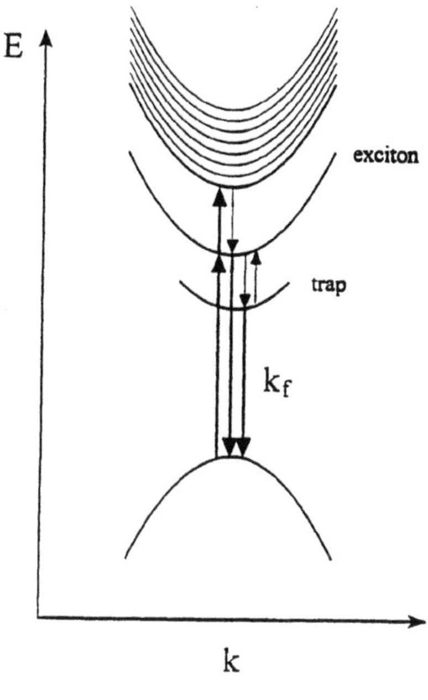

Fig.7

In contrast to the case of the dye J-aggregate (Fig.6) the excitonic transitons in nm-semiconductor particles can be described by the creation of electron and hole quasi-particles. Direct optical transitions and some non-radiative transitions are illustrated in Fig.7 for a nm-semiconductor particle. It is well documented that the separation between the hole and electron states widens when the size of the semiconductor particle is decreased to a diameter smaller than the excitonic radius in the bulk material [1]. The excitonic transitions at the low energy side of the absorption spectrum have been identified and assigned by the groups of Bawendi [22] and of Alivisatos [23] for the case of well-prepared CdSe colloidal particles. These excitonic peaks are immediatlely followed by a steeply increasing absorption unlike the case of the J-aggregate shown in Fig.6. The assignment of the transitions following at higher energies is still incomplete. In some cases they may be derived from the crowded transitions

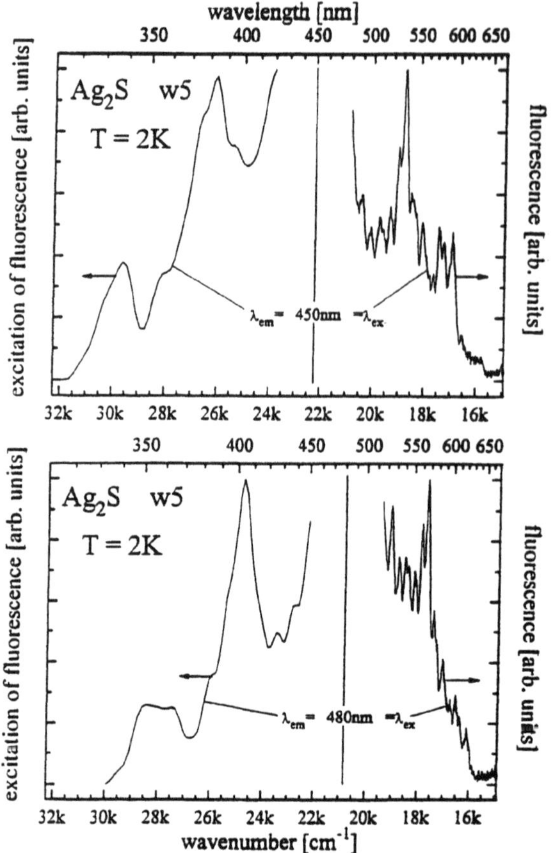

Fig.8

seen already in the molecular constituents. Alivisatos et al. [23] reported recently that relaxation of these higher states leads only with small probability to the population of the lower excitonic states. Fluorescence of nm-particles originates usually from trapped state for either electron or hole, or both. Excitons relax in all cases in less than one ps into shallow trap states [24]. Excitonic transitions can be completely hidden in the room temperature absorption spectrum of nm-particles. Fig.8 shows luminescence spectra and excitation spectra of luminescence for Ag_2S particles prepared in inverse micelles that do not show any structure in the absorption spectra at room temperature. The average diameter of the particles was determined with TEM in the range of 20 Å with a halfwidth of about 10 Å for the distribution. As can be seen in Fig.8 pronounced excitonic peaks in the excitation spectra of the luminescence followed the excitation wavelength when the latter was moved through the distribution of the particle sizes. For larger particles this shift was observed over a much wider range than shown in Fig.8. At higher photon energies the the excitation spectra show a low yield for populating the light emitting state. an observation also reported by Alivisatos et al.

[23] for the CdSe particles. The phonon structure seen in the luminescence spectra on the right hand side of Fig.8 agree with the energies reported for bulk phonons in the stable monoclinic crystal modification of Ag_2S. Lattice parameters determined from X-ray scattering agree with the crystal structure of the stable bulk phase. The main peak in the luminescence spectrum is shifted by about 3500 cm^{-1} to the red with respect to the excitation wavelength. This was confirmed also with time-resolved luminsecence measurements. The latter revealed a lifetime in the range of 2 ns for the light emitting state at 2 K that decreased by about a factor of 2 at room temperature. The peak in the fluorescence spectrum shifted in the same way through the distribution of particles sizes when the excitation wavelength was shifted as already noted above for the excitonic peaks in the excitation spectrum of the luminescence. Thus, the excitonic levels of the light absorbing state and of the light emitting state are linked and both shift blue with decreasing particles size. Light emission in this system is attributed to a trap state that mixes with and shifts with the excitonic state. The two excitonic peaks in Fig.8 are attributed to spin-orbit splitting. AgBr nm-particles are of considerable interest for photographic emulsions. Some more details of the optical and redox behavior of these particles will be published elsewhere.

Recently it was reported [25] that the approximate range of the reduction potential of nm-ZnO particles can be determined from electrical and optical signals measured in dependence on the electrochemical potential imposed on the ZnO nm-particles at room temperature. The particles were deposited as a thin film on a conductive support, and the latter was mounted as an electrode in an electrochemical cell. The reduction potential deduced from the onset of pronounced electron injection into the deposited film was found to shift by up to 0.4 V towards more negative potential, when the average diameter of the particle distributions was decreased from 250 Å to to 37 Å. A similarly determined shift in the reduction potential by 0.2 V with respect to bulk anatase for a thin film composed of anatase TiO_2 particles was attributed to the 10 to 20 Å particle diameter [26]. However, studies of oxide semiconductor surfaces, e.g. ZnO, in ultra-high-vacuum showed that the ionisation energy and electron affinity can shift by up to 1 V depending on the preparation and ambient conditions [27. Therefore, it would be important to establish the spread in the above electrochemical data in different laboratories for samples prepared with wet-chemistry methods. Detailed information on the charging behavior of nm-semiconductor quantum dots has been obtained with electrical measurements at He-temperature [28], where the dots were created with split-gates formed on AlGaAs-GaAs heterostructures. Assuming a linear relationship with the reciprocal radius the change in the potential was found in the range of 25 mV for an increase by one electron in the charge on a quantum dot of 40 Å radius and a dielectric constant of 13 corresponding to GaAs. Already the first atomic layer in direct contact with the nm-particle can strongly influence the spatial distribution of the excess charge in the nm-particle. Also the presence or absence of a counter charge in the particle's vicinity can influence greatly the screening of the excess electron and thus the potential seen by the approaching electron. The influence of just one atomic surface layer on the charge distribution of a nm-particle can be seen clearly in the pronounced shift in the peak position and in the increase to 0.9 eV halfwidth for the surface plasmon peak of small silver particles, e.g. 20 Å diameter, when sulfur atoms are attached to the surface [29]. A similar effect had been observed earlier when small Ag-particles were immersed in a CO matrix [30].

4. Experimental results on electron injection by excited dye molecules and simple considerations concerning electron transport through a colloidal TiO_2 electrode

Fig.9 gives an artist's view of the tip of a finger protruding into the solution as part of the TiO_2 electrode prepared via deposition of colloidal particles on the surface of a conducting support. The finger is in contact with a redox couple, Red and Red^+ added in the form of the iodide/iodine couple, that regenerates the oxidized dye species D^+ and defines the redox

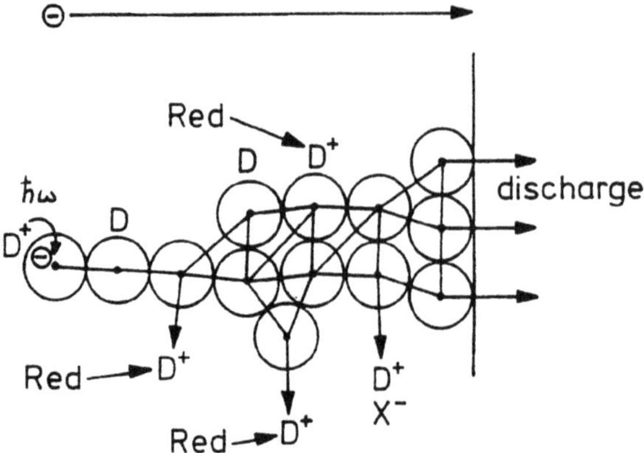

Fig.9

potential in the system in the dark. Injection of excess electrons by excited dye molecules adsorbed at the surface of the colloidal electrode is identified by the corresponding photocurrent. A clearer picture of the injection process can be obtained when applying a dye molecule D with negligible singlet-triplet conversion and with a pronounced vibrational structure in the absorption and fluorescence spectrum. For this purpose the aromatic molecule perylene was employed. However, a modified chromophore was employed where four bulky spacer groups, tetra-butyl, were covalently attached to the four corners of the perylene molecule to form TTB-perylene. This elliminate the complicated role that dye aggregates in the adsorbed layer usually play in such a system [16]. This TTB-perylene molecule retains the complete vibrational structure of the monomer luminescence even when deposited in the form of a microcrystalline powder [31], due to the distance to any molecular neighbor or surface imposed by the bulky tetra-butyl groups. The upper curve in Fig.10 shows the vibrational structure of the perylene moiety in the excitation spectrum of the dye-sensitized photocurrent in a 4 µm thick colloidal TiO_2 electrode kindly donated by P. Liskas of Prof. Grätzel's lab. The lower curve in Fig.10 shows the time-resolved fluorescence spectrum of the adsorbed TTB-perylene layer. Reabsorption in the thick dye layer elliminated almost completely the first vibrational peak in the fluorescence spectrum and distorted also the second peak. The decay behavior was fitted with 3 time constants where the amplitude of the fast component was 25 times larger than those for the slower components. The fast fluorescence component had a time constant shorter than 20 ps, the value of 20ps was the apparent value delivered by the least square fit employing a convolution with the apparatus response function of 37 ps halfwidth. The fast time constant was independent of temperature between 300 K and 150 K. The slower time constants were 1.5 and 4 ns, i.e. the same that were obtained when the TTB-perylene adsorption layer was formed on a quartz plate where electron transfer quenching does not occur. The fast time constant was attributed to electron transfer quenching of the excited singlet state of TTB-perylene. The appearance of the slow fluorescence decay constants can be understood from the experimental result shown in Fig.11.

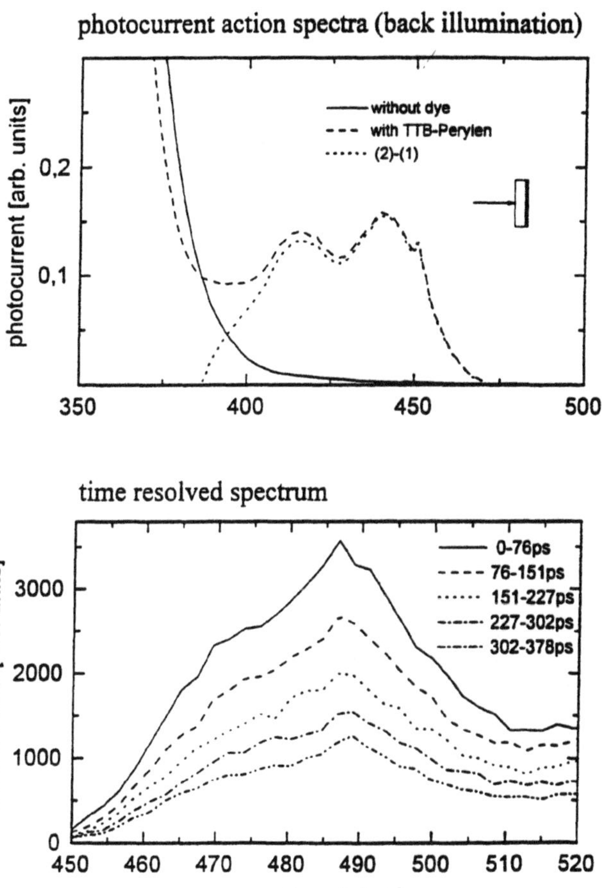

Fig.10

The fluorescence decay curve changed when the 10^4 times attenuated beam of a frequency doubled Ti-saphire laser was moved over the dye covered colloidal TiO_2 electrode. The laser beam encountered active spots with fast fluorescence decay and inactive spots with slower fluorescence decay. Clearly, the surface properties of this electrode vary from spot to spot and there are dye species on the electrode surface that decay without being affected by electron transfer to a TiO_2 colloid. With the attenuated laser beam dye bleaching, as shown in Fig.4 for the photographic system, was not observed in the present system, though dye bleaching could be provoked also in the present system when applying a high laser power. It is clear from Figs.10 and 11 that the ps time scale expected for electron transfer on the basis of the results discussed under point 2 above is clearly occurring in the present system at well-behaved spots of the complicated colloical TiO_2 electrode prepared with wet-chemistry methods. It has been

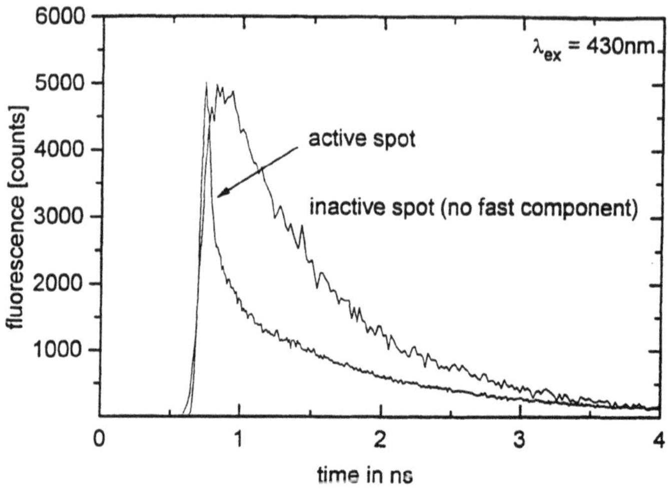

Fig.11

shown before [32, 16] that the fluorescence decay of rigid dye molecules on the surface of the TiO_2 electrode is fastest at very low dye coverage and becomes slower with increasing dye coverage. This observation suggests that dye adsorption is stronger at surface sites that permit fast electron transfer compared to other adsorption sites, from where electron transfer into the TiO_2 electrode is slow.

The transport of injected electrons from the place of adsorbed dye molecules through the fingers of the colloidal TiO_2 electrode and finally capture at the planar back electrode was investigated by analyzing the rise of the photocurrent in the ms range [33]. Electron injection occurred as depicted in Fig.9 via electron injection from the excited state of adsorbed dye molecules. The rise of the photocurrent occurred after a pseudo-stationary light source was switched-on with a rise time of a few μs. Transport of electrons was also measured as photocurrent transients in the μs to ms range [34]. In the latter case electron-hole pairs were generated with a short laser pulse in the collodial TiO_2 electrode. In the latter case an ITO layer and a Pt layer at the opposite sides of the colloidal TiO_2 layer were employed as solid contacts. Both types of transients were measured in dependence on the applied bias voltage. The two different measuring techniques gave the same qualitative picture for electron transport through the colloidal electrode. The transport was dispersive due to multiple trapping of the injected electrons during their transit to the back electrode. The occupation of deep traps, with e.g. 0.5 eV trap depth, was saturated at a high enough electron injection level, i.e. at a high enough light intensity. Saturated occupation of deep traps led to faster transport, sine it was controlled only by electron trapping in a distribution of shallow traps. Systematic studies in dependence on the doping level of the colloidal electrode have not yet been carried out. The potential distribution seen by the injected electrons when moving about in the colloidal electrode is not known for the case of photocurrent flow in the absence of an applied voltage. It appears reasonable to assume that already the fairly high dielectric constant of anatase together with the fs-response of a polar solvent to the creation of an excess charge [35], and also the at least ns-response of a high concentration of background electrolyte will screen the charge of the excess electron as long as the fingers of the colloidal electrode are sufficiently thin. The completely screened excess electron can move about between the same energy levels of identical neighboring colloidal particles. Transport of the excess electron in such a finger of the colloidal electrode is illustrated in Fig.9 with the simple picture of a random walker moving about in a nework of identical electronic levels. The excess electron can either disappear irreversibly into the increasingly thicker extension of the electrode finger towards the back contact or recombine with redox species as illustrated in Fig.9. Direct transfer via a deep trap level is also shown as a possibility at the third colloidal particle from the left. To obtain maximum screening and reside at the minimum of the potential the excess electron is expected to reside mostly in the center of a colloidal particle of the electrode finger. The excess electron can be destroyed via electron transfer when residing on a colloidal particle with either an oxidized dye species adsorbed, including the oxidized parent dye molecule, or an oxidized species of the added redox couple residing on the surface of the particle. Repeated encounters of the excess

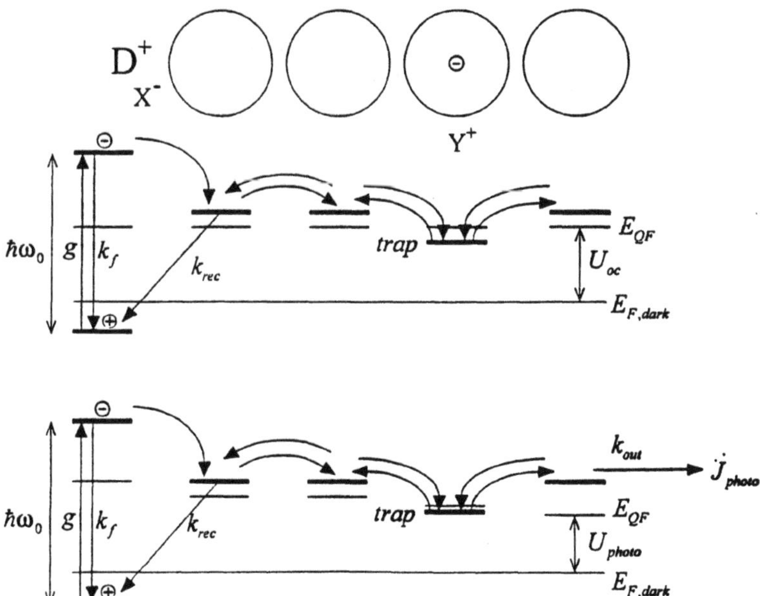

Fig. 12

electron with both these oxidized redox species can occur before the excess electron is destroyed, since the time constants for both these electron transfer reactions are expected to fall into the ns range, or even μs to ms range, due to a small Franck-Condon arising from the large downhill free energy difference for only one electronic acceptor level, corresponding to the so called inverted Marcus region. Irreversible escape with rate constant k_{out} is assumed at some distance away from the location of the oxidized dye, since firstly the electrons are collected irreversibly at the back electrode and secondly the number of steps leading away from the tip of the finger is assumed to increase towards the back electrode due to an increasing thickness of the electrode finger. The latter geometry can be assumed since the colloidal layer is formed on top of the back electrode as solid substrate. The simplest case is illustrated in Fig. 12. It is straightforward to calculate the photocurrent yield and open circuit voltage for this situation [36]. The yield for electron escape is high when both, the rate constant for the jumps between adjacent colloids and the rate constant k_{out} at the exit, are greater than the rate constant for recombination at the entrace. The current yield is simply controlled by only the branching ratio of the rate constants for the forward step and the recombination step at the entrance, if a downhill potential difference greater than $k_B T$ exists between adjacent colloids in the chain. Information on these rate constants is not yet available for such complicated collodial networks. To prepare for a better understanding of transport in such collodial systems it appears necessary that collodial networks are being built, where the interfacial layer between the colloidal particles can be controlled and the doping level can be varied systematically.

References

[1] Brus, L.E. (1984) J. Chem. Phys. **80**, 4403

[2] Nazeeruddin, M., Kay, A., Rodicio, I., Humphry-Baker, R., Muller, E., Liska, P., Vlachopolous, N., and Grätzel M. (1993) J. Am. Chem. Soc. **115**, 6382

[3] Hagfeldt, A., Vlachopolous, N., and Grätzel, M. (1994) J. Electrochem. Soc. **141**, L82

[4] Bello, J. M., Stokes, D. L., and Vo-Dinh, T. (1989) Anal. Chem. **61**, 1779

[5] Chumanov, G., Sokolov, K., Gregory, B. W., and Cotton, Th. M. (1995) J. Phys. Chem. **99**, 9466

[6] Colvin, V. L., Schlamp, M. C., and Alivisatos, A. P. (1994) Nature **370**, 354

[7] Dabbousi, B. O., Bawendi, M. G., Onitsuka, O., Rubner, M. F. (1995) Appl. Phys. Lett. **66**, 1316

[8] Vogel, R., Pohl, K., and Weller, H. (1990) Chem. Phys. Lett. **174**, 241

[9] Linsebigler, A. L., Lu, G., and Yates , J. T. (1995) Chem. Rev. **95**, 735

[10] Kubo, R. (1952) Phys. Rev. **86**, 929

[11] Jortner, J. J. (1976) Chem. Phys. **64**, 4860

[12] Huang, K.; Rhys, A. (1950) Proc. R. Soc. Lond. **A204**, 406

[13] Price, P. J. (1981) Annals of Physics **133**, 217

[14] Marcus, R. A. (1963) J. Phys. Chem. **67**, 853

[15] Sakata, Y., Tsue, H., O'Neil, M. P., Wiederrecht, G. P., and Wasielewski, M. R. (1994) J. Am. Ch,em. Soc. **116**, 6904

[16] Willig, F. (1995) Surface Electron Transfer Processes, chapter 5, VCH, New York

[17] Gadzuk, J. W. (1967) Surface Science **6**, 133; ibid 159

[18] Trösken, B. and Willig, F. unpublished results

[19] Briggs, J. S. and Herzenberg, A. (1971) J. Mol. Physics **21**, 865; Fidder, H., Terpstra, J., and Wiersma, D. A. (1991) J. Chem. Phys. **94**, 6895

[20] Feldmann, J., Nunnenkamp, J., Peter, G., Göbel, E., Kuhl, J., Ploog, J., Dawson, P., Foxon, C. T. (1990) Phys. Rev. **B42**, 5819

[21] Kasha, M. Rawls, H. R., El-Bayoumi, M. A. (1965) Pure Appl. Chem. **11**, 371

[22] Norris, D. J., Sacra, A., Murray, C. B., and Bawendi, M. G. (1994) Phys. Rev. Lett. **72**, 2612 and of Alivisatos

[23] Hoheisel, W. Colvin, V. L., Johnson, C. S., Alivisatos, A. P. (1994) J. Chem. Phys. **101**, 8455

[24] Shank, C. V., Schönlein, R. W., Mittlemann, D. M., and Shiang, J. J. (1993) Ultrafast Phenomena VIII, 438, Springer, Berlin

[25] Hoyer, P., Eichberger, R. and Weller, H. (1993) Ber. Bunsensges. Phys. Chem. **97**, 630

[26] Kavan, L., Stoto, T., and Grätzel, M. (1993) J. Phys. Chem. **97**, 9493

[27] Jacobi, K., Zwicker, G., and Gutman, A. (1984) Surface Sci. **141**, 109

[28] Kouwenhoven, L. P., van der Vaart, N. C., Johnson, A. T., Harmans, C. J., Williamson, J. G., and Staring, A. A. (1991) Advances in Solid State Physics, 329, Vieweg, Braunschweig

[29] Pileni, M.-P., unpublished result

[30] Charlé, K.-P., Frank, F., and Schulze, W. (1984) Ber. Bunsenges. Phys. Chem. **88**, 350.

[31] Mahrt, J. and Willig, F. unpublished results

[32] Kietzmann, R., Willig, F., Weller, H., Vogel, R., Nath, D. N., Eichberger, R., Liska, P., and Lehnert, J. (1991) Mol. Cryst. Liqu. Cryst. **194**, 169

[33] Schwarzburg, K. and Willig, F. (1991) Appl. Phys. Lett. **58**, 2520. There is a misprint in the caption to Fig.3 of this paper. It should read $k_{es} = 10^4\ s^{-1}$.

[34] Könenkamp, R., Henninger, R., and Hoyer, P., (1993) J. Phys. Chem. **97**, 7328

[35] Knödler, D., Dieterich, W. Lonsky, C., and Nitzan, A. (1995) J. Chem. Phys. **102**, 465

[36] Willig, F., Charlé, K.-P., and Van der Auweraer, M. (1986) Mol. Cryst. Liqu. Cryst. **137**, 329

MORPHOLOGICAL AND SURFACE CHEMICAL CHARACTERIZATION OF FINE ZrO_2 PARTICLES FOR CERAMIC APPLICATIONS

C. MORTERRA and G. CERRATO
Department of Inorganic, Physical and Materials Chemistry, University of Turin
via Pietro Giuria, 7 - 10125 - Turin, Italy

1. Introduction

In the last decade a great deal of work has been devoted to the characterization of Yttria-stabilized tetragonal ZrO_2 (YSTZ), because of its large use as a ceramic material. Many efforts have been made to evaluate some macro- and mesoscopic characteristics of YSTZ, such as those obtained by mechanical tests (e. g., shrinkage, sinterability, K_{IC}, specific surface area SSA, etc), whereas only little attention has been focused so far on the evolution of the nanostructural and surface reactivity features of the starting powders. It is normally observed that YSTZ precursors, that have been preliminarly treated in different conditions, exhibit quite different sintering behaviours.

It is thus thought that the role of the pre-calcining conditions of the powders may play an important role in determining the characteristics of the final sintered bodies. Moreover, the final mechanical properties of YSTZ might be expected to be closely correlated to the physicochemical features (surface terminations, structure, morphology of the crystallites, etc) of the micro- or nanocrystalline starting materials [1]. As the meso-macroscopic characterization of YSTZ allows to discriminate among the various preparations only at the end of the ceramic production cycle, we tried to find out if any significant differences are already present on a nanostructural scale in the starting materials, and to do so two of the techniques typical of surface science have been resorted to. We have used High Resolution Transmission Electron Miscoscopy (HRTEM) of the powders, and *in situ* Fourier Transform Infra-Red Spectroscopy (FTIRS) of the starting materials as well as of the gas-solid adsorption of probe molecules thereon. Side infomation, deriving from X-ray diffraction (XRD) and Scanning Electron Microscopy (SEM), has been also taken into account in order to improve the comprehension of the above mentioned characteristics.

It has been noted that:

i) depending on the Yttria content, YSTZ powders may exhibit rather different ceramic properties. In fact, low % materials yield compacted/sintered bodies of very poor quality, whereas higher Y_2O_3 percentages yield sintered bodies of better quality.

ii) depending on the conditions of the preliminary treatment of YSTZ powders, (i. e., on the initial firing temperature $\mathbf{T_c}$), compacted and sintered materials may lead to *very* different ceramic products; in particular, low $\mathbf{T_c}$ materials exhibit much worse sintering charactestics (like the formation of hard agglomerates leading to an easier fracture) than high $\mathbf{T_c}$ specimens.

E. Pelizzetti (ed.), Fine Particles Science and Technology, 609–622.

We have thus two open questions: i) why do low-Y_2O_3 content YSTZ preparations exhibit a much worse sintering behaviour than higher-Y_2O_3 content preparations (even if it is known that Y_2O_3 content should be kept as low as possible, compatibly with the stabilization of the tetragonal ZrO_2 phase); ii) why is a relatively high $\mathbf{T_c}$ treatment needed to give YSTZ a good sintering behaviour (even if it is known that, when using microcrystalline ceramic precursors, the crystallites size should be kept as small as possible).
This contribution is aimed at answering these two questions, by exploring some nano-scale features of YSTZ microcrystalline preparations.

2. Experimental

2.1. MATERIALS

Two YSTZ preparations, differing by the amount of Y_2O_3 added (2 and 3 %-mol, respectively), were prepared by a sol-gel method described elsewhere [1]; the resulting powders were then calcined for 30 min at a chosen temperature ($\mathbf{T_c}$; °C) in the 450-1000°C range. The samples are therefore designated by the symbols TO2-$\mathbf{T_c}$ and TO3-$\mathbf{T_c}$, respectively. BET specific surface areas range in the 110-25 $m^2\ g^{-1}$ interval.
Also a commercial 3%-mol YSTZ sample (TSK-Zirconia, lot N. Z306341P) has been considered as a reference material; it is manufatured by Toyo Soda Co., Tokyo, Japan, and shows BET SSA of 15.5 $m^2\ g^{-1}$.

2.2. TECHNIQUES

X-ray diffraction patterns were obtained with a Philips PW 1830 using Co K_α radiation in a conventional Bragg-Brentano geometry.
High Resolution Transmission Electron Microscopy was performed with a Jeol JEM 2000 EX microscopy (200 kV acceleration, LaB_6 filament), equipped with top-entry stage; all samples were first ultrasonically dispersed in n-heptane, and then deposited onto Cu holey grids coated with amorphous carbon.
FTIR spectra were obtained at 2 cm^{-1} resolution using a Bruker 113v spectrophotometer, equipped with MCT cryodetector; all samples, in the form of self-supporting pellets ($\approx$ 20 mg cm^{-2}), were activated at 400°C in a quartz cell connected to a vacuum line (final pressure $< 10^{-5}$ Torr) in order to get rid of most of the abundant hydrated layer present at the surface of oxides when exposed to the atmosphere.

3. Results and Discussion

3.1. X-RAY DIFFRACTION DATA

Figure 1 reports segments of the X-Ray diffraction pattern of all the TOn-$\mathbf{T_c}$ powders examined.
As a function of $\mathbf{T_c}$, XRD peaks modify their structure: in fact, as the specific surface area decreases upon increasing the calcining temperature, all peaks intensify and

sharpen, indicating a larger-range crystalline order. In particular, on passing from T_c = 600°C to T_c = 900°C, all peaks become definitely much sharper, while little amount of monoclinic phase starts being present. It is worthwhile noting that, up to this T_c, no structural differences are evidenced between TO2 and TO3 samples.
Only after the calcination treatment at 1000°C, the TO2 sample shows an amount of monoclinic phase that is still quite low, but ≈ 3 times larger than in the corresponding TO3 preparation.

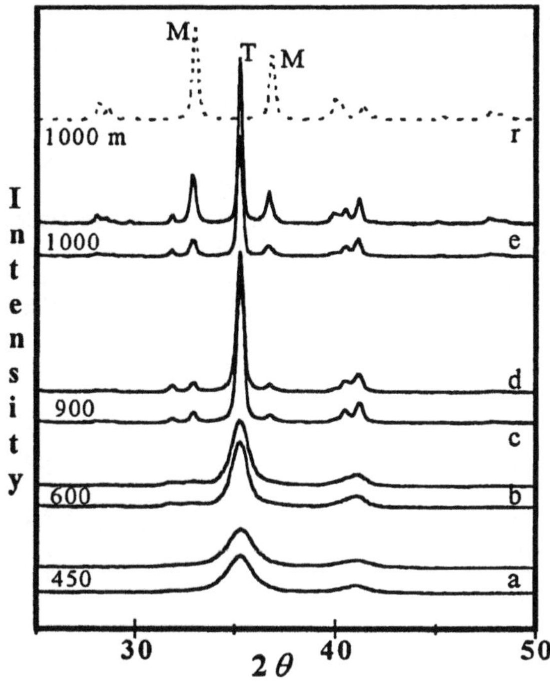

Figure 1. XRD pattern of various YSTZ materials. (T_c firing temperatures are reported in °C). Upper and lower curves refer to TO2 and TO3 materials, respectively. Curve r is of a reference monoclinic ZrO_2 sample.

3.2. HRTEM INVESTIGATION

In order to obtain a better comprehension of the sintering behaviour of the different YSTZ specimens HRTE microscopy has been employed. In fact, this technique gives information on the structure of the materials and on the morphology of the relevant particles.
Figure 2 shows that some morphological features of YSTZ vary with continuity on passing from low T_c (see, for instance, sections a and b), to high T_c (sections c-f).
The starting TOn-450 materials are made up of tiny roundish nano-particles: the frequent patterns of interference fringes in fig. 2a indicate that the system is crystalline, and the round contours suggest a very high incidence of crystallographically defective

situations (steps, high-index planes, etc). Fig. 2b shows that, for $T_c = 600°C$, the crystals size has increased, but the overall morphology of the crystallites is still the same. For T_c < 900°C, no appreciable differences are noted between TO2 and TO3 systems.

The system TO3-900 presents definitely larger crystallites size and somewhat sharper edges, but the overall morphology of the ultimate particles is not so different from that of the starting material. Rather extended flat patches of low-index crystal planes (likely ascribable to the (111) and/or the (100) planes of tetragonal ZrO_2) are now constituting the most frequent top terminations of the crystallites (fig. 2c).

Fig. 2d shows HRTEM images of a TO2-900 specimen. It is interesting to note that an amorphous overlayer (5-15 nm thickness) is present around all particles. The good cristallinity achieved by the material (as monitored by the XRD pattern trace of fig. 1d) is still observable with some difficulty through this layer. XPS data relative to this sample [2] indicate that there is a surface enrichment in Y_2O_3 and, consequently, we can conclude that an yttria-rich phase has been segregated, leaving the solid solution with ZrO_2. Crystallites of ZrO_2 with low (or null) Y_2O_3 content must thus exist, and this agrees with the formation of some monoclinic ZrO_2-phase.

Some major morphological and dimensional differences are evidenced for the TO3 material that has been calcined first at 600°C and, then, at 1000°C: in fact, as shown in fig. 2f, the average crystallites size is ca twice as large as in the case of the ZrO_2 system calcined directly at 1000°C (fig. 2c), while crystallinity has not noticeably improved.

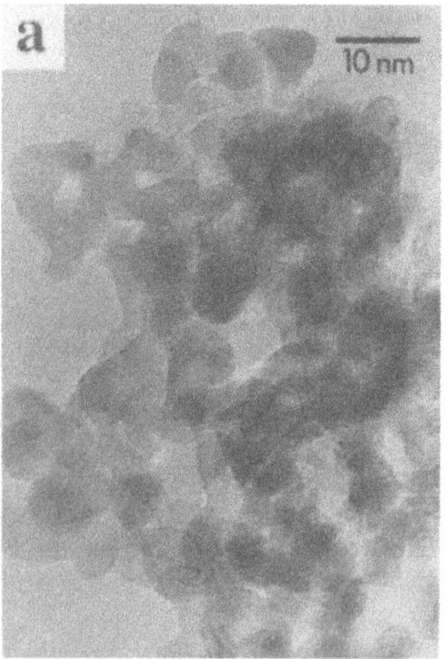

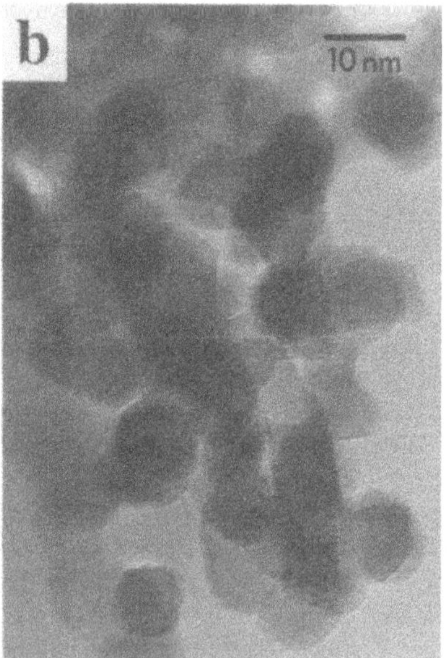

Figure 2. HRTEM images of various YSTZ systems. A) TO3-450; b) TO3-600; c) TO3-900; d) TO2-900; e) TO3-1000; f) TO3-600-1000.

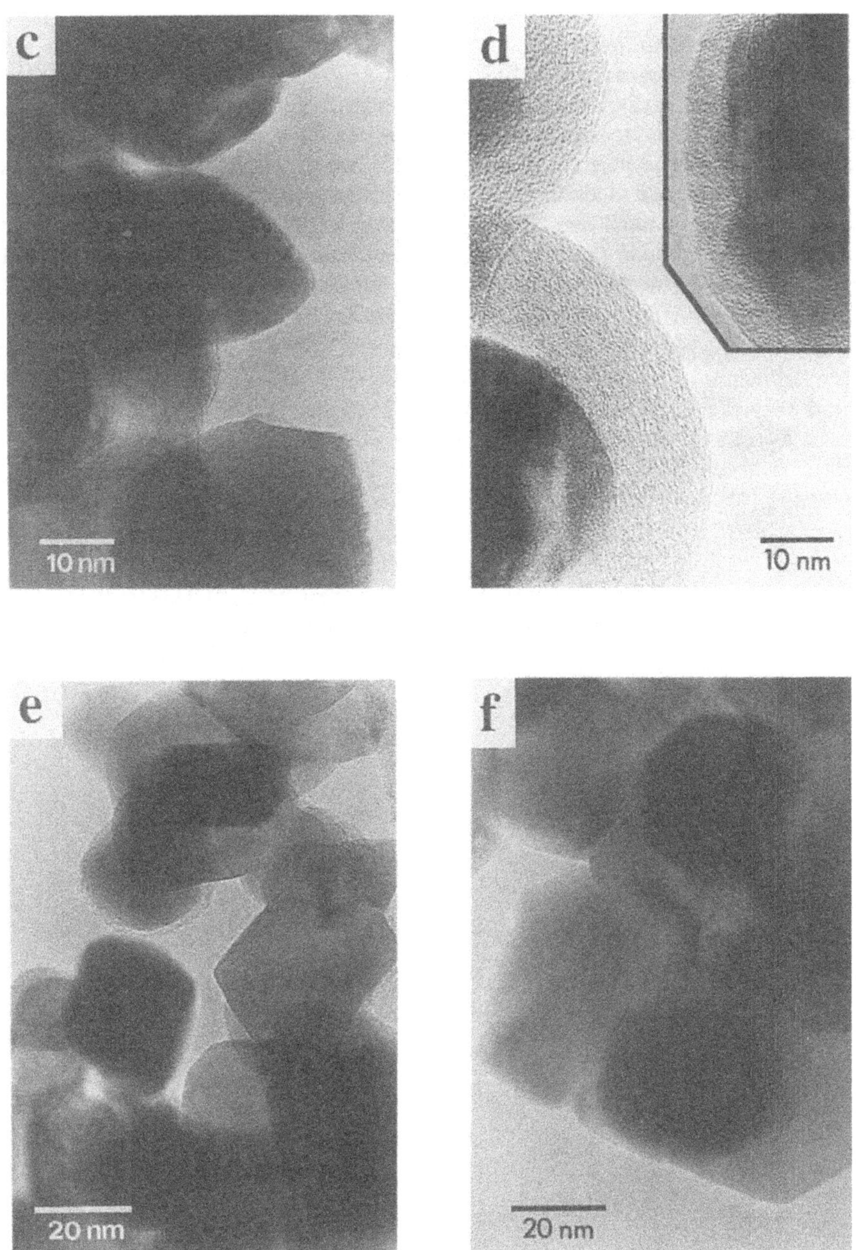

Figure 2. Continued.

3.3. FTIR INVESTIGATIONS

3.3.1 *Surface Functionalities*

In situ vibrational spectroscopy of the different YSTZ samples has been resorted to in order to investigate surface functionalities, essentially H_2O molecules and OH groups. These species are generated at the surface of most oxidic systems by the exposure to the atmosphere of truncated portions of crystallites. When the oxide is thermally activated *in vacuo*, so as to achieve a medium grade of dehydration (400-500°C), water molecules are eliminated fast, while most free OH groups remain at the surface and yield information about both the coordination state and the structural configuration of surface cations they belong to. According to the literature [3], three OH species, absorbing in the 3800-3550 cm^{-1} range, are structurally possible at the surface of ZrO_2, bonded respectively to one (higher ν, 3790-3760 cm^{-1} range), two (medium ν, 3750-3700 cm^{-1} range) and three surface Zr^{4+} cations (lower ν, 3700-3640 cm^{-1} range).

Figure 3 reports the spectral pattern in the OH stretching region relative to some YSTZ systems. Spectra a and b (TOn-450 and TOn-600; in this stage, no differences are observed between TO2 and TO3 systems) refer to the materials calcined at low temperature (450-600°C). Only one component is observable, absorbing at $\approx$ 3660 cm^{-1} and ascribable to tri-bridged OH groups; as far as the ν_{OH} frequency is concerned, these OH groups are fairly acidic, in that their ν_{OH} frequency lies at the lowest end of the spectral range expected for tribridged OH groups. Increasing T_c of TO3 up to 900°C (fig. 3c) leads to the appearence of minor, though not negligible, OH components absorbing at higher ν.

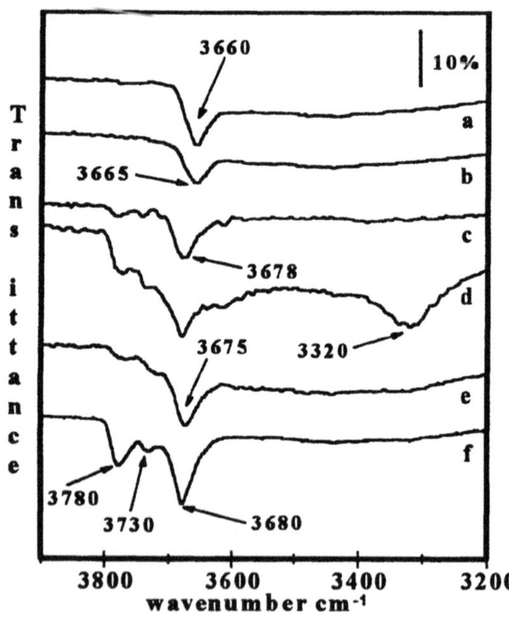

Figure 3. (Transmittance) FTIR spectra in the ν_{OH} region of some YSTZ systems characterized by different T_c. a)TO3-450; b)TO3-600; c)TO3-900; d) TO2-900; e)TO3-1000; f)TO3-600/1000°C.

This means that the surface termination of the crystallites has been modified after these thermal treatments to a very little extent. More important, the ν_{OH} frequency of the still dominating tribridged OH component has shifted towards higher wavenumbers of some 15-20 cm^{-1}. Within the spectral range of existence of the same OH species, a shift of ≈ 20 cm^{-1} is very high, and indicates that the acidity of tribridged OH groups at the surface of the TO3 system has decreased a lot.
The interpretation is that, for low T_c temperatures, tribridged OH groups of the TOn systems are mainly localized in the highly energetic (highly defective) side terminations of the crystallites, whereas for $T_c \geq 900°C$ most of the highly energetic terminations have been annealed, and tribridged OH groups are now located in the extended patches of regular low-index crystal planes.
When $T_c \geq 900°C$, dramatic differences are evidenced for a TO2 system, (see curve d in fig. 3). Three OH components are present with a very different intensity ratio; in particular, a strong OH component is present at higher ν (terminal OH groups), most likely indicative of the presence of monoclinic ZrO_2. Moreover, a broad and intense new OH band is now present at ≈ 3320 cm^{-1}, i. e., in a position typical of OH groups that are no longer free but are perturbed by ineractions of the H-bonding type. The band at 3320 cm^{-1} is likely to be due to the ν_{OH} mode of OH groups that are now trapped within the amorphous overlayer that coats all the particles. This hypothesis is consistent with the HRTEM data, which indicated that all TO2-900 crystallites became coated by a new segregated phase.
Important differences are also present in the OH spectrum of the TO3-600-1000 material (fig. 3f): in fact, if we compare the behaviour of this system with that of the TO3-1000 system (fig. 3e), we observe that not only the size of the crystallites became different in the two systems, but also the termination of the crystallites became different, as demonstrated by the much more abundant presence of (more reactive) mono- and bibridged OH groups.

3.3.2 *Adsorption of CO*
Gas-solid adsorption of probe molecules, like for instance carbon monoxide (CO), is a well established technique in surface chemistry to test surface reactivity [4,5]. CO is a soft Lewis base that, in the case of d^0 systems, interacts mildly and reversibly with surface coordinatively unsaturated (c.u.s.) cations to give carbonyl species [6,7], similar to those described in homogeneous chemistry. In particular, CO interaction is very different from that attained by other probe molecules (like, for instance NO), because CO gives information about the surface"as is", i. e., as this were virtually unperturbed by the test molecule [8,9].
In the case of ZrO_2, CO forms only linear (σ) carbonyl species, since this system possesses no d electrons able to give π back donation towards CO; this fact is reflected by an increase of the force constant of the CO molecule and, consequently, of the frequency of the CO stretching; in fact, upon adsorption, there is a marked blue-shift ($\Delta\nu > 35$ cm^{-1}) with respect to the ν_{CO} frequency of the free gaseous molecule (2143 cm^{-1})
CO adsorption can be carried out both at room temperature (RT) and at liquid nitrogen temperature (-196°C). At RT, carbon monoxide yields very low coverages and interacts only with the strongest fraction of surface Zr^{4+} cations, i. e., with sites located in crystallographically defective configurations (e. g., corners, edges, terraces), and, at

most, with a small fraction of Zr^{4+} sites located in defective situations within exteded (ordered) patches of low-index crystal planes. At liquid nitrogen temperature, on the contrary, CO interacts with the whole surface: in this way, we obtain information also about those sites that at RT form CO complexes with too short an average life to be detected by FTIR.

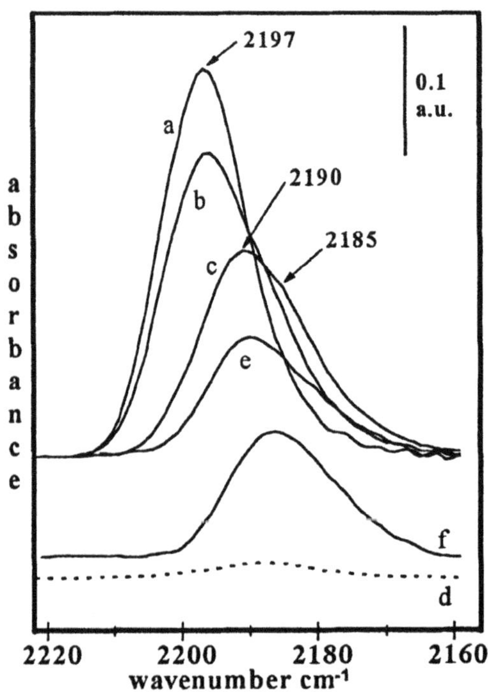

Figure 4. (Absorbance) FTIR spectra of 100 Torr CO adsorption at 25°C onto various YSTZ specimens treated at increasing T_c. a) TO3-450; b) TO3-600; c) TO3-900; d) TO2-900; e) TO3-1000; f) TO3-600-1000

CO adsorption at RT. In the following we will report data on CO adsorption carried out at RT onto ZrO_2 specimens activated *in vacuo* at 400°C, unless otherwise specified.
Figure 4 shows the spectral features of CO uptake onto various YSTZ specimens. It was noted that there is no difference between TO2 and TO3 specimens, if T_c < 900°C. Also, very little differences exist between the samples fired at the lowest T_c temperatures (i. e., 450 and 600°C; see curves a and b in fig. 4, respectively): in both spectra one single broad CO band is present, characterized by ν_{CO} at ≈ 2197 cm^{-1}, indicative of a very high Lewis acidity. In the TO3-600 system the CO band just starts showing a slight asymmetry on the low frequency side. When T_c is raised up to 1000°C (see, for instance, curves c and e in fig. 4), the high ν component (2197 cm^{-1}) is selectively eliminated, and only the low ν component is now evident (2190-2180 cm^{-1}). This means that, on the starting materials, very defective configurations are most abundant, as also put into evidence by the TEM data (sections a-b in fig. 2), while on raising T_c the defective

terminations are selectively annealed, and flat portions of extended crystals planes now form the most favourable termination of the crystallites.

Almost the same behaviour is evidenced for the TO3-600-1000°C material (curve f in fig. 4): in fact, the only CO component lies at ≈ 2185 cm^{-1}, indicating interaction of CO with Zr^{4+} cations located on defective configurations positioned on extended crystal patches of ZrO_2 as indicated by the lower ν_{CO}. On the basis of these data, the 1000°C and the 600-1000°C specimens should look very similar, i. e., formed by extended and ordered ZrO_2 crystals: this is only partly true, as shown by the TEM images (section e and f in fig. 2), since the latter system exhibits a much lower specific surface area and tends to be more aggregated with respect to the former material.

Unlike TO3, CO adsorption onto TO2-900 or TO2-1000 specimens leads to the formation of a negligible CO component [6] (see curve d in fig. 4), implying that no highly reactive Zr^{4+} cations are available at the surface of this material to interact with the probe molecule: this observation is in agreement with the formation of the amorphous overlayer coating the surface of the crystallites, as evidenced by TEM images (section d in fig. 2).

CO Adsorption at -196°C. When the adsorption is carried out at liquid nitrogen temperature, namely -196°C, the spectral features become more complicated: in fact, all surface sites can now interact with CO, to form several carbonyl complexes, as shown in figures 5-7.

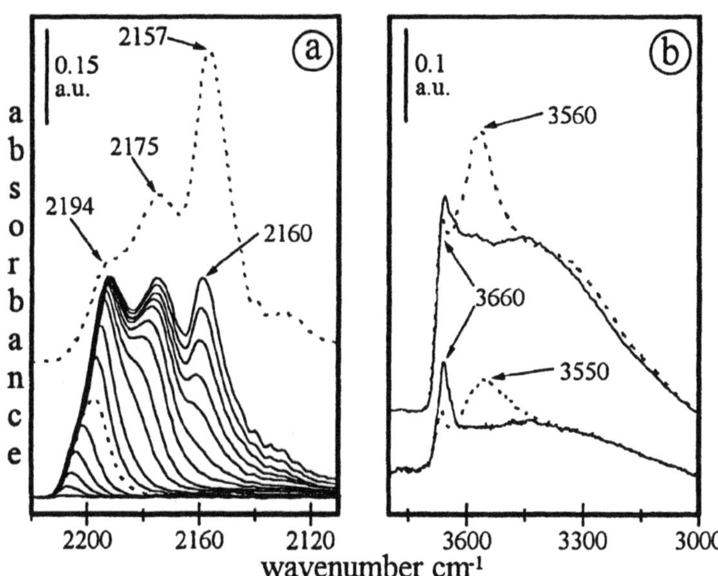

Figure 5. Section **a**: CO adsorption isotherm at -196°C onto TO3-450 systems. (CO pressure in the 50-1x10^{-2} Torr range). The upper broken-line curve refers to the (50 Torr CO) adsorption onto the same specimen, but in a more hydrated situation). Section **b**: the ν_{OH} region of the same ZrO_2 systems of section a, before (solid lines) and after (broken lines) the admittance of 50 Torr CO. (Upper spectra refer to the more hydrated specimen, lower spectra to the sample standard activated at 400°C).

This is because CO adsorption at RT is able to evidence only part of the Zr^{4+} cations, and above all those located in highly defective crystallographic configurations, i. e., the most energetical sites: for instance, see the lower broken-line spectrum in section **a** of fig. 5, that refers to the fraction of CO that could be detected at room temperature for the TO3-450 specimen.

The lower part of section a in figure 5 reports the adsorptive behaviour of the system TO3-450 activated in standard conditions (400°C *in vacuo*). It can be observed that three main CO species form in these conditions: the first band envelope, positioned at high frequency (ν_{CO} in the 2205-2190 cm^{-1} range), is ascribable to CO interacting with Zr^{4+} cations located in crystallographically defective configurations (note that this is the only CO species that we could observe in the RT CO adsorption: see curve a in fig.4) [7]. The weak central envelope, $\nu_{CO} \approx 2175$ cm^{-1}, refers to CO adsorption onto Zr^{4+} cations located on flat portions of extended patches of low-index crystal planes (this component is totally missing at RT). The last CO component, lying at still lower frequency ($\nu_{CO} \approx$ 2160 cm^{-1}), is the weakest CO species formed at the surface of this system, and is present only for high CO coverages: it is ascribed to CO molecules interacting by H-bonding with surface OH groups [10,11].The assignment of the latter species is demonstrated by: i) the different relative intensity assumed by this component on varying the activation treatment (see, for instance, the broken-line spectrum in sect. **a** of fig. 5, relative to a more hydrated TO3-450 specimen); ii) the perturbation brought about on the OH band after CO uptake at high CO pressures, and shown in section **b** of figure 5.

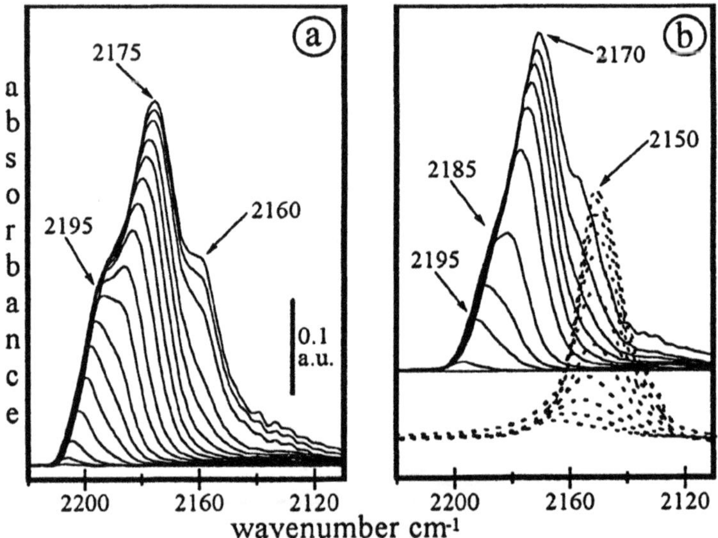

Figure 6. CO adsorption isotherms at -196°C onto high T_c ZrO_2 systems. Section **a**: CO onto TO3-600. Section **b**: CO onto TO3-1000 (upper solid-line curves), and CO onto TO2-1000 (lower broken-line curves). (CO pressure in the 50-1x10^{-2} Torr range)

On passing to the TO3-600°C system, the spectral features of CO adsorption undergo some important changes, as shown in section **a** of figure 6. First, there is a general increase in intensity; secondly, the ratio between CO species formed on Zr^{4+} located in crystallographically defective configurations ($\nu_{CO} \geq 2195$ cm^{-1}) and those formed on Zr^{4+} located in ordered crystallographic situations (ν_{CO} at 2195-2175 cm^{-1}) has evolved in favour of the latter ones. These species are by far the most important at the surface of this system: the singleton ν_{CO} frequency is positioned at $\approx$ 2195 cm^{-1}, while, at high CO coverages, this band undergoes a downwards shift of $\approx$ 20 cm^{-1} due to strong mutual CO-CO interactions.

This mechanism of electronic effects transmission *through the solid* is well established in surface chemistry [12,13], and can be explained by: i) the incipient sinterization achieved by the ZrO_2 crystallites, which involves the formation of extended patches of regular crystal planes; ii) almost uninterrupted dipole-dipole interactions on these large size crystallites. HRTEM images have already shown that, in these conditions, there is a fair increase of the crystallites size (e. g., compare section a and b in fig. 2), and the present observation implies a selective annealing of the defective configurations upon increasing T_c. Note that, up to this T_c temperature, the behaviour of TO2 and TO3 specimens is perfectly equivalent, and, consequently, for the sake of brevity, only the spectral features relative to the TO3 system have been reported.

Section **b** of fig. 6 shows that, when CO adsorption is carried out onto TO3-900 or TO3-1000, the spectral component relative to CO species interacting with coordinatively unsaturated Zr^{4+} cations located in crystallographically defective configurations is (almost) totally missing. In fact, this species reaches saturation fast, and its singleton band is very weak and positioned below 2200 cm^{-1}, i. e., at lower ν, than in the case of TO3-600.

The main component now formed (ν_{CO} of the singleton band at 2195 cm^{-1}) is then the species ascribable to the interaction of CO with coordinatively unsaturated Zr^{4+} cations located in extended patches of low-index crystal planes. This band shows a stronger downwards shift ($\Delta\nu \approx$ 25 cm^{-1}) than in the case of TO3-600, and this fact implies a stronger and longer-range CO-CO dipole-dipole coupling taking place at the surface of the larger size crystallites. This means also that the TO3-1000 system (and TO3-900, too) has evolved towards an even more ordered crystallographic configuration of the particles, as confirmed by the HRTEM images (lower SSA *vs* larger size of the crystallites, as shown in the HRTEM images, fig. 2c and 2e).

Major differences are observed in the case of the TO2-1000 system: the dotted-line spectra reported in section **b** of fig. 6 show that both CO species on crystal defects and in regular crystal planes are missing, as the spectrum is characterized by one single CO component, absorbing at extremely low ν ($\approx$ 2150 cm^{-1}), and thus ascribable to CO interacting via H-bonding with OH groups. This indicates that the amorphous overlayer coating the ZrO_2 particles, reported in the HRTEM images (see fig. 2d), is preventing CO from reaching all exposed Zr^{4+} surface sites. It is also indicating that the amorphous overlayer is probably rich in OH groups [2].

In order to test the validity of the analytical method here proposed, i. e., obtaining the possibility of information on the structure and surface reactivity of a nanocrystalline system by the gas-solid adsorption of CO at -196°C, we have examined a different YSTZ preparation of commercial origin.

The reference commercial YSTZ system presents a somewhat different behaviour towards CO (see figure 7). In fact, three CO components are formed at the surface of this material. The first band, absorbing at low ν ($\approx$ 2150 cm^{-1}), is evidenced only at high CO coverages, meaning that it represent a very weak perturbation, such as that ascribable to the interaction of CO with surface OH groups by H-bonding [10]. The band is sharp and very strong, suggesting for this material a very high OH population. In the inset to fig. 7 the solid-line spectrum refers to the OH pattern of the reference YSTZ material: it is worthwhile noting that the intensity of the OH spectrum is much higher than in the case of TO3-1000 (broken-line curve in the inset), and this observation confirms the above mentioned hypothesis about the interaction between CO and surface hydroxy groups and about the surface OH concentration of this system.

The high ν OH band, which is ascribed to monobridged OH groups [3], is very abundant, if compared to the same (very weak) envelope present in TO3-1000, and could imply a different surface reactivity for this reference system.

No CO component, absorbing at $\nu > 2200\ cm^{-1}$ and due, if present, to CO σ coordinated to Zr^{4+} located in crystallographically defective configurations, is evidenced in fig. 7. This fact is in agreement with: i) a high T_c (reported by the manufacturer to be ≈ 900°C); ii) very low SSA (≈ 15 $m^2\ g^{-1}$).

The medium wavemunbers CO component, ascribed to adsorption onto cus Zr^{4+} centres located on flat low-index crystal planes, shows a peculiar behaviour: the singleton band is only one, centred at ≈ 2195 cm^{-1}, like in the cae of TO3-1000, but the downwards shift with CO coverage exhibits two patterns: one pattern tends to ≈ 2178 cm^{-1} ($\Delta\nu \approx 17\ cm^{-1}$), indicative of relatively weak dipole-dipole interactions and thus of small patches of regular crystal planes. The second pattern tends to 2165 cm^{-1} ($\Delta\nu \approx 30\ cm^{-1}$), indicative of very strong dipole-dipole interactions and thus of very large patches of regular crystal planes.

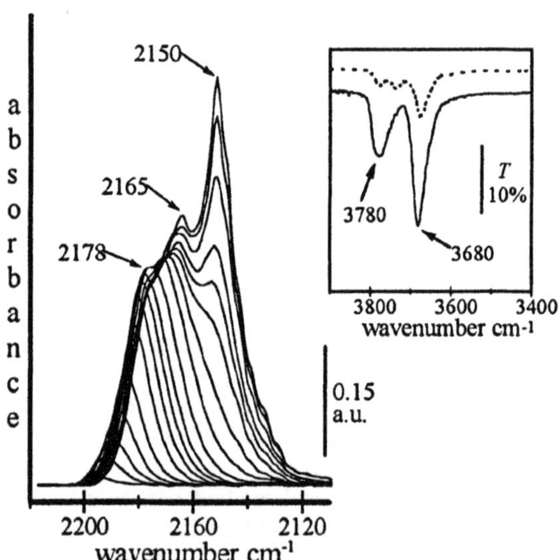

Figure 7. CO adsorption at -196°C onto a reference commercial YSTZ system. (P_{CO} are in the 50-1x10^{-2} Torr range). Inset: OH transmittance spectra (3900-3400 cm^{-1}) relative to the reference ZrO_2 sample (solid-line curve) and to TO3-1000 (broken-line curve).

A HRTEM study, reported elsewhere [2,6] showed that indeed the TZ-3Y preparation by Toyo Soda has a double morphology: there are huge aggregates of relatively small particles, in which the regular pattern of interference fringes are small, and there are large crystallites (30-50 nm size) in which the temination is along large portions of regular low-index crystal planes.

4. Conclusions

1) Most of the large-scale (macroscopic) properties of materials can be sought for among the nano-scale features of the ultimate particles that the materials are made. These features can be understood, provided the proper analytical tools are employed. In the present work, by the use of HRTEM, we could understand why high Y_2O_3 loadings are needed for and adequate sintering behaviour: systems with low Y_2O_3 contents are phase unstable, and segregate amorphous phases when brought to relatively high temperatures. By the use of IR spectroscopy we could understand why high temperatures of preliminary firing are needed for an adequate sintering behaviour: materials fired at low $\mathbf{T_c}$ are too reactive (highly acidic OH groups, highly energetic Lewis surface centres), whereas firing at $\mathbf{T_c} \geq 900°C$ anneals the highly energetic centres associated with crystallographically defective configurations.

2) Not all the analytical techniques are equivalent, although many techniques are complementary. In the present case, we saw that the HRTEM technique was determinant for answering the first of the two questions we had to answer, but was largely insufficient to answer satisfactorily the second question. In fact, according to the microscopic approach, two systems like TO3-600 (whose sintering behaviour is bad) and TO3-1000 (whose sintering behaviour is good) are different only on a quantitative ground: they mainly differ by the average size of the crystallites, but the overall morphology of the particles is the same. According to the spectroscopic approach, we could understand that the two mentioned systems are actually qualitatively different, in that they are systems possessing largely different surface reactivity. In particular, the spectroscopic study of gas-phase adsorption phenomena can give, on a molecular scale , an actual physical and chemical meaning to the term *surface reactivity*.

5. Acknowledgements

Thanks are due to Dr. L. Ferroni, Turin Polytechnic, who provided the various ZrO_2 powders and carried out mechanical tests, and to Dr. M. Baricco, University of Turin, who obtained and interpreted the XRD data.

6. References

1. Cerrato, G., Ferroni, L., Montanaro, L. and Morterra, C. (1995) Correlation between nanostructure and sintering properties of zirconia powders, *Proc. 8th World Ceramic Congress "Ceramics: Charting the Future"*, P. Vincenzini (ed.), Techna Srl, Faenza (Italy), 1419-1426.

2. Morterra, C., Cerrato, G., Ferroni, L. and Montanaro, L. (1994) Surface characterization of yttria-stabilized tetragonal ZrO_2 Part 1. Structural, morphological, and surface hydration features, *Mater. Chem. Phys.* **37**, 243-257.
3. Tsyganenko, A. A. and Filimonov, V. N. (1973) Infrared spectra of surface hydroxyl groups and crystalline structure of oxides *J. Mol. Struct.* **19**, 579-589.
4. Little, L. H. (1966) *Infrared Spectra of Adsorbed Species*, Academic Press, London.
5. Davydov, A. A. (1984) *Infrared Spectroscopy of Adsorbed Species on the Surface of Transition Metal Oxides*, Wiley, New York.
6. Morterra, C., Cerrato, G., Ferroni, L., Negro, A. and Montanaro, L. (1993) Surface characterization of tetragonal ZrO_2, *Appl. Surf. Sci.* **65/66**, 257-264.
7. Morterra, C., Cerrato, G., Bolis, V., Lamberti, C., Ferroni, L. and Montanaro, L. (1995) Surface characterization of yttria-stabilized tetragonal ZrO_2. Part 2. Adsorption of CO, *J. Chem. Soc. Faraday Trans.* **91(1)**, 113-123.
8. Bolis, V., Fubini, B., Giamello, E. and Reller, A. (1989) Effect of form on the surface reactivity of differently prepared zinc oxides, *J.Chem. Soc. Faraday Trans.* **85(4)**, 855.
9. Bolis, V., Fubini, B., Garrone, E. and Morterra, C. (1989) Thermodynamic and vibrational characterization of CO adsorption on variously pretreated anatase, *J.Chem. Soc. Faraday Trans.* **85(6)**, 1383-1395.
10. Zaki, M. I. and Knözinger, H. (1987) Carbon monoxide - A low temperature infrared probe for the characterization of hydroxyl group properties on metal oxide surfaces, *Mater. Chem. Phys.* **17**, 201-215.
11. (a) Morterra, C., Orio, L., Bolis, V. and Ugliengo, P. (1991) Bi-dimensional CO clusters at the surface of polycrystalline monoclinic ZrO_2, *Mater. Chem. Phys.* **29**, 457-466. (b) Morterra, C., Orio, L., Fubini, B., Bolis, V. and Williams, T. B. (1991) A FTIR and HRTEM study of some morphological and adsorptive properties of monoclinic ZrO_2 microcrystals, *Surf. Sci.*, **251/252**, 540-545.
12. (a) Spoto, G., Morterra, C., Marchese, L., Orio, L. and Zecchina, A. (1990) The morphology of TiO_2 microcrystals and their adsorptive properties towards CO: a HRTEM and FTIR study, *Vacuum* **41**, 37. (b) Cerrato, G., Marchese, L. and Morterra, C. (1993) Structural and morphological modifications of sintering microcrystalline TiO_2: an XRD, HRTEM and FTIR study, *Appl. Surf. Sci.* **70/71**, 200-205.
13. Morterra, C., Cerrato, G. and Ferroni, L. (1995) Surface characterization of yttria-stabilized tetragonal ZrO_2 Part 3. CO_2 adsorption and the CO_2-CO interaction, *J. Chem. Soc. Faraday Trans.* **91(1)**, 125-132.

Use of Chemistry as a Powerful Means for Synthesis and Processing of Nano Particles

H. Schmidt, C. Lesniak, T. Schiestel,
Institut für Neue Materialien gem. GmbH, D-66 123 Saarbrücken, Germany

1. Introduction

The chemical synthesis of small particulate systems has become an interesting area of scientific activities since the potential of nano-scale systems has been demonstrated by the early work of Gleiter [1] and Henglein [2]. Colloidal systems in liquid phases with particle sizes in the lower nano range, however, have been known for centuries, and commercial applications of SiO_2 sols are known since the middle of the last century. It is a matter of fact that during each precipitation process from solution a growth reaction takes place and a nano size range is passed through. If the precipitation process is carried out far away from the point of zero charge (p. z. c.), the particles will be charged and prevented from agglomeration as long as the conditions for stability (e.g. critical distance, particle size versus precipitation) are not violated. This leads to stable sols, the concentration of which has to be kept rather low, as described in [3] and the references cited therein. Approaching the p. z. c., the interaction mechanisms get activated and gels (strong interaction) or precipitates (slow interaction) are formed. In fig. 1 the differences of charge stabilized and sterically stabilized systems during upconcentration are shown. In the charge stabilized systems, due to a strong interaction of surface groupings which are necessary to build up surface charges, in general, aggregates to be formed, are

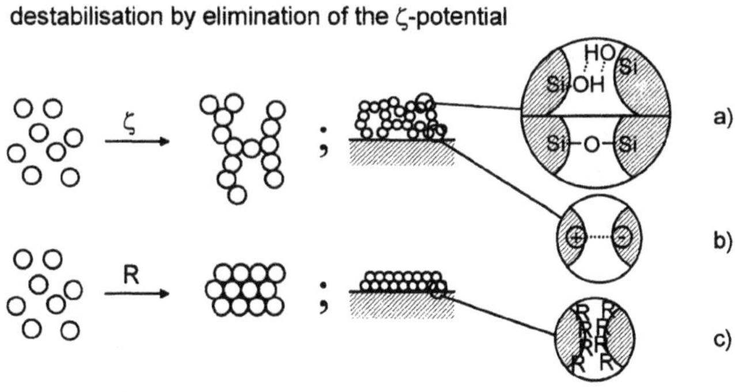

Fig. 1: Models for small particle interactions and destabilization processes of reactive and non-reactive surfaces; a: hydrogen bonds; b: dipole interactions; c: van der Waal's forces.

E. Pelizzetti (ed.), Fine Particles Science and Technology, 623–642.

rather uncontrolled, and irregularly structured gels or precipitates are formed. As a consequence, in gels the green density is rather low (≤ 20 % by volume). If the surface of the small particles, however, is covered by unreactive groupings, only weak van der Waal's forces are active and it should be possible to obtain more regularly packed solids. Typical interaction mechanisms are hydrogen bonds (in solution), dipole-dipole interactions, hydrophobic interactions, van der Waal's forces or even the formation of chemical bonds.

In oxidic systems, which, in general, are covered by hydroxyls, particle aggregation very often is accompanied by neck formation and aggregates are formed which cannot be redispersed (hard agglomerates). If further processing to monoliths or thick films is required, these agglomerated systems are difficult to be shaped defect-free, and defect structures leading to problems in densification and defect-containing microstructures are produced (pin hole formation, pore formation, stress formation).

For these reasons, it is very important to develop solutions for the control of the surface reactivity of small particles, which is not only interesting for the control of the particle-to-particle interaction but also to generate desired surface reactivities for example for further reactions or to generate desired charges for ξ-potential tailoring, e.g. polymerisation, polycondensation.

The aim of this paper is to show principle synthesis routes for particles with tailored surface reactivities and to give examples for the preparation of powders and colloids and examples for processing them to materials already in or close to application.

2. General considerations and results

Appropriate surface modification cannot only be of interest in the case of the control of the particle surface reactivity, but also to be used as a means for controlling the growth reaction by itself. In the growing system, after nucleation has taken place, the growth reaction is determined by the decrease of the free energy. In equations (1) and (2), the nucleation frequency is expressed as a function of the free energy of nucleation, the free energy of diffusion and the surface free energy σ [4, 5].

$$I = A \cdot e^{\frac{-(\Delta G_n + \Delta G_D)}{kT}} \qquad (1)$$

$$A = 2\, n_\upsilon \cdot \upsilon^{1/3} \frac{kT}{h} \sqrt{\frac{\sigma}{kT}} \qquad (2)$$

ΔG_n: free energy of nucleation
ΔG_D: free energy of diffusion
n_υ: number of species able to nucleate
υ: volume of a single species
σ: surface free energy

If the volume υ changes are small and the diffusion rate is high compared to the nucleation step, nucleation and growth are mainly dependent on the surface free energy σ, which depends on the surface area of the growing particles. this leads to the concept of introducing agents interfering with the nucleation process and interfering with the growing surface.

In fig. 2, free energy dependence of a nucleating and growing system is schematically shown as a function of the particle radius. After exceeding the critical nucleation radius, the free energy starts to decrease, and the system grows. If strongly bonded components are added to the surface an energy barrier can be introduced which leads to a stop of the growth process.

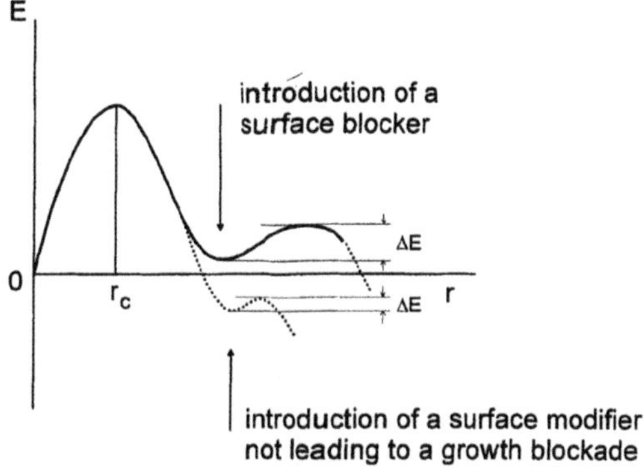

Fig. 2: Model of the growth reaction as a function of the surface free energy and the effect of surface additives; r_c: critical radius for nucleation.

This model suggests the possibility to control growth and particle size by use of appropriate surface additives during a growth or precipitation process. In order to carry out surface modification, the bonds between the modifying molecule and the surface are of importance. Furthermore, by appropriate choice of this modifying molecules, e.g. by using bi- or multifunctional molecules, the surface properties of particles can be varied either with respect to its surface charge or its surface reactivity. In this paper, it is demonstrated how the described principles can be used for the synthesis of a nano-scale particulate system and its processing to materials like ceramics and inorganic-organic composites.

3. Synthesis of Small Particles with Narrow Size Distributions

From the principles pointed out in the introduction, a concept of small particle synthesis has been developed. In this concept, several assumptions are incorporated. Firstly, it is assumed that bonds with sufficient strength can be built up to a growing particle's surface.

Secondly, it is assumed that each component adsorbed on the particle's surface requires a specific area and that an optimum coverage exists leading to a minimum of the surface free energy. This then leads to the conclusion that for components not interrupting the growth process but bond to the surface with sufficient strength, the particle size is controlled by the concentration of these "ligands" in the system. The schematics for this model is shown in fig. 3.

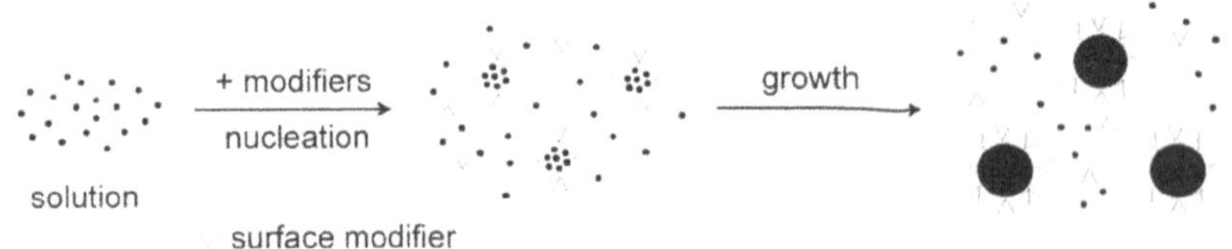

Fig. 3: General principles for the synthesis of small particulate systems by control of the surface reactivity.

For demonstrating the bond strength, investigations have been carried out with hydrolysed alkoxides after complexation. In sol-gel reactions, alkoxides are interesting starting compounds since they are soluble in a variety of solvents, e.g. alcohols, in which acids or bases also can be dissolved in order to adjust appropriate pH. Reactive alkoxides can form complexes with a variety of components like carboxylic acids (carboxylate complex formation [6], ß-diketones [6 - 8], or even etheralcohols [9]. Examples are titanium alkoxides, zirconium alkoxides or aluminum alkoxides. The question arises whether this complex forming ability can be transferred from the ionic state in solution to ions incorporated in nano-scale particles. In principle, surface ions in a crystalline lattice are in a very unsaturated state, and for these reasons it should be possible to coordinate these unsaturated entities by saturating ligands. In fig. 4, an example is given, resulting from a reaction of $Al(OR)_3$ and a β-dicetone (ethyl aceto acetate) by hydrolysis and condensation. The particle size as determined by photon correlation spectroscopy after hydrolysis is about 2 nm (± 10 %) and was identified as boehmite.

As one can see from the shift of the carbonyl frequencies, there is no difference between the non-hydrolysed and hydrolysed complexed form, leading to the conclusion that the ß-diketone complex is still intact on the nanoparticles.

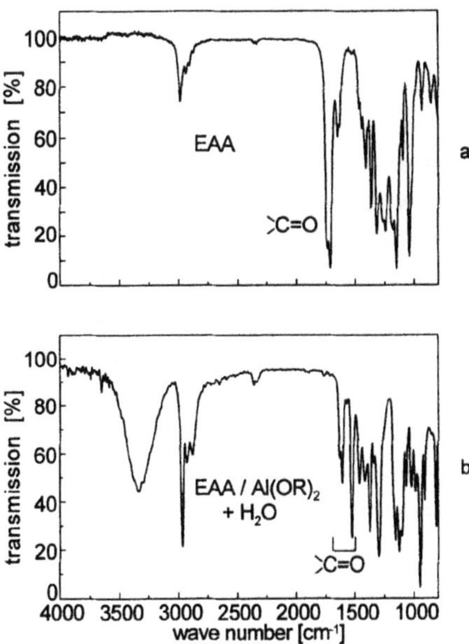

Fig. 4: IR spectra of ethylaceto acetate (a), after complexing with Al(i-prop)$_3$ (b) and after subsequent hydrolysis (b).

In order to find out how far the ligand concentration can be used for tailoring the particle size in a wider range, experiments with Zr alkoxides and two different types of ligands have been carried out. In fig. 5, two examples are given with a nucleating and growing system in presence of β-diketones or alkoxylic acids. The particle size was determined as a function of the ligand concentration. A $y = 1/x$ relation should be expected if the ligand concentration is the size controlling parameter.

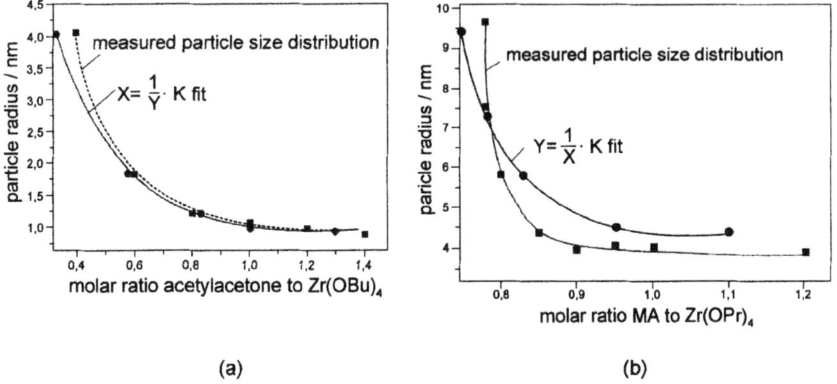

Fig. 5: Zr alkoxides reacted with acetyl acetone (a) and methacrylic acid (MA) (b) and subsequently hydrolysed by addition of H_2O; dependence of particle size on the ratio alkoxide: complex former as determined by photon correlation spectroscopy (after [10]).

The figures show the fitting of the experimental curve with a calculated curve and as one can see in the case of the β-diketones, where a strong bond can be assumed, the theoretical curve fits very well to the experimental curve. In the case of a carboxylic acid, where an adsorption - desorption equilibium has to be taken into consideration, the fit is not quite as well. Summarising it is to say that assuming a surface free energy driven model, particle size follows a rather simple function, depending on the ligand concentration, but anyway calibration curves for different systems should be possible. For the synthesis of metal colloids, the described approach is depending on the presence of components having sufficient affinity to the metal surface as well as being able to stabilize the metal ion. In the case of Au and Ag, amines were found to be able to play this ambivalent role. For this reason, the soluble metal salts (e.g. nitrates or chlorides) were reacted with γ-aminopropyl triethoxysilanes or γ-aminoethyl aminopropyl triethoxysilane (DIAMO) to form amino complexes followed by a photochemical reduction, the details of which were published elsewhere [11], and the particle sizes were determined as a function of the complex former ratio. In fig. 6, curves are shown.

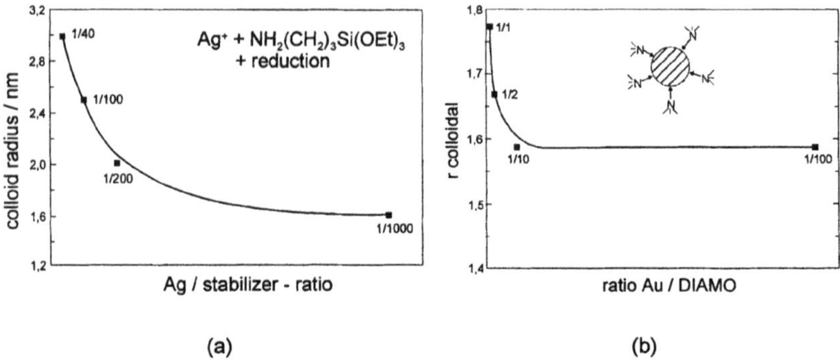

Fig. 6: Dependence of particle size of metal colloids obtained from complexed metal salts in MeOH by photochemical reaction. a: Ag/γ-aminopropyl triethoxysilane; b: Au/DIAMO [12].

In this case, rather high concentrations of complex formers had to be used in order to obtain these curves. This is explained by a relatively unfavorable absorption desorption equilibrium of the amines on the colloid surface. Silanes have been used in order to generate a "second" reactivity and to use these precursors for further sol-gel processing. The particle size distributions obtained by these methods have been determined by TEM and are relatively narrow within a range of ± 1.5 nm deviation from the average.

The results show that the described principle can be employed for ceramic as well as for metal particles. The use of bifunctional ligands (e.g. organofunctional silanes or functionalized acids) is of special interest since they allow to add special reactivities to the particles. In the following, some composite developments are described, using two different preparation routes, the preparation of particles within and separately from the matrix system.

4. Nanocomposite Development

4.1. In-situ grown ZrO_2 containing systems

Based on the considerations shown above, several materials developments have been carried out. Beside the fact, that the results have been used for the preparation of pure ceramic nano scale systems [13], the synthesis of composites with nanoparticles is of interest for various reasons, for example, for the preparation of transparent composites, if particle sizes are used below the Rayleigh scattering limit.

Two routes have been developed for the composite synthesis, the so-called in-situ route, that means the generation of nanoparticles within a reactive system which later on can be reacted to the composite matrix or the seperate preparation route where nanoparticles are prepared outside of the matrix systems and then modified to obtain an appropriate surface to be introduced into the matrix system.

In the first example, the in-situ synthesis of a zirconia nanoparticle containing system is described. The basic principle of the in situ route is shown in fig. 7. Little experimental details are shown elsewhere [14, 15]. As matrix former, γ-methacryloxypropyl triethoxy-silane (MPTS) was used in order to form an inorganic-organic composite matrix by combination of polymerization and polycondensation of silanol groupings.

Methacrylic acid was used as complex former, since the double bond of MA should be able to react with other monomeric acrylates, and various types of polymerization reactions should be accessable (e.g. radical polymerization, thermally or photochemically initiated), and photoresist properties for micropatterning should be possible.

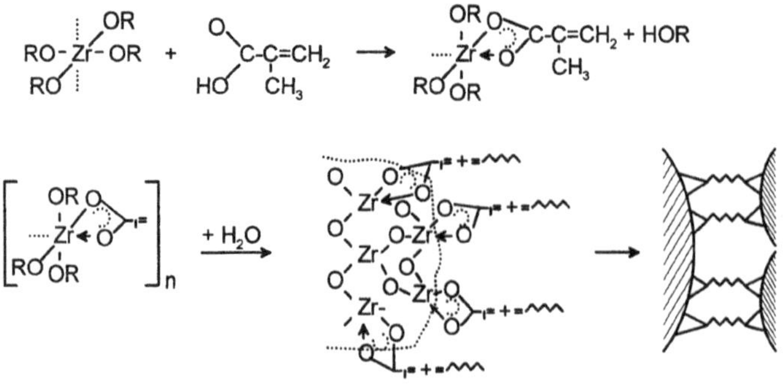

Fig. 7: Model of the ZrO_2 nanoparticle containing in situ grown composite; /\/\/: methacrylate type of monomers.

In order to obtain sufficient transparency, the particle size has been established at about 2 nm by appropriate choice of the Zr:MA ratio. The material was synthesized by complexing zirconium alkoxides by methacrylic acid [16], hydrolysing the system with an understoichiometric concentration of water and mixing it with a prehydrolysed methacryloxic group containing silane. After polymerizing the system at temperatures of around 100 °C with a photoinitiator, a clear transparent composite was obtained. In order to follow the reaction, the carbonyl frequency of the methacrylic acid was monitored by IR spectroscopy and by electron scattering measurements the crystallinity of the system was investigated. In fig. 8, the IR spectrum of the carbonyl frequencies of the system and the electron diffraction patterns are shown [17].

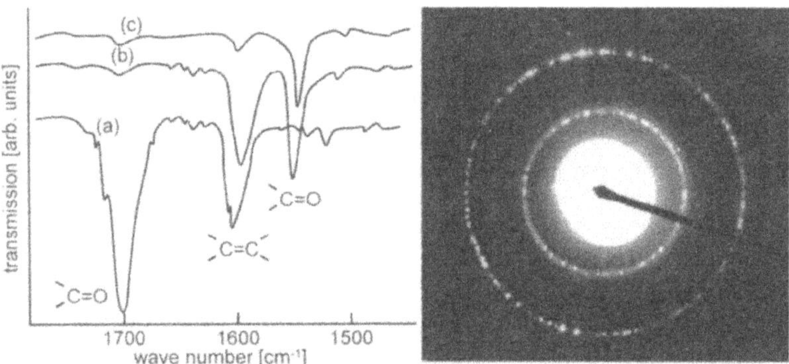

Fig. 8: IR spectrum of the carbonyl frequency (a: methacrylic acid; b: zirconia alkoxide complex with methacrylic acid, hydrolized and mixed with a methacryloxy group containing silane; c: polymerized film after 2 weeks of water exposure.); electron diffraction pattern: after photocuring and heat treatment at 120 °C. The pattern can be attributed to a mainly monoclinic phase.

Fig. 8 shows that a crystalline phase is formed at extremely low temperatures through the in-situ growth process.

ZrO_2 also could be detected in HTEM investigations [18]. Films prepared from this system are highly transparent and have been used for patterning processes, for wave-guides diffractive gratings and micro-lens arrays [19]. Moreover, and this seems to be of importance, in the polymerized system the bonds between the zirconia particles and the methacrylic acid are still present.

This finding leads to the question about the mobility of the particles in the polymerizing system during a photopolymerization process. As already shown elsewhere [19], it could be shown that this type of system could be polymerized to high degrees of conversion including the double bonds of the MA bond to ZrO_2. For generating a "driving force" for a difference of double bonds, a gradient electromagnetic field has been established, obtained by the interference of two laser beams (two-wave mixing process). As it is already well known from the photopolymerization state of the art with photo masks or two wave mixed laser beams with polymers, partial polymerization can be introduced into films and diffractive grating patterns are obtain due to the polymerization shrinkage in the photopolymerized areas. However, if the full area is irradiated after the first step the pattern disappears due to the polymerization of the rest of the volume. The question arises whether in the multicomponent system with comoponents having a higher refractive index (in this case the zirconia) a diffusion process of zirconia particles can be initiated to establish permanent refractive index changes. In order to prove this hypothesis, the two-wave mixing irradiation process has been used. After the partial polymerization, the film was fully irradiated by a UV lamp and completely polymerized. Micro Raman and analysis showed that in the areas polymerized in the first step in the two wave mixing process show an increased concentration of zirconia compared to the areas irradiated in the second step only. The basic process is shown in figure 9 and the Raman patterns in fig. 10 which show a higher concentration of ZrO_2 within the firstly irradiated

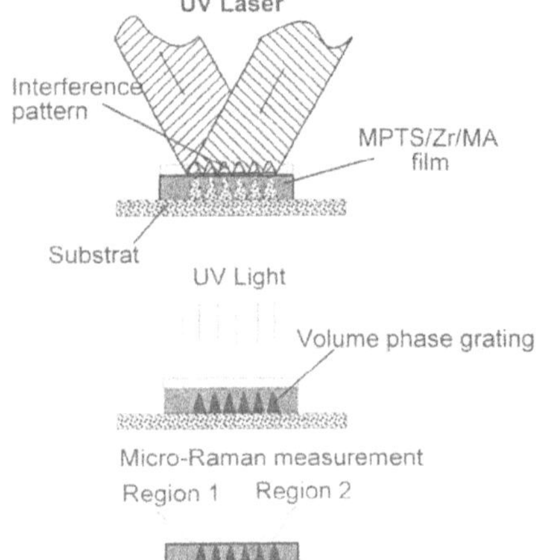

Fig. 9: Schematics of the two-wave mixing process for the fabrication of a volume diffraction pattern based on the composition ZrO_2:MA:methacryloxy propyl trimethoxy silane (MPTS) = 1:1:1 (after [15]); region 1: irradiated, region 2: non-irradiated zone.

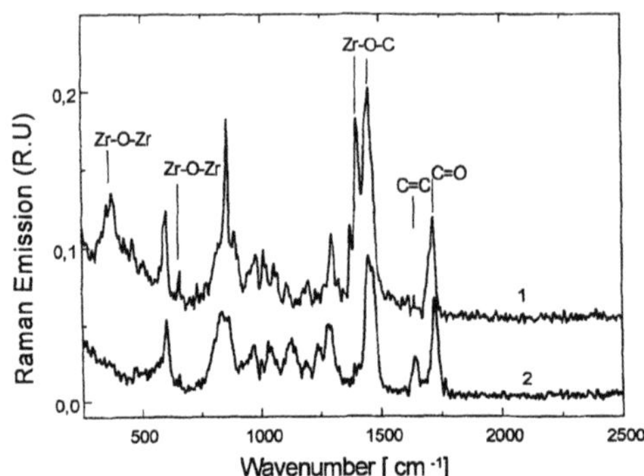

Fig. 10: Micro Raman spectra of region 1 and 2.

ions and which confirms the hypothesis that during the first irradiation step zirconia particles diffuse into the polymerizing areas, driven by a chemical potential gradient. The differences in refractive indices to be obtained by this route have been determined and reach from 0.001 to 0.03, depending on ZrO_2 contents [18]. In the second polymerization step the concentration gradients are completely immobilized, so that a stable system is shown. In fig. 11, the diffractive grating produced by this process is shown. This process has also been used for the development of micro Fresnel lenses by a different two wave mixing process [19].

The investigations have shown that it is possible to generate in-situ nanocomposites which can be photopolymerized and, moreover, that through gradient electromagnetic fields a diffusion of zirconia particles can be introduced. The diffusivity of the zirconia particles is linked to the methacrylic acid groups bonded to the surface which move into areas with lower content of double bonds due to the consumption of them by the polymerization step.

Fig: 11: Diffractive grating after full irradiation.

4.2 Fe Oxide Containing Nanocomposites

It is well known that FeO_x nanoparticles can be produced by precipitation from Fe^{2+}/Fe^{3+} containing aqueous solutions [20], but they form agglomerates. These agglomerates can build up strong aggregates, caused by condensation reactions of the OH-groups at the surface of the nanoparticles. If isolated nanoparticles should be prepared, it is necessary to destroy the agglomerates and to stabilize the individual particles [21 - 24]. For this reason, thermal ultrasonic energy has been used to destroy the agglomerates, but the lifetime of the individual particles is only very short due to a rapid reagglomeration rate.

To prevent the immediate reagglomeration, a new approach has been developed, based on the ultrasonic energy deagglomeration of iron oxide particles in the presence of reactive surface modifiers. In order to establish an appropriate surface reactivity, bifunctional compounds have been used, where one of the functional groups was supposed to interact with the iron oxide surface and the other to stabilize the particles by building up a surface potential and to generate desired surface reactivity. For this reason, γ-aminopropyl triethoxy silane was used. The schematics of the reaction is shown in fig. 12. The amino silane is hydrolysed in the aqueous solution, leading to an adsorption of the $\equiv$SiOH groupings on the =FeOH group containing particles as soon as they are separated. The amino grouping of the silane is building up a ξ potential preventing re-agglomeration. The isoelectric point of the suspension is shifted from pH 6 (uncoated Fe oxide system) to pH 9 (after coating), showing clearly the effect of the coating process.

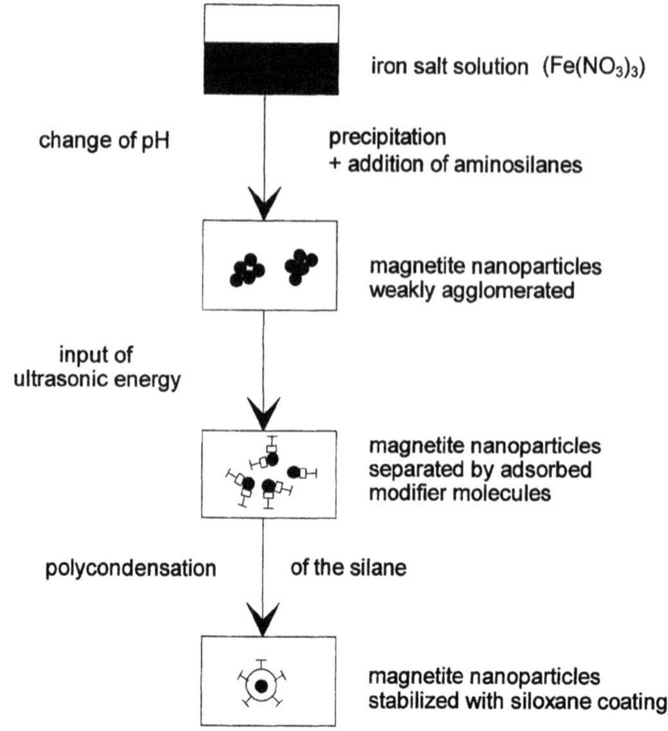

Fig. 12: Reaction scheme for the synthesis of modified magnetite.

The process has to take place over 48 hrs of elevated temperature (70 - 90 °C) to be completed. After this time, the coating is irreversible. This is interpreted by a subsequent polycondensation of ≡Si-OH + HO-Si≡ → ≡Si-O-Si≡ of the silanes adsorbed or bonded to the Fe oxide particles surface.

The effect of the ultrasonic treatment with and without the addition of γ-aminopropyl triethoxysilane is shown in fig. 13. The particle size distribution was determined by photon correlation spectroscopy and by ultracentrifugation.

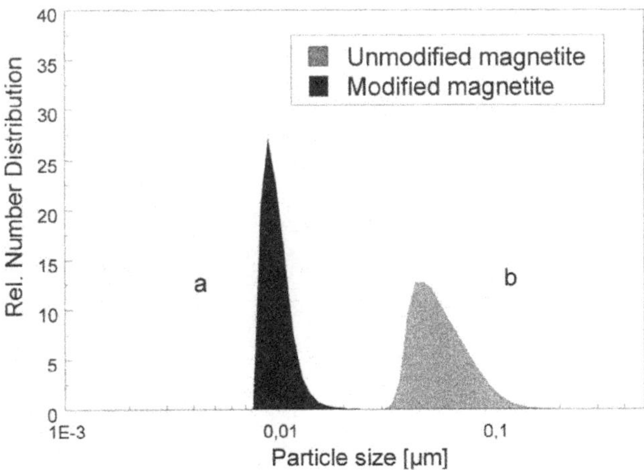

Fig. 13: Particle size distribution of precipitated Fe oxide particles, a: in presence of aminosilanes, b: without aminosilanes.

One clearly can see the effect of the surface modifier on particle size and distribution.

The aqueous suspension can be transformed into dry powders by removal of the water and completely redispersed. In order to synthesize polymer-ceramic nanocomposites, the powder was dispersed in ethanol and a commercially available diepoxide (Degacure K 126) was dissolved in the suspension. Thin films of about 1 - 2 µm were prepared by dip coating of microscopic slides and cured by UV light using a photo initiator (Degacure K 385). The solid films were investigated by HRTEM [25]. In fig. 14 the process is depicted, and fig. 15 shows the HRTEM image of the resulting nanocomposite.

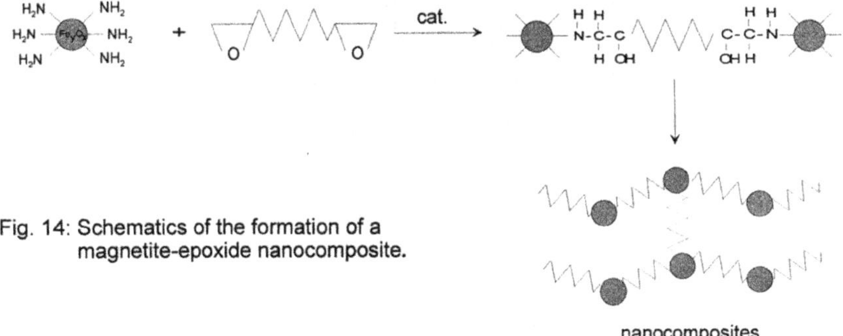

Fig. 14: Schematics of the formation of a magnetite-epoxide nanocomposite.

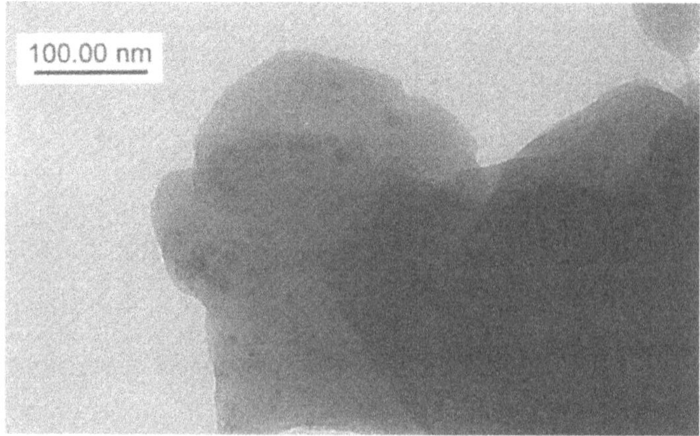

Fig. 15: HRTEM micrograph of a magnetite-epoxide nanocomposite.

The micrograph clearly shows the completely deagglomerated magnetite particles within the epoxide matrix. The films are superparamagnetit with magnetisation limits close to bulk magnetite and do not show any hysteresis. The question whether magnetite or maghemite is the dominating phase could not be clearly answered, neither by Moessbauer spectroscopy nor by HRTEM. In fig. 16 the HRTEM micrograph of a Fe oxide nanoparticle within the nanocomposite is shown. The lattice parameters within the possible preciseness fit to maghemite as well as to magnetite.

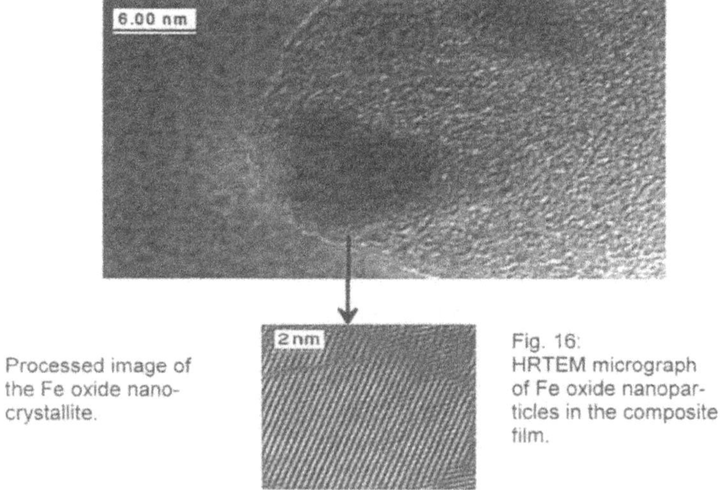

Fig. 16: HRTEM micrograph of Fe oxide nanoparticles in the composite film.

The films show a superparamagnetic behavior without any hysteresis at room temperature and a saturation magnetisation of about 80 % of a bulk magnetite phase (calculated on the iron oxide content as basis). The results show that the so-called separate preparation route, that means the preparation of nanoparticles separately from the matrix system using surface modifiers can be used to fabricate nanocomposites similar to those prepared by the in situ route. Using this approach, it was possible to synthesize completely deagglomerated amino group modified magnetite particles in the range between 6 and 20 nm by varying the processing parameters (concentration and pH). Investigations on chemical stability showed that the amino groups are surprisingly stably bonded to the surface (stable between pH 3 and 11). Amino groups are important since they easily can be reacted with biological molecules, organic monomers or sol-gel precursors to be incorporated into various types of matrices.

4.3 Metal Colloid Containing Systems

As shown in chapter 2 metal colloids can be formed using the surface reactivity control approach, too. It is well known that noble metal colloids show quantum size effects leading to non-linear optical properties (high third-order susceptibilities) based on the plasmon resonance of hot electrons. The question arises how far these effects can be influenced by the surface properties of these colloids, since the χ^3 value is strongly affected by the so-called local electric field factor f [11] ($\chi^3 \sim f^4$), and f depends strongly on the dielectric function of the surrounding media. To investigate this question, two different types of nanocomposites have been prepared. The general preparation route is shown in fig. 17. For this reason, amino group containing silanes and alternatively SH-group containing silanes have been used for the stabilisation of the colloids during and after the photochemical reduction step [11]. In this case, the $-NH_2$ and -SH groupings are supposed to act as bonding agents to the particle surface or as complex fomers. In fig. 18 the HRTEM micrograph of a Au particle after reduction is shown.

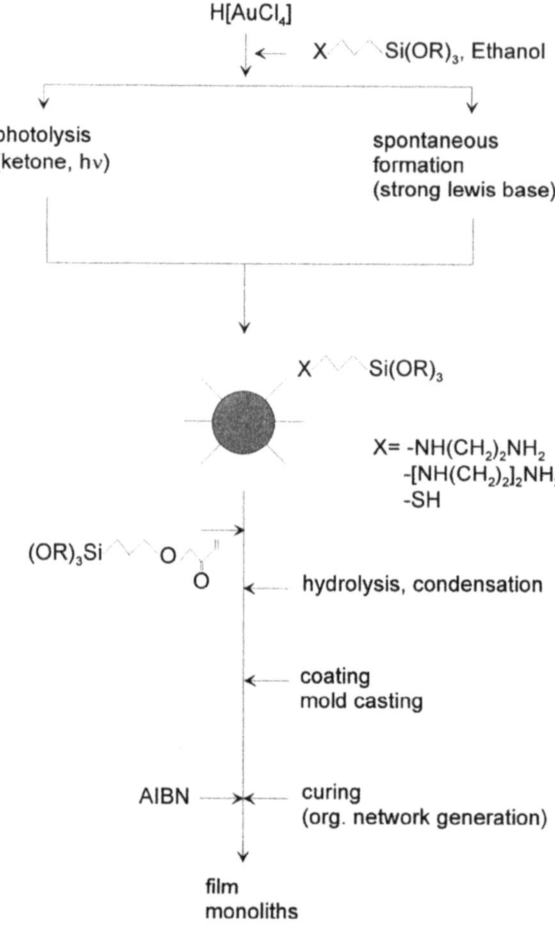

Fig. 17: Preparation scheme of Au containing nanocomposites; ABIN: azobis isobutyronitrile.

Twin formation is typical. The average size depends on the complex former concentration and was chosen to be about 7 - 9 nm. In combination with methacryloxy group containing silanes a polymerization/polycondensation process was carried out on films of about 3 µm in thickness obtained by dip coating of microscopic slides. These films have been photocured by adding AIBN as photoinitiator. In fig. 19 a Au particle in a composite matrix prepared from amino grouping covered Au particles and a methacryloxy silane, located at the film surface is shown. One can recognize an interface structure different from the bulk structure. In this case, the particle size is around 12 nm.

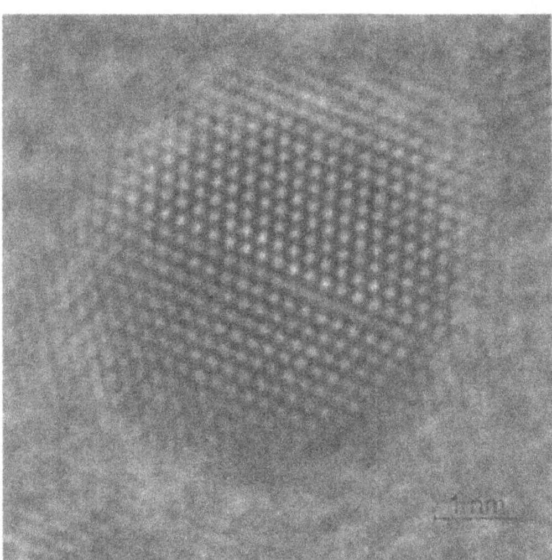

Fig. 18: Au colloid prepared by amino group control.

The non-linear properties have been determined by a four-wave mixing process, generating a self-induced diffractive grating. The χ^3 value then can be calculated from the diffraction efficiency [11].

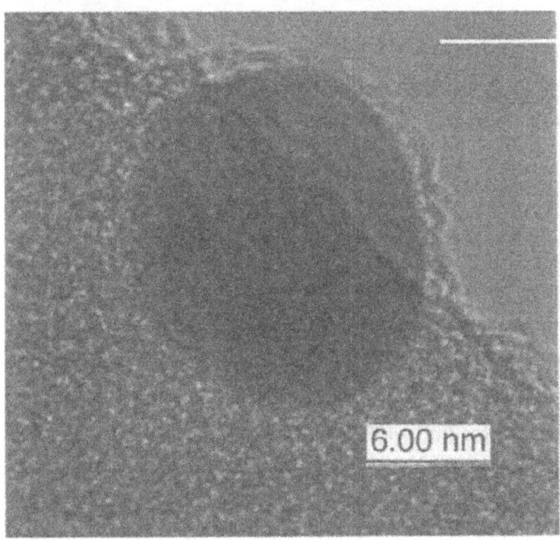

Fig. 19: Au particle in a composite matrix after polymerization of the methacrylic groupings.

Au colloid in Ormocer coating (amino silane ligand)

In table 1 the results of the χ^3 measurements of various composites and after firing them are shown.

Table 1: χ^3 values of Au particles of about 10 nm in different matrices; χ^3_m = value calculated for the pure metal.

Matrix	Ligand	$\chi^{(3)}_m$ [esu]
Composite	NH_2-	2.0 10^{-6}
	SH-	9.0 10^{-7}
SiO_2*	NH_2-	7.0 10^{-7}
	SH-	<1 10^{-7}

*after burning out the organics

The results show that very high χ^3 values were obtained in the amino ligand case. For SH ligands, the values are distinctly lower. If the organics are burnt out, the χ^3 values drop sharply. In the case of NH_2, the drop is far lower than in the case of SH. Whereas the amino groups are burnt out completely, leaving back a pure SiO_2 matrix, SH leads to sulfate formation entrapped in the system. The SO_4^{2-} ion as a ligand seems to interfere in an undesired way with respect to high χ^3 values.

The results show that the interface generated in Au colloid systems affects the χ^3 value strongly and is able to influence it by two orders of magnitude. In the described system, electron donating ligands seem to be able to increase the χ^3 value substantially.

5. Conclusion

The described investigations have shown that the synthesis and processing of new materials by chemistry opens interesting perspectives with respect to composite materials. The most interesting part is the possibility of the surface chemistry of the nucleating and growing particulate systems as well as the control of their further reactivity. Control of surface thermodynamics was used for the synthesis of nanoparticles with controlled sizes, and a simple model has been developed for the correlation between particle size and surface modifier concentration in cases of strong surface-to-modifier interaction. Using bifunctional ligands, it is possible to generate highly reactive nanoparticles within a reactive matrix and to fabricate in situ composites.

The results show that various types of nanocomposites can be obtained using sol-gel processing.

6. Acknowledgements

The authors want to thank Dr. T. Krajewski for the HRTEM micrographs, the Minister for Education, Research and Technology of the Federal government, the European Community, the State of Saarland and Deutsche Forschungsgemeinschaft for the financial support.

7. References

[1] H. Gleiter,
Nanocrystalline Materials. Pergamon Press, Oxford (1989).

[2] A. Henglein,
Topics in Current Chemistry: Mechanisms of reactions on colloidal microelectrodes and size quantization effects, 143 (1988), 115.

[3] C. J. Brinker and G. W. Scherer,
Sol-Gel Science, Academic Press Inc., Boston (1990).

[4] J. W. Christian,
The Theory of Transformations in Metals and Alloys. Pergamon Press, London, 1975.

[5] A. C. Zettlemoyer, ed.,
Nucleation, Marcel Dekker, New York, 1969.

[6] D. C. Bradley, R. G. Mehrotra and D. P. Gaur,
Metal Alkoxides, Academic Press, London, 1978.

[7] R. Naß and H. Schmidt,
in: Ceramic Powder Processing Science; eds.: H. Hausner, G. L. Messing, S. Hirano. Deutsche Keramische Gesellschaft e. V., Köln, 1989, 69.

[8] C. Sanchez and M. In.
J. Non-Cryst. Solids 147&148 (1992) 1.

[9] R. Kasemann, E. Geiter, H. Schmidt, E. Arpac, G. Wagner and V. Gerhard,
Verfahren zur Herstellung von Zusammensetzungen auf der Basis von epoxidgruppenhaltigen Silanen.
10.11.1993, German Offen 43 38 361 A1.

[10] H. Schmidt,
in: Proceedings 8th Intl. Workshop on Glasses and Ceramics from Gels, ed.: R. M. Almeida, J. Sol-Gel Science and Technology (in print).

[11] M. Mennig, M. Schmitt, U. Becker, G. Jung and H. Schmidt,
in: SPIE Vol. 2288 "Sol-Gel Optics III", ed.: J. D. Mackenzie. SPIE, Bellingham/ Washington, 1994, 130.

[12] M. Schmitt, M. Mennig, private communication.

[13] H. Schmidt and R. Naß,
in: Proc. Austceram '94, eds.: C. C. Sorell and A. J. Ruys. International Ceramic Monographs, Vol. 1, No. 2, Australasian Ceramic Society, Sydney, 1994, 1065.
[14] H. Schmidt, R. Kasemann, T. Burkhart, G. Wagner, E. Arpac and E. Geiter,
in: ACS Symposium Series No. 585: Hybrid Organic-Inorganic Composites, eds.: J. E. Mark, C. Y.-C. Lee and P. A. Bianconi, American Chemical Society, Washington, 1995, 331.
[15] C. Becker, M. Zahnhausen, H. Krug and H. Schmidt,
in: Ceramic Transactions Vol. 55: Sol-Gel Science and Technology, eds.: E. Pope, S. Sakka and L. Klein. American Ceramic Society, 1995, 299.
[16] M. Popall, H. Meyer, H. Schmidt and J. Schulz,
Mat. Res. Soc. Symp. Proc. 180 (1990), 995.
[17] F. Tiefensee, Ph. D. Thesis, University of Saarland, Saarbrücken, 1994.
[18] H. Schmidt, H. Krug, P. Oliveira,
private communication.
[19] H. Schmidt,
in: Ceramic Transactions, Vol. 55: Sol-Gel Science and Technology. Eds.: E. Pope, S. Sakka and L. Klein, American Ceramic Society, 1995, 253.
[20] T. Wood and H. Dislich,
in: Ceramic Transactions Vol. 55: Sol-Gel Science and Technology, eds.: E. Pope, S. Sakka and L. Klein. American Ceramic Society, 1995, 3.
[21] M. Kresse, R. Lawaczeck and D. Pfefferer,
EP 0 516 252 A2.
[22] H. Pilgrim,
EP 0 284 549 A2.
[23] K. Osseo-Asare and F. J. Arriagada,
in: Ceramic Transactions, Vol. 12, Ceramic Powder Science III, eds.: G. L. Messing, S. Hirano and H. Hausner, Am. Ceram. Soc., Ohio 1990.
[24] H. Schmidt, R. Naß and D. Burgard,
in: Proceedings Eurogel 1992, Dec. 2. - 4., Colmar.
[25] T. Krajewski, private communication.

THE USE OF NANOPARTICLES IN ENVIRONMENTAL APPLICATIONS

W.A. Zeltner and M.A. Anderson
University of Wisconsin - Madison
Water Chemistry Program
660 North Park Street
Madison, WI 53706
USA

1. Introduction

Growing public concern with environmental contaminants is producing increased interest in developing novel methods of treating contaminated media. A wide variety of contaminants present potential problems, including organics (e.g., volatile organic compounds, pesticides, polychlorinated biphenyls and derivatives such as dioxin, etc.), heavy metals, and radioactive species, in matrices such as ground waters, surface waters, soils, sediments, and air. Two of the commonly employed treatment methods, adsorption (primarily on activated carbon) and air stripping, simply transfer the contaminant from one medium to another. Incineration is capable of converting organic contaminants into more innocuous forms but does not enjoy strong public support. Thus, the development of novel methods for environmental remediation is a very active area of research at present. In addition, processes such as separations are being improved because these processes have direct applications in the treatment of hazardous wastes and offer significant energy savings over other processes presently in use in industry. The use of nanoparticles offers significant advantages in many of these processes. In this paper, we will discuss potential applications of nanoparticles in several areas with environmental implications. Applications involving adsorption and catalysis will only be summarized. Applications to photocatalysis, separations, and energy storage devices, areas in which we have active research programs, will be presented in more detail. We will not discuss possible applications in biotreatment because of our lack of expertise in this area.

2. Adsorption

The most advanced treatment scheme in widespread use that employs physical processes for treating environmental contaminants is adsorption on granular activated carbon (GAC) [1]. Although this process does immobilize the contaminant, some method must be found for disposing of or treating the adsorbent once its adsorptive capacity is spent.

E. Pelizzetti (ed.), Fine Particles Science and Technology, 643–656.

Disposal of the adsorbent (e.g., in landfills) only transfers the problem to another medium. For heavy metals and radioactive species, the preferred approach for treating these contaminants after adsorption involves desorbing these materials in a manner that allows them to be reused (i.e., recycled). Similar approaches are being considered for handling industrial waste streams that contain organic contaminants, with the ultimate goal being to achieve zero discharge processes in industrial applications [1].

Granular activated carbon is most commonly employed in treatment processes that rely on adsorption because GAC has a high surface area and thus a high adsorptive capacity, is relatively inexpensive, and is capable of adsorbing many different types of contaminants. However, there are many situations in which one desires to adsorb selected compounds under specific conditions and then change those conditions to desorb the compounds. For example, many hydrous oxides can adsorb heavy metals such as lead under neutral to alkaline conditions and then release the adsorbed metals when exposed to acidic conditions. There is thus a need for materials with the capability of selectively adsorbing only certain desired compounds. It is likely that the effectiveness of such selective adsorbents would be improved by fabricating them in nanoparticulate forms because of the resulting increase in effective surface area.

3. Catalysis

While adsorption in some form appears to be the most viable method of treating media that contain heavy metal and/or radioactive contaminants, oxidation processes can be employed for treating organic contaminants. Various oxidation and/or reduction processes are also being developed to treat inorganic gases such as ammonia, sulfur dioxide, and the various nitrogen oxides. Applications of these processes that would employ nanoparticles would primarily involve the use of nanoparticles as catalysts.

Heterogeneous catalytic oxidation of toxic organic chemicals requires the use of solid catalysts that are characterized by both high reactivity and high selectivity for deep oxidation products (CO_2, H_2O, etc.). Catalytic oxidation occurs at a lower reaction temperature than the corresponding homogeneous process (i.e., incineration in various forms) and has several advantages over other destruction techniques. It is mechanically simple and is advantageous when on-site destruction of the contaminant is required for reasons of safety or economics. Applications of catalytic oxidation techniques to environmental remediation are under study [2-5]. It has been demonstrated that many volatile organic compounds (VOCs), including halogenated VOCs, can be efficiently degraded to CO_2, H_2O and, in the case of chlorinated hydrocarbons, HCl, over a variety of metallic (e.g., ruthenium, platinum, and palladium) and metal oxide (e.g., manganese oxide, copper oxide, ferric oxide, titanium dioxide and cobalt (III) oxide) catalysts.

The effectiveness of these conventional oxidation catalysts should be significantly improved by using highly divided materials which are composed of nanometer sized particles (<10 nm). These materials, exhibiting high specific surface areas and large pore volumes, can be obtained through sol-gel chemical methods [6-9]. For many catalytic reactions (e.g., hydrogenation of pyrene [10] and methylation of phenol [11]), dramatic improvements in reaction rates were observed when the characteristic dimensions of the catalytic particles were in the nanometer range. (The catalytic activities increased by factors of ca. 100 and 10, respectively). In other cases, nanometer sized particles of MgO exhibited quite high reactivities for the destructive adsorption of organophosphorus compounds and in other catalytic processes [12,13]. If the improved performance noted above can be carried over into commercially available nanoparticle catalysts, significant environmental benefits would result. Treatment methods involving the complete oxidation of organic contaminants would be improved. In addition, significant energy savings would result if the efficiencies of the catalysts employed in numerous industrial processes (e.g., hydrogenation, dehydrogenation, partial oxidation, cracking, isomerization, etc.) were improved. This latter benefit may potentially be more important than the former.

4. Photocatalysis

While we are interested in expanding our research activities to include studies of the effects of employing nanoparticle oxides in selected heterogeneous catalytic reactions, this work is only in its early stages. However, much of our present research focuses on a somewhat different approach to catalysis - the study of photocatalytic oxidation directed primarily towards the complete oxidation of organic contaminants in both aqueous and gas phase systems. Because photocatalytic oxidation is a technique that can completely oxidize many organic compounds, photocatalysis is most likely to be used for environmental remediation. This use, however, can involve more than just organic contaminants in liquid waste streams. Specific applications under study include the use of photocatalysis to treat indoor air problems involving sensitivity to both organic molecules and microbes in homes and offices and to purify air in industrial settings such as dry cleaning establishments. At present, commercial interest in photocatalytic oxidation technology centers more on applications in gaseous systems rather than aqueous systems.

4.1. UNDERLYING PRINCIPLES OF PHOTOCATALYTIC OXIDATION

Semiconducting materials contain electrons that are confined to relatively narrow energy bands. The band of highest energy that contains an abundance of electrons is the valence band, while the band lying above the valence band, the conduction band, contains very few electrons. The difference in energies between the highest energy of the valence band and the lowest energy of the conduction band is termed the band gap energy. Illumination of semiconductors with radiation of energies greater than the band gap energy promotes electrons from the valence band to the conduction band, leaving behind positive holes. (UV light with a wavelength of 390 nm or less has sufficient energy for this purpose when one employs titania photocatalysts.) If these photogenerated electrons and holes reach the surface of the semiconductor before they recombine, they can then react there with adsorbed species (e.g., VOCs adsorbed from the surrounding medium).

The relevant literature contains both detailed discussions of this process and its applications in electrochemistry and photochemistry [14-17] and extensive reviews of photochemistry over semiconductor surfaces [18-20].

Although photocatalysis can occur over many different semiconductors, the photocatalyst most commonly utilized is titania because titania is stable under most reaction conditions. However, it must be noted that UV radiation is necessary to activate titania photocatalysts. Numerous studies indicate that illuminating the anatase form of TiO_2 with near UV radiation causes the complete oxidation of many organic species [21-26]. Photocatalytic degradation technology based on the use of semiconductor particles was employed to completely oxidize chlorinated organic compounds that constitute environmental hazards [27-30]. Complete photo-oxidation of organic compounds was achieved using a UV-illuminated TiO_2 film coated on silica gel or on the inside of a tubular glass reactor [25,31]. Most studies of photocatalytic oxidation are performed in aqueous systems, but photodegradation of gaseous contaminants has been demonstrated [32-36].

4.2 PREPARATION OF PHOTOCATALYSTS

Most reported photocatalysis studies employ commercially available titania powders as the catalyst. The most commonly used material is the P25 titania provided by Degussa. This photocatalyst combines several useful properties (low cost, high specific surface area, slow recombination of photogenerated electron-hole pairs [20], etc.). In our studies, we employ sol-gel techniques to prepare titania-based materials that can serve as possible photocatalysts. By employing a dialysis step during processing, we have been able to prepare highly porous (50-55% porosity) titania particles that display photocatalytic activity [34,37]. These particles can be made in various size ranges up to 2 mm in size by suitable grinding and sieving. However, these particles are highly aggregated collections of titania nanoparticles. Size estimates based on the widths of the X-ray diffraction peaks of these materials indicate that the primary particles that make up these aggregates are only ca. 8 nm in diameter.

An advantage of sol-gel processing is that it provides researchers considerable flexibility in the preparation of potential catalytic and photocatalytic materials. For example, several routes are available for preparing mixed oxide systems (e.g., cohydrolysis of mixtures of metal alkoxides, partial hydrolysis of a metal alkoxide which is then mixed with another metal alkoxide for final hydrolysis, mixing of nanoparticle sols that are prepared separately, etc.) [6,38]. Thus, numerous potential photocatalytic materials could be studied. An additional approach that can improve the performance of the catalyst is to modify it by depositing other materials on its surface. In gas phase photoreactors, we have increased the rates of oxidation and photooxidation of selected VOCs on our titania and titania-based mixed oxide catalysts by depositing islands of platinum on the surfaces of these catalysts. The high surface areas that are characteristic of these nanoparticulate catalysts (100-500 m^2/g depending on the firing temperature and composition of the catalyst) appear to enhance this deposition process and the resulting activity of the catalyst.

4.3. PERFORMANCE OF TITANIA-BASED PHOTO/CATALYSTS

An example of the behavior of these various photocatalysts is given by their use in a bench-scale photoreactor for the complete oxidation of a gaseous stream of benzene (ca. 500 ppmv) in air [39]. A fluorescent UV bulb was used to illuminate a reactor that was packed with either pure titania or platinized (0.1 wt% Pt) titania particles. A Hewlett-Packard gas chromatograph, equipped with both a flame ionization detector (used for quantifying levels of benzene) and a thermal conductivity detector (used for quantifying levels of carbon dioxide), was employed to determine concentrations of gases in both the feed and effluent streams. The gas stream was not recirculated through the reactor.

Figure 1 illustrates the effect of the reaction temperature on the conversion of benzene (defined as the percent of benzene removed from the feed stream under steady-state conditions) in this reaction system for both types of photocatalyst under UV illumination and without UV illumination [39]. Without illumination, no conversion of benzene is observed until the reaction temperature exceeds 70°C when the pure titania catalyst is used. The limited reaction that occurs above 70°C is due solely to conventional heterogeneous catalytic processes occurring on the surface of the titania. It appeared that the primary removal process was adsorption of some degradation product on the catalyst, as the color of the catalyst was observed to change and no organic compounds other than unreacted benzene were detected in the effluent stream. Only a limited amount of oxidation occurred, since only a small amount of carbon dioxide was detected in the effluent even at the highest operating temperature. Under UV illumination, however, significant photocatalytic activity was observed, with the conversion of benzene approaching 90% at a reaction temperature of 140°C. At this temperature, the amount of carbon dioxide detected was about 70% of that expected if all of the benzene that reacted was completely oxidized.

The effect of platinizing this catalyst (i.e., depositing small islands of Pt on the surface of the material) is clearly illustrated in this figure. Without illumination, there is no advantage gained by employing the platinized catalyst until the reaction temperature exceeds 90°C. Above this temperature, the conversion of benzene increases rapidly, reaching complete conversion at a reaction temperature of 130°C. The improved performance of this platinized titania at temperatures above 90°C is due to heterogeneous catalytic reactions occurring on the deposited platinum. Platinum is known to dissociatively adsorb numerous gases, and it is likely that this mechanism is the initial step in the overall oxidation reaction. However, complete oxidation of benzene does not occur even at 130°C, as only some 75% of the theoretical amount of carbon dioxide is detected in the effluent under these conditions. Under UV illumination, however, the additional photocatalytic reactions that occur on the platinized catalyst lead to complete conversion of benzene at reaction temperatures of 70°C or higher, and essentially complete oxidation of the benzene feed is obtained by operating at 110°C.

The ability to prepare nanoparticulate titania-based photo/catalysts that are readily platinized allows one to combine both conventional heterogeneous catalytic processes and photocatalytic processes in one reactor. The resulting system offers great promise for the remediation of undesired VOCs in a variety of settings.

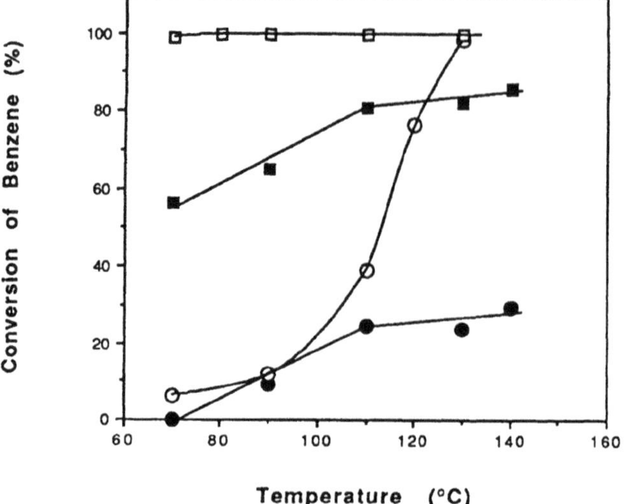

Figure 1. Effect of reaction temperature on the conversion of benzene under UV illumination (■,□) and without UV illumination (●,○). Closed symbols represent data obtained for the pure titania catalyst, while open symbols represent data obtained for the platinized titania catalyst.

5. Separations

5.1. GENERAL CONSIDERATIONS

Another area in which the successful application of nanoparticles would have important consequences is the development of inorganic nanoparticulate filtration media for separations. Because separations involving filtration are generally not very energy intensive, such separations offer considerable cost savings over other separation processes (e.g., distillation or crystallization for separations involving liquids, pressure-swing adsorption for the separation of gases). Markets for reverse osmosis (removal of low molecular weight ions and molecules from water) and gas phase separations presently utilize membranes that are fabricated from organic polymers. These polymeric membranes generally have limited chemical and thermal stabilities. Thus, fouling of these membranes becomes a serious problem because few methods are available to treat the membranes without degrading their performance. Reverse osmosis separations are conducted under high pressure so that a reasonable flow of purified water through the membrane is obtained. However, the high pressures employed compress polymeric membranes, thus closing off their pores and minimizing the flow of water through them. The passage of material through polymeric membranes is even more of a challenge for gas phase separations, as the mechanism for these separations involves differential solubility within the membrane of the gases being separated. As a result, the separation is achieved because of the difference in the rates of transport of the gases through the membrane. Since polymeric membranes for gas phase separations are not porous, the fluxes of gases through these materials are extremely low.

Ceramic membranes that employ inorganic nanoparticles for the separation layer should overcome many of these problems. Inorganic membranes in general will have much greater thermal and chemical stability than corresponding organic polymeric membranes, especially when exposed to organic solvents and chlorinated water. While certain inorganic membranes may degrade when exposed to strongly acidic or alkaline solutions, this problem also affects most polymeric membranes. Thus, it should be much easier to remove foulants and to sterilize inorganic membranes than to perform these processes with organic membranes. Inorganic membranes should also be much more resistant to high pressures than organic membranes, so inorganic membranes should eventually play a large role in reverse osmosis applications. If the surfaces of inorganic membranes can be suitably modified so that different gases pass through such membranes at different rates, then those membranes could also be employed for gas phase separations. Since it should be possible to prepare inorganic membranes with open pore structures, fluxes of gases through such membranes may well be several orders of magnitude larger than fluxes of gases through the presently available non-porous polymeric membranes.

While the advantages of employing ceramic membranes for reverse osmosis and gas phase separations are clear, the need to employ nanoparticulate materials in fabricating the separation membrane has led to problems that are limiting the implementation of these membranes in commercial processes. A significant challenge is the need to deposit an ultra-thin (i.e., sub-micron thickness) flaw-free porous ceramic membrane either on or within a porous support. In practice, the smallest pore diameters that are available in commercial ceramic membranes are ca. 4 nm. However, such pore diameters are too large for applications involving reverse osmosis and gas phase separations. Ceramic membranes with smaller pore diameters have been found to be extremely susceptible to the formation of cracks and pin-holes during fabrication. Several research approaches to circumvent this problem are under active investigation.

An example of the performance of ceramic membranes for separations in aqueous systems is presented in Figure 2 [40]. Data for our laboratory units were obtained by performing permselectivity experiments using aqueous solutions of polyethylene glycol of different molecular weights. Rejections for other membranes, including a commercially available ceramic membrane unit, have also been plotted to provide a reference. Our titania membrane is fabricated by dip coating a commercially obtained Anotec Anodisc (electrolytically prepared alumina supports with pore diameters of ca. 0.02 μm) with a sol that consists of titania particles (ca. 5 nm diameter at a concentration of ca. 0.05 M) suspended in amyl alcohol [41]. Gas permeability measurements conducted using this titania membrane indicate that it is essentially flaw-free [42]. However, Anotec supports are quite fragile. We will need to cast flaw-free membranes on sturdy porous supports in order to generate commercial interest in developing ceramic membranes based on our nanoparticle technology.

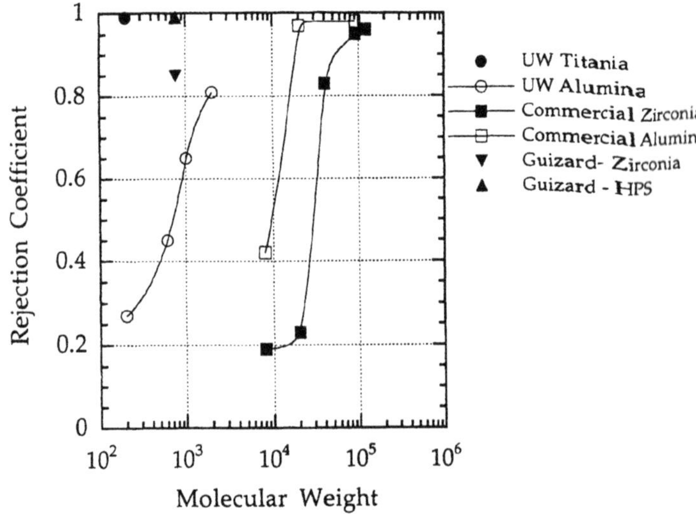

Figure 2. Rejection as a function of molecular weight of the test compound dissolved in water for various ceramic membranes. Data obtained for our ceramic membranes are labeled UW [41,43]. Data labeled Guizard were given in ref. 44.

5.2. CERAMIC MEMBRANE REACTORS

A specific application of ceramic membranes for gas phase separations that is receiving considerable interest is their use in membrane reactors [45]. Many chemicals are synthesized commercially at high temperature and/or high pressure under conditions that are equilibrium limited (i.e., the reaction cannot proceed to completion). If a ceramic membrane can be prepared that can withstand the required reaction conditions and can allow one of the reaction products to be removed from the reactor while retaining the other reactants and products, the equilibrium limitations on performing these reactions can be removed with a noticeable improvement in the performance of the process. A reaction of great interest for such an application is the non-oxidative dehydrogenation of ethylbenzene to styrene and hydrogen [46]. Availability of a membrane with pore sizes small enough to allow passage of hydrogen while retaining ethylbenzene and styrene that is also stable at typical operating temperatures (ca. 600°C) would lead to considerable industrial interest. An additional advantage of membrane reactors is that, because of the removal of equilibrium limitations, reactions in membrane reactors can be performed at or near ambient pressure.

A major problem in applying nanoparticulate membranes for such high temperature applications as membrane reactors is the tendency of nanoparticles to sinter readily due to the high surface energies of these materials. As a result, the thermal stabilities of nanoparticulate membranes may not be good enough to allow the use of pure oxides in membranes at temperatures of 600°C or so. However, we have found that materials prepared from mixed oxide nanoparticles are thermally stable at temperatures of 100-150°C higher than materials prepared from the corresponding pure oxide nanoparticles [38]. Such nanoparticle mixed oxide membranes may prove feasible for use in ceramic membrane reactors that can be operated at relatively high temperatures.

6. Energy Storage and Power Devices

A final application of nanoparticles that has implications for the environment is the use of nanoparticulate oxides to fabricate thin-film power devices such as ultra-thin batteries, ultracapacitors, and fuel cells. Our initial research in this area led to a thin-film battery based upon the Zn/KOH/MnO_2 redox system [47]. In this battery we employed an Al_2O_3 thin film microporous separator between the anode and the cathode. In this system the size of the electrodes can be greatly reduced due to the fact that each electrode layer is less than 2 μm thick while the separator is itself of the same dimensions. This means that there is little I^2R loss in these thin film systems. Furthermore, since the separator is a ceramic membrane, it is highly resistant to puncturing that results from the growth of dendrites from the zinc electrode as the battery is recharged. This dendritic growth would otherwise tend to short-out the thin film battery. Since both electrodes as well as the separator can be produced by dip-coating of colloidal nanoparticulate suspensions, the economics of battery fabrication should be able to be reduced substantially. The power and energy density characteristics of our thin-film battery appear to be far superior to present alkaline cells. We are now in the process of further characterizing this battery with respect to rechargeability and are also examining other electrochemical couples for use in this application.

Ultracapacitors represent another energy storage application for nanoparticulate oxides. An ultracapacitor can be defined as a device which stores its energy in the electrical double layer around each of the nanoparticles and also by Faradaic processes involving either the proton or other potential determining ions which adsorb on the surface of the oxide [48] . Since capacitance scales inversely with distance of charge separation (i.e., distance between the electrodes) and directly with surface area, nanoparticulate ultracapacitors can utilize the short distances that are characteristic of the electrical double layer (Angstroms) for energy storage as well as the large surface areas of the nanoparticulate oxide films.

We have fabricated an electrochemical capacitor by coating a nickel foil with a NiO_x colloidal suspension prepared using sol-gel techniques [49]. The power and energy density characteristics of this capacitor are compared with those of other electrochemical systems in Figure 3. Note that ultracapacitors fall between conventional capacitors and batteries with respect to energy and power densities. Note also that our capacitor has a much higher energy storage density than the commercial RuO_2 capacitor fabricated by Pinnacle. However, the Pinnacle system has a higher power density. It is believed that the reason for the better power densities of the Pinnacle units is that Pinnacle employs a stacked bi-polar arrangement of many cells while the device that we have fabricated contains only one cell.

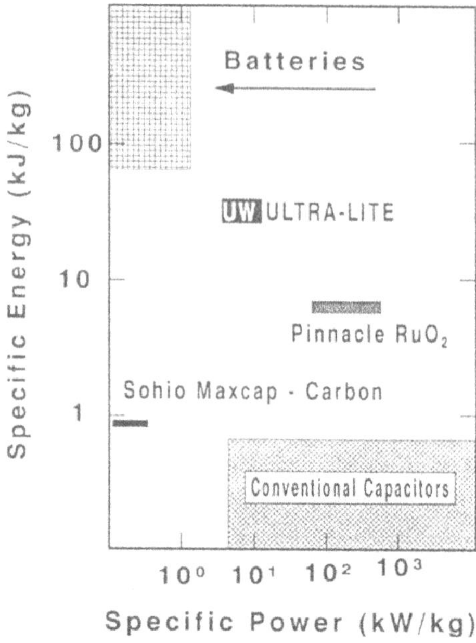

Figure 3. Ranges of specific energy vs. specific power measured for several batteries and capacitors, including our NiO capacitor (UW Ultra-lite), a commercial carbon-based ultracapacitor (Sohio Maxcap), and a commercial RuO_2 ultracapacitor (Pinnacle RuO_2) [48,49].

7. Conclusions

Clearly, nanoparticles have an important role to play in environmental applications. The development of nanoparticle-based catalysts and ceramic membranes promises major improvements in environmental remediation, development and implementation of zero discharge processes for industry, and energy savings in a variety of industrial applications that utilize separations. The use of nanoparticle technology in thin-film energy storage devices will also lead to significant improvements in the performance of batteries and capacitors.

8. References

1. LaGrega, M.D., Buckingham, P.L., and Evans, J.C. (1994) *Hazardous Waste Management*, McGraw-Hill, Inc., New York.

2. Shaw, H., Wang, Y., Yu, T.C., and Cerkanowicz, A.E. (1993) Catalytic oxidation of trichloroethylene and methylene chloride, in D.W. Tedder and F.G. Pohland (eds.), *Emerging Technologies for Hazardous Waste Management III*, ACS Symposium Series No. 518, American Chemical Society, Washington, DC, pp. 358-379.

3. Barresi, A.A. and Baldi, G. (1994) Deep catalytic oxidation of aromatic hydrocarbon mixtures: reciprocal inhibition effects and kinetics, *Ind. Eng. Chem. Res.* **33**, 2964-2974.

4. Gandhi, H.S. and Shelef, M. (1991) Effects of sulfur on noble metal automotive catalysts, *Appl. Catal.* **77(2)**, 175-186.

5. Papenmeier, D.M. and Rossin, J.A. (1994) Catalytic oxidation of dichloromethane, chloroform, and their binary mixtures over a platinum alumina catalyst, *Ind. Eng. Chem. Res.* **33**, 3094-3103.

6. Brinker, C.J. and Scherer, G.W. (1990) *Sol-Gel Science: The Physics and Chemistry of Sol-Gel Processing*, Academic Press, New York.

7. Xu, Q. and Anderson, M.A. (1994) Sol-gel route to synthesis of microporous ceramic membranes: preparation and characterization of microporous TiO_2 and ZrO_2 xerogels, *J. Amer. Ceram. Soc.* **77**(7), 1939-1945.

8. Ward, D.A. and Ko, E.I. (1995) Preparing catalytic materials by the sol-gel method, *Ind. Eng. Chem. Res.* **34(2)**, 421-433.

9. Pajonk, G.M. (1991) Aerogel catalysts, *Appl. Catal.* **72**, 217-266.

10. Wilcoxon, J.P., Martino, A., Baughmann, R.L., Klavetter, E., and Sylwester, A.P. (1993) Synthesis of transition metal clusters and their catalytic and optical properties, *Mat. Res. Soc. Symp. Proc.* **286**, 131-136.

11. Tleimat-Manzalji, R., Manzalji, T., and Pajonk, G.M. (1992) Aerogels and xerogels for catalytic applications, *J. Non-Cryst. Solids* **147** & **148**, 744-747.

12. Li, Y.-X. and Klabunde, K.J. (1991) Nanoscale metal oxide particles as chemical reagents. Destructive adsorption of a chemical agent simulant, dimethyl methylphosphonate, on heat-treated magnesium oxide, *Langmuir* **7**, 1388-1393.

13. Hoq, M.F., Nieves, I., and Klabunde, K.J. (1990) Mechanistic studies of hydrocarbon CH/deuterium exchange over thermally activated magnesium oxide, *J. Catal.* **123(2)**, 349-363.

14. Nozik, A.J. (1978) Photoelectrochemistry: applications to solar energy conversion, *Annu. Rev. Phys. Chem.* **29**, 189-222.

15. Bard, A.J. and Faulkner, L.R. (1980) *Electrochemical Methods: Fundamentals and Applications*, John Wiley & Sons, New York.

16. Gerischer, H. (1985) Semiconductor electrodes and their interaction with light, in M. Schiavello (ed.), *Photoelectrochemistry, Photocatalysis and Photoreactors: Fundamentals and Developments*, NATO ASI Series C, Vol. 146, D. Reidel Publishing Co., Boston, MA, pp. 39-106.

17. Memming, R. (1988) Photoelectrochemical solar energy conversion, in *Topics in Current Chemistry*, Vol. 143, Springer-Verlag, Berlin-Heidelberg, pp. 79-112.

18. Kamat, P.V. (1993) Photochemistry on nonreactive and reactive (semiconductor) surfaces, *Chem. Rev.* **93**, 267-300.

19. Fox, M.A. and Dulay, M.T. (1993) Heterogeneous photocatalysis, *Chem. Rev.* **93**, 341-357.

20. Hoffmann, M.R., Martin, S.T., Choi, W., and Bahnemann, D.W. (1995) Environmental applications of semiconductor photocatalysis, *Chem. Rev.* **95**, 69-96.

21. Bard, A.J. (1979) Photoelectrochemistry and Heterogeneous Photocatalysis at Semiconductors, *J. Photochem.* **10(1)**, 59-75.

22. Barbeni, M., Pramauro, E., Pelizzetti, E., Borgarello, E., and Serpone, N. (1985) Photodegradation of pentachlorophenol catalyzed by semiconductor particles, *Chemosphere* **14(2)**, 195-208.

23. Fox, M.A. (1992) Photocatalysis: decontamination with sunlight, *Chemtech* **22(11)**, 680-685.

24. Low, G.K.-C., McEvoy, S.R., and Matthews, R.W. (1991) Formation of nitrate and ammonium ions in titanium dioxide mediated photocatalytic degradation of organic compounds containing nitrogen atoms, *Environ. Sci. Technol.* **25**, 460-467.

25. Matthews, R.W. (1988) Kinetics of photocatalytic oxidation of organic solutes over titanium dioxide, *J. Catal.* **111**, 264-272.

26. Tunesi, S. and Anderson, M.A. (1987) Photocatalysis of 3,4-DCB in TiO_2 aqueous suspensions: effects of temperature and light intensity: CIR-FTIR interfacial analysis, *Chemosphere* **16(7)**, 1447-1456.

27. Borgarello, A., Serpone, N., Barbeni, M., Minero, C., Pelizzetti, E., and Pramauro, E. (1986) Putting photocatalysis to work, in E. Pelizzetti and N. Serpone (eds.), *Homogeneous and Heterogeneous Photocatalysis*, NATO ASI Series C, Vol. 174, Kluwer Academic Publishers, Dordrecht, pp. 673-689.

28. Ahmed, S. and Ollis, D.F. (1984) Solar photoassisted catalytic decomposition of the chlorinated hydrocarbons trichloroethylene and trichloromethane, *Solar Energy* **32(5)**, 597-601.

29. Hisanaga, T., Harada, K., and Tanaka, K. (1990) Photocatalytic degradation of organochlorine compounds in suspended titanium dioxide, *J. Photochem. Photobiol. A: Chem.* **54(1)**, 113-118.

30. Turchi, C.S. and Ollis, D.F. (1990) Photocatalytic degradation of organic water contaminants: mechanisms involving hydroxyl radical attack, *J. Catal.* **122(1)**, 178-192.

31. Matthews, R.W. (1988) An adsorption water purifier with in situ photocatalytic regeneration, *J. Catal.* **113(2)**, 549-555.

32. Dibble, L.A. and Raupp, G.B. (1990) Kinetics of the gas-solid heterogeneous photocatalytic oxidation of trichloroethylene by near UV illuminated titanium dioxide, *Catal. Lett.* **4**, 345-354.

33. Peral, J. and Ollis, D.F. (1992) Heterogeneous photocatalytic oxidation of gas-phase organics for air purification: acetone, 1-butanol, butyraldehyde, formaldehyde, and m-xylene oxidation, *J. Catal.* **136**, 554-565.

34. Anderson, M.A., Yamazaki-Nishida, S., and Cervera-March, S. (1993) Photodegradation of trichloroethylene in the gas phase using TiO_2 porous ceramic membrane, in D.F. Ollis and H. Al-Ekabi (eds.), *Photocatalytic Purification and Treatment of Water and Air*, Elsevier, New York, pp. 405-420.

35. Yamazaki-Nishida, S., Nagano, K.J., Phillips, L.A., Cervera-March, S., and Anderson, M.A. (1993) Photocatalytic degradation of trichloroethylene in the gas phase using titanium dioxide pellets, *J. Photochem. Photobiol. A: Chem.* **70**, 95-99.

36. Obee, T.N. and Brown, R.T. (1995) TiO_2 photocatalysis for indoor air applications: effects of humidity and trace contaminant levels on the oxidation rates of formaldehyde, toluene, and 1,3-butadiene, *Environ. Sci. Technol.* **29**, 1223-1231.

37. Xu, Q. and Anderson, M.A. (1991) Synthesis of porosity controlled ceramic membranes, *J. Mater. Res.* **6(5)**, 1073-1081.

38. Xu, Q. and Anderson, M.A. (1993) Sol-gel route to synthesis of microporous ceramic membranes: thermal stability of TiO_2-ZrO_2 mixed oxides, *J. Amer. Ceram. Soc.* **76(8)**, 2093-2097.

39. Fu, X., Zeltner, W.A., and Anderson, M.A. (1995) The gas phase photocatalytic mineralization of benzene on porous titania-based catalysts, accepted for publication in *Appl. Catal. B: Environ.*

40. Peterson, R.A., Webster, E.T., Niezyniecki, G.M., Anderson, M.A., and Hill, C.G., Jr. (1995) Ceramic membranes for novel separations, *Separat. Sci. Technol.* **30(7-9)**, 1689-1709.

41. Peterson, R.A., Anderson, M.A., and Hill, C.G., Jr. (1993) Application of ceramic membranes to reverse osmosis, *Separat. Sci. Technol.* **28(1-3)**, 327-342.

42. Peterson, R.A., Anderson, M.A., and Hill, C.G., Jr. (1994) Development of TiO_2 membranes for gas phase nanofiltration, *J. Membrane Sci.* **94**, 103-109.

43. Gieselmann, M.J., Anderson, M.A., Moosemiller, M.D., and Hill, C.G., Jr. (1988) Physico-chemical properties of supported and unsupported γ-Al_2O_3 and TiO_2 ceramic membranes, *Separat. Sci. Technol.* **23(12-13)**, 1695-1714.

44. Guizard, C., Julbe, A., Larbot, A., and Cot, L. (1992) Nanostructures in sol-gel derived materials: application to the elaboration of nanofiltration membranes, *J. Alloys Compounds* **188**, 8-13.

45. Saracco, G. and Specchia, V. (1994) Catalytic inorganic-membrane reactors: present experience and future opportunities, *Catal. Rev. - Sci. Eng.* **36(2)**, 305-384.

46. Tiscareño-Lechuga, F., Hill, C.G., Jr., and Anderson, M.A. (1993) Experimental studies of the non-oxidative dehydrogenation of ethylbenzene using a membrane reactor, *Appl. Catal. A: General* **96**, 33-51.

47. Stadniychuk, H.P., Anderson, M.A., and Chapman, T.W. (1995) Sol-gel-derived thin-film manganese dioxide cathodes, submitted to *J. Electrochem. Soc.*

48. Raistrick, I.D. (1992) Electrochemical capacitors, in J. McHardy and F. Ludwig (eds.), *Electrochemistry of Semiconductors and Electronics: Processes and Devices*, Noyes Publications, Park Ridge, NJ, pp. 297-355.

49. Liu, K.-C. and Anderson, M.A. (1995) Porous nickel oxide films for electrochemical capacitors, submitted to *Mater. Res. Soc. Symp. Proc.*

SUBNANOSECOND CHARACTERISTICS AND PHOTOPHYSICS OF NANOSIZED TiO_2 PARTICULATES FROM R_{part} = 10 A TO 134 A: *MEANING FOR HETEROGENEOUS PHOTOCATALYSIS*

N. SERPONE,[a] D. LAWLESS,[a] and E. PELIZZETTI[b]

[a] *Center for Fast Laser Spectroscopy, Department of Chemistry & Biochemistry, Concordia University, Montreal, Que., CANADA.*
[b] *Dipartimento di Chimica Analitica, Universita di Torino, 10125 Torino, ITALIA.*

1. Abstract

Three nanosized TiO_2 clusters (R_{part} = 10 A, 65 A, and 134 A), prepared by the arrested hydrolysis of $TiCl_4$, had their photophysical properties assessed by absorption and photoluminescence spectroscopy and their subnanosecond characteristics determined by fast laser spectroscopy using 355 nm, 30-ps pulse excitation in the time window -50 ps to 10 ns. Direct and indirect electron transitions between energy levels in the Brillouin zone were observed and assigned for this indirect gap semiconductor. It is argued that for colloidal particles of TiO_2 with $2R_{part}$ = 20 A or greater there is no display of a size quantization effect. The subnanosecond transient spectra are composites of spectra of trapped electrons and trapped holes with trapping occurring in subpicosecond time (< 1-10 ps); as well, nearly 90% of the charge carriers have recombined by 1 ns. This has important consequences in heterogeneous photocatalyzed oxidations in the debate as to whether these oxidations implicate directly the valence band holes or the trapped holes, inferred earlier to be surface-bound •OH radicals. Quantum yields of these photoredox processes are therefore expected to be less than 10%.

2. Introduction

Nanosized semiconductor clusters (quantum dots) and particles have received considerable attention owing to their optical properties and their potential applications in fast optoelectronics (see refs in [1]) and as photocatalysts for solar

E. Pelizzetti (ed.), Fine Particles Science and Technology, 657–673.

energy conversion [2,3] and for detoxification of polluted environmental ecosystems [4-6]. Ultrasmall crystallites (clusters) composed of a few molecular units maintain their discrete HOMOs and LUMOs [7]. These clusters are said to be quantized (Q particles) because of the confinement of the charge carriers in potential wells of small dimensions [8-10]. Size quantization in semiconductor materials causes large changes in optical properties. Accumulation of charge carriers in a small reaction volume as exemplified by ultrasmall semiconductor particulates enhances the effective redox potentials of the photogenerated electrons and holes, and consequently enhances their photoredox chemistry with electron acceptors and donors (not otherwise possible with bulk-type semiconductor materials) and particularly so when the particles are quantized [11].

Around the time of our first report on the picosecond spectral properties and photophysical events in titania colloids ($2R_{part}$ = 120 A) [12], a number of other reports concerning the photophysics of TiO_2 sols appeared in the literature [9,13-17]. They have shown that particle size has a pronounced effect on spectral properties and on charge carrier dynamics on semiconductor particulates [18]. More recent reports [19-24] have noted that extremely small TiO_2 colloid particles in the size range about 10 to 120 A show exciton confinement and thus Q-size effects in this range. Systematic studies of the dependence of the photophysical properties of TiO_2 particles on particle size have not kept pace with studies in heterogeneous photocatalysis where TiO_2 plays a very significant role (see e.g., refs. [4-6] and the many references therein).

We have recently undertaken a systematic study [25] of the photophysical characteristics of titania particles of colloidal dimensions ($2R_{part}$ = 21 A, 133 A, and 267 A); a full report of this study is presented elsewhere [26,27]. We show that there are no quantum size effects on the spectral properties of these nanosized TiO_2 specimen. In fact our results suggest that what has been taken to be Q-size effects in the past [19-24] from observations of spectral blue shifts may well be due to direct band transitions for an otherwise indirect bandgap semiconductor such as titania. Analogous observations of direct transitions in TiO_2 and in other indirect gap semiconductors (e.g., $SrTiO_3$, $Sr_{0.5}Ba_{0.5}Nb_2O_6$, WO_3, and Fe_2O_3) were reported long ago by physicists [28-30] and more recently by Wold and coworkers [31]. The particle sizes examined in our studies [25-27] seem to affect charge carriers relaxation dynamics; however, they appear not to be related to exciton confinement.

3. Considerations of Photophysical Events in a Small Semiconductor Particle

Figure 1 illustrates a typical spherically-shaped small TiO_2 particle, taken here as a model semiconductor system, with a definition of the valence (VB) and conduction (CB) bands together with some (by no means all) photophysical and photochemical processes within and on the particle surface. Subsequent to photogeneration of CB electrons and VB holes by bandgap excitation (3.2 eV; anatase) of these titania particles, the charge carriers undergo a plethora of photophysical events. We must first recognize that colloidal semiconductor particulates comprise a high density of lattice defect sites and surface states, the nature of which is highly dependent on the

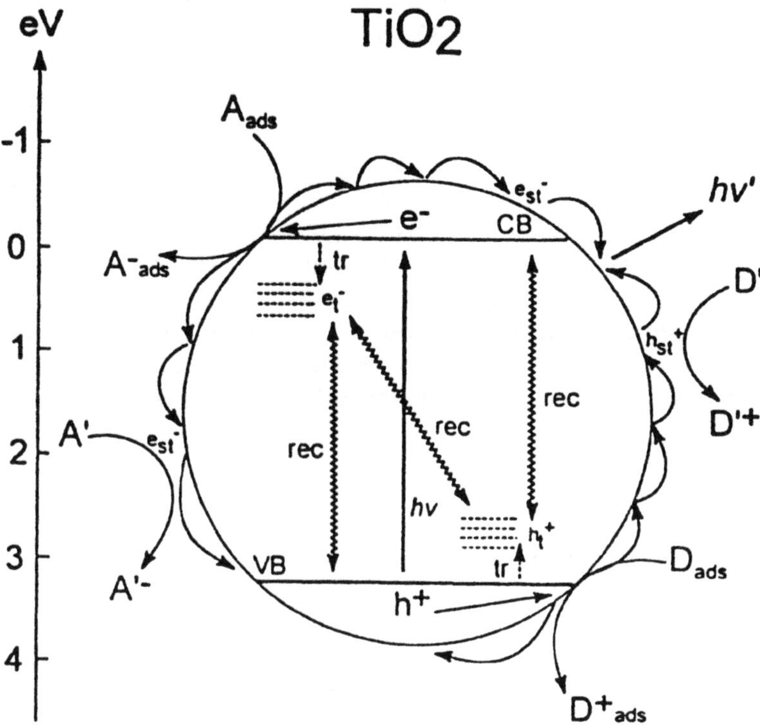

Figure 1.- Graphic illustrating the various, but non-exhaustive, series of photophysical events in a small semiconductor particle.

method(s) of preparation. We expect these defects to have some influence on the photophysical processes. As illustrated in Figure 1, charge carriers can rapidly migrate to the surface, where they are trapped in competition with trapping, detrapping, and radiative and nonradiative recombination (eqns 1- 4).

$$TiO_2 + h\nu \rightarrow TiO_2 (e^-_{cb} + h^+_{vb}) \tag{1}$$

$$e^-_{CB} + traps \rightarrow e^-_t \quad and \quad h^+_{VB} + traps \rightarrow h^+_t \tag{2}$$

$$e^-_{CB} / e^-_t + h^+_{vb} / h^+_t \rightarrow h\nu' ; h\nu'' ; ... \tag{3}$$

$$e^-_{st} + h^+_{st} \rightarrow h\nu' + h\nu''.... \tag{4}$$

In addition, the CB electrons and VB holes upon migration to the surface may reduce a pre-adsorbed acceptor A and oxidize a pre-adsorbed donor D, respectively. Where D donors are the surface chemisorbed OH^- or H_2O species, the oxidized donor is the •OH radical, and where A is adsorbed molecular O_2 the reduced donor is the superoxide radical anion, O_2^-•. We have long held the view, supported by past [32] and recent work [33], that h^+_{st} is the surface-bound •OH radical at ambient temperature and that •OH is the principal oxidizing entity in photooxidations. The lifetime of this surface-bound •OH is a few microseconds [34], long enough for an electron donor D' to be oxidized to D'^+ or to form the •OH-adduct, D'•OH. Another feature depicted in Figure 1 is the notion that charge carriers migrate along the particle surface where they are trapped and recombine in competition with redox chemistry. Such migration is reminiscent of the spillover effect with hydrogen atoms in catalysis. The extent of migration will depend on the depth of the surface traps for electrons and holes.

4. Experimental Methods

Three nanosized colloids ($2R_{part}$ = 21 A, 133 A, and 267 A) were preparted using the arrested hydrolysis (and dialysis) of titanium tetrachloride at various temperatures from -20°C to 24°C. Photoluminescence studies were done with a Perkin-Elmer MPF-44B spectrofluorimeter and absorption spectra were recorded on a Shimadzu UV-265 spectrophotometer. Powder X-ray diffraction studies indicated the only crystalline phase in these colloids was anatase. Subnanosecond transient absorption and emission studies were carried out in the Center for Fast Laser Spectroscopy of Concordia University using the pump-and-probe technique with the system depicted in Figure 2. Excitation was at 355 nm with a 30-ps, 2.5 mJ pulse. Details of the general procedure may be found elsewhere [35,36].

5. Exciton Confinement in TiO_2 Particles

For a size-quantized particle, the increase in bandgap energy (ΔE_{BG}, eqn 5) varies

[37] with the radius of the particle, R_{part}, the reduced effective mass of the exciton, μ, and the dielectric constant of the semiconductor ($\epsilon = 184$ for TiO_2 [19]); $\mu^{-1} = m_e^{-1} + m_h^{-1}$, where m are effective masses of the charge carriers. For TiO_2, μ has been estimated as $1.63m_e$ by Kormann et al [19] and as $2.72m_e$ by Kuskinski and coworkers [38]. Clearly, ΔE_{BG} depends critically on the effective masses of electrons

$$\Delta E_{BG} = h^2 / 8 \ \mu R^2_{part} - 1.8 e^2 / \epsilon R_{part} \quad (5)$$

and holes; for TiO_2 m_e has been taken by some [39-41] to range from 5 to $13m_e$ whereas others [42] took m_e as $30m_e$ and m_h as $3m_e$ (where m_e is the electron rest

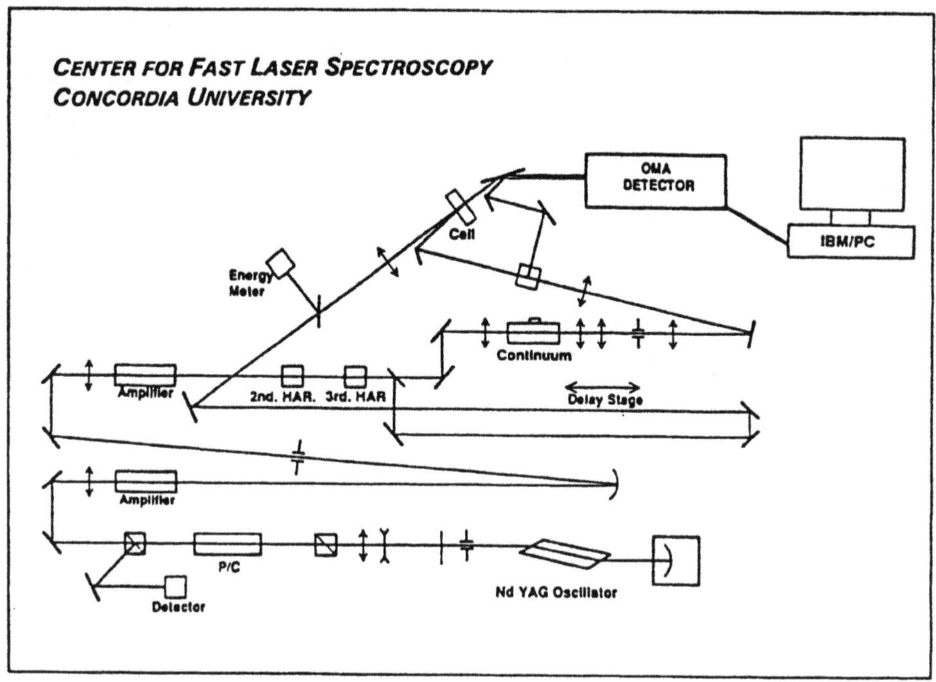

Figure 2.- Schematic of the picosecond laser system used in these studies, at the Center for Fast Laser Spectroscopy (Concordia University).

mass). Exciton radii R_{exc} between 7.5 and 19 A were estimated [19] for extremely small particles of titania from $R_{exc} = \epsilon a_o m_e / \mu$ [43] where a_o is the Bohr radius of the hydrogen atom (0.53 A). Gratzel [42] estimated an exciton radius of ~ 3 A if contributions from valence band holes were neglected. On the basis of the above discussion, TiO_2 particles should then display a Q-size threshold *either* at $2R_{part} \leq 15$

to 38 A by some estimates *or* at $2R_{part} \leq 6$ A by others. Note how predictions of thresholds of Q-size effects depend critically on the magnitude of $\mathbf{m_e}$ and $\mathbf{m_h}$.

A few reports in the last decade have described exciton confinement in small TiO_2 particles. Anpo and coworkers [22] prepared particles of titania with size $2R_{part}$ = 55 to 2000A for rutile and 38 to 530A for anatase, and noted that these crystallites display size-quantization even for such large sizes. For a 120-A rutile particle E_{BG} increased by 67 meV relative to bulk rutile bandgap (~ 3.03 eV); for anatase E_{BG} increased by 156 meV (vs. ~ 3.18 eV). As well, quantum yields of photocatalytic activity of TiO_2 increased with the magnitude of the blue shift of the effective bandgap [22]. Kormann and coworkers [19,24] indicated that small particles of TiO_2, prepared by the arrested hydrolysis of either $TiCl_4$ or $Ti(i\text{-}PrO)_4$ showed size quantization when $2R_{part}$ = 20 to 40 A. The particles exhibited a 150 - 170 meV blue shift of the UV absorption edge (bandgap) under anaerobic illumination, assigned to exciton confinement by trapped electrons as Ti^{3+}. Others [21] suggested this shift from the bandgap of bulk anatase (~ 3.2 eV), also exhibited by R_{part} = 25 to 100 A particles [14,44-48], to be due [21] to a Burstein-Moss effect (the band-filling mechanism [43]. For TiO_2particles of $2R_{part}$ = 10 to 20 A, Kavan et al [21] reported a bandgap shift of ~ 100 meV, also attributed to a quantum size effect; they used low intensity illumination to minimize the significance of the Burstein-Moss shift. Joselevich and Willner [23] observed the onset of the absorption edge in TiO_2 (particles prepared in a water-in-oil microemulsion) to be 3.6 eV (345 nm); the blue shift of 400 meV if anatase or 600 meV if rutile). The estimated size of the particles (eqn 5) was $2R_{part}$ = 15 A (if anatase) and 12 A (if rutile); dilution from 0.5 g L^{-1} to 0.1 g L^{-1} had no effect on the absorption edge.

The expression for ΔE_{BG} (eqn 5) has often been used to calculate the size of small semiconductor particulates and later compared to sizes from transmission electron microscopy. The reasonable congruence notwithstanding, there are other plausible explanations for the observed blue shifts in bandgap. We have addressed this issue in our systematic studies on how size affects semiconductor photophysics [25,34]. We examine these alternatives below.

6. Absorption and Photoluminescence of TiO_2 Specimen

Absorption spectra of the three TiO_2 specimen examined are exhibited in Figure 3 for a concentration of ~15 g L^{-1}. The three spectra are identical to the absorption spectrum of bulk anatase TiO_2 and are in accord with the spectra of the $2R_{part}$ = 120A colloidal sols reported earlier [19,20]; the inset shows the spectral changes for the

133-A specimen at concentrations ranging from 0.5 to 15 g L^{-1}. The diffuse reflectance spectra of the three dried specimen were identical to the spectrum in Figure 3. For our sols, no aggregation took place at the higher concentrations. At a concentration of 15 g L^{-1} the absorption coefficient was α ~1.9 x 10^3 cm^{-1} at 355 nm and ~ 2.3 x 10^2 cm^{-1} at 375 nm. Treatment of these coefficients α by the relationship for indirect gap semiconductors, $\alpha_{(h\nu)} \sim (h\nu - E_{BG})^2$ [43], gave two linear relationships with intercepts at ~ 2.96 eV and ~ 3.22 eV. This behavior makes us infer that the three TiO_2 specimen examined are not size-quantized and that size quantization must occur at sizes $2R_{part}$ less than ~20 A.

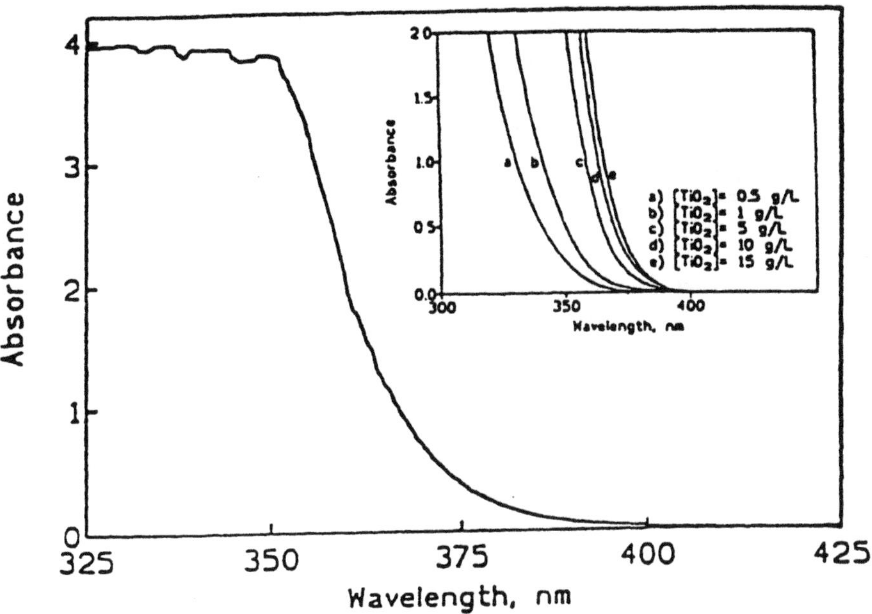

Figure 3.- Absorption spectra of the three TiO_2 specimen at a concentration of 15 g L^{-1}. Inset shows the concentration dependent spectra of the 133-A sol.

To examine photoluminescence of the three sol specimen of TiO_2 by the conventional right-angle technique, the highly concentrated sols (15 g L^{-1}) were diluted a thousand fold (0.015 g L^{-1}) to minimize inner filter effects; to obtain an absorption spectrum for the $2R_{part}$ = 267 A specimen nearly identical to those of the other two sols ($2R_{part}$ = 21 A and 133 A) necessitated dilution to ~ 0.30 g L^{-1}. This was rather curious since at the 15 g L^{-1} sols the spectra were identical. Figure 4 illustrates the absorption spectra and the resulting luminescence spectral envelope of the 21-A particles (excitation wavelength, 270 nm); the dashed lines represent a qualitative delineation of the various emission components. The absorption edge in these spectra occurs at

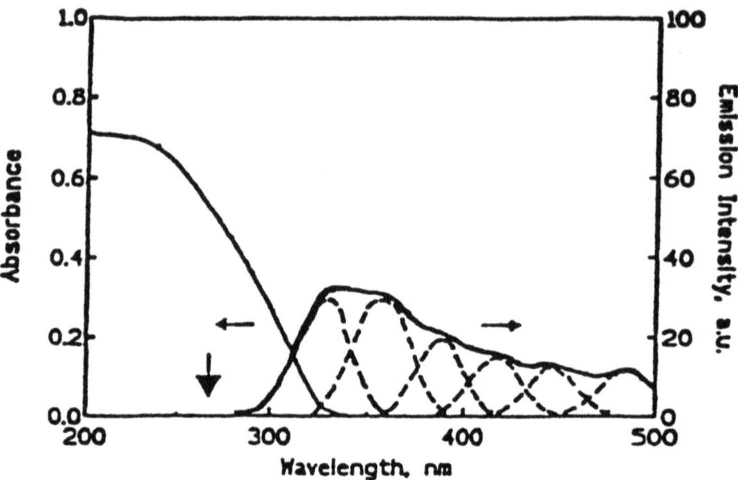

Figure 4.- Absorption spectra and luminescence envelope for the 23-A TiO_2 particles at a concentration of 0.015 g L^{-1}.

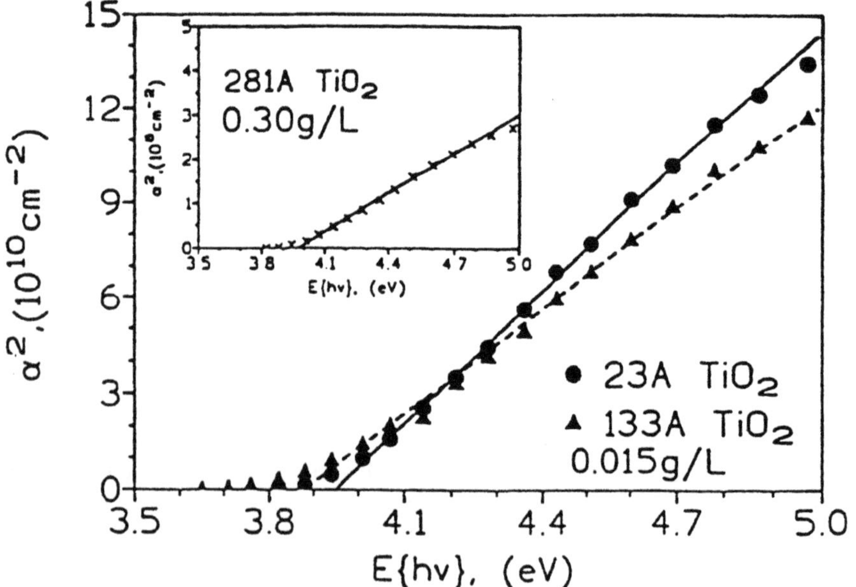

Figure 5.- Plots of α^2 versus E(h v) for all three TiO_2 sols.

~ 3.64 ± 0.04 eV which accords with an earlier observation [31] and is ascribed to a direct transition (see below).

An attempt at describing the absorption coefficients by the $\alpha_{(h\nu)} \sim (h\nu - E_{BG})^2$ relationship for indirect bandgap semiconductors failed to provide the expected linear correlation over the appropriate range of wavelengths. Linear correlation could only be had from plots of the relation $\alpha_{(h\nu)} \sim (h\nu - E_{BG})^{1/2}$ [43] that applies to direct bandgap semiconductors. Figure 5 portrays the resulting treatment for the three sol specimen. Good linearity is evident at wavelengths below 340 nm. We infer that the stronger absorption features in Figure 3 are best described as allowed direct band transitions in an otherwise indirect bandgap semiconductor. The respective absorption threshold for one of the direct transition is 3.93 ± 0.05 eV (all three sols). The emission envelope confirms and provides additional transition energies. Table 1 summarizes transition energies and corresponding assignments [26,27]. The present results are in accord with data reported earlier [29-31,49-53]. Some of the allowed indirect and direct electron transitions (solid arrows) based on the theoretical calculations by Daude and coworkers [28] are depicted in Figure 6.

Table 1.- Direct and Indirect Transitions in the three TiO_2 sols

Transition[a]	Theoretical[a] (eV)	Energies[b] (eV)
$X_{1a} \rightarrow \Gamma_{1b}$	2.91	2.96
$X_{2b} \rightarrow \Gamma_{1b}$	3.05	2.96
$\Gamma_3 \rightarrow X_{1b}$	3.19	3.22
$X_{1a} \rightarrow X_{1b}$	3.45	3.61-3.68
$X_{2b} \rightarrow X_{1b}$	3.59	3.61-3.68
$\Gamma_{5'a} \rightarrow \Gamma_{1b}$	4.05	3.88-3.98
$\Gamma_{2'} \rightarrow \Gamma_{1b}$	4.3	---

[a] Reference [28]. [b] From absorption only.

Emission bands at wavelengths longer than 400 nm (i.e., at energies smaller than ~ 3.0 eV) implicate intragap surface states, shallow traps, at energies less than ~ 1.0 eV below the conduction band. We observed these states at energies 0.41 to 0.64 eV below the conduction band. Some of these surface states may implicate oxygen vacancies [54].

7. Subnanosecond Transient Emission of TiO_2 Sols

Radiative charge carrier recombination was examined by transient emission spectroscopy in the time window 0 to 7 ns with excitation at 355 nm. With our experimental setup only emissions at wavelengths longer than 400 nm could be monitored by the streak camera system. For all three sols (15 g L^{-1}) luminescence decay traces followed good single exponential kinetics. Emission decay times ranged from ~70 to ~90 ps for the 21-A and 267-A colloids; the decay time for the 133-A

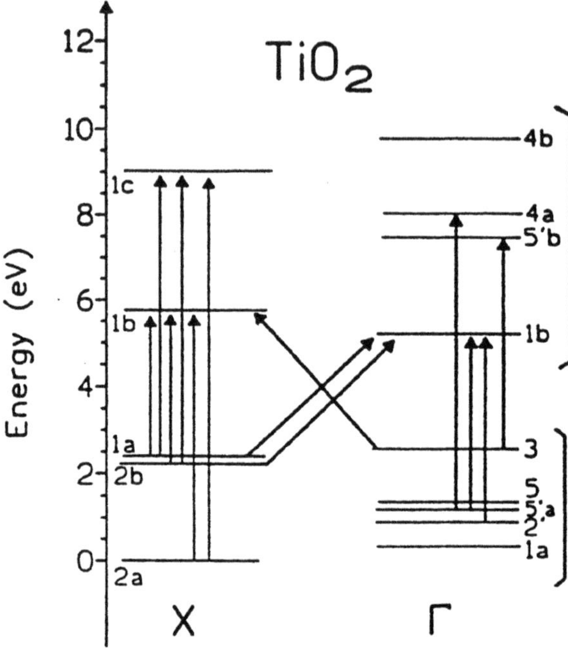

Figure 6.- Energy level diagram of the lowest energy allowed indirect and direct electron transitions for TiO_2 (adapted from ref. [28]).

sol was ~420 ps. These decay times and the corresponding emission intensities infer that the luminescence quantum yield is largest for the 133-A TiO_2 sol and less but equal to each other for the 21-A and 267-A colloidal sols.

8. Subnanosecond Transient Absorption of TiO_2 Specimen

The temporal evolution of the transient spectra of one of the sols examined, the 21-A TiO_2 specimen, at various delay times after excitation with the 30-ps, 2.5-mJ laser

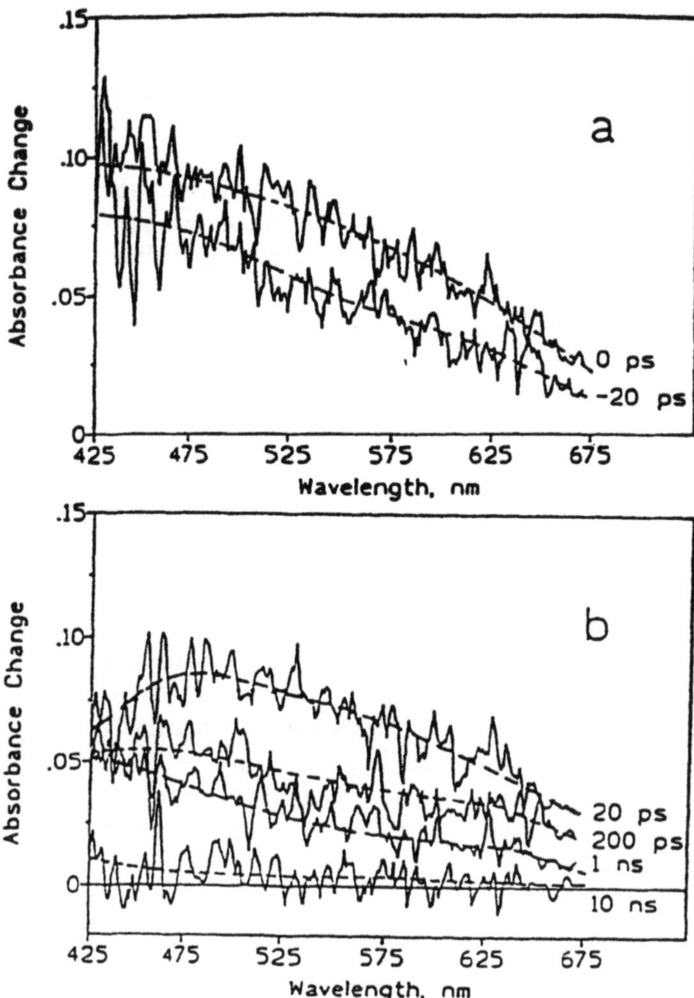

Figure 7.- Transient absorption spectra at various delay time intervals after 30-ps pulse excitation of 21-A TiO_2 colloidal particles in aqueous media (pH 2.7 and 15 g L^{-1}).

pulse is illustrated in Figure 7. Subsequent to electron/hole pair formation and charge carrier separation, in competition with recombination, a fraction of these carriers are trapped at lattice sites and some migrate to the surface in about 0.05 to ~ 10 ps, depending on size, where they also get trapped to give Ti^{3+} for the electron and $\{Ti^{4+}$-O^{2-}-$Ti^{4+}\}$-•OH for the hole [34]. The 20-ps spectrum shows that the transients were fully developed by the end of the laser pulse. Earlier work established that the absorption spectrum of the surface-trapped hole [13-16] on TiO_2 reveals a broad band at ~ 420 nm, whereas surface trapped electrons showed broad absorption in the wavelength range ~ 500 to ~ 650 nm depending on conditions [13-16]. From the evolution of the spectra, we infer that they are composite spectra of spectra of trapped holes and trapped electrons with trapping occurring in less than ~ 1-10 ps. By 10 ns, there is little transient absorption observable except a hint at the lowest wavelengths monitored (around 425-450 nm)..

A recent femtosecond spectroscopic study by Colombo and coworkers [55] on 2R = 20 A TiO_2 colloids indicates that trapping of electrons takes place in less than ~500 fs; the fluence-dependent decay kinetics of trapped electrons showed a fast and a slow component. A global fit gave an electron/hole pair lifetime on the nanocluster of 23 ± 5 ps. In our study [27], the 23-A colloid also showed biphasic decay kinetics for the transients with the fast component being 240 ± 50 ps and the slower component, 10 ± 5 ns. The inference that the charge carriers are trapped in such a short time has significant consequences in heterogeneous photocatalysis in the debate of redox chemistry being initiated by "free" valence band holes, h^+_{VB}, and conduction band electrons, e^-_{CB}, *OR* by surface-trapped carriers. Another interesting observation in the femtosecond study relates to the notion of the extent of recombination in the short time. The authors [55] concluded that by 30 ps about 50% of the charge carriers have recombined. Our own studies [25,27] carried out during the period 1990-1992 indicate that by 1 ns nearly 90% of all photogenerated electrons and holes have recombined; the extent of this recombination at shorter time is summarized in Table 2. This calls immediate attention to the realization that the quantum yield of formation of surface redox species that initiate the redox chemistry in heterogeneous photocatalysis will likely be less than ~ 10%. In this regard, this conclusion is in accord with a recent estimate of the quantum yield of ~ 4% for the formation of •OH radicals when TiO_2 is illuminated at appropriate wavelengths [56].

Table 2.- Summary of Data of TiO_2 Specimen

Size of TiO_2 ($2R_{part}$ in A)	Average e^-/h^+ pairs per particle	Number of Photons per particle[a]	% e^-/h^+ recomb'n by 20 ps	% e^-/h^+ recomb'n by 1 ns
21	0.2	0.9	78	94
133	157	225	30	86
267	1504	1811	17	91

[a] Laser fluence ~ 1.42 x 10^{17} photons cm^{-2}

9. Concluding remarks

In this paper we have attempted to show that what has been taken in the past and to this day by various authors (see e.g., refs. [57,58]) of having observed size quantization on TiO_2 particles with 2R ~ 20 A may in fact not be a quantum size effect from observations of spectroscopic blue shifts in the effective bandgap of this semiconductor. Predictions of size quantization rely very heavily on the "actual" magnitude of the effective masses of the photogenerated electrons and holes. The present study confirms earlier observations that in fact some of the shifts may simply reflect observations of direct electron transtion in an otherwise indirect bandgap semiconductor such as TiO_2.

Another feature of our study is the observation, at least in colloidal sols of TiO_2, that the charge carriers have undergone almost ~ 90% recombination by about 1 ns which leads us to believe that the quantum yields of any surface photoredox reaction cannot be greater than ~ 10%. Moreover, our transient absorption spectroscopic studies suggest that the charge carriers have been trapped by the end of the laser pulse of 30 ps, leading to the additional conclusion that any photoredox chemistry occurring at the particle surface emanates from trapped electrons and trapped holes and not from so-called *free* valence band holes, h^+_{VB}, and conduction band electrons, e^-_{CB}.

10. Acknowledgements

Support of our work by the Natural Sciences and Engineering Research Council of Canada, by the North Atlantic Treaty Organization (CRG-089746), and by the France-Quebec Collaborative Exchange Programme is gratefully appreciated.

11. References

[1]. *See for example references* in Rajh, T., Micic, O.I.,.Lawless D., and Serpone, N., (1992), Semiconductor Photophysics. 7. Photoluminescence and picosecond charge carrier dynamics in CdS quantum dots confined in a silicate glass, *J.Phys.Chem.*, **96**, 4633-4641.

[2] *See* Kamat, P.V., and Dimitrijevic, N.M., (1990), Colloidal Semiconductors as Photocatalysts for Solar Energy Conversion, *Solar Energy,* **44**, 83-98, and references therein.

[3] *See* Kamat, P.V. (1994) Interfacial Charge Transfer Processes in Colloidal

Semiconductor Systems, *Prog. Reaction Kinetics,* 19, 277-316, and references therein.
[4] Bahnemann, D., Cunningham, J., Fox, M.A., Pelizzetti, E., Pichat, P., and Serpone, N., (1994) ,Photocatalytic Treatment of Waters, in *"Aquatic and Surface Photochemistry"*,Crosby, D., Helz, G., and Zepp, R., Eds., Lewis Publishers, Boca Raton, FL, 387-398.
[5] Ollis, D.F. and Al-Ekabi, H., eds., (1993), *"Photocatalytic Treatment of Water and Air"*, Elsevier Science Publishers, Amsterdam.
[6] Ollis, D.F., Pelizzetti, E., and Serpone, N., (1991), Destruction of Water Contaminants, *Environ.Sci.Technol.,* **25,** 1523-1529.
[7]. Brus, L., (1994), Luminescence of Silicon Materials: chains, sheets, nanocrystals, nanowires, microcrystals and porous silicon, *J.Phys.Chem.*, **98**, 3575-3581.
[8]. Henglein, A., (1987), Q-particles: size quantization effects in colloidal semiconductors, *Prog.Colloid Polym.Sci.,* **73**, 1-4.
[9]. Henglein, A., (1988), Mechanism of reactions on colloidal micro-electrodes and size quantization effects, *Top.Curr.Chem.,* **143**, 113-180.
[10]. Henglein, A., (1989), Small Particle Research: physicochemical properties of extremely small colloidal metal and semiconductor particles, *Chem.Rev.,* **89**, 1861-1873.
[11]. Micic, O.I., Rajh, T., Nedeljkovic, J.M., and Connor, M.I., (1993), Enhanced redox chemistry in quantized semiconductor colloids, *Isr.J.Chem.*, **33**, 59-65.
[12]. Rothenberger, G., Moser, J., Gratzel, M., Serpone, N., and Sharma, D.K., (1985), Charge carrier trapping and recombination dynamics in small semiconductor particles, *J.Am.Chem.Soc.*, *107*, 8054-8059.
[13]. Henglein, A. (1982), Colloidal TiO_2 catalyzed photo- and radiation chemical processes in aqueous solution, *Ber.Bunsenges.Phys.Chem.*, **86**, 241-246.
[14]. Duonghong, D., Ramsden, J., and Gratzel, M., (1982), Dynamics of interfacial electron transfer processes in colloidal semiconductor systems, *J.Am.Chem.Soc., 104,* 2977-2985.
[15]. Bahnemann, D., Henglein, A., Lillie, J., Spanhel, L., (1984), Flash photolysis observation of the absorption spectra of trapped positive holes and electrons in colloidal TiO_2, *J.Phys.Chem., 88*, 709-711.
[16]. Bahnemann, D., Henglein, A., Spanhel, L., (1984), Detection of the intermediates of colloidal TiO_2-catalyzed photoreactions, *Faraday Disc. Chem.Soc., 78*, 151-157.
[17]. Howe, R.F., and Gratzel, M., (1985), EPR observation of trapped electrons in colloidal TiO_2, *J.Phys.Chem., 89*, 4495.
[18]. *See for example*, Micic, O.I., Meglic, M., Lawless, D., Sharma, D.K., and Serpone, N., (1990), Semiconductor photophysics. 5. Charge carrier trapping in

ultrasmall silver iodide particles and kinetics of formation of silver atom clusters, *Langmuir*, *6*, 487-492.
[19]. Kormann, C., Bahnemann, D. W., Hoffmann, M. R., (1988), Preparation and characterization of quantum-size titanium dioxide colloids, *J. Phys. Chem.*, *92*, 5196-5201.
[20]. Bahnemann, D.W., (1993), Ultrasmall metal oxide particles: preparation, photophysical characterization and photocatalytic properties, *Isr.J. Chem.*, *33*, 115-125.
[21]. Kavan, L., Stoto, T., Gratzel, M., Fitzmaurice, D., and Shklover, V., (1993), Quantum size effects in nanocrystalline semiconducting TiO_2 layers prepared by anodic oxidative hydrolysis of $TiCl_3$, *J.Phys.Chem.*, *97*, 9493-9498.
[22]. Anpo, M., Shima, T., Kodama, S., and Kubokawa, Y., (1987), Photocatalytic hydrogenation of CH_3CCH with H_2O on small-particle TiO_2: size quantization effects and reaction intermediates, *J.Phys.Chem.*, *91*, 4305-4317.
[23]. Joselevich, E., and Willner, I., (1994), Photosensitization of quantum-size TiO_2 particles in water-in-oil microemulsions, *J.Phys.Chem.*, *98*, 7628-7635.
[24]. Choi, W., Termin, A., and Hoffmann, M.R., (1994), The role of metal ion dopants in quantum-sized TiO_2: correlation between photo-reactivity and charge carrier recombination dynamics, *J.Phys.Chem.*, *98*, 13669-13679.
[25]. Lawless, D., (1993), Photophysical Studies on Ultra-Small Semiconductor Particles: CdS quantum dots, doped and undoped TiO_2, and Silver Halides, *Ph.D. Dissertation*, Concordia University, Montreal, Canada.
[26]. Serpone, N. and Lawless, D., (1995), Size Effects on the Photophysical Properties of Colloidal TiO_2 Particulates, *J.Phys.Chem.*, submitted for publication.
[27] Serpone, N., Lawless, D., and Pelizzetti, P., (1995), Subnanosecond relaxation dynamics of TiO_2 particulates and their significance in heterogeneous photocatalysis, *J.Phys.Chem.*, submitted for publication.
[28]. Daude, N.; Gout, C.; Jouanin, C., (1977), Electronic band structure of titanium dioxide, *Phys.Rev.B*, *15*, 3229-3235.
[29]. Frova, A., Brody, P.J., and Chen, Y.S., (1967), *Phys.Rev.*, **157**,157.
[30]. Cardona, M., and Harbeke, G., (1965), *Phys.Rev.*, **137**, 1467.
[31]. Koffyberg, F.P., Dwight, K., and Wold, A., (1979), Interband Transitions of Semiconducting Oxides Determined from Photoelectrolysis Spectra, *Solid State Commun.*, **30**, 433-437.
[32]. *see for example*, Serpone, N., Pelizzetti, E., and Hidaka, H., (1993), Heterogeneous Photocatalysis: issues, questions, some answers and some successes, in *Photochemical and Photoelectrochemical Conversion and Storage of Solar Energy*, Z.W. Tian and Y. Cao, eds., International Academic Publishers, Beijing, China, 33-73.
[33]. Micic, O.I., Zhang, Y., Cromack, K.R., Trifunac, A.D., and Thurnauer, M.C.,

(1993), Trapped holes on TiO_2 colloids studied by electron paramagnetic resonance, *J.Phys.Chem.*, **97**, 7277-7783.
[34]. Lawless, D., Serpone, N., and Meisel, D., (1991), Role of OH radicals and trapped holes in photocatalysis. A pulse radiolysis study, *J.Phys.Chem.*, **95**, 5166-5170.
[35]. Ponterini, G., Serpone, N., Bergkamp, M.A., and Netzel, T.L., (1983), Comparison of radiationless decay processes in osmium and platinum porphyrins, *J.Am. Chem.Soc.*, **105**, 4639-4645.
[36]. Serpone, N., Sharma, D.K., Moser, J., and Gratzel, M., (1987), Reduction of acceptor relay species by conduction band electrons of colloidal titanium dioxide: light-induced charge separation in the picosecond time domain, *Helv.Chim.Acta,* **136**, 47-51.
[37]. Brus, L., (1986), Electronic wavefunctions in semiconductor clusters. Experiment and theory, *J. Phys. Chem.*, *90,* 2555-2560.
[38]. Kuskinski, J.J., Gomez-Jahn, L.A., Faran, K.J., Gracewski, S.M., and Miller, D., (1989), *J.Chem.Phys.*, *90*, 1253.
[39]. Acket, G.A., and Volger, J., (1966) *Physica (Amsterdam)*, *32*, 1680.
[40]. Pascual, J., Camassel, J., and Mathieu, H., (1978). *Phys.Rev.B: Condens.Matter*, *18*, 5606.
[41]. Agekyan, V.T., Berezhnaya, A.A., Lutsenko, V.V., and Stepanov, Y.A., (1980), *Sov.Phys.Solid State, 22,* 6.
[42]. Gratzel, M., (1989), *"Heterogeneous Photochemical Electron Transfer"*, CRC Press, Boca Raton, FL.
[43]. Pankove, J.I., (1975), "*Optical Processes in Semiconductors*", Dover, New York.
[44]. Fox, M.A., (1987), Selective formation of organic compounds by photoelectrosynthesis at semiconductor particles, *Top.Curr.Chem.*, *142*, 72-79.
[45]. Matthews, R.W., (1987), Solar-electric water purification using photocatalytic oxidation with TiO_2 as a stationary phase, *Solar Energy, 38*, 405-413.
[46]. Ollis, D.F., (1985), Contaminant degradation in water. Heterogeneous photocatalysis degrades halogenated hydrocarbon contaminants, *Environ.Sci.Technol.*, *19,* 480-484.
[47]. Nishimoto S.-I., Ohtani, B., Shirai, H., Adzuma, S., and Kagiya, T., (1985), *Polymn.Commun.*, *26*, 292.
[48]. Moser, J., and Gratzel, M., (1984), Photosensitized electron injection in colloidal semiconductors, *J.Am.Chem.Soc.*, *106,* 6557-6564.
[49]. Arntz, F.; Yacoby, Y., *Phys.Rev.Lett.*, **1966,** *17*, 857.
[50]. Vos, K.; Krusemeyer, *Solid State Commun.*, **1975,** *15*, 949.
[51]. Ghosh, A.K.; Wakim, F.G.; Adiss Jr., P.R., (1969), Photoelectronic processes in rutile, *Phys.Rev.*, *184,* 979-988.

[52]. Hashimoto, K., Hiramoto, M., and Sakata, T., (1988), Photoluminescence of TiO_2 powder and its relation with photocatalytic reactions, in "*Photoelectrochemistry and Electrosynthesis on Semiconducting Materials*", Ginley, D.S., Nozik, A., Armstrong, N., Honda, K., Fujishima, A., and Sakata, T., Eds., The Electrochemical Society, Inc., Pennington, N.J., *88-14*, 395-400.
[53]. Forss, L. and Schubnell, M., (1993), Temperature dependence of the luminescence of TiO_2 powder, *Appl.Phys.*, *B56*, 363-366.
[54]. Lu, G., Linsebigler, A., and Yates, Jr., J.T., (1994), Ti^{3+} defect sites on TiO_2(110): production and chemical detection of active sites, *J.Phys.Chem.*, *98,* 11733-11738.
[55]. Colombo Jr., D.P., Roussel, K.A., Saeh, J., Skinner, D.E., Cavaleri, J.J. and Bowman, R.M., (1995), Femtosecond study of the intensity dependence of electron-hole dynamics in TiO_2 nanoclusters, *Chem.Phys. Letters*, *232,* 207-214.
[56]. Bolton, J.R., *Proceedings First International Conference on Advanced Oxidation Technologies*, London, Ontario, Canada, June 1994.
[57]. Martin, S.T., Herrmann, H., Choi, W. and Hoffmann, M.R., (1994), Time-resolved microwave conductivity. Part I. TiO_2 photoreactivity and size quantization, *J.Chem.Soc.Faraday Trans.*, *90*, 3315-3322.
[58]. Martin, S.C., Morrison, C.L. and Hoffmann, M.R., (1994), Photochemical mechanism of size-quantized vanadium-doped TiO_2 particles, J.Phys.Chem., 98, 13695-13704.

FROM NANO-SIZED PARTICLES TO COMMERCIAL PRODUCTS:

The Search for Novel Photocatalysts

D. BOCKELMANN, M. LINDNER, D. BAHNEMANN
Institut für Solarenergieforschung GmbH
Sokelantstaße 5
D-30165 Hannover
Germany

Abstract

Colloidal Ti/Fe mixed oxide nanoparticles were synthesized in aqueous solution with the Fe^{3+} content varying from 0.05 up to 50 at% and particle diameters between 4 and 6 nm. The absorption spectra of these particles could be shifted considerably into the visible part of the solar spectrum with increasing content of ferric ions. The photophysical properties of the newly synthesized particles were examined in detail. Steady state illumination experiments showed a reversible bleaching of the absorption spectrum in the presence of a hole scavenger indicating that a negative excess charge could be stored. The origin of the bleaching was studied in detail using time-resolved flash photolysis measurements. The photocatalytic activity of the Ti/Fe mixed oxide particles was compared to that of pure colloidal TiO_2 by studying the photooxidation of dichloroacetate (DCA) in acidic and alkaline media. It could be demonstrated that the incorporation of iron into TiO_2 leads to an enhancement of the degradation efficiency, especially in acidic solutions. However, a release of Fe(II) was observed during the course of these detoxification reactions evincing that the newly synthesized catalyst partially photocorrodes. Strategies are suggested to suppress this cathodic process.

Furthermore, a commercially available TiO_2 powder consisting of 5 nm anatase particles forming a micrometer-sized secondary structure has been studied. The photocatalytic activity of this novel material, namely Hombikat UV 100 from Sachtleben Chemie GmbH, has been investigated and compared with that of a well-studied material. Laboratory experiments using once again DCA as the model pollutant have been carried out showing a very high quantum yield reaching almost unity at low light intensities. The effect of substrate and catalyst concentration has been thoroughly investigated. The inhibiting effect of inorganic anions such as chloride, nitrate, sulphate and phosphate was reversible upon washing the catalyst powder with pure water or diluted base, respectively. Models are proposed to explain these results. The effect of the photo-deposition of small noble metal islands onto the surface of these particles on their photocatalytic activity has also been studied in detail.

E. Pelizzetti (ed.), Fine Particles Science and Technology, 675–689.

1. Introduction

Persistent organic chemicals are present as pollutants in waste water effluent from industrial manufacturers, dry cleaning facilities or even normal households. They can be found in ground water wells and surface waters where they have to be removed to achieve drinking water quality [1, 2]. Therefore many processes have been proposed over the years and are currently being employed to destroy these toxins.

The so-called photocatalytic detoxification has been discussed as an alternative method for clean-up of polluted water in the scientific literature since 1976 [3]. Lately, considerable public attention has been focussed on this possibility of combining heterogeneous catalysis with solar technologies to achieve the mineralization of toxins present in water [4–7]. Several reviews have recently been published discussing the underlying reaction mechanisms of photocatalytic detoxification and illustrating examples of successful laboratory and field studies [8-11]. The energetic requirements for the illumination source appear to have by far the greatest impact on the overall cost of the method. Depending upon the desired application it could, for example, be ideal to use the sun instead of an electrically powered lamp. Anatase, titanium dioxide, the material with the highest photocatalytic detoxification efficiency, is a wide bandgap semiconductor ($E_g \approx 3.2$ eV [12]). Thus, only light below 400 nm (*i.e.*, 5% of the solar energy reaching the surface of the earth) can be absorbed to form e^-/h^+ pairs [13] which are a prerequisite for the process. For solar applications other materials have to be found or developed which exhibit similar efficiencies as anatase, but possess spectral properties more closely adapted to the terrestrial solar spectrum. Here we present results of our investigation on the synthesis and characterization of photocatalysts which are able to absorb the visible part of the solar spectrum to a greater extent with the simultaneous improvement of their photocatalytic detoxification properties. For a detailed understanding of photocatalytical processes it is important to have a certain knowledge about the photophysical and photochemical properties of the photocatalyst itself. Most of the photocatalytic detoxification experiments performed worldwide used suspensions of semiconductor particles. On the other hand, colloidal particles can be synthesized with diameters considerably smaller than the wavelength of the incident light, so their aqueous solutions are completely transparent. Thus it is possible to obtain absorption spectra of the material and therefore the kinetics of processes taking place on the surface can be recorded, if they are reflected in a change of the absorption. Colloidal solutions of mixed oxide semiconductor particles, namely Ti(IV)/Fe(III) mixed oxides, have been synthesized to serve both purposes, *i.e.*, they can absorb photons in the visible and due to their small size light scattering can be neglected in this wavelength regime.

In recent years various TiO_2 powders have been tested as photocatalysts for the degradation of different model compounds. Since the quantum efficiencies obtained with these catalysts were mostly distinctly below 10% and often smaller than 1% even under optimal laboratory conditions, it is evident that novel materials have to be developed, in particular, to increase the overall efficiency of solar detoxification processes. Here we also present results of a laboratory study using a newly developed anatase powder (Hombikat UV 100 from Sachtleben Chemie) as the photocatalyst and comparing its efficiency to "traditional" materials. The investigations have been carried out using dichloroacetate which is one of the model compounds extensively studied in our laboratory.

2. Experimental Procedures

The Ti/Fe-mixed oxide colloids with different iron content were prepared as powders, which are stable at room temperature for several months and can be resuspended in water or water/ethanol mixtures yielding transparent colloidal solutions.

For the preparation of Ti/Fe-mixed oxide colloids freshly distilled $TiCl_4$ cooled to –20°C was added slowly to cold (≈ 0°C) freshly prepared $FeCl_3$-solution under vigorous stirring . The concentration of the $FeCl_3$-solution was varied in order to get colloids with different iron content. The resulting colloidal suspension was stable for several hours at temperatures below 5 °C. To increase the stability of the colloids and to facilitate powder formation during evaporation of the solvent, the ionic strength was subsequently reduced by dialyzing against pure water until a final pH between 2.5 - 3 was reached. Aliquots of 200 ml were dried with the aid of a rotary evaporator (25 mbar, 30 °C). The residue was dried under higher vacuum (1 mbar) for one minute, resulting in crystalline yellowish to brownish powders.

Detailed photophysical and photochemical investigations of colloidal Ti/Fe mixed oxide suspensions were performed using a flash photolysis apparatus for time-resolved measurements. The flow cell had an optical path length of 6 cm and a thickness of 0.7 mm. The Xenon flash lamp and the cell were located at the two focal points of an elliptical cavity. The lamp was triggered by a spark-gap using a 4 μF/10 kV capacitance. A 100 W quartz halogen lamp was used as analyzing light source. Signals were detected with a photomultiplier EMI 9558 and recorded with a transient recorder DL905 (Data Lab., England) with a time resolution of ≈2 ms. The excitation flash had a half-time of 9 μs and an energy of approximately 100 J/pulse. All flash photolysis experiments were performed in 20% methanol/water solutions with 0.16 g/l colloidal Ti/Fe mixed oxides. The solutions also contained 10^{-3} mole/L $HClO_4$ to adjust the pH-value to ≈2.3. For experiments in alkaline solution (pH≈11.3) an appropriate amount of 1 mole/L NaOH was added quickly to the colloidal solution. Solutions were deaerated with Argon for about one hour.

The photocatalytic activity of all semiconductor powders was tested and compared by detoxification measurements using dichloroacetic acid (DCA) as the probe molecule. DCA is a relatively strong organic acid (pK_A = 1.48) and thus nearly exclusively present in its anionic form over the entire pH range studied in this paper. According to the stoichiometry observed previously [14], the oxidation of one dichloroacetate ion leads to the formation of one proton. Therefore we used a pH-stat technique [15] which allows the in-situ measurement of H^+_{aq} formed during the photolysis experiments with extremely high sensitivity. The photochemical reactor was made of quartz glass and filled with 50 mL of the aqueous suspension containing the respective photocatalyst. These suspensions were thermostated, vigorously stirred using a magnetic stirrer, and continuously purged with molecular oxygen to ensure a constant O_2 concentration throughout the experiment. Illuminations were carried out with a mercury doped high pressure Xe-lamp (OSRAM HBO 500 or XBO 450). Short-wavelength UV-radiation was eliminated by a 320 nm band pass filter and IR radiation was suppressed by a 10 cm water filter. Actinometry was performed using Aberchrome 540 [16] in order to determine the light intensity in the wavelength region between 320 and 400 nm.

A Dionex 4500i ion chromatograph equipped with a HPIC-AS4A separator column was used to determine carbonate, chloride, and dichloroacetic acid concentrations. Detailed conditions are given in the literature [17]. Absorption spectra of the colloids were measured using a Bruins Omega 10 UV-VIS spectrophotometer. Chemicals and solvents were of reagent grade and used without further purification. The water employed in all preparations was purified by a Milli–Q/RO system (Millipore) resulting in a resistivity > 18 MΩcm.

3. Results and Discussion

3.1. TI(IV)/FE(III) MIXED OXIDE COLLOIDS

3.1.1. *Photophysical Properties*

Using an X-ray-fluorescence analysis of the dried powders we were able to show that the proportion of titanium to iron in the synthesized Ti(IV)/Fe(III) mixed oxide particles corresponds to the estimations based upon the stoichiometric ratio used for the preparation within an uncertainty of approximately ±2%. Particle size analysis with a transmission electron microscope (TEM) showed that freshly prepared Ti/Fe-colloids are between 4 and 6 nm in diameter.

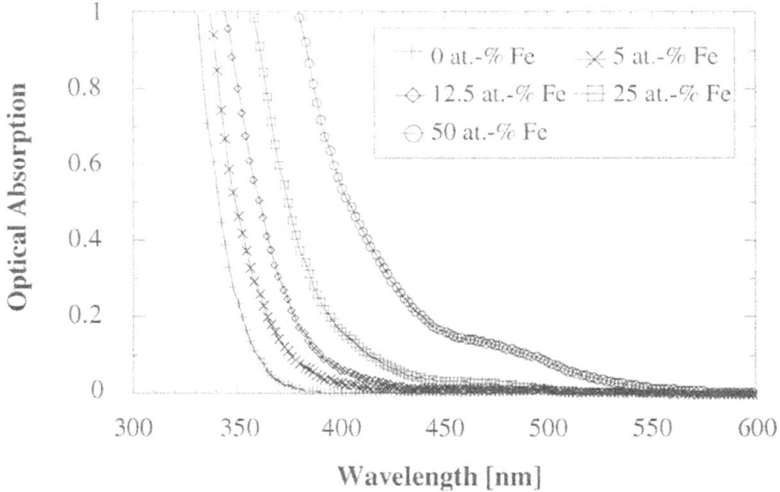

Figure 1. Absorption spectra of various Ti(IV)/Fe(III) mixed oxide colloids; 0.5 g/L colloid, pH 2.6, d(opt)=1 cm, 294K

Absorption spectra of some Ti/Fe mixed oxides are shown in Figure 1. It can be seen that the absorption edge is shifted towards the visible region of the spectrum as the iron content is increased. Additionally, a new absorption shoulder around 480 nm appears in the spectra when the particles contain more than 5 at% iron. Colloids containing 25 at% or more iron exhibit a significant absorption up to 550 nm. This can be interpreted as a decrease of the effective bandgap by more than 1 eV as compared with pure TiO_2. It has been shown that the absorption spectra of ultrasmall colloidal semiconductor particles can often be treated using mathematical models originally derived for amorphous semiconductors [13]. We have therefore calculated the absorption coefficient α for the colloidal particles at each wavelength λ via

$$\alpha = \frac{2.303 * \rho * 1000}{c(coll) * d(opt)} * A(\lambda) \qquad (1)$$

with $A(\lambda)$: measured absorption at given wavelength,
c(coll): concentration of colloid [g/L],
d(opt): optical pathlength [cm],
ρ: density of the material [g/cm^3].

The respective data are plotted in Figure 2 in the form of $\ln(\alpha)=f(\text{photon energy } E_\lambda)$. As shown by Kormann *et al* [13], the bandgap energy for a given nano-cristalline semiconductor material can be derived from this plot by taking the energy values at $\ln(\alpha)=6.5$. We employed this method to determine the bandgap energies for the different mixed oxide particles. The respective bandgap energies for all materials prepared during this study are listed in Table 1.

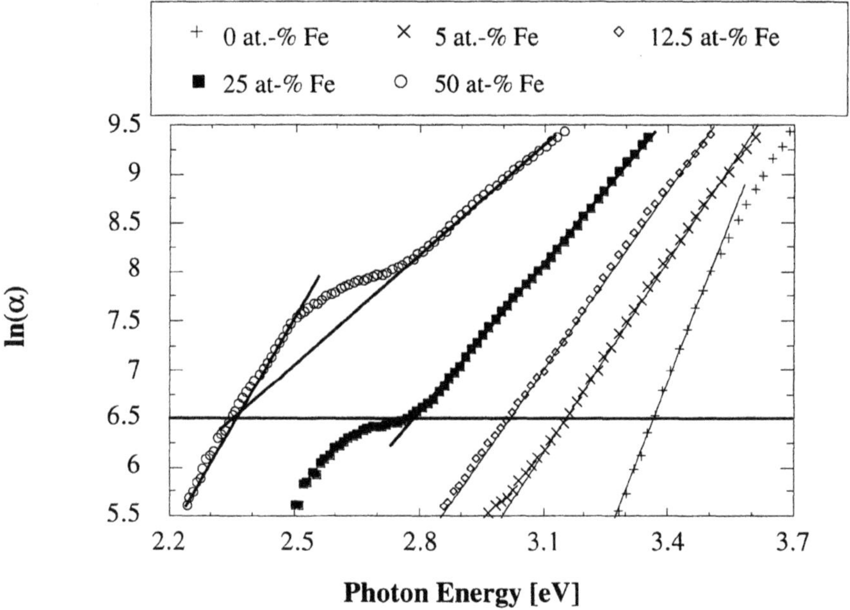

Figure 2. Plot of ln(α) *vs* photon energy (exp. details see Fig. 1)

Obviously, an increasing amount of iron can be utilized to shift the absorption properties of TiO_2 into the visible part of the (solar) spectrum.

TABLE 1: Bandgap energies of various Ti(IV)/Fe(III) mixed oxides determined from plots ln(α) *vs* photon energy (*cf.* Fig. 2)

Ti/Fe-Coll. [at.-% Fe]	Bandgap E_g [eV]	λ_g [nm]
0	3.35	370
0.1	3.33	372
0.25	3.2	388
0.5	3.29	377
2.5	3.21	386
5	3.15	394
12.5	3.01	411
25	2.79	444
50	2.33	532

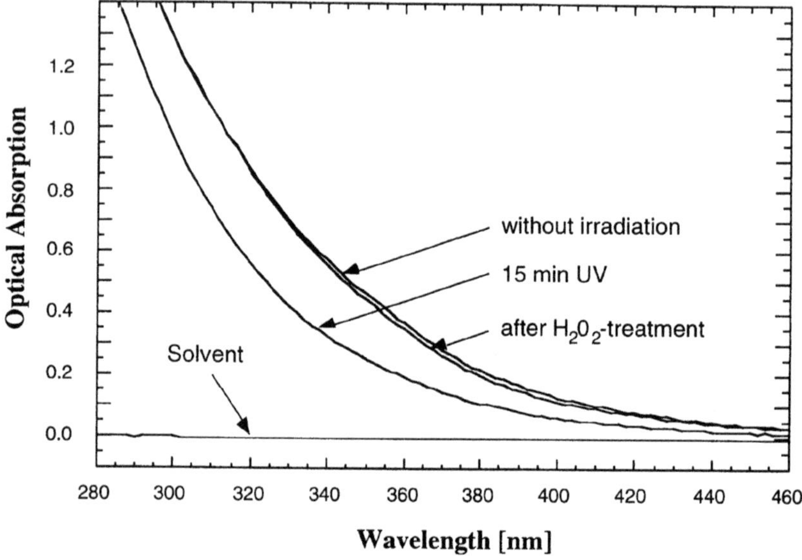

Figure 3. Absorption spectra before and after the UV-illumination of 2g/L Ti/Fe-mixed oxide colloid (50 at% Fe), pH 2.3, 1 mole/L ethanol, I_{hv}= 350 µmole/Ls, subsequent addition of 6 mmole/L H_2O_2

Changes in the spectral characteristics of the mixed oxide colloid containing 50 % Fe^{3+} which are observed upon illumination with near UV light (λ_{ex}=360 nm) in the presence of ethanol (acting as a hole-scavenger) are shown in Figure 3. The absorption onset is shifted significantly towards shorter wavelengths following the UV-irradiation of 15 minutes. Subsequent addition of an oxidant (e. g. H_2O_2) leads to complete recovery of the spectrum. In pure quantum-sized TiO_2-colloids a similar effect has already been reported. A slightly blue-shifted absorption spectrum has been observed after anoxic UV-illumination [13]. Also it has been reported that such an effect can be observed when quantum-size ZnO or CdS particles contain excess negative charges [18-21].

These spectral blue-shifts have been explained by an increased confinement of the exciton arising from trapped negative charges. In the case of quantum-sized TiO_2 the formation of Ti^{3+} has been verified by its spectral absorption around 700 nm. The lifetime of trapped electrons in TiO_2 can be very long (*i.e.*, hours or even days) when oxidants (e. g. molecular oxygen) are absent. Addition of O_2 leads to a fast removal of the charge carriers and consequently to a reversal of the spectral shift [13].

However, in the case of the Ti/Fe-colloids the bleaching of the absorption spectrum can be observed even in the presence of dioxygen while no absorption band characteristic of Ti^{3+}-centers is formed. Therefore, we envisage that photogenerated electrons are not mainly captured at Ti^{4+}-centers but can in fact reduce Fe^{3+}-centers within the microcrystals forming Fe^{2+}. Recovery of the spectrum is observed only for oxidants with an oxidation potential which is more positive than that of molecular oxygen, such as H_2O_2 (*cf.* Fig. 3).

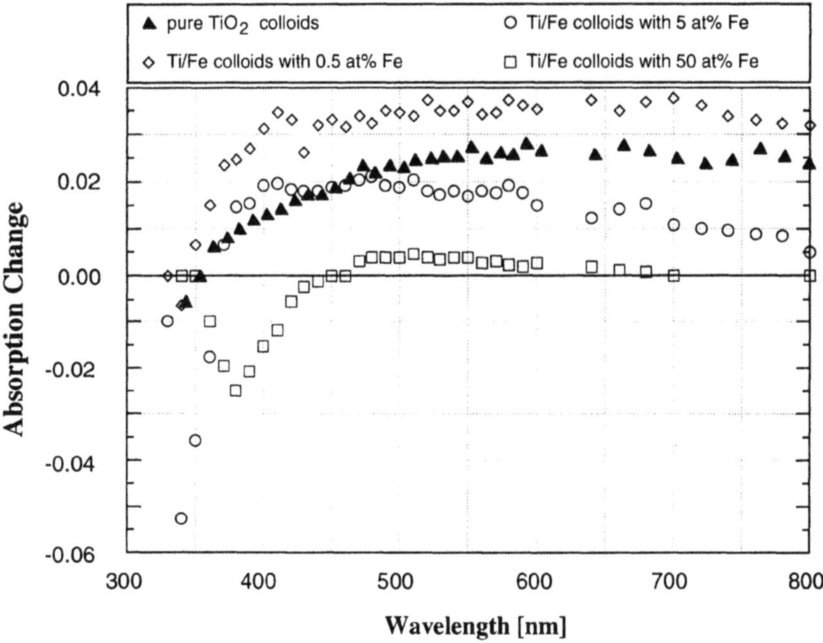

Figure 4. Transient absorption spectra observed 2 ms after flash-illumination of 0.16 g/L Ti/Fe-mixed oxide colloids containing no, 0.5, 5 or 50 at% Fe^{3+}, pH 2.3, 4.9 mole/L methanol, 5 mmole/L $HClO_4$, Ar–sat.

For a more detailed understanding of this bleaching effect time-resolved flash-photolysis studies have been carried out. Figure 4 shows the transient absorption spectra observed 2 ms after the UV-flash for representative materials at pH 2.3 with methanol acting as the hole scavenger. Pure TiO_2 colloids exhibit a broad transient absorption starting at 350 nm and peaking around 600 nm. Similar absorption changes have previously reported by Kormann *et al* and have been attributed to the formation of Ti^{3+}-centers [13].

As shown in Fig. 4, a similar broad transient absorption band is observed after the flash-illumination of Ti/Fe mixed oxide particles with an iron content of 0.5 at% exhibiting, however, an increased intensity as compared with pure TiO_2. Colloids with iron contents of 5 at% exhibit a transient absorption with a broad band starting above 400 nm. This indicates that excess photogenerated electrons in these mixed oxide semiconductor particles might have a different chemical environment than the Ti^{3+}-centers observed in pure TiO_2. The latter's absorption band around 600 nm disappears with increasing iron content and simultaneously a bleaching of the Ti/Fe-colloid spectrum occurs in the wavelength range between 330 and 370 nm.

It is intriguing to note that Ti/Fe-colloids containing 50 at% of iron exhibit a strong bleaching effect in the whole wavelength range between 350 and 450 nm. Above 450 nm an increased spectral intensity observed 2 ms after the UV-flash indicates that the photogenerated electrons are initially located at Ti^{3+}-centers. 200 ms after the flash the transient spectrum has significantly changed (to ensure a better clarity of the presentation these results are not shown in Fig. 4). Bleaching is now encountered for wavelengths up to 550 nm while no positive absorption change is observed above 550 nm compared to the original absorption spectrum before the UV-flash. Here the transient spectrum due to Ti^{3+} is no longer exhibited. In agreement with the steady-state UV-illumination experiments (*cf.* Fig. 3) we explain this bleaching effect by a trapping of the electrons at energetically more favored iron centers. Bleaching effects have been observed for all Ti/Fe-colloids with their extent being dependent on the amount of iron incorporated in the TiO_2-lattice.

Flash photolysis experiments have also been carried out in alkaline solution at pH 11. In this case the surface of the colloidal particles is negatively charged. Nevertheless, the transient spectrum of pure TiO_2 colloids was found to be almost identical to the spectrum recorded in acidic solution. Thus the surface charge seemingly does not influence the trapping of conduction band electrons at titanium-centers. However, a significant change in the shape of the transient absorption spectra was observed in alkaline solutions for Ti/Fe-colloids containing 0.5 and 5 at% iron. Here an absorption peak with a pronounced maximum at 400 nm could be seen. The most surprising difference between the transient absorption spectra of the mixed oxide colloids containing 50 at% iron is that at pH 11 there is no bleaching effect at all. The transient spectrum changes with time but a broad band indicating Ti^{3+} still remains in the range of 400...650 nm. This observation led to the conclusion that in alkaline solution the photogenerated electrons in the Ti/Fe-colloids are not localized at iron centers forming Fe^{2+} but rather in the form of Ti^{3+}. This is in agreement with the chemical behavior of iron in alkaline homogeneous systems where Fe^{2+}-ions are not thermodynamically stable.

3.1.2. *Detoxification properties*

The photophysical properties of the 50 at% Fe-colloids are promising with respect to the goal of employing a wider range of wavelengths of the incident sunlight as compared to pure TiO_2. Therefore, we have compared the efficiency of all colloidal mixed oxide semiconductor particles for the photocatalytic degradation of the model compound dichloroacetic acid ($Cl_2HCCOOH$, DCA). The photocatalytic activity (expressed as photonic efficiency ζ, *i.e.*, the rate of DCA degradation divided by the incident UV photon flux) of the colloids is shown in Figure 5 for particles with various iron contents at pH≈2.6 and pH≈11.3.

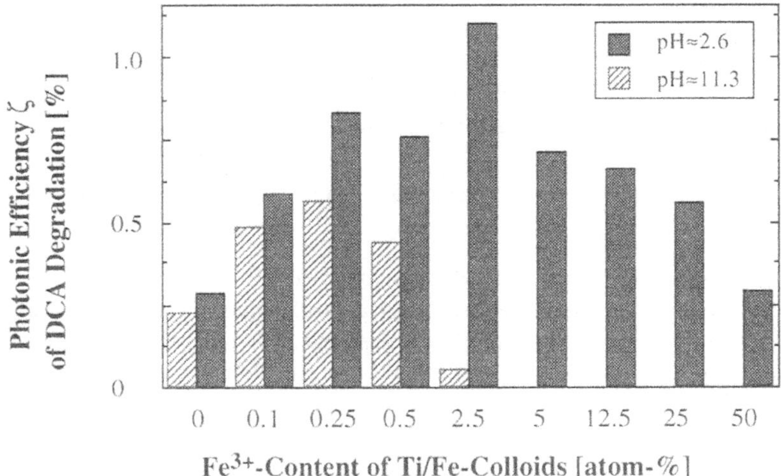

Figure 5. Photonic efficiency ζ of the photocatalytic degradation of dichloroacetic acid with 0.5 g/L Ti/Fe mixed oxide colloids of different iron content, 2.5 mmole/L DCA, O_2-sat.

The data presented in this figure show that at pH≈2.6 the efficiency of the destruction of dichloroacetic acid obtained with the Ti/Fe-mixed oxide particles is always higher than for pure TiO_2 colloids. Also the yield increases strongly at low iron contents reaching a maximum at 2.5 at% Fe^{3+}. At this point the photonic efficiency is almost four times higher than for pure TiO_2 colloids. Therefore, we assume that in acidic solution small iron contents inhibit the recombination of photogenerated charge carriers. At pH≈11.3 the photonic efficiencies of the destruction of dichloroacetic acid are generally smaller than at pH≈2.3. This effect might be due to the pH-dependence of the adsorption of dichloroacetic acid on the TiO_2 surface. In acidic solution the metal oxide surface is positively charged whereas dichloroacetic acid is deprotonated (pK_A=1.48) and therefore negatively charged and should thus be well adsorbed.

It should also be noted that at pH≈11.3 the efficiency of the detoxification of dichloroacetic acid increases only slightly up to 0.25 at% Fe^{3+} and subsequently decreases with increasing iron content. Already at a content of 5 at% iron there is no detectable detoxification in alkaline solutions. A possible explanation for this observation is that the recombination of charge carriers is the predominant process in alkaline solution. We suppose that the trapping of the photogenerated electrons at surface states forming Ti^{3+} or Fe^{2+} suppresses the recombination. Considering the flash photolytic measurements described in the previous chapter it is concluded that in alkaline solution the iron on the TiO_2 surface cannot be reduced to Fe^{2+} by the photogenerated electrons. Moreover, the amount of titanium-atoms on the surface is getting smaller with increasing iron content compared to pure TiO_2. Thus, the probability for the photogenerated electrons to be trapped at Ti-surface states (Ti^{3+}) will also be reduced resulting in an enhanced recombination.

3.2. NOVEL TITANIUM DIOXIDE PHOTOCATALYSTS

3.2.1. *Concentration Dependencies*

Dichloroacetic acid (DCA) which was used as the model compound in all experiments described below can be completely mineralized photocatalytically upon UV illumination of TiO_2 powders obeying the following stoichiometry [14]:

$$CHCl_2COO^- + O_2 \xrightarrow[TiO_2]{h\nu} CO_2 + H^+ + 2\,Cl^- \qquad (2)$$

Both photocatalysts employed in this part of the study, *i.e.*, the novel material Hombikat UV 100 supplied by Sachtleben Chemie and a standard material (in the following called "TiO_2-powder") which consists of 75% anatase and 25% rutile with a surface area of 50 m^2/g, are able to degrade dichloroacetate upon bandgap illumination as is evident from Figure 6.

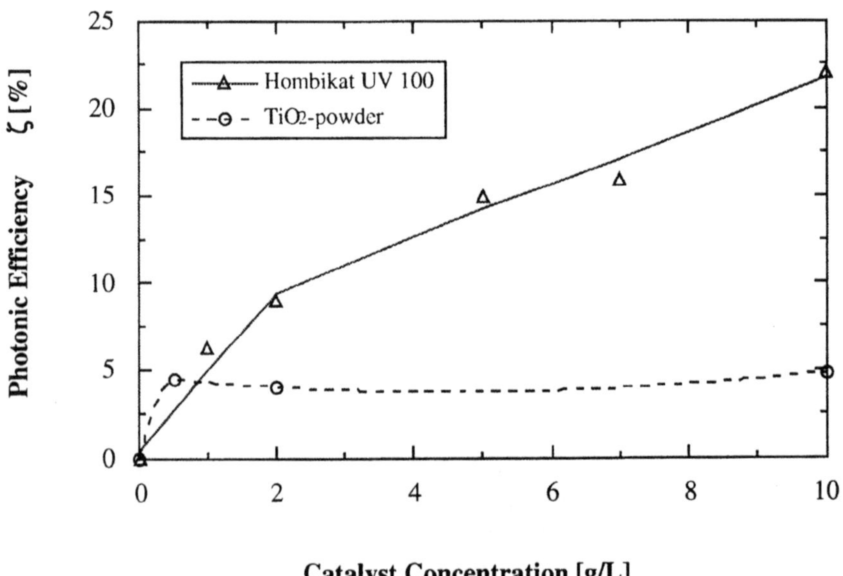

Figure 6. pH3, 1 mmole/L DCA, 293K, 10 mmole/L KNO_3, $I_{h\nu}$= 12 mmole/Ls, O_2-sat.

The effect of the total catalyst concentration used in the experiments on the degradation efficiency is shown in this figure. While a maximum photonic efficiency ζ of approx. 4% (based upon the initial number of photons below 400 nm available in the photoreactor) is reached already with 0.5 g/l TiO_2-powder, ζ considerably increases up to a catalyst concentration of 5 g/l when Hombikat UV 100 is used instead. Indeed, following a plateau region with a photonic efficiency around 15% between 5 and 7 g of catalyst per liter, an increase of the Hombikat UV 100 concentration to 10 g/l even results in a DCA degradation efficiency of 22%. The respective catalyst concentration at the plateau value, *i.e.*, 0.5 g/l for TiO_2-powder and 5.0 g/l for Hombikat UV 100, has been used in all the following experiments described within this paper.

Hombikat UV 100 is a material which consists of 100% anatase with very small primary particles (approx. 5 nm diameter) forming secondary agglomerates with 1 μm diameter and a surface area of >250 m^2/g. It can therefore be envisaged that the increased photocatalytic activity of this material can be explained by its very high surface area which should be beneficial for the adsorption of pollutant molecules. However, the different dependence of the photonic efficiency on the catalyst concentration (illustrated in Fig. 6) is difficult to explain. Recent results suggest that with the reactor configuration used in this study the maximum possible number of incident photons is already absorbed at 0.5 g/l TiO_2-powder, while the different absorptivity of the "quantum-sized" 5 nm-particles forming the secondary aggregates of Hombikat UV 100 results in a blue-shifted absorption spectrum in comparison with TiO_2-powder at identical catalyst concentration. Additionally, Hombikat UV 100 exhibits a much more pronounced light scattering than TiO_2-powder. Therefore, much higher concentrations of Hombikat UV 100 are necessary to achieve an increased light absorption and consequently an increased activity of this material. Thus, it was reasonable to choose different catalyst concentrations as explained above to compare the materials' efficiency.

Using these different catalyst concentrations a series of experiments have been carried out employing various substrate concentrations. While the efficiency of the photocatalytic DCA degradation increases considerably when the substrate concentration is increased to about 1 mmole/L, ζ reaches a saturation level above this concentration. Both photocatalysts exhibit qualitatively identical behavior, however, the maximum photonic efficiency which could be attained with Hombikat UV 100 (16%) is about four times higher than that of TiO_2-powder (4%).

3.2.2. *Catalyst Lifetime*

The photocatalytic properties of Hombikat UV 100 have been investigated in more detail. Figure 7 shows that the photonic efficiency of the photocatalytic DCA degradation is considerably decreased upon the addition of various "inert" salts, e.g., Na_2SO_4, NaCl, and Na_3PO_4. In addition to the salt concentration shown on the x-axis all solutions except those marked with the open triangle (KNO_3) contained 10 mmole/L KNO_3 to ensure constant ionic strength during a photocatalytic experiment. It is interesting to note that ζ increases from 15% in the presence of 10 mmole/L NO_3^- to 23% in the absence of any salts. While the addition of 10 mmole/L NaCl results in an efficiency decrease of 50%, an inhibition of more than 90% is noted when the same concentrations of phosphate or sulphate are added. It is important to note that the addition of the above mentioned salts has almost no effect on the degradation efficiency when the experiments were carried out at pH 11 instead of pH 3 (optimum pH–value for DCA degradation). The results shown in Fig. 7 are in good agreement with those published by Abdullah *et al* [22] who reported a strong inhibiting effect of chloride, sulphate and phosphate when they studied the photocatalytic degradation of salicylic acid, aniline and ethanol at pH 4.1. Since we observed that the choice of the cation does not effect the apparent efficiency, the crucial role of the various anions will be discussed below.

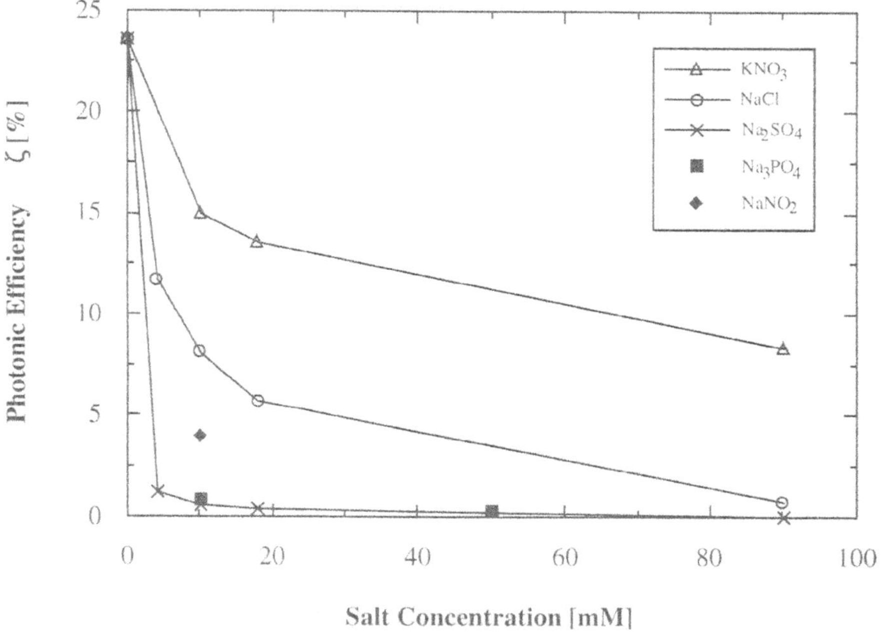

Figure 7. 5g/l Hombikat UV100, 1 mmole/L DCA, pH3, 293K, 10 mmole/L KNO_3 (except for —△—)

Two different mechanisms are proposed to explain the effect of anion addition on the observed photocatalytic activity. Chloride or nitrite ions can act as hole scavengers following the reaction sequence (3) - (5).

$$TiO_2 \xrightarrow[\lambda_{ex} < 400\ nm]{h\nu} TiO_2\,(h_{VB}^+, e_{CB}^-) \qquad (3)$$

$$Cl^-/NO_2^- + h_{VB}^+ \longrightarrow Cl^\bullet/NO_2^\bullet + H^+ \qquad (4)$$

$$Cl^\bullet + Cl^- \longrightarrow Cl_2^{\bullet -} \qquad (5)$$

This is a typical example for a competitive inhibition. The reaction of DCA with the holes

$$h_{VB}^{+} + CHCl_2COO^- \longrightarrow CHCl_2COO^{\bullet} \qquad (6)$$

will have to compete with reaction (4) with chloride and nitrite ions being at least as well adsorbed on the particle surface as dichloroacetate. While chlorine atoms will be forming the respective radical anions almost instantaneously (*cf.* reaction (5)), the radicals formed in reaction (6) will decarboxylate, add O_2 and subsequently form the products given in eq. (2).

Alternatively, surface sites normally available at the TiO_2/water interface for the adsorption and electron transfer from DCA can be blocked by multivalent anions such as sulphate or phosphate which are not readily oxidizable but yet very effective inhibitors for the detoxification process (*cf.* Fig. 7). A monovalent anion such as nitrate will also more likely act as a geometrical rather than a kinetic inhibitor but will be less effective since it does not form very strong complexes on metal oxide surfaces. It should be pointed out that all of these inhibitions were found to be completely reversible. While a few wash cycles of the catalyst powder with pure water were sufficient to fully restore its photocatalytic activity when chloride or nitrate were present, dilute base was required to eliminate inhibiting traces of sulphate or phosphate from the catalyst's surface.

Batch experiments have been conducted with the same powder as photocatalyst but adding 1mmole/L of dichloroacetate after the same amount had been degraded in a preceding run. The photocatalytic activity of Hombikat UV 100 decreased from ζ=15% to 7% after 9 subsequent batch experiments, that of TiO_2-powder from ζ=4% to 1% within 6 experiments. Washing the catalyst powders with distilled water between the individual experiments completely restored their photocatalytic activity. This inhibiting effect can readily be explained by the formation of 2 mmole/L chloride ions upon the degradation of 1 mmole/L DCA (*cf.* reaction (2)). In fact, the reduction in the degradation rate can be predicted accurately from the data shown in Fig. 7 for the effect of Cl^-.

The inhibiting effect of chloride ions which are formed during the photocatalytic degradation of toxic chlorinated hydrocarbons presents a serious problem since the latter class of non-biodegradable compounds is among the most interesting groups of pollutants which could be mineralized photocatalytically. Fortunately, this inhibition is completely reversible, *i.e.*, interferences caused by chloride ions can simply be washed away with pure water. The effectiveness of photocatalytic systems will therefore strongly depend on the development of a sophisticated photoreactor design avoiding batch reactor conditions as much as possible.

3.2.2. *Catalyst Platinization*

Even though Hombikat UV 100 already showed increased photocatalytic properties for the degradation of the model pollutant DCA in comparison to TiO_2-powder, we have carried out additional experiments with a platinized modification of this material. Using 0.1 mole/L methanol as a hole scavenger and the appropriate concentrations of H_2PtCl_6 to yield the desired content of Platinum, aqueous suspensions of Hombikat UV 100 were illuminated to prepare platinized powders which were washed thoroughly with distilled water (3 times) to remove any organic residues and subsequently dried by centrifugation. Figure 8 shows that the photocatalytic activity of the Hombikat UV 100 for the degradation of DCA can indeed be considerably increased by this platinization, *i.e.*, when Platinum islands are present on its surface. While a Pt-content of only 0.1 wt.% already resulted in an effective doubling of the DCA degradation rate, the maximum photonic efficiency (50 %) was reached with 0.5 wt.% Pt.

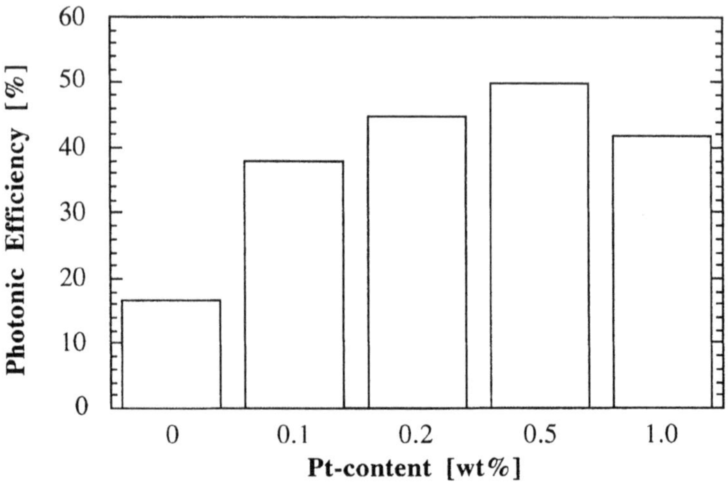

Figure 8. 5 g/L Hombikat UV 100, 1 mmole/L DCA, 293 K, pH 3, 10 mmole/L KNO_3, I_{hv}= 5 μmole/Ls, O_2–sat., var. Pt-concentrations (wt-%)

It is a well known effect of Platinum deposits on semiconductor surfaces to act as efficient catalysts for electron transfer reactions. It is therefore proposed that the reaction of the electrons formed upon the illumination of the photocatalyst powder (e_{CB}^- , *cf.* reaction (3)) with molecular oxygen is rate limiting when DCA is reacting with the holes (reaction (6)). The tiny Pt-deposits on the TiO_2 surface seem to catalyze this reaction effectively and therefore a considerable increase in ζ is observed. We have, in fact, observed in ongoing studies that the situation can be quite different when the same photocatalyst materials are used for the degradation of other model compounds. No increase in the photonic efficiency could, for example, be observed with platinized Hombikat UV 100 as compared with the non-treated material when 4-chlorophenol was the model pollutant. In the latter case, the oxidative step involving the holes appears to be rate limiting which can also be concluded from the overall ζ which was more than one order of magnitude smaller than that for the DCA mineralization. It would only be through the catalysis of this oxidation step that the overall efficiency could be increased which, however, requires other catalyst materials than Platinum. Alternatively, additional oxidants might be effective to increase the overall efficiency of these processes.

4. Acknowledgement

Parts of the investigations presented here were financially supported by the Sachtleben Chemie GmbH. In particular, we thank Dr.Griebler and Mr. Hirthe (Sachtleben Chemie GmbH) for their continuous support and interest in our work. The flash photolysis experiments described were performed at the Heyrovsky Institute of the Czech Academy of Sciences in Prague and we would like to thank Dr. Jirkovsky for his collaboration in this project.

5. References

1. Häfner, M. (1989) Wasser ist kein Naturprodukt mehr, *Natur* **10**, 20-23.
2. Dieter, H. H. (1989) Gebt dem Grenzwert eine Chance! Pestizide im Trinkwasser, *Wechselwirkung* **43**, 4-9.
3. Carey, J. H., Lawrence, J., and Tosine, H. M. (1976) Photodechlorination of PCB's in the presence of titanium dioxide in aqueous suspensions, *Bull. Environ. Contam. Toxicol.* **16**, 697-701.
4. Robin, E. (1989), Sunlight found to help break down organic pollutants, *San Francisco Examiner* **May 20**, D-3.
5. Fettwell, J. (1989) Semiconductors help the sun to clean water, *New Sci.* **June 10**, 36.
6. Thornton, J. (1989) Sunlight destroys hazardous waste, *SERI S&T In Review*, 8.
7. Hecht, J. (1990) Sunlight gives toxic waste a tanning, *New Sci.* **April 14**, 28.
8. Serpone, N. and Pelizzetti, E. (1989) *Photocatalysis : Fundamentals and Applications*, J. Wiley & Sons, New York.
9. Pelizzetti, E., Minero, C., and Maurino, C. (1990) The role of colloidal particles in the photodegradation of organic compounds of environmental concern in aquatic systems, *Adv. Coll. Interface. Sci.* **32**, 271-316.
10. Bahnemann, D., Bockelmann, D., and Goslich, R. (1991) Mechanistic studies of water detoxification in illuminated TiO_2 suspensions, *Solar Energy Materials* **24**, 564-583.
11. Hoffmann, M.R., Martin, S.T., Choi, W., and Bahnemann, D. W. (1995), Environmental applications of semiconductor photocatalysis, *Chem. Rev.* **95**, 69-96.
12. Gerischer, H. (1979) Solar photoelectrolysis with semiconductor electrodes, *Topics Appl. Phys.* **31**, 115-171.
13. Kormann, C., Bahnemann, D. W., and Hoffmann, M. R. (1988) Preparation and characterization of quantum size titanium dioxide (TiO_2), *J. Phys. Chem.* **92**, 5196-5201.
14. Hilgendorff, M. (1991) *Mechanistische Untersuchungen zum Photokatalytischen Abbau von Dichloressigsäure*, Diplomarbeit, Universität Hannover,.
15. Bockelmann, D. (1989) *Photokatalytischer Abbau Halogenierter Kohlenwasserstoffe an Halbleiteroberflächen*, Diplomarbeit, TU Clausthal,.
16. Heller, H. G. and Langan, J. R. (1981) A new reusable chemical actinometer, *EPA Newslett.*, 71-73.
17. Weiß, J. (1983) Einführung in die Ionenchromatographie: Grundlagen, Instrumentation und Anwendung, Teil 1, *Chemie in Labor und Betrieb*, 34, 293-297.
18. Koch, U., Fojtik, A., Weller, H., and Henglein, A. (1985) Photochemistry of semiconductor colloids: preparation of extremely small ZnO particles, fluorescence phenomena and size quantization effects, *Chem. Phys. Lett.* **122**, 507-510.
19. Henglein, A., Kumar, A., Janata, E., and Weller, H. (1986) Photochemistry and radiation chemistry of semiconductor colloids: reaction of hydrated electrons with CdS and non-linear optical effects, *Chem. Phys. Lett.* **132**, 133-136.
20. Spanhel, L., Weller, H., Fojtik, A., and Henglein, A. (1987) Photochemistry of semiconductor colloids: strong luminescing CdS and CdS-Ag_2S particles, *Ber. Bunsenges. Phys. Chem.* **91**, 88-94.
21. Bahnemann, D. W., Kormann, C., and Hoffmann, M. R. (1987) Preparation and characterization of quantum size zinc oxide: a detailed spectroscopic study, *J. Phys. Chem.* **91**, 3789-3798.
22. Abdullah, M., Low, G. K.-C., and Matthews, R. W. (1990) Effects of common inorganic anions on rates of photocatalytic oxidation of organic carbon over illuminated titanium dioxide, *J. Phys. Chem.* **94**, 6820-6825.

NANOSCALE METAL OXIDES AS DESTRUCTIVE ADSORBENTS. NEW SURFACE CHEMISTRY AND ENVIRONMENTAL APPLICATIONS

K.J. KLABUNDE, D.G. PARK, J.V. STARK, O. KOPER, S. DECKER, Y. JIANG and I. LAGADIC
Department of Chemistry
Kansas State University
Manhattan, Kansas 66506
USA

ABSTRACT. An aerogel procedure combined with hypercritical drying has yielded magnesium oxide and calcium oxide in ultrahigh surface area forms. These nanoparticles of MgO and CaO possess intrinsically higher surface reactivities, and serve as destructive adsorbents for a variety of toxic substances, including organophosphorus compounds, and chlorocarbons. They also serve to adsorb large amounts of gases very strongly, such as CO_2, SO_2, SO_3, and HX. A second generation of even more effective destructive adsorbents has been prepared by depositing a monolayer of transition metal oxide on the MgO or CaO nanoparticles, for example $[Fe_2O_3]MgO$, [NiO]CaO, [ZnO]MgO, and others. As a test reaction $CCl_4 + [Fe_2O_3]MgO \rightarrow [FeCl_x]MgCl_2 + CO_2$ was employed. This gas-solid reaction was facilitated and enhanced by two things: (1) Unusual morphology of nanoscale MgO, probably because of exposure of {111} crystal faces and high concentrations of edge sites and defect sites, and (2) the presence of the thin layer of Fe_2O_3 (or other transition metal oxide), which allows a catalytic O^{2-}/Cl^- solid state ion/ion exchange to take place. The reaction proceeded to almost stoichiometric proportions when Fe_2O_3 was present, which indicates that the surface Fe_2O_3-$FeCl_x$ layer is mobile and dynamic, allowing continual O^{2-}/Cl^- exchange deeper into the nanoparticle. Morphological studies were aided by Atomic Force Microscopy experiments, which will also be discussed.

1. Introduction

The safe destruction of hazardous materials requires the development of new processes that preferably operate at temperature well below incineration temperatures, do not involve large volumes of hot flow gases, and produce nontoxic useable byproducts.[1] One such approach is the use of destructive adsorbents, which depend on the high reactivity and reaction capacity of high surface area metal oxide nanoparticles.[2] Destructive adsorbents made from benign materials such as metal oxides of Mg, Ca, and Fe have been used successfully to destroy chlorocarbons as well as organophosphorus compounds. (eqn. 1 and 2):[3]

E. Pelizzetti (ed.), Fine Particles Science and Technology, 691–706.

$$(CH_3O)_2\overset{\overset{O}{||}}{P}-CH_3 \xrightarrow{MgO} CH_3O\text{-}\overset{\overset{O}{||}}{P}\text{-}CH_{3\,ads} + CH_3O_{ads}$$

$$CH_3O_{ads} + (CH_3O)_2\overset{\overset{O}{||}}{P}CH_3 \longrightarrow CH_3O\text{-}PCH_{3\,ads} + HO\text{-}\overset{\overset{O}{||}}{C}\text{-}H + CH_3OH \quad (1)$$

$$CCl_{4(g)} + 2\,CaO_{(s)} \longrightarrow CO_{2(g)} + 2\,CaCl_{2(s)} \quad (2)$$

In attempts to maximize the capacities of MgO and CaO as destructive adsorbents for chlorocarbons, we have employed aerogel-hypercritical drying procedures to produce nanoscale MgO and CaO particles. In further attempts to enhance reactivities and capacities, we have applied overlayers of transition metal oxides on these nanoparticles, the reasoning being that these overlayers should give a thermodynamic advantage that could possibly lead to more efficient use of the inner layers of the MgO(CaO) in the bulk of the nanocrystals. The work of Berty and coworkers[4] showing an unusual and positive effect in chloride capture by Na_2CO_3 during oxidative decomposition of chlorocarbons also encouraged as to pursue this avenue. In fact, we did find a substantial beneficial effect of Fe_2O_3 and a preliminary report has appeared.[5] Basically, the presence of an Fe_2O_3 overlayer was found to catalytically enhance O^{2-}/Cl^- solid state exchange. Although O^{2-}/F^- exchange on Al_2O_3 has been observed by Hedge and Barteau[6], our results appear to be the first examples of a catalytic enhancement of this process.[7]

Herein are reported further experiments on nanoparticle surface chemistry with environmental applications, in particular data on capacities for nanoparticle adsorption of acid gases, and on destruction of a model chlorocarbon (CCl_4) by nanoparticle destructive adsorbents.

2. Experimental Section

Materials: Magnesium turnings were purchased from Aldrich, $Fe(acac)_3$ from Strem, commercial MgO (calcined magnesia) from Fisher Scientific, and commercial CaO from Fisher. All other chemicals were reagent grade. Methanol (MeOH), toluene, and tetrahydrofuran (THF) were dried over $Mg(OMe)_2$, CaH_2, and sodium benzophenone, respectively.

Synthesis of Destructive Adsorbents: A series of MgO and $[Fe_2O_3]$MgO samples were prepared of varying surface area and homogeneity:

CM-MgO; a commercial sample of MgO was pretreated under vacuum at 500°C. This procedure was carried out by slowly heating several grams of powdered sample under dynamic vacuum (about 1 x 10^{-6} Torr). This was carried out in a Pyrex Schlenk type reaction tube that fit into a cylindrical furnace. The heat up took several hours, and the sample was maintained at 500°C overnight.

CP-MgO; a "conventionally prepared" sample was prepared by first boiling several grams of commercial MgO in 500 mL distilled water overnight (with magnetic stirring in a 1L flask equipped with a reflux condenser). After cooling the slurry was filtered and the filter cake dried in an oven at 120°C. The dried powder was broken into pieces and then heat treated up to 500°C under vacuum as described for CM-MgO.

AP-MgO; these samples were prepared from an aerogel with hypercritical drying in an autoclave.[2] Briefly, in a 3-necked 2 L creased round bottom flask equipped with a mechanical stirrer, water cooled condenser, argon inlet with a 3-way stopcock, was placed 300 mL of toluene. In another flask, 2.4 g of Mg turnings were allowed to react with 100 mL of MeOH under Ar. The resultant 1 M solution of $Mg(OMe)_2$ was added dropwise to the toluene with vigorous stirring under Ar. Then 4 mL of water was added dropwise from a syringe over a 30 min. period. This solution was stirred at room temperature under Ar overnight. The resultant slightly milky solution (sol) was placed in an autoclave, heated, vented, and thus converted to a $Mg(OH)_2$ aerogel as previously described.[2] This white fine powder was heat treated under vacuum up to 500°C as described for CM-MgO.

Iron Oxide Coated MgO designated [Fe_2O_3]MgO; the Fe_2O_3 surface layer was prepared by placing the heat treated MgO sample in a Schlenk reaction tube under Ar. It was treated with a 0.2 M $Fe(acac)_3$ solution in THF (prepared by stirring 6.2 g of $Fe(acac)_3$ in 87.5 mL of THF) overnight at room temperature. During this time, much of the $Fe(acac)_3$ adsorbed onto the MgO surface. The solid was filtered off under Ar and washed with pure THF until the filtrate had no color. The sample was dried under vacuum and the heat treated at 500°C as described above. Thus, were prepared [Fe_2O_3]CM-MgO, [Fe_2O_3]CP-MgO, and [Fe_2O_3]AP-MgO.

Composite Mixture of Fe_2O_3 and MgO designated Fe_2O_3/MgO; In a 3-necked 2L creased round bottom flask equipped with a mechanical stirrer, water condenser, and Ar inlet, 150 mL of a 1 M $Mg(OMe)_2$ methanol solution was rapidly mixed with 18.8 mL of 0.2 M $Fe(acac)_3$ in THF. The deep red color of the $Fe(acac)_3$ solution turned a light greenish yellow immediately upon mixing. Into this solution 450 mL of toluene was added. With vigorous stirring 5.5 mL of water was added dropwise using a syringe over a 30 min period. Stirring was continued overnight. The resultant light greenish, hazy sol was converted into a brown aerogel using the autoclave, further dried at 120°C in an oven, and then heat treated to 500°C under vacuum as described above. (see Scheme 1)

Gas Adsorption on AP-MgO and CP-MgO: Loaded on a spring balance, 80 mg samples of AP-$Mg(OH)_2$ and CP-$Mg(OH)_2$ were slowly heated under dynamic vacuum to temperature 500°C (or lower temperature such as 25, 120, 250, 300 or 400°C) and maintained for 3 hours. The sample was converted to AP-MgO or CP-MgO by dehydration during this period, and it was then cooled to room temperature over a 1 hour period. At this time the sample was isolated from the vacuum pump, and 20 torr of SO_2 or CO_2 was introduced, allowing 15 minutes contact time, followed by 100 minutes of evacuation, the change in spring position, thus calibrated weight difference, was measured before and after heating and before and after evacuation. Temperature effects on SO_2 (or CO_2) desorption were carried out by evacuation at certain elevated temperature (100, 200, 300, 400 and 500°C for ½ hour intervals, and weight changes recorded.

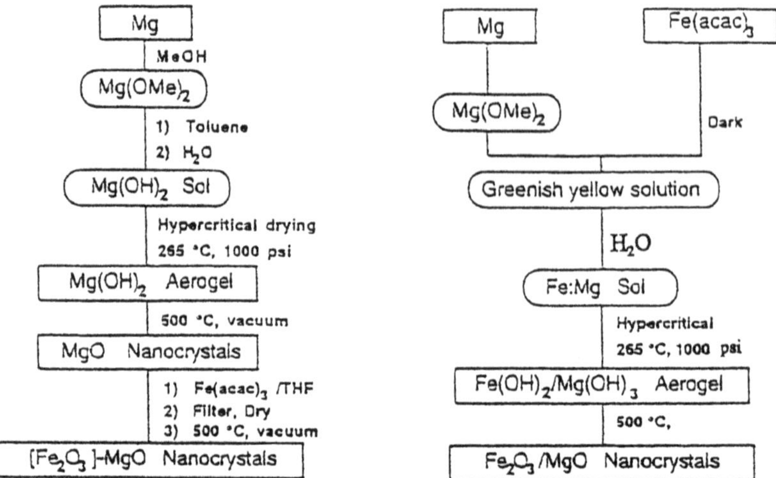

Scheme 1. Schematic outline of the preparation of nanocrystalline samples: (a) iron oxide coated MgO[Fe_2O3]Ap-MgO) NS (B) composite mixture of Fe_2O_3 and MgO.

Instrumental Analyses: FTIR spectra were recorded on a Bio-rad 3240 SPC. Infrared samples were prepared by mixing about 1 mg of sample with pre-dried KBr powder with a mortar and pestle. The mixture was pressed into a transparent pellet. These manipulations were carried out in air. X-Ray diffraction (XRD) spectra were collected on a Scintag XDS 2000 x-ray diffractometer. Surface areas were measured by BET N_2 with a Micrometries Flowsorb II 2300. Elemental analysis were obtained by Galbraith Laboratories.

Gas chromatographic studies were carried out on a GOW MAC model 580. A U-tube reactor was introduced between the injection port and the column as shown in Figure 1. Either one of two U-tube reactor types was employed, a quartz reactor or a stainless steel reactor. Into the reactor tube was placed 0.10 g of the oxide sample and held in place with glass wool plugs. Helium flowed through the reactor at 30 cc/min while it was heated to 425°C. Then CCl_4 (dried over molecular sieves) was injected in 1 microliter pulses with 7 minute intervals. The injections were continued until a constant GC profile (mix of products) was observed. Then the pulse sizes were increased to 2 microliters. Similar experiments were performed with water saturated CCl_4.

Atomic Force Microscopy (AFM) was carried out on Wyco Scanning Probe Microscope Model M-30. The AFM measurements on powdery samples presented serious technical problems caused by mobility of the particles. Therefore, measurements on pressed pellets (0.1 g powder pressed at 86,000 kPa) were also obtained. In order to check the validity of the images, scans in several directions were taken on each pellet.

Atomic Force Microscopy (AFM) was carried out on Wyco Scanning Probe Microscope Model M-30. The AFM measurements on powdery samples presented serious technical problems caused by mobility of the particles. Therefore, measurements on pressed pellets (0.1 g powder pressed at 86,000 kPa) were also obtained. In order to check the validity of the images, scans in several directions were taken on each pellet.

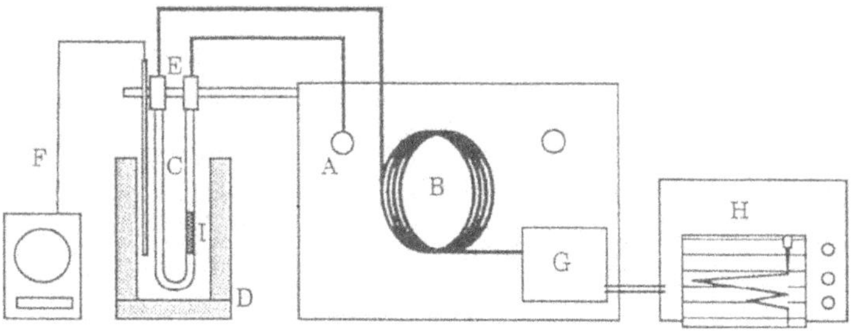

Figure 1 Schematic diagram of the gas chromatographic experimental setup used in monitoring the generation of the decomposition products. A) injection port, B) GC column, C) reactor, D) cylindrical furnace, E) connector, F) thermocouple and temeperature controller, G) GC detector, H) recorder, I) destructive adsorbent sample.

3. Results and Discussion

3.1. LOW PRESSURE (20 TORR) ADSORPTION OF SO_2 AND CO_2 ON AP-MgO AND CP-MgO

Pretreatment Temperature. The amount of SO_2 adsorbed at room temperature on AP-MgO and CP-MgO was measured as SO_2/nm^2 at various pre-heat treatment temperatures of 25, 120, 250, 300, 400 and 500 °C for the samples. The results are shown in Figure 2. The pre-evacuation amounts include physically and chemically adsorbed SO_2 species, while the post-evacuation results show only the chemically adsorbed species. Since these results are already corrected for surface area, they are in SO_2/nm^2, it is clearly shown that more SO_2 is adsorbed on AP-MgO than on CP-MgO per surface unit. Thus it unmistakingly implies that AP-MgO must have intrinsic characteristics and perhaps different or more numerous active sites that allow higher reactivity with SO_2. Comparing pre-evacuation results with post-evacuation results, it shows that most of the absorption is chemisorption. For both samples, the adsorbing ability increases with heat treatment temperature. Noting that the samples began as $Mg(OH)_2$ and were dehydrated with heat treatment, an explanation for this was drawn. Dehydration opened sites for adsorption formerly occupied by water, hydroxyls and hydrogen atoms. For AP-MgO, additional sites formerly occupied by methoxy groups and other carbon species were also formed at 500 °C.

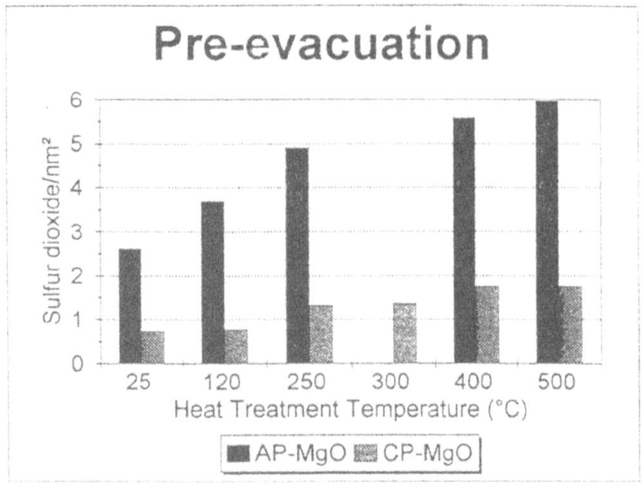

Figure 2 Number of SO_2/nm_2 adsorbed on AP-MgO and CP-MgO heat treated at various temperatures.

A comparison gas, CO_2, was also adsorbed on AP-MgO and CP-MgO heat treated at 500 °C. Once again, pre-evacuation and post-evacuation results, showing total adsorbed species and chemisorbed species respectively, demonstrated that most of the CO_2 is chemisorbed on both MgO samples. Surface area corrected results show that significantly more CO_2 was adsorbed on AP-MgO than on CP-MgO. It can also be noted that SO_2, being a stronger acid gas, adsorbed three times more than CO_2. Therefore, it appears that CO_2 is more specific than SO_2 and probes all basic sites whatever their strength.

3.1.2 Desorption of SO_2. AP-MgO heat treated at 500 °C and exposed to SO_2 was systematically heated under dynamic vacuum on a spring balance to observe the desorption of SO_2. The desorption of SO_2 did not appreciably occur until temperatures of 300 °C and greater were met. By 500 °C, 0.90 SO_2/nm^2 remained on the surface and 0.63 SO_2/nm^2 after overnight treatment.

3.2 IRON OXIDE COATED NANOPARTICLES OF MgO

3.2.1 Preparation and Characterization

Earlier studies showed that the nanocrystalline AP-MgO contained some residual surface methoxy groups, but that these are eliminated by 500°C treatment.[8] Some isolated surface-OH groups remain as well as some unidentate carbonate species.[9] In the current study these carbonate groups exhibited bands a 1634, 1475, 1430, 1073 and 865 cm^{-1}.

Iron oxide coated samples were also analyzed by FTIR. The AP-MgO (after 500°C treatment and cooling to room temperature) was treated with $Fe(acac)_3$ in THF; see Scheme

1. The sample after drying at 120°C showed bands characteristic of $Fe(acac)_3$; 1610(s), 1520(s), 1462(s), 1409(s), 1264(s), 1194(w), 1020(s), and 921(s) but various were bands shifted from 5-40 cm^{-1} from free $Fe(acac)_3$; lit: 1575(sh), 1570(s), 1555(sh), 1535(sh), 1525(s), 1425(s), 1490(s), 1465(s), 1285(s), 1190(w), 1025(s), 930(s). These results suggest a strong adsorption process but with at least some acac ligands in tact and probably still attached to Fe^{3+}. After another 500°C treatment all the acac ligands were eliminated, leaving light grey $[Fe_2O_3]$CM-MgO, dark grey $[Fe_2O_3]$CP-MgO, and dark brown $[Fe_2O_3]$AP-MgO. Table I shows elemental analyses for these samples, and earlier work showed that the iron was on the surface of the particles.[5]

For comparison, a more homogeneous mixture of iron and magnesium oxides was prepared by a modified aerogel procedure (Scheme 1). Solutions of $Mg(OMe)_2$ and $Fe(acac)_3$ were mixed and allowed to react. A rapid color change took place from deep red to light green indicating that ligand exchange must have occurred:

$$Fe(acac)_3 + Mg(OMe)_2 \longrightarrow Fe(OMe)_3 + Mg(acac)_2$$

This reaction was followed by UV-vis; the $Fe(acac)_3$ in THF showed three major adsorptions at 285, 355, and 440 nm, and upon mixing a new peak at 295 nm replaced the originals, and this latter peak is due to $Mg(acac)_2$. Upon hydrolysis it would be expected that $Mg(OH)_2$ and $Fe(OH)_3$ would form a gel together. Indeed, within an hour the solution became viscous. After the autoclave drying step the aerogel was a brown solid of very low density. Investigation by FTIR showed residual methoxy groups but no acac, indicating that all the acac was lost in the hydrolysis step. The final 500°C dehydration step yielded a dark brown solid that was converted to a powder form by brief grinding, designated Fe_2O_3/MgO.

Surface areas of these sample are shown in Table I. The over-coating of Fe_2O_3 did not change surface areas, indicating that the layer was deposited with conservation of surface morphology. As expected, the AP-MgO samples exhibited higher surface areas and smaller crystallites. Interestingly, the Fe_2O_3/MgO sample was surprisingly small in area even though prepared by the aerogel method. Apparently the presence of iron ion during the gelation process caused polymer formation with more cross-linking thus yielding a solid aerogel with a comparatively lower BET surface area.

TABLE I. Elemental analysis results, surface area, crystallite size calculated from XRD, and breakthrough points of the MgO and Fe_2O_3 coated MgO samples.

	CM-MgO	CP-MgO	AP-MgO	[Fe]CM-MgO	FeCP/ MgO	[Fe]AP- MgO	Fe_2O_3/ MgO
SA (m^2/g)	36	241	311	37	241	325	194
Breakthrough (μL)	0	0	0	0	5	12	7
Crystallite (Å)				280	77	54	
CO_2/CCl_4 Constant ratio	0.180	0.110	0.080	0.110	0.180	0.330	0.150
Carbon (%)	<0.5	1.13	1.91	0.54	2.40	4.29	2.37
Hydrogen	<0.5	1.61	2.18	0.50	1.25	2.15	1.90
Magnesium	65.02	53.46	47.69	60.68	51.75	44.02	50.02
Iron				<0.15	0.88	1.43	2.49

Complete elemental analyses were obtained for the samples (oxygen by difference), and calculate
d empirical formulas are shown in Table II. Interestingly CM-MgO and $[Fe_2O_3]$CM-MgO showed deficiency of oxygen, whereas the other samples showed an overall surplus of oxygen. This surplus can probably be accounted for as surface carbonate: FTIR showed peaks at 1634, 1475, 1430, 1073 and 865 cm^{-1} suggesting the presence of unidentate CO_3.[9] If it is assumed that CO_3 groups possess an H species and the remainder of H is in the form of surface -OH, fractional empirical formulas can be calculated. Actually, it is likely that some of the carbon is in the form of graphitic material, but we have not yet determined this in a quantitative way.

TABLE II. Formulas of MgO and Fe_2O_3 coated MgO, calculated on the bases of the elemental analysis and

	Empirical Formula	Proposed Components
CM-MgO	$MgO_{0.82}$	$MgO_{0.82}$
CP-MgO	$MgO_{1.24}\ C_{0.43}\ H_{0.73}$	$MgO_{0.43}(CO_3H)_{0.043}(OH)_{0.68}$
AP-MgO	$MgO_{1.536}\ C_{0.81}\ H_{1.10}$	$MgO_{0.27}\ (CO_3H)_{0.081}\ (OH)_{1.02}$
$[Fe_2O_3]$CM-MgO	$MgO_{0.95}\ C_{0.018}\ H_{0.20}\ Fe_{7.17\,E\text{-}4}$	$MgO_{0.72}\ (CO_3H)_{0.018}\ (OH)_{0.18}\ (Fe_2O_3)_{7.17*E\text{-}4}$
$[Fe_2O_3]$CP-MgO	$MgO_{1.28}\ C_{0.094}\ H_{0.58}\ Fe_{7.403*E\text{-}3}$	$MgO_{0.49}\ (CO_3H)_{0.094}\ (OH)_{0.49}$ $(Fe_2O_3)_{7.40*E\text{-}3}$
$[Fe_2O_3]$AP-MgO	$MgO_{1.66}\ C_{0.19}\ H_{1.17}\ Fe_{0.0141}$	$MgO_{0.048}\ (CO_3H)_{0.20}\ (OH)_{0.98}\ (Fe_2O_3)_{0.014}$
Fe_2O_3/MgO	$MgO_{1.31}\ C_{0.096}\ H_{0.91}\ Fe_{0.022}$	$MgO_{0.14}\ (CO_3H)_{0.096}\ (OH)_{0.82}\ (Fe_2O_3)_{0.022}$

The trends are interesting. As the samples became higher in surface area, more surface carbonates and hydroxides were found. It is likely that the carbonates were formed during the preparation, especially the initial drying step. Likewise, some hydroxide was probably taken up during handling of the hygroscopic samples. Also note that iron content increased with surface area, and this is reasonable based on the method of preparation where a competitive adsorption process for $Fe(acac)_3$ was employed.

In the XRD these surface -OH, carbonate, and iron oxide species were not observable, as expected. The broadness of the peaks confirms the very small crystallite size of about 4 nm. Although only the MgO diffraction pattern was observed, the presence of the surface Fe_2O_3 caused a shift in the d spacing of 0.15 Å as the MgO crystallite size decreased. This suggests a change in the bulk properties as surfacial iron oxide species become more important; that is, the surface/bulk ratio becomes larger. Thus, the presence of Fe_2O_3 perhaps causes strain in the very small MgO crystallites.

3.2.2. Destructive Adsorption of Carbon Tetrachloride

3.2.2.1 MgO without Iron. The main point we hope to convey in this section is that the presence of iron dramatically changes the capacity for destructive adsorption as judged by the following reaction (unbalanced):

$$CCl_4\,(g) + MgO(s) \longrightarrow CO_2\,(g) + MgCl_2\,(s)$$

Two numbers are of particular interest: (1) the breakthrough number; which is the number of injections of CCl_4 over the MgO bed before any CCl_4 exists the bed, and this is a measure of initial decomposition capacity, and (2) constant decomposition ratio; after initial decomposition, a reasonably constant composition of CO_2/CCl_4 exiting the bed was observed up until exhaustion.

The decomposition of CCl_4 over MgO samples (without iron) was carried out in a quartz U-tube reactor.[10] The loading of MgO was 0.10 g and each injection (pulse) of CCl_4 was 1 μl. The breakthrough number was only about 1 (% decomposition = 96% AP-MgO, 92% CP-MgO, and 81% CM-MgO for this one pulse). Soon thereafter a constant decomposition ratio was established, but at a fairly low value. In short, under the conditions of these experiments, the MgO samples were only marginally effective in CCl_4 decomposition, and this indicates that the MgO surface was being transformed into a protective $MgCl_2$ coating:

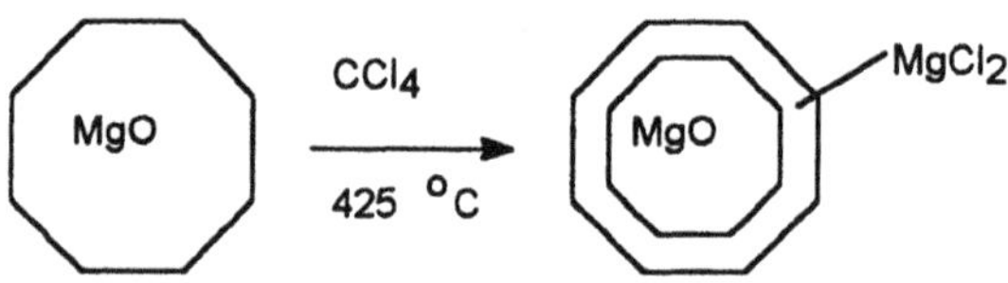

During this process surface carbonate was eliminated, according to FTIR studies of the used samples.

3.2.2.2 MgO Doped with Fe_2O_3 In the presence of iron the results were dramatically different. Figure 3 summarizes the results for reactions in a quartz reactor tube.

Initial decomposition studies gave breakthrough numbers ranging up to 15, and the effectiveness in this regard was in the order $[Fe_2O_3]$AP-MgO > Fe_2O_3/MgO ≈ $[Fe_2O_3]$CP-MgO > AP-MgO > CP-MgO > $[Fe_2O_3]$CM-MgO > CM-MgO. In addition many more pulses of CCl_4 were required to reach a constant decomposition ratio for the iron containing samples compared with their iron free counterparts. Indeed, it took more than 100 μl of CCl_4 to reach a reasonably constant decomposition ratio for $[Fe_2O_3]$AP-MgO. And, as previously reported, both initial and constant decomposition ratios could be improved by increasing the interval time between injections.

It was further deduced that the amount of Fe_2O_3 present had an effect on the constant decomposition ratio. Thus, the iron content for surface coated samples was $[Fe_2O_3]$AP-MgO > $[Fe_2O_3]$CP-MgO > $[Fe_2O_3]$CM-MgO. The amount of iron taken up in the preparative scheme should be surface area dependent since adsorption of $Fe(acac)_3$

from THF solution was the determining factor. However, note that the intimate mixture (possibly homogeneous) of Fe_2O_3/MgO was less effective in the CCl_4 decomposition, and yet possessed the highest iron content.

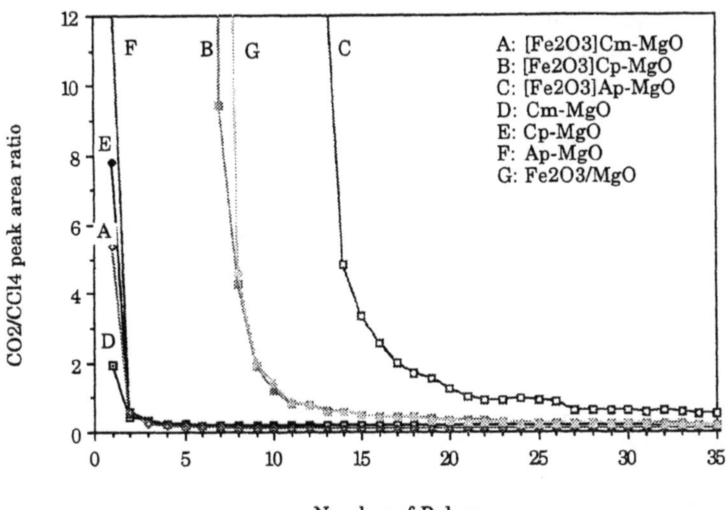

Figure 3 Decomposition of ccl4 over a series of [Fe2O3]MgO nanocrystalline samples represented by the peak area ratio of the decomposition product, CO2, to the unreacted Ccl4.

These results clearly show that surface iron is necessary for effective destructive adsorption of CCl_4. Before we try to rationalize these findings in mechanistic way, it will be helpful to add a few other pieces of data.

3.2.2.3 Effect of Water When water saturated CCl_4 was injected over the samples, breakthrough numbers did not change. Therefore, unlike the destructive adsorption of organophosphorus compounds on MgO,[10] water has no apparent effect on CCl_4 decomposition under the conditions employed.

3.2.2.4. Effect of Reactor Material (Quartz or Stainless Steel) In our earlier studies, a stainless steel u-tube reactor was employed [5] and even higher breakthrough numbers were realized. Thus, we undertook studies under exactly the same conditions except reactor material (quartz or stainless steel). When $[Fe_2O_3]$ AP-MgO in quartz was examined, breakthrough occurred at 13 μl, whereas in stainless steel breakthrough occurred at around 70 μl. In addition, product mix was different with stainless steel. For example, C_2Cl_4 became a major product after only 6 μl. In contrast, in a quartz reactor only very small amounts of C_2Cl_4 were formed. Also, in the stainless steel reactor, using

$[Fe_2O_3]MgO$, even more complete conversion of $MgO \longrightarrow MgCl_2$ was observed as monitored by XRD. Finally, when no $[Fe_2O_3]MgO$ was present (an empty reactor tube) no decomposition was observed in quartz, whereas a small amount of C_2Cl_4 and Cl_2 eluted from stainless steel. These results strongly suggest a predissociative mechanism in stainless steel, and this is discussed later.

3.2.2.5 Solid Products (MgO vs. $MgCl_2$) XRD spectra were taken of solid residues after 0.10 g $[Fe_2O_3]MgO$ samples had been treated with 120 μl of CCl_4 (note that 120 μl is the theoretical maximum for a stoichiometric $2\ MgO + CCl_4 \longrightarrow 2\ MgCl_2 + CO_2$ reaction). The extent of conversion to $MgCl_2$ was roughly in the order $[Fe_2O_3]$AP-MgO > $[Fe_2O_3]$ CP-MgO > $[Fe_2O_3]$ CM-MgO >> MgO samples without Fe_2O_3. In fact for the $[Fe_2O_3]$AP-MgO almost all of the MgO had been converted.

3.2.2.6 Atomic Force Microscopy Imaging the MgO and $[Fe_2O_3]MgO$ particles is a difficult problem due to their mobility, caused by electrostatic interaction between the particles and the probe tip, and the relatively low electron density of MgO. However, if surface reactivity is related to morphology, it is important to do so. We report here some preliminary results using a Scanning Probe Microscope in the Atomic Force mode. In order to anchor the small particles, a variety of techniques were attempted, including embedding in epoxy and other types of glue. However, with this technique we had difficulty differentiating the background from embedded particles. In the end we found that simply by pressing the pure samples at 86,000 kPa: the particles were partially "ceramitized" and could be studied. Scanning was performed on several sections of each pellet. The images obtained for CP-MgO showed a collection of smooth platelets about 200 x 200 x 50 nm. Several measurements over different sections of the sample indicated that the platelets have edge angles larger than 90° and the middle sections of the platelets appear thinner than the edges. The platelets were stacked together in one direction, probably caused by the action of the pressing.

In contrast, the AP-MgO sample imaged as a collection of spheres of about 80 nm diameters. No directional orientation was observed. The spheres showed a roughened surface. (Figures 4 and 5).

The features observed for these two samples were consistent with earlier TEM and SEM images.[2,8] Thus, the CM- and CP-MgO samples are more crystalline and usually in the form of platelets, whereas the AP-MgO samples are spheroidal. These different morphologies suggest preferential crystal growth along different crystalline facets, resulting in different crystalline planes being exposed.[11]

Figure 4 An AFM micrograph of AP-MgO

Figure 5 An AFM micrograph of CP-MgO

When cubic MgO grows into a cube, the exposed surface consists solely of {100} faces, in which Mg^{2+} and O^{2-} ions are alternately arranged as in structure I shown below. When the crystal takes on a spheroidal shape (such as a cubooctahedron or octahedron), the corners of the cube are missing (cleaved off), exposing {111} faces.[8,11] In an extreme case of the octahedron, the surface is solely {111} faces. The {111} faces consist of either a layer of exposed Mg^{2+} ions or O^{2-} ions, as shown in structures IIa and IIb.

I	IIa	IIb
Mg O Mg O	Mg Mg Mg Mg	O O O O
O Mg O Mg	Mg Mg Mg Mg	O O O O
Mg O Mg O	Mg Mg Mg Mg	O O O O
O Mg O Mg	Mg Mg Mg Mg	O O O O

The plate-like morphology of the CM- and CP-MgO suggests that a large fraction of {100} faces are exposed, whereas the spheroidal AP-MgO would have much more {111} faces exposed. The {111} faces have more tendency to develop (or possess) surfacial defects in order to disperse charge density. For example, anion or cation vacancies would be more prevalent. Thus, a roughened surface would be more likely. It would be expected that these two strikingly different surfaces would exhibit different surface reactivities; if fact defective {111} surfaces should be much more reactive, and this helps explain why AP-MgO is more effective as a destructive adsorbent, even in the absence of the Fe_2O_3 overlayer.

When the Fe_2O_3 overlayer was present, the overall plate-like and spheroidal morphologies did not change. The roughness of the AP-MgO sample (with Fe_2O_3) remained. Further AFM investigation at high resolution is needed in order to ascertain more surface detail.

4. Conclusions

The adsorption of acid gases, and destructive adsorption of CCl_4 on MgO is enhanced by two things:(1) spheroidal morphology of AP-MgO samples, and this is probably due to the presence of larger amounts of exposed {111} crystalline facets possessing large concentrations of defect sites. (2) a thin overlayer of Fe_2O_3 greatly enhances the capacity of the sample for CCl_4 destruction. This is apparently due to a catalytic effect that enhances O^{2-}/Cl^- exchange. Since this O^{2-}/Cl^- exchange continues into the interior of the MgO particle, it is proposed that the presence of the Fe_2O_3-$FeCl_x$ outer layer is morphologically rough and probably mobile such that no protective $MgCl_2$ coating can form. Therefore, O^{2-}/Cl^- exchange continues in a catalytic sense (Figure 6).

							Cl		Cl		
O	Fe	O	Fe	O	CCl_4	O	Fe		Fe	O	
Mg	O	Mg	O	Mg	------->	Mg	O	Mg	O	Mg	exchange ----------->
O	Mg	O	Mg	O	$-O^{2-}$	O	Mg	O	Mg	O	
					$+2Cl^-$						

					CCl_4		Cl		Cl		
O	Fe	O	Fe	O	------->	O	Fe		Fe	O	
Mg	Cl	Mg	Cl	Mg	$-O^{2-}$	Mg	Cl	Mg	Cl	Mg	exchange ----------->
O	Mg	O	Mg	O	$+2Cl^-$	O	Mg	O	Mg	O	
	O						O				

O	Fe	O	Fe	O
Mg	O	Mg	O	Mg
Cl	Mg	Cl	Mg	O
	Cl		Cl	

Figure 6 Proposed O2-/Cl⁻ exchange sequence that is enhanced by Fe2O3 overcoating.

5. Acknowledgements

The support of the Army Research Office and the Hazardous Substance Research Center of Kansas State University (administered by EPA and DOE) are acknowledged with gratitude.

6. References

1. a. Musick, J.R., Williams, F.W. (1974), Catalytic decomposition of halogenated hydrocarbons over hopcalite catalyst, *Ind. Eng. Chem. Res.*, **13**, 175-179.
   b. Pope, D., Walker, D.S., Moss, R. L. (1976) Evaluation of platinum - honeycomb catalysts for the destruction oxidation of low concentrations of odorous compounds in air, *Atmos. Environ.*, **12**, 1921-1927.
   c. Manning, M.P. (1984) *Hazardous Waste*, **1**, 41-65.
   d. Klabunde, K.J. (1984) *Centerpoint*, Vol. 2, No. 1, pg. **1**, publ. by Hazardous Subst. Res. Centers/S and SW, Atlanta, GA 30332-0837.
2. Utamapanya, S., Klabunde, K.J., Schulup, J.R. (1991) Nanoscale metal oxide particles/clusters as chemical reagents. Synthesis and properties of ultrahigh surface area magnesium hydroxide and magnesium oxide, *Chem. Mater.*, **3**, 175-181.
3. a. Koper, O., Li, Y.X., Klabunde, K.J. (1993) Destructive adsorption of chlorinated hydrocarbons on ultrafine (nanoscale) particles of calcium oxide, *Chem. Mater.*, **5**, 500-505.
   b. Li, Y.X., Klabunde, K.J. (1991) Nanoscale metal oxide particles as chemical reagents. Destructive adsorption of a chemical agent simulant, dimethyl methylphosphonate on heat-treated magnesium oxide, *Langmuir*, **7**, 1388-1393.
4. a. Berty, J.M. June 4, 1991 U.S. Patent 5,021,383, to Berty Reaction Engineers.
   b. Stenger, H.G. Jr.; Buzan, G.E., Berty, J.M. (1993) Chlorine capture by catalyst-sorbents for the oxidation of air pollutants, *Appl. Catal. B.*, **2**, 117-30.
5. a. Klabunde, K.J., Khaleel, A., Park, D.G., Overlayer of iron oxide on nanoscale magnesium oxide. Chlorocarbon destruction by catalytic solid state ion/ion exchange at 400°C, *High Temp. Sci.*, in press.
   b. Klabunde, K.J., Khaleel, A., U.S. Patent pending.
6. Hedge, R. I., Barteau, M.A. (1989), Preparation, characterization, and activity of fluorinated aluminas for halogen exchange, **120**, *J. Catal.*, 387-400.
7. Li, Y.X., Klabunde, K.J. (1992) Heterophasic isotope exchange in nanoscale metal oxide particles. Lattice oxygen and surface OH groups with water vapor (D_2O and $H_2{}^{18}O$), *Chem. Mater.*, **4**, 611-615 reports surprisingly facile $^{16}O^{2-}/^{18}O^{2-}$ exchange in nanoscale MgO particles reacting with $H_2{}^{18}O$.
8. Itoh, H., Utamapanya, S., Stark, J.V., Klabunde, K.J., Schlup, J.R. (1993) Nanoscale metal oxide particles as chemical reagents. Intrinsic effects of particle size on hydroxyl content and on reactivity and acid/base properties of ultrafine magnesium oxide, *Chem. Mater.*, **5**, 71-77.
9. a. Grigorec, A.I., Turova, N.Ya. (1965) Infrared spectra of berylium, magnesium and other alkaline earth metal alkoxides, *Dokl. Akad. Nauk SSSR*, **162**, 98-101.
   b. Gregg, S.J., Ramsey, J.D. (1970) Adsorption of carbon dioxide by magnesia studied by use of infrared and isotherm measurements, *J. Chem. Soc., (A)*, **2784-2787**.
   c. Sadtler, High Resolution Spectra of Inorganics and Related Compounds, Sadtler Res. Lab., Inc. Y333.
10. Li, Y.X., Koper, O., Atteya, M., Klabunde, K.J. (1992) Adsorption and decomposition of organophosphorus compounds on nanoscale metal oxides particles. In situ GC-MS studies of pulsed microreactions over magnesium oxide, *Chem. Mater.*, **4**, 323-330.
11. West, A.R. (1984) "Solid State Chemistry and it's Applications. John Wiley, New York, Chap. 1.

PHOTOTRANSFORMATIONS OF ATRAZINE OVER DIFFERENT METAL OXIDE PARTICLES

V. Maurino[a], C.Minero[a], E.Pelizzetti[a*] and **N. Serpone[b]**

[a]Dipartimento di Chimica Analitica, University of Torino, Italy
[b]Dept. of Chemistry, Concordia University, Montreal, Canada

1. ABSTRACT

The photocatalytic degradation of atrazine has been examined under simulated solar light irradiation of various semiconductors powders (TiO_2, ZnO, WO_3, SnO_2, and Fe_2O_3) and different titania specimens (nanometer-size colloids, micrometer particles and millimeter spherical beads). Three major different degradation pathways operate in the photocatalyzed process: (a) dehalogenation, (b) dealkylation, and (c) deamination; the final product is cyanuric acid. Comparison of the results suggests that, amongst the different semiconductors, TiO_2 is the most efficient. Titania, ZnO and WO_3 degrade atrazine through the generation of active species which mainly promote the alkyl chain oxidation. Fe_2O_3 and SnO_2 behave similarly to Al_2O_3 (an insulator) through a slow surface-assisted hydrolysis at the chlorine substituent position. The activity and selectivity of titania definitely depends on the differently preparative modes used to make the titania specimens. The temporal distribution of intermediates changes significantly as a consequence of the manner by which the organic probe and the catalyst interact.

2. INTRODUCTION

Interfaces between metal oxides and the aqueous phase play a crucial role in the transformations of organic and inorganic molecules. An adsorbed molecule shows chemical and physical properties that are different from those in homogeneous phase. Under light exposure, reactions can take place that may involve electronically excited surface

E. Pelizzetti (ed.), Fine Particles Science and Technology, 707–718.

species or reactive intermediates that were formed at the metal oxide particle surface by semiconducting-type mechanisms. The latter case, which is now referred to as *heterogeneous photocatalysis*, has gained increased attention as a possible technique for the detoxification of drinking waters and waste waters polluted with organic contaminants.[1,2]

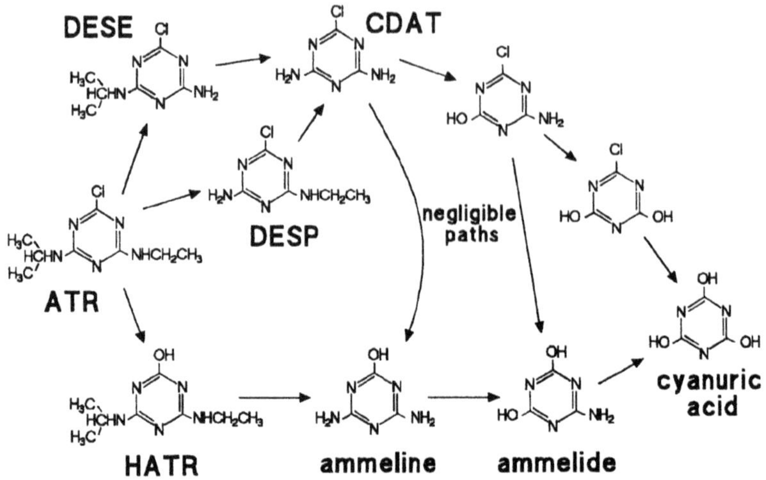

Scheme. Main reaction pathways for atrazine disappearance.

The search of efficient and selective metal oxide semiconductors and investigation of various alternarive preparation techniques are being actively pursued to acquire a better understanding of the underlying mechanistic steps in the various processes and to obtain a series of photocatalysts suitable for practical applications.[3-8]

Atrazine is a widely used herbicide often found in drinking waters at $\mu g\ L^{-1}$ levels. Since the photocatalytic treatment is very efficient even at such low concentration levels, the treatment process and the degradation mechanism of atrazine have been investigated extensively in our laboratory.[9,10] The principal degradation pathways under abiotic conditions are illustrated in the Scheme.

Atrazine is a propitiously good probe molecule to test the activity and the selectivity of various photocatalysts for many reasons. First, as indicated in the Scheme, there exist three competing degradative routes: (i) chlorine substitution by a hydroxyl group, (ii) dealkylation, and (iii) amino group substitution also by a hydroxyl group. Secondly, the s-triazine ring is integrally retained to the end-product of the transformation: cyanuric acid. And, thirdly, although the atrazine structure seems complex, few key intermediates are possible; they are increasingly more stable toward further transformation on going from the initial substrate to the final product. This has permitted a careful and detailed examination of the kinetic profiles.

The chosen semiconductor systems (in powder form) were TiO_2, ZnO, WO_3, SnO_2, and Fe_2O_3. The insulator system, Al_2O_3, was also examined for comparison. In so far as examining the influence that surface characteristics of the different specimens of the same semiconductor species might have on the photooxidative process of a test probe, we focused much of our attention on differently prepared titania colloidal and particulate systems.

3. PHOTOCATALYTIC MECHANISM

Photocatalysis over irradiated semiconductor dispersions (e.g. TiO_2 particles) provides an attractive method that leads to highly effective and spatially controlled oxidations and reductions of organic substrates.[1]

The primary photochemical act, following the near-UV light absorption by TiO_2 (λ<380nm for the anatase polymorph) is the generation of electron/hole pairs in the bulk semiconductor, eq.(1):

$$TiO_2 + h\nu\ (>E_{bg}) \longrightarrow e^-_{CB} + h^+_{VB} \quad (1)$$

$$e^-_{CB} + h^+_{VB} \longrightarrow \text{heat and/or } h\nu' \quad (2)$$

The charge carriers can either recombine according to eq.(2) or migrate rapidly to the surface where they ultimately trapped (the electron in a surface Ti(III), e^-_T, and the hole as a surface hydroxyl radical species, Ti(IV)-O•H, h^+_T). Henceforth, e^-_{CB} and e^-_T or h^+_{VB} and h^+_T will be referred simply as e^- and h^+, respectively, when this differentiation is unnecessary. If electron acceptors (Ox_1) and/or electron donors (Red_2) are present at the surface (ads), interfacial electron transfers may occur according to reactions (3)-(4).

$$e^-_{CB}\ (\text{or } e^-_T) + Ox_{1,ads} \longrightarrow Red_{1,ads} \quad (3)$$

$$h^+_{VB}\ (\text{or } h^+_T) + Red_{2,ads} \longrightarrow Ox_{2,ads} \quad (4)$$

Other possible reactions in these primary steps are the back reactions, eqs.(5) and (6):

$$e^-_{CB} \text{ (or } e^-_T) + Ox_{2,ads} \longrightarrow Red_{2,ads} \quad (5)$$

$$h^+_{VB} \text{ (or } h^+_T) + Red_{1,ads} \longrightarrow Ox_{1,ads} \quad (6)$$

Under some circumstances, a radical species formed by oxidation by h^+ can inject an electron into the conduction band of the semiconductor giving rise to the now well known doubling (of photocurrent) effect phenomenon.

The reactions (3) and (4) can generate different (radical) species depending on the type of primary active species (either e^-_{CB} or e^-_T, and either h^+_{VB} or h^+_T) and on the ensuing chemical reactions: notably, redox reactions, reaction with solvent and oxygen, and dimerization reactions, among others. In aerated systems, oxygen acts as an efficient electron scavenger. Thus:

$$O_{2(ads)} + e^-_{CB} \text{ (or } e^-_T) \longrightarrow O_2^-{}_{ads} \quad (7)$$

In the presence of organic compounds a series of oxidation/reduction processes, both initiated through eq.(3) and (4) and involving such other species as e.g. O_2, H_2O_2, and $O_2^{\bullet -}$, may lead to their complete degradation to CO_2 and inorganic anions.[2]

4. DIFFERENT METAL OXIDES

The principal characteristics and kinetic results germane to the conversion of atrazine to cyanuric acid are shown in Figures 1 and 2 and are collected in Table I.[11] Several properties play a role in the photocatalytic process efficiency:

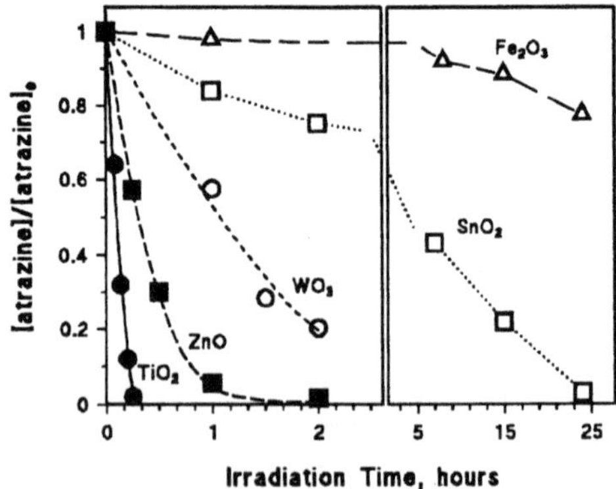

Figure 1. Atrazine disappearance as a function of irradation time in the presence of different metal oxides. Conditions: initial atrazine 25 mg L^{-1}, TiO_2 and Fe_2O_3 0.5 g L^{-1}, other oxides 1 g L^{-1}, light of λ>340 nm, pH 5.6. From ref.11, with permission.

(i) The surface area of the catalyst, the fraction of the irradiated surface (because of possible presence of pores), and the quantum yield of light absorption by the photocatalyst.[12]

(ii) The lifetimes of the active species (electron/holes pairs, radicals), that is the rate of their generation at the surface and their rate of recombination.[13] For a photocatalyst to be highly efficient, the interfacial concentration of the photogenerated charge carriers should be as high as possible. The presence of recombination centers or dopants that act as inhibitors affects the process. The hydroxyl group density at the metal oxide surface [14] and the extent of O_2 adsorption [15,16] are also important factors.

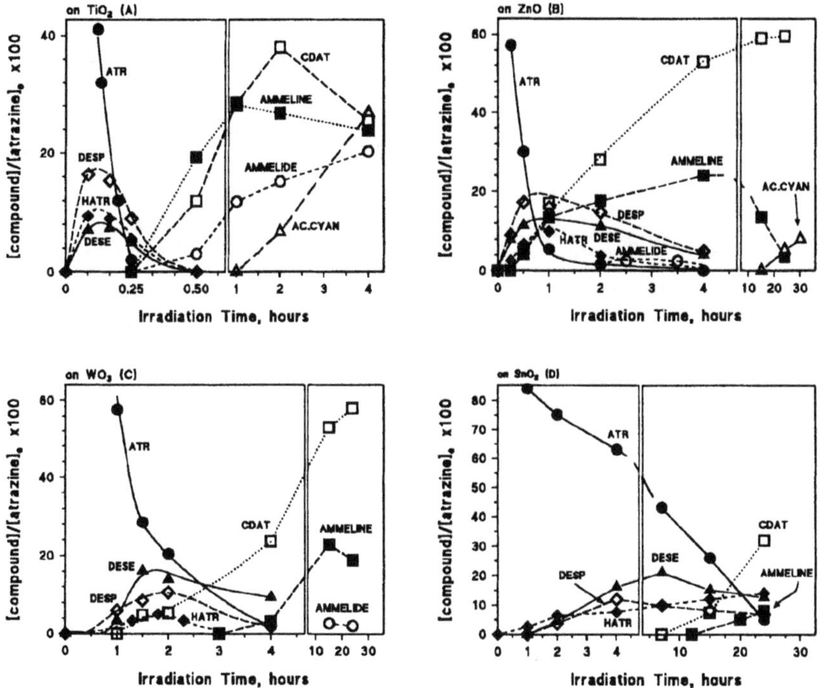

Figure 2. Atrazine degradation and intermediate evolution in the presence of TiO_2 (0.5 g L^{-1}, A), ZnO ((1.0 g L^{-1}, B), WO_3 (1.0 g L^{-1}, C) and SnO_2 (1.0 g L^{-1}, D). Acronyms refer to compounds depicted in the Scheme. Other conditions as in Fig.1. Adapted from ref.11, with permission.

(iii) Molecular positioning and molecular energy levels for interfacial chemical catalysis should be as favourable as possible to reduce the activated barrier and to drive the desired reactions; this will depend on the specific composition and the surface structure of the metal oxide.[17,18]

(iv) The nature and the extent of adsorption of the organic substrate can vary depending on the surface chemical composition and on the adsorption of anions, solvent and species other than oxygen as electron scavengers. These species can play a crucial role. They can compete for the sites where the degradative process originates, thus reducing the overall process efficiency.[19,20]

Table I. Data on metal oxides and key results concerning atrazine degradation under irradiation at wavelength > 340 nm.

entry	**added oxide**	**supplier**	**surface area (b) $m^2 g^{-1}$**	**load $g L^{-1}$**	**kapp (c) min^{-1}**	**comments (for acronyms see Scheme)**
1(a)	none	-	-	-	$(3.6 \pm 0.8) \times 10^{-5}$	HATR and CDAT formed at ca. the same time during the early degradation stages (up to 400 h).
2	TiO_2	Degussa	55	0.5	$(1.4 \pm 0.3) \times 10^{-1}$	lateral alkyl chain oxidation accounts for the large extent of degradation.
3	ZnO	C.Erba	15	1.0	$(4.0 \pm 0.2) \times 10^{-2}$	intermediate distribution similar to TiO_2. 5% conversion to cyanuric acid after 24 h.
4	WO_3	Ventron	1	1.0	$(1.2 \pm 0.2) \times 10^{-2}$	*ibid.* Only traces of cyanuric acid after 24 h.
5	SnO_2	BDH	8	1.0	$(2.1 \pm 0.2) \times 10^{-3}$	CDAT becomes relevant at high irradiation times. No cyanuric acid after 24 h.
6	α-Fe_2O_3	home prepared	150	0.5	$(1.6 \pm 0.1) \times 10^{-4}$	traces of HATR initially, traces of dealkylated compounds after 24 h
7	Al_2O_3	Degussa	100	1.0	$(1.7 \pm 0.2) \times 10^{-4}$	initially HATR is the largest predominant intermediate.

(a) ref.20

(b) BET technique.

(c) pseudo first order degradation rate constant k from fits to $C=C_0 exp(-kt)$, where t is the irradiation time and C_0 the initial concentration of atrazine.

The different distributions of intermediates observed when using different metal oxide photocatalysts infer a different selectivity toward specific reaction mechanisms. In a homogeneous solution, as well as in a heterogeneous medium in which atrazine is adsorbed on Al_2O_3, the most relevant degradation pathway is chlorine substitution (see Figure 3). By contrast, dealkylation that implicates hydrogen abstraction and decarboxylation mediated by h^+ or bound hydroxyl radicals, largely prevails in the presence of TiO_2, ZnO and WO_3. Amongst these semiconductor oxides, different distributions of intermediates are observed because of factors (iii) and (iv) above (note, for example, the intermediate compounds CDAT, ammeline and ammelide).

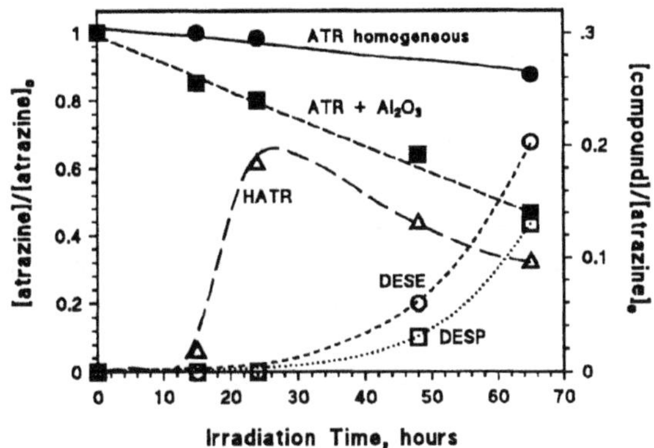

Figure 3. Atrazine degradation in homogeneous solution (full circles) and atrazine degradation (full squares) plus intermediate evolution in the presence of Al_2O_3 (1 g L^{-1}). Acronyms refer to compounds depicted in the Scheme. Other conditions as in Fig.1. Adapted with permission from ref.11.

Table II. Data on different TiO_2 specimens and key results concerning atrazine degradation under irradiation at wavelength > 340 nm.

entry	TiO_2 specimen	cristallinity	surface area (a) m^2 g^{-1}	diameter μm	loading g L^{-1}	kapp (b) min^{-1}
A	Degussa P25	anatase 80%	55	0.3	1.0	0.18 ± 0.01
B	AGIP powder	anatase 90%	43	0.05-0.17	2.0	0.34 ± 0.01
C	AGIP micro sphere	anatase 90%	84	1000	2.9	$(6.8 \pm 0.2)x10^{-2}$
D	colloidal	-	-	0.05	1.0	$(7.6 \pm 1.2)x10^{-2}$

(a) BET technique.

(b) pseudo first order degradation rate constant k from fits to $C=C_0exp(-kt)$, where t is the irradiation time and C_0 the initial concentration of atrazine.

The behaviour of Fe_2O_3 and, in part, of SnO_2 is worth noting. Based on the trend reported in Table I, α-Fe_2O_3 does not behave as a photocatalyst, at least under the conditions used and with respect to atrazine as the probe molecule. This is in accord with earlier suggestions to the effect that α-Fe_2O_3 reacts only in the presence of strong reducing agents or multidentate ligands.[22] The photocatalytic activity of SnO_2 is low under otherwise similar experimental conditions. The distribution of intermediates lies between the distribution observed for Al_2O_3 and the one noted for the semiconductors TiO_2, ZnO and WO_3. This result can be attributed to the selected wavelength of the incident light, which is close to the band gap requirements for excitation.[23]

5. DIFFERENT TiO_2 MATERIALS

Nanometer-size colloids, micrometer-size powders prepared by sol-gel techniques and millimeter-size spherical bead specimens of TiO_2 were tested toward atrazine degradation in order to explore their photocatalytic activity and selectivity.[24] The specimen characteristics are reported in Table II. Some results are given in Table II and illustrated in Figures 4 and 5.

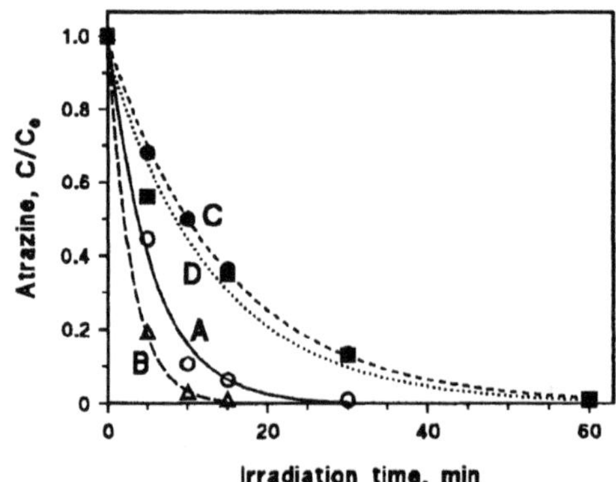

Figure 4. Degradation of atrazine in the presence of different TiO_2 specimens (see Table II for notation). Initial atrazine concentration 25 mg L^{-1}. Loading of TiO_2: (A) 1 g L^{-1}, (B) 2.0 g L^{-1}, (C) 2.9 g L^{-1}, (D) 1 g L^{-1}. Other conditions as in Fig.1. Adapted with permission from ref.24.

The photooxidative degradation process for atrazine has been investigated in the presence of two types of TiO_2 particles that show good photocatalytic activity (samples A and B), TiO_2 beads (sample C) and TiO_2

colloids (sample D). In the presence of particles A and B, the rates of disappearance of ATR are quite similar (comparable half-lives, $t_{1/2}$~ 2 to 3 minutes), whereas a slightly lower degradation rate is observed for colloids ($t_{1/2}$≈ 10 minutes) and the spheres ($t_{1/2}$≈ 8 to 12 minutes). It is remarkable that after several repeated cycles, the activity of the beads remains unchanged as far as atrazine disappearance is concerned.[24]

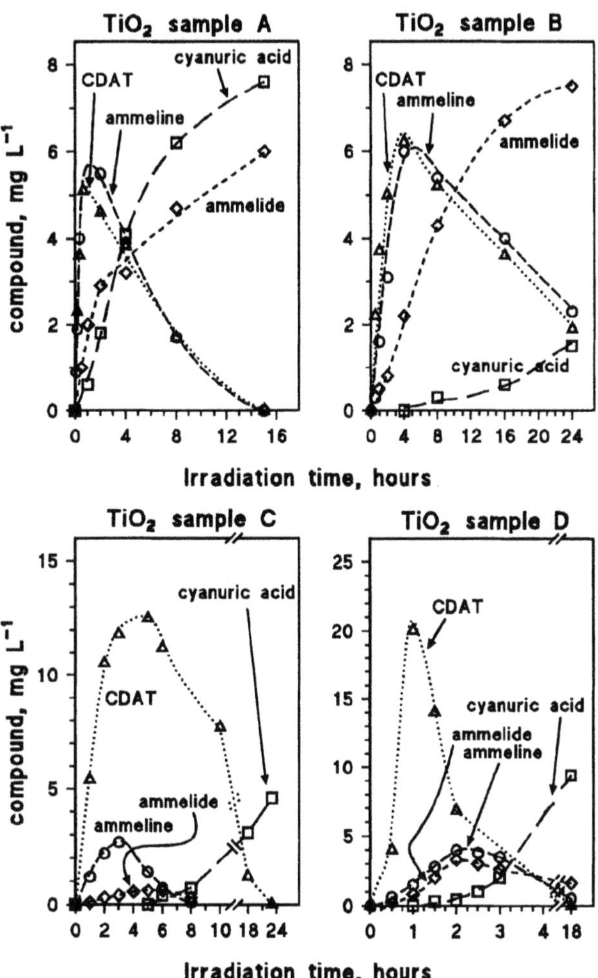

Figure 5. Time evolution of intermediates in the photocatalytic degradation of atrazine in the presence of different TiO_2 specimens (see Table II for notation and Scheme for acronyms). Other conditions as in Fig.4. Adapted with permission from ref.24.

Although the same intermediates and final product (cyanuric acid) are obtained, different trends are observed for the rates of formation and the quantity of intermediates.

A similarity in the shapes of the plots is evident for the powders, but formation of the intermediates and cyanuric acid is sligthly slower with catalyst B relative to catalyst A. The formation of intermediate CDAT is predominant with colloidal TiO_2 and with TiO_2 spheres. The direct photocatalytic transformation of intermediate CDAT yields only small amounts of ammeline and ammelide. It appears that with these two catalyst samples, the dehalogenation pathway is strongly attenuated (Figure 5).

In the photo-oxidation of phenol,[24] the different TiO_2 specimens used showed small but significant variations in activity; however, we found no evidence for variations in product selectivity for the various materials owing to the rather similar nature of the intermediates formed (hydroquinone, catechol, benzoquinone, pyrogallol) and the rather analogous interaction(s) of these with the active oxidizing species ($^{\bullet}OH$ radicals and probably also $O_2^{\bullet -}$).

When atrazine is the test probe, there is a dramatic variation in activity, and a more significant variation in the selectivity as witnessed by the differences in the temporal distributions of intermediate species and product formed (Figure 5) for the different preparations of TiO_2 materials. This is a consequence of the different surface characteristics of the catalyst particles, together with the different modes of interaction between the test probe and the catalyst: (a) dehalogenation, (b) dealkylation, and (c) deamination.

A final point is worth noting. For a given specimen, the differences observed in the activity and selectivity between a virgin sample used for the first time and with one used in subsequent cycles calls attention to (and should alert workers in this area) the different surface properties of catalysts from different batches and to comparisons often made between results from different laboratories. Although this has been suspected in the past, the present results demonstrate the variations.

6. CONCLUSIONS

In the presently investigated systems, when the surface area is considered, TiO_2 still remains the most active material. The half-life $t_{1/2}$ for atrazine is 5 minutes for TiO_2 (55 m^2g^{-1}) and 20 minutes for ZnO (15 m^2g^{-1}) at the same loading (0.5 g L^{-1}). Of greater importance is the process efficiency based on the formation of cyanuric acid. At about 15 hours of irradiation, the yield of cyanuric acid is 52% of the stoichiometric value in the presence of TiO_2; it is only 5% after 24 hours when ZnO is used.

The present report has illustrated several salient features of heterogeneous photocatalysis, which were earlier either implied or suspected: (i) a wide range of photoreactivity is observed within different TiO_2 specimens; (ii) such differences are also observable for specimens with the same crystal structure; (iii) all the investigated specimens lead to the same final products (e.g., CO_2 for phenol [24] and cyanuric acid for atrazine); (iv) in the case of atrazine as the test molecule, a different distribution of intermediate products as well as different rates of formation and disappearance of intermediates are noted, depending on the surface characteristics of the photocatalyst; (v) the different surface/substrate interactions seem to affect the degradation pathways and to influence directly the predominance of dehalogenation relative to alkyl chain oxidation in the case of atrazine; (vi) the reproducibility of the degradation process over several cycles has been demonstrated for TiO_2 beads, although the behavior of a virgin specimen may vary with respect to a specimen that was aged for a given process insofar as the temporal distribution of intermediates and the distribution of rates are concerned (this may be important in the design of active materials to be used in large volume photoreactors); (vii) colloidal TiO_2 is photoactive towards the degradation of organic substrates. This is relevant in fundamental studies of primary events and transient formation [25], as well as for application of colloidal semiconductors in the decontamination of pollutants that are strongly adsorbed on inert materials (e.g., soils) [26].

7. ACKNOWLEDGEMENTS

Generous financial support for this work was received from the European Economic Community, CIEMAT, MURST and CNR Rome. We are also grateful to the North Atlantic Treaty Organization for an exchange grant (NATO Grant No. 890746). We also thank the American Chemical Society and Gauthier-Villars for permission to reproduce Figures and Tables.

8. REFERENCES

1. N. Serpone and E. Pelizzetti, eds., "*Photocatalysis. Fundamentals and Applications*", Wiley, NewYork 1989

2 D. Bahnemann, J. Cunningham, M. A. Fox, E. Pelizzetti, P. Pichat and N. Serpone, in "*Aquatic and Surface Chemistry*", G. R. Helz, R. G. Zepp and D. G. Crosby, eds., Lewis, Boca Rabin, 1994, p. 261

3. L. Zang, C. Y. Lin and X. M. Reu, *J. Chem. Soc. Faraday Trans.*, 1995, **91**, 917.

4. A. M. Mills and S. Morris, *J. Photochem. Photobiol. A: Chem.*, 1993, **71**, 285.

5. V. Chhabra, V. Pillati, B .K. Mishra, A. Morrone and D. O. Shah, *Langmuir*, 1995, **11**, 3307.

6. Y. Xu and C. H. Langford, *J. Phys. Chem.*, 1995, **99**, 11501

7. N. Takeda, T. Torimoto, S. Sampath, S. Kuwabata and H. Yoneyama, *J. Phys. Chem.*, 1995, **99**, 9986.

8. C. Anderson amd A. J. Bard, *J. Phys. Chem.*, 1995, **99**, 9982.

9. E. Pelizzetti, V. Maurino, C. Minero, V. Carlin, E. Pramauro, O. Zerbinati and M. L. Tosato, *Environ. Sci. Technol.*, 1990, **24**, 1559

10. E.Pelizzetti, C. Minero, M. Vincenti, E. Pramauro, V. Carlin and M. Dolci, *Chemosphere*, 1992, **24**, 89

11. E. Pelizzetti, C. Minero and V. Carlin, *New J. Chem.*, 1993, **17**, 315

12. R. Terzian, N. Serpone, C. Minero, E. Pelizzetti and H. Hidaka, *J. Photochem. Photobiol. A: Chem.,* 1990, **55**, 243.

13. J. M. Warmann, M. P. de Haas, P. Pichat and N. Serpone, *J. Phys. Chem.,* 1991, **95**, 8858.

14. Y. Oosawa and M. Gratzel, *J. Chem. Soc. Faraday I,* 1988, **84**, 157.

15. A. Sclafani, L. Palmisano and M. Schiavello, *J. Phys. Chem.*, 1990, **94**, 829.

16. A. Sclafani, L. Palmisano and E. Davì, *New.J. Chem.*, 1990, **58**, 361.

17. G. Hodes and M. Gratzel, *Nouv. J. Chem.*, 1984, **8**, 509.

18. H. Gerisher, *Angew. Chem., Int. Ed. Engl.*, 1988, **27**, 63.

19. J. Cunningham and S. Srijaranai, *J. Photochem. Photobiol. A: Chem.* 1991, **58**, 361.

20. C. Minero, E. Pramauro, E. Pelizzetti, M. Dolci and A. Marchesini, *Chemosphere*, 1992, **24**, 1597

21. P. Pichat in "*Photocatalysis*", N. Serpone and E. Pelizzetti eds., Wiley, New York, 1989.

22. C. Kormann, D. W. Banhemann and M. R. Hoffmann, *J. Photochem. Photobiol. A: Chem.*, 1989, **48**, 683.

23. K. Rajeshwar, P. Sing and J. DuBow, *Electrochim. Acta,* 1978, **23**, 1117.

24. E. Pelizzetti, C. Minero, E. Borgarello, L. Tinucci and N. Serpone, *Langmuir*, 1993, **9**, 2995.

25. M. T. Nenadovic, Z. I. Saponjic, O. I. Micic, C. Minero, E. Pelizzetti, submitted.

26. E. Pelizzetti, C. Minero, V. Carlin, E. Borgarello, *Chemosphere*, 1992, **25**, 343.

NANOCRYSTALLINE ELECTRONIC JUNCTIONS

M.GRAETZEL
Institut de Chimie Physique, Ecole polytechnique fédérale de Lausanne
CH-1015 Lausanne, Switzerland

1. General Features of Nanocrystalline Electronic Junctions

Nanocrystalline semiconductor films are constituted by a network of mesoscopic oxide particles, such as TiO_2, ZnO, Nb_2O_5 which are interconnected to allow for electronic conduction to take place. The pores between the particles are filled with a semiconducting or a conducting medium, such as a hole transmitter or an electrolyte, forming a junction of extremely large contact area. The salient features of such mesoporous films are:

i) an extremely large internal surface area, the roughness factors being in excess of 1000 for a film thickness of 8 microns;
ii) the ease of charge carrier percolation across the nanoparticle network making this huge surface electronically addressable;
iii) a very rapid and highly efficient interfacial charge transfer between the oxide and redox active species anchored to the particle surface. and
iv) the rapid intercalation and release of Li^+ ions into such films.

The unique optical and electronic features of such junctions have already found a number of important applications including photovoltaic injection cells and electrochromic displays, as well as photochromic switches and intercalation batteries. Particularly

E. Pelizzetti (ed.), Fine Particles Science and Technology, 719–732.

intriguing is the observation of close to 100 % conversion of photons in electric current made with films that are derivatized by charge transfer sensitzers adsorbed onto the surface of the oxide. This has led to the development of a new type of photovoltaic cell (1,2) which will be described in more detail in my lecture. Figure 1 gives an overview of the applications foreseen for such mesoscopic oxide semiconductor films. Figure 2 illustrates the morphology of nanocrystalline TiO_2 (anatase) layer deposited on conducting (TCO) glass. The mean particle diameter of the oxide is 15 nm in this case. The size can be adjusted by varying the conditions of the sol gel-type process used for film preparation.

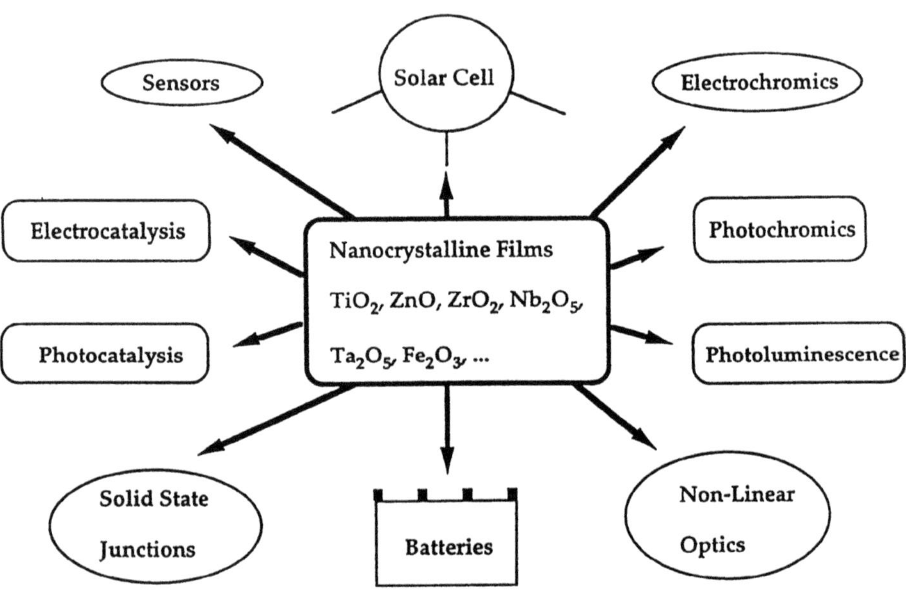

Figure 1. Summary of applications of the new nanocrystalline electronic junctions.

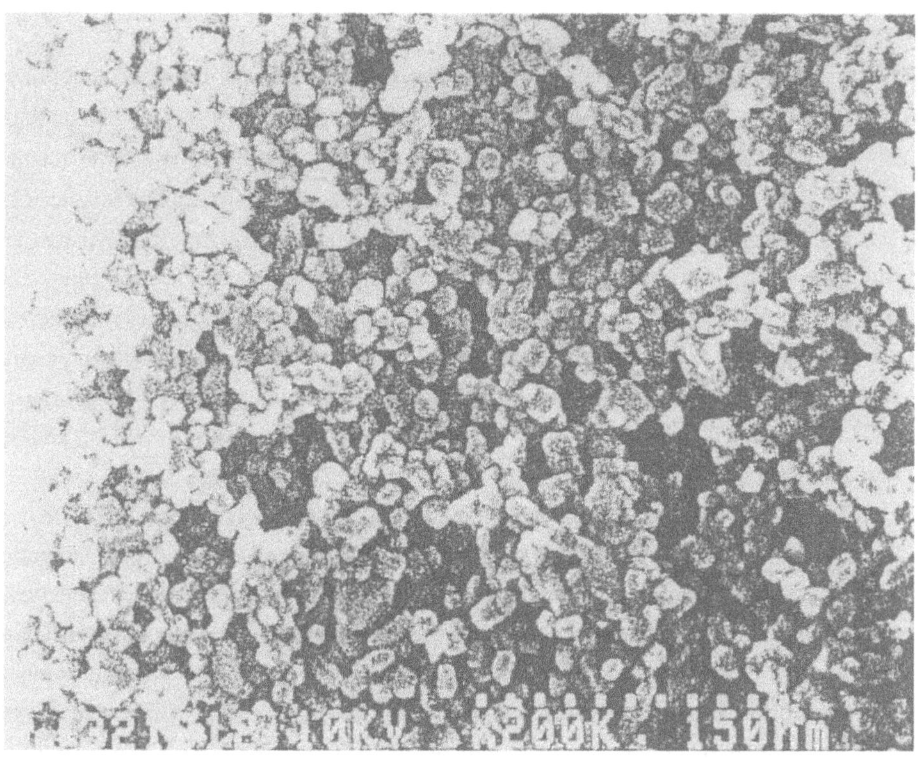

Figure 2. Scanning electron micrograph of a nanocrystalline TiO_2 film

2) The Principle of Operation of the Nanocrystalline Photovoltaic Injection Devices

The fundamental processes involved in any photovoltaic conversion process are:

i) the absorption of sunlight

ii) the generation of electric charges by light, and

iii) the collection of charge carriers to produce electricity

The incident monochromatic photon to current conversion efficiency or "external quantum yield" of such a device is then given by the equation:

$$\eta_i \text{ (l)} = \text{LHE } (\lambda) \times \phi_{inj} \times \eta_e \quad (1)$$

where η_i (λ) expresses the ratio of the measured electric current to the incident photon flux for a given wavelength, LHE is the light harvesting efficiency, ϕ_{inj} is the quantum yield for charge injection into the oxide and η_e is the charge collection efficiency. Conventional solar cells convert light into electricity by exploiting the photovoltaic effect that exists at semiconductor junctions. They are thus closely related to transistors and integrated circuits. The semiconductor performs two processes simultaneously: absorption of light, and the separation of the electric charges (electrons and "holes") which are formed as a consequence of that absorption. However, to avoid the premature recombination of electrons and holes, the semiconductors employed must be highly pure and defect-free. The fabrication of this type of cell presents numerous difficulties, preventing the use of such devices for electricity production on a large industrial scale.

In contrast, the solar cell developed in our group at the Swiss Federal Institute of Technology operates on a different principle, whereby the processes of light absorption and charge separation are differentiated, Figure 3. Light absorption is performed by a monolayer of dye (S) adsorbed chemically at the semiconductor surface. After having been excited (S*) by a photon of light, the dye - usually a transition metal complex whose molecular properties are specifically engineered for the task - is able to inject an electron into the conduction band of the oxide semiconductor. The back reaction is intercepted by transferring the positive charge from the dye (S^+) to a redox mediator R/R^+ present in the electrolyte with which the cell is filled, and thence to the counter electrode. Via this last electron transfer, in which the mediator is returned to its reduced state, the circuit is closed. The system operates as a regenerative electrochemical cell that converts light into electricity without inducing any permanent chemical transformation. The maximum voltage DV that such a device could deliver corresponds to the difference between the redox potential of the mediator and the Fermi level of the semiconductor. The electrolyte containing the mediator could be replaced by a p-type conductor or a hole transmitting solid, such as the amorphous organic compounds used in electroluminescence devices. This is an attractive option which is presently being explored in our laboratory.

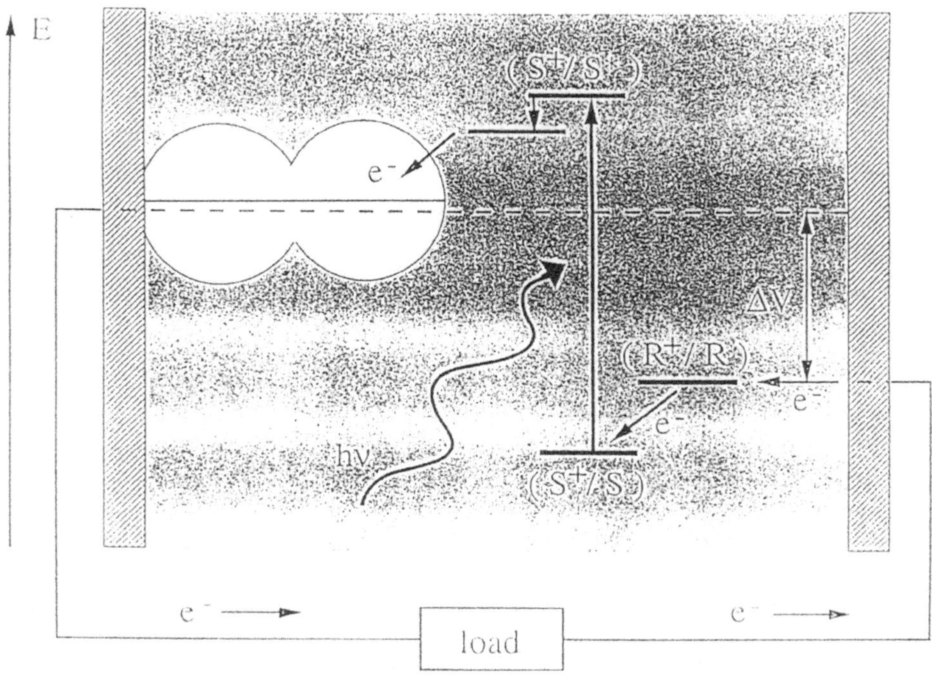

Figure 3. Operation priniciple of the nanocrystalline injection solar cell

The absorption of light by a monolayer of dye is weak due to the fact that the area occupied by one molecule is at much larger than its optical cross section for light capture. A respectable photovoltaic efficiency cannot therefore be obtained by the use of a flat semiconductor surface but rather by use of a porous, nanostructured film of very high surface roughness. When light penetrates the photosensitized, semiconductor "sponge", it crosses hundreds of adsorbed dye monolayers. The mesoporous structure thus fulfills a function similar to the thylakoids in green leaves which are stacked in order to enhance light harvesting.

In contrast to chlorophyll which is continuously regenerated by the leaf, the sensitizer in the nanocrystalline cell must be selected to satisfy the high stability requirements encountered in practical applications. A photovoltaic device must remain serviceable for 20 years without significant loss of performance corresponding to a number of 50 - 100 million turnovers for the dye. Our work has focused on the molecular engineering of suitable ruthenium compounds which are known for their excellent stability. *Cis*-di(thiocyanato)bis(2,2'-bipyridyl)-4,4'-dicarboxylate)-ruthenium(II), **I**, was found to perform as an outstanding solar light absorber and charge-transfer sensitizer [2], unmatched by any other dyestuff known so far. This sensitizer satisfies also the stability requirements for practical applications. The carboxylate groups serve to attach the Ru complex to the surface of the oxide and to establish good electronic coupling between the p* orbital of the electronically excited complex and the 3d wavefunction manifold of the TiO_2 film. Using **I** in conjunction with nanocrystalline TiO_2 films, solar cells have been fabricated for which all three factors in equation (1) are close to unity.

3) Conversion Efficiencies

A graph which presents the monochromatic current output as a function of the wavelength of the incident light is known as a "photocurrent action spectrum". Figure 4 shows such action spectra for three ruthenium complexes, illustrating the very high efficiency of current generation, exceeding 75% obtained with these complexes. When corrected for the inevitable reflection and absorption losses in the conducting glass serving to support the nanocrystalline film, yields of practically 100 percent are obtain. Historically, RuL_3 (L = 2,2'-bipyridyl-4,4'-dicarboxylate) was the first efficient and stable charge transfer sensitizer to be used in conjunction with high surface area TiO_2 films. In a long term experiment carried out during 1988 it sustained 9 months of intense illumination without loss of performance. However, the visible light absorption of this sensitizer is insufficient for solar light conversion. A significant improvement of the light harvesting was achieved with the trimeric complex of ruthenium whose two peripheral ruthenium moieties were designed to serve as antennas [3]. However, the most successful charge transfer

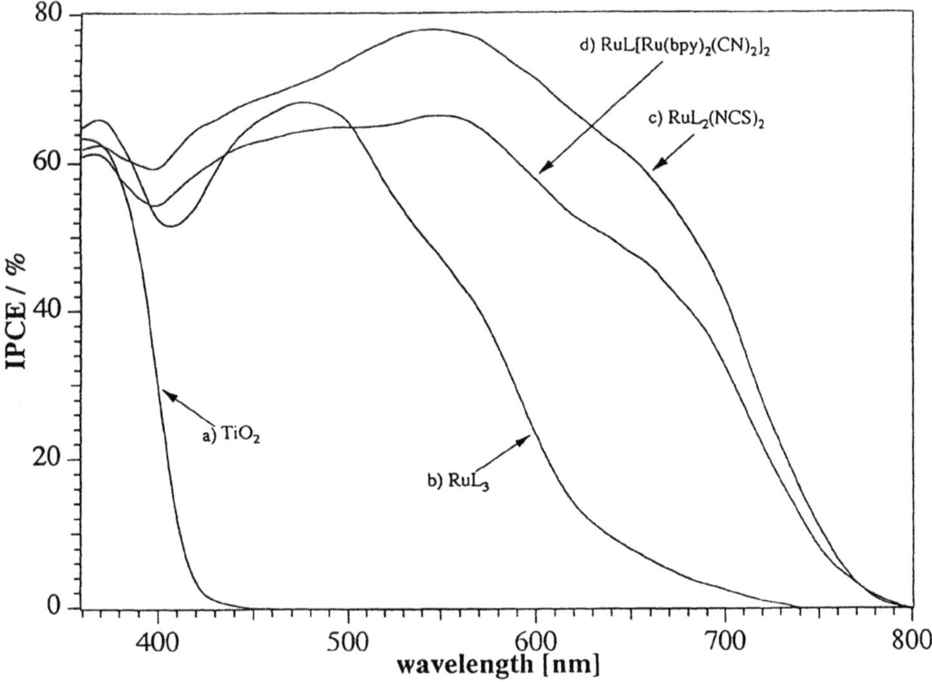

Figure 4. Photocurrent action spectrum obtained with three sensitizers and TiO_2 alone

sensitizer investigated so far is without any doubt *cis*-dithiocyanatobis(2,2'-bipyridyl-4,4'-dicarboxylate)-ruthenium(II). This achieves close to quantitative photon to electron conversion over the whole visible range (2). Even at 700 nm current generation is still 40 -60 percent efficient depending on the thickness of the film.

The overall efficiency (h_{global}) of the photovoltaic cell can easily be calculated from the integral photocurrent density (i_{ph}), the open-circuit photovoltage (V_{oc}), the fill factor of the cell (ff) and the intensity of the incident light (I_s).

$$(2) \qquad \eta_{global} = i_{ph} \times V_{oc} \times ff/I_s$$

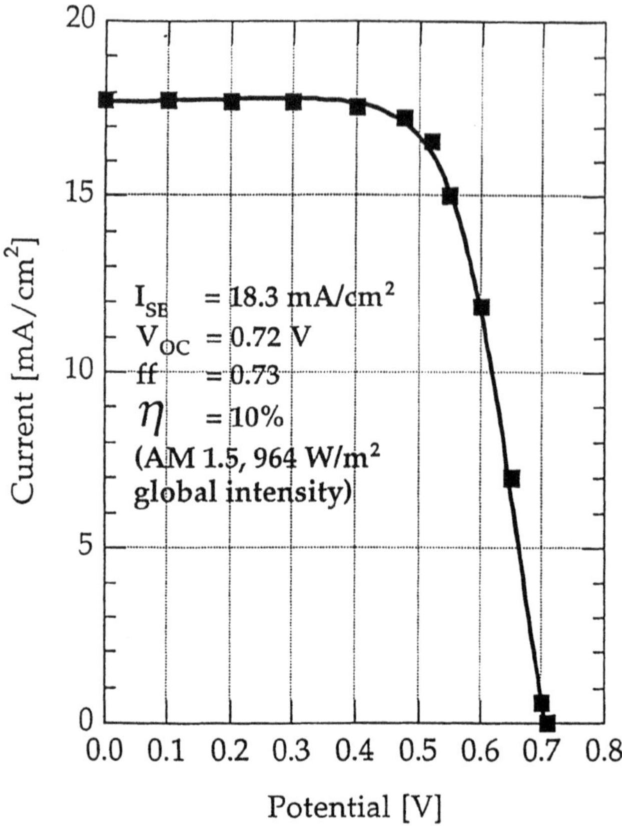

Figure 5. Photocurrent voltage characteristic of a nanocrystalline injection cell

Figure 5 illustrates the current-voltage characteristic of a nanocrystalline cell sensitized by **I**. The photocurrent measured at 96.4 mW/cm^2 of simulated AM 1.5 global solar intensity was 18.3 mA/cm^2 the open circuit voltage was 0.72 V, and the fill factor was 0.73 yielding for the overall efficiency of the cell a value of 10%. This is significantly below the value of 33 percent corresponding to the upper limit for a quantum converter that is based on a single light absorber. The main reason for the difference is the mismatch in the redox level of the dye and that of the iodide/triiodide redox system used as electrolyte, leading to a voltage loss of 0.7 V. Adjusting the redox levels to reduce this loss to a more reasonable figure of 0.3 V would allow to double the overall conversion

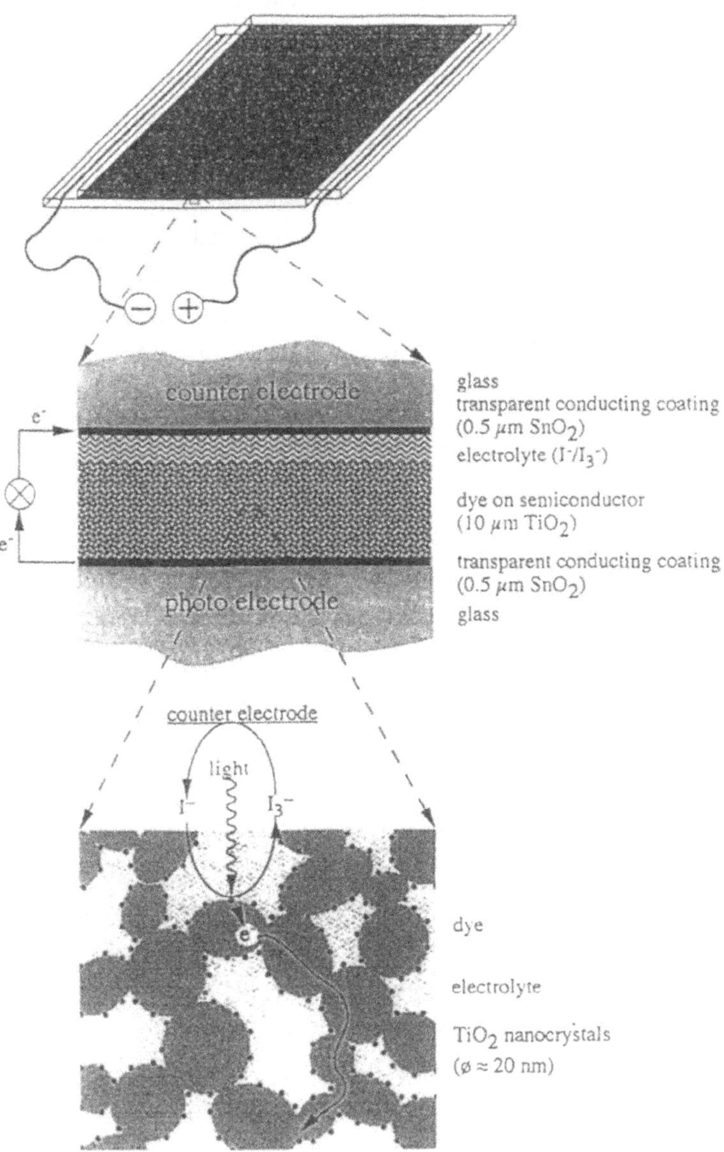

Figure 6. **Layout of the nanocrystalline injection solar cell**

efficiency from 10 to 20 percent. A further improvement of the cell efficiency by ca 30 % should be possible through better light harvesting in the 700 to 800 nm range where the absorption of **I** is relatively weak. Thus it is feasable to reach with nanocrystalline materials efficiencies around 25 % which falls in the same range as that observed with top quality, single crystal GaAs solar cells.

An advantage of the nanocrystalline solar cell with respect to solid state devices is ta temperature increase form from 20 to 60 °C has practically no effect on its conversion efficiency. In contrast conventional silicon cells exhibit a significant decline by over the same temperature range amounting to ca 20 percent. Since the tempreraturе of a solar cell will reach about 60 °C under full sunlight this feature of the new cell is particularly attractive for power generation under natural conditions. At this stage industrial companies are already involved in production of such cells which will reach the market quite soon.

4) Stability, Cost and Environmental Compatibility of the New Solar Cell

Meanwhile, the development and testing of the first cell module for practical applications has begun. The lay out of the module is presented in Fig. 6. The cell consists of two glass plates, which are coated with a transparent conducting oxide (TCO) layer. The nanocrystalline titanium dioxide film deposited on one plate functions as a support for the ruthenium complex acting as charge transfer sensitizer. On illumination, this injects an electron into the titanium dioxide conduction band. The electrons pass over the collector electrode into the external current where they perform work. They are then returned to the cell via the counter electrode. The sensitizer film is separated from the counter electrode by the electrolyte.containing the redox couple, for example triiodide/iodide, whose role is to transport electrons from the counter electrode to the sensitizer layer. A small amount of platinum (5-10 mg/cm^2) is deposited onto the counterelectrode to catalyze the cathodic reduction of triiodide to iodide.

Upon long time illumination, complex **I** sustained more than 5×10^7 redox cycles without noticeable loss of performance corresponding to at least 12 years of continuous operation in natural sunlight. By contrast, practically all organic dyes tested so far underwent photo-bleaching after less than 10^4 redox cycles. This clearly outlines the exceptionally stable operation of the newly developed transition metal charge transfer sensitizers which is of great advantage for the practical application of these devices. The reason for this amazing stability is the very rapid deactivation of the excited triplet state of via charge injection into the TiO_2 which occurs in the femtosecond time domain. This is many orders of magnitude faster than any other competing channel of excited state deactivation including those leading to chemical transformation of the dye.

Stability test on sealed cells have progressed significantly over the last few years. These tests are very important, since - apart from the sensitizer - other components of the device, such as the redox electrolyte or the sealing, may fail under long term illumination. Indeed, a problem emerged with electrolytes based on cyclic carbonates, such as propylene- or ethylene carbonate, which were found to undergo thermally activated decarboxylation in the presence of TiO_2 rendering these solvents unsuitable for practical usage. They were therefore replaced by ionic liquids which are harmless and nonvolatile and do not exhibit this undesirable property. More recently, room temperture molten salts based on imidazolium iodides and triflates have revealed very attractive stability features. Thus, fully assembled cells showed no decline in photovoltaic performance, i.e. photocurrent, photovoltage and fill factor, when submitted to accellerated aging performed in a sun test (AM 1) chamber at 44°C and 85°C for at least 2300 hours and 1000 hours respectively. Direct excitation of electron hole pairs in the anatase by l > 380 nm light was avoided in these experiments by using a polycarbonate protective film.

It may be argued that the presence of ruthenium renders the price of the sensitizer too high for commercial exploitation or that there is insufficient supply of it. However, the required amount of ruthenium is only 10^{-3} moles/m^2 of ruthenium complex corresponding to the small investment of ca US $ 0.05/m^2 for the noble metal. The world trade in ruthenium reached 10 tons in 1994 mainly due to its use as dimensionally stable anode in electrochemical chlorine production. One ton of ruthenium alone incorporated in the charge transfer sensitizer **I** could provide one gigawatt of electric power under full sunlight. This is more than twice the total photovoltaic capacity presently installed worldwide. Thus the cost and supply of ruthenium based sensitizers is of no real concern here. The price determining factor for this photovoltaic technology is undoubtedly the conducting glass. Several industrial sources, including Asea Brown Boveri, Strategies Unlimited and Research Triangle Institute have performed a cost analysis which yields 0.6 US$/pW and 2000 US$/kWp for module and total systems cost, respectively. Similar values were reported by Smestad[4].

Apart from efficiency and stability any future photovoltaic technology will be valued according to its environmental and human compatibility. There is great concern about the adverse environmental effects and acute toxicity of CdTe or $CuInSe_2$ which are being considered for practical development as thin solar cells. Such concerns are unjustified for our nanocrystalline device. Titanium dioxide is a harmless environmentally friendly material, remarkable for its very high stability. It occurs in nature as ilmenite, and is used in quantity as a white pigment and as an additive in toothpaste. Worldwide annual production is in excess of 1 million tons. Similarly ruthenium has been used without adverse health effects as an additive for bone implants. Quite aside form its intrinsic merits as a photovoltaic device, the sensitized nanocrystalline photovoltaic device will undoubtedly promote the acceptance of alternative energy technologies, not least by setting new standards of convenience and economy for the photovoltaic industry as a whole.

5) Nanocrystalline Intercalation Batteries

The diffuse and intermittant nature of sunlight renders necessary the storage of solar energy in electrical or chemical form. There is presently a thrive for improved batteries effecting electric energy storage. In this context the so-called "rocking chair" batteries deserve particular attention. Electric power generation is associated with the migration of lithium ions from one host , i.e. TiO_2 , constituting the anode, to another host electrode, i.e. NiO_2/CoO_2 or MnO_2 , constituting the anode. The voltage delivered by the device is simply the difference of the chemical potential of lithium in the two host materials. It was discovered in our joint research with Dr. Exnar from Renata Corporation and Dr. Randin from Asulab that the intercalation of lithium is greatly facilitated if the oxides employed have a nanocrystalline morphology as compared to bulk electrodes, Figure 6. Thus, for TiO_2 the capacity for Li^+ uptake is increased by a factor of roughly two and the intercalation and the release of the ions is much faster than for bulk materials. The reason for this behavior is that the diffusion time for lithium ions in the host oxide is dramatically shortened by using oxide particles of mesoscopic dimensions as electrodes. A standard size R921 coin cell has been developed supplying 4 to 4.5 mAh corresponding to 50 mAh/g capacity which compares well with the rocking chair battery having a carbon anode. These findings provide a very promising basis for the development of a new type of rechargeable battery [5].

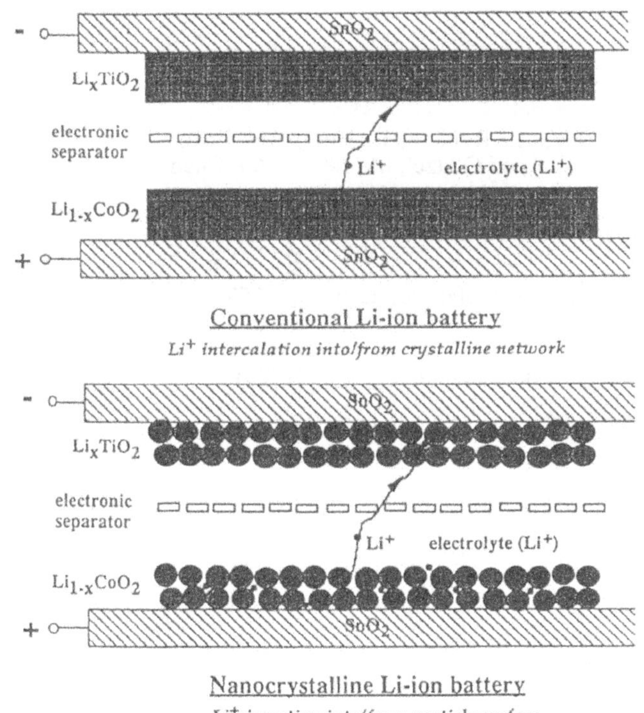

Figure 7. Nanocrystalline and conventional rocking chair batteries

In conclusion it appears that nanocrystalline electronic junctions involving transition metal oxides form not only the heart of a new photovoltaic device but offer also attractive perspectives for the storage of electrical energy that is generated by sunlight.

6. Acknowledgement

Support of this work by the Institut für Angwandte Photovoltaic (INAP), Gelsenkirchen, Germany and the Commission pour l'encouragement de la recherche scientifique (CERS) is gratefully acknowledged.

7. References

1) O'Regan,B. and Grätzel,M.(1991) Nature (London) 335, 737.

2) Nazeeruddin,M.K.; Kay, A.; Rodicio, I.; Humphrey-Baker, R.; Müller, E.; Liska, P.; Vlachopoulos, N.and Grätzel, M.(1993) J:Am Chem Soc. 115, 6382

3) Amadelli, R.; Argazzi, R.; Bignozzi, C. A, and Scandola, F.(1990) J.Am. Chem. Soc. 112 , 7029

4) Smestad, G. (1994) Sol Energy Mat. Sol. Cells 32, 259.

5) Huang, S.Y.; Kavan,K.; Exnar, I.and Grätzel, M.(1995) J.Electrochem.Soc.142, L142

THE FORMATION AND PROPERTIES OF NANO-SIZED SEMICONDUCTOR PARTICLES IN LANGMUIR-BLODGETT FILMS

[a]Franz Grieser, [b]D. Neil Furlong, [b]Robert S. Urquhart and [b]David J. Elliot
[a]Advanced Mineral Products Research Centre, School of Chemistry, University of Melbourne, Parkville, 3052. Australia. [b]CSIRO Division of Chemicals and Polymers, Private Bag 10, Rosebank MDC, Clayton, 3169. Australia

1. Introduction

The *insitu* formation of Q-state semiconductor particles in Langmuir-Blodgett (LB) films is but one of many ways these particles can be synthesised. LB films not only act as a matrix that limits the size of the particles that are formed, but also act as a "vehicle" by which these particles can be deposited onto substrates of interest. LB films themselves may contain molecules of specific functional value that may complement certain aspects of the inorganic particles imbedded in the layered structure. In addition, LB films can be constructed consisting of as little as one monolayer, very nearly the lower limit of miniaturisation. This is a very attractive feature considering the commercial markets available for ultra small devices.

This review summaries our present state of knowledge of the formation of these particles in LB films, the optical characteristics of the particles, the type of photoelectrochemical cells that can be made utilising these particle-films, and possible future applications of ultra thin films containing inorganic nanocrystallites.

2. Q-state particle formation in LB films

The deposition of ordered metal ion-fatty acid films onto selected substrates using the LB technique is a convenient method for creating either mono- or multi-layer thin organic films. The technique involves spreading a solution of a long chain fatty acid, such as arachidic acid (ArH), onto an aqueous subphase containing metal ions. If the subphase is at an appropriate pH, ion exchange occurs between the protons of the carboxylic group of the fatty acid and the metal ions in solution, leaving a monolayer of the metal alkyl carboxylate at the air-water interface. This

E. Pelizzetti (ed.), Fine Particles Science and Technology, 733–744.

layer is usually compressed to achieve close packing of the monolayer before transfer is initiated by either passing a substrate through the monolayer or by horizontally contacting a substrate with the air-water monolayer. By repeating this process a number of times, a multilayer film can be constructed.

Once such an LB film has been formed it is now possible to react the metal ion in the film with a suitable complexing agent. This can be achieved by either exposing the film to a reactive gas, e.g., H_2S, or by dipping the film into an aqueous solution containing a complexing ion, e.g., I^-. The interesting result that is obtained by doing either of the above, is the formation of Q-state particles in the LB matrix. Where information is provided, the particles are usually in the size range of 2-3 nm diameter. A number of different Q-state semiconductor particles in LB films, have been made to date and a summary is given in Table 1.

TABLE 1: Type and absorption onset of Q-state semiconductor particles in LB films

Type	Absorption onset	References
CdS	450 nm	1, 2, 3
CdSe	600 nm	3
CdTe	740 nm	3
HgS	375 nm	4, 5
CuS	--	6, 7, 8
PbS	640 nm	7, 9, 10
CoS	--	11
CdI_2	--	12,13
$CdBr_2$	--	13
$CdCl_2$	--	13
PdS	550 nm	14
PtS_2	485 nm	14

It has also been suggested that sheets of semiconductor material is formed between the fatty acid bilayers of the LB films. However, this is usually deduced indirectly from ellipsometry and X-ray or electron diffraction data. The one exception seems to be the cadmium dihalides[13]. Direct microscopy of the metal sulfides in the films always reveals particles which are of the size expected from the wavelength of the absorption spectrum onset of the films[15, 16].

Our work has primarily focused on Cd chalcogenides, predominantly in arachidic acid films. The reaction that occurs in the LB film when exposed to a chalcogenide gas , e.g. H_2S can be written as:

$$[CH_3(CH_2)_{18}CO_2^-]_2Cd^{2+} + H_2S \rightarrow 2CH_3(CH_2)_{18}CO_2H + CdS \quad ....(1)$$

The reaction as shown is essentially quantitative, after about an hour of exposure of a multilayer film (up to 40 layers) to an atmosphere of H_2S[15,17].

A close examination of this reaction by grazing angle FTIR spectroscopy and using a Quartz Crystal Microbalance (QCM) to measure the weight change of the film with reaction time, supports reaction (1)[15,18]. The IR spectra clearly show a decrease in the $\nu_s(COO^-)$ band at 1433 cm^{-1} with a growth in signals at 1708 cm^{-1} and 3130 cm^{-1}, due to the carbonyl stretch $\nu(CO)$ and hydroxyl stretch $\nu(OH)$ of a protonated carboxylic acid, respectively, with time of H_2S exposure. In addition, IR spectra show a series of progression bands below 1400 cm^{-1} typical of ArH. The spectra also reveal that even after extensive exposure to H_2S some of the $\nu_s(COO^-)$ band remains. Our results using a QCM suggest that the amount of arachidate remaining after gas exposure corresponds to about a monolayer covering of the particles in the film[19]. This is an important result because it provides evidence that one of the factors that limits the growth of particles in the film is "surfactant capping" of the particle surface.

Reaction (1) is then the first step in the creation of molecules of CdS; these then must diffuse in the LB lattice to some trapping site from which particle growth develops. What these sites are is still not known, although it is not unreasonable to suspect high energy grain boundary sites. Limits on particle growth must then be dependent on surfactant adsorption which limits the rate molecular CdS adds onto a nascent nuclei. This is quite similar to what is seen in bulk solution particle formation. Another unknown in the mechanism of particle growth in the films is whether or not the molecular CdS, that must be formed in the hydrophilic microdomains of the head group regions within the LB matrix, can diffuse through the alkyl bilayers. We are currently working on experiments to throw more light on this possibility.

The answer to what primarily controls the particle size in an LB film is not yet known. Our experiments in varying the amount of Cd^{2+} in the film, the alkyl chain length of the fatty acid and the temperature over the range of about 2°C and 30°C, showed no effect. Some effect was however noted when the composition of the film was changed[20]. The work of Moriguchi et al.[21] put this last observation onto a more quantitative basis. They found that systematically changing the head group size brought about a systematic change in the bandgap of the CdS produced. The biggest bandgap was observed for the smallest headgroup area.

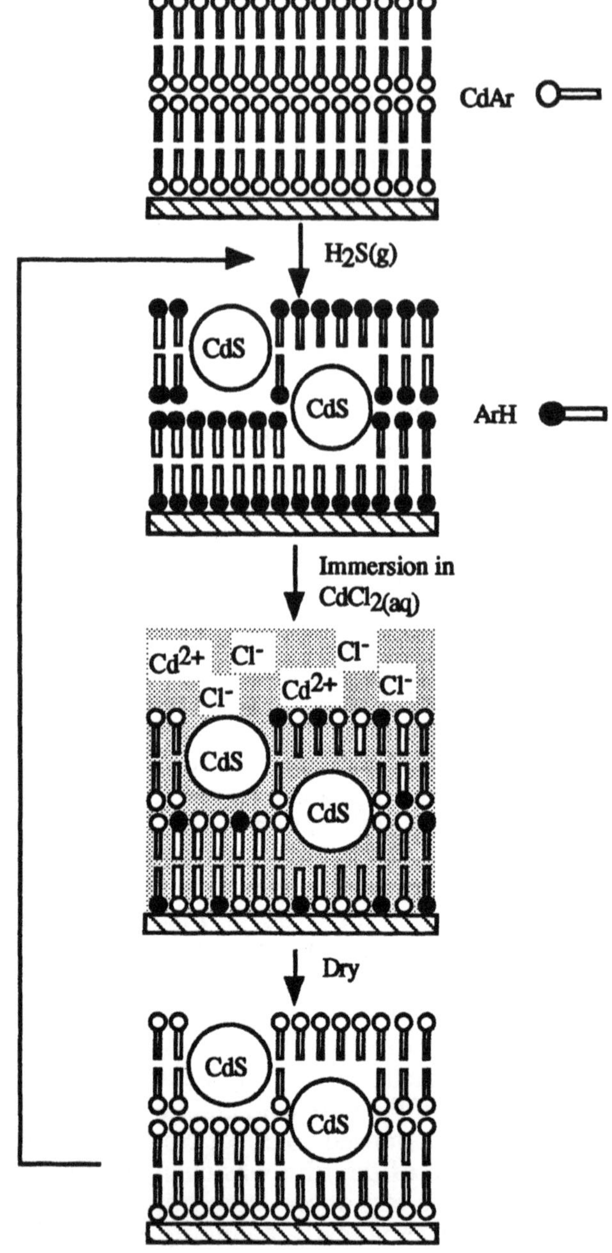

Figure 1. A schematic diagram showing the sequence by which CdAr LB films are converted to ArH films containing Q-state CdS particles.

3. Growth of Q-state particles in LB films

Once the particles have been formed in the LB film they remain stable for months showing no signs of growth, a feature which can be easily monitored via their absorption spectra. The particles can however be used as "seeds" onto which more semiconductor material can be deposited. The method by which this is achieved is shown schematically in Figure 1.

If the LB-particle film is immersed into a solution containing metal ions that ion exchange with the fatty acid protons, and this film is then reexposed to H_2S gas, particle growth occurs[2, 16, 20]. This is clearly seen from a red shift in the absorption spectrum onset and from the particle size histograms of films exposed to a different number of immersion-gas exposure cycles, Figure 2. Particles can be grown until bulk semiconductor properties are spectroscopically observed[20].

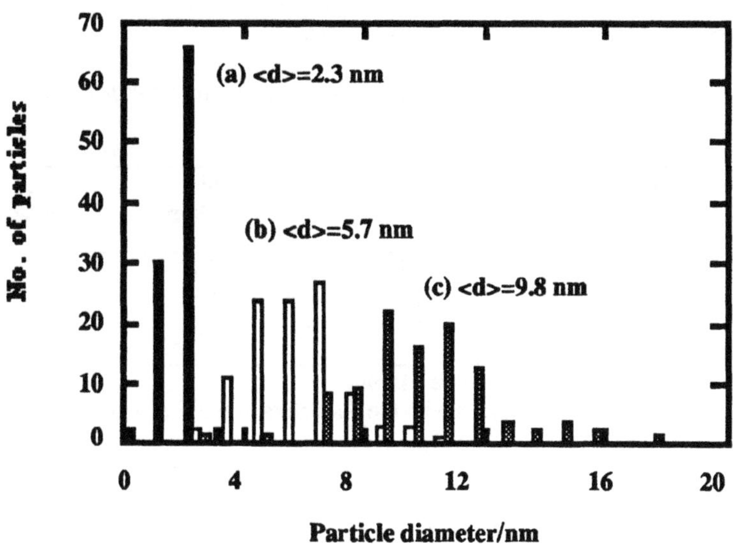

Figure 2. CdS particle size distribution obtained from AFM images of samples for which the LB film had undergone a cycle of being dipped into an aqueous Cd^{2+} solution and then exposed to H_2S gas.(a) a CdAr film exposed to H_2S; (b) 3 dipping and gas exposure cycles; (c) 5 dipping and gas exposure cycles.

4. Core-shell Q-state particles in LB films

The manipulation used to grow the particles in the film illustrates that even though the LB matrix limits the growth of the particles in the film it does not seem to prevent further chemical reaction of the semiconductor particles. Another example that illustrates the reactivity of the particles is the conversion of the outer layer(s) of CdS colloids to CdSe by exposing a CdS-LB film to H_2Se. We have recently shown that exposing different sized CdS particles in an arachidic acid film to H_2Se yields the same optical absorption spectrum with virtually the same absorption onset wavelength, irrespective of the initial CdS particle size[22]. Table 2 presents some of the parameters measured for these systems. An XPS elemental analysis of the 5.7 nm diameter particles gave their composition as $CdS_{0.4}Se_{0.6}$. Taking the case that Se^{2-} only replaces S^{2-} at the surface of the particle, the stoichiometry means that between 1-2 layers of CdSe exist at the surface of the particle, i.e., a core shell particle is formed with a shell thickness of about 0.7 nm. Evidence that such a thin layer is present on a core of CdS, is obtained when comparing the absorption onsets in Table 2 with the data of Figure 3.

TABLE 2. Absorption onset wavelengths (nm) of CdS/LB films exposed to $H_2Se(g)$

diameter(nm)	initial bandgap	final bandgap	Pure CdSe bandgap(b)
2.3	450	520	550
5.7	480	510	700
9.8	510	510	>700

(a) bandgap energies are ±10 nm
(b) for CdSe particles with the diameters in the first column. Bandgaps were calculated using Fig.3

As can be seen from this information, the CdSe film must act as an equivalent particle of <1 nm radius.

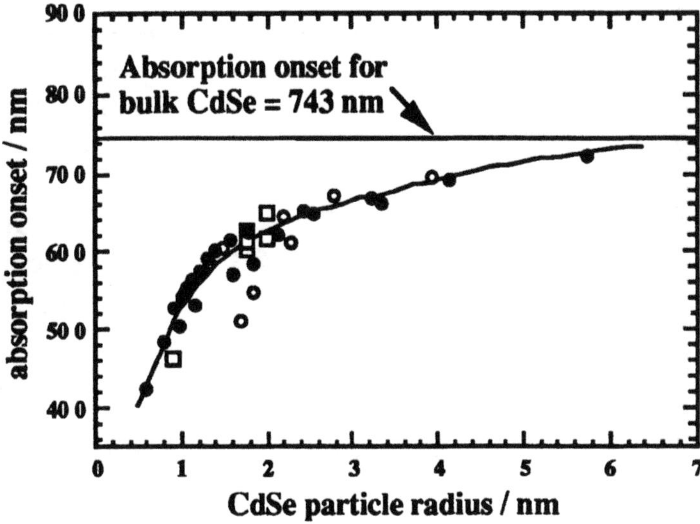

Figure 3. Absorption onset versus particle radius for Q-state CdSe particles in a variety of systems. Data taken from references 23 and 24.

An experiment that we haven't yet carried out but one which would lead to the formation of an "onion ring" like Q-state particle would be to take a film containing the CdS:CdSe material and process this through the cycle shown in Figure 1. This in principle, would produce a CdS:CdSe:CdS "sandwich" particle. Continuing the cycle and alternating the gas between H_2S and H_2Se would then produce a multilayered particle.

5. Photoelectrochemical (PEC) cells using Q-state particles in LB films

We have recently constructed a PEC cell consisting of a layered glass/ITO-CdS-LB plate as the photo anode and have examined its photoelectrochemical characteristics. What we had anticipated was a high photovoltage for small Q-state CdS particles which would progressively approach that of a macroscopic CdS electrode as the particle size increased. This expectation was simply based on a decreasing bandgap energy (hence conduction band potential) with increasing particle size. The experimental observations were in fact in the opposite order as can be seen in Figure 4. The observed behaviour can be qualitatively explained by considering Figure 5.

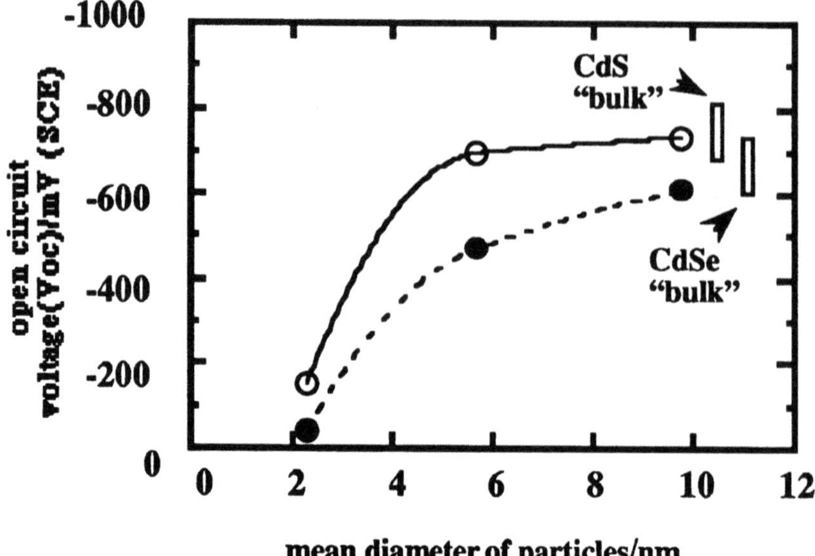

Figure 4. The open circuit voltages as a function of CdS (o) and $CdS_xSe_{(1-x)}$ (•) particle size in 19 layer arachidic acid LB films.

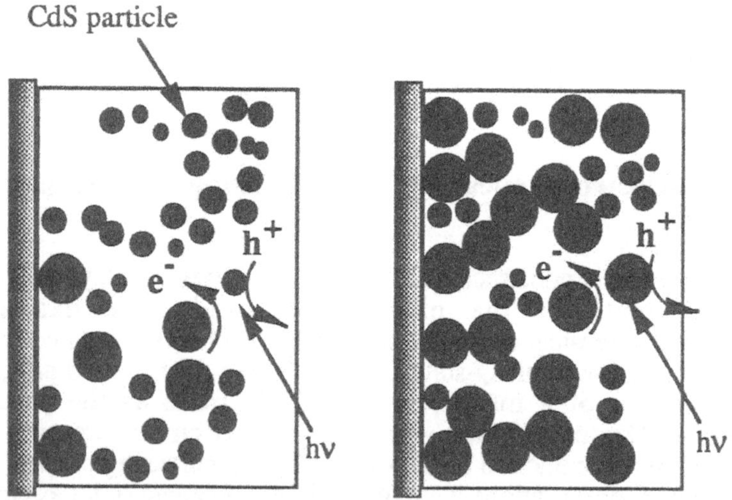

Figure 5. Detail of two different sized Q-state CdS particles in a LB film showing that the larger sized particles are more interconnected than the small sized particles.

It can be assumed that the photogenerated hole in the semiconductor is almost immediately scavenged by the electrolyte solution due to the relatively high concentration of reductant. (The electrolyte was 1.0 mol dm^{-3} Na_2SO_3). On the other hand the photo generated electron must move through the resistive LB film to reach the conductive ITO layer to complete the electrical circuit. This process results in an energy loss in the electron and hence a relatively lower effective photovoltage. As the CdS particle size is increased more particles may interlink, forming lower resistance conduits, hence resulting in less energy being lost in the electron path.

This explanation in the V_{OC} behaviour with particle size has some support in the trends observed for the CdS:CdSe system. For the latter system, the absorption characteristics suggest that the conduction band energy is the same for all the 3 different particle size films examined (see Table 2). In this case one would expect the measured V_{OC} potentials to be the same for the 3 films. However, as can be seen in Figure 4, the variation in the V_{OC} with increasing particle size in the film follows the same trend as observed for the CdS system. Therefore, the increase in the V_{OC} must be due to a less resistive pathway for the electron, stemming from, as suggested above, the increasing interconnection of the particles in the film.

What remains to be explored is the photoelectrical behaviour in more conductive films to see if the high potentials of the smallest Q-state particles can be realized. We are currently looking at impregnating the LB films with polypyrrole to achieve this goal.

6. Future Applications

The three main features of the LB-particle films that make them attractive systems to explore can be summarised as follows:

(a) Their limiting thinness lends itself well to miniaturisation applications particularly in areas involving chemical and biochemical sensors.

(b) The LB matrix, which in some ways can be viewed as an "organic swelling clay", provides a unique reaction zone which allows a host of particle types to be synthesised over a range of particle sizes. Indeed, it may be possible to control the morphology of the inorganic materials synthesised, creating not just spherical particles but also "sheets" or 'nano-wires'. To date however, the only convincing evidence that non- spherical particle can be formed in an LB film comes from the Cd halide group[13].

(c) The organic film itself provides an added dimension in increasing the functional qualities of the inorganic-organic film structure. This feature has yet to be taken advantage of, but it is not hard to see how certain functional groups could alter the characteristics of the film to take advantage of the properties of the inorganic crystallites imbedded in the matrix.

This review has concentrated exclusively on semiconductor particles in fatty acid films because these system have been paid the most attention. It is worth noting that because LB films are not restricted to just fatty acids advantage can be taken of other functional head groups to incorporate a host of water soluble complexes. To illustrate this, we have recently examined the uptake of $AuCl_4^-$ and Pt(II) and Pd(II) complexes into dimethyldioctadecylammonium bromide films[14]. Reduction of the complexes has produced 3 nm diameter metal particles. We have also been able to react $TiCl_4$ with an arachidic acid film which when hydrolysed produced TiO_2 particles having a band gap of about 340 nm[25]. Films produced with mixed particles is something that has not been done, but can be expected to be relatively easily constructed.

Finally, an area that has yet to be investigated is the use of particle-films as nonlinear optical components[26]. Since the particles can be manipulated both in size and composition and the film material can also be varied it may be anticipated that these type of systems will find some use in the laser optics field.

7. Acknowledgements

We thank the Australian Research Council for support and we acknowledge the input from our collaborators, Thomas Gengenbach, Nobuo Kimizuki, Toyoki Kunitake, Catherine Hoffman, John Rabolt, Kentaro Tanaka, Yoshio Okahata, Norman Geddes, Herman Mansur, Simon Biggs, and David Scoberg.

8. References

1. Smotkin, E.S., Lee, C., Bard, A.J., Campion, A., Fox, M.A., Mallouk, T.E.,Webber, S.E., and White, J.M. (1988) Size quantization effects in CdS films formed by a LB technique, *Chem. Phys. Lett.* **152,** 265-268.
2. Moriguchi, I., Tanaka, I., Teraoka, Y., and Kagawa, S. (1991) Production and growth of size quantized CdS in hydrophilic interlayer of LB films, *J. Chem. Soc., Chem. Commun.* 1401-2.
3. Scoberg, D.J., Grieser, F., and Furlong, D.N. (1991) Control of size and composition of chalcogenide colloids in LB films, *J. Chem. Soc., Chem. Commun.* 515-517.
4. Zylberajch, C., Ruaudel-Teixier, A., and Barraud, A. (1989) Properties of inserted mercury sulfided single layers in a LB matrix, *Thin Solid Films* **179,** 9-14.
5. Elliot, D.J., Furlong, D.N., Gengenbach, T., and Grieser, F. (1995) Preparation of HgS in LB films of mercury(II)-arachidate and mercury(II)-behenate, *Colloids and Surfaces* (in press)
6. Leloup, J., Ruaudel-Teixier, A., and Barraud, A. (1992) Electrical properties of copper sulfide semiconductors inserted in LB films, *Thin Solid Films* **210/211,** 407-409.

7. Ruaudel-Teixier, A., Leloup, J., and Barraud, A. (1986) Insertion compounds in LB films, *Mol. Cryst., Liq. Cryst.* **134**, 347-354.
8. Chen, H., Chai, X., Wei, Q., Jiang, Y., and Li, T. (1989) X-ray photoelectron spectroscopy study of the synthesis of copper sulfide in LB films, *Thin Solid Films* **178**, 535-540.
9. Peng, X., Guan, S., Chai, X., Jiang, Y., and Li, T. (1992) Preparation and structure of Q-state lead sulfide monolayers in metastable stearic acid LB films, *J. Phys. Chem.* **96**, 3170-3174.
10. Peng, X., Lu, R., Zhao, Y., Qu, L., Chen, H., and Li, T. (1994) Control of distance and size of inorganic nanoparticles by organic matrices in ordered LB monolayers, *J. Phys. Chem.* **98**, 7052-7055.
11. Luo, X., Zhang, Z., and Liang, Y. (1994) Structure of cobalt stearate and cobalt stearic acid LB films, *Langmuir* **10**, 3213-3216.
12. (a)Pike, J.K., Byrd, H., Morrone, A.A., and Talham, D.R. (1993) Template-directed synthesis: orientated CdI_2 prepared in a LB film, *J. Am. Chem. Soc.* **115**, 8497-8488.(b)Pike, J.K., Byrd, H., Talham, D.R., and Morrone, A.A. (1994) X-ray photoelectron spectroscopy, attenuated total reflectance FTIR spectroscopy and transmission electron diffraction studies of orientated CdI_2 prepared in a LB template, *Thin Solid Films* **243**, 510-514.
13. Pike, J.K., Byrd, H., Morrone, A.A., and Talham, D.R. (1994) Orientated cadmium dihalide particles prepared in LB films, *Chem. Mater.* **6**, 1757-1765.
14. Elliot, D.J., Furlong, D.N., Gengenbach, T.R., Grieser, F., Urquhart, R.S., Hoffman, C.L., and Rabolt, J.F. (1995) Reactions of complex ions of platinum and palladium in LB films of dimethyldioctadecylammonium-chlorometallates, *Colloids and Surfaces* (in press)
15. Urquhart, R., Furlong, D.N., Mansur, H., Grieser, F., Tanaka, K., and Okahata, Y. (1994) Quartz Crystal Microbalance and UV-vis absorption study of Q-state CdS particle formation in cadmium arachidate LB films, *Langmuir* **10**, 899-904.
16 Mansur, H.S., Grieser, F., Marychurch, M.S., Biggs, S., Urquhart, R.S., and Furlong, D.N. (1995) Photoelectrochemical properties of Q-state CdS particles in arachidic acid LB films, *J. Chem. Soc. Faraday Trans.* **91**, 665-672.
17. Facci, P., Erokhin, V., Tronin, A., and Nicolini, C. (1994) Formation of ultrathin semiconductor films by CdS nanostructure aggregation, *J. Phys. Chem.* **98**, 13323-13327.
18. Urquhart, R.S., Furlong, D.N., Gengenbach, T., Geddes, N.J., Rabolt, J.F., and Grieser, F. (1995) FTIR study of the changes in orientation and chemistry of LB films of cadmium arachidate during the formation of Q-state CdS particles, *J. Phys. Chem. (in press).*
19. Furlong, D.N., Urquhart, R., Grieser, F., Tanaka, K., and Okahata, Y. (1993) Size-quantised cadmium sulfide particles in LB films:film thermal stability, *J. Chem. Soc. Faraday Trans.* **89**, 2031-2035.
20. Grieser, F., Furlong, D.N., Scoberg, D., Ichinose, I., Kimizuka, N., and Kunitake, T. (1992) Size-quantised semiconductor cadmium chalcogenide particles in LB films, *J. Chem. Soc. Faraday Trans.* **88**, 2207-2214.
21. Moriguchi, I., Teraoka, Y., and Kagawa, S. (1994) Precise control of structure and property of inorganic materials with the assistance of molecular assembly. Third Australia/Japan Symposium *"Self-assembly in Colloid Science"*, Abstract Book, pp74-75, Kyushu University, Fukuoka, Japan.

22 Mansur, H.S., Grieser, F., Urquhart, R.S., and Furlong, D.N. (1995) The photoelectrochemical behaviour of Q-state $CdS_xSe_{(1-x)}$ particles in arachidic acid LB films, *J. Chem. Soc. Faraday Trans.*(in press)
23. Urquhart, R.S., Furlong, D.N., Gengenbach, T., Geddes, N.J., and Grieser, F. (1995) Studies of the formation and growth of Q-state cadmium selenide particles in cadmium arachidate LB films, *Langmuir* **11**, 1127-1133.
24. Murray, C.B., Norris, D.J., and Bawendi, M.G. (1993) Synthesis and characterization of nearly monodisperse CdE (E = S, Se, Te) semiconductor nanocrystallites, *J. Am. Chem. Soc.* **115**, 8706-8715.
25. Furlong, D.N., Grieser, F., Urquhart, R.S., Matsuura, K. and Okahata, Y. (1995) The fabrication of titania Q-state particles in LB films, J. Chem. Soc., Chem. Commun. (in press).
26. Furlong, D.N. and Grieser, F. (1992) Surfactant/particle films for non-linear optics, *Chemistry in Australia*, **59**, 617-619.

HYDROLYSIS AND CONDENSATION OF OCTADECYLTRIMETHOXY AND OCTADECYLDIMETHYLMETHOXY SILANES IN MONOLAYERS WITH AND WITHOUT SURFACTANTS.

J. SJÖBLOM, G. STAKKESTAD[⊗] AND M. LEHTISYRJÄ[⊗⊗]
Department of Chemistry
University of Bergen
Allegt.41
N-5007 Bergen
Norway

and

S. E. FRIBERG
Center for Advanced Materials Processing
Clarkson University
Potsdam, New York 13699-5814
U.S.A.

Langmuir films of octadecyltrimethoxy silane (C_{18}TMS) and octadecyldimethylmethoxy silane (C_{18}DMMS) were studied at different pHs and salinities with and without surfactant in the aqueous subphase. The surfactant used in this study are n-dodecylsulphate sodium salt (SDS) and hexadecyltrimethyl-ammoniumbromide (HTAB). C_{18}TMS is shown to give a fully condensed monolayer consisting of two-dimensional surface polymers within 1 hours at pH 1. At pH 11 the same process takes 2 hours and 14 hours at pH 5.6. At pH 1 there is no observable influence of the surfactants added. At pH 5.6 a compression of the film will take place upon addition of the HTAB and SDS, while at pH 11 HTAB gives a strong compression and SDS a minor increase in area at low interfacial pressures. At the higher pHs the kinetics of condensation are so slow that a surfactant monomer - silane monomer interaction will dictate the shape of the π-A isotherms. For C_{18}DMMS no fully condensed surface polymers will form at any pH, due to a low surface charge density of the films or alternatively less charged monomers the interaction with HTAB is low.

[⊗] Present address: Prototech AS, P. O. Box 12, Fantoftvegen 38, N-5036 Fantoft, Bergen, Norway.
[⊗⊗] Present address: Eteläinen Hesperiank, 12 A 12, SF-00100 Helsinki, Finland.

E. Pelizzetti (ed.), Fine Particles Science and Technology, 745–759.

1. Introduction.

Langmuir films [1] of different amphiphiles, like long chain fatty acids, alcohols, esters, amines and amides, have traditionally been used to study different molecular transitions at the liquid/vapour phase boundary [2]. The kinetics of polymerisation reactions in monolayers may also be studied by means of this technique [3]. The Langmuir-Blodgett technique [4] makes it possible to transfer the monolayer from the water surface onto a solid substrate, and a well-oriented monomolecular film is formed. Further analysis of the films is now possible by means of microscopy and spectroscopy [5], for instance FT-IR and ESCA [6]. Technical applications of the deposited films can be found within electronics, optics and sensor technology [5].

The sol/gel technique utilises reactions of organosilanes to form silane polymers either as a gel at low pH values or as particles at high pH [7]. Up to now hydrolysis and condensation reactions of alkoxide silanes have been mapped mostly by means of Si^{29}NMR in bulk solutions [8-12]. The Langmuir technique offers a possibility to study the hydrolysis and condensation reactions of organosilanes at the water/air interface. By using the knowledge obtained from the surface-monolayer studies, it can be possible to get more detailed information that contributes to the understanding of the reaction steps in the complicated sol/gel process. Formation and use of highly organised mono- and multilayer films of alkyltrichloro silanes have been the focus of recent investigations [13, 14].

In this paper the hydrolysis and condensation reactions of alkylmethoxy silanes are studied by means of Langmuir films with and without surface active agents in the bulk phase. The choice of the film-forming agents and the surfactants gives a variety of possibilities with regard to hydrolysis and condensation reactions in the film and the molecular interactions with the surfactants. Octadecyltrimethoxysilane (C_{18}TMS) can undergo condensation reactions at three different sites (Q_3) while octadecyldimethylmethoxysilane (C_{18}DMMS) can just undergo one possible condensation reaction (Q_1). Hence, in the first case a polymeric two-dimensional film can form, while in the second case maximally condensed dimers. The hydrolysis and condensation are influenced by pH, electrolyte, temperature, etc.. Kinetic studies performed by the Langmuir technique reveal that the condensation for C_{18}TMS is complete in 1 hour at pH 1, at 2 hours at pH 11 while it takes 14 hours to complete the reaction at pH 5.6 [15].

A charged surfactant will display different interaction forms with monolayers depending on the surface charges. For opposite charges on the monolayer and the surfactant the electrostatic interaction will be strong and is expected to result in an uptake of the surfactant monomer into the film. This is observed as altered packing conditions in the monolayer. For equal charges in the monolayer and on the surfactant the interaction pattern will be depending on the surface charge density and the van der Waals interactions.

In this study we have altered the pH in order to vary the charge conditions at the silane surface. Further we have added both negatively and positively charged surfactants in order to study different kinds of surfactant monolayer interactions.

2. Experimental.

2.1. MATERIALS.

Octadecyltrimethoxysilane (C_{18}TMS, purity 90%), octadecyl-dimethylmethoxysilane (C_{18}DMMS, purity 90%) and hexadecyltrimethylammoniumbromide (HTAB, purity > 98%) were all supplied by Fluka Chemicals. Aluminiumnitrate nonahydrate ($Al(NO_3)_3 \cdot 9H_2O$, purity >98.5%) and n-dodecylsulphate sodium salt (SDS, purity >99%) were supplied by Merck. The water used in all experiments was first purified by distillation. The final purification was undertaken by Seralpur Pro 90 water purification system (Seral), which include five filter stages: one activated charcoal long, two ion exchanger long one adsorbent resin long, and 0.2 μm sterile filter. The pH of the subphase was adjusted by adding HCl (suprapur, Merck) or NaOH (suprapur, Merck) to the purified water. The spreading solvent was redistilled toluene (p.a.) from Merck. All other chemicals were used as supplied.

2.2. THE SURFACE BALANCE EXPERIMENT.

Surface pressure-area isotherms as well as monolayer relaxation isotherms were recorded with a KSV 5000 Langmuir trough double-barrier system (KSV Chemicals, Helsinki Finland). The trough was made of Teflon and the barrires were made of Delrin. The temperature was kept at 294 ± 0.5 K. The surface area could be determined with an accuracy of 0.5 $Å^2$ molecule^{-1}.

Surface pressure-area isotherms and area relaxation measurements were recorded with a maximum barrier speed of 6 mm min^{-1} and a maximum increase in surface pressure of 1 mN m^{-1} min^{-1}. The purity of the surface was monitored before each run by moving the barrier across the pure liquid surface. The experiment was continued when the surface pressure was less than 0.15 mN m^{-1}. In the final step the isotherms were measured on freshly spread monolayers after allowing the spreading solvent to evaporate for 15 min.

The influence of surfactant on the film material at different pHs was studied by compressing the film on a subphase containing HTAB or SDS at fix concentrations. By adding $Al(NO_3)_3$ into the subphase at pH 2 and 4 the salt effect was studied.

The kinetics of the hydrolysis and condensation reactions were studied by compressing the film molecules at pH 5.6 and 11 to a surface pressure of 20 mN m^{-1} and left for 2, 14 and 42 hours (at a constant surface pressure), before the compression was completed. This experiment will be referred to as the two-step compression.

3. Results.

3.1. SURFACE PRESSURE-AREA ISOTHERMS.

Surface pressure-area (π-A) isotherms of the C_{18}TMS and C_{18}DMMS films with and without surfactant in the subphase are shown in figures 1-4.

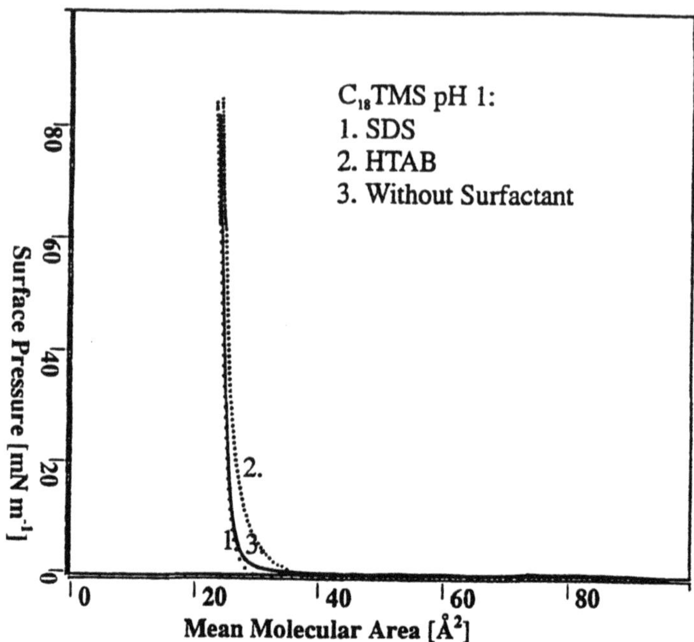

Figure 1: π-A isotherms of C_{18}TMS films at pH 1 with and without surfactants in the subphase.

At pH 1 the C_{18}TMS films display basically a solid condensed state with molecular areas of ~34 Å^2 for pure water, ~39 Å^2 for HTAB and ~31 Å^2 for SDS subsolutions. The surface pressure in the C_{18}DMMS film at this pH, both with and without HTAB in the subsolution, does not start to increase before a molecular area of ~23 Å^2 is reached.

Addition of surfactant to a C_{18}DMMS film at pH 5.6 gives a significant contraction and the film becomes less compressible. The C_{18}TMS film at pH 5.6 with HTAB in the subsolution seems to be more compressible than without surfactant. When SDS is added to the subsolution the film is contracted and becomes less compressible.

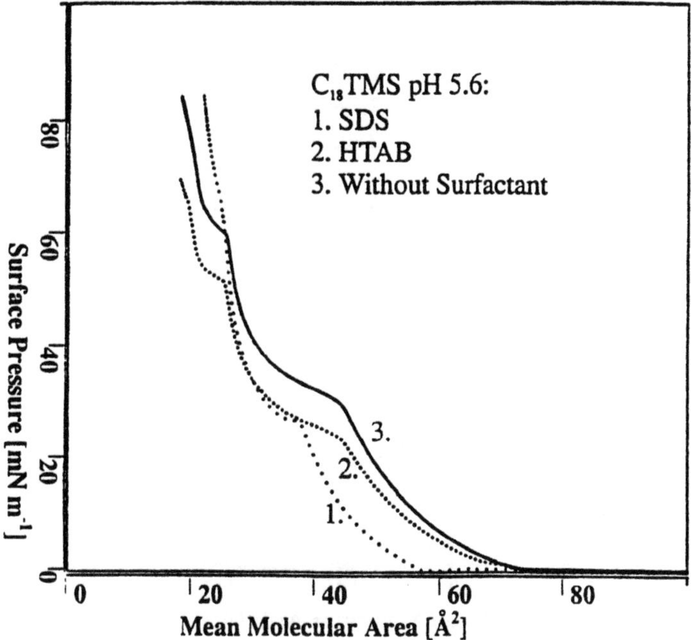

Figure 2: π-A isotherms of C_{18}TMS films at pH 5.6 with and without surfactants in the subphase.

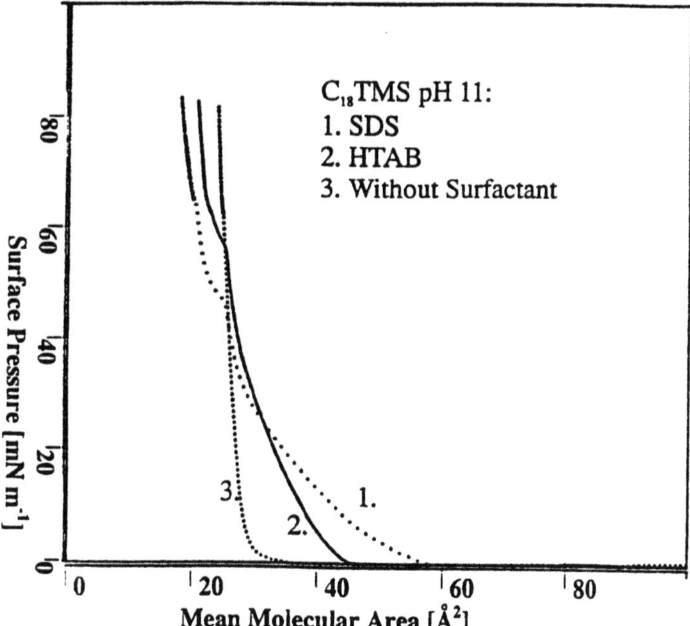

Figure 3: π-A isotherms of C_{18}TMS films at pH 11 with and without surfactants in the subphase.

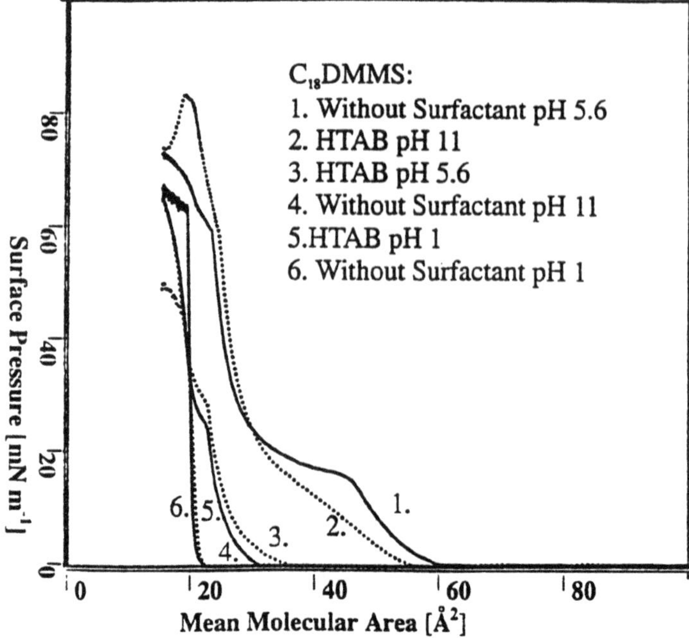

Figure 4: π-A isotherms of C_{18}DMMS films at pH 1, 5.6 and 11 with and without surfactants in the subphase.

At pH 11 addition of SDS to the C_{18}TMS subphase makes the film more compressible and the surface pressure starts to rise at molecular areas higher than without surfactant in the system. HTAB contracts the film molecules and makes the film less compressible. For the C_{18}DMMS film at pH 11 addition of HTAB increases the area where the surface pressure starts to rise and the film seems to be more compressible than without HTAB.

TABLE 1. Molecular areas A_0 and A_{40} (Å^2) from the isotherms of C_{18}TMS and C_{18}DMMS at pH 1, 5.6 and 11 with and without surfactant in the subphase. Estimated error for the areas ± 0.05 Å^2.

Film material	Area	pH 1	pH 5.6	pH 11
C_{18}TMS	A_0	33.46	72.84	45.33
	A_{40}	25.04	30.51	27.51
C_{18}TMS+HTAB	A_0	38.98	74.07	34.92
	A_{40}	25.75	27.34	26.10
C_{18}TMS+SDS	A_0	30.86	56.61	57.50
	A_{40}	24.69	27.87	26.10
C_{18}DMMS	A_0	22.57	61.38	31.39
	A_{40}	19.75	25.22	19.22
C_{18}DMMS+HTAB	A_0	22.57	35.80	55.91
	A_{40}	19.40	19.22	26.46

Table 1 summarises the limiting areas, A_0, and the molecular areas at a surface pressure of 40 mN m^{-1}, A_{40}. A_0 is determined by extrapolating the expanded or condensed part of the isotherm to zero surface pressure.

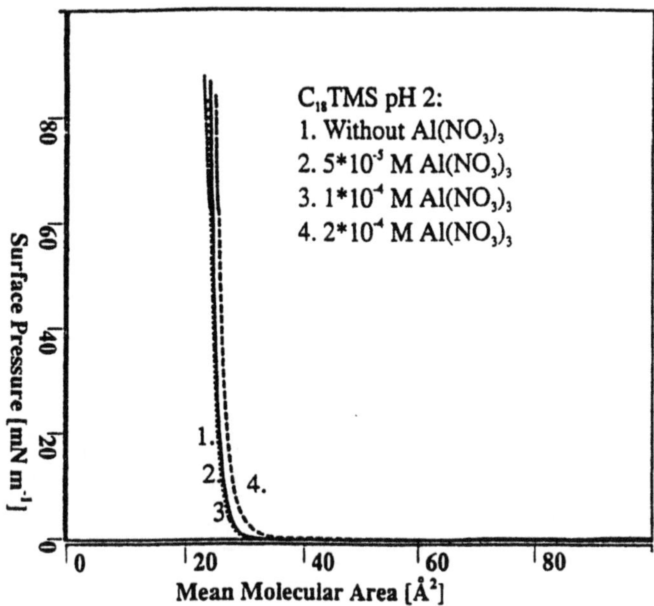

Figure 5: π-A isotherms of C_{18}TMS films at pH 2 with and without $Al(NO_3)_3$ in the subphase.

The salt effect was studied only for C_{18}TMS at pH 2 and pH 4. The concentration of $Al(NO_3)_3$ in the subphase was 5*10^{-5}, 1*10^{-4} and 2*10^{-4} M. Figure 5 and 6 show the results from these experiments. At pH 2 there is no remarkable difference with or without salt in the subphase. In contrast, at pH 4 there is an obvious reduction in molecular area upon $Al(NO_3)_3$ addition. Table 2 summarises A_0 and A_{40} for the isotherms with and without salt added to the subphase.

TABLE 2. Molecular areas A_0 and A_{40} (Å^2) from the isotherms of C_{18}TMS at pH 2 and 4 with and without $Al(NO_3)_3$ in the subphase. Estimated error for the areas ± 0.05 Å^2.

Film material	Area	pH 2	pH 2 with $Al(NO_3)_3$			pH 4	pH 4 with $Al(NO_3)_3$		
			5'10^{-5}M	1*10^{-4}M	2*10^{-4} M		5*10^{-5}M	1*10^{-4}M	2*10^{-4}M
C_{18}TMS	A_0	31.31	30.51	30.16	33.33	42.77	39.15	36.60	34.48
	A_{40}	24.96	24.60	24.51	26.66	31.66	28.48	28.84	28.31

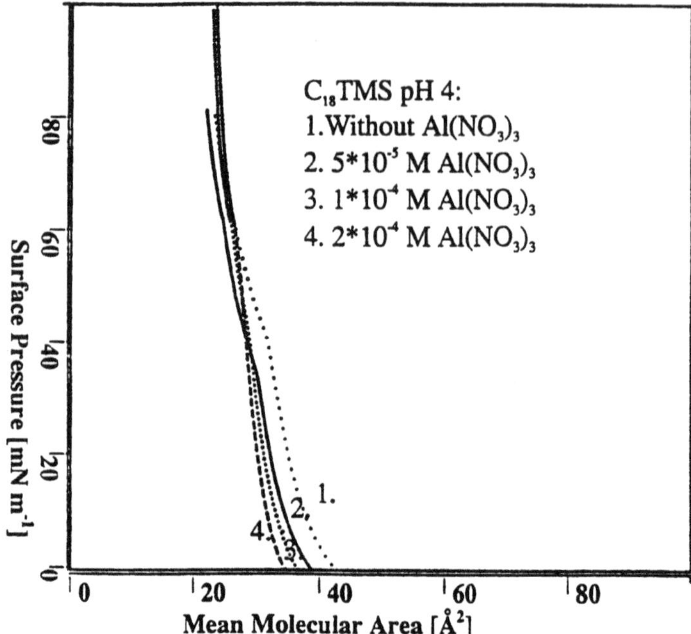

Figure 6: π-A isotherms of C_{18}TMS films at pH 4 with and without $Al(NO_3)_3$ in the subphase.

3.2. KINETICS.

The time dependence on the hydrolysis and condensation reactions was studied with the two-step compression procedure at pH 5.6 and 11. The concentration of HTAB was $5*10^{-8}$ M, in the case when surfactant is added.

Figure 7 shows π-A isotherms for C_{18}TMS films at pH 5.6 with and without HTAB, when the reaction time at a constant surface pressure of 20 mN m^{-1} was 14 hours. Without HTAB the film is fully condensed after 14 hours. Addition of HTAB seems to prevent the condensation in an effective way.

Figure 8 shows π-A isotherms for C_{18}TMS films at pH 11 with and without HTAB, when the reaction time at a constant surface pressure of 20 mN m^{-1} was 2 hours. After 2 hours the film seems to be fully condensed, regardless of the presence of surfactant.

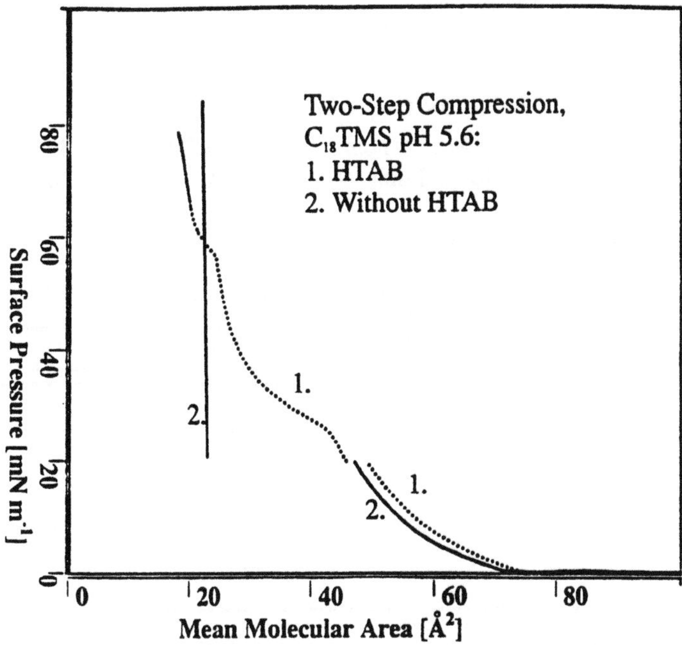

Figure 7: π-A isotherms of C_{18}TMS films at pH 5.6 with and without HTAB in the subphase. The films are compressed to a surface pressure of 20 mN m and left for

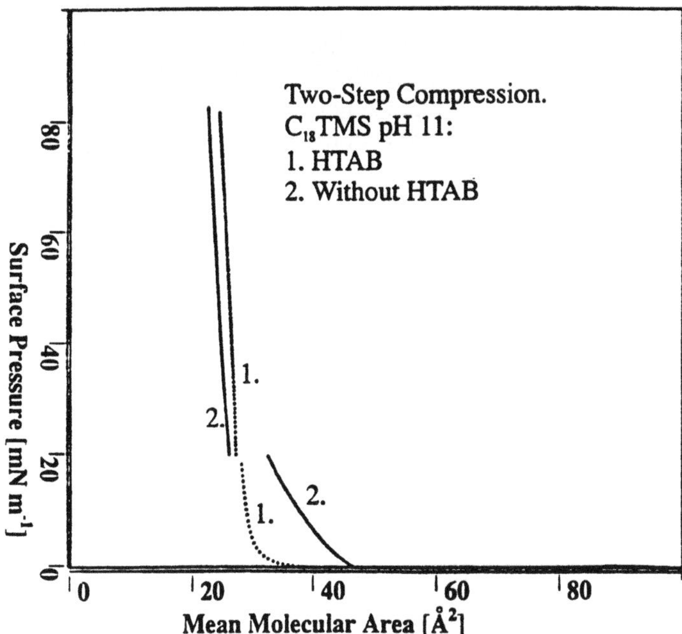

Figure 8: π-A isotherms of C_{18}TMS films at pH 11 with and without HTAB in the subphase. The films are compressed to a surface pressure of 20 mN m and left for 2 hours at a constant surface pressure.

In figure 9 the result of the two-step compression procedure is displayed for the C_{18}DMMS film at pH 5.6 without surfactant. After 42 hours time the film is still not fully condensed.

Figure 10 shows π-A isotherms for C_{18}DMMS films at pH 11 with and without HTAB, when the reaction time was 2 hours. Without HTAB the film is fully condensed after 22 hours, while addition of surfactant seems to hamper the condensation effectively.

The values of A_{40} for the kinetic reactions are summarised in table 3.

TABLE 3. Comparision of the molecular area A_{40} from the isotherms of C_{18}TMS at pH 5.6 and 11 when submitted to direct compression or the two-step compression procedure.

Film material	Area	pH 5.6 direct comp.	pH 5.6 after 14 h.	pH 11 direct comp.	pH 11 after 2 h.
C_{18}TMS	A_{40}	30.51	22.66	27.51	24.69
C_{18}TMS+HTAB	A_{40}	27.34	27.87	26.10	26.46
CDMMS	A_{40}	25.22	22.13	19.22	18.17
CDMS+HTAB	A_{40}	19.22	-	26.46	23.54

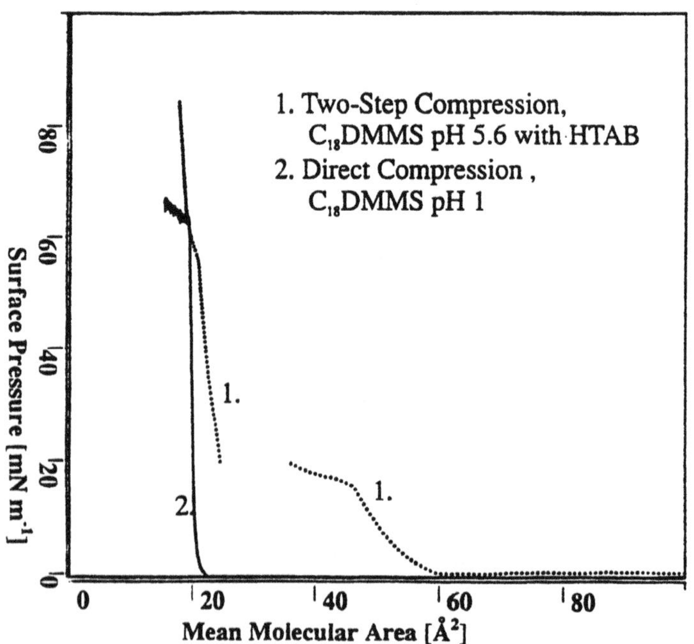

Figure 9: π-A isotherms of C_{18}DMMS film at pH 5.6 with HTAB in the subphase. The film is compressed to a surface pressure of 20 mN m and left for 42 hours at a constant surface pressure. π-A isotherm of C_{18}DMMS at pH 1, direct pression, is inserted for comparision.

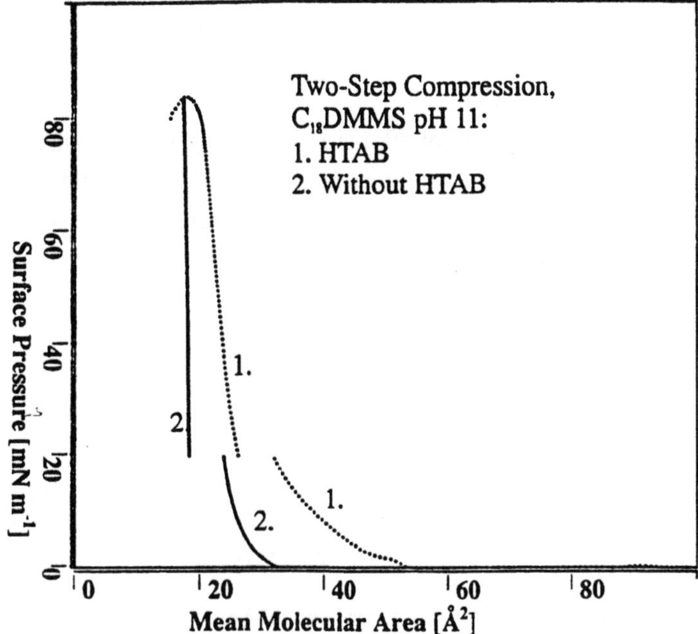

Figure 10: π-A isotherms of C_{18}DMMS films at pH 11 with and without HTAB in the subphase. The films are compressed to a surface pressure of 20 mN m^{-1} and left for 22 hours at a constant surface pressure.

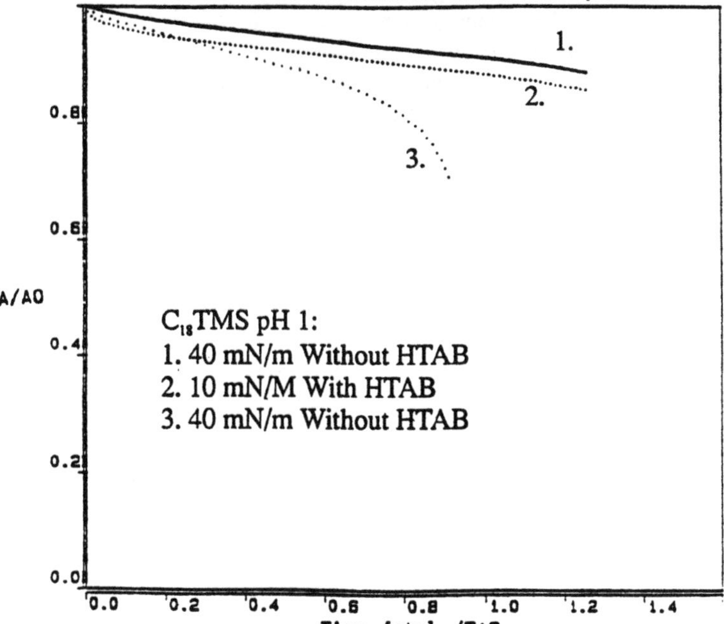

Figure 11: Constant surface pressure area relaxation curves for C_{18}TMS films at pH 1 with and without HTAB in the subphase.

3.3. STABILITY TEST.

Constant surface pressure area curves for C_{18}TMS films at pH 1 with and without surfactant are shown in figure 11.

At a constant surface pressure of 40 mN m^{-1} the C_{18}TMS film without HTAB is stable for more than 20 hours. At this surface pressure the C_{18}TMS film with HTAB is stable for about 12 hours, but then the relaxation time increases. At a surface pressure of 10 mN m^{-1} with HTAB the stability of the C_{18}TMS film exceeds 20 hours.

4. Discussion.

4.1. THE INFLUENCE OF SURFACTANT AT DIFFERENT pHs.

When a monolayer is spread upon a dilute solution of appropriate components, two competing interaction processes can take place; i.e. there can be a penetration of bulk components into the film or a dissolution of the film components into the dilute subphase [17-19].

The chain cohesion in the insoluble monolayer will indicate the level of penetration. With a strong cohesion in the monolayer, the penetration ability of the guest molecule is usually reduced. The uptake in the film is very dependent on the molecular structure of the penetrating species and of the subsolution. The properties of mixed films are mainly dependent on the extent of the hydrophobic chain and polar group interactions between the film components. If the association is weak, one component may be dissolved into the subphase upon film compression, with strong interactions the mixed monolayer generally possesses characteristics entirely different from those of the pure components.

The surfactants HTAB and SDS have different amount of CH_2 groups in the hydrocarbon chain, but which is more important, HTAB is a cationic and SDS is an anionic surfactant. So the main difference between these compounds is the nature of the ionic group. The bulk concentration of both surfactant was chosen to $5*10^{-8}$ M which is far below cmc value. This guarantees an interaction between surfactant monomers and the monolayer.

It is known that the polymerisation level in pure octadecyltrimethoxy silane films increases from pH 5.6 $\rightarrow$ 11 $\rightarrow$ 1 [15]. At pH 1 the monolayer is little compressible and is in a condensed phase. There does not seem to be any remarkable differences between curves with or without surfactant, (figure 1). This reflects that the surface film is not charged, since otherwise there would have been a strong electrostatic interaction. The surfactant can also take place in a replacement process where the condensed C_{18}TMS film might be completely or partly substituted by the surfactant. However, the π-A isotherms in figure 2 show that the C_{18}TMS film is intact and no significant replacement of the original film components has taken place. Hence, the original

C_{18}TMS film will be very stable, compressed, and does not allow any appreciable penetration of surfactant to take place.

At pH 5.6 however there are clear differences. In direct compression the C_{18}TMS film can not reach the fully condensed state. The interactions will take place between surface monomers and surfactant molecules instead of surface polymers and surfactant molecules. When C_{18}TMS is compressed upon subsolutions containing $5*10^{-8}$ M SDS, the monolayer is contracted and becomes less compressible, when compared to C_{18}TMS monolayer upon pure water. A possible explanation to this is a complexation between the C_{18}TMS and SDS which results in a polar water-soluble complex [20]. HTAB in the subphase expands the monolayer upon penetration due to the longer hydrocarbon chain. Even if a higher surface pressure is needed, it is still possible to compress the monolayer to smaller specific areas as compared to the pure C_{18}TMS film. This can be explained analogous for SDS, but because of the longer hydrocarbon chain, the complex is more hydrophobic. As a consequence larger forces are required to press the complex from the film into the subsolution.

At pH 11 there is a profound effect of both SDS and HTAB on the monolayer properties of the C_{18}TMS. In the presence of HTAB there is a distinct contraction of the film molecules and the compressibility of the film decreases. Obviously the negatively charged C_{18}TMS film will attract HTAB molecules from the bulk and complexation will take place. As a consequence of this complexation the electrostatic nature of the film is neutralised and it is possible to compress the film components further. This state of the film is further verified by the influence of SDS. An addition of SDS will result in an increased repulsion at low surface pressures. This can easily be understood from a partial penetration of the SDS molecules into the C_{18}TMS film with an increased repulsion as a consequence. Obviously the SDS can due to a strong cohesive attraction overcome the negative surface charge and reorganise the film.

The only surfactant used in the C_{18}DMMS studies was cationic HTAB. The reason for this is that C_{18}DMMS is not able to fully condense in the direct compression even without any surfactant.

At pH 1 there is no difference between isotherms with and without HTAB. At pH 5.6 and 11, however, the differences seem to be clear. Addition of HTAB seems to clearly decrease the compressibility at pH 5.6. Instead of decreasing, at pH 11 the compressibility seems to increase due to the influence of HTAB. The result at pH 5.6 can possibly be explained in a way similar to that of C_{18}TMS.

The result at pH 11 is, however, totally different from that of C_{18}TMS. A plausible explanation can be different ratios between alkylsilane molecules and surfactant molecules in these films. The ratio between silane molecules in the film and surfactant molecules was 5:1 for C_{18}TMS systems and 8:5 for C_{18}DMMS systems. So, in the C_{18}DMMS monolayers there is an excess of negative charge, and that is probably the reason for increasing repulsion with HTAB at pH 11.

4.2. INFLUENCE OF $Al(NO_3)_3$ ON C_{18}TMS.

The salt effect was studied only for C_{18}TMS at pH 2 and pH 4. The concentration of $Al(NO_3)_3$ in the subphase was $5*10^{-5}$, $1*10^{-4}$ and $2*10^{-4}$ M. There is no remarkable difference between curves with and without salt at pH 2. At pH 4 there seems to be a lower repulsion, when $Al(NO_3)_3$ is added into subphase. The molecular area reduction is proportional to salt concentration. The explanation for decreasing compressibility with $Al(NO_3)_3$ at pH 4 is the state of Al and its interaction with the monolayer film. Below pH 2 the Al-aqua complexes are in the form $[Al(H_2O)_x]^{3+}$ while at pH 4 the complexation will also give $[Al(H_2O)_x(OH)]^{2+}$. At the same time the silane film have a higher surface charge density at pH 4 which emphasizes more importance of an electrostatic interaction at the higher pH.

4.3. KINETICS.

It is already known that C_{18}TMS needs more time to polymerise at pH 5.6 than at pH 1 [15]. When the C_{18}TMS film has been spread for 14 hours at constant surface pressure (20 mN m^{-1}) at pH 5.6 without surfactant, it is fully condensed. Addition of HTAB seems to prevent the condensation in an effective way. The reason why HTAB seems to hamper the condensation can be explained by a complexation between the C_{18}TMS and HTAB molecules which results in a polar water-soluble complex.

At pH 11 the C_{18}TMS film seems to be fully condensed after 2 hours reaction time, regardless of the presence of surfactant. The cationic surfactant is not effective enough to prevent the condensation reaction at this pH and at the studied reaction times.

For the C_{18}DMMS film at pH 5.6 even 42 hours was not enough to give a fully condensed state. At pH 11 the C_{18}DMMS film is able to condense when the reaction time is more than 22 hours. This shows that the HTAB surfactant is able to hamper the condensation.

4.4. STABILITY AT pH 1.

Monolayer instability at surface pressures below film fracture can be investigated by maintaining a constant surface pressure and observing the decrease in surface area with time [20-25]. Instability due to rearrangement of the molecules in the film [20-25], evaporation [26], dissolution of the film molecules in the subphase [27, 28], or transformation of the monlayers into the stable bulk phase [22, 25] can be followed from the shape of the isotherms.

Relaxation due to structural rearrangements within the monolayer is characterised by a small initial area loss followed by a slower relaxation that asymptotically reaches a constant value. If the film dissolves into the subphase, one expects a continuously decreasing area at constant pressure.

At a constant surface pressure of 40 mN m^{-1} the C_{18}TMS film without HTAB is stable for more than 20 hours. At the same surface pressure with HTAB in the subsolution the C_{18}TMS film is stable for 12 hours before the relaxation starts to increase. This can be explained by a collapse due to nucleation and growth of nuclei [22, 25, 29].

5. Acknowledgements.

Gro Stakkestad acknowledges a PhD grant from the Research Council of Norway (NFR). The Langmuir-Blodgett equipment was also financed by NFR.

6. References

1. Langmuir, I. (1917) *J. American Soc.*, **39**, 1848.
2. Dervichian, D. G. (1939) *J. Ch. Phys.*, **7**, 931.
3. Beredijk, N.; Burlant, W. J. (1970) *J. Polymer Sci.*, **8**, 2807.
4. Blodgett, K. B. (1935) *J. Am. Chem. Soc.*, **57**, 1007.
5. Roberts, G. (1991) *Langmuir-Blodgett Films*, Plenum Press, New York,.
6. Siegbahn, K. (1982) *Science*, **217**, 111.
7. Iler, R. K. (1979) *The Chemistry of Silica*, Wiley, New York.
8. Vega, A. J.; Scherer, G. W. (1989) *J. Non-Cryst. Solids*, **111**, 153.
9. Zerda, T. W.; Artaki, I.; Jones, S. M. (1986) *J. Non-Cryst. Solids*, **81**.
10. Keltz, L. W.; Effingen, N. J.; Melpunder, S. M. (1986) *J. Non-Cryst. Solids*, **83**, 353.
11. Jones, S. M.; Friberg, S. E. (1992) *J. Disp. Sci. Techn.*, **13**, 6.
12. Seile, M. H.; Sjöblom, J.; Friberg, S. E.; Young, T.; Yang, C. C. (1992) *Prog. Colloid Polymer Sci.*, **88**, 42.
13. Barton, S. W.; Goudot, A.; Rondelez, F. (1991) *Langmuir*, **7**, 1029.
14. Samha, H.; DeArmond, M. K. (1993) *Langmuir*, **9**, 1927.
15. Sjöblom, J., Stakkestad, G., Ebeltoft, H., Friberg, S. E.and Claesson, P. (1995) *Langmuir*, in Press.
16. Munger, G. and Leblanc, R. M. (1980) *Rev. Sci. Instrum.*, **51**, 710.
17. Schulman, J. H. and Stenhagen, E. (1937) *Proc. R. Soc.*, London, **B**, **122**, 29.
18. Schulman, J. H. and Stenhagen, E. (1938) *Proc. R. Soc.*, London, **B**, **126**, 356.
19. Adam, N. K., Askew, F. A. and Pankhurst K. G. A.(1939) *Proc. R. Soc.*, London, **A 170**, 485.
20. Gaines, Jr. G. L. (1966) *Insoluble Monolayers at Liquid-Gas Interfaces*, Wiley-Interscience, New York.
21. Pezron, E., Claesson, P. M., Berg, J. M. and Vollhardt, D. (1990) *J. Coll. Int. Sci.*, 138, 245.
22. Smith, R. D. and Berg, J. C. (1980) *J. Coll. Int. Sci.*, **74**, 273.
23. Honig, E. P., Hengst, J. H. and den Engelsen, D. (1973) *J. Coll. Int. Sci.*, **45**, 92.
24. Neuman, R. D. (1976) *J. Coll. Int. Sci.*, **56**, 505.
25. Xu, S., Miyano, K. and Abraham, B. M. (1928) *J. Coll. Int. Sci.*, **89**, 581.
26. Brooks, J. H., Alexander, A. E. (1962), in La Mer, V. K. (ed.) *Retardation of monolayers*; Academic Press, *New York, pp.* 245.
27. Minassian-Saraga, L. T. (1956) *J. Coll. Int. Sci.*, **11**, 398.
28. Minassian-Saraga, L. T. (1955) *J. Chim. Phys. Phys.-Chim. Biol.*, **52**, 181.
29.Gabrielli, G. Guarini, G. G. T. (1976) *J. Coll. Int. Sci.*, **54**, 424.

PREPARATION AND CHARACTERIZATION OF NANO-SIZED CAROTENOID HYDROSOLS

D. HORN and E. LÜDDECKE

BASF-Aktiengesellschaft, Department of Polymer and Solid State Physics
67056 Ludwigshafen/Rhein, Germany

Abstract

Recent developments in the preparation of nano-sized beta-carotene hydrosols by polymer- and surfactant-controlled precipitation and phase separation techniques are described.

As found with electron microscopy and quasielastic light scattering dispersions are obtained with core-shell-type particles of the size range between 15 nm and 250 nm. From electrophoretic mobility measurements it is concluded that colloidal stabilization of the gelatin-coated nanoparticles is governed by steric effects. The stability of the beta-carotene loaded mixed micelles with particle sizes of ~ 20 nm is based on enthalpic stabilization effects of the ethoxylated fatty acid surfactant used as solubilizer.

Concomitant with changes in particle size and state the UV/VIS-spectra of the different preparations exhibit characteristic shifts in respect with the electronic absorption spectra of dilute molecular solutions. The red shifts observed with crystalline dispersions and the blue shifts found in precipitates and solubilizates are interpreted as caused by process-controlled head-to-tail and card-pack aggregation structures, respectively.

The corresponding control of colour shade and strength and the enhanced bioavailability of the obtained formulations provide the basis for numerous new applications for carotenoids in the pigmentation of foodstuffs and feed and in pharmaceutical preparations.

E. Pelizzetti (ed.), Fine Particles Science and Technology, 761–775.

1. Introduction

Carotenoids form a class of dyes which are widely found in nature [1]. They impart a characteristic yellow to red colour in plants and marine animals. The most important example of these dyes is beta-carotene which is used as a colourant for foodstuffs and beverages. The chemical structures and UV/VIS-spectra of all-trans and cis isomers dissolved in cyclohexane are shown in Figure 1.

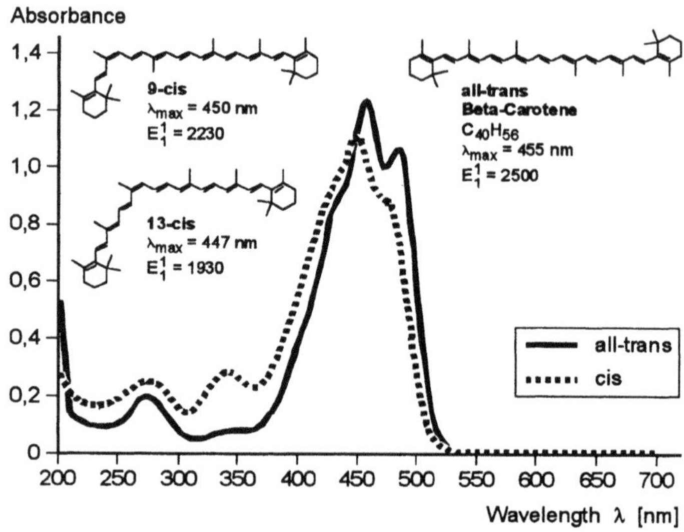

Figure 1. Chemical structures and UV/VIS absorption spectra of beta-carotene isomers.

Moreover, due to its pro-vitamin A activity is beta-carotene of particular interest in pharmaceutical formulations and as a health food additive [1].

All carotenoids are insoluble in water and only poorly soluble in fats and oils. This limited solubility hampers the direct application of the coarse crystalline products obtained from synthesis since only low colour yields can be obtained or - as in aqueous systems - colouration is even impossible. Similarly is the biological resorption of coarse beta-carotene during gastro intestinal passage seriously hampered and the pro vitamin A activity cannot be made available.

Accordingly, there have been many attempts to improve the colour yield and to increase the bioavailability of carotenoids by special formulation techniques with emphasis on reducing the particle size [2].

In the following, with beta-carotene as an example, some more recent developments in technologies are described that transform the coarse raw material into nano-sized micronizates [3]. The achieved colloidal state of carotenoids not only provides the basis of their applications in aqueous systems with high potential in bioavailability but also opens up a wide variety in colour shade as controlled by particle size and state.

2. Fundamentals of the Preparation of Carotenoid Hydrosols

In principle two approaches may be taken to transform the coarse raw material into micronizates.

As is common practice in pigment technology, grinding processes yield preparations with particle sizes in the micrometer size range [2]. In the presence of an edible oil one thus obtains formulations readily to be used for colouring fats and oils since the solubility of carotenoids in oils, though low, suffices for colouring purposes.

For applications in aqueous systems, however, micron-sized carotenoid dispersions exhibit only a fraction of their colouring potential and are only poorly bioavailable [3]. As it turns out, the desired colouring characteristics and bioavailability levels require micronization techniques that yield preparations with particle size distributions in the 10 - 100 nm size range. Micronization technologies that meet this target are in general based on the principle of size-controlled precipitation from the solute state. The number of solvents with technologically useful solubilities at room temperature is, however, limited to toxicologically inacceptable and water-incompatible halogenated or aromatic hydrocarbons.

The newest developments based on this principle relate to the use of water-miscible and toxicologically safe solvents like alcohols and ketones at elevated temperatures [3]. As shown in Figure 2 the solubility of beta-carotene in acetone increases dramatically with rising temperature. At 175°C solubilities of 80 g/l can be extrapolated with an Arrhenius plot establishing a technologically useful level. At elevated temperatures, however, trans → cis isomerization of beta-carotene is a detrimental side reaction leading to changes in colour shade (Figure 1) and reduced pro-vitamin A activity. In order to prevent losses of the active ingredient during processing the residence time in the solute state at elevated temperatures has to be strictly limited, e.g. to < 0.5 seconds at 170°C (Figure 2). These limiting constraints in temperature and residence time define the technical fundamentals of a novel micronization process.

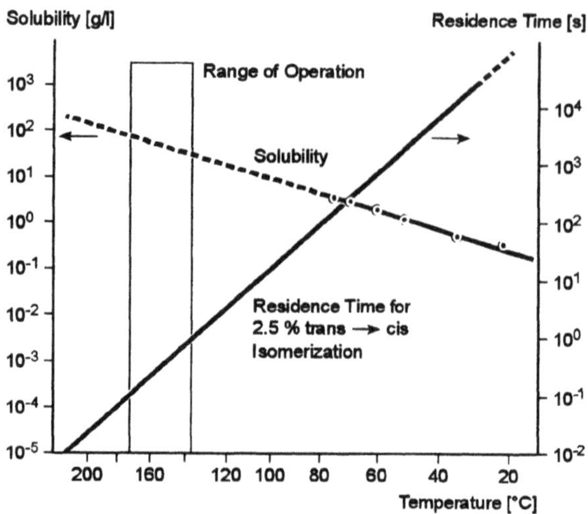

Figure 2. Arrhenius plot of solubility and tolerable residence times of beta-carotene in acetone.

Figure 3 shows the scheme of the process. A suspension of the coarse carotenoid in the selected solvent, e.g. an alcohol, is pumped into mixing chamber I where it is rapidly mixed with an appropriate volume of the plain solvent at elevated temperature. The residence time, the temperature of mixing, and the final concentration of the active ingredient are readily controlled by the delivery rate of each pump. Obviously, if the process is run at temperatures above the boiling point of the solvent, proper pressure control of the process has to be applied. With the required combination of solvent temperature and residence time in order to prevent major isomerization, a molecular solution can be obtained by feeding a coarse carotenoid suspension.

After passing mixing chamber I the molecular solution is fed into mixing chamber II where an aqueous solution of a stabilizing polymer and other additives, if required, is added under turbulent mixing conditions. Due to the rapid temperature drop and change of the solvent character, nucleation and growth of carotenoid particles is induced in the presence of the polymer stabilizer, e.g. gelatin. The spontaneous adsorption of the protective polymer onto the growing particles controls their growth and yields a colloidally stable hydrosol which is readily transformed into a dry powder by conventional techniques, e.g. spray-drying. With proper choice of the stabilizer and other water-soluble additives, e.g. sugar, a redispersable powder is obtained that readily dissolves in water reestablishing the hydrosol state of the active ingredient with a particle size distribution similar to the originally obtained micronizate.

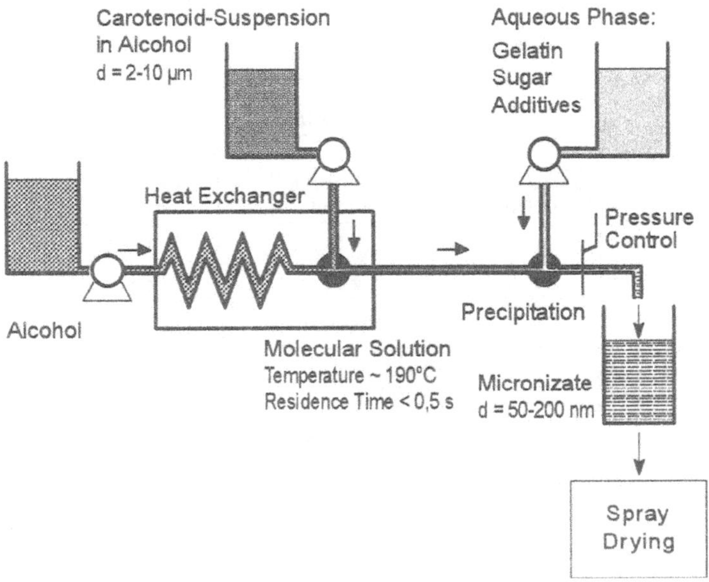

Figure 3. Scheme of the micronization process for carotenoids by controlled precipitation.

With the same technology but by adding an edible oil into the process, e.g. by adding the oil to the coarse suspension, an emulsion type micronizate can be obtained where the active ingredient is present in the final product dissolved in nano-sized oil droplets. The carotenoid loading capacity of the oil droplets increases with the degree of trans → cis isomerization which can be controlled by the temperature/residence time profile during processing.

Still another variety of the process yields micellar solutions of carotenoids with particle sizes in the range of 15 - 25 nm. In this case the coarse raw material is dissolved in a nonionic surfactant, e.g. an ethoxylated fatty acid, at elevated temperature and exposed to water in mixing chamber II whereby a micellar solution of the active ingredient is obtained. The particle size increases with the degree of loading, as will be shown later. Due to their narrow size distributions in the 20 nm range these formulations can readily be used for injection purposes.

A schematic depiction of the three described micronization processes is shown in Figure 4.

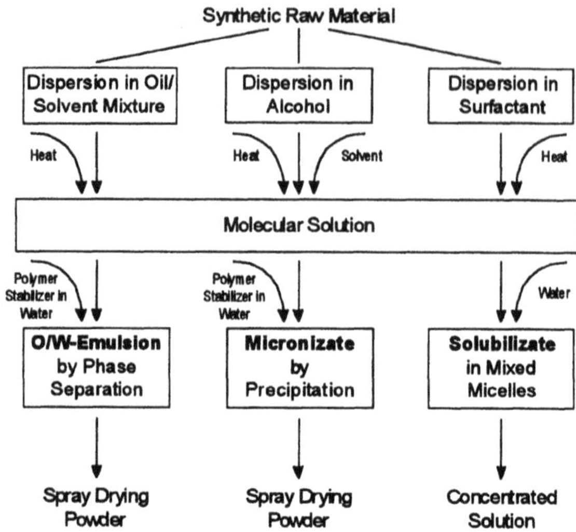

Figure 4. Schematic depiction of various methods for preparing water-compatible carotenoid nano-particles.

3. Characterization of Particle Size, Charge and Colour

3.1. SIZE DETERMINATION AND QUALITY CONTROL BY QUASIELASTIC LIGHT SCATTERING

Qualitatively the characterization of the obtained micronizates is readily possible by electron microscopy. A typical result is shown in Figure 5. By means of specific staining procedures the core/shell-morphology of the hydrosol particles can be elucidated. With OsO_4-treatment the carotenoid core is specifically enhanced in contrast. The complementary contrast enhancement of the gelatin adsorption layer is achieved with UO_2-acetate staining.

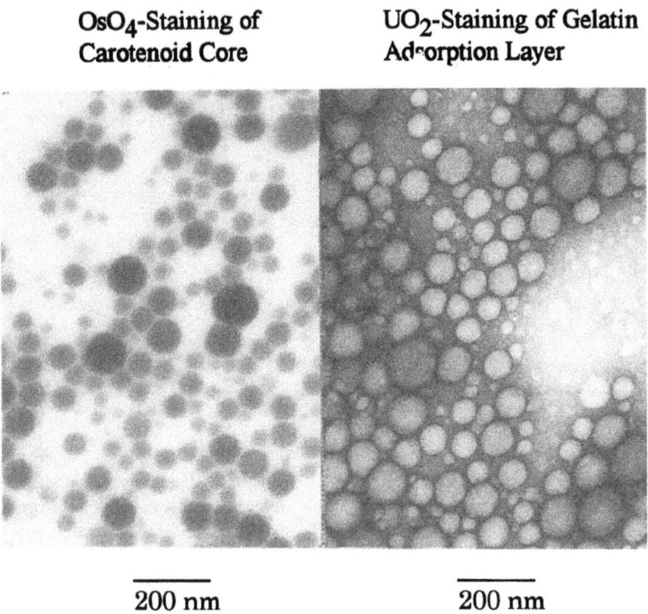

Figure 5. Electron micrographs of beta-carotene micronizates. The core-shell structure is revealed by specific staining.

Quantitative particle size distributions are readily accessible by standard quasielastic light scattering (QELS) as a method of choice [4]. Recent developments in correlation analysis considering the angular dependence of scattered light intensity [5] enable the characterization of polydisperse systems as typically obtained in micronization. According to the principles involved in QELS-size-analysis, the hydrodynamic diameter of the colloidal objects is determined including the polymer adsorption layer. The adsorption layer thickness can be determined by comparing the diameter values of the naked particles with those after polymer adsorption. In a model experiment polymer-free micronizates were obtained in different size ranges by changing the running conditions of the process and with anionic surfactants used as stabilizers. After QELS-size-determination (Ar^+ ion laser, Model 165-06, Spectra Physics; Correlator K 7025, Malvern Instr.) these preparations were exposed to gelatin solutions with increasing concentrations. The results of size determinations before and after gelatin adsorption are shown in Figures 6 and 7. As shown in Figure 6 the mean particle size increases by about 80 nm accounting for an adsorption layer thickness of ~40 nm. Of particular interest, however, is the fact that the observed layer thickness increases with increasing polymer to beta-carotene ratio indicating a multilayer adsorption pattern (Figure 7). This corresponds with the finding that at a given polymer to pigment ratio the observed layer thickness decreases with decreasing primary particle size. A discussion of the respective adsorption isotherms is beyond the scope of this article. They can readily be determined by polyelectrolyte titration [6].

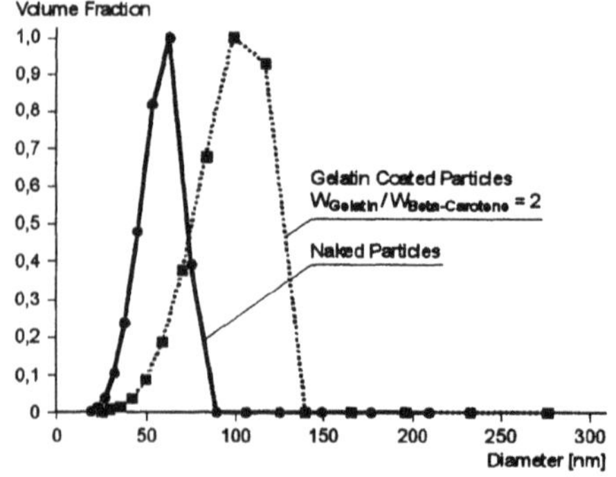

Figure 6. Particle size distribution from QELS studies of beta-carotene micronizate before and after gelatin adsorption.

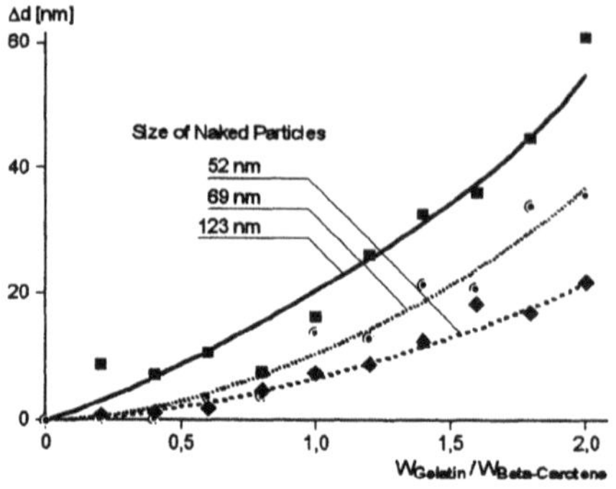

Δd [nm]
Size of Naked Particles
52 nm
69 nm
123 nm
$W_{Gelatin}/W_{Beta-Carotene}$

Figure 7. Thickness of gelatin (Type B 200) adsorption layers in beta-carotene micronizates as a function of gelatin/carotene-ratio.

As will be outlined later, strict control of particle size during production is required to assure product quality as related to colour shade and strength and to bioavailability. Standard QELS, however, requires high dilution of the sample and is, therefore, not readily adaptable as an in situ method at solids contents of practical relevance. This disadvantage can be overcome by employing a fiber-optic version of QELS recently developed in our laboratories [7]. Fiber-optic quasielastic light scattering (FOQELS) is applicable even to highly turbid media like latex dispersions with weight fractions up to 50% [8]. Distances between the reactor and the measuring device of several hundred meters can be bridged by the optical fibers. The method is therefore ideally suited for on-line applications even in explosion hazardous environments [9].

QELS is also readily suited for characterizing the micellar solubilizates described before. Figure 8 depicts the dependence of micelle size on beta-carotene loading. Solutol® HS 15 (ethoxylated 12-hydroxy-stearic acid) as a nonionic surfactant forms micelles of approximately 12 nm in diameter with narrow size distribution. With increasing loading of the micelles with beta-carotene the micellar size increases up to about 25 nm thus defining the limit above which no stable solubilizates can be formed.

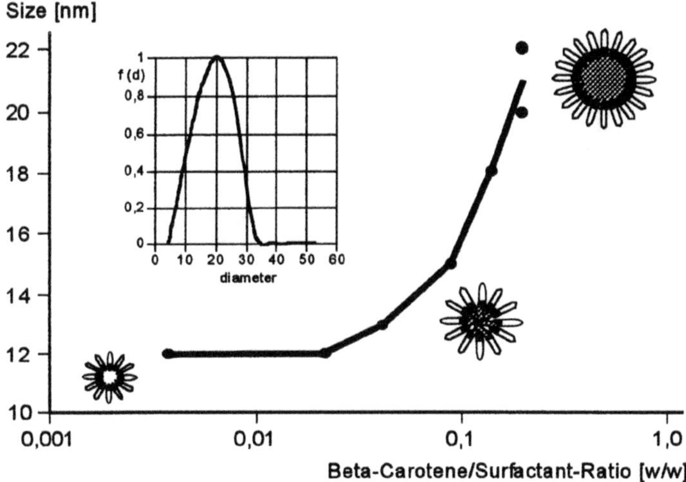

Figure 8. Increase of micellar size (ethoxylated 12-hydroxy stearic acid) with increasing degree of loading with beta-carotene.

3.2. CONTROL OF SURFACE CHARGE AND COLLOIDAL STABILITY BY POLYMER ADSORPTION

Both application features, colour strength/shade and bioavailability, are not only controlled by the primary particle size but also by the state of agglomeration. Hence, colloidal stability has to be maintained during processing of the micronizates and under the conditions of final application. The proper selection of the stabilizing polymers is, therefore, of major concern. For food, feed and pharmaceutical applications various proteins and polysaccharides are accepted as stabilizers [10]. Both the adsorption behaviour and the performance as colloidal stabilizers [11] is largely controlled by their polyelectrolyte character. Their charge density, its dependence on pH and the determination of adsorption isotherms are aspects that are readily accessible by polyelectrolyte titration [6]. For in-process control purposes the determination of the electrophoretic mobility by microelectrophoresis is a convenient method with direct relevance towards the prediction of colloidal stability. Respective mobility data obtained from beta-carotene micronizates at different pH-values and with various polymer stabilizers are shown in Figure 9 (Electrokinetics Analyzer® Model 3000, PenKem Inc.).

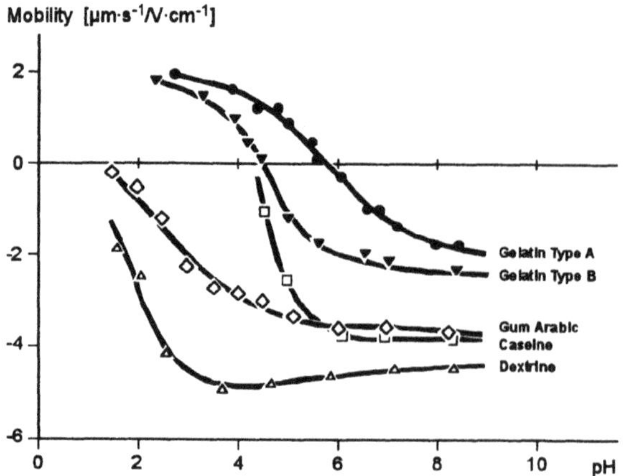

Figure 9. Electrophoretic mobilities of beta-carotene micronizates with various polymer stabilizers.

In the case of polysaccharides, the charge density of the adsorbed polymers is determined by the pH-dependent degree of dissociation of carboxyl groups. Accordingly, in the acidic pH-range the negative electrophoretic mobility approaches values close to zero. The observed colloidal instabilities of polysaccharide-stabilized micronizates in acidic environments and their sensitivity to water hardness ions like Mg^{++} and Ca^{++} points evidence towards an electrostatics dominated stabilization mechanism according to the DLVO-theory [12]. On the other hand, micronizates stabilized with proteins exhibit at specific pH-values isoelectric points (i.e.p.) which are not necessarily identical with the i.e.p. values of the free polymers [13], apparently due to conformational changes during adsorption. Obviously, colloidal stability is again expected to improve with increasing values of the electrophoretic mobility. It is noteworthy, however, that gelatin-stabilized hydrosols exhibit at room temperature even at pH-values close to the i.e.p. good colloidal stability indicating that steric effects are dominating the stabilizing action of gelatin. The performance of gelatin-based formulations is, therefore, quite insensitive to environmental conditions. With caseine as a protective polymer, however, close to the i.e.p. spontaneous agglomeration accurs and stable conditions are only observed at pH levels well above the i.e.p.

The stability of the solubilizates is governed by steric effects as typical for micellar systems based on ethoxylated nonionic surfactants where hydration of the ethoxylate shell provides the thermodynamic basis of enthalpic stabilization [12]. Electrokinetic effects are of no concern in those systems.

3.3. COLOUR SHADE AND STRENGTH AS DETERMINED BY PARTICLE SIZE AND STATE

The electronic absorption spectra of organic pigments are in general determined by their molecular structure and by intermolecular interactions in the solid and solute state, respectively. Particle size effects due to Mie scattering and perhaps quantum size effects if nano-sized particles are involved may further modify the absorption spectra. Figure 10 shows UV/VIS spectra of 5 ppm beta-carotene exhibiting characteristic shifts apparently due to changes in the supramolecular pattern of interaction in solubilizates and micronizates as compared with the spectrum of the molecular solution in cyclohexane ($1^1A_g \rightarrow 1^1B_u$, $S_o \rightarrow S_2$, -Transition).

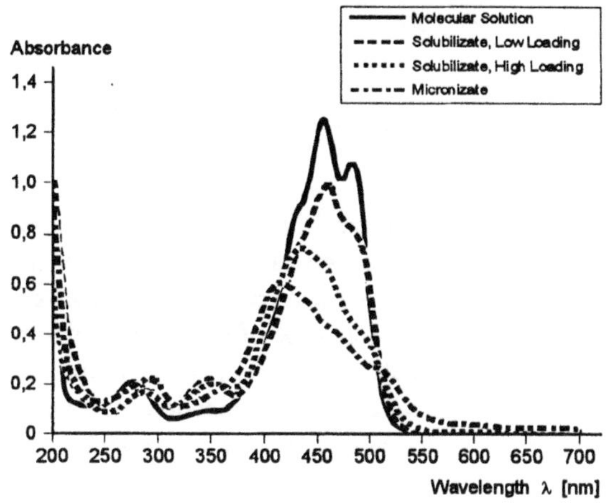

Figure 10. UV/VIS-spectra of 5 ppm beta-carotene. The blue shift indicates card-pack aggregation in solubilizates and micronizates.

It is interesting to note that changes in the absorption spectra of micronizates are also obtained if by altering the process parameters changes in the mean particle size are induced, as shown in Figure 11. All micronizate spectra are characterized by a blue shift as compared to the spectrum of molecular and micellar solutions (Figure 10). In contrast to that observation exhibit suspensions of crystalline beta-carotene red shifted absorption spectra. In Figure 12 an experimental spectrum of a beta-carotene suspension obtained by milling the coarse raw material into the 1 μm size range is compared to calculated spectra based on experimental data from reflection studies of single crystals [14].

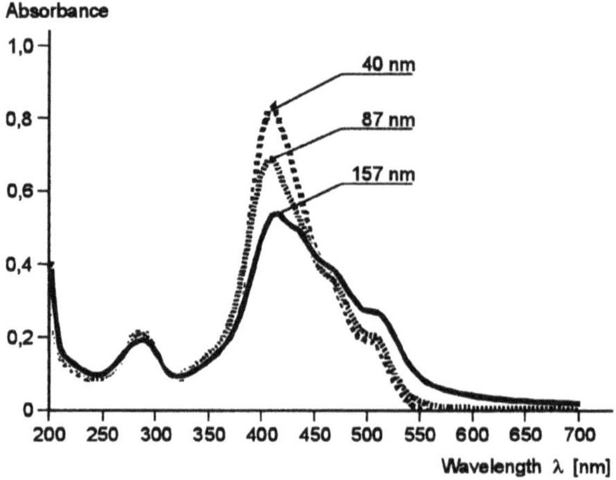

Figure 11. UV/VIS-spectra of 5 ppm beta-carotene micronizates. Dependence on particle size.

For particle sizes of 50 nm and 200 nm Mie-scattering-calculations yield absorption spectra, considering monodisperse spherical particles, which closely resemble the spectra obtained from the crystalline suspension. The observed red shift is typical for "head-to-tail" aggregation [15] and this type of stacking occurs in the beta-carotene unit cell containing two molecules [16]. The observation of a blue shift in preparations obtained by micellization or micronization indicates a "card-pack" type aggregation pattern [17, 18] as being typical for these formulations.

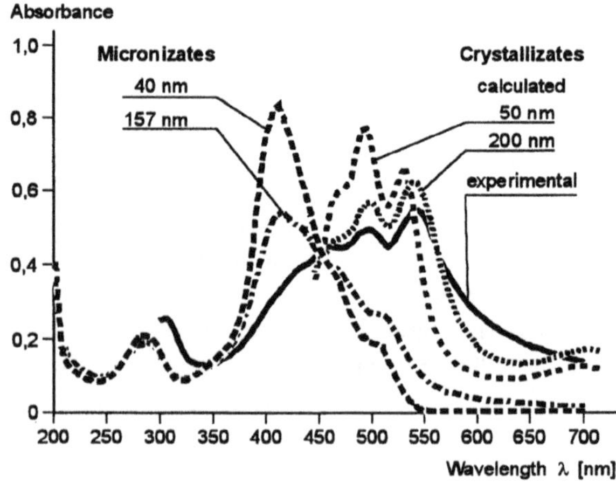

Figure 12. UV/VIS-spectra of beta-carotene micronizates and crystallizates. Dependence on particle size and state.

Obviously, there is a wide variety of packing structures possible depending on the experimental conditions during precipitation. In the case of solubilizates the packing structure may depend on the degree of loading.

A random card-pack type aggregation pattern is also in accord with the X-ray diffraction results indicating an amorphous solid state of the micronizates. Whether the increasing blue shift with decreasing particle size of the precipitates (157 nm → 40 nm) is due to process-induced systematic changes in the packing structure or due to superimposed quantum size effects is at present unclear.

4. Concluding Remarks

The described processes of controlled precipitation and phase separation, respectively, allow for the transformation of the coarse crystalline synthetic carotenoids into nano-sized colloidally stable hydrosols. In the size range between 25 and 250 nm these preparations develop their full potential in colour strength and bioavailability thus establishing a large variety of technological and pharmaceutical applications. Of particular interest are the concomitant changes in the structure of the electronic absorption bands leading to characteristic changes in colour shade. This observation resembles similar findings with different "spectroscopic forms of carotenoids" discovered in biological systems. They are believed to be related to their function as light harvesting pigments in photosynthesis [19] and as photo-protecting agents against molecular oxygen [20]. It appears that proper control of the structure of molecular aggregates is also operative in biological systems [15] besides the formation of specific carotenoid-protein complexes [21] whose structures have not yet been elucidated. It remains to be a challenging perspective to exploit this latter principle for technological applications.

Whether the enhanced biological resorption during gastro-intestinal passage of beta-carotene micronizates of particle sizes < 200 nm, as observed in plasma level studies with carotene-depleted calves [3], is simply a matter of size or related to specific aggregation structures is at present unclear due to our very limited knowledge of the resorption mechanism.

Acknowledgements

The authors are expressing their thanks to Dr. Thomas Beuermann for calculating the absorption spectra of monodisperse beta-carotene nanospheres on the basis of spectral data from single crystals, to Dr. Chi Wu for the dynamic light scattering investigation of the adsorption layer thickness of gelatin on beta-carotene nanoparticles, to Dr. Walter Heckmann for the electron micrographs, and to Dr. Peter Hoppe for the bioavailability investigations.

References

1. Bauernfeind, J.C., Brubacker, G.B., Kläui, H.M., and Marusich, W.C. (1971) Use of carotenoids, in O. Isler (ed.), *Carotenoids*, Birkhäuser Verlag, Basel/Stuttgart, pp. 743-770.

2. Kläui, H.M., Bauernfeind, J.C. (1981) Application Forms, in J.C. Bauernfeind (ed.), *Carotenoids as Colourants and Vitamin A Precursors*, Academic Press, Boston, pp. 86-102.

3. Horn, D. (1989) Preparation and characterization of microdisperse bioavailable carotenoid hydrosols, *Angew. Makromol. Chem.* **166/167,** 139-153.

4. Chu, B. (1991) *Laser Light Scattering*, Academic Press, Boston.

5. Unterforsthuber, K., Lilge, D., Lüddecke, E., Horn, D., and Wu, C.(1992), Determination of particle size distribution by an analysis of intensity constrained multi-angle photon correlation spectroscopy data, *Proc. 5. European Symp. Particle Characterization*, Nürnberg, pp. 70-81.

6. Horn, D. and Heuck, C.C. (1983) Charge determination of proteins with polyelectrolyte titration, *J. Biol. Chem.* **258,** 1665-1670.

7. Auweter, H. and Horn, D. (1985) Fiber-optical quasi-elastic light scattering of concentrated dispersions, *J. Colloid Interface Sci.* **105**, 399-409.

8. Wiese, H. and Horn, D. (1991) Single-mode fibers in fiber-optic quasielastic light scattering: A study of the dynamics of concentrated latex dispersions, *J. Chem. Phys.* **94**, 6429-6443.

9. Wiese, H. and Horn, D. (1993) Fiber-optic quasielastic light scattering in concentrated dispersions: The on-line process control of carotenoid micronization, *Ber. Bunsenges. Phys. Chem.* **97**, 1589-1597.

10. Dickinson, E. and Stainsby, G. (1982) *Colloids in Food*, Applied Science, London.

11. Dickinson, E. (1986) Mixed proteinaceous emulsifiers: review of competitive protein adsorption and their relationship to food colloid stabilization, *Food Hydrocolloids* **1**, 3-23.

12. Hiemenz, P.C. (1977) *Principles of Colloid and Surface Chemistry*, Marcel Dekker, New York/Basel.

13. Maxey, C.R. and Palmer, M.R. (1976) The isoelectric point distribution of gelatin, in R.J. Cox (ed.), *Photographic Gelatin*, Academic Press, London, pp. 27-36.

14. Gaier, K. (1990) Diploma Thesis, *Optische Spektroskopie an Beta-Carotin-Einkristallen*, Universität Stuttgart.

15. Wloch, E., Wieckowski, S., and Turek, A.M. (1987) Spectroscopic characteristics of the long-wavelength absorbing form of beta-carotene, *Photosynthetica* **21**, 2-8.

16. Sterling, C. (1964) Crystal-structure analysis of beta-carotene, *Acta Cryst.* **17**, 1224-1228.

17. Kasha, M. (1963) Energy transfer mechanisms and the molecular exciton model for molecular aggregates, *Radiation Res.* **20**, 55-71.

18. Kasha, M., Rawls, H.R., and El-Bayoumi, M.A. (1965) The exciton model in molecular spectroscopy, *Pure Appl. Chem.* **11**, 371-392.

19. Krinski, N.I. (1971) Function, in [1], pp. 669-716.

20. Krinski, N.I. (1979) Carotenoid Protection against oxidation, *Pure Appl. Chem.* **51**, 649-660.

21. Nishimura, M. and Takamatsu, K. (1960) Studies on a carotene protein complex isolated from green leaves, *Plant Cell Physiol.* **1**, 305-309.

BIORECEPTOR FERROFLUIDS: NOVEL CHARACTERISTICS AND THEIR UTILITY IN MEDICAL APPLICATIONS

P. A. LIBERTI, J. N. CHIARAPPA, A. C. HOVSEPIAN, C. G. RAO
Immunicon Corporation
1310 Masons Mill II
Huntingdon Valley, PA 19006, USA

Abstract:

Aqueous stably coated magnetic material to which bioligands can be coupled has been synthesized. Sizes can be controlled in the synthesis and via magnetic fractionation. Materials from 50 - 150nm exhibit properties of ferrofluids and remain stable colloids; larger materials show particle-like behavior. The discovery that these materials can be separated in magnetic gradients one order of magnitude lower than previously believed leads to the hypothesis that in magnetic fields they form north-south pole chains decreasing the Stokes drag. Magnetic "chaining" has been found to be related to the size and magnetic susceptibility of crystalline cores, degree and thickness of coating, and is inhibited by charge repulsion. Chaining has been utilized to prepare relatively monodispersed materials. It allows the use of lower magnetic open field gradients and makes possible the use and design of inexpensive disposable separator chambers containing no ferromagnetic elements. These materials form stable phases which can be moved and/or controlled by magnetic gradients. Phasing has been used to obtain uniform magnetic collection in continuous flow separators. Under the influence of magnetic gradients, these materials expel non-targeted entities such as cells. This may account for the ability to remove massive numbers of cells in a mixture while retaining small numbers of non-targeted cells. In immunoassay these materials provide significant advantage as regards assay time and performance. Because so little material can be used compared with large magnetic beads, ferrofluid can be collected in magnetic monolayers which allows washing away of unwanted signal without resuspension and signal readout in this bound state. The superior performance of the bioreceptor ferrofluids combined with their fluidics makes them ideal for engineering automated, high throughput analyzers and processors.

E. Pelizzetti (ed.), Fine Particles Science and Technology, 777–790.

Introduction.

Ferrofluids, often described as magnetic liquids or stable magnetic colloids, were discovered by Elmore more than 50 years ago. These materials are single crystals (≅ 10nM) or small crystal clusters of magnetite coated with surfactant, typically detergent, rendering them stable colloids. In solution they have the thermodynamic properties of solutes. Ferrofluids have several unusual properties. For example, their apparent solution density or viscosity is related to applied magnetic field gradient. They have been used in a variety of industrial applications such as in magnetic seals and bearings and for density separation of minerals.

In the early 80's, ferrofluid-like materials were synthesized that employed aqueous soluble polymers or proteins as surfactants [Molday and MacKenzie (1982), Owen and Sykes (1984), Owen and Liberti (1987)]. These materials are produced by titrating Fe^{+2} and Fe^{+3} with base in the presence of polymer or protein to form magnetite and other less magnetic oxides of iron. From electron microscopy, the dispersed phase appears as flocs composed of iron oxide crystals embedded in polymer or protein clusters. Because these materials have magnetic properties that can be used for separations and because they would be expected to react with diffusion limited kinetics, ligands such as antibody, protein A or Streptavidin have been coupled to their polymeric coatings and they have been used in a variety of bioseparations such as in cell isolation and immunoassays. Separations have been accomplished using high magnetic gradients (100 to 150 kGauss/cm) induced on magnetic stainless steel wool packed in columns placed in uniform external fields. Materials so collected are eluted by turning down the field or removing the column from the field.

Unfortunately, these materials suffer from several shortcomings, which are: (1) Resultant flocs have magnetite crystals on their surfaces which are difficult to coat completely and the polarity of surface iron atoms leads to non-specific binding, particularly to mammalian cells; (2) The size and dispersions of these flocs are hard to control; (3) The presence of polymer or protein during synthesis interferes with the formation of magnetite and consequently magnetics are compromised such that gradients higher than might be needed are required for separation. Internal gradient magnetic devices such as stainless steel columns also present issues because they can clog, they are difficult and expensive to manufacture in a disposable form, they are not easy to position in and out of permanent magnetic external field devices or must be used with expensive electromagnets.

This paper reports on (1) the synthesis and properties of a highly magnetic polymer/protein coated ferrofluid, (2) the discovery that these and similar materials form North-South "chains" in magnetic fields which results in the ability to separate them in relatively low gradient magnetic fields, (3) the discovery that they act as phases retaining solute molecules and even cells in the ferrofluid phase while separation is taking place and (4) the ionic strength colloidal stability of these materials. Their performance in immunoassays and in cell separations is also reported.

Synthesis and properties of protein/polymer coated ferrofluid.

As depicted in Fig. 1, aqueous suspensions of magnetite crystals are peptized and dispersed into crystal agglomerates by sonication. Next follows a process which places a layer of protein or polymer onto the surfaces of the agglomerates.

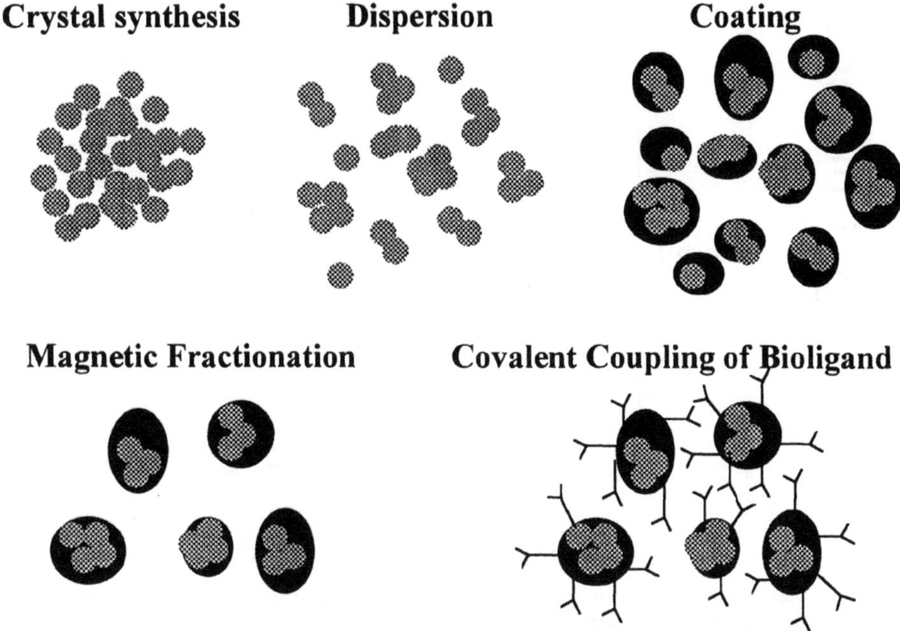

Figure 1. Synthesis of bioactive ferrofluids. Synthesis involves preparation of magnetite crystals (Step 1); dispersion of magnetite into small agglomerates by sonication (Step 2); complete covering of agglomerate surfaces with a protein or a polymer (Step 3); magnetic fractionation of coated agglomerates to get uniform distribution (Step 4) and final Step 5 is covalent coupling of bioligand to agglomerate surfaces.

It is noteworthy that by separating the synthesis of iron oxide crystals from the coating reaction it is possible to produce essentially pure magnetite of defined crystal size and type in contrast with the production of gamma ferric oxide and other less magnetic crystals in those methods where synthesis of the crystals and coating is done simultaneously. Although the process of Fig.1 can be done in a single step where dispersal into agglomerate and the coating reaction is done simultaneously, both reactions compete resulting in sub-optimally dispersed crystal agglomerates.

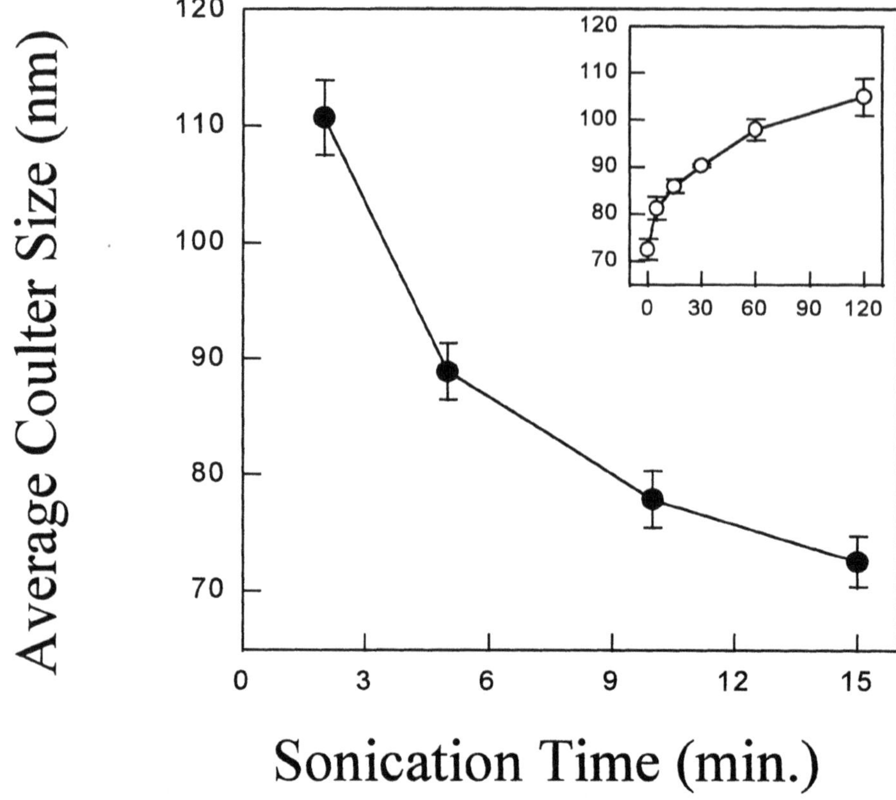

Figure 2. Dispersion of magnetic crystals slurries into agglomerates as a function of sonication time. Resuspended magnetite in proprietary buffer is sonicated in flask cooled to -4°C using 475 watt Sonic Dismembrator. At intervals, dispersed magnetite is removed and diluted in proprietary buffer to measure average size by photon correlation spectroscopy on Coulter N4SD sub-micron particle analyzer. Insert shows growth of dispersed magnetite post sonication as a function of time.

Figure 2 shows the results of dispersion of magnetite slurries into crystal agglomerates as a function of sonication time. The insert shows the growth in size of the agglomerates as a function of time post sonication. Size growth under these conditions is relatively slow and can be totally inhibited by initiation of the coating reaction. The mean size of coated product is a function of duration and power of sonication and, more importantly, is related to nature of the oxide crystals. Figure 3 shows data which indicates that the ability to disperse magnetite into agglomerates is a function of crystal size. In experiments where iron oxide crystals doped with manganese were synthesized

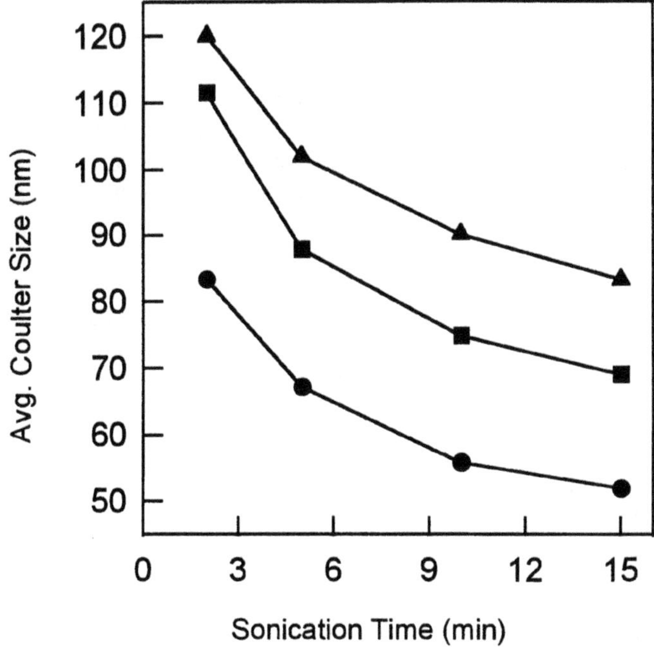

Figure 3. Dispersion of magnetite preparations of different crystal size and character.

to produce a series of crystals of decreasing magnetic susceptibility, a relationship between susceptibility and mean dispersion size was found. Coated materials as small as 30nm could be produced (Tang and Liberti, unpublished results) using the lowest susceptibility crystals. Thus it appears that magnetic moment of crystals significantly affects the ability to disperse them.

TABLE I. *Colloidal salt stability of ferrofluids as a function of protein/polymer coating.*

Coat Protein µg BSA/mg Fe	Size (nm) in 0.75M NaCl Photon correlation spectroscopy (Coulter N4 SD)		
	0 hour	**1 hour**	**25 hour**
73.4	91	settled	settled
117	93	settled	settled
174	103	249	settled
196	113	119	194
198	126	129	136

At neutral pH, dispersed agglomerates of magnetite crystals form large aggregates with time as do poorly coated agglomerates. Size growth of dispersed magnetite is markedly increased by addition of neutral salts or by magnetic separation. For the applications we have targeted, salt stability and the ability to readily disperse magnetically collected material is critical. Bovine or human serum albumins (BSA or HSA) have been found to be excellent coating materials for biological applications. Table I lists data showing that coating increased quantities of BSA onto dispersed crystal agglomerates results in materials which show no size growth (are colloidally stabile) in 0.75 M NaCl. Thus the well coated materials assume the properties of BSA, a highly hydrophilic protein. These well coated ferrofluids can repeatedly be magnetically collected and redisposed without size growth. Figure 4 shows that as coating increases non-specific binding to cells also decreases. Cells have large quantities of sialic acid on their surfaces which imparts them with their negative charge. Most likely that is responsible for binding to the positively polarized iron of bare crystal "spots" on incompletely coated ferrofluids.

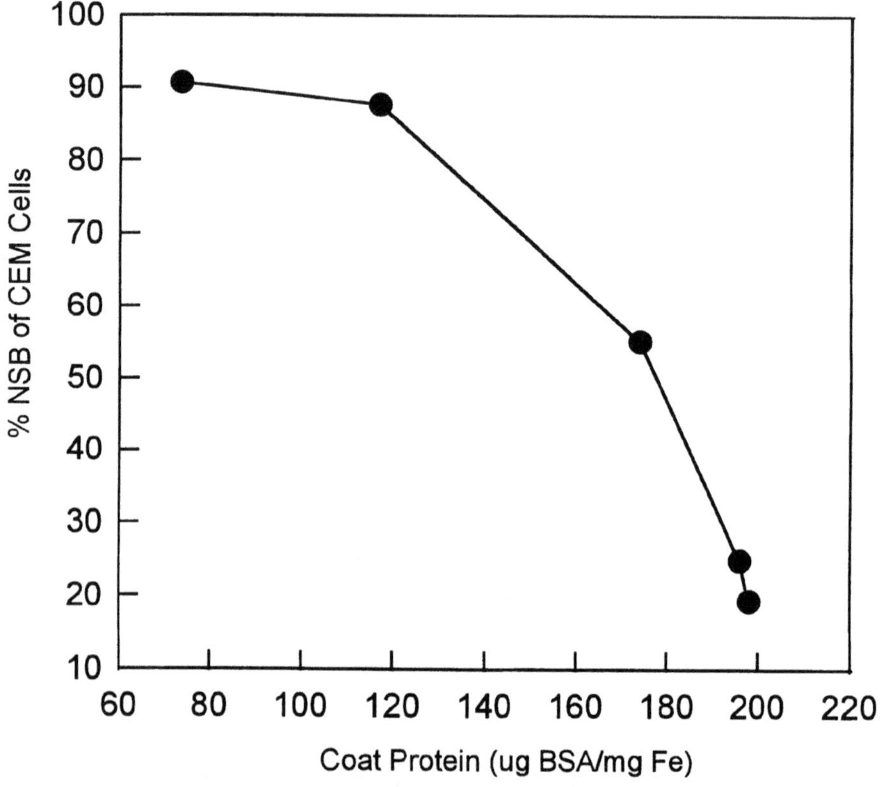

Figure 4. % Non-specific binding (NSB) of ferrofluid to CEM cells as a function of protein/polymer coating. % NSB is measured with a radioactive difference assay. CEM cells are labeled with ^{51}Cr. 160µl of labeled cells (2.5×10^6 cells/ml) are mixed with 160µl of ferrofluid at 20 µg Fe/ml. The mixture is incubated for 5 minutes and ferrofluid is separated with a 5 minute magnetic collection.

Magnetic properties and separation.

By combining the Lorentz equation for the force on a magnetic particle of radius, **R**, with the Stokes law viscous drag for spheres, the velocity, **V**, of a ferrofluid particle in a magnetic field gradient, **∇H** is given by

$$V = 2\,\mu_o M_s R^2\, \nabla H / 9\,\eta \qquad (1)$$

where μ_o, M_s and η are permeability of vacuum, magnetic saturation and viscosity, respectively. From eq. (1), it can be calculated that a 50nm sphere, half filled with magnetite (the approximate density of the crystalline cores of the flocs, mentioned above, or these materials) and coated with a 5nm layer of protein would have a velocity of 0.18 and 0.018 cm/min in a 100 and 10 kGauss/cm gradient field, respectively. These values are consistent with the notion that reasonable velocities can only be achieved in very high gradient fields and verify the rationale for using high gradient magnetic separation technology employing internal gradient devices (>100kGauss/cm) introduced by those who pioneered the use of these materials in bioseparations. On the other hand, we have discovered that very effective separation takes place in fields as low as 8 kGauss/cm for such particles and that is entirely inconsistent with equation (1). For the materials described here which have crystal agglomerate cores in contrast to mixed polymer/crystal cores of the Molday or Owen process, separations can be done in gradient fields of 2 kGauss/cm for 80nm magnetic core preparations. Gradient fields of this magnitude enable the design of open field magnetic gradient devices which can use permanent magnets and have allowed us to perform separations in disposable chambers containing no internal ferromagnetic elements.

TABLE II. *Factors affecting N-S pole chaining of aqueous ferrofluids*

Factor	Supporting Observation
Core Mass and Susceptibility	a. Repeated limited separation results in preferential collection of larger core particles.
	b. Larger cores of lower susceptibility do not collect well
Distance of Closest Approach of Cores	a. Degree of Coating
	b. Thickness of Coating (multi-layers)
	c. Repulsive Charge (collection affected by ionic strength)

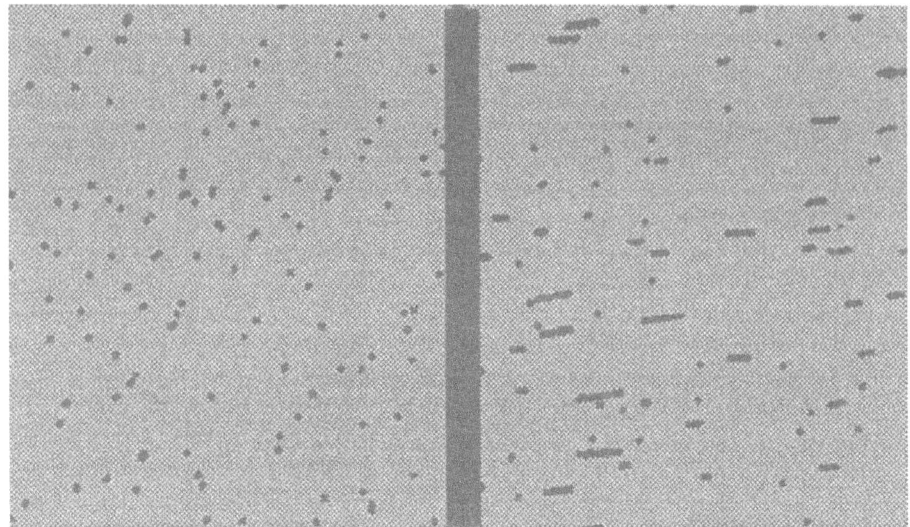

Figure 5. Dynal particles viewed under microscope. Left without magnetic field. Right in the magnetic field.

To explain the discrepancy between theory and experiment as regards magnetic gradient requirement, we theorize that these materials as well as those prepared by simultaneous crystal synthesis and coating form north-south pole magnetic "chains" under the influence of magnetic fields. Figure 5 is a photograph of an observation first made by Ugelstad (personal communication) on magnetic chaining of 5 μ Dynal beads under the influence of a magnetic field, hence there is some basis for our concept. If ferrofluids form chains, then the Stokes drag diminishes dramatically and the use of lower than expected gradient fields for separation is reasonable.

Table II lists the parameters which appear to effect chaining, viz., magnetic dipole of the ferrofluid particle (size and susceptibility) and factors which affect the closeness with which magnetic cores can approach each other such as repulsive charge or the coupling of macromolecules onto coating protein or polymer. Figure 6 shows the effect on separation of low ionic strength. Since size data in the different buffers shows no change nor would the magnitude of the change in size due to charge/hydration layer be expected to affect fractional drag, the observed effect is believed to be due to magnetic chaining.

As might be expected the materials produced in our process have a significant size dispersion. From a wide body of experience, we have utilized concepts consistent with chaining to obtain a relatively narrow dispersion. For magnetic separation of a monodispersed ferrofluid, material remaining unseparated as the separation proceeds should be of the same size and be identical to the mean of the bulk. Further, identical materials should have a characteristic half time of separation. Preparations whose size differ by less than 10% of the mean size can be made by repeated magnetic separation under very defined conditions of gradient, separation chamber geometry and time, discarding in each cycle material which does not collect. Since magnetic chaining is

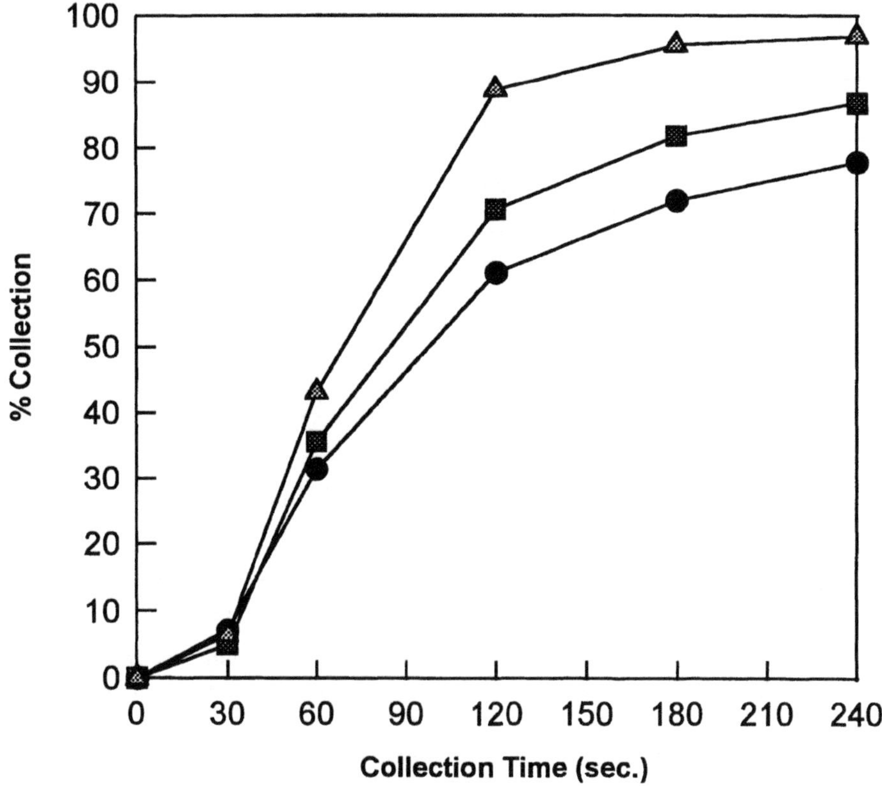

Figure 6. Effect of ionic strength on magnetic collection. Ferrofluids are diluted to a final concentration of 0.25mg Fe/ml in proprietary buffer with/without NaCl to a final NaCl concentration of (●) 0.0mM, (■) 2.0mM, and (▲)100mM, placed in microtiter wells and separated in a 2kGauss/cm magnetic collection device. Collection is determined by the absorbance of supernatant at 490nm.

related to size, the supernatant size for each successive cycle should increase in size and, of course, decrease in mass of material left behind. Table III lists the data from such an experiment as well as magnetic clearance of the final product. Note that when 90% of the material has magnetically cleared the supernatant size is essentially identical with that of the starting material. Even though these size measurements are not terribly sensitive to dispersion, observing supernatant size as a function of amount separated is analogous to determining polymer dispersion by fractionation thereof and determining mole fraction and molecular weights. Hence, it allows us to examine dispersion.

TABLE III. *Magnetic fractionation of BSA ferrofluid and collection kinetics of fractionated ferrofluid.*

Fractionation

# of Cycles	Size (nm)	
	Pellet	Supernatant
1	126	58
2	132	71
3	130	79
4	132	87
5	133	91
6	133	97

Collection Kinetics

Collection Time (sec.)	% Collected	Pellet Size (nm)	Supernatant Size (nm)
30	69	139	132
60	88	139	132
90	99	135	131
120	100	138	
240	100	134	

We have observed two other properties of these materials related to their magnetics the first of which we refer to as "ferrophasing." Figure 7 shows a photograph (left side) of buffer layered over two concentrations of ferrofluid. Layering is facilitated by placing a magnet under the tube during the process. The right side of the photograph shows that in a horizontal magnetic gradient the ferrofluid layers reorientate. This process of reorientation can be repeated many times without loss of the integrity of the layers. In a large vessel, a bolus of ferrofluid can be moved around the periphery of the vessel in a similar fashion. In one flow through separator design we use this phenomenon to place ferrofluid to be separated uniformly along the gradient surface and effectively prevent buildup on the feed end.

Figure 8 is a photograph of the resultant of another ferrofluid phenomenon as it relates to cell separation. When mammalian cells are placed in a microtiter cup and the ferrofluid separated in a radial magnetic gradient, it is observed that as the ferrofluid separates cells will move radially with the ferrophase. At the point when approximately 2/3's of the ferrofluid is collected, cells are ejected from the phase back into the center of the cup. We believe this ejection is a consequence of increase in apparent density of the ferrophase induced by the gradient field so that cells are "buoyed" out of the phase. The ability, as shown below, to massively remove a major population of cells leaving behind rare, untargeted cells of interest, may be a consequence of the latter being ejected from the collecting phase.

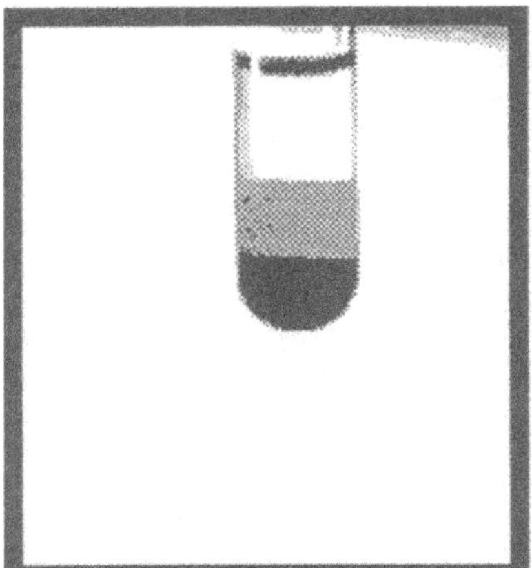

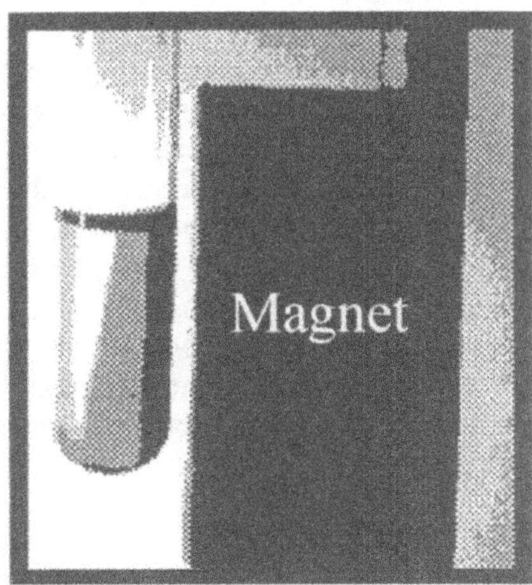

Figure 7. Photograph of Ferrophasing. The layer of two concentrations of ferrofluid and buffer is facilitated by forming layer on a magnetic field (left). When magnet is moved to the side of the tube, ferrofluid layers move as separate phases without mixing (right).

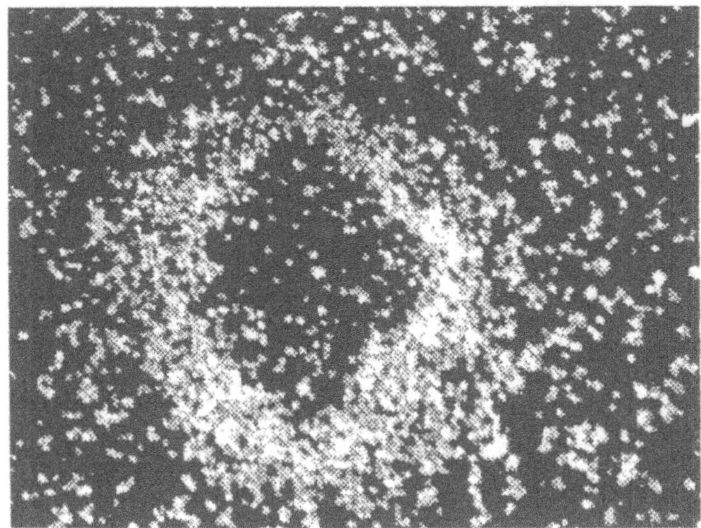

Figure 8. Characteristic resultant pattern of non-targeted cells after ferrofluid separation in radial magnetic field.

Application to rare cell isolation and detection.

The occurrence of tumor cells in blood early in disease or fetal cells in maternal blood early in gestation are events which occur with frequencies of one cell per 10^5 or 10^6. Conventional technologies can either not be used to detect such frequencies or are not economically feasible. Figure 9 shows the results of 10 experiments where 70 to 100 tumor cells were spiked into 10mls of blood and the majority population of cells (leukocytes) depleted by labeling these cells with monoclonal antibody (MoAb), CD45, and next with ferrofluid directed to this antibody. Better than 99% of this population is removed, while at least 70% of the spiked cells are left in solution. The frequency or occurrence of the latter which before the depletion could not be detected can now be readily observed by microscopy or flow cytometry.

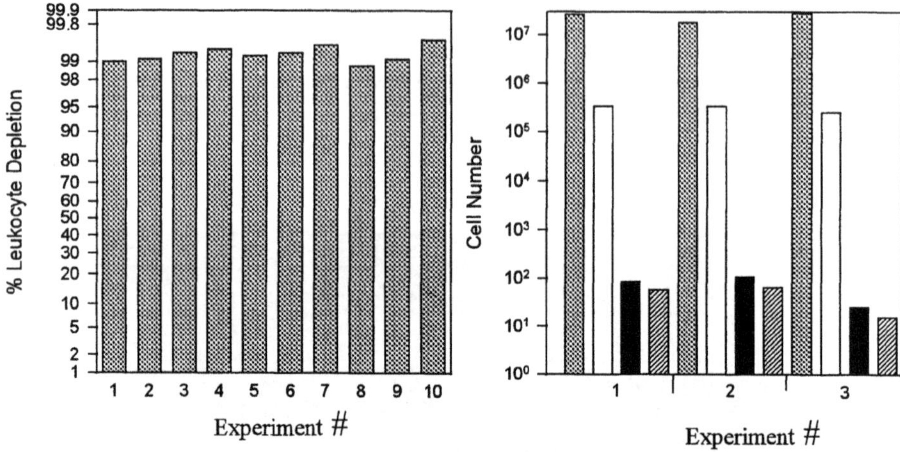

Figure 9. Concentration of non-leukocytes by bulk depletion of peripheral blood leukocytes. Low-density fractions of peripheral blood of 10 normal donors were labeled with CD45 followed by ferrofluids. The labeled cells were separated from the non labeled cells in an internal field magnetic separator, and the non-captured cells were counted. Panel A of the figure illustrates the percentage of cells depleted in each of the experiments. Panel B shows three experiments with excellent recovery of target cells when small quantities of tumor cells are spiked into the blood before cell processing,

Starting cell number — *Cell number after depletion.* — *Number of spiked tumor cells* — *Number of tumor cells after leukocyte depletion.*

Application to immunoassay.

Figure 10 depicts the steps for a ferrofluid based sandwich type immunoassay utilizing biotinylated MoAb and streptavidin conjugated ferrofluid. After ferrofluid captures the biotin-MoAb, it is separated onto the inside walls of a microtiter cup with a radial gradient magnetic field.

TABLE IV. *Signal to noise ratio for ferrofluid based TSH immunoassay*

TSH (uIU/ml)	Signal	Average	S.D.	%C.V.	Signal/Noise
0	2.65 2.74	2.69	0.063	2.3	-
0.05	6.77 6.43	6.55	0.02	3.2	2.4
0.13	16.2 17.2	16.7	0.7	4.2	6.2
0.75	160 165	162.5	3.5	2.1	60
4	991 991	991	0	0	368
20	4527 4484	4505	30.4	0.67	1674
110	17368; 17921	17644	391	2.2	6559

80μl of TSH analyte is incubated with 40μl of antibody mixture containing biotinylated antibody and signal antibody. After 10 minutes, 40μl of Streptavidin ferrofluid (0.5μg) is added and incubated for another 3 minutes. Ferrofluid is then separated (2minutes) followed by 4 washes (4x 300μl) to remove unbound label antibody. 200μl of chemiluminescence signal reagent is added to the ferrofluid and signal is read at 5 minutes.

The advantages of employing ferrofluid vs. large magnetic particles are in kinetics of reaction: 5 - 20 fold faster than 1 - 100 μ magnetic particles and in mass of separation material added (0.5 μg vs. 100 - 300 μg for particles). Because of the smallness of the mass of ferrofluid added, the collected material can actually be monolayered. Unlike magnetic particles, this monolayer can be washed free of contaminates without resuspension. Further chemiluminescent or colorimetric signal can be read directly from the monolayered ferrofluid. Results for a Thyroid Stimulating Hormone (TSH) assay are shown in Table IV. The signal to noise ratio for this system exceeds commercial systems which are considered ultra sensitive and which require incubation times of 30-60 minutes compared to the total time of 15 minutes for the ferrofluid system. The advantages for automation inherent for liquid reagents such as these compared with magnetic particle slurries is readily apparent.

In summary, we have developed biologically active novel and stable magnetic colloids -- ferrofluids -- which have many unusual characteristics, and we have discovered how to use them in bioseparations (immunoassay and cell separations) in a simple and highly efficient manner. The systems developed are ideal for automation.

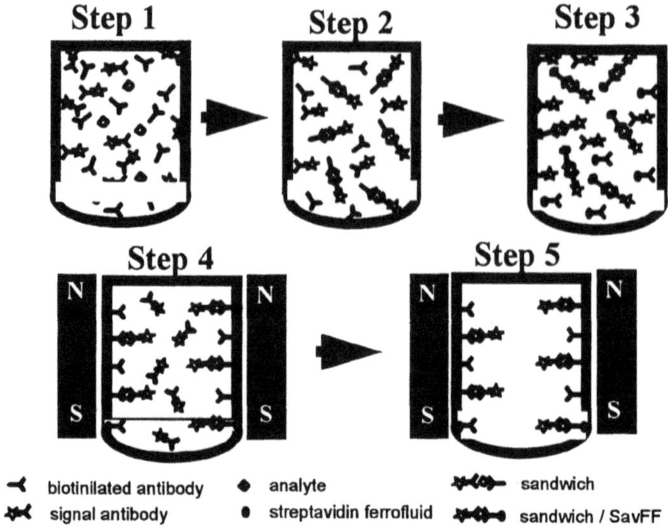

Figure 10. Schematic representation of steps in ferrofluid based sandwich immunoassay. Step 1 shows incubation of biotinylated capture MoAb and signal MoAb with analyte. Step 2 shows sandwiches which link capture and signal MoAbs through analyte. Ferrofluid binds capture MoAb in Step 3 and is separated in Step 4 where it is magnetically immobilized for washing and detection of signal (Step 5).

The authors acknowledge Dhanesh Gohel, Brian Feeley, Steve Gross, Richard Horan, Yuzhou Wang, David Chianese and Leon Terstappen for many contributions to this work and Tammy Henderson and Ellen Turk in the preparation of this manuscript.

References

1. Elmore, W. C. et al., *Physical Review*, 54, 1092, 1938
2. Molday, R.S. and MacKenzie, D. (1982), *J. Immunol. Meth.* 52, 353.
3. Owen,C.S. and Liberti, P.A. (1987), *In Cell Separation: Methods And Selected Applications,* Pretlow and Pretlow, Eds., 4, 259.
4. Owen, C.S. and Sykes, N.L. (1984), *J. J. Immunol. Meth.* 73, 41.

SOME PROPERTIES OF HYDROPHILIC LATICES: FROM HAIRY PARTICLES TO MICROGELS

B. VINCENT
School of Chemistry, University of Bristol,
Cantock's Close, Bristol BS8 1TS, U.K.

1. Introduction

The surfaces of most latex particles prepared in water are naturally hydrophobic, the classical example being polystyrene latex. (Kinetic) stability to coagulation is due to the interparticle electrostatic forces, associated with the terminal charge groups on the polymer chains, arising from the initiator used; such groups tend to migrate to the particle/water interface. Thermodynamic *stability* (as opposed to *metastability*), may be achieved if the interparticle van der Waals forces are sufficiently reduced as to be negligible compared to kT. Swelling of the latex particles by the medium (water) can achieve this. Such swelling may be confined to the interfacial region ("hairy particles) or extend essentially across the body of the latex particles (microgels).

"Hairy" particles may be prepared in a number of ways, e.g. (i) by post-grafting a water-soluble polymer onto the latex particles; (ii) by copolymerising relatively small amounts of a water-soluble monomer (e.g. acrylic acid) or a macromonomer (e.g. polyethylene glycol methacrylate), *in situ*, usually towards the end of the latex polymerisation. Microgel particles, on the other hand, are normally prepared (i) by incorporating significantly higher amounts of a water-soluble monomer at the beginning of the polymerisation, or (ii) by the emulsion polymerisation of the monomer of a polymer which is in soluble at the (normally high) temperature of the polymerisation, but which becomes water-soluble at the room temperature. Incorporation of a small percentage of a cross-linking monomer is necessary in this case.

E. Pelizzetti (ed.), Fine Particles Science and Technology, 791–813.

Hydrophilic latices of this type do not normally coagulate at the relatively low concentrations of added electrolyte required to induce rapid (diffusion controlled) coagulation of *hydrophobic* latex particles. Flocculation may be induced, however, if the solvency conditions are altered to sufficiently reduce the swelling of the water-miscible polymer, at the surface ("hairy" particles) or throughout the bulk of the particles (microgels). Alternatively, flocculation may be brought about by the addition of a controlled amount of a water-soluble polymer or polyelectrolyte. Two regimes may be distinguished here: (i) the addition of relatively small amounts of a polymer which associates with the latex surface polymer, giving rise to *bridging* flocculation; (ii) the addition of larger quantities of a *non*-associating polymer, giving rise to *depletion* flocculation.

In this paper some examples will be described which illustrate these stability/flocculation aspects, both for "hairy" particles (polystyrene with grafted polyethylene glycol chains) and for microgel particles (poly[N-isopropylacrylamide], cross-linked with bisacrylamide).

2. Bridging and Depletion Flocculation

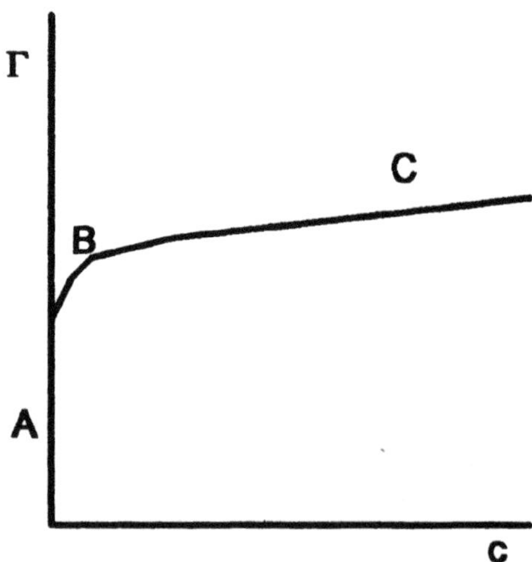

Figure 1. Schematic polymer adsorption isotherm. Γ is the amount of polymer adsorbed per unit area; c is the equilibrium concentration of free polymer in solution.

Somewhat paradoxically, the addition of a polymer (or polyelectrolyte) to a colloidal dispersion may lead to either enhanced stability or to induced flocculation of the particles. This situation can be understood in terms of the adsorption isotherm for the polymer on the particles concerned. A typical case is illustrated schematically in Figure 1 above. Three "regions" of the adsorption isotherm may be distinguished, as illustrated. Region B is where steric repulsion between the particles, leading to enhanced stabilisation, occurs. Regions A and C, on the other hand, may lead to flocculation: bridging flocculation in region A; depletion flocculation in region C, i.e. where there is a large excess of free polymer in solutions.

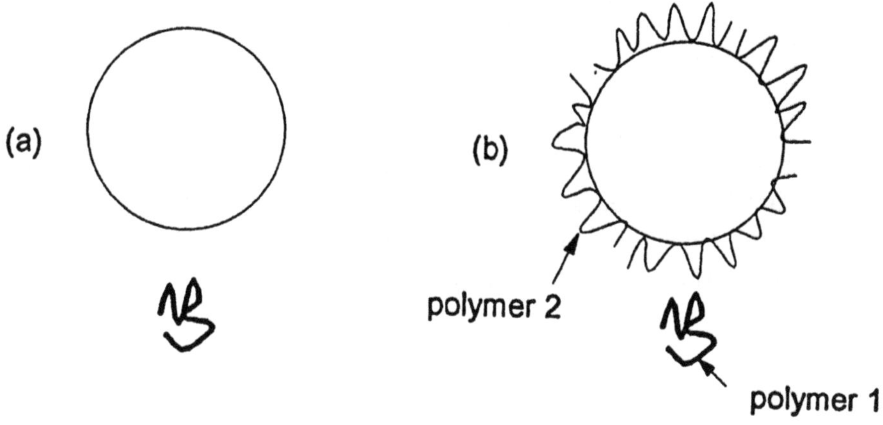

Figure 2. Situations in which depletion flocculation may occur: (a) bare particles with non-adsorbing polymer; (b)polymer-covered particles with polymer in solution (polymers 1 and 2 may be of the same type or different).

Figure 2 illustrates two cases where depletion flocculation may occur. Case (a) may be considered as "hard" particles and case (b) "soft" particles. Which category hairy particles and microgel particles fall in depends, as this paper will illustrate, on their degree of coverage or selling, respectively.

3. Hairy particles

3.1 INTRODUCTION

Whether a polymer adsorbs onto a surface or not depends on the χ_S parameter i.e. the net surface/segment-surface/solvent interaction parameter. There exists a critical value for χ_S (χ_{SC}) only above which adsorption occurs [1]. It is possible to vary χ_S systematically by varying the solution conditions. In this study we do this for the following system: aqueous dispersions of hairy particles, *viz* polystyrene (PS) latex particles, carrying terminally-anchored poly(ethylene oxide)(PEO) chains (i.e. PS-g-PEO particles), to which poly(acrylic acid) (PAA) is added. PAA has a pK_a value of 5.5 [2]. Hence by varying the solution pH, the degree of ionisation of the acid hydrogen on the PAA chains may be varied. At low pH values, there is little ionisation, and the PAA chains can be expected to "adsorb" onto the PEO chains of the latex particles (through H-bonding), whereas at high pH values no such adsorption is expected. This situation reflects the well-established behaviour of mixtures of aqueous PEO and PAA solutions: coacervation is observed at low pH values, but not at high pH values [3]. Figure 3 shows the pH dependence of the precipitate yield for mixtures of various poly(carboxylic acids) and PEO [4].

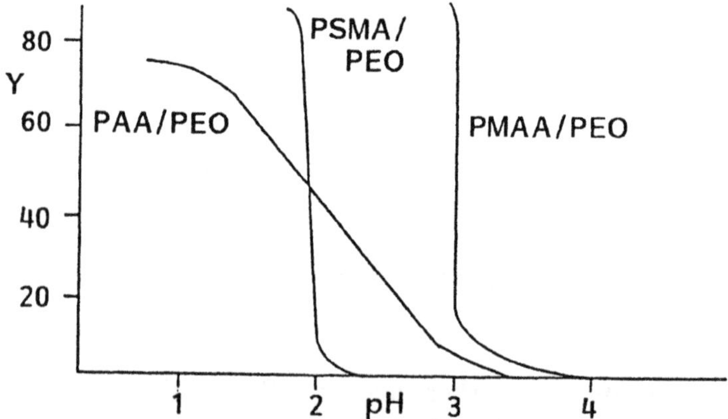

Figure 3. pH-dependence of precipitate yield (Y%) for various poly(carboxylic acids) + PEO: PAA = poly(acrylic acid); PMMA = poly(methacrylic acid); PSMA = a copolymer of styrene and maleic acid. [4]

For the PS-g-PEO latex plus PAA systems, therefore, bridging flocculation may be anticipated at *low* pH values. Indeed, Evans and Napper [5] have already reported such behaviour. Moreover, a similar recipe may be used in practice in "electrocoat" paint systems [6]. At *high* pH values, on the other hand, one might expect to see depletion flocculation. This is indeed what is seen.

3.2 MATERIALS

3.2.1 *Polystyrene-g-(Poly(ethylene oxide) Particles*

A PS-g-PEO latex was prepared using the method described by Cowell and Vincent [7] and by Bromley [8]. This involves the dispersion copolymerisation of styrene and PEO-methacrylate, together with a small quantity of divinylbenzene as a cross-linker, using α-azoisobutyronitrile as the initiator, in aqueous media. The mean-particle diameter of the latex

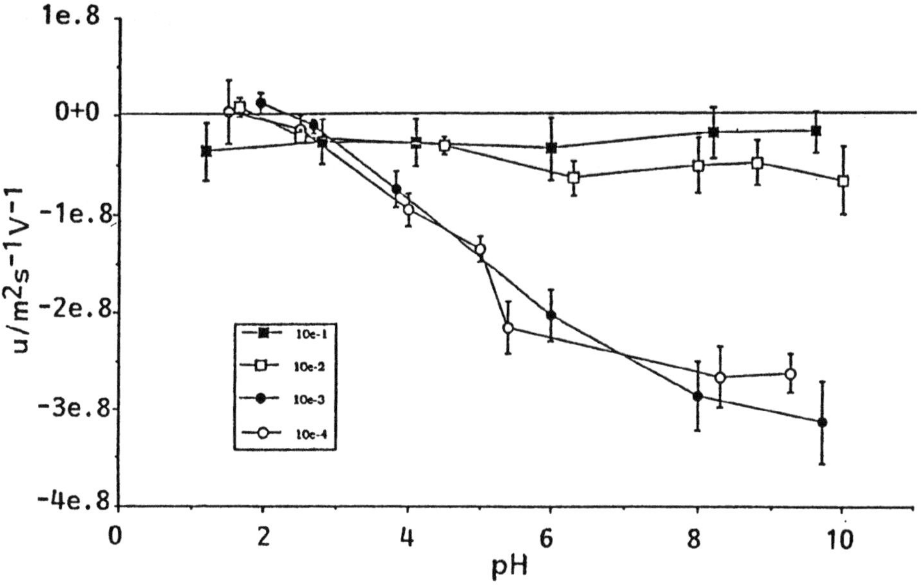

Figure 4. Electrophoretic mobility (u) of PS-g-PEO particles, as a function of pH and NaCl concentration (see inset)

particles was found from transmission electron microscopy to be 500 ± 37 nm. The PEO chains had an M_n value of 2000. Cosgrove and Ryan [9] have reported a value of 3.6 nm for the thickness (δ) of the hydrated PEO sheath on similar latex particles. The critical flocculation temperature of the PS-g-PEO latex in 0.26 mol dm^{-3} aqueous $MgSO_4$ solution was found to be 58 ± 1°C. This value is identical to the value reported previously by Bromley [8] for a similar latex, and is close to the theta-temperature for PEO in 0.26 mol dm^{-3} $MgSO_4$ solution [10].

In order to check for any possible electrostatic interactions between the latex particles, their electrophoretic mobility was monitored as a function of pH and NaCl concentration, using a *Pen Kem 3000* apparatus. The results are shown in Figure 4. As may be seen, the particles become increasingly negatively charged with increasing pH, particularly at low ionic strengths; this effect was found to be reversible. The maximum mobility value at pH 9 (-3×10^{-8} m^2 s^{-1} V^{-1}) in 10^{-4} mol dm^{-3} NaCl solution, corresponds to a zeta potential of -46 mV. The charge is probably due to the presence of weak carboxylic acid groups at the polystyrene core surface, arising from the initiator.

3.2.2 *Poly(acrylic acid)*

The preparation of monodisperse samples of poly(acrylic acid) (PAA) based on the anionic polymerisation, and subsequent hydrolysis of t-butyl acrylate, has been described by Miller and Raubut [11] and by Leon, Vincent and Cawdery [12]. Because GPC analysis of aqueous polyelectrolyte solutions is not straightforward, the GPC analyses were carried out on the parent poly(t-butylacrylate) polymers. ^{1}H n.m.r. analysis of the PAA samples showed that complete hydrolysis of the parent polymers had been achieved. Thus, it was straightforward to calculate the (number-average) molecular weights (M_n) of the PAA samples produced, and the polydispersity index (p.i.). The p.i. values for the range of PAA samples produced all lie in the range 1.1 to 1.2 (based on the parent polymer). Two samples were subsequently used in the work reported in this paper: PAA 1 having an M_n value of 48,000 and PAA 4 having a M_n value of 111,000.

It should be noted that the PAA samples were prepared with a terminal diphenylmethylene group (a 1:1 adduct of t-butyl lithium and diphenylethylene was used as the initiating species for the anionic polymerisation of t-butylacrylate). This u.v. chromophore enables the concentration of the PAA samples in solution to be readily determined using u.v. absorption (peak at 215 nm).

3.3 RESULTS

3.3.1 *Adsorption isotherms for PAA onto PS-g-PEO latex*

The adsorption isotherm for PAA 1 onto the PS-g-PEO latex particles was determined as a function of pH, at a fixed background NaCl concentration (10^{-2} mol dm^{-3}). These were established by monitoring the PAA concentration, prior to and after, equilibration of latex plus PAA solutions, using the u.v. method referred to in §3.2.2. The results are shown in Fig. 5.

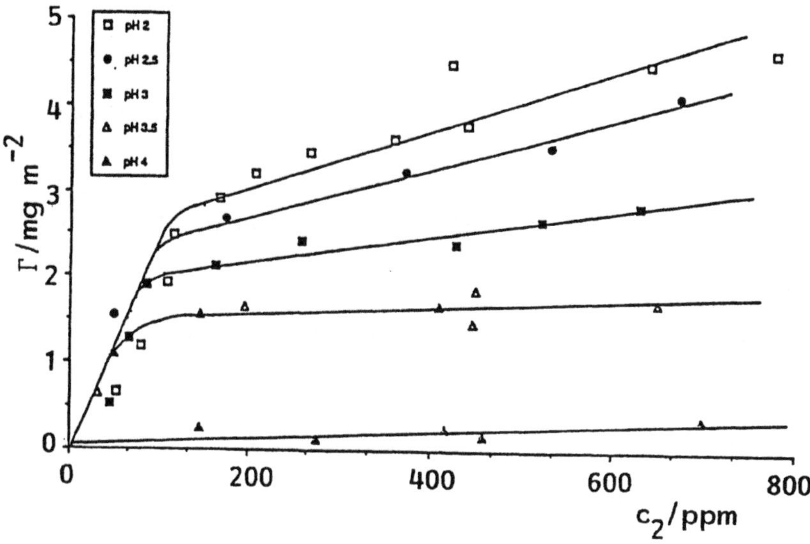

Figure 5. Additional amount (Γ) as a function of equilibrium PAA concentration (c_2) at various pH values. [NaCl]=10^{-2} mol dm^{-3}.

As may be seen, the adsorbed amount decreases, at fixed PAA concentration, with increasing pH, until at pH 4.0 very little adsorption occurs. These results reflect the PAA/PEO coacervation behaviour shown in Fig 3. Clearly, at pH > 4 only a small fraction of the carboxylic acid groups of the

PAA chains remain undissociated, so that H-bonding with the PEO chains on the latex particles (i.e. χ_s) is greatly reduced.

With increasing NaCl concentration, it was found that the PAA adsorbed amount, at a given pH value, increased slightly (~ 20% over the NaCl concentration range 10^{-4} to 10^{-1} mol dm^{-3}). This presumably reflects the compaction of the chains as the intra-chain repulsion is increasingly screened with increasing NaCl concentration.

3.3.2 *Stability Studies*

The following solutions were prepared, at the required pH values:

1) PS-g-PEP latex of known particle volume fraction and at twice the required final NaCl concentration.
2) PAA solution
3) Water

x g of (2) was then mixed with y g of (3) (x+y=1) in a small sample tube, before adding 1 g of (1). The final PAA concentrations are, therefore, quoted as weight fractions (W_2) (rather than volume fractions, ϕ_2). In a given experiment a series of stoppered tubes was prepared containing mixtures for which all parameters were kept constant, except that the PAA concentration was systematically increased.

The tubes were allowed to stand at ambient temperature for 24 h and evidence for flocculation assessed visually. This was quite straightforward and reproducible for latex volume fractions in the range, ϕ=5 x 10^{-4} to 10^{-1}. In some cases turbidity measurements were also made. Good agreement with the direct visual assessment method was obtained. The PAA weight fraction where flocculation is first observed and that where restabilisation is observed, are designate $W^{\dagger}_2$ and $W^{\dagger\dagger}_2$, respectively.

Figure 6 shows a "stability map" for the PS-g-PEO latex (particle volume fraction ϕ-5 x 10^{-3}) plus PAA (M=111,000), in a background NaCl concentration of 10^{-4} mol dm^{-3}. This takes the form of a PAA wt. fraction (W_2) verus pH graph. Two instability regions may be identified: one below pH 2.8, the other above pH 3.5. In the low pH region, where PAA adsorption occurs, "fluffy" flocs form very quickly and seemingly irreversibly, over a certain range of polymer wt. fractions ($W^{\dagger}_2 < W_2 < W^{\dagger\dagger}_2$. This would seem to be bridging flocculation. In the high pH region, on the other hand, where (virtually) no PAA adsorption occurs, much weaker flocculation occurs, again over a certain PAA wt. fraction range. In

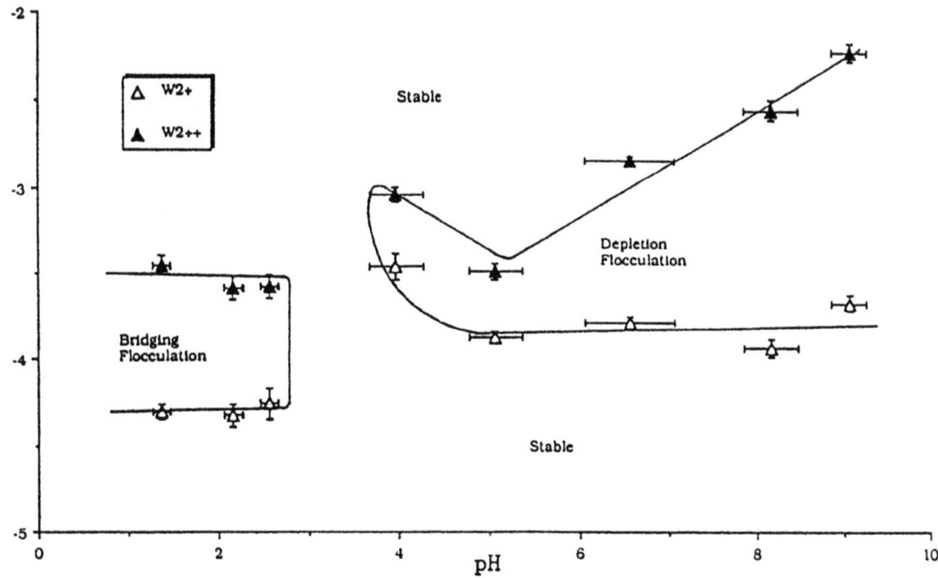

Figure 6. Stability map for PS-g-PEO latex plus PAA (M=111,000) $\phi=5\times10^{-3}$, [NaCl]=10^{-4} mol dm^{-3}.

this case flocculation occurs much more slowly (~ 24 h), and the system effectively forms two co-existing colloidal "phases": a low volume fraction ("gas-like") phase in equilibrium with a condensed ("solid-like") phase. This is typical of depletion flocculation behaviour, as described previously, for example by Vincent et al. (1986). No flocculation was observed over the pH range 2.8 to 3.5

Figure 7 shows a similar stability map to that shown in Figure 6, except that the background NaCl concentration has been raised to 10^{-1} mol dm^{-3}. Again, two instability regions occur, one at low pH, the other at higher pH values. There seems to be very little effect of salt concentration on the bridging flocculation region at low pH, but there is a much stronger effect on the depletion flocculation region at high pH. The onset of depletion flocculation occurs at a higher minimum pH value: ~6 at 10^{-1} mol dm^{-3} NaCl. A much higher value of $W_2^{\dagger}$ is observed at the higher NaCl concentration. Also no restabilisation could be achieved by the highest PAA

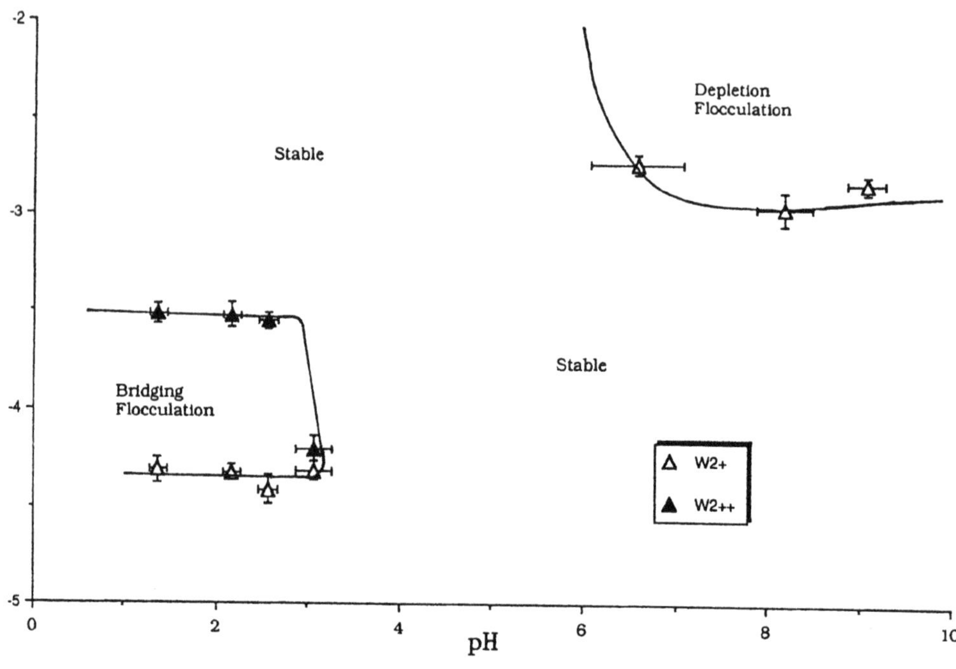

Figure 7. Stability map for PS-g-PEO latex plus PAA (M=111,000). ϕ=5 x 10^{-3}, [NaCl] = 10^{-1} mol dm^{-3}.

concentration (W_2 ~ 0.05) which could be studied before the high viscosity of the solution made the experiments intractable.

In all the experiments described so far the volume fraction (ϕ) of the latex particles was fixed (at 5 x 10^{-3}). The effect of varying ϕ, over the range from 5 x 10^{-4} to 10^{-1}, on $W_2^{\dagger}$ and $W_2^{\dagger\dagger}$ was investigated for both the bridging and depletion flocculation regions (at 10^{-4} mol dm^{-3} NaCl). In the case of bridging flocculation it was found that both $W_2^{\dagger}$ and $W_2^{\dagger\dagger}$ increased steadily with increasing ϕ. In the case of depletion flocculation, however, both $W_2^{\dagger}$ and $W_2^{\dagger\dagger}$ *decreased with increasing* ϕ.

The effect of varying the molecular weight of the PAA used was also investigated. In the case of bridging flocculation, at low pH values, flocculation was only observed with PAA 4 (M.W. 111,000). No flocculation was seen with the lower M.W. PAA, sample 1 (M.W.=48,000), over the NaCl concentration range 10^{-4} to 10^{-1} mol dm^{-3}. A "cut-off", minimum molecular weight of added polymer, for the onset of depletion flocculation, has been reported for other systems, for example, for dispersions of silica particles, carrying terminally-anchored polymethylmethacrylate chains, in dioxan, to which free polymethylmethacrylate was added [13].

3.4 DISCUSSION

3.4.1 *Bridging Flocculation*

It is clear that the flocculation behaviour observed at low pH's is bridging flocculation, and reflects the adsorption of PAA onto the PS-g-PEO particles at coverages less than saturation (Figure 5). The $W_2^{\dagger}$ and $W_2^{\dagger\dagger}$ "boundaries" shown in the stability maps (Figures 6 and 7) are not strictly "thermodynamic" boundaries in this case; rather, they are "kinetic" boundaries. They define the range of W_2 values where the rate of flocculation is sufficiently fast that flocs become visible over a timescale of minutes.

Since the PS-g-PEO particles have only small electrophoretic mobilities in this pH range (Figure 4), corresponding to a maximum value for the zeta potential of ~ 5 mV, electrostatic repulsion between the particles can be ignored. It is not too surprising, therefore, that little effect of NaCl concentration on $W_2^{\dagger}$ and $W_2^{\dagger\dagger}$ increase with increasing particle volume fraction, ϕ imply reflects the increasing particle surface area, and therefore the higher initial PAA concentration required to achieve a similar level of coverage. No quantitative analysis of the bridging flocculation results is attempted here.

3.4.2 *Depletion Flocculation*

In the high pH flocculation regimes (Figures 6 and 7), where weak depletion flocculation is occurring, the systems effectively phase separate into two co-existing colloidal phases. This is a phenomenon which is often observed in the case of depletion flocculation [14]. Further evidence for this is the fact that $W_2^{\dagger}$ and $W_2^{\dagger\dagger}$ in this case, *decrease* with increasing particle volume fraction (i.e. the opposite trend to that observed in the case of bridging flocculation). This behaviour has also been observed before, for example, by Cowell and Vincent [7], who studied the effect of adding a *neutral* polymer, poly(ethylene oxide) to PS-g-PEO particles similar to those used here.

In order to try to interpret, at least in a semi-quantitative way, the trends shown in Figures 6 and 7 for the depletion flocculation behaviour at high pH values, the total interparticle pair interaction energy (V) was calculated as a function of particle separation (h). V(h) is given by,

$$V(h) = V_A(h) + V_E(h) + V_{dep}(h) \qquad (1)$$

where $V_A(h)$ is the van der Waals interaction energy between the particles; $V_E(h)$ is the electrostatic (electrical double layer) interaction energy, associated with the particle charge at high pH (Figure 4), and $V_{dep;}(h)$ is the depletion interaction energy.

$V_A(h)$ is negligible if the assumption is made that the Hamaker constant of the (hydrated) PEO sheath around the PS particle cores is the same as that of the medium

$V_E(h)$ may be calculated using the following standard expression,

$$V_E = 2\pi\,\varepsilon a\,\varsigma^2\,\ell n\left[1+\exp(\kappa h)\right] \qquad (2)$$

In eqn. (2), ε is the permittivity of the medium: a is the particle radius; ζ is the zeta potential of the particles, calculated from the experimental electrophoretic mobility values, using the O'Brien and White formulation [15]. In calculating ζ and κ (the Debye-Hückel screening parameter), consideration has to be given to the value of the electrolyte concentration between the particles. When two particles approach closely then PAA chains are effectively excluded from the region between the particles (the basis of the depletion interaction). Hence, a Donnan salt effect is set up, such that the electrolyte concentration between the particles is greater than in the bulk solution containing the polyelectrolyte chains. The standard correction term for the Donnan effect across a semi-permeable membrane was applied in calculating the electrolyte concentration between the particles. With regard to the depletion interaction itself, Fleer, Scheutjens and Vincent [16] (SFV) showed that, for dispersions of hard-sphere particles (which to a first approximation the PS-g-PEO particles may be considered to be), then $V_{dep}(h)$ is given by,

$$V_{dep} = 2\,\pi a\,\Pi\,\Delta^2\left(1-\frac{h}{2\Delta}\right) \qquad (\Delta \ll a) \qquad (3)$$

or when h = 0 (particle contact),

$$V_{dep} = 2\,\pi a \Pi \Delta^2 \tag{4}$$

where Π is the osmotic pressure of the bulk polymer solution, and Δ is the effective depletion layer thickness for polymer chains at a non-adsorbing surface ($\chi^s < \chi_c{}^s$). At low polymer concentrations the SFV theory predicts that $\Delta = 1.4\ r_g$, where r_g is the radius of gyration of the polymer coils in dilute solution. Thus, evaluation of $V_{dep}(h)$, as a function of bulk PAA weight fraction (W_2) or volume fraction (ϕ_2), requires knowledge of $\Pi(\phi_2)$, and $\Delta(\phi_2)$. The former quantity can be obtained directly from osmotic pressure measurements. Figure 8 shows the $\Pi(W_2)$ data obtained in this work for PAA solutions, at different NaCl concentrations. These were obtained using a Knauer membrane osmometer with cellulose acetate membrane.

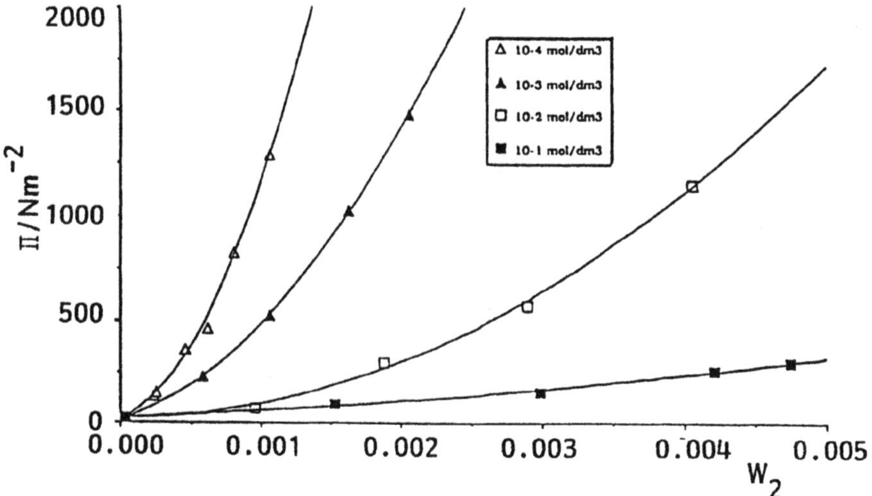

Figure 8. Osmotic pressure (Π) as a function of weight fraction (W_2) of PAA 4, in various NaCl concentrations at pH9. T = 25 ± 0.1°C.

As expected, for a given W_2 value, Π decreases with increasing electrolyte concentration, due to the screening of the intramolecular interactions within the polyelectrolyte chains.

With regard to Δ, Vincent [17] has derived the following analytical expression for $\Delta(\phi_2)$ for *neutral* random coil polymers.

$$\frac{\Delta}{\Delta_0^2} - \frac{1}{\Delta} = \frac{\Pi}{2kT}\left(\frac{v_2}{\phi_2}\right)^{2/3} \tag{5}$$

where Δ_0 is the value of Δ in the limit of zero polymer concentration and v_2 is the molar volume of the polymer chains. However, *polyelectrolyte* chains only approach random coil behaviour at *high* background electrolyte concentrations. At *low* background electrolyte concentration the behaviour is more rod-like. For the case of *rigid rods*, Auvray [18] has suggested that Δ is *independent* of ϕ_2.

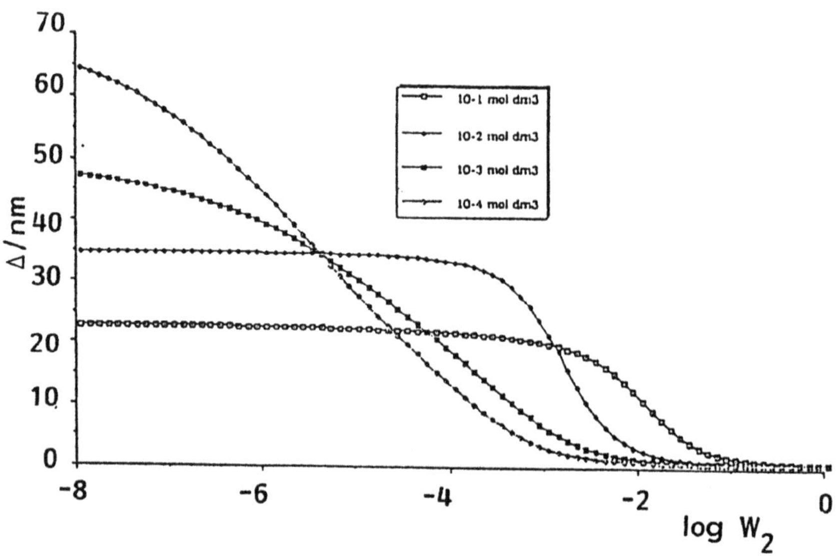

Figure 9. Apparent depletion-layer thickness (Δ) as a function of PAA weight fraction (W_2) for PAA 4, for various NaCl concentrations.

Figure 9 shows the dependence of Δ on W_2 for PAA 4, calculated at various electrolyte concentrations according to the theory of Vincent [17], i.e. assuming random-coil behaviour. The value for Δ was taken to be 1.4 r_g following the SFV theory (see earlier). Values for r_g were taken from the literature for sodium polyacrylate in NaBr [19]. Total V(h) curves are shown in Figures 10 and 11 below for the PS-g-PEO particles in PAA 4 solutions at pH9, at NaCl concentrations of 10^{-1} mol dm^{-3} and 10^{-2} mol dm^{-3}, respectively. In both cases, four W_2 values were selected, scanning the range between the observed $W_2^{\dagger}$ and $W_2^{\dagger\dagger}$ values.

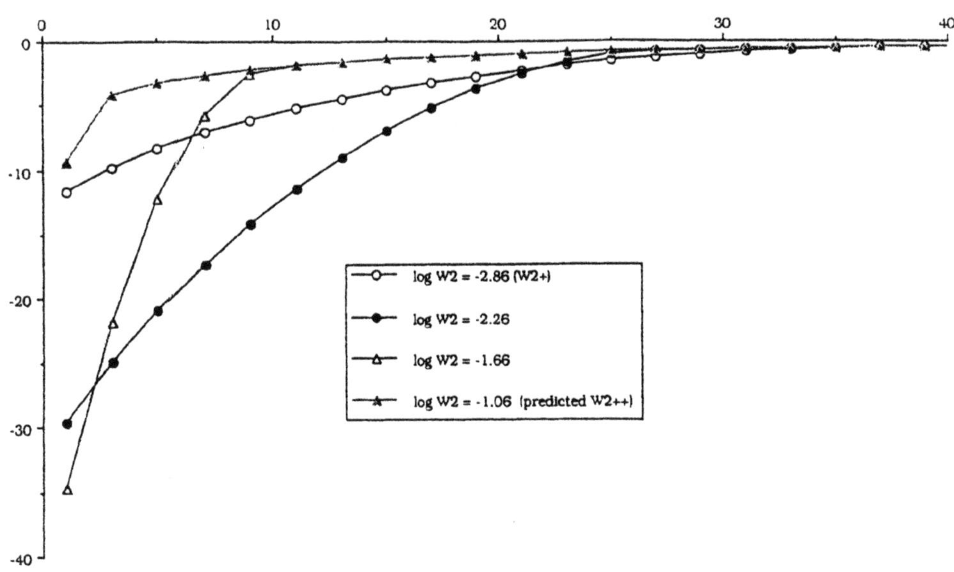

Figure 10. V/kT versus h/nm plots for PS-g-PEO particles in the presence of PAA 4 at pH 9 and 10^{-1} mol dm^{-3} NaCl, for the various PAA weight fractions indicated.

As can be seen, at both 10^{-1} and 10^{-2} mol dm^{-3} NaCl, $V(h \to o)$ is smallest at $W_2^{\dagger}$ and $W_2^{\dagger\dagger}$, and greater at intermediate PAA concentrations. Thus, at least semi-quantitatively, the theoretical approach, outlined above, seems to account reasonably well for the experimental observations at these higher NaCl concentrations. When, however, these equations were applied at 10^{-3} or 10^{-4} mol dm^{-3} NaCl solutions, the corresponding V(h) curves showed *no* net attraction, clearly in contrast to the experimental observations (Figures 6 and 7). The most probable explanation is that the use of eqn. (5) for $\Delta(\phi)$ is no longer valid, i.e. the random-coil approximation breaks down, at these low electrolyte concentrations.

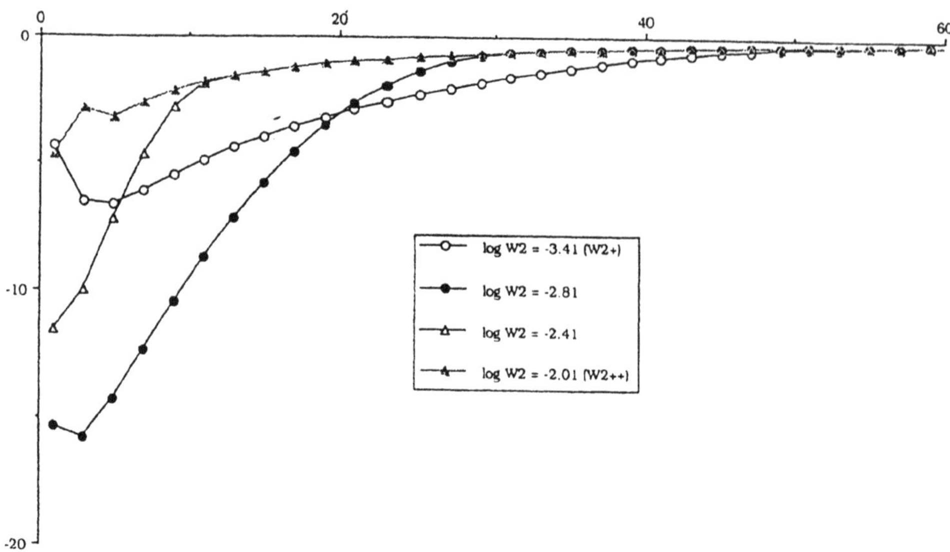

Figure 11. As for Figure 10, but for 10^{-2} mol dm^{-3} NaCl.

Using an alternative approach, Böhmer *et al.* [20] extended the Scheutjens-Fleer, self-consistent, mean-field lattice model to the polyelectrolyte depletion situation. Some computated data from this approach are shown in Figure 12, for polyelectrolyte chains in an athermal solvent, at an uncharged, planar, non-adsorbing ($\chi_S = 0$) surface. Only relatively *high* background electrolyte concentrations (> 10^{-2} mol dm^{-3}) are considered.

Figure 12 (a) should be compared to Figure 9, where Δ has been calculated using eqn. (5) from the analytical theory. The general form of the curves in the two figures is similar. However, because of the mean-field assumptions in the Scheutjens-Fleer model, one difference is that Δ_o

(a)

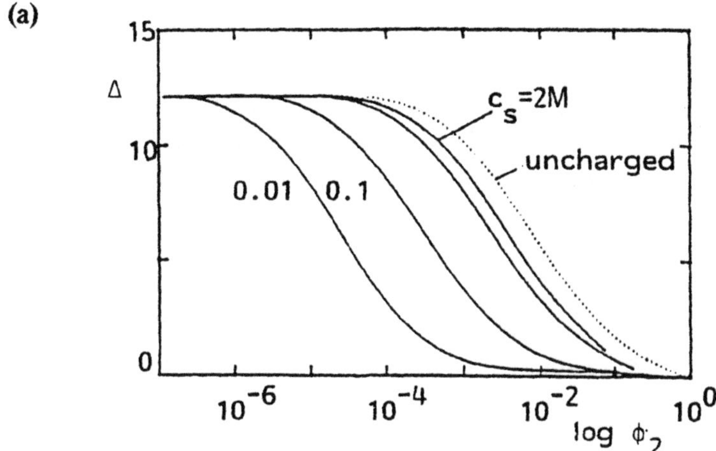

(b)

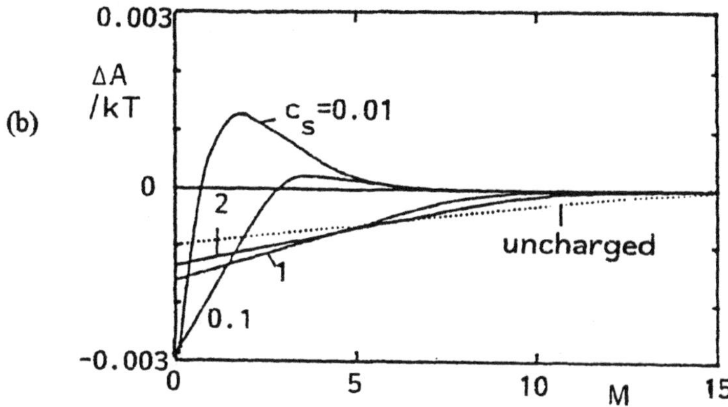

Figure 12 *(a)* Depletion layer thickness Δ (lattice units) for a fully -charge polyelectrolyte versus bulk volume fraction (ϕ_2);
(b) Interactionfree energy ΔA (kT/segment area) versus plate separation M (lattice units), for $\phi_2 = 10^{-2}$.
In both cases the polyelectrolyte chain has 500 segments, and $\chi = \chi_s = 0$; also the plate(s) are uncharged. The full curves represent four different 1:1 electrolyte concentrations as indicated; the dotted curve is for an uncharged polymer, for comparison.

(limiting value of Δ at low ϕ_p) in Figure 12 (a) appears to be independent of electrolyte concentration, which is clearly unreasonable.

The form of the $\Delta A(M)$ plots given in Figure 12 (b) may be compared with those for V(h) given in Figures 10 and 11, calculated using the analytical theory for 10^{-1} and 10^{-2} mol dm^{-3} NaCl solution, respectively. One striking difference between the curves of Böhmer et al. (Figure 12 (b), and those shown in Figures 10 and 11, is the presence of a *maximum* in the interaction energy curves in the former case, at 10^{-1} and 10^{-2} mol dm^{-3} electrolyte. The origin of the repulsive contribution is related to the work required to displace the polyelectrolyte chains into bulk solution, against the Donnan (electrostatic) potential difference set up across the depletion zone between the particles and the bulk polyelectrolyte solution. This potential difference is greater the lower the background electrolyte concentration. The presence of an energy barrier in the interaction curve implies some *kinetic* control of the depletion flocculation behaviour seen at low background electrolyte concentrations. However, we have no direct evidence for this; even in 10^{-4} mol dm^{-3} NaCl the flocculation seems to be thermodynamically controlled. Clearly, this is an aspect which needs further investigation; force-balance measure would be useful here, if they can be made sufficiently sensitive.

4. MICROGELS

4.1 INTRODUCTION

In a previous publication [21] the temperature-controlled, reversible flocculation of poly(N-isopropylacrylamide) (Poly(NIPAM) microgel particles, on the addition of simple electrolytes such as NaCl was described. The particles were prepared by the method developed by Pelton and Chibante [22]. At 25°C the dispersion was stable to the addition of electrolyte to high concentrations. This is because the particles are swollen with water at this temperature, so that their Hamaker constant is close to that of the medium. Hence, the van der Waals attractive force between the particles is negligibly small. At 40°C, however, the particles shrink, to become effectively hard spheres, as a result of an increase in the Flory interaction parameter (χ) for the poly(NIPAM)/water system. Aggregation of the dispersion can now take place in the presence of sufficient electrolyte, as a result of the increase in the van der Waals attractive forces. On cooling the flocculated dispersions from 40°C to 25°C, the particles redisperse as a result of reswelling. In this paper, the effect of adding a *polyelectrolyte* sodium poly(styrene sulphonate), in contrast to the simple electrolytes studied earlier [21], to a poly(NIPAM) dispersion is described. The idea was to see if the depletion interaction induced by the presence of the polyelectrolyte, would lead to aggregation of the poly(NIPAM) particles, even at 25°C.

4.2 MATERIALS

Poly(NIPAM) particles were prepared by the free radical polymerisation of NIPAM in water at 70°C, using persulphate initiator, in the presence of N/N-methylenebisacrylamide, following the procedure described by Pelton and Chibante [22]. Following dialysis against distilled water, transmission electron micrographs showed the particles to be monodispersed spheres, having a mean diameter 460 ± 12 nm. Sodium poly(styrene sulphonate) having M_W = 46000 and M_W/M_n = 1.05, was purchased from Polymer Laboratories, U.K.

4.3 RESULTS

4.3.1 *Temperature-induced, reversible swelling of poly(NIPAM) particles.*

Figure 13 shows how the particle size of the poly(NIPAM) particles, as measured using dynamic light scattering, varies with temperature. The instrument used was a Coherent Krypton-ion laser (λ = 530.9 nm) coupled to a temperature-controlled cell and a type 7027 correlator. As may be seen there is virtually a two-fold change in diameter (i.e. 8-fold change in particle volume) between 25°C and 50°C, for these particular poly(NIPAM) particles.

4.3.2 *Stability Studies*

The stability/flocculation behaviour of the poly(NIPAM) dispersions was assessed by determining the "n" value for the dispersions, using a standard u.v./visible spectrometer. It is defined as -dlog (turbidity)/dlog (wavelength).

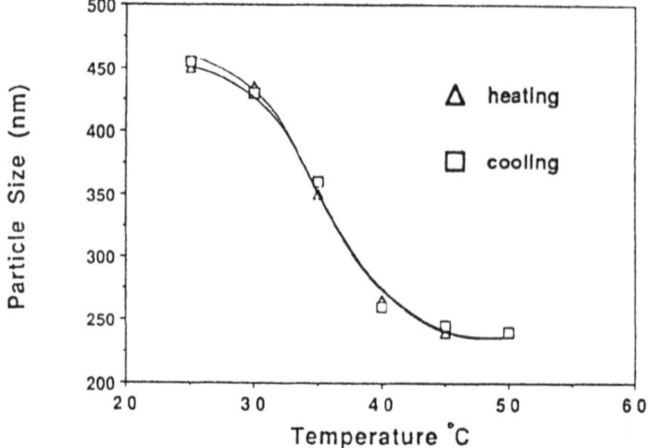

Figure 13. **Particle diameter versus temperature for the poly(NIPAM) particles in water.**

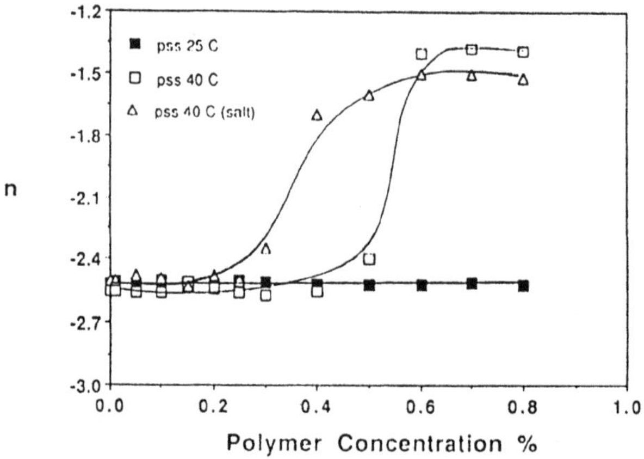

Figure 14. **A plot of u (see text versus concentration of sodium poly(stryrene sulphonate), at 25° and 40° in water at pH7, and at 40°C in 10^{-2} mol dm^{-3} NaCl.**

Figure 14 shows how u varies with increasing concentration of sodium poly(styrene sulphonate) under various conditions. It is clear that the poly(NIPAM) particle remains *stable* at 25°C, i.e. in the swollen state, even up to the highest polyelectrolyte concentrations effectively attainable (i.e. prior to the viscosity of the solution becoming intractibly high). However, on heating to 40°C flocculation could be induced on addition of the polyelectrolyte. Furthermore, the minimum concentration of polyelectrolyte required to induce flocculation, is reduced in the presence of a background concentration of 10^{-2} mol dm^{-3} NaCl.

4.4 DISCUSSION

It is clear that microgel particles, in the swollen state, are resistant to depletion flocculation by polyelectrolytes even at relatively high concentrations, where the ionic strength effect reduces the electrical double layer repulsion between the particles. This presumably has to do with the "softness" of the microgel/solution interface. Milling *et al.* [23] have studied the effect of the surface coverage of poly(methylmethacrylate) [PMMA] chains, grafted onto silica particles, on the stability of dispersions of such particles in non-aqueous media, on the addition of free PMMA chains. They showed that the minimum amount of free PMMA required to induce flocculation of the particles goes through a *maximum* with increasing coverage of grafted chains. The decrease beyond the maximum, is attributed to the decreasing "softness" (i.e. penetrability) of the grafted chains by the free chains with increasing coverage, at high coverages approaching maximum packing of the grafted chains. Penetration of the grafted chains by the free chains reduces the depletion interaction [24]. A similar situation probably pertains in the present case: some peripheral interpenetration, at high ionic strengths, of free polyelectrolyte chains into the surface regions of the swollen microgel particles (consisting of "dangling" polymer chains), is very likely, hence reducing any depletion effect. Increasing the temperature to 40°, makes the surface of the microgel particles "hard" again, leading to significant depletion effects. However, sufficient polyelectrolyte still needs to be added to overcome the electrostatic repulsion between the surface charge groups on the microgel particle and those on the polyelectrolyte. This point is supported by the fact that the addition of NaCl reduces the minimum concentration of polyelectrolyte required to induce flocculation at 40°C (Figure 4). Snowden *et al* [25] showed a similar effect. They used poly(NIPAM) particles containing surface - COOH groups (rather than the $-OSO_3^-$ groups in the present work). They showed that (partial) methaylation of the -COOH groups significantly reduce the surface charge density, and hence the electrostatic repulsion between the microgel particles. A much lower concentration of added sodium poly(styrene sulphurate) was needed, at pH7, to induce flocculation of the carboxylated microgel particles compared to the much higher charged microgel particles.

5. CONCLUSIONS

Both bridging and depletion flocculation may be observed in the same system, if some mechanism exists for altering *in situ*, the χ_S parameter for the particle/polymer solution interface. This has been demonstrated for hydrophilic, PS-g-PEO latex particles in the presence of poly(acrylic acid), where changes in pH lead to changes in χ_S: adsorption occurs at low pH, depletion at high pH values. In the case of the PS-g-PEO particles, the packing of the grafted PEO chains around the PS core is sufficiently high, that these particles behave as "pseudo hard-spheres". The depletion interaction may be modelled, reasonably successfully, on this basis, provided the electrostatic interactions between the (weakly charged) PS-g-PEO particles and the polyelectrolyte chains are properly included. However, the model used here only works at intermediate (and higher) background electrolyte concentrations (> 10^{-2} mol dm^{-3}). Where the polyelectrolyte chains can be reasonably modelled on random coils. At lower electrolyte concentrations, better models for the polyelectrolyte chains need to be incorporated. With poly(NIPAM) microgel particles, in the swollen state at room temperature, the packing of the bound chains is just too low to allow depletion effects to occur on adding polyelectrolyte, even at reasonably high concentrations. However, at high temperatures when the particles contract back to "hard spheres", depletion effects may be seen.

Acknowledgements

I should like to thank Nick Cawdery and Martin Snowden who carried out the experimental work reported here.

6. REFERENCES

1. Cohen Stuart, M.A., Cosgrove, T. and Vincent, B. (1986) *Advances in Colloid and Interface Sci.* **24**, 143.
2. Kawaguchi, Y. and Nagasawa, A. (1969), *J.Phys.Chem.*, **73**, 4382.
3. Bailey, F.E.J. (1964) *J.Polymer Sci.* **192**, 845.
4. Ikawa, T., Abe, K., Honda, K. and Tsuchida, E. (1975) *J.Polymer Sci.* **A13**, 1505.

5. Evans, R. and Napper, D.H. (1973) *Nature* **246**, 34.
6. Doroszkowski, A. (1986) *Colloids and Surfaces* **17**, 13.
7. Cowell, C. and Vincent, B. (1982) *J.Colloid Interface Sci.* **87**, 518.
8. Bromley, C. (1989) *Colloids and Surfaces* **17**, 1.
9. Cosgrove, T. and Ryan, K. (1990) *Langmuir* **6**, 136.
10. Boucher, E.A. and Hines, P.D. (1976) *J.Polymer Sci. (Physics)* **14**, 2241.
11. Miller, M.L. and Raubut, C.A. (1958) *J.Amer.Chem.Soc.* **80**, 4115.
12. Leon, A., Vincent, B. and Cawdery, N. (1994) *Colloid and Polymer Sci.* **272**, 427.
13. Milling, A., Vincent, B., Emmett, S. and Jones, D.A.R. (1991) *Colloids and Surfaces* **57**, 185.
14. Emmett, S. and Vincent, B. (1990) *Phase Transitions* **21**, 197.
15. O'Brien, R.W. and White L.R. (19778) J.Chem.Soc.Faraday Trans.II **74** 1607.
16. Fleer, G.J., Scheutjens, J.M.H.M. and Vincent, B. (1984) Polymer Adsorption on Dispersion Stability, in E.D. Goddard and B. Vincent (eds.), *Amer .Chem. Soc.* **240,** 245.
17. Vincent, B. (1990) *Colloids and Surfaces* **50**, 241.
18. Auvray, L. (1981) *J.PhysiqueI* **42**, 49.
19. Takahashi, A. and Nagasawa (1964) *J.Amer.Chem.Soc.* **86,** 543.
20. Böhmer, M.R., Evers, O.A. and Scheutjens, J.M.H.M. (1990) *Macromolecules* **93**, 2288.
21. Snowden, M.J. and Vincent, B. (1992) *J.Chem.Soc.Chem.Comm.* 1103.
22. Pelton, R.H. and Chibante, P. (1986) *Colloids and Surfaces* **20** 247.
23. Milling, A., Vincent, B., Emmett, S. and Jones, D.A.R. (1991 *Colloids and Surfaces* **51**, 185.
24. Jones, D.A.R. and Vincent, B. (1989) *Colloids and Surfaces* **42**, 113.
25. Snowden, M.J., Marston, N.J. and Vincent, B. (1994) *Colloid and Polymer Sci.* **272**, 1273.

PREPARATION OF COMPOSITE PARTICLES WITH MULTILAYERS BY HETEROCOAGULATION AND ENCAPSULATION POLYMERIZATION

Kunio Furusawa, Kazuomi Nagashima, and Chikai Anzai

Department of Chemistry, University of Tsukuba, Tennodai-1, Tsukuba, Ibaraki 305, Japan

Abstract: Firstly, the heterocoagulation behaviour of amphoteric polymer latices(2a=250 nm) on spherical silica(2a=240~1600 nm) was investigated as a function of pH, particle size ratio, and of electrolyte and polymer concentrations in the medium. On the basis of these data, subsequent experiments to prepare composite particles with multilayers comprised of usual anionic latices (2a = 180 ~ 900 nm), NiO.ZnO.Fe_2O_3 particles(2a=20 nm) and polystyrene layer, have been carried out. It is found that regular heterocoagulates which are made up of many magnetic particles regularly adhering to latex surface, were generated under the suitable medium condition. Furthermore, the best encapsulation condition of the heterocoagulates with polystyrene via emulsion polymerization, was surveyed. The encapsulation of the heterocoagulates was greatly promoted by the existence of adsorption layer of Sodium Oleate molecules on their surfaces and made uniform composite particles with multilayers. It is easily realized that the total sizes of the present composite particles with multilayers, can be controlled freely by selecting a latex particle with different size as the core.

Key Words; Composite Particles/ Heterocoagulates/ Encapsulation/ Multilayer Composite/ Polymer Latices/ Adsorption Layer/ Sodium Oleate

1. Introduction

The exceptional uniform shape and wide diameter range of latex particles are of interest because their combination with other dispersion systems should extend their application

E. Pelizzetti (ed.), Fine Particles Science and Technology, 815–827.

in the field of academic investigation as well as industrial material development. Especially, on the synthesis of composite particles comprised of organic and inorganic substances, the adoption of monodispersed polymer latices as their one component largely expands the variety of composite particles, because many techniques to prepare latex samples with different sizes and structure have already been developed.

In this paper, firstly we report on the preparation and properties of heterocoagulates, generated from mixture of amphoteric polymer particles with a constant size(2a=250 nm) and monodispersed spherical silica with a wide range of particle diameters(2a=240 ~ 1590 nm)(1). Also, for controlling the composition of organic and inorganic materials in the heterocoagulates, the effect of addition of electrolyte or aqueous polymer to the medium has been examined(2).

In the synthesis of inorganic-organic composite particles, the establishment of a method to control the orientation of each element, should be regard as an important factor, because the characteristic of the composite particles largely depends on the local orientation of the component particles. Simultaneously, since the function of the composite particles is frequently determined by their size and uniformity, the development of a way to fix the particle diameter in a narrow size distribution is also indispensable. Aiming at the systematic resolution of these problems, the synthesis method of magnetic composite particles was investigsted adopting the concept of heterocoagulation theory; heterocoagulates were prepared from magnetic ultra-fine powder ($NiO.ZnO.Fe_2O_3$,2a=20 nm) and four types of monodispersed negative latices with different particle diameter(2a=180~980 nm). In addition, the optimum condition for encapsulating the heterocoagulates by a polystyrene layer via emulsion polymerization, were surveyed(3).

2. Heterocoagulation Behavior of Polymer Latices with Spherical Silica

A most important parameter controlling the morphology of heterocoagulates generated from the mixture of amphoteric latices and silica particles is the surface charge of each single particle. A stable suspension consisting of regular composite particles has been prepared only from a medium controlled at a definite pH(pH=5~6), where the ζ-potentials of the two components are in contrast with each other and they carry charges of opposite sign(1). The next important parameter to control is the particle size ratio of the component particles when they are mixed in the vessel. Figure 1 shows schematic pictures of the morphology of heterocoagulates of different silica particles and amphoteric latices(4). Figure 2 shows an optical micrograph of the real heterocoagulate generated from different silica samples and amphoteric latex system, where the other conditions (e.g., the particle number ratio,N_{silica}/N_{latex} = 1/300; the medium pH= 5.6) have kept constant. It may be seen that at a particle size ratio($r=D_{silica}/D_{latex}$) higher than 3, the suspension is composed of uniform heterocoagulate particles and each hetero-coagulate is still undergoing Brownian motion as an isolated unit. The insert of Fig.2-A shows a scanning electron micrograph of the heterocoagulate. It is apparent that heterocoagulate takes a raspberry shape with one silica particle in the core. In contrast to this, the heterocoagulates generated at a particle size ratio lower than r=3 (Fig.2-C and 2-D),are composed of large irregular aggregates and regular coagulates were hardly formed at any medium pH and particle number ratio investigated.

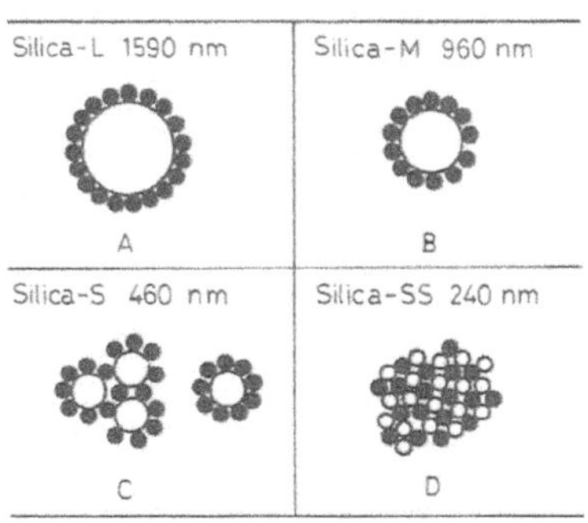

Fig. 1. Schematic pictures showing the morphology coagulate particles formed from different silica samp

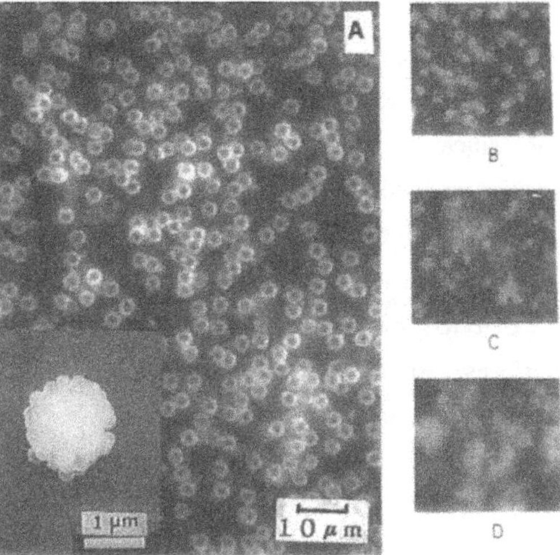

Fig.2, Optical micrograph showing the heterocoagulates prepared from different silica samples: A, Silica-L; B, Silica-M; C, Silica-S; D, Silica-SS.

3. Effects of Electrolyte and Water soluble Polymer

It is interesting to analyse the heterocoagulation behaviors from the concept of the adhesion isotherm for the amphoteric latices on the silica sample. Figure 3 shows some typical isotherms for the latices on Silica-L(2a=1590 nm) at various K_2SO_4 concentrations, where all the systems were controlled at pH5.2. It is evident that the isotherms are all of the very high affinity type with a well-defined plateau and the plateau value increases with increasing the K_2SO_4 concentration within the range from 10^{-5} ~ 10^{-2} mol dm^{-3}. This means that in this concentration range, adhesion proceeds in the way characteristic of mono-layer adhesion. This may be due to the strong blocking effect of adhering particles . However, in K_2SO_4 aqueous solutions more concentrated than $2x10^{-2}$ mol.dm^{-3}, no reproducible isotherm could be obtained under any conditions tested, and only some irregular aggregates were generated in the course of the experiment.

The effect of water-soluble polymer has been studied using hydroxypropyl cellulose(HPC) which has a lower critical solution temperature(LCST) at about 45°C(5). Firstly, HPC adsorption on the single latex and silica particles has been examined at 45 °C. It is found that HPC-M (Mw =30.3x10^4) adsorbed fairly well on the silica surface and showed a maximum adsorption at 0.5x10^{-1}%wt. residual concentration of HPC. However, no adsorption of HPC can be detected on the amphoteric latex surfaces. Figure 4 shows adhesion isotherms of amphoteric latices on the silica surfaces in a solution containing 0.6 mg ml^{-1} HPC with different molecular weights. It is apparent that the isotherm obtained in the HPC-L (Mw=5.3x10^4) aqueous solutions is of the high affinity type and the adsorption layer of HPC-L has no effect on the latex adhesion. In contrast to this, the isotherms obtained in the HPC-M(Mw=30.3x10^4) and HPC-H(Mw=92.5x10^4) aqueous solutions show curved shapes with low maximum adhesions; the effect of an HPC-H layer is especially remarkable.

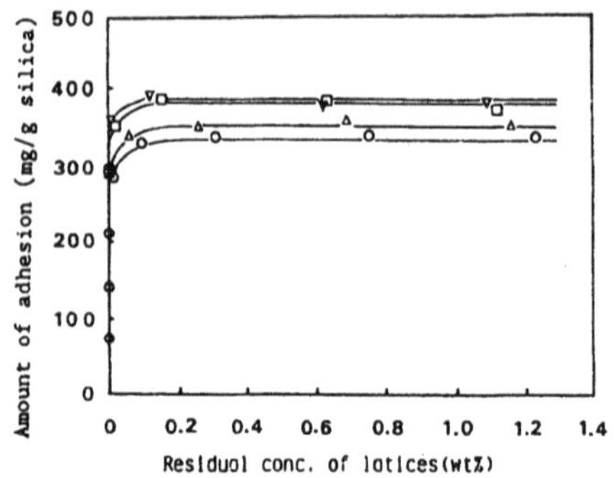

Fig. 3. Adhesion isotherms of amphoteric latices onto silica at various K_2SO_4 concentrations: ▽, $1.46 \cdot 10^{-2}$ M; □, $1.46 \cdot 10^{-3}$ M; △, $1.46 \cdot 10^{-4}$ M; ○, 0 M.

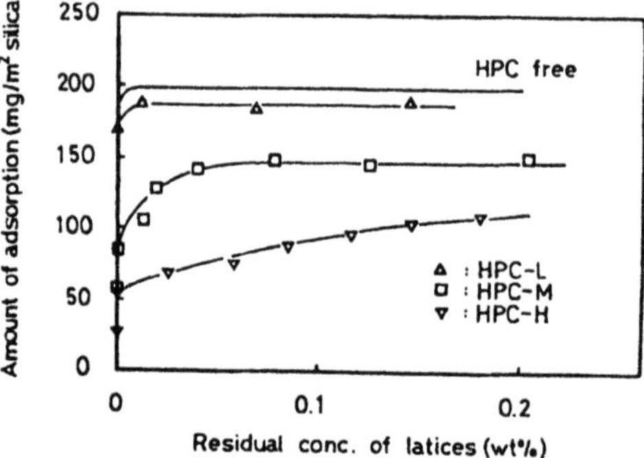

Fig. 4. Adsorption isotherms of latex particles on Silica-L: Effect of molecular weight of HPC (25℃, HPC conc. 0.6 mg/ml).

All these results indicate that these additives(electrolyte and polymer) are useful to control the composition of organic and inorganic compounds in the regular composite particles prepared by the heterocoagulation technique.

4.Analyses of Heterocoagulation Behaviour using Heterocoagulation Theory

The heterocoagulation between amphoteric latices and silica particles will proceed at the same time as the homocoagulation between latex particles and silica particles themselves. The interaction potential energy between these identical particles(two equal spheres of radius a) is given by

$$V = \frac{8R^2T^2\gamma^2\epsilon a}{Z^2F^2}\exp(-2\kappa h_0) - \frac{A}{6} \times \left[\frac{a^2}{2h_0(h_0+2a)} + \frac{a^2}{2(h_0+a)^2} + \ln\frac{h_0(h_0+2a)}{(h_0+a)^2}\right] \quad (1)$$

where

$$\kappa = \sqrt{8\pi n^0 Z^2 e^2/\epsilon kT} \qquad \gamma = \frac{\exp(Ze\psi/2kT)-1}{\exp(Ze\psi/2kT)+1}$$

In the above equations, R is the gas constant, $2h_o$ is the shortest distance of separation between two particles, ε is the dielectric constant of the medium(78.5 for water at 20 C), T is the absolute temperature of the suspension, Z is the valency of the ion, n_o is the ionic density, e is the elementary charge, ψ is the surface potential of the particle(here, the ζ-potenial measured at the appropriate salt concentration), and A is the Hamaker constant which is reported as $0.3\text{x}10^{-20}$ J for latices in water and $1\text{x}10^{-20}$ J for latices in water(6)

The potential energy of interaction between two dissimilar spherical particles is given as follows(7)

$$V = \frac{\epsilon a_1 a_2(\psi_1^2+\psi_2^2)}{4(a_1+a_2)}\left\{\frac{2\psi_1\psi_2}{\psi_1^2+\psi_2^2}\ln\frac{1+\exp(-2\kappa h_0)}{1-\exp(-2\kappa h_0)} + \ln[1-\exp(-4\kappa h_0)]\right\} - \frac{A_{12/3}a_1a_2}{6h_0(a_1+a_2)} \quad (2)$$

where, a_1, a_2 and ψ_1, ψ_2 are radii and surface potentials of the respective dissimilar particles and for ψ_1 and ψ_2, the ζ-potentials obtained in the same salt solutions were used. $A_{12/3}$ in equ(2) has been calculated from the following equation, using A_{33} = $4.83\text{x}10^{-20}$ J for water in vacuum, $A_{11}=14.8\text{x}10^{-20}$J for silica in vacuum and A_{22} = $11.5\text{x}10^{-20}$ J for latices in vacuum,

$$A_{12/3} = (A_{11} - A_{33})(A_{22} - A_{33}) \quad (3)$$

In Figs 5 and 6, the total interactio potential curves between silica-silica, latex-latex and silica- latex particles in $1\text{x}10_{-3}$ mol dm_{-3} and and $1\text{x}10_{-2}$ mol dm_{-3} K_2SO_4 aqueous solutions are pictured in the same drawing. From Fig.5, it is clear that in $1\text{x}10^{-3}$ mol dm^{-3} K_2SO_4 solutions, strong attraction energy between latex and silica particles appears and heterocoagulation will occur predominantly in comparison with the other two homocoagulations. In this salt concentration, the interaction potential between latex

- latex particles shows a high enoughbarrier to prevent mutual coagulation. However, in $1x10^{-2}$ mol dm^{-3} K_2SO_4 solution, the height of the potential barrier between latex-latex particles becomes about 10 kT which is the lower limit needed to keep the colloid stable.

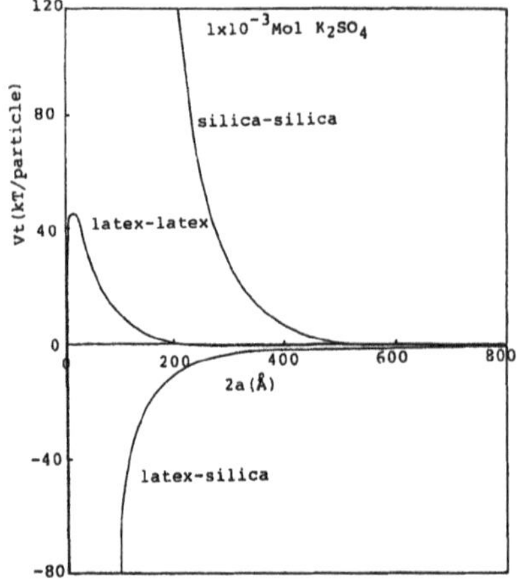

Fig.5, The total interaction potential curves between silica-silica, latex-latex and latex-silica particles in $1x10^{-3}$ M K_2SO_4.

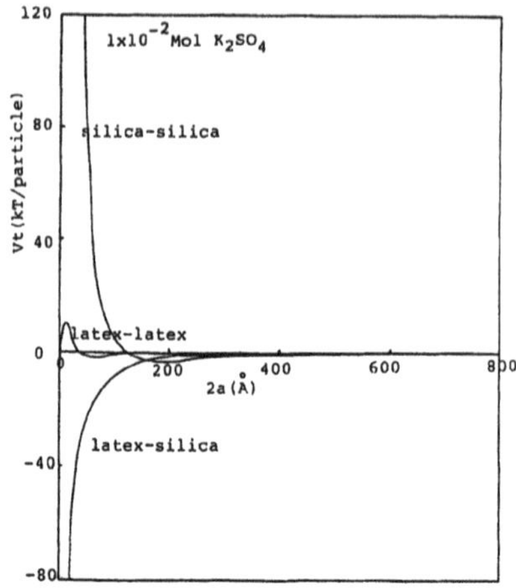

Fig.6, The total interaction potential curves between silica-silica, latex-latex and latex-silica particles in $1x10^{-2}$ M K_2SO_4.

This analysis indicates that the fractional coverage of latex particles on the large silica surface increases with increasing K_2SO_4 concentration, but the concentration of $1x10^{-2}$ mol dm^{-3} K_2SO_4 will be an upper limit to make a stable hetero-coagulation, because above $1x10^{-2}$ mol dm^{-3} K_2SO_4, some aggregates of the latex particles will be generated in the course of the experiments. It is shown in Fig.3 that the same behaviour could be observed in the present system.

5. Process to Control the Total Size and Component Distribution of Multi-layer Composite Particles

A new synthetic process to prepare composite particles with multi-layers, compresed of usual polymer latices, ultra-fine magnetic particles, and a polystyrene layer was examined under various solution conditions.

As an inorganic material, the magnetic ultrafine powder $NiO.ZnO.Fe_2O_3$ offered by Sumitomo Chemical Co.Ltd. Japan, was used without further purification. On the other hand, four different polystyrene latices were employed as organic species. The latex-L is a standard sample offered by Japan Synthetic Rubber Co. Ltd. The other three latex samples, the latex-M, the latex-S, and the latex-SS were synthesized by surfactant-free emulsion polymerization in our laboratory(7). All of these latex samples were used after sufficient dialysis in purified water and subsequent ion exchange treatment with resin grains. The diameter and functional group of each sample are listed in Table 1.

The procedure to prepare the heterocoagulates is as follows. To begin with, 200 ml of pure water was added to a beaker containing 0.25 g of the magnetic powder and sonicated for 20 min with a 600 W homogenizer(T-180, Ultrasonic Co. Ltd.) at the best tuning. Meanwhile, the original latex dispersion was diluted with purified water to control the particle number ratio to magnetic particles at 1/4500. These two suspensions were quickly blended and ultrasonic emission was used for another 3 min to promote complete mixing. Next, the pH of the blend suspension was adjusted with nitric acid to pH2.5 to insure that each particle acquuired the opposite charge to one another(see Fig.7). After the suspension was kept for 24 h. at room temperature, the heterocoagulates, comprised of the latex-M(core) and the magdnetic particles(shell), accumulated at the bottom of the vessel. The supernatant was replaced by purified water to remove the residual magnetic particles which did not adhere to the latex surface The product was repeatedly washed in purified water and kept in a stock bottle.

Table 1. Diameter and functional groups of samples

Sample	Diameter (nm)	Functional group
Latex-L	900	$-COO^-$
Latex-M	600	$-COO^-$
Latex-S	530	$-SO_3^-$
Latex-SS	180	$-SO_3^-$
$NiO \cdot ZnO \cdot Fe_2O_3$	20	–

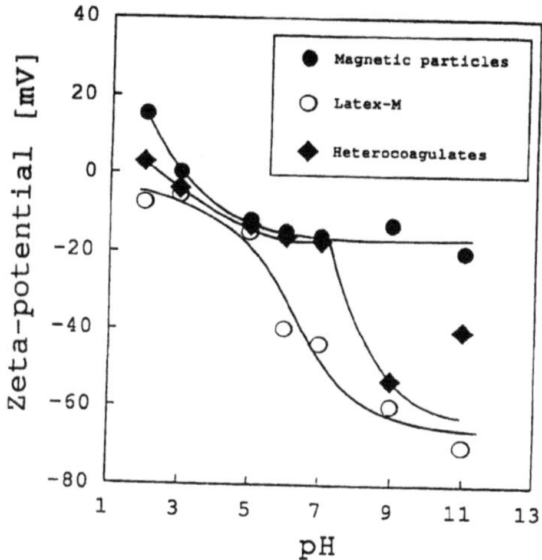

Fig.7 Zeta-potential of $NiO \cdot ZnO \cdot Fe_2O_3$ particles (●), latex-M (○) and heterocoagulates (◆) as a function of pH at 10^{-3} M KNO_3.

Figure 8 indicates the scanning electron micrographs which show four kinds of heterocoagulates, provided from the suspensions of latices with different sizes and magnetic particles. As seen, the heterocoagulates generated from the Latex-L, the

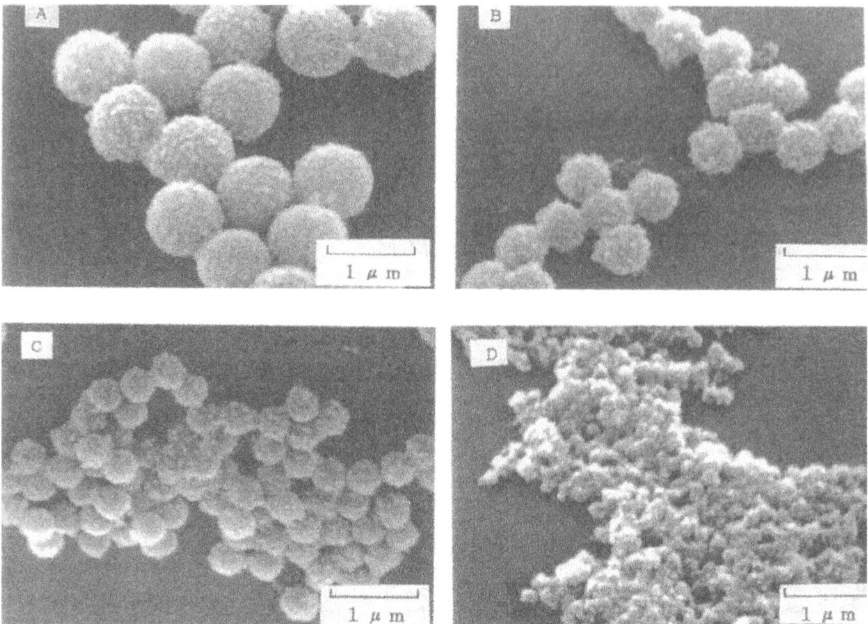

Fig.8, Electron micrographs showing the heterocoagulates prepared from different latex samples: A, Latex-L; B, Latex-M; C, Latex-S; D, Latex-SS.

Latex-M, and Latex-S are comprised of the uniform composite particles and all the heterocoagulates are separated as an isolated unit having an even surface. In contrast with these particles, the heterocoagulates generated from the Latex -SS are made of large irregular aggregates and the regular heterocoagulates were difficult to recognize under any mixing conditions investigated. Furthermore, it should be noted that the heterocoagulates took over the uniformity as the original latex particles under the suitable mixing conditions. The composite particles were formed by the heterocoagulation of the magnetic particles onto the large latex particles with a constant layer, and they exactly maintain the uniformity of the latex particles after the heterocoagulation with the magnetic particles.

In Fig.7, the ζ-potential profile of the latex-M particles and the magnetic particles are presented as a function of medium pH, along with that of the heterocoagulates prepared by mixing their original suspensions at pH2.5. The heterocoagulates display zero ζ-potential at pH2.5 due to the charge neutralization between the two component particles. This suggests that the rapid sedimentation of the products after the heterocoagulation is attributed to the homoaggregation of the products rather than to the increase of their diameter or density. However, the dynamic light scattering measurement and the transmission electron microscopy confirm that the aggregated products are re-dispersed into single particles by moderate ultrasonication without any desorption of the magnetic particles from the surfaces of the composite particles. Furthermore, Fig. 7 indicates that the ζ-potential of the heterocoagulates approaches that of the original magnetic particles in the alkaline medium. This behavior suggests that the magnetic particles, which cover the surface of the core latices in the acid and neutral pH systems, desorb from their surface in the alkaline system where the pH is above 9. This implies that the interaction between the surfacesa of the two component particles changes from lateral electrostatic attraction into electrostatic repulsion according to the increase of medium pH.

Figure 9 shows the thermograms of the magnetic particles, the latex-M particles, and the heterocoagulates prepared from these two kinds of particles. The rate of mass decrease for the magnetic particles was 2.0 to 2.3 % during the temperature transition from 100℃ to 600℃, while the result of the latex-M was complete decomposition. The thermogram curve of the heterocoagulates exhibits a slow steady mass decrease below 200 ℃、 a noticeable loss around 350 ℃, and a plateau above 400 ℃. The residual amount above 400 ℃is considered to be the weight of magnetic particles which did not decompose. As a result, the composition of the heterocoagulates is calculated to be 60.7% latex-M and 39.3% $NiO.ZnO.Fe_2O_3$ particles by weight percent. The application of the model produced by Hansen et al(8)(which assumes that the latex-M and $NiO.ZnO.Fe_2O_3$ are sphere particles with diameters of 600 nm and 20 nm respectively) calculated the fractional coverage of the heterocoagulates(θ) to be ca. 1.0.

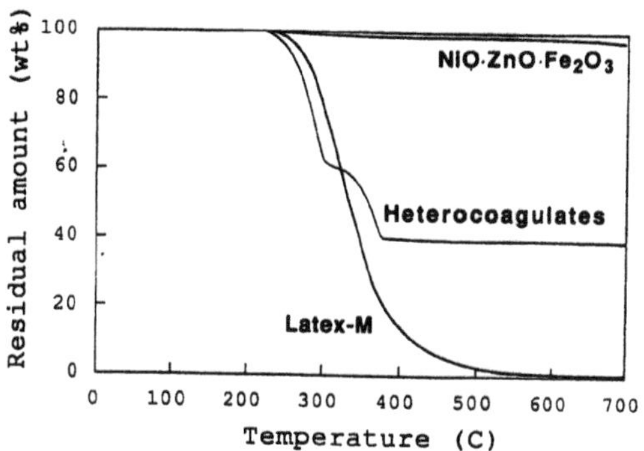

Fig. 9 Thermograms of latex-M, $NiO \cdot ZnO \cdot Fe_2O_3$ particles, and heterocoagulates.

6. Encapsulation of Heterocoagulates.

Concerning the heterocoagulates, both desorption of the magnetic particles and dissolution of ion species decrease their value as bioreactors because they inactivate biological functions. Hence, the optimal condition of encapsulation polymerization, in which the heterocoagulate surface is covered with a thin polystyrene layer, was surveyed. The conditions for the emulsion polymerization and their results are shown in Table 2 and 3. After 24 h on dialysis, a fixed amount of the hetero- coagulates prepared from the latex-M was used as the seeds to examine the polymerization conditions. Approximately 0.2g of the seed heterocoagulates was dispersed into 180 ml of water using the homogenizer, before it was moved into a special 200 ml vial with a screw cap. After that, distilled water containing a definite amount of potassium persulfate($K_2S_2O_8$) and sodium oleate($C_{17}H_{33}COONa$) was carefully addedd to the vial. The pH of the sodium oleate aqueous solution was primarily adjusted to pH6.0 with nitric acid, because the magnetic particles on the heterocoagulates desorb from the latex surface in the alkaline medium (Fig.7). This vial was set on the rotating disk of polymerization equipment and polymerized for 36 hr at 60 C. Encapsulation behavior of the heterocoagulates was examined for the eight polymerization conditions set up with regard to the amount of styrene monomer, the kind of initiator, and the presence of a surfactant. An initiator which provides a positive terminal group, azo-bis(isobutylamidine hydrochloric acid) was employed, while potassium persulfate was selected as a negative-terminal initiator. Moreover, sodium oleate aqueous solution was adopted to promote hydrophobilicity of the magnetic particles on the heterocoagulate surfaces. As seen in the line of product on Table 2, it is considered that the best condition in the list is the system No.3 where ideal encapsulation was virtually accomplished. Under the conditions of the system No.1, No.2, No.5, and No.6, bridging flocculation among the seed particles or the generation of independent latex particles occurred in the course of the polymerization. Similarly, in the system No.4, No.7, and No.8, the flocculation and the new particle generation were

Table 2 Conditions for encapsulation polymerization of heterocoagulates (temp. 60 °C, scale 200 ml, stirring spee time of polymerization 36 h)

Sample number	1	2	3	4	5	6	7
Seed composite [g]	0.2	0.2	0.2	0.2	0.2	0.2	0.2
Styrene monomer [g]	0.2	1.0	0.2	1.0	0.2	1.0	0.2
KPS [g]	10^{-3}	10^{-2}	10^{-3}	10^{-2}	–	–	–
AIBA·2HCl [g]	–	–	–	–	10^{-3}	10^{-2}	10^{-3}
Sodium oleate [g]	–	–	0.06	0.06	–	–	0.06
PRODUCT	×	×	⊙	×	×	×	△

KPS; potassium persulfate
AIBA·2HCl; azo-bis(isobutylamidine hydrochloric acid)
⊙; good encapsulation, △; encapsulation and new particle generation,
×; flocculation or new particle generation, or both.

Table 3. Conditions for encapsulation polymerization of heterocoagulates. (temp. 60 °C, scale 200 ml, stirring spe time of polymerization 36 h)

Sample number	9	10	11	12	13	14
Seed composite [g]	0.2	0.2	0.2	0.2	0.2	0.2
Styrene monomer [g]	0.2	0.2	0.2	0.2	0.2	0.2
KPS [g]	10^{-3}	10^{-3}	10^{-3}	10^{-3}	10^{-3}	10^{-3}
Sodium oleate [g]	0.003	0.006	0.03	0.06	0.12	0.30
PRODUCT	×	×	⊙	⊙	△	×

⊙; good encapsulation, △; encapsulation and partially no reaction, ×; flocculation or decomposition.

observed although a part of the seed particles were encapsulated. As a result, it can be considered that sodium oleate is indispensable for the encapsulation of the hetero-coagulates, and that potassium persulfate is a suitable initiator for this system. Furthermore, as pointed out in Table 3, under similar conditions to the system No.3(the same amount of seed, potassium persulfate, and styrene monomer), the encapsulation polymerization was carried out at various sodium oleate concentration ranges from 0.003 to 0.60 g/200 ml. Below the concentration of 0.01 g/200 ml, the seed particles aggregated soon after the vials were equipped, while above 0.1 g/200 ml, desorption of the magnetic particles from the seed particle surface occurred. Consequently, the concentration of sodium oleate has to be adjusted in the range from 0.01 to 0.1 g/200 ml for the encapsulation. This fact suggests that the adsorption state of oleate ion onto the seed particle surface largely affects the process of encapsulation polymerization.

Figure 10 is a transmission electron micrograph of the encapsulated composite particles whose surface was covered with a thin polymer layer, and it shows smoother surface than that of the seed particles (Fig.8).

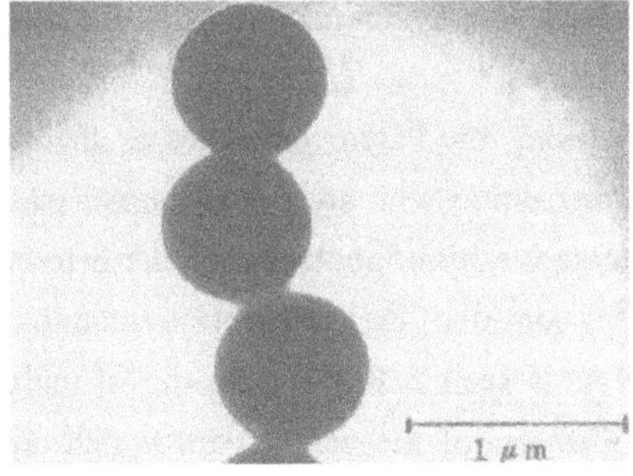

Fig.10,Microphotograph showing the multilayer composite particles after encapsulation polymerization.

7. Conclusion

Figure 11 shows the preparation process of multilayer magnetic composite particles.

1) The heterocoagulates were produced from the magnetic ultra-fine powder and the polymer latices with various sizes and functional groups. In this case, the particle size ratio and the ζ– potential of the two components should be noted in order to prepare the

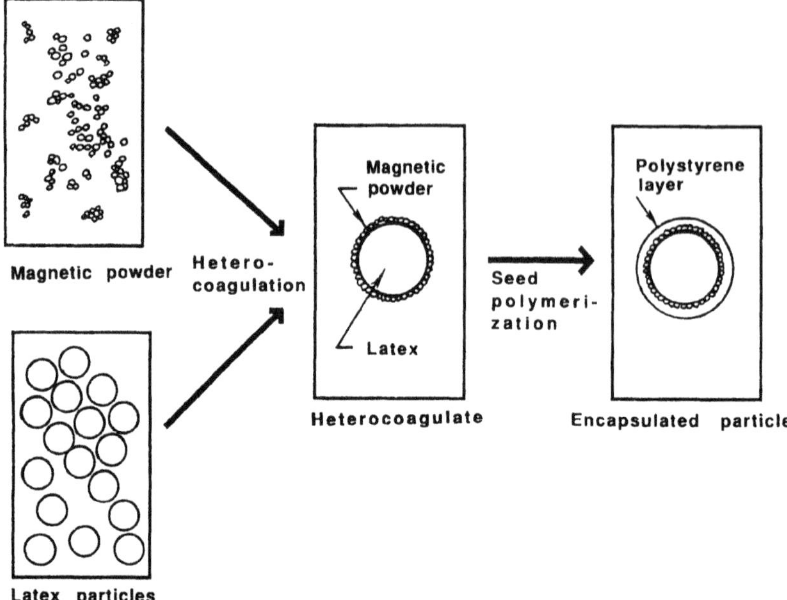

Fig.11, Schematic picture showing the process of synthesizing the multilayer composite particles.

heterocoagulates whose fractional coverage is ca. 100 %. The prepared heterocoagulates take over the uniformity of the latex particles adopted as the core. However, in the alkaline system above pH9, the magnetic particles desorb from the surface of the core latices.

2) Using the heterocoagulates as the seeds, the optimum condition of encapsulation polymerization with styrene monomer was surveyed. In this case, it is indispensable to promote the hydrophobicity of the heterocoagulate surface with oleate ions. With the proper control of their adsorption state, the encapsulation is promoted efficiently.

3) As is seen in Fig.11, the size of multilayer composite particles can be controlled by the diameter of the core latices which are applied primarily. This method realizes the localization of the magnetic particles on the outside of the composite particles covered with a thin polystyrene layer, therefore, they can give full function to the surrounding system.

References

1) Furusawa, K. and Anzai, C.(1987) Preparation of composite fine particles by heterocoagulation, Colloid & Polymer Science, **265**,882~ 888.

2) Furusawa, K and Anzai, C.(1992) Heterocoagulation behaviour of polymer latices with spherical silica, Colloids and Surfaces, **63**, 103~111.

3) Furusawa, K.,Nagashima, K. and Anzai, C.(1994), Synthetic process to control the total size and component distribution of multilayer magnetic composite particles, Colloid & Polymer Science, **274**,1104~1110.

4) Harding,R.D.(1972), Heterocoagulation in mixed dispersion - Effect of particle size, size ratio, relative concentration, and surface potential of colloidal components, Journal of Colloid and Interface Science,**40**, 164 ~ 173.

5) Furusawa, K. and Tagawa, T.(1985), Adsorption behavior of water soluble polymer with lower critical solution temperature, Colloid & Polymer Science,**263**, 353 ~ 360.

6) Visser, J.(1972) On hamaker constants: A comparison between hamaker constants and Lifshitz- Van der Waals constants, Advance of Colloid and Interface Science, **3**, 331 ~ 363.

7) Furusawa, K. Norde, W., and Lyklema, J.(1972), A method for preparing surfactant-free polystyrene latices of high surface charge, Kolloid Zeitschrift und Zeitschrift fur Polymere, **250**, 908 ~ 909.

8) Hansen,F.K., and Matijevic, E.(1980), Heterocoagulation, part 5, adsorption of a carboxylated polymer latex on monodispersed hydrated metal oxides, Journal of Chemical Society, Faraday Transition-1, **76**, 1240 ~ 1262.

SUBJECT INDEX

GPSR Compliance
The European Union's (EU) General Product Safety Regulation (GPSR) is a set of rules that requires consumer products to be safe and our obligations to ensure this.

If you have any concerns about our products, you can contact us on

ProductSafety@springernature.com

In case Publisher is established outside the EU, the EU authorized representative is:

Springer Nature Customer Service Center GmbH
Europaplatz 3
69115 Heidelberg, Germany

www.ingramcontent.com/pod-product-compliance
Ingram Content Group UK Ltd.
Pitfield, Milton Keynes, MK11 3LW, UK
UKHW021903190726
13853UKWH00003B/1396

* 9 7 8 9 4 0 0 9 0 2 6 0 2 *